Elementary Multivariate Analysis for the Behavioral Sciences
Applications of Basic Structure

Elementary Multivariate Analysis for the Behavioral Sciences

Applications of Basic Structure

Clifford E. Lunneborg
Robert D. Abbott
University of Washington

North-Holland
New York • Amsterdam • Oxford

Elsevier Science Publishing Co., Inc.
52 Vanderbilt Avenue, New York, New York 10017

Distributors outside the United States and Canada:

Elsevier Science Publishers B.V.
P.O. Box 211, 1000 AE Amsterdam, The Netherlands

Library of Congress Cataloging in Publication Data

Lunneborg, Clifford E.
Elementary multivariate analysis for the behavioral sciences.

Includes bibliographies and index.
1. Multivariate analysis. I. Abbott, Robert D. II. Title.
QA278.L86 1983 519.5'35 83-4016
ISBN 0-444-00753-9

Manufactured in the United States of America

Contents

Preface

Multivariate analysis models the structure of or relations among a set of measures or the relations between two (or more) sets of measures. Multiple measures are essential to research in the behavioral sciences: Behavior is complexly determined, and a single independent variable is seldom the object of isolated study. Similarly, behavioral science constructs have a richness that precludes their being well represented by single indicators or outcome measures.

In this book we introduce the student of behavioral sciences to the better known multivariate approaches. Clearly, it is important that the student gain experience with the currently popular techniques of data analysis. It is equally important, we believe, that he or she be provided a grounding in the bases for those techniques. Such grounding provides the best assurance that, over a professional lifetime, the student will intelligently use available procedures and assimilate new techniques as they are introduced. Knowing how to "run SPSS" is one thing. Knowing why to do so is something else. We strive to stimulate the latter understanding.

The book's title refers to elementary multivariate analysis, to indicate our interest in presenting the elements of multivariate analysis. The organization of multiple and multivariate observations into matrices, the transformation of these observations, and some understanding of how multivariate observations may be distributed and how they may be represented in a sample will be necessary whatever particular analysis is chosen.

In presenting and organizing these elements we are most privileged to draw upon an approach we owe to our mentor, Paul Horst: the representation of matrices in terms of their basic structures. It is a unifying concept. Not only data and data-derived matrices, but also those used to achieve the goal-directed transformations at the heart of multivariate analysis, possess basic structures; we use these to advantage to link data, transformations, and results.

We intend the book for use at the senior or graduate level, and the materials have been used for a number of years in such courses. We intentionally include topics to permit use of the book with students whose statistical background consists of a single introductory statistics course and whose mathematical training extends no further than high school algebra. The 15 chapters have been organized into four parts, making it easier for the instructor to assign topics in keeping with the mathematical and statistical background of students and the length of the course. In a one-quarter course some selection of advanced topics may be necessary.

Part I covers the tools used in building multivariate analyses. Chapter 1 familiarizes students with the organization of multivariate observations into data matrices and reviews the characteristics of behavioral science observations. The second chapter develops the elementary matrix manipulations—transposition, addition, and multiplication—and illustrates these with computations of a basic statistical nature. In the third chapter students confront the sometimes arbitrary nature of behavioral science observations, study some techniques for standardization, and are introduced to the critical concept of the rank of a matrix. The latter leads to the definition of a regular inverse for certain matrices.

Basic structure as a unique representation, basis, or transformation of the data matrix (the singular value decomposition) is introduced in Chapter 4. A powering approach is used to show how the basic structure might be found, and the properties of the basic structure of a number of specialized statistical matrices are discussed. Chapters 5 and 6 discuss techniques for transforming matrices of multivariate observations to achieve the goals of multivariate analysis. Properties of linear, quadratic, and spherizing transformations are presented in Chapter 5, and Chapter 6 reviews elementary differential calculus, illustrates the use of Lagrangian multipliers in optimizing transformations, and introduces a symbolic matrix calculus useful in finding multivariate solutions.

Students with a background in linear algebra can quickly review Chapters 2–5, while those with coursework in introductory calculus may need only to study that part of Chapter 6 dealing with symbolic matrix differentiation. The matrix algebra and calculus chapters have been written in sufficient detail that the reader without previous mathematical training can study these independently, without class time being spent on what may be for most students a review.

Part II develops systematically the general linear model (GLM) for the univariate criterion. In Chapter 7, starting with the sample-bound prediction problem, we introduce the technique of linear constraints and then coordinate the ideas of multiple and partial correlation with this sample-bound model, for which a least squares minimization is appropriate. We follow this in Chapter 8 with a survey of the basis for normal linear regression, our linear model for inference. Structural and sampling aspects of the model are discussed. Maximum likelihood is used as a technique for estimating unknown population values and for developing tests of hypotheses, usually phrased as linear constraints on these unknown populations. Chapter 9 develops some specific forms or applications of the general linear model. We present the fixed effects analysis of variance and the analysis of covariance from a GLM perspective and then examine a number of types of applications of multiple regression.

The GLM is the most influential data analytic philosophy in the behavioral sciences. Many students in a multivariate analysis course (perhaps nearly all students) will have already encountered it either as multiple regression or as the analysis of variance, if not in a more general form. Our goals in presenting it here are twofold: First, to let the student see that these familiar data analytic procedures have a common statistical basis and to see what that basis is. Second, when developed in this fashion, the GLM provides a way of linking the newer, more sophisticated, or more complicated multivariate analyses to be introduced in Part III with the familiar. The amount of time the instructor gives to Part II will depend upon how comfortable students are with a linear regression approach to data analysis. As in Part I, we give sufficient breadth of coverage to the simpler linear models that the less experienced reader can develop an understanding of their basis and range of application before going on to consider the multivariate extensions.

Part III develops and applies the multivariate general linear model (MGLM), the basis of classical continuous multivariate analysis. Chapter 10 introduces the multivariate normal distribution and uses the concepts of basic structure to establish the generalized variance and Wishart matrices as extensions of the univariate criterion variance and quadratic forms of the criterion, essential to both estimation and hypothesis testing. In Chapter 11 the MGLM is presented and its application demonstrated in Hotelling's T^2, the multivariate analysis of variance and covariance, profile analysis, and multivariate linear regression. Continuity with the univariate GLM is stressed, and the two major hypothesis testing approaches, likelihood ratio and union–intersection, are discussed. Chapter 12 illustrates the application of the MGLM, drawing upon data sets presented earlier in the text to provide examples of multivariate analysis of variance, multivariate linear regression, etc. The linear discriminant function is developed and its similarity to the MGLM highlighted. Following an example of the discriminant function, we use the likelihood ratio principle to show how tests of the homogeneity of variance–covariance matrices can be developed.

Part IV provides an overview of multivariate techniques that addresses questions other than those relating to the mean structure of measured multivariate attributes. In Chapter 13 we take up the analysis of categorical data. Adopting a log linear approach, we focus on the discrimination information techniques of Gokhale and Kullback. Their orientation permits us to point up similarities to the GLM in the use of design matrices and linear constraints to phrase and test hypotheses or competing models. The goal is to suggest that the same kinds of questions asked of measured attributes may also be asked of categorical data. We return in Chapter 14 to the discussion of basic structure, now in the form of principal components analysis. Our discussion of the analysis of interdependence among attributes moves from the determination and application of principal components to the role of unobserved or latent attributes in factor analysis. Latent attributes are also featured in the final chapter, in which their emerging role in dependency models, even "causal" models, is discussed. We use some applications of Joreskog's LISREL approach to illustrate these ideas.

Elementary Multivariate Analysis for the Behavioral Sciences

Applications of Basic Structure

PART I
BASIC TOOLS OF MULTIVARIATE DATA ANALYSIS

When researchers choose to use empirical approaches to investigate relations among theoretical constructs, they must resolve a series of questions, judging the degree to which relations among the data are supportive of a priori theoretical expectations. Part I presents basic tools that, when mastered, permit the researcher to address those questions.

In Chapter 1 we discuss fundamental issues in the collection of data: Which measurement procedures are used to assess attributes of interest? How are individuals or other observational units (entities) selected for inclusion in the study? How should data be arranged so that they "fit" the analyses?

Because an investigation in the behavioral sciences typically involves many attributes, we orient the reader in Chapter 2 to a multivariate approach that features a matrix representation of observations. Important manipulations of the data matrix are discussed. As the analysis of relations among attributes frequently involves covariance or correlation matrices, we organize our discussion of matrix operations around the calculation and determination of the properties of these derived matrices.

Most research in the behavioral sciences involves multivariate data based on measurement operations that are somewhat arbitrary. As a result, the researcher is free to make certain decisions about scaling the resulting observations. Frequently, this takes the form of standardizing the observations, a decision that may cause a loss of some of the information in the original data. In Chapter 3 we take up transformations of the data matrix and introduce the rank and inverse of a matrix.

More often than not, there are interdependences among the attributes, making our task more difficult. To provide a framework within which to work with such interrelations, we introduce in Chapter 4 the basic structure representation of a data matrix and derived matrices. The basic structure representation is a nonarbitrary transformation of the data into orthogonal variates. Such a transformation is a powerful way of dealing with multivariate observations, and the orientation developed in Chapter 4 is fundamental to the rest of our discussions.

The use of basic structure also greatly simplifies our presentation of the statistical underpinning of multivariate analyses. In Chapter 5 we explore the basic structure characteristics of key multivariate transformations. Multivariate analyses consist at heart of transforming the data matrix in a fashion that permits us to judge more easily the degree of consistency between the observed data and theoretical expectations. To understand such analyses, we must understand the nature of these transformations.

Multivariate transformations are chosen because they are optimal in some sense; often we seek to make some criterion function take its smallest or largest value. In Chapter 6 we review some elementary differential calculus and extend the principles of optimization to the use of Lagrangian multipliers for cases where a "side condition" must be satisfied under optimization. We also introduce a symbolic matrix calculus that allows us to summarize the results of simultaneously differentiating with respect to a collection of variables.

The first six chapters, then, discuss processes and procedures fundamental to statistical analyses of multivariate data. These basics are necessary to the understanding of the multivariate techniques developed and illustrated in Parts II–IV.

Chapter 1
The Data Matrix

Introduction

This chapter introduces concepts the authors think are basic to empirical research involving observations on multiple variables. Each reader, however experienced, will need at least to scan this chapter's contents before studying later portions of the book.

The analysis of *multivariate* data is usefully approached from a *matrix* orientation. We first discuss writing a matrix of data. As it is frequently important to distinguish observations that are presumed explanatory from those to be explained, we distinguish between and illustrate explanatory or *predictor* attributes and response or *criterion* attributes.

We next take up the question of the *measurement* properties of data, distinguishing between measured and categorical observations. Then we consider how such observations are sampled. In the experimental tradition the researcher manipulates the independent or predictor variables and observes the values assumed by the dependent or criterion measures, whereas in the survey sampling tradition we observe values of both predictors and criteria as they occur. Implications of type of measurement and sampling strategy are discussed.

Finally, we introduce research examples, in the form of data matrices, that we shall use throughout the book to illustrate multivariate analyses.

1.1 Writing the Data Matrix

Multivariate analyses of the kind discussed in this book will always involve many observations—many measurements. Because of the number of these observations and the relatedness of the observations among themselves, we must be very systematic in the way we record those observations. The data matrix provides a way of ensuring this systematic arrangement.

Table 1.1 Annual Appropriations in Support of Higher Education by State, in the United States

State	Income[a]	High school completions[a, b]	Geographic region	Appropriations[a]
Alabama	5227.84	0.01568	6	76.26
Alaska	11,800.00	0.01869	8	143.03
Arizona	6130.98	0.01366	7	90.90
Arkansas	5043.65	0.01406	6	50.15
California	7356.39	0.01467	8	98.15
Colorado	6650.64	0.01595	7	75.72
Connecticut	7415.80	0.01538	1	44.24
Delaware	7155.32	0.01658	5	73.30
Florida	6267.00	0.01150	5	50.80
Georgia	5653.42	0.01358	5	50.02
Hawaii	7337.28	0.01669	8	98.58
Idaho	5882.35	0.01652	7	78.47
Illinois	7411.73	0.01463	3	64.85
Indiana	6191.37	0.01529	3	55.42
Iowa	6269.70	0.01797	4	70.79
Kansas	6563.88	0.01780	4	80.70
Kentucky	5481.08	0.01451	6	55.35
Louisiana	5526.04	0.01645	6	52.87
Maine	5444.13	0.01643	1	42.41
Maryland	6961.41	0.01534	5	57.84
Massachusetts	6603.45	0.01503	1	35.93
Michigan	6759.73	0.01572	3	65.05
Minnesota	6254.79	0.01838	4	64.03
Mississippi	4604.13	0.01420	6	67.38
Missouri	5966.08	0.01423	4	48.25
Montana	5850.34	0.01782	7	63.27
Nebraska	6156.83	0.01653	4	69.35
Nevada	7076.78	0.01152	7	65.79
New Hampshire	6064.36	0.01510	1	27.85
New Jersey	7394.27	0.01539	2	41.63
New Mexico	5525.85	0.01889	7	71.12
New York	7012.31	0.01497	2	87.15
North Carolina	5556.59	0.01329	5	71.01
North Dakota	5965.46	0.01900	4	78.81
Ohio	6389.12	0.01576	3	43.94
Oklahoma	5832.41	0.01528	6	48.21
Oregon	6443.07	0.01540	8	82.04
Pennsylvania	6455.42	0.01630	2	55.61
Rhode Island	6296.69	0.01580	1	51.01
South Carolina	5280.17	0.01584	5	77.80
South Dakota	5131.96	0.02009	4	52.35
Tennessee	5473.48	0.01429	6	41.54
Texas	6431.54	0.01305	6	71.39
Utah	5626.60	0.01679	7	74.85
Vermont	5531.92	0.01723	1	42.77
Virginia	6499.59	0.01469	5	56.48
Washington	7077.16	0.01683	8	89.21
West Virginia	5527.64	0.01614	5	57.57
Wisconsin	6176.08	0.01706	3	87.76
Wyoming	7072.42	0.01699	7	105.29

(*Source: Chronicle of Higher Education*, 25 October 1977, 8 March 1970.)

[a]Per capita. [b]Public and nonpublic.

Observations will typically have two characteristics or dimensions. Each observation will be *on some entity*—an individual, a census tract, a classroom—and it will be *of some characteristic or attribute* of that entity—an airline pilot's reaction time, the average number of years of education of heads of households in a tract, the number of pupils in the classroom who exceeded the national median score on an arithmetic achievement test. Our observations are never of the entity itself, but of some attribute of the entity. In multivariate analysis, observations are of several attributes of the same entity.

As a convention, this entity–attribute structure will be used to arrange the observations into a rectangular array called the data matrix. Table 1.1 shows data related to the annual spending in support of higher education for each state in the United States. In this book we shall frequently consider this data matrix, which will be referred to as the Higher Education Study. The numerical values in this table satisfy the two rules for a data matrix. Each row corresponds to a particular entity. All of the entries in the sixth row, for example, relate to the state of Colorado. Each column corresponds to a particular attribute. Entries in the first column of figures are all per capita incomes. The intersection of the sixth row and the first column, then, contains the per capita income of the state of Colorado—$6650.64.

In a data matrix, *entities are rows* and *attributes are columns*. The data matrix is thus an array of numbers that have been systematically arranged. Almost universally, the observations contributing to a multivariate analysis will be based on a larger number of entities than attributes. As a result, the data matrix will be a *vertical matrix*; that is, it will have more rows than columns. Later we shall describe other matrices, *derived* from data matrices, that are horizontal or square.

The data matrix of the Higher Education Study in Table 1.1 contains observations on four attributes (per capita personal income, per capita high school completions, geographic region, and dollars appropriated for higher education) on each of 50 entities—the 50 states of the United States. In describing any matrix, it will be important to state its dimensions. We shall follow the convention of mentioning first the number of its rows and then the number of its columns. The data matrix of our Higher Education Study, then, is a "fifty-by-four" matrix, written 50×4. It has 50 rows and 4 columns.

1.2 Predictors and Criteria

The social and behavioral sciences, no less than the natural sciences, are concerned with the relations between attributes. Some attributes are hypothesized to respond to, or to be affected by other attributes. Different disciplines have different traditions for naming these two classes of attributes: the dependent variable responds to the independent variable; an explanatory variable affects a response variable; predictors change criteria. We shall use the terms *predictor* and *criterion*.

However, there is nothing to an attribute itself that makes it either a predictor or a criterion. Such labeling is a function of the specific research question. Annual personal income, for example, is a criterion if the researcher is interested in how income varies as a function of education and a predictor if the research asks whether the poor regard present taxation practices as fairer than do the rich.

In the data matrix the predictor–criterion distinction is frequently emphasized by arranging the columns of the matrix so that the predictor attributes are *partitioned* from the criterion observations. This means that the predictors are collected in the left (or right) columns of the matrix and the criteria in the remaining partition. In analyzing the Higher Education Study we shall be interested in the extent to which the per capita amount of money appropriated by a state government for higher education is a function of the state's per capita income, per capita high school completions, and geographic location. When writing the data matrix, the criterion was segregated from the other attributes by reserving the final column for the amount of dollars appropriated per capita for higher education. As the data on a criterion are often collected later in time than predictor data, it is more usual to record the criterion observations *after* (to the right of) the predictors in the data matrix.

Data matrices may also be partitioned by rows. This is normally a result of the data collection or sampling design. Although something of an oversimplification, consider at the moment only three sampling plans: exhaustive, random predictor sampling, and fixed predictor sampling.

Inspection of the Higher Education Study data matrix indicates that the data resulted from *exhaustive sampling* because all 50 states were included. We have no good convention for how the rows of such a data matrix should be arranged. The arrangement of the states in Table 1.1 in alphabetical order makes it easy for one to find out how much Iowa spent on higher education, so long as the names of the states are *appended* to the matrix. The data matrix, however, is just a 50×4 array of numbers, and to locate Iowa in the data matrix would require that we work out that Iowa is, alphabetically, 15th in the list of states. Given the intended analysis of the Higher Education data matrix, an alternative arrangement of rows might order the states from largest to smallest per capita appropriation. The first row of the reordered matrix, listing the observations for Alaska, would then read

$$11{,}800.00 \quad 0.01869 \quad 8 \quad 143.03$$

and the last row, listing the observations for New Hampshire, would be

$$6064.36 \quad 0.01510 \quad 1 \quad 27.85$$

Random predictor sampling will refer to designs in which entities were chosen independently and at random from the population of relevant

Table 1.2 Precollege Test Scores and Earned Grade Point Averages for 50 University Students

Test scores			Grade point averages		
vocabulary	quantitative skills	spatial ability	English	mathematics	physics
61	39	38	2.59	1.48	2.80
46	38	43	1.53	1.88	2.50
44	62	48	2.76	2.16	2.35
57	49	38	2.67	4.00	3.53
60	66	57	3.00	3.61	3.75
49	54	43	2.50	1.98	2.16
57	73	55	3.31	3.33	3.60
63	64	65	3.04	3.82	4.00
64	54	55	2.64	2.88	1.90
42	47	45	2.48	3.02	2.43
51	60	38	2.35	3.37	3.75
55	57	46	3.16	2.45	2.85
42	38	57	1.93	2.16	2.75
63	68	56	2.33	2.70	2.02
54	53	56	3.19	2.55	2.40
45	47	51	2.64	4.00	2.79
62	54	60	2.85	3.11	2.56
36	47	39	1.68	2.32	1.49
62	68	62	3.51	4.00	4.00
46	59	38	2.12	3.07	3.51
51	59	53	3.13	3.19	3.46
58	61	51	2.80	2.12	2.21
57	51	69	2.40	2.32	2.51
59	62	57	4.00	4.00	4.00
55	62	47	3.65	3.66	4.00
62	46	60	3.30	3.18	3.95
77	69	62	2.81	3.68	3.85
41	59	50	2.31	2.08	2.00
63	57	58	2.70	2.14	2.25
53	55	56	2.69	3.46	2.60
28	23	50	1.26	2.19	2.12
56	50	60	2.82	2.58	2.70
54	57	44	2.59	1.77	2.06
40	58	45	2.13	1.58	2.52
52	63	46	3.70	2.01	1.36
53	54	75	2.93	2.90	2.37
52	63	64	2.32	2.20	2.04
70	60	57	2.58	2.67	3.37
66	52	52	2.72	3.79	2.04
66	63	79	3.68	3.06	3.23
51	45	56	2.44	1.33	1.86
61	57	59	4.00	3.48	2.54
59	49	43	3.55	3.19	3.24
42	37	34	1.74	1.41	2.10
61	51	53	3.77	3.45	2.30
51	59	57	2.10	2.23	2.61
64	62	46	3.71	3.23	2.43
42	50	54	2.92	2.92	3.32
64	54	55	3.45	2.95	1.98
47	59	61	2.93	3.83	4.00

entities and predictor and criterion attributes were then observed for the sampled entities. Table 1.2 presents data for such a sampling design. This research will be referred to as the College Prediction Study. The population of relevant entities for this College Prediction Study consisted of all students entering the University of Washington directly from high school in the fall of 1976 who subsequently completed at least two courses each in English, mathematics, and physics during their first two university years. Students were then sampled randomly and without replacement from that population until 50 were chosen.

Six attributes (three predictors and three criteria) were observed for each of the selected students. The predictors were scores from three tests, of

Table 1.3 Level of Aspiration, Career Centrality, and Career Choice Certainty as a Function of Career Education and Socioeconomic Status (SES)

Career education	SES	Level of aspiration	Career centrality	Choice certainty
1	3	9	1	6
1	3	3	1	2
1	3	6	1	6
1	3	7	1	7
1	3	9	1	6
1	3	5	1	7
1	3	6	1	5
1	3	6	1	4
1	3	4	0	5
1	3	4	1	4
1	2	8	1	6
1	2	9	1	6
1	2	7	1	5
1	2	2	0	4
1	2	5	1	5
1	2	6	0	5
1	2	6	1	7
1	2	6	1	7
1	2	2	0	2
1	2	5	1	5
1	1	4	0	7
1	1	6	0	4
1	1	6	1	4
1	1	6	0	1
1	1	6	0	4
1	1	7	0	6
1	1	7	0	4
1	1	5	1	2
1	1	6	0	4
1	1	7	1	7

vocabulary, quantitative skills, and spatial ability, taken by the students while in high school. The criteria were grade point averages or mean grades received in each of the three areas, English, mathematics, or physics. For the entire population of high school juniors writing the tests, the mean score of each test is set at 50 and the standard deviation at 10. University grades can range from 0.0 (low) to 4.0 (high) in steps of 0.1.

The data matrix for the College Prediction Study has been written with predictors to the left and criteria to the right. How have the rows been arranged? In random predictor sampling the rule will be to write the rows of the matrix to correspond to the order in which the entities were sampled. That is, the first row contains observations on the first randomly chosen

Table 1.3 (*continued*)

Career education	SES	Level of aspiration	Career centrality	Choice certainty
0	3	9	1	3
0	3	6	1	3
0	3	7	1	2
0	3	7	1	1
0	3	4	0	1
0	3	6	1	3
0	3	9	1	1
0	3	6	1	1
0	3	6	1	1
0	3	4	0	1
0	2	5	1	1
0	2	3	0	2
0	2	2	0	1
0	2	4	0	1
0	2	2	0	1
0	2	2	0	1
0	2	1	0	1
0	2	4	1	2
0	2	3	0	5
0	2	5	1	1
0	1	1	0	1
0	1	3	0	1
0	1	1	0	1
0	1	1	0	1
0	1	3	0	2
0	1	2	1	4
0	1	3	1	3
0	1	1	0	1
0	1	2	0	1
0	1	1	0	3

student, and the 50th row contains data on the last student chosen. An advantage of this ordering is that it permits the researcher to later ask whether there was any bias in the order of selection of the sample.

In *fixed predictor sampling*, the researcher has controlled or fixed the distribution of observations on at least one of the predictors in the study. For example, if subjects are to be selected to ensure equal numbers of men and women respondents to a survey, the distribution of gender will be fixed in the sample. Table 1.3 contains hypothetical observations for what will be referred to as the Women's Career Education Study. This study was interested in certain possible effects of participation in a high school career education course for women. Additionally, the investigator wanted to know whether such effects were differential and depended upon the socioeconomic status (SES) of the woman's family.

Because sampling at random from a *single* population (all women students in the high school) could produce a sample in which the predictors—whether the course was taken or not and whether the student's family was of high, medium, or low SES—were not adequately represented, the experimenter decided to select *subsamples* randomly from different populations. In the example, ten women were chosen randomly and without replacement from each of the following six populations of students:

1. High SES, completed career education for women
2. Medium SES, completed career education for women
3. Low SES, completed career education for women
4. High SES, did not take career education for women
5. Medium SES, did not take career education for women
6. Low SES, did not take career education for women

Such a sampling scheme *fixed* the distribution of the scores on the predictors [1 and 0 for women who, respectively, took (and completed) and did not take career education; 1, 2, and 3 for students from, respectively, low-, medium-, and high-SES families]. On those randomly sampled (from *different* populations), these three criteria were observed: level of aspiration (status of chosen career scored from 1—low—to 10—high), career centrality (1 if the woman reported her career was expected to be the center of her adult life, 0 otherwise), and career choice certainty (scaled 1–7 from very uncertain to very certain). The experimenter knew in advance how many 1s and 0s there would be in column 1 of the data matrix, but not how many 1s and 0s there would be in column 5. Thus the researcher used fixed predictor sampling.

When the distribution of one or more predictors is fixed in the sampling process, it is customary to write (or partition) the rows of the matrix to reflect that. This is accomplished by using successive rows for entities that share the same values of the fixed predictors. Thus for the Women's Career Education Study, the 60 rows of the data matrix have been arranged so that the first 10 rows correspond to the sample from the first population described above, the second 10 rows were from the second population, and so on to the last set of 10 rows. The data matrix for the Women's Career Education Study has thus been partitioned vertically between predictor and criteria attributes and horizontally between the six populations of students according to the values of the two fixed predictors.

1.3 Measured and Categorical Observations

Much has been written about the status of the scales of measurement that behavioral scientists employ. There may be a gradation of levels of measurement and, associated with that gradation, a hierarchy of statistical and mathematical operations that are not permissible, given the level of measurement. Because this is not a book on scaling or measurement, we shall neither recreate nor review all of the arguments that have been offered. Rather, the reader is asked to consider only the following question about the observations of any given attribute: "Do I believe there is an *interval* in these observations?"

What does it mean for there to be an interval? We believe it means, "I am comfortable with the result of *averaging* several observations of the attribute." Some attributes are observed in *natural intervals*. The four partners in a law firm earned \$42,000, \$36,000, \$51,000, and \$39,000 last year. An air traffic controller responded to five successive radar blips in 720, 860, 750, 910, and 840 msec. Three overweight professors lost 8, 6, and 10 kg by running at noon rather than taking lunch at the Faculty Club during February and March. On the average, our lawyers earned \$42,000/yr, the air traffic controller responded to an aircraft entering a monitoring sector in 816 msec, and, huffing and puffing, the academicians lost an average of 8 kg over 2 months circling the track rather than the buffet. Because we know, or at least we think we do, what a dollar, a second, and a kilogram are, we are comfortable averaging them. Through custom and usage, they define natural intervals.

Although natural intervals are encountered in the behavioral sciences, there are other measurements researchers employ that they want to assume are interval. An *assumed interval* does not have the force of custom behind it and may never have it. Consider the data in the College Prediction Study. Neither the spatial ability test score nor the course grade in English 306 is observed, apparently, in natural intervals. Nevertheless, the question "How does the grade point average in English change as scores on the spatial ability test increase?" assumes intervals in the observation of both attributes.

As behavioral researchers, we are not troubled by assuming intervals in this example. Measurement and scaling experts, however, would be. This difference in viewpoint is understandable if we remember to separate two questions that might be asked of a measure. In one, we are concerned only with the relation between the numbers that are test scores and the numbers that are grade point averages. Differences exist in both sets of numbers; there are differences of 2 test score points between scores of 42 and 44 and between scores of 58 and 60. In the analysis of the relation of the scores to grades, these two differences will be treated as equal. The same can be said for two pairs of grade point averages each separated by, say, 0.25 units. A second question, one that was not asked in this example but is of concern in some other contexts, is whether the same difference in the underlying nonobservable trait of spatial ability results in the 2-point differences between 42 and 44 and between 58 and 60 on the test of spatial ability. That is a *scaling* question. In this book we shall be concerned primarily with the relations between observables. More often than not, those observables will make it reasonable to assume an interval of measurement.

However, such a "weak" position on measurement should not be interpreted to mean that all alternative measures or scalings of an attribute are equally good. The researcher is encouraged to use the best available measure in each instance, taking into account all the evidence that has accumulated about the alternatives. Use of a poor scale can only make the researcher's task harder.

There are instances, however, in which the numbers in our data matrix cannot be regarded as having an interval basis. Usually these instances will be quite obvious. The term *measured* will be reserved for observations with an assumed or natural interval. Observations that are not considered interval will be termed *categorical*. An example has already been presented in which the differences between the numbers assigned or the averages of those numbers cannot be given any meaning. In the Higher Education Study, the geographic region to which each state belonged was coded with an integer from 1 to 8. The average of a 6 (South Central states) and an 8 (Pacific states) certainly cannot be interpreted as a 7 (Mountain states). Nor would we wish to claim that the difference between Delaware (5) and Illinois (3) was "geographically" equal to the difference between Alabama (6) and Alaska (8). The assignment of a number does not serve to *measure* a state's geographic region; it merely *categorizes* it. States assigned the same number belong to the same category; those assigned different numbers belong to different geographic categories.

Three types of categories will be distinguished: natural, ordered, and assigned. Consonant with natural intervals, there are *natural categories*. They are also defined by custom and usage. Marital status, historically, has been categorized as married, single, or divorced. Researchers have no way of deciding that the difference between married and single is less than, greater than, or equal to the difference between single and divorced. We just agree to attach one of these three different labels to each relevant entity when marital status is observed.

The geographic regions of the United States are also a set of natural categories. Note that 4, 6, or 10 categories instead of 8 could have been used. What is important is that states are assigned to a category only if they are contiguous and that the categorizing rule provides for assigning *all* possible entities uniquely to a category.

Ordered categories come very close to measurement. In the Women's Career Education Study the three categories of socioeconomic status can clearly be ordered. Why do they provide ordered categories of SES and not a measurement of SES? It would be simple enough to say that it is because we do not believe we know how much bigger or smaller the difference between low and medium is than the difference between medium and high. However, such an argument based upon scaling uncertainty, could also be made for many observations that we wish to assume have an interval basis. The difference is that when the number of ordered categories is small enough (or, equivalently, the measurement is coarse enough), we can afford (both conceptually and computationally) not to commit ourselves to an interval. We can wait and see how large the difference between low and medium is when compared with the difference between medium and high. However, such an argument suggests two questions. First, how many categories can be ordered before it becomes more reasonable to treat the observations as measurements? Somewhat arbitrarily, the convention of five categories as an upper limit will be adopted here. For more than five ordered categories the researcher has almost certainly given some attention to the measurement properties of the scaling. Second, do ordered categories give researchers the best of both worlds? Not necessarily, since researchers may have to accept very coarse measurements in instances where much finer measurements are possible.

Finally, there are categories that are created by the researcher for a particular study. These are *assigned categories*. Table 1.4 introduces a data matrix for a hypothetical study, the Pilot Response Time Study. In this study, 48 active pilots were randomly chosen from the roster of a major airline. The 48 pilots were then randomly divided into four groups of 12 and each pilot's time to respond, by pushing a switch, to each of a series of 50 stimuli was measured. Pilots assigned to the four groups received different treatments:

Group 1: visual stimuli to left visual field
Group 2: visual stimuli to right visual field
Group 3: aural stimuli to left ear
Group 4: aural stimuli to right ear

The predictor attribute, termed *stimulus condition group* and coded 1, 2, 3, or 4, is made up of assigned categories. The experimenter chose, for this study, to include exactly these four stimulus conditions. They do not exhaust the set of possible stimulus conditions. Bichannel, bimodal, or

Table 1.4 Mean Response Times, Trials 1–25 and 26–50, Commercial Pilots Receiving Stimuli Directed to the Left (Group 1) or Right (Group 2) Visual Field or to the Left (Group 3) or Right (Group 4) Ear

Group	Mean response times (msec)		Intelligence
	trials 1–25	trials 26–50	
1	828	632	100
1	843	669	110
1	766	719	120
1	1146	857	115
1	781	690	115
1	593	553	115
1	463	530	130
1	832	738	120
1	1032	817	110
1	898	859	115
1	850	710	110
1	833	755	120
2	642	531	115
2	696	624	120
2	709	589	115
2	506	499	120
2	589	626	120
2	914	788	110
2	977	723	110
2	670	583	115
2	890	708	115
2	801	674	115
2	682	631	105
2	635	519	125
3	925	759	110
3	708	625	110
3	660	672	130
3	724	747	120
3	727	673	120
3	871	853	115
3	1012	838	105
3	721	673	115
3	750	665	125
3	712	658	115
3	989	834	115
3	1125	872	105
4	622	592	130
4	581	529	120
4	809	619	110
4	901	750	115
4	788	688	115
4	694	618	120
4	900	795	115
4	872	794	100
4	622	622	120
4	782	710	115
4	795	616	115
4	808	663	105

Table 1.5 Higher Education Data by Geographic Region

State	Income	High school completions	Geographic region	Appropriations
Alabama	5227.84	0.01568	00000100	76.26
Alaska	11800.00	0.01869	00000001	143.03
Arizona	6130.98	0.01366	00000010	90.90
Arkansas	5043.65	0.01406	00000100	50.15
California	7356.39	0.01467	00000001	98.15
Colorado	6650.64	0.01595	00000010	75.72
Connecticut	7415.80	0.01538	10000000	44.24
Delaware	7155.32	0.01658	00001000	73.30
Florida	6267.00	0.01150	00001000	50.80
Georgia	5653.42	0.01358	00001000	50.02
Hawaii	7337.28	0.01669	00000001	98.58
Idaho	5882.35	0.01652	00000010	78.47
Illinois	7411.73	0.01463	00100000	64.85
Indiana	6191.37	0.01529	00100000	55.42
Iowa	6269.70	0.01797	00010000	70.79
Kansas	6563.88	0.01780	00010000	80.70
Kentucky	5481.08	0.01451	00000100	55.35
Louisiana	5526.04	0.01645	00000100	52.87
Maine	5444.13	0.01643	10000000	42.41
Maryland	6961.41	0.01534	00001000	57.84
Massachusetts	6603.45	0.01503	10000000	35.93
Michigan	6759.73	0.01572	00100000	65.05
Minnesota	6254.79	0.01838	00010000	64.03
Mississippi	4604.13	0.01420	00000100	67.38
Missouri	5966.08	0.01423	00010000	48.25
Montana	5850.34	0.01782	00000010	63.27
Nebraska	6156.83	0.01653	00010000	69.35
Nevada	7076.78	0.01152	00000010	65.79
New Hampshire	6064.36	0.01510	10000000	27.85
New Jersey	7394.27	0.01539	01000000	41.63
New Mexico	5525.85	0.01889	00000010	71.12
New York	7012.31	0.01497	01000000	87.15
North Carolina	5556.59	0.01329	00001000	71.01
North Dakota	5965.46	0.01900	00010000	78.81
Ohio	6389.12	0.01576	00100000	43.94
Oklahoma	5832.41	0.01528	00000100	48.21
Oregon	6443.07	0.01540	00000001	82.04
Pennsylvania	6455.42	0.01630	01000000	55.61
Rhode Island	6296.69	0.01580	10000000	51.01
South Carolina	5280.17	0.01584	00001000	77.80
South Dakota	5131.96	0.02009	00010000	52.35
Tennessee	5473.48	0.01429	00000100	41.54
Texas	6431.54	0.01305	00000100	71.39
Utah	5626.60	0.01679	00000010	74.85
Vermont	5531.92	0.01723	10000000	42.77
Virginia	6499.59	0.01469	00001000	56.48
Washington	7077.16	0.01683	00000001	89.21
West Virginia	5527.64	0.01614	00001000	57.57
Wisconsin	6176.08	0.01706	00100000	87.76
Wyoming	7072.42	0.01699	00000010	105.29

tactile stimuli would provide different conditions, as would visua stimuli of different intensities or waveforms. The stimulus con treatment categories, were thus an arbitrary choice by the researc purposes of this study.

Observations of an attribute, then, provide either a measurem attribute or a categorization of it, depending on whether the ma *differences* between the numbers assigned different entities can b ed. Having made this distinction, we now introduce a convention categorical observations that will permit the researcher to treat data as though they consisted solely of observations on measure

This coding method will be illustrated with the attribute of gen one high school student is labeled male and another female, rese more likely to say they are categorizing the students by gender they are measuring their gender. This would be true even if th record not "male" and "female" but the numbers 0 (if male) female). However, in another sense, the researcher in arran numbers would have measured gender. Whenever an attribute is that only two categories may be observed, assigning distinct n each of the two categories results in the measurement of that attr

Although this seems in conflict with what we said earlier, rec interval, a difference between two numeric observations, need o consistent meaning before the title "measurement" is accorde numbers. Because there is only one way a difference can occu observations based on a two-category attribute, interpretatio difference is certain. In coding gender, for example, a difference only mean that one observation is male and the other female. averages can be interpreted. Given the above coding scheme, if th of ten observations was 0.6, the ten observations would have contained exactly four males and six females.

These arguments would be the same whatever two distinctive were used to code male and female. If, for example, we had used rather than 0 and 1, a difference of 45 would always be found b male and female observation (and could only be found in that com and the average of ten observations would be 29 only if it consiste males and six females—$(4 \times 2) + (6 \times 47) = 290$. The 0–1 co some advantages. An attribute so coded can be interpreted as pres absent (0) in an entity, and averages can be interpreted directl proportion of observations possessing the attribute. Consequentl the 0–1 coding system for all two-category attributes.

Based upon this reasoning, if all categorical observations coul placed with 0–1 coded attributes, a data matrix would then cont observations on measured attributes. A simple coding that allo searcher to make exactly such replacements is to replace any containing a $k \geqslant 2$ category attribute with k new columns, each a p

0s and 1s. The first of the new columns would have 1s only where the k-category column entry was the smallest of its k values, the second 0–1 column would have 1s only where the next to the smallest of the k values occurred, and the third through the kth new columns would have 1s similarly positioned.

Table 1.5 is a restatement of the data matrix for the Higher Education Study with the eight-category regional attribute replaced by 0–1 coded regional attributes. A 1 in column 3 of the data matrix indicates the state is in the New England region; a 0 in that column indicates the state is not in New England. Column 3 is thus a measure of the attribute "in New England."

Similarly, Table 1.6 is a new data matrix for the Women's Career Education Study with the three-category socioeconomic status attribute replaced by three measured attributes: low SES, medium SES, and high SES. Finally, Table 1.7 is a revision of the data matrix for the Pilot Response Time Study. Here the 0–1 attributes denote the experimental conditions under which response times were collected.

1.4 Selecting vs Sampling Observations

In Section 1.2 we made a distinction between fixed and random predictor sampling. This distinction is partly one between two scientific traditions, sometimes described oversimply as adopting laboratory (experimental) or survey (observational) designs. In the classic experimental design, the researcher controls observations on the predictor attributes by actively manipulating the treatments. The Pilot Response Time Study provides a good example of such a design. The four treatment conditions were carefully constructed by the experimenter, and by randomly assigning equal numbers of experimental participants to the four conditions, the statistical analysis of the outcome focuses on contrasts among the testing conditions. The experimenter has selected the stimulus conditions to be observed and fixed the frequency with which each will occur. In so distributing predictor observations, the first four columns to the data matrix in Table 1.7 are fixed.

Fixing the distribution of one or more predictor attributes can and does occur outside the laboratory as well. The Women's Career Education Study described earlier provided an example. Table 1.8 reports data for a second nonexperimental but fixed predictor study, referred to as the Census Tract Study. Based on the 1970 census, this study investigated income and fertility data for a set of 32 census tracts in the city of Seattle. The tracts were systematically chosen to provide eight tracts meeting each of the following descriptions:

1. High education, high mobility
2. High education, low mobility

Table 1.6 Women's Career Education Data Matrix with SES Attributes

Career	Low SES	Medium SES	High SES	Level of aspiration	Career centrality	Career certainty
1	0	0	1	9	1	6
1	0	0	1	3	1	2
1	0	0	1	6	1	6
1	0	0	1	7	1	7
1	0	0	1	9	1	6
1	0	0	1	5	1	7
1	0	0	1	6	1	5
1	0	0	1	6	1	4
1	0	0	1	4	0	5
1	0	0	1	4	1	4
1	0	1	0	5	1	5
1	0	1	0	8	1	6
1	0	1	0	9	1	6
1	0	1	0	7	1	5
1	0	1	0	2	0	4
1	0	1	0	5	1	5
1	0	1	0	6	0	5
1	0	1	0	6	1	7
1	0	1	0	6	1	7
1	0	1	0	2	0	2
1	1	0	0	4	0	7
1	1	0	0	6	0	4
1	1	0	0	6	1	4
1	1	0	0	6	0	1
1	1	0	0	6	0	4
1	1	0	0	7	0	6
1	1	0	0	7	0	4
1	1	0	0	5	1	2
1	1	0	0	6	0	4
1	1	0	0	7	1	7

3. Low education, high mobility
4. Low education, low mobility

The values of the predictor attributes were fixed in advance of collecting any data. Eight tracts were selected at random from each of the four populations described and income and fertility data then recorded for each selected tract. Observations on the two criteria were sampled (a different random selection of 32 tracts would likely yield a different set of fertility and income figures) while the predictor observations were fixed. The researcher decided in advance to include eight tracts with each combination of education and mobility scores.

Another example of a *random* predictor design that we explore in this book is the High School Aspiration Study. In this study, high school

Table 1.6 (*continued*)

Career	Low SES	Medium SES	High SES	Level of aspiration	Career centrality	Career certainty
0	0	0	1	9	1	3
0	0	0	1	6	1	3
0	0	0	1	7	1	2
0	0	0	1	7	1	1
0	0	0	1	4	0	1
0	0	0	1	6	1	3
0	0	0	1	9	1	1
0	0	0	1	6	1	1
0	0	0	1	6	1	1
0	0	0	1	4	0	1
0	0	1	0	5	1	1
0	0	1	0	3	0	2
0	0	1	0	2	0	1
0	0	1	0	4	0	1
0	0	1	0	2	0	1
0	0	1	0	2	0	1
0	0	1	0	1	0	1
0	0	1	0	4	1	2
0	0	1	0	3	0	5
0	0	1	0	5	1	1
0	1	0	0	1	0	1
0	1	0	0	3	0	1
0	1	0	0	1	0	1
0	1	0	0	1	0	1
0	1	0	0	3	0	2
0	1	0	0	2	1	4
0	1	0	0	3	1	3
0	1	0	0	1	0	1
0	1	0	0	2	0	1
0	1	0	0	1	0	3

students were sampled at random from Washington state schools and the following set of attributes assessed: family background (father's education, mother's education, family size, sibling rank), ability (verbal ability, quantitative ability), academic performance (high school grade point average), and educational aspiration (post-high school educational goal). The division of this set of attributes between predictors and criteria will be a function of the specific question asked of the sample. Table 1.9 reproduces a small part of the data matrix for this study. Given the design of this study, we cannot know in advance of data gathering the distribution of scores in any column.

The distinction between fixed and random predictor samplings is sometimes relevant to phrasing behavioral science research questions. Frequently, fixed predictor designs are associated with research questions of the form, "*Is there a difference*... in response times to visually and to aurally pre-

Table 1.7 Pilot Response Time Data Matrix with 0–1 Attributes

Type of stimulus (1 = visual; 0 = auditory)	Source of stimulus (1 = left; 0 = right)	Mean response times (msec)	
		trials 1–25	trials 26–50
1	1	828	632
1	1	843	669
1	1	766	719
1	1	1146	857
1	1	781	690
1	1	593	553
1	1	463	530
1	1	832	736
1	1	1032	817
1	1	898	859
1	1	850	710
1	1	833	755
1	0	642	531
1	0	696	624
1	0	709	589
1	0	506	499
1	0	589	626
1	0	914	788
1	0	977	723
1	0	670	583
1	0	890	708
1	0	801	674
1	0	682	631
1	0	635	519
0	1	925	759
0	1	708	625
0	1	660	672
0	1	724	747
0	1	727	673
0	1	871	853
0	1	1012	838
0	1	721	673
0	1	750	665
0	1	712	658
0	1	989	834
0	1	1125	872
0	0	622	592
0	0	581	529
0	0	809	619
0	0	901	750
0	0	788	688
0	0	694	618
0	0	900	795
0	0	872	794
0	0	622	622
0	0	782	710
0	0	795	616
0	0	808	663

Table 1.8 Census Tract Data Matrix

Education[a]	Mobility[b]	Income[c]	Fertility[d]
1	1	103	27
1	1	94	14
1	1	95	17
1	1	110	12
1	1	112	15
1	1	113	5
1	1	106	18
1	1	105	11
1	0	128	39
1	0	117	30
1	0	129	41
1	0	117	34
1	0	129	28
1	0	128	33
1	0	114	29
1	0	107	24
0	1	83	15
0	1	91	13
0	1	75	2
0	1	71	1
0	1	74	1
0	1	85	2
0	1	72	4
0	1	89	31
0	0	89	26
0	0	88	22
0	0	95	22
0	0	70	29
0	0	91	41
0	0	93	45
0	0	92	27
0	0	88	22

[a]Percent of professional and technical households in 1977 (0: low, < 8%; 1: high, > 10%).
[b]Mobility rate k for all households (moves/yr), 1977 (0: low, < 0.45; 1: high, > 0.64).
[c]Income index 1977.
[d]Percent of households with children, 1977.

Table 1.9 Sample Rows of the High School Aspiration Data Matrix[a]

FE	ME	FS	SR	VA	QA	HS	EG
2	2	3	1	51	50	2.84	1
1	2	2	0	48	51	2.92	0
3	3	1	1	55	60	3.50	3
⋮	⋮	⋮	⋮	⋮	⋮	⋮	⋮

[a]Attributes: FE, father's education; ME, mother's education; FS, family size; SR, sibling ranks; VA, verbal ability test score; QA, quantitative ability test score; HS, high school grade point average; EG, post-high-school educational goal.

sented stimuli?" Random predictor designs, on the other hand, are often stimulated by such questions as, "*How strong is the relation* between tested academic ability and grades earned in university courses?" We shall not make too much of the distinction between the two questions. Indeed, the failure to measure the strength of an experimental effect or to test for significance an obtained correlation between survey items is to be regretted.

1.5 Additional Data Sets Discussed in This Book

A wide range of types of data is used in multivariate analyses, and many types of information are sought from the analyses. We have already described some of the data sets we work with in this book. We now describe additional data sets spanning the range of applications covered in Parts II–IV.

Throughout the book, and especially in Chapter 12 on linear discriminant function analysis, we refer to data from the Vocational Interest Study. In this study, university graduates in each of 22 majors or disciplines responded to the Vocational Interest Inventory (VII) 5–7 years prior to graduation. The VII provides scores on eight vocational interest areas: service, business, organizational, technical, outdoors, science, general cultural, and arts and entertainment. The criteria in this research were categories such as *did* or *did not graduate in political science*.

In Chapter 12, we illustrate multivariate linear regression by analyzing data from the Preemployment Study. Here, data were obtained on a set of preemployment measures (work history, education, age, aptitudes, interests, and values) and a set of postemployment measures (time to attain job skills, competence, job productivity, supervisory ratings, and self-ratings of job satisfaction). Researchers wanted to know how many different linkages there were between the two sets of data, how strong each linkage was, and whether the stronger relations supported any psychological or sociological interpretation.

In Chapter 13 we illustrate log linear models of analysis based on the First Job Satisfaction Study. In this study data were collected from a group of psychology majors on their university performance and their selection of a first postcollege job. The graduates' satisfaction with their first job is related to a set of categorical attributes such as their choice of an arts or science degree program, the selection of a job related or unrelated to psychology, and the gender (male or female) of the graduate. Noteworthy in this example and its analysis is that all attributes are categorical. The relations among the different frequencies in each of the cross-classified "cells" are what the researchers seek to describe.

Data from a classic study that we shall call the Primary Mental Abilities Study are used to illustrate research concerned with the interdependence of attributes rather than the dependence of one partition of the attributes (criteria) on another (predictors). In this study, data from 145 seventh- and eighth-grade students were obtained on 12 psychological tests. Questions researchers have asked about this data set include the following: Can the correlations or covariances among this battery of tests be efficiently described by studying the performance of and contributors to composites formed from the battery? How many composites of attributes are needed to provide a good description of the pattern of relations observed among the tests?

Among these examples, the types of both data and multivariate analysis differ. An initial reaction might be that the type of data determines the type of analysis. This is seldom true in multivariate analysis.

1.6 Partitioned Matrices and Computer Analyses

Section 1.2 may have left the reader with the impression that only an aesthetic or traditional interest exists in how the columns or rows of a data matrix are arranged. While the conventions do make it easier to consider a data matrix visually, there are additional, more compelling, reasons for such preferences.

Many if not most real-world multivariate analyses involve data matrices that are too large to consider visually all at once. Rather than invalidating our conventions for expressing a data matrix, this makes such conventions more important. Researchers are almost certainly dependent upon the computer for computational solutions. Indeed, most researchers work with standard, packaged computer programs for multivariate analyses. Your work as a researcher will be less time consuming and less error prone if you are able to arrange your data to take advantage of the power and flexibility of computer program packages. How you order your data will be important.

For example, one of the more frequently used statistical computing packages, SPSS (statistical package for the social sciences), has a feature that permits analyses to be conducted on some or all of the *subgroups* in a data file. Subgroups are nothing more than sets of sequential rows of a data matrix. Consequently it will be advantageous to have observations for entities belonging to the same group in adjoining records or punched cards.

Additionally, describing in terms of a computer program the sets of attributes to be employed in an analysis is simplified if each set occupies consecutive columns in the data matrix. For example, in utilizing the SPSS correlation subroutine to obtain Pearson correlations between the three predictors and the three criteria in the College Prediction Study, we could take advantage of the order in which we arranged the six attributes to direct the computer to produce the correlations with the command

```
PEARSON CORR    VOCAB TO SPACE WITH ENGL TO PHYS
```

Table 1.10 College Prediction Study Intercorrelation Matrix

	VOCAB	QUAN	SPA	English GPA	Mathematics GPA	Physics GPA
Vocabulary score (VOCAB)	1.0000	0.5094	0.4146	0.5941	0.4079	0.2707
Quantitative score (QUAN)	0.5094	1.0000	0.3184	0.5176	0.3845	0.3184
Spatial score (SPA)	0.4146	0.3184	1.0000	0.3309	0.2773	0.1895
English GPA	0.5941	0.5176	0.3309	1.0000	0.5446	0.3309
Mathematics GPA	0.4079	0.3845	0.2733	0.5446	1.0000	0.6677
Physics GPA	0.2707	0.3184	0.1895	0.3309	0.6677	1.0000

The result might look like this:

```
          ENGL     MATH     PHYS
VOCAB     .5941    .4079    .2707
QUANT     .5176    .3845    .3184
SPACE     .3309    .2772    .1895
```

As we shall see in developing a matrix approach to multivariate analysis, partitioning the data matrix leads to a corresponding partitioning of matrices derived from the data matrix. Partitioning these derived matrices not only makes computation easier but also makes reading or interpretation of those matrices more illuminating.

Consider the result of asking that all the attributes be intercorrelated. The result of an SPSS command like

```
PEARSON CORR    VOCAB TO PHYS
```

would be a 6×6 matrix as given in Table 1.10. The important thing to note is that the order of the attributes in the data matrix was preserved both in the rows and in the columns of the intercorrelation matrix. The rows and columns of the intercorrelation matrix may now be partitioned between predictors and criteria in the same fashion as the data matrix was partitioned. This partitioning results in four submatrices. The upper left quadrant contains correlations among the predictors. Correlations among the three criteria are in the lower right quadrant, and in the upper right (as well as in the lower left) are correlations between predictors and criteria. If we partition the data matrix, that partitioning will frequently carry through to derived matrices, such as the correlation matrix.

1.7 Summary

This chapter considered several issues basic to the analysis of multivariate data in the behavioral sciences. Researchers rely on empirical observation to answer questions about relations among attributes. Such observations are best arranged in a two-dimensional array, the data matrix, with entities as rows and attributes as columns. Often the attributes will be partitioned into explanatory or predictor variables and response or criteria variables. When generalizing beyond the entities in a study, the researcher must consider whether the sampling method leads to fixed or random predictors.

This chapter has considered distinctions between measured and categorical observations and methods for recording categorical observations as measured variables. These methods are extremely important and allow one to treat the data matrix as if it contained only measured attributes.

The chapter has also introduced several data sets that will be referred to when we consider different multivariate analyses later in the book and has examined the interrelation between the matrix representation of data and its subsequent computer analysis.

Exercises

The following are several attributes that might be associated with members of a human population: age, extent of education, occupation, intelligence, place of residence, career goal, gender, ethnicity, and work history.

1. Select five attributes from the list and describe for each a *measure* of that attribute. Which provide *natural* and which *assumed* intervals?

2. For the same five attributes, describe how observations of each might yield categories. Which provide *natural*, *ordered*, and *assigned* categories?

3. Select three attributes from the list, choosing two to be predictors and the third a criterion. Describe a study, involving no more than 20 entities, in which you would control the distribution of the two predictors. What population are you drawing from, what is your sampling plan, what is your research question, and what do your observations look like? (Which attributes are measured? Which yield categories? What range of numbers is associated with each?)

4. Write the data matrix that might result if you were to conduct the study. If you have one or more categorical attributes in your design, write first the data matrix using the numeric designations for categories you worked out in Exercise 3. Then rewrite the data matrix developing a set of 0–1 attributes to replace each categorical attribute.

5. Repeat Exercise 3, this time letting the criterion attribute be a predictor and selecting a new criterion from among your two predictors in Exercise 3. (*Hint*:

You will probably want to think of a different population of potential entities from that used in Exercise 3.) Also, for this second study, assume both predictors are to be randomly sampled. Finally, replace each attribute that was measured in Exercise 3 with a set of categorical observations and replace each categorical attribute of Exercise 3 with a measure. Describe the new study as you did the one in Exercise 3.

6. As in Exercise 4, write a possible data matrix for your study in Exercise 5.

Additional Reading

The following is a good reference to some of the problems of measurement in the behavioral sciences.

Jones, L. V. (1971). The nature of measurement, in *Educational Measurement*, 2nd ed. (R. L. Thorndike, ed.). Washington, DC: American Council on Education.

Problems of data collection and organization are addressed in the next reference.

Kerlinger, F. N. (1973). *Foundations of Behavioral Research*, 2nd ed. New York: Holt.

The final reference focuses on errors of measurement.

Blalock, H. M. (1974). *Measurement in the Social Sciences*. Chicago: Aldine.

Chapter 2

Working with Matrices

Introduction

In general, matrices of raw or untransformed data are useful only for archival purposes. In order to answer the questions that led to the collection of observations, data matrices must be manipulated, yielding *derived* matrices or other summaries, for example, means, deviation scores, variances, or correlations.

In this chapter we (a) introduce terminology used to refer to data and derived matrices; (b) define and illustrate some basic manipulations of matrices—transposition, addition, subtraction, and multiplication; (c) demonstrate some frequently encountered matrix products; and (d) illustrate the importance of sequencing matrix operations.

The reader who has some experience with linear or matrix arithmetic may need only to scan quickly this chapter to identify any peculiarities of our notation. We give sufficient detail, however, that the complete novice may learn enough "matrix algebra" to follow the discussion of multivariate analysis.

2.1 Types of Matrices

A matrix, for our purposes, is a two-dimensional array of real-valued numbers. Matrices, then, can have different dimensions. The dimensions of a matrix are identified by describing the number of rows and the number of columns in that matrix. If the number of rows to a matrix exceeds the number of columns, the matrix is described as *vertical*; if the number of columns exceeds the number of rows, the matrix is *horizontal*; and if the number of rows and columns are the same, the matrix is *square*. Matrices **A**,

B′, and **C** are, respectively, vertical, horizontal, and square:

$$\mathbf{A} = \begin{bmatrix} 1 & 0 & 0 \\ 1 & 0 & 0 \\ 0 & 1 & 0 \\ 0 & 1 & 0 \\ 0 & 0 & 1 \\ 0 & 0 & 1 \end{bmatrix}$$

$$\mathbf{B}' = \begin{bmatrix} 15 & 12 & 16 & 10 \\ 5 & 5 & 5 & 5 \\ 3.0 & 2.4 & 3.2 & 2.0 \end{bmatrix}$$

$$\mathbf{C} = \begin{bmatrix} 1.0 & 0.5 \\ 0.5 & 1.0 \end{bmatrix}$$

Matrices will be named with single uppercase letters whenever possible.

2.1.1 Matrix Dimensions

The dimensions of a matrix are given by mentioning first the number of rows, then the number of columns. Thus, **A** is described as "six by three" or as 6×3; **B′** is 3×4; and **C** is 2×2; or, equivalently, **C** is a square matrix of order 2. If we know a matrix is square, only one integer is needed to give its dimensions. The dimensions of a matrix are written as a parenthesized subscript to the name of the matrix: $\mathbf{A}_{(6\times 3)}$, $\mathbf{B}'_{(3\times 4)}$, and $\mathbf{C}_{(2\times 2)}$ or, since it is square, $\mathbf{C}_{(2)}$. Knowing the dimensions of a matrix will tell us what operations can be performed with that matrix. Because the second of our matrices is horizontal it has been given the special name **B′**, read "**B** transpose." (In Section 2.1.2 we explain this naming convention.)

The numbers that make up a matrix are called entries to or *elements* of the matrix. We single out or identify a particular element of a matrix with a nonparenthesized subscript notation: A_{ij} is the entry at the intersection of the ith row and the jth column of matrix **A**. The first subscript identifies the row and the second the column.

Matrix **A** is part of the data matrix (Table 1.3) in the Women's Career Study and indicates the SES of six women (low SES = 1 in column 1, medium SES = 1 in column 2, and high SES = 1 in column 3). $A_{21} = 1$, $A_{43} = 0$, and $A_{62} = 0$ indicate, respectively, that woman 2 is from a low-SES family, woman 4 is not from a high-SES family, and woman 6 is not from a medium-SES family. Note that A_{34}, for example, would be without meaning as there is no fourth column to **A**.

Matrix **B′** was derived or calculated from the data matrix (Table 1.2) of the College Prediction Study. The first row of **B′** contains the sums of the grades earned by five students in their courses in, from left to right, history, mathematics, English, and chemistry. The second row records the number of students contributing to each sum (here, always five), and the third row contains the averages resulting from dividing the entry in the first row by the entry in the second row. When we speak of matrix $\mathbf{B}'$ as derived from a data matrix, we can imagine that the first row was obtained by adding up five entries (rows) for each of the four columns in a 5×4 data matrix. In the subscript notation, we may say that the sum of the mathematics grades is $B'_{12} = 12$, the number of students contributing chemistry grades is $B'_{24} = 5$, and the average grade earned in English is $B'_{33} = 3.2$.

$\mathbf{C}$ is also a derived matrix. Assume we started with a 50×2 matrix resulting from recording years of education (range 9–20) and occupational status ($1 =$ lowest to $10 =$ highest) in columns 1 and 2. Our entities were 50 randomly chosen North Central high school freshmen from the year 1947, and years of education and occupational status were obtained 30 years later. $\mathbf{C}$ might be the resulting set of correlations among those attributes. Although the derivation of a correlation matrix from a data matrix will be described in Section 2.6, we can describe here two features of the result. The intercorrelation matrix will be square, of the same order as the number of columns (attributes) in the data matrix, and the sequence of the attributes across the columns of the data matrix will be preserved in the sequence of rows (and columns) of the intercorrelation matrix. Thus, for our $\mathbf{C}$ matrix, C_{11} gives the correlation of the first attribute (educational attainment) with the first attribute and is 1.0, as it should be. C_{12} will be the correlation between attribute 1 (educational attainment) and attribute 2 (occupational status) and is 0.5. Note that $C_{12} = C_{21}$; the correlation of attributes 1 and 2 is identical to the correlation of attributes 2 and 1. When a square matrix has this latter property, it is said to be a *symmetric* matrix.

2.1.2 Matrix Transposition

Working with matrices will frequently necessitate changing a vertical matrix into a horizontal one and vice versa. This is done so often, in fact, that a special name and notational convention is adopted for the process. Every matrix is said to have a *transpose form*, which is produced by interchanging the rows and columns in the original matrix—that is, by writing the rows of the original matrix as columns and the columns of the original matrix as rows. What was the first row becomes the first column and what was the jth column becomes the jth row. The transpose of matrix $\mathbf{A}$ is written

$$\mathbf{A}' = \begin{bmatrix} 1 & 1 & 0 & 0 & 0 & 0 \\ 0 & 0 & 1 & 1 & 0 & 0 \\ 0 & 0 & 0 & 0 & 1 & 1 \end{bmatrix}$$

Since $\mathbf{A}$ is 6×3, its transpose must be 3×6. To show symbolically that this matrix is the transpose of the matrix $\mathbf{A}$, it is labeled $\mathbf{A}'$, where the prime indicates the operation of *transposition*. An important property of transposition is illustrated by the following matrix equation:

$$(\mathbf{A}')' = \mathbf{A} \tag{2.1.1}$$

The left side of this equation directs us to transpose the matrix $\mathbf{A}'$ (find the transpose of $\mathbf{A}$ transpose), and the right-hand side ensures that the result will be the original matrix $\mathbf{A}$. In words, Eq. (2.1.1) states that the transpose of a transpose of a matrix is the original matrix. If the columns of $\mathbf{A}'$ are written as rows and its rows as columns, the resultant matrix is $\mathbf{A}$.

There are, then, two forms of any matrix, its natural form and its *transposed* form. We shall adopt the convention that a vertical matrix is in natural form and, consequently, a horizontal matrix is in transposed form. Henceforth an unprimed capital letter denotes a vertical (or square) matrix, and a primed capital letter denotes a horizontal (or square) matrix.

The transpose of a vertical matrix is horizontal, the transpose of a horizontal matrix is vertical, and the transpose of a square matrix is another square matrix of the same order. The square matrix $\mathbf{C}$ was called a symmetric matrix. The defining property of a symmetric matrix is given by the results of transposition: $\mathbf{C}' = \mathbf{C}$.

When one or both of the two dimensions of a matrix is 1, we give the matrix a specific name. Consider these three examples:

$$\mathbf{r}' = (1, 2, 3, 4, 5, 6), \qquad \mathbf{c} = \begin{bmatrix} 0.9 \\ -0.6 \\ 0.2 \end{bmatrix}, \qquad s = \sqrt{3.0}$$

The matrix $\mathbf{r}'$ is 1×6 and is called a *row vector* of order 6; $\mathbf{c}$ is 3×1 and is, consequently, a *column vector* of order 3. s is termed a *scalar*, a matrix that is 1×1.

Lowercase letters will be used to name vectors and scalars. Note that the transposition convention is followed in naming vectors: The natural order of a vector will be taken to be the column form, and a row vector will be considered to have been transposed and will be named with a prime. One may think of a scalar as a 1×1 symmetric matrix since $s' = s$.

Earlier we noted that a matrix was a two-dimensional array of numbers or scalar quantities. A matrix may also be regarded as made up of vectors. That is, we can think of the rows and columns of a matrix as vectors. This will be useful, for example, in describing data matrices. In data matrices a column corresponds to an attribute and a row to an entity. Within a data matrix, then, specified attributes or entities can be referred to as column or row vectors.

To do so, however, we further refine our notation and utilize a dot convention. Recall now that A_{ij} indicated the jth element from the left, in the ith row (or equivalently, the ith element, from the top, in the jth column) of matrix $\mathbf{A}$. To name row vectors and column vectors embedded in matrices, we use the following notation:

$\mathbf{A}_{i.}$: the ith *row* of matrix **A**, written as a *column* vector

$\mathbf{A}'_{i.}$: the ith *row* of matrix **A**, written as a *row* vector

$\mathbf{A}_{.j}$: the jth *column* of matrix **A**, written as a *column* vector

$\mathbf{A}'_{.j}$: the jth *column* of matrix **A**, written as a *row* vector

Three things are important in using this row and column notation: (1) the subscript, $i.$ or $.j$, tells whether a row or a column is to be extracted from a matrix; (2) the row or column is *always* extracted from the natural order matrix **A**, never from **A**′; and (3) the prime notation indicates that the extracted row or column is to be written as a row vector.

Let us illustrate these rules with the following matrix:

$$\mathbf{A} = \begin{bmatrix} 1 & 0 & 0 \\ 1 & 0 & 0 \\ 0 & 1 & 0 \\ 0 & 1 & 0 \\ 0 & 0 & 1 \\ 0 & 0 & 1 \end{bmatrix}$$

From this matrix and the preceding rules we may write

$$\mathbf{A}_{2.} = \begin{bmatrix} 1 \\ 0 \\ 0 \end{bmatrix}, \qquad \mathbf{A}'_{5.} = (0, 0, 1)$$

$$\mathbf{A}_{.1} = \begin{bmatrix} 1 \\ 1 \\ 0 \\ 0 \\ 0 \\ 0 \end{bmatrix}, \qquad \mathbf{A}'_{.3} = (0, 0, 0, 0, 1, 1)$$

Not only does this notation permit us to write a vector of observations corresponding to a selected attribute or entity, it allows us to describe the entire data matrix as a collection of attribute, or entity, vectors. This symbolic convenience will be important when we perform arithmetic with matrices.

For now, notice that the matrix **A** can be expressed as a set of three column vectors placed side by side,

$$\mathbf{A} = (\mathbf{A}_{.1} \quad \mathbf{A}_{.2} \quad \mathbf{A}_{.3})$$

or as a set of six row vectors stacked up,

$$\mathbf{A} = \begin{bmatrix} \mathbf{A}'_{1.} \\ \mathbf{A}'_{2.} \\ \mathbf{A}'_{3.} \\ \mathbf{A}'_{4.} \\ \mathbf{A}'_{5.} \\ \mathbf{A}'_{6.} \end{bmatrix}$$

The transpose of a matrix can also be expressed as an orderly arrangement of the rows or columns making up the natural form of the matrix. Thus the matrix $\mathbf{A}'$ can be represented either as

$$\mathbf{A}' = (\mathbf{A}_{1.} \quad \mathbf{A}_{2.} \quad \mathbf{A}_{3.} \quad \mathbf{A}_{4.} \quad \mathbf{A}_{5.} \quad \mathbf{A}_{6.})$$

or as

$$\mathbf{A}' = \begin{bmatrix} \mathbf{A}'_{.1} \\ \mathbf{A}'_{.2} \\ \mathbf{A}'_{.3} \end{bmatrix}$$

The first representation states that the columns of the transpose matrix are the rows of the natural form of the matrix column vectors. Similarly, the second expression restates the transposition rule that the rows of $\mathbf{A}'$ are the columns of $\mathbf{A}$.

Although this "dot-and-transpose" notation appears clumsy, the reader will find it easier with practice. More important, it will prevent us from mistakenly treating a row vector as a column vector or a horizontal matrix as a vertical matrix. In working with matrices, dimensions and orientation must be taken into account.

2.2 Special Matrices and Vectors

Multivariate analysis starts with a data matrix and then, to answer questions about that matrix, derives matrices from it. The matrices commonly derived from the data matrix and the matrices used to operate on the data matrix in computing the derived matrices may have special properties. We introduce some of these special matrices here. Later we discuss how they enter into multivariate computations.

A symmetric matrix has been defined as a square matrix whose natural and transpose forms are indistinguishable. Symmetric matrices are frequently encountered as matrices of intercorrelations, as in Table 1.10, or as variance–covariance matrices. Table 2.1 shows the variance–covariance matrix for the Vocational Interest Study described in Section 1.5. Note that the rows, from top to bottom, and the columns, from left to right, are labeled in exactly the same way.

Table 2.1 Variance–Covariance Matrix of Vocational Interest Inventory Scale Scores[a]

	SER	BUS	ORG	TEC	OUT	SCI	CUL	ART
Service (SER)	100	−16	−19	−34	−7	−3	4	−14
Business (BUS)	−16	100	17	0	−28	−44	−10	−15
Organization (ORG)	−19	17	100	−11	−46	−26	11	−22
Technical (TEC)	−34	0	−11	100	−400	−5	−38	−8
Outdoors (OUT)	−7	−28	−46	−4	100	7	−27	−11
Science (SCI)	−3	−44	−26	−5	7	100	−18	−19
General cultural (CUL)	4	−10	11	−38	−27	−18	100	−12
Arts and entertainment (ART)	−14	−15	−22	−8	−11	−19	−12	100

[a]Scale scores for the high school norm group for the Vocational Interest Study.

A *variance–covariance matrix* has the following property:

$$S_{ij} = \begin{cases} \text{variance of attribute } i, & i = j \\ \text{covariance between attributes } i \text{ and } j, & i \neq j \end{cases}$$

Variances occupy the *diagonal* elements of the square matrix (those extending from the upper left corner down to the lower right corner), while covariances occupy the *off-diagonal* (or remaining) positions in the matrix.

Those off-diagonal elements of a square matrix above and to the right of the diagonal are called *supradiagonal* elements, while those below and to the left of the diagonal are termed *infradiagonal.*

If the variance–covariance matrix in Table 2.1 is named **S**, then

$S_{33} = 100$ is the variance of scores on the organization scale of the Vocational Interest Inventory

$S_{56} = 7$ is the covariance between the outdoors and science scales

$S_{71} = 4$ is the covariance between the general cultural and service scales

Another definition of a symmetric matrix **S** is

$$S_{ij} = S_{ji} \qquad \text{for all} \quad i \text{ and } j \tag{2.2.1}$$

Note that all of the diagonal elements of Table 2.1 are identically 100. These data are for the high school normative sample, for which the scales were standardized so that each has a mean of 50 and a standard deviation of 10.

If all off-diagonal elements in a square matrix are identically zero, so that the only nonzero elements are contained in the diagonal, we call the matrix a *diagonal* matrix.

$$\mathbf{D} = \begin{bmatrix} 0.7955 & 0 & 0 \\ 0 & 0.8729 & 0 \\ 0 & 0 & 0.8729 \end{bmatrix}$$

is a diagonal matrix. The diagonal elements in this case are the standard deviations of the grade point averages in English, mathematics, and physics courses, in order, from the College Prediction Study. For a diagonal matrix, then,

$$D_{ij} = 0 \quad \text{if} \quad i \neq j \tag{2.2.2}$$

Quite often we obtain diagonal matrices from other matrices. For example, in the present instance we might well have first computed a variance–covariance matrix among the three achievement measures:

$$\mathbf{C} = \begin{bmatrix} 0.6328 & 0.3782 & 0.2898 \\ 0.3782 & 0.7619 & 0.5088 \\ 0.2898 & 0.5088 & 0.7620 \end{bmatrix}$$

Then, we could have extracted the diagonal elements from this matrix to form a diagonal matrix:

$$\mathbf{D_C} = \begin{bmatrix} 0.6328 & 0 & 0 \\ 0 & 0.7619 & 0 \\ 0 & 0 & 0.7620 \end{bmatrix}$$

The name of this new matrix $\mathbf{D_C}$ summarizes its derivation: **D** indicates that it is a diagonal matrix and the subscript **C** that $\mathbf{D_C}$ takes as its diagonal elements the diagonal elements of the square matrix **C**.

The diagonal elements of $\mathbf{D_C}$ would be variances, since **C** was a variance–covariance matrix. The matrix **D** was formed by replacing the diagonal elements of $\mathbf{D_C}$ with their positive square roots:

$$\mathbf{D_C^{1/2}} = \begin{bmatrix} 0.7955 & 0 & 0 \\ 0 & 0.8729 & 0 \\ 0 & 0 & 0.8729 \end{bmatrix}$$

The superscript $\frac{1}{2}$ indicates the power to which the extracted diagonal elements were raised. $\mathbf{D_C^{1/2}}$ is a diagonal matrix whose diagonal elements are the square roots of the diagonal elements obtained from a square matrix. For another example,

$$\mathbf{D_C^{2}} = \begin{bmatrix} 0.4004 & 0 & 0 \\ 0 & 0.5805 & 0 \\ 0 & 0 & 0.5806 \end{bmatrix}$$

has diagonal elements equal to the squares of the diagonal elements of **C**, and

$$\mathbf{D_C^{-1/2}} = \begin{bmatrix} 1.2571 & 0 & 0 \\ 0 & 1.1461 & 0 \\ 0 & 0 & 1.1461 \end{bmatrix}$$

has diagonal elements equal to the reciprocals of the square roots of the diagonal elements of **C**: $1/(0.6328)^{1/2} = 1.2571$. (Fractional powers are roots, and negative powers are reciprocals of powered values.) In the last example we could have omitted the off-diagonal zeros in writing out the matrices. Since by the naming convention the matrix is diagonal, only the diagonal elements need be indicated.

If all elements in a diagonal matrix are equal, it is customary to call the matrix a *scalar* matrix. For the **S** matrix of Table 2.1,

$$\mathbf{D}_{\mathbf{S}}^{1/2} = \begin{bmatrix} 10 & & & & & & & \\ & 10 & & & 0 & & & \\ & & 10 & & & & & \\ & & & 10 & & & & \\ & & & & 10 & & & \\ & & 0 & & & 10 & & \\ & & & & & & 10 & \\ & & & & & & & 10 \end{bmatrix}$$

which is a scalar matrix $\mathbf{D}_{10}$ with all diagonal elements equal to 10. **D** with a subscript number will represent a scalar matrix.

For our work, the most important scalar matrix is the *identity* matrix, for which all diagonal elements are equal to 1. It is always written **I**. Identity matrices may be of different dimensions, which we can note by a subscript; for example,

$$\mathbf{I}_{(5)} = \begin{bmatrix} 1 & & & & \\ & 1 & & 0 & \\ & & 1 & & \\ & 0 & & 1 & \\ & & & & 1 \end{bmatrix}, \qquad \mathbf{I}_{(3)} = \begin{bmatrix} 1 & 0 & 0 \\ 0 & 1 & 0 \\ 0 & 0 & 1 \end{bmatrix}$$

We shall find that, just as multiplying by 1 does not change the value of a scalar expression, multiplying by the identity matrix leaves any matrix unchanged.

A matrix each of whose elements is either 1 or 0 is called a *binary* matrix. The identity matrix is a binary matrix. If we rearrange the rows or columns of an identity matrix, we produce a *permutation* matrix:

$$\pi = \begin{bmatrix} 0 & 1 & 0 \\ 1 & 0 & 0 \\ 0 & 0 & 1 \end{bmatrix}$$

The identifying characteristic of a permutation matrix π is that it has exactly a single 1 in each row, a single 1 in each column, and no other nonzero elements. We shall find in Section 2.5.7 that permutation matrices can be used to change the order of the rows or columns in a matrix.

If a binary matrix has but a single nonzero entry, it is called an *elementary* matrix; for example,

$$\mathbf{E}_{21} = \begin{bmatrix} 0 & 0 \\ 1 & 0 \\ 0 & 0 \end{bmatrix}$$

Here we have an exception to our subscripting convention. $\mathbf{E}_{21}$ does not refer to the element at the intersection of the second row and first column of a matrix $\mathbf{E}$. Rather, when the context makes it clear that $\mathbf{E}$ is an elementary matrix, the subscript tells us which matrix element is nonzero.

If all entries in a matrix are zero, that matrix is called the *null* matrix written $\mathbf{0}$. Null matrices, can have different dimensions:

$$\mathbf{0} = \begin{bmatrix} 0 & 0 & 0 & 0 \\ 0 & 0 & 0 & 0 \\ 0 & 0 & 0 & 0 \\ 0 & 0 & 0 & 0 \end{bmatrix}, \qquad \mathbf{0}' = \begin{bmatrix} 0 & 0 & 0 & 0 & 0 \\ 0 & 0 & 0 & 0 & 0 \end{bmatrix}$$

As the identity matrix is the "matrix one," the null matrix operates as a "matrix zero."

The null matrix may be a vector, in which case we call it a *null vector*, although we use the same symbol, $\mathbf{0}$. A vector with a single 1 and zeros elsewhere is an *elementary* vector; for example,

$$\mathbf{e}_3' = (0, 0, 1, 0), \qquad \mathbf{e}_2 = \begin{bmatrix} 0 \\ 1 \\ 0 \\ 0 \end{bmatrix}$$

As in the case of the elementary matrix, the subscript to the elementary vector denotes the position of the 1 in the vector.

If all the entries in a vector are 1s, we call it a *unit* vector, written $\mathbf{1}$:

$$\mathbf{1}' = (1, 1, 1, 1), \qquad \mathbf{1} = \begin{bmatrix} 1 \\ 1 \\ 1 \end{bmatrix}$$

A vector all of whose elements are equal is a *scalar vector*. The null and unit vectors are scalar vectors, as are

$$\mathbf{2}' = (2, 2, 2, 2) \qquad \text{and} \qquad \mathbf{5} = \begin{bmatrix} 5 \\ 5 \\ 5 \end{bmatrix}$$

These special matrices and vectors are useful because they facilitate the symbolic representation and description of many computations important in multivariate analysis.

2.3 Addition and Subtraction of Matrices

In scalar algebra any two quantities may be added or subtracted, with the result some third scalar. The result may defy interpretation, as when a man's height (190 cm) is added to his weight (79 kg), but we do have a resultant quantity (269)!

Not every pair of matrices may be added, however—they must have the same number of rows and the same number of columns. Their sum will be a third matrix the same size as the matrices being added, and it is obtained by adding elements that occupy the same position in the two matrices. Symbolically,

$$\mathbf{A} + \mathbf{B} = \mathbf{C}$$

if and only if **A** has both the same number of rows and columns as **B**. Given this *additive conformability*, **C** will be such that

$$C_{ij} = A_{ij} + B_{ij} \qquad \text{for all} \quad i \text{ and } j \tag{2.3.1}$$

Consider an example we have drawn from the Preemployment Study. As part of the preemployment assessment battery, applicants completed a job values test consisting of 20 items. Each item could be answered either in the *intrinsic* direction (applicant values achievement, creativity, social service) or the *extrinsic* direction (applicant values salary, job stability, working conditions). For nine applicants chosen at random from the study, matrix **A** gives the frequency with which the ten odd-numbered items from the test were marked in the intrinsic (first column) or extrinsic (second column) direction; matrix **B** is similarly defined, but is based on responses to the ten *even-numbered* items:

$$\mathbf{A} = \begin{bmatrix} 7 & 3 \\ 8 & 2 \\ 0 & 10 \\ 2 & 8 \\ 1 & 9 \\ 5 & 5 \\ 6 & 4 \\ 1 & 9 \\ 6 & 4 \end{bmatrix}, \qquad \mathbf{B} = \begin{bmatrix} 8 & 2 \\ 10 & 0 \\ 1 & 9 \\ 0 & 10 \\ 1 & 9 \\ 4 & 6 \\ 7 & 3 \\ 3 & 7 \\ 4 & 6 \end{bmatrix}$$

Matrices **A** and **B** are the same size and can be added:

$$\underset{\mathbf{A}+\mathbf{B}}{\begin{bmatrix} 7+8 & 3+2 \\ 8+10 & 2+0 \\ 0+1 & 10+9 \\ 2+0 & 8+10 \\ 1+1 & 9+9 \\ 5+4 & 5+6 \\ 6+7 & 4+3 \\ 1+3 & 9+7 \\ 6+4 & 4+6 \end{bmatrix}} = \underset{\mathbf{C}}{\begin{bmatrix} 15 & 5 \\ 18 & 2 \\ 1 & 19 \\ 2 & 18 \\ 2 & 18 \\ 9 & 11 \\ 13 & 7 \\ 4 & 16 \\ 10 & 10 \end{bmatrix}}$$

The sum matrix **C** can be given an interpretation *if* the identification of attributes (columns) and of entities (rows) was consistent from **A** to **B**. That is, because we had the same nine job applicants, ordered in the same way, and the same two job values, ordered in the same way, for both **B** and **A**, we can read total intrinsic and extrinsic interest scores from **C**.

Equation (2.3.1) defines matrix addition as element-by-element addition, relying upon our original definition of a matrix as a rectangular array of numbers. Because a matrix also can be expressed as a set of row or column vectors, an alternative but equivalent definition of matrix addition can be formulated. If the two data matrices **A** and **B** are each expressed as a set of two column vectors,

$$\mathbf{A} = (\mathbf{A}_{.1} \quad \mathbf{A}_{.2}) \qquad \text{and} \qquad \mathbf{B} = (\mathbf{B}_{.1} \quad \mathbf{B}_{.2})$$

then the result of adding **A** and **B** may be symbolized

$$\begin{aligned} \mathbf{A} + \mathbf{B} &= (\mathbf{A}_{.1} \quad \mathbf{A}_{.2}) + (\mathbf{B}_{.1} \quad \mathbf{B}_{.2}) = (\mathbf{A}_{.1} + \mathbf{B}_{.1} \quad \mathbf{A}_{.2} + \mathbf{B}_{.2}) \\ &= (\mathbf{C}_{.1} \quad \mathbf{C}_{.2}) = \mathbf{C} \end{aligned} \tag{2.3.2}$$

That is, we may add the corresponding vector components of two matrices (adding corresponding columns or corresponding rows) to obtain the matrix sum.

Alternatively, now that we know how to carry out the actual additions, we can add, *symbolically* at least, even more extensive partitions of a matrix. We illustrate this property by extending the Preemployment Study example. Assume that the first five applicants (the first five rows of **A** and **B**) contain data for women and that the remaining four rows contain data for men. We can use this information to partition **A** and **B** as follows:

$$\mathbf{A} = \begin{bmatrix} 7 & 3 \\ 8 & 2 \\ 0 & 10 \\ 2 & 8 \\ 1 & 9 \\ \hline 5 & 5 \\ 6 & 4 \\ 1 & 9 \\ 6 & 4 \end{bmatrix} = \begin{bmatrix} \mathbf{F} \\ \hline \mathbf{G} \end{bmatrix}$$

$$\mathbf{B} = \begin{bmatrix} 8 & 2 \\ 10 & 0 \\ 1 & 9 \\ 0 & 10 \\ 1 & 9 \\ \hline 4 & 6 \\ 7 & 3 \\ 3 & 7 \\ 4 & 6 \end{bmatrix} = \begin{bmatrix} \mathbf{J} \\ \hline \mathbf{K} \end{bmatrix}$$

Each of the 9 × 2 matrices has been partitioned into an upper 5 × 2 matrix (**F**, **J**) and a lower 4 × 2 matrix (**G**, **K**). The sum of **A** and **B** can be written

$$\mathbf{A} + \mathbf{B} = \begin{bmatrix} \mathbf{F} \\ \hline \mathbf{G} \end{bmatrix} + \begin{bmatrix} \mathbf{J} \\ \hline \mathbf{K} \end{bmatrix} = \begin{bmatrix} \mathbf{F} + \mathbf{J} \\ \hline \mathbf{G} + \mathbf{K} \end{bmatrix} = \begin{bmatrix} \mathbf{M} \\ \hline \mathbf{N} \end{bmatrix} = \mathbf{C} \qquad (2.3.3)$$

Matrices **M** and **N** would be total score matrices for the female and male applicants, respectively.

These examples illustrate that two additively conformable matrices can each be partitioned and their sum represented symbolically by the sums of those partitions, provided that we partition each of the two matrices in the same way. That is, the partitions themselves must be additively conformable. To calculate the sum of two matrices, we must work with their scalar elements. However, symbolic representations of matrix addition, as in (2.3.2) and (2.3.3), will prove helpful with data matrices that are naturally partitioned into predictor and criterion attributes or groups of entities.

Matrix *subtraction*, as might be expected, follows the same rules as matrix addition. The difference matrix **W** = **U** − **V** is defined only if **U** and **V** have the same dimensions. The matrix **W** will then be of the same dimensions as

U or **V** and

$$W_{ij} = U_{ij} - V_{ij} \qquad \text{for all} \quad i \text{ and } j \tag{2.3.4}$$

Two matrices are *equal*, or identical, if they are the same size and if the result of subtracting one from the other is the null matrix. That is, to be equal, two matrices must be identical element by element.

More than two matrices all the same size may be summed, and the result is independent of the order in which they are summed:

$$(\mathbf{A} + \mathbf{B}) + \mathbf{C} = \mathbf{A} + (\mathbf{B} + \mathbf{C}) = (\mathbf{A} + \mathbf{C}) + \mathbf{B} \tag{2.3.5}$$

2.4 Matrix Products

In the preceding section we saw that not every pair of matrices can be added. To be added, two matrices had to have the same dimensions. Not every pair of matrices will form a product, either. Dimensions will again be important. Furthermore, the order of multiplication is critical. The products **AB** and **BA** when defined, may well be different. This contrasts, of course, with scalar multiplication.

Because order is important in matrix multiplication a special terminology must be adopted. Assuming for the moment that the product **AB** = **C** exists, **A** is called the *prefactor* in the multiplication and **B** the *postfactor*. Equivalently, we say that the matrix **C** results from the *postmultiplication of* **A** *by* **B** or from the *premultiplication of* **B** *by* **A**.

With this terminology, a simple rule determines whether a matrix product is defined or not and says something about the dimensions of the product:

Matrix Product. A product **AB** = **C** exists only if the number of columns in the prefactor **A** is identical to the number of rows in the postfactor **B**. The matrix product **AB** = **C** will have as many rows as in the prefactor **A** and as many columns as the postfactor **B**.

Symbolically,

$$\mathbf{A}_{(n\times m)}\mathbf{B}_{(m\times p)} = \mathbf{C}_{(n\times p)} \tag{2.4.1}$$

Note that if n and p are different integers, the product **BA** is not defined. **B**, now the prefactor, does *not* have as many columns (p) as the postfactor **A** has rows (n).

Consider some examples, based on the following matrices:

$$\mathbf{e}_2' = (0 \quad 1 \quad 0), \qquad \mathbf{1}' = (1 \quad 1 \quad 1 \quad 1), \qquad \mathbf{u}' = (3 \quad 6 \quad 9)$$

$$\mathbf{v}' = (1 \quad 2 \quad 3 \quad 4), \qquad s = (5)$$

$$\mathbf{F}' = \begin{bmatrix} 1 & 7 & 2 & 4 \\ 1 & 1 & 9 & 3 \\ 0 & 4 & 4 & 8 \end{bmatrix}, \qquad \mathbf{G} = \begin{bmatrix} 1 & 0 & 1 \\ 3 & 2 & 4 \\ 6 & 2 & 4 \end{bmatrix}, \qquad \mathbf{H} = \begin{bmatrix} 1 & 2 & 2 \\ 3 & 4 & 5 \\ 2 & 4 & 2 \end{bmatrix}$$

The product $\mathbf{e}_2\mathbf{1}'$,

$$\begin{bmatrix} 0 \\ 1 \\ 0 \end{bmatrix}(1 \quad 1 \quad 1 \quad 1) = \begin{bmatrix} \times & \times & \times & \times \\ \times & \times & \times & \times \\ \times & \times & \times & \times \end{bmatrix}$$

exists bccausc thc number of columns in the prefactor matches the number of rows in the postfactor—both are 1. Furthermore, the product is a matrix with three rows (the number of rows in the prefactor) and four columns (the number of columns in the postfactor). We do not yet know *what* the elements of the product matrix are.

The product $\mathbf{1}'\mathbf{e}_2$,

$$(1 \quad 1 \quad 1 \quad 1)\begin{bmatrix} 0 \\ 1 \\ 0 \end{bmatrix}$$

is not defined because the number of columns in the prefactor does not equal the number of rows in the postmultiplier. However, the product $\mathbf{1}'\mathbf{v}$,

$$(1 \quad 1 \quad 1 \quad 1)\begin{bmatrix} 1 \\ 2 \\ 3 \\ 4 \end{bmatrix} = (\times)$$

is defined and must be a scalar, a matrix with one row (the number of rows in $\mathbf{1}'$) and one column (the number of columns in $\mathbf{v}$).

The products $s\mathbf{u}'$ and $\mathbf{u}s$,

$$(5)(3 \quad 6 \quad 9) = (\times \quad \times \quad \times) \qquad \text{and} \qquad \begin{bmatrix} 3 \\ 6 \\ 9 \end{bmatrix}(5) = \begin{bmatrix} \times \\ \times \\ \times \end{bmatrix}$$

both exist, but not $\mathbf{u}'s$ or $s\mathbf{u}$:

$$(3 \quad 6 \quad 9)(5) \qquad \text{or} \qquad (5)\begin{bmatrix} 3 \\ 6 \\ 9 \end{bmatrix}$$

The matrix $\mathbf{G}$ may be premultiplied by $\mathbf{e}_2'$,

$$(0 \quad 1 \quad 0)\begin{bmatrix} 1 & 0 & 1 \\ 3 & 2 & 4 \\ 6 & 2 & 4 \end{bmatrix} = (\times \quad \times \quad \times)$$

but $\mathbf{G}\mathbf{e}_2'$,

$$\begin{bmatrix} 1 & 0 & 1 \\ 3 & 2 & 4 \\ 6 & 2 & 4 \end{bmatrix}(1 \quad 0 \quad 1)$$

does not exist. The products

$$\mathbf{GF}' = \begin{bmatrix} 1 & 0 & 1 \\ 3 & 2 & 4 \\ 6 & 2 & 4 \end{bmatrix} \begin{bmatrix} 1 & 7 & 2 & 4 \\ 1 & 1 & 9 & 3 \\ 0 & 4 & 4 & 8 \end{bmatrix} = \begin{bmatrix} \times & \times & \times & \times \\ \times & \times & \times & \times \\ \times & \times & \times & \times \end{bmatrix}$$

and

$$\mathbf{FG}' = \begin{bmatrix} 1 & 1 & 0 \\ 7 & 1 & 4 \\ 2 & 9 & 4 \\ 4 & 3 & 8 \end{bmatrix} \begin{bmatrix} 1 & 3 & 6 \\ 0 & 2 & 2 \\ 1 & 4 & 4 \end{bmatrix} = \begin{bmatrix} \times & \times & \times \\ \times & \times & \times \\ \times & \times & \times \\ \times & \times & \times \end{bmatrix}$$

are defined, but we cannot compute the product $\mathbf{F}'\mathbf{G}$. The products $\mathbf{GH}$ and $\mathbf{HG}$ are both 3×3 matrices:

$$\begin{bmatrix} 1 & 0 & 1 \\ 3 & 2 & 4 \\ 6 & 2 & 4 \end{bmatrix} \begin{bmatrix} 1 & 2 & 2 \\ 3 & 4 & 5 \\ 2 & 4 & 2 \end{bmatrix} = \begin{bmatrix} \times & \times & \times \\ \times & \times & \times \\ \times & \times & \times \end{bmatrix}$$

and

$$\begin{bmatrix} 1 & 2 & 2 \\ 3 & 4 & 5 \\ 2 & 4 & 2 \end{bmatrix} \begin{bmatrix} 1 & 0 & 1 \\ 3 & 2 & 4 \\ 6 & 2 & 4 \end{bmatrix} = \begin{bmatrix} \times & \times & \times \\ \times & \times & \times \\ \times & \times & \times \end{bmatrix}$$

Finally, s cannot be pre- or postmultiplied by any of the matrices **F**, **G**, or **H**.

What, however, are the values of the several $\times$s entered in these matrices? Matrix multiplication is sometimes referred to as *row-by-column multiplication* because of the following rule: In the matrix product $\mathbf{A}_{(n \times m)}\mathbf{B}_{(m \times p)} = \mathbf{C}_{(n \times p)}$, each element of the product **C** is computed from the elements of the matrices **A** and **B** by

$$C_{ij} = \sum_{k=1}^{m} (A_{ik} B_{kj}) \tag{2.4.2}$$

where m is the *common order* of **A** and **B**, the number of columns in the prefactor and the number of rows in the postmultiplier.

The Product Matrix. The element at the intersection of the ith row and the jth column of the product matrix is computed by forming the sum of products of the elements of the ith row of the prefactor and the jth column of the postfactor.

This rule expressed in Eq. (2.4.2) indicates at once two things. First, the number of columns in the prefactor must equal the number of rows in the postfactor. (The sum of products is *over that common order*. There must be the same number of elements in a row of the prefactor as there are elements in a column of the postmultiplier.) Second, the product has rows and columns corresponding in number to the number of premultiplier rows and postfactor columns. [Equation (2.4.2) yields a product element for each possible pairing of a row of the premultiplier with a column of the postmultiplier. With n rows in the prefactor and p columns in the postfactor, the np elements computed by Eq. (2.4.2) can be systematically arranged in an $n \times p$ matrix.]

Now that we have our row-by-column rule, consider the numerical examples introduced earlier in this section. Although a single rule covers all possibilities, it is customary to discuss separately vector-by-vector, vector-by-matrix, matrix-by-matrix, and scalar-by-matrix (or vector) products.

2.4.1 Matrix Products and Vectors

There are two types of vector-by-vector products. The *major product* of two vectors is a matrix and results from postmultiplying a column vector by a row vector. One example is the product $\mathbf{e}_2\mathbf{1}'$,

$$\begin{bmatrix} 0 \\ 1 \\ 0 \end{bmatrix} (1 \quad 1 \quad 1 \quad 1) = \begin{bmatrix} 0 \times 1 & 0 \times 1 & 0 \times 1 & 0 \times 1 \\ 1 \times 1 & 1 \times 1 & 1 \times 1 & 1 \times 1 \\ 0 \times 1 & 0 \times 1 & 0 \times 1 & 0 \times 1 \end{bmatrix}$$

$$= \begin{bmatrix} 0 & 0 & 0 & 0 \\ 1 & 1 & 1 & 1 \\ 0 & 0 & 0 & 0 \end{bmatrix}$$

As the common order is 1, the product matrix consists, in each element, of the product of an element of the column prefactor multiplied by an element of the row postfactor. The major product of any two vectors is found by postmultiplying the one written as a column by the second written as a row.

If a major product $\mathbf{1e}_2'$ is computed, the result is

$$\begin{bmatrix} 1 \\ 1 \\ 1 \\ 1 \end{bmatrix} (0 \quad 1 \quad 0) = \begin{bmatrix} 0 & 1 & 0 \\ 0 & 1 & 0 \\ 0 & 1 & 0 \\ 0 & 1 & 0 \end{bmatrix}$$

Our interest in this outcome is that this 4×3 matrix is the transpose of the 3×4 matrix first obtained. That is, the two results can be linked by the equation

$$\mathbf{1e}_2' = (\mathbf{e}_2\mathbf{1}')'$$

This relation is an instance of a frequently applied rule:

Transpose Product Rule. The transpose of a product is equal to the product of the transposes of the factors taken in reverse order. Symbolically,

$$(\mathbf{AB})' = \mathbf{B}'\mathbf{A}' \tag{2.4.3}$$

B, the original postfactor, becomes in its transposed form the prefactor. Applying this rule in the preceding instance,

$$(\mathbf{e}_2\mathbf{1}')' = (\mathbf{1}')'\mathbf{e}_2' = \mathbf{1}\mathbf{e}_2'$$

$$(\mathbf{AB})' = (\mathbf{B})'\mathbf{A}'$$

we obtain the result with which we started.

The second type of vector-by-vector product is the *minor product*. The minor product is a scalar and results from postmultiplying a row vector by a column vector. Our numerical example was the product

$$\mathbf{1}'\mathbf{v} = (1 \quad 1 \quad 1 \quad 1)\begin{bmatrix} 1 \\ 2 \\ 3 \\ 4 \end{bmatrix} = (1 \times 1) + (1 \times 2) + (1 \times 3) + (1 \times 4)$$

$$= 10$$

With a single row in the prefactor and a single column in the postmultiplier, application of the row-by-column rule calls for computing a single sum of products between the two vectors.

As the transpose of a scalar is that same scalar, application of the product transpose rule suggests that the product

$$(\mathbf{1}'\mathbf{v})' = \mathbf{v}'(\mathbf{1}')' = \mathbf{v}'\mathbf{1}$$

ought also to equal 10.

Note that while a major product can be computed between any two vectors, a minor product can be obtained only if the two vectors have the same number of elements. We have seen the conditions under which row and column vectors may be multiplied; two column vectors may never be multiplied, nor may two row vectors.

A matrix may be premultiplied by a row vector or postmultiplied by a column vector so long as the vector satisfies the common order rule. As numerical examples,

$$\mathbf{e}_2'\mathbf{G} = (0 \quad 1 \quad 0)\begin{bmatrix} 1 & 0 & 1 \\ 3 & 2 & 4 \\ 6 & 2 & 4 \end{bmatrix}$$

$$= ((0 \times 1) + (1 \times 3) + (0 \times 6), (0 \times 0) + (1 \times 2) + (0 \times 2),$$

$$(0 \times 1) + (1 \times 4) + (0 \times 4))$$

$$= (3 \quad 2 \quad 4)$$

and

$$\mathbf{F'1} = \begin{bmatrix} 1 & 7 & 2 & 4 \\ 1 & 1 & 9 & 3 \\ 0 & 4 & 4 & 8 \end{bmatrix} \begin{bmatrix} 1 \\ 1 \\ 1 \\ 1 \end{bmatrix}$$

$$= \begin{bmatrix} (1 \times 1) + (7 \times 1) + (2 \times 1) + (4 \times 1) \\ (1 \times 1) + (1 \times 1) + (9 \times 1) + (3 \times 1) \\ (0 \times 1) + (4 \times 1) + (4 \times 1) + (8 \times 1) \end{bmatrix} = \begin{bmatrix} 14 \\ 14 \\ 16 \end{bmatrix}$$

Premultiplying a matrix by a row vector produces a row vector with as many elements as there were columns in the matrix. Postmultiplying a matrix by a column vector yields a column vector with number of elements equal to the number of rows in the matrix. A matrix may not be premultiplied by a column vector, nor may it be postmultiplied by a row vector.

Applying the transpose product rule to the vector-by-matrix products yields

$$(\mathbf{e}_2'\mathbf{G})' = \mathbf{G}'\mathbf{e}_2 = \begin{bmatrix} 3 \\ 2 \\ 4 \end{bmatrix}$$

and

$$(\mathbf{F'1})' = \mathbf{1'F} = (14 \quad 14 \quad 16)$$

Computing the matrix product

$$\begin{bmatrix} 1 & 0 & 1 \\ 3 & 2 & 4 \\ 6 & 2 & 4 \end{bmatrix} \begin{bmatrix} 1 & 2 & 2 \\ 3 & 4 & 5 \\ 2 & 4 & 2 \end{bmatrix} = \begin{bmatrix} (1, 0, 1)\begin{pmatrix} 1 \\ 3 \\ 2 \end{pmatrix} & (1, 0, 1)\begin{pmatrix} 2 \\ 4 \\ 4 \end{pmatrix} & (1, 0, 1)\begin{pmatrix} 2 \\ 5 \\ 2 \end{pmatrix} \\ (3, 2, 4)\begin{pmatrix} 1 \\ 3 \\ 2 \end{pmatrix} & (3, 2, 4)\begin{pmatrix} 2 \\ 4 \\ 4 \end{pmatrix} & (3, 2, 4)\begin{pmatrix} 2 \\ 5 \\ 2 \end{pmatrix} \\ (6, 2, 4)\begin{pmatrix} 1 \\ 3 \\ 2 \end{pmatrix} & (6, 2, 4)\begin{pmatrix} 2 \\ 4 \\ 4 \end{pmatrix} & (6, 2, 4)\begin{pmatrix} 2 \\ 5 \\ 2 \end{pmatrix} \end{bmatrix}$$

yields

$$\mathbf{GH} = \begin{bmatrix} 3 & 6 & 4 \\ 17 & 30 & 24 \\ 20 & 36 & 30 \end{bmatrix}$$

a result the reader will want to verify.

As additional practice, we compute the product **HG**:

$$\mathbf{HG} = \begin{bmatrix} 1 & 2 & 2 \\ 3 & 4 & 5 \\ 2 & 4 & 2 \end{bmatrix} \begin{bmatrix} 1 & 0 & 1 \\ 3 & 2 & 4 \\ 6 & 2 & 4 \end{bmatrix}$$

$$= \begin{bmatrix} (1,2,2)\begin{pmatrix}1\\3\\6\end{pmatrix} & (1,2,2)\begin{pmatrix}0\\2\\2\end{pmatrix} & (1,2,2)\begin{pmatrix}1\\4\\4\end{pmatrix} \\ (3,4,5)\begin{pmatrix}1\\3\\6\end{pmatrix} & (3,4,5)\begin{pmatrix}0\\2\\2\end{pmatrix} & (3,4,5)\begin{pmatrix}1\\4\\4\end{pmatrix} \\ (2,4,2)\begin{pmatrix}1\\3\\6\end{pmatrix} & (2,4,2)\begin{pmatrix}0\\2\\2\end{pmatrix} & (2,4,2)\begin{pmatrix}1\\4\\4\end{pmatrix} \end{bmatrix}$$

$$= \begin{bmatrix} 19 & 8 & 17 \\ 45 & 18 & 39 \\ 26 & 12 & 26 \end{bmatrix}$$

Note, as indicated earlier, that **GH** and **HG** are quite different matrices.

For practice, apply the transpose product rule:

$$(\mathbf{GH})' = \mathbf{H}'\mathbf{G}' = \begin{bmatrix} 3 & 17 & 20 \\ 6 & 30 & 36 \\ 4 & 24 & 30 \end{bmatrix}, \qquad (\mathbf{HG})' = \mathbf{G}'\mathbf{H}' = \begin{bmatrix} 19 & 45 & 26 \\ 8 & 18 & 12 \\ 17 & 39 & 26 \end{bmatrix}$$

2.4.2 *Partitioned Products*

In computing **GH** (and **HG**) we in effect *partitioned* the prefactor into a number of row vectors and the postfactor into a number of column vectors:

$$\mathbf{G} = \begin{bmatrix} \mathbf{r}' \\ \mathbf{t}' \\ \mathbf{w}' \end{bmatrix} \quad \text{and} \quad \mathbf{H} = (\mathbf{x} \quad \mathbf{y} \quad \mathbf{z})$$

and then obtained the product as a systematic arrangement of minor products of vectors:

$$\mathbf{GH} = \begin{bmatrix} \mathbf{r}'\mathbf{x} & \mathbf{r}'\mathbf{y} & \mathbf{r}'\mathbf{z} \\ \mathbf{t}'\mathbf{x} & \mathbf{t}'\mathbf{y} & \mathbf{t}'\mathbf{z} \\ \mathbf{w}'\mathbf{x} & \mathbf{w}'\mathbf{y} & \mathbf{w}'\mathbf{z} \end{bmatrix}$$

The dot-and-transpose notation of Section 2.1 could have been used to symbolize the rows of **G** and columns of **H**. An alternative notation was used to allow us to introduce the first of two important rules on the product of partitioned matrices.

If the matrix product $\mathbf{AB} = \mathbf{C}$ is defined, we may partition the prefactor horizontally into two or more submatrices:

$$\mathbf{A} = \begin{bmatrix} \mathbf{U} \\ \mathbf{V} \\ \mathbf{W} \end{bmatrix}$$

partition the postfactor vertically:

$$\mathbf{B} = (\mathbf{X} \quad \mathbf{Y} \quad \mathbf{Z})$$

and then represent the product $\mathbf{AB} = \mathbf{C}$ as the partitioned matrix

$$\begin{bmatrix} \mathbf{U} \\ \mathbf{V} \\ \mathbf{W} \end{bmatrix} (\mathbf{X} \quad \mathbf{Y} \quad \mathbf{Z}) = \begin{bmatrix} \mathbf{UX} & \mathbf{UY} & \mathbf{UZ} \\ \mathbf{VX} & \mathbf{VY} & \mathbf{VZ} \\ \mathbf{WX} & \mathbf{WY} & \mathbf{WZ} \end{bmatrix} \tag{2.4.4}$$

Further, the partitioning of $\mathbf{B}$ can be quite arbitrary, independent of the partitioning of $\mathbf{A}$. A possible arbitrary partition of the two factors is not to partition one of them at all, in which case the rule gives, for an unpartitioned postfactor,

$$\mathbf{AB} = \begin{bmatrix} \mathbf{U} \\ \mathbf{V} \\ \mathbf{W} \end{bmatrix} \mathbf{B} = \begin{bmatrix} \mathbf{UB} \\ \mathbf{VB} \\ \mathbf{WB} \end{bmatrix} = \mathbf{C}$$

Let us apply this partitioned product rule to the numeric example $\mathbf{GH}$, taking the partition of $\mathbf{G}$ and of $\mathbf{H}$ to be

$$\mathbf{G} = \left[\begin{array}{ccc} 1 & 0 & 1 \\ \hline 3 & 2 & 4 \\ 6 & 2 & 4 \end{array}\right] = \left[\begin{array}{c} \mathbf{u}' \\ \hline \mathbf{V}' \end{array}\right], \qquad \mathbf{H} = \left[\begin{array}{cc|c} 1 & 2 & 2 \\ 3 & 4 & 5 \\ 2 & 4 & 2 \end{array}\right] = (\mathbf{X} \mid \mathbf{y})$$

The product $\mathbf{GH}$ may be represented symbolically

$$\mathbf{GH} = \left[\begin{array}{c} \mathbf{u}' \\ \hline \mathbf{V}' \end{array}\right] (\mathbf{X} \mid \mathbf{y}) = \left[\begin{array}{c|c} \mathbf{u}'\mathbf{X} & \mathbf{u}'\mathbf{y} \\ \hline \mathbf{V}'\mathbf{X} & \mathbf{V}'\mathbf{y} \end{array}\right]$$

which we can expand numerically to yield

$$\mathbf{GH} = \left[\begin{array}{c|c} \mathbf{u}'\mathbf{X} & \mathbf{u}'\mathbf{y} \\ \hline \mathbf{V}'\mathbf{X} & \mathbf{V}'\mathbf{y} \end{array}\right] = \left[\begin{array}{c|c} (1 \quad 0 \quad 1)\begin{pmatrix} 1 & 2 \\ 3 & 4 \\ 2 & 4 \end{pmatrix} & (1 \quad 0 \quad 1)\begin{pmatrix} 2 \\ 5 \\ 2 \end{pmatrix} \\ \hline \begin{pmatrix} 3 & 2 & 4 \\ 6 & 2 & 4 \end{pmatrix}\begin{pmatrix} 1 & 2 \\ 3 & 4 \\ 2 & 4 \end{pmatrix} & \begin{pmatrix} 3 & 2 & 4 \\ 6 & 2 & 4 \end{pmatrix}\begin{pmatrix} 2 \\ 5 \\ 2 \end{pmatrix} \end{array}\right]$$

or

$$\mathbf{GH} = \left[\begin{array}{cc|c} 3 & 6 & 4 \\ \hline 17 & 30 & 24 \\ 20 & 36 & 30 \end{array}\right]$$

The product corresponds to the earlier result.

The value of this partitioning rule (and the one to follow) is not computational but symbolic. In Chapter 1 we pointed out that data matrices, more often than not, are partitioned—vertically between predictors and criteria, for example, and horizontally between groups of research entities. Because research questions may depend on these partitionings of the data, it will be helpful to be able to trace the effects of the partitioning on the results.

A second rule on the product of partitioned matrices states that if the matrix product $\mathbf{AB} = \mathbf{C}$ is defined and the prefactor and postfactor have each been symmetrically partitioned, vertically and horizontally, respectively,

$$\mathbf{A} = (\mathbf{U} \quad \mathbf{V} \quad \mathbf{W}) \qquad \text{and} \qquad \mathbf{B} = \begin{bmatrix} \mathbf{X} \\ \mathbf{Y} \\ \mathbf{Z} \end{bmatrix}$$

then the overall product may be represented as a sum of matrix products:

$$\mathbf{AB} = (\mathbf{U} \quad \mathbf{V} \quad \mathbf{W}) \begin{bmatrix} \mathbf{X} \\ \hline \mathbf{Y} \\ \hline \mathbf{Z} \end{bmatrix} = (\mathbf{UX} + \mathbf{VY} + \mathbf{WZ}) \tag{2.4.5}$$

Because the product $\mathbf{AB}$ is defined, we know that the number of columns in $\mathbf{A}$ is equal to the number of rows in $\mathbf{B}$. Symmetrically partitioning the two factors divides the columns of $\mathbf{A}$ in exactly the same way that it separates the rows of $\mathbf{B}$: $\mathbf{U}$ will have as many columns as $\mathbf{X}$ has rows, $\mathbf{V}$ as many columns as $\mathbf{Y}$ has rows, and $\mathbf{W}$ as many columns as $\mathbf{Z}$ has rows. This is essential, since in Eq. (2.4.5) the products $\mathbf{UX}$, $\mathbf{VY}$, and $\mathbf{WZ}$ must all be defined.

This second rule may be illustrated by recomputing the product $\mathbf{HG}$ after partitioning the two matrices as

$$\mathbf{H} = (\mathbf{H}_{.1} \quad \mathbf{H}_{.2} \quad \mathbf{H}_{.3}) = \left[\begin{pmatrix} 1 \\ 3 \\ 2 \end{pmatrix} \begin{pmatrix} 2 \\ 4 \\ 4 \end{pmatrix} \begin{pmatrix} 2 \\ 5 \\ 2 \end{pmatrix}\right]$$

and as

$$\mathbf{G} = \begin{bmatrix} \mathbf{G}'_{1.} \\ \mathbf{G}'_{2.} \\ \mathbf{G}'_{3.} \end{bmatrix} = \begin{bmatrix} (1, 0, 1) \\ (3, 2, 4) \\ (6, 2, 4) \end{bmatrix}$$

using the dot-and-transpose notation. **H** and **G** have been symmetrically partitioned. Applying the partitioned product rule of Eq. (2.4.5), we can write the product

$$\mathbf{HG} = (\mathbf{H}_{.1} \quad \mathbf{H}_{.2} \quad \mathbf{H}_{.3}) \begin{bmatrix} \mathbf{G}'_{1.} \\ \mathbf{G}'_{2.} \\ \mathbf{G}'_{3.} \end{bmatrix} = (\mathbf{H}_{.1}\mathbf{G}'_{1.} + \mathbf{H}_{.2}\mathbf{G}'_{2.} + \mathbf{H}_{.3}\mathbf{G}'_{3.})$$

and then substitute the numerical vectors

$$\mathbf{HG} = \begin{bmatrix} 1 \\ 3 \\ 2 \end{bmatrix} (1, 0, 1) + \begin{bmatrix} 2 \\ 4 \\ 4 \end{bmatrix} (3, 2, 4) + \begin{bmatrix} 2 \\ 5 \\ 2 \end{bmatrix} (6, 2, 4)$$

to obtain

$$\mathbf{HG} = \begin{bmatrix} 1 & 0 & 1 \\ 3 & 0 & 3 \\ 2 & 0 & 2 \end{bmatrix} + \begin{bmatrix} 6 & 4 & 8 \\ 12 & 8 & 16 \\ 12 & 8 & 16 \end{bmatrix} + \begin{bmatrix} 12 & 4 & 8 \\ 30 & 10 & 20 \\ 12 & 4 & 8 \end{bmatrix}$$

or finally

$$\mathbf{HG} = \begin{bmatrix} 7 & 4 & 9 \\ 15 & 8 & 19 \\ 14 & 8 & 18 \end{bmatrix} + \begin{bmatrix} 12 & 4 & 8 \\ 30 & 10 & 20 \\ 12 & 4 & 8 \end{bmatrix} = \begin{bmatrix} 19 & 8 & 17 \\ 45 & 18 & 39 \\ 26 & 12 & 26 \end{bmatrix}$$

which is the result we obtained earlier, calculating row-by-column products.

2.4.3 Scaling

From time to time we perform such *scaling* operations as, say, to multiply every element in a matrix **X** by 50 or to divide each element in a vector **v** by N, which might, for example, be the number of cases. Assume that

$$\mathbf{v} = \begin{bmatrix} 14 \\ 12 \\ 10 \end{bmatrix} \quad \text{and} \quad \mathbf{X} = \begin{bmatrix} -10 & 20 \\ 8 & -5 \\ 12 & 10 \end{bmatrix}$$

By our rule on the common order of products, it would be perfectly proper to write

$$\mathbf{v}(1/N) = \begin{bmatrix} 14 \\ 12 \\ 10 \end{bmatrix} (1/N) = \begin{bmatrix} 14/N \\ 12/N \\ 10/N \end{bmatrix}$$

but the only way to secure the desired scaling of the **X** matrix is to introduce a *scalar matrix* (hence the name) to form one of the factors.

$$\mathbf{D}_{50}\mathbf{X} = \begin{bmatrix} 50 & 0 & 0 \\ 0 & 50 & 0 \\ 0 & 0 & 50 \end{bmatrix} \begin{bmatrix} -10 & 20 \\ 8 & -5 \\ 12 & 10 \end{bmatrix} = \begin{bmatrix} -500 & 1000 \\ 400 & -250 \\ 600 & 500 \end{bmatrix}$$

or

$$\mathbf{XD}_{50} = \begin{bmatrix} -10 & 20 \\ 8 & -5 \\ 12 & 10 \end{bmatrix} \begin{bmatrix} 50 & 0 \\ 0 & 50 \end{bmatrix} = \begin{bmatrix} 500 & 1000 \\ 400 & -250 \\ 600 & 500 \end{bmatrix}$$

Both notations, $\mathbf{v}(1/N)$ and $\mathbf{D}_{50}\mathbf{X}$, seem rather clumsy, and since a major reason for introducing matrix notation and convention is to facilitate understanding through the simplification of multivariate expressions, we propose to abandon a purist stance. We shall write (50)**X** for the multiplication of every element in a matrix **X** by 50; and when we want to divide each element in a vector **v** by N, we shall write $(1/N)\mathbf{v}$.

2.5 Special Matrix Products

There is a small collection of special vector-by-matrix and matrix-by-matrix products that occur frequently in multivariate analysis. They provide the matrix techniques for computing sums, sums of squares and cross products, and other statistical quantities of recurring interest. In this section we introduce these special products.

2.5.1 Product Moments

In multivariate analysis, a matrix or vector is often multiplied by itself or, more precisely, by its transpose. We term the result a *product moment*.

Consider again the intrinsic scores earned on the odd-numbered items of the job values test by the five women applicants in the Preemployment Study who contributed data to matrix **A** of Section 2.3. Let a vector **u** represent those scores:

$$\mathbf{u}' = (7, 8, 0, 2, 1)$$

The *minor product moment of a vector* is the result of premultiplying a vector by its transpose. Recall that the natural form of a vector is the column form and the transpose form is the row. For the vector **u**,

$$\mathbf{u}'\mathbf{u} = (7, 8, 0, 2, 1)\begin{bmatrix} 7 \\ 8 \\ 0 \\ 2 \\ 1 \end{bmatrix} = 118$$

Thus *the minor product moment of a vector is a scalar whose value is the sum of the squares of the elements of the vector*. If **a** is a vector, $\mathbf{a}'\mathbf{a}$ is the sum of squares of that vector. In the form of a scalar equation,

$$\mathbf{u}'\mathbf{u} = \sum_{i=1}^{m} u_i^2 \qquad (2.5.1)$$

The *major product moment of a vector* results from postmultiplying a vector by its transpose:

$$\mathbf{uu}' = \begin{bmatrix} 7 \\ 8 \\ 0 \\ 2 \\ 1 \end{bmatrix} (7, 8, 0, 2, 1) = \begin{bmatrix} 49 & 56 & 0 & 14 & 7 \\ 56 & 64 & 0 & 16 & 8 \\ 0 & 0 & 0 & 0 & 0 \\ 14 & 16 & 0 & 4 & 2 \\ 7 & 8 & 0 & 2 & 1 \end{bmatrix}$$

The major product moment of a vector is a symmetric matrix. The diagonal elements of $\mathbf{uu}'$ are the squares of the elements $\mathbf{u}$ and the off-diagonals of the matrix are products of pairs of elements of $\mathbf{u}$. This result may be summarized as

$$(\mathbf{uu}')_{ij} = u_i u_j \tag{2.5.2}$$

where u_i and u_j are the ith and jth elements of the vector $\mathbf{u}$ and $(\mathbf{uu}')_{ij}$ denotes the ijth element of the matrix $\mathbf{uu}'$, the element at the intersection of the ith row and the jth column.

If the vector $\mathbf{u}$ is *augmented* by a second vector $\mathbf{v}$ carrying the intrinsic scores for these same women applicants but based now on the even-numbered test items (from matrix $\mathbf{B}$ of Section 2.3), we can write a matrix $\mathbf{X}$, where

$$\mathbf{X}' = \begin{bmatrix} \mathbf{u}' \\ \mathbf{v}' \end{bmatrix} = \begin{bmatrix} 7 & 8 & 0 & 2 & 1 \\ 8 & 10 & 1 & 0 & 1 \end{bmatrix}$$

The *minor product moment of a matrix*, like that of a vector, is the result of premultiplying the matrix by its transpose:

$$\mathbf{X}'\mathbf{X} = \begin{bmatrix} \mathbf{u}' \\ \mathbf{v}' \end{bmatrix} (\mathbf{u}, \mathbf{v}) = \begin{bmatrix} \mathbf{u}'\mathbf{u} & \mathbf{u}'\mathbf{v} \\ \mathbf{v}'\mathbf{u} & \mathbf{v}'\mathbf{v} \end{bmatrix} = \begin{bmatrix} 118 & 137 \\ 137 & 166 \end{bmatrix}$$

We have taken advantage of the first partitioned product rule by initially partitioning the prefactor horizontally and the postfactor vertically. $\mathbf{u}'\mathbf{u}$ and $\mathbf{v}'\mathbf{v}$ are sums of squares of the elements of $\mathbf{v}$ and of $\mathbf{u}$, and $\mathbf{u}'\mathbf{v}$ is the sum of products between the two vectors. Note that $\mathbf{v}'\mathbf{u}$ is identical to $\mathbf{u}'\mathbf{v}$. This is consistent with the transpose product rule,

$$(\mathbf{u}'\mathbf{v})' = \mathbf{v}'\mathbf{u}$$

since $\mathbf{u}'\mathbf{v}$ is a scalar.

Quite generally the minor product moment of a matrix is a symmetric matrix with elements defined by

$$(\mathbf{X}'\mathbf{X})_{ij} = \mathbf{X}'_{.i}\mathbf{X}_{.j} \tag{2.5.3}$$

The diagonal elements of $\mathbf{X}'\mathbf{X}$ are the sums of squares of the respective columns of $\mathbf{X}$, and the off-diagonal elements are sums of products between columns. Thus *the minor product moment of a data matrix contains the sums of squares of the observations on each attribute down the diagonal and the sums of products between attributes in the off-diagonal positions.*

To obtain the *major product moment of a matrix*, we postmultiply a matrix by its transpose. Thus

$$\mathbf{XX}' = \begin{bmatrix} 7 & 8 \\ 8 & 10 \\ 0 & 1 \\ 2 & 0 \\ 1 & 1 \end{bmatrix} \begin{bmatrix} 7 & 8 & 0 & 2 & 1 \\ 8 & 10 & 1 & 0 & 1 \end{bmatrix} = \begin{bmatrix} 113 & 136 & 8 & 14 & 15 \\ 136 & 164 & 10 & 16 & 18 \\ 8 & 10 & 1 & 0 & 1 \\ 14 & 16 & 0 & 4 & 2 \\ 15 & 18 & 1 & 2 & 2 \end{bmatrix}$$

The major product moment of a matrix is also symmetric, with elements defined by

$$(\mathbf{XX}')_{ij} = \mathbf{X}'_{i.}\mathbf{X}_{j.} \qquad (2.5.4)$$

The diagonal elements of $\mathbf{XX}'$ are sums of squares over the rows of $\mathbf{X}$ ($7^2 + 8^2 = 113$, $8^2 + 10^2 = 164$), and the off-diagonal elements are the sums of products between rows [$(7 \times 8) + (8 \times 10) = 136$, $(7 \times 0) + (8 \times 1) = 8$]. Thus *the major product moment of a data matrix has sums of squares of the observations for each entity down the diagonal and sums of products between entities in the off-diagonal positions.*

Every square matrix has associated with it a scalar value called its *trace* and symbolized tr($\mathbf{S}$), where $\mathbf{S}$ is the square matrix. It is computed for a $p \times p$ matrix as

$$\text{tr}(\mathbf{S}) = \sum_{i=1}^{p} S_{ii} \qquad (2.5.5)$$

That is, the trace of $\mathbf{S}$ is the sum of the elements along its diagonal. Because they are each symmetric, the minor and major product moment matrices are also square and have traces.

For the minor product moment of $\mathbf{X}$,

$$\text{tr}(\mathbf{X}'\mathbf{X}) = 118 + 166 = 284$$

and for the major product moment of $\mathbf{X}$,

$$\text{tr}(\mathbf{XX}') = 113 + 164 + 1 + 4 + 2 = 284$$

This equality of the traces will hold for any matrix $\mathbf{X}$:

$$\text{tr}(\mathbf{X}'\mathbf{X}) = \text{tr}(\mathbf{XX}') \qquad (2.5.6)$$

Why is this true? What interpretation can we give tr($\mathbf{X}'\mathbf{X}$) and tr($\mathbf{XX}'$)? Recall that the diagonal elements of $\mathbf{X}'\mathbf{X}$ are sums of squares down the columns of $\mathbf{X}$. If we sum those diagonal elements, we shall have the sum of squares of all the elements of $\mathbf{X}$. As the diagonal elements of $\mathbf{XX}'$ are sums of squares across the rows of $\mathbf{X}$, summing those ought also to give the overall sum of squares. For an $n \times m$ matrix $\mathbf{X}$,

$$\operatorname{tr}(\mathbf{X}'\mathbf{X}) = \operatorname{tr}(\mathbf{X}\mathbf{X}') = \sum_{i=1}^{n} \sum_{j=1}^{m} x_{ij}^2 \tag{2.5.7}$$

2.5.2 Unit Vector Products

The unit vector, introduced in Section 2.2, is sometimes called the *summing vector* for its role in vector-by-matrix multiplication. For example, we can premultiply

$$\mathbf{X} = \begin{bmatrix} 7 & 9 \\ 8 & 10 \\ 0 & 1 \\ 2 & 0 \\ 1 & 1 \end{bmatrix}$$

by the conformable unit vector. Because $\mathbf{X}$ is the postfactor, the premultiplier must have five columns, matching the number of rows in $\mathbf{X}$. The premultiplying unit vector, then, is a row with five entries:

$$\mathbf{1}'\mathbf{X} = (1, 1, 1, 1, 1)\begin{bmatrix} 7 & 8 \\ 8 & 10 \\ 0 & 1 \\ 2 & 0 \\ 1 & 1 \end{bmatrix} = (18, 20)$$

Premultiplying a matrix by the row unit vector produces a row vector of column sums. The sums of the elements in the two columns of $\mathbf{X}$ are 18 and 20. If $\mathbf{X}$ is a data matrix, $\mathbf{1}'\mathbf{X}$ sums, separately, the observations on each attribute.

We can also postmultiply $\mathbf{X}$ by a unit vector. Since $\mathbf{X}$ is now the prefactor, the postmultiplier must have rows to match the number of $\mathbf{X}$ columns. We can write

$$\mathbf{X1} = \begin{bmatrix} 7 & 8 \\ 8 & 10 \\ 0 & 1 \\ 2 & 0 \\ 1 & 1 \end{bmatrix}\begin{bmatrix} 1 \\ 1 \end{bmatrix} = \begin{bmatrix} 15 \\ 18 \\ 1 \\ 2 \\ 2 \end{bmatrix}$$

Postmultiplying a matrix by the column unit vector produces a column of row sums. If the matrix is a data matrix, the effect is to sum the observations for each entity.

A matrix can be both pre- and postmultiplied by unit vectors. It would be more correct to say we can *first* premultiply and *then* postmultiply (or first postmultiply and then premultiply), as we do not know how to pre- and postmultiply simultaneously. The result will be the same whether we first pre- or postmultiply:

$$\mathbf{1}'\mathbf{X}\mathbf{1} = (\mathbf{1}'\mathbf{X})\mathbf{1} = \mathbf{1}'(\mathbf{X}\mathbf{1})$$

For our numerical example,

$$(\mathbf{1}'\mathbf{X})\mathbf{1} = (18 \quad 20)\begin{bmatrix} 1 \\ 1 \end{bmatrix} = (38)$$

and

$$\mathbf{1}'(\mathbf{X}\mathbf{1}) = (1 \quad 1 \quad 1 \quad 1 \quad 1)\begin{bmatrix} 15 \\ 18 \\ 1 \\ 2 \\ 2 \end{bmatrix} = (38)$$

This *triple product* is equal to the sum of the elements in the matrix **X**. That is, if **X** is $n \times m$,

$$\mathbf{1}'\mathbf{X}\mathbf{1} = \sum_{i=1}^{n} \sum_{j=1}^{m} X_{ij} \tag{2.5.8}$$

Forming a triple product is possible only when adjacent factors match in their common orders. The triple product has the number of rows of the first factor and the number of columns of the last factor:

$$\mathbf{A}_{(n\times m)}\mathbf{B}_{(m\times p)}\mathbf{C}_{(p\times q)} = \mathbf{P}_{(n\times q)} \tag{2.5.9}$$

In general, it makes no difference whether we compute **AB** or **BC** first:

$$\mathbf{A}(\mathbf{BC}) = (\mathbf{AB})\mathbf{C} = \mathbf{ABC} \tag{2.5.10}$$

2.5.3 Elementary Vector Products

Another special case of vector-by-matrix multiplication is that of multiplying by *elementary* $\mathbf{e}_i$ vectors. For example, we can premultiply **X** by the $\mathbf{e}_3$ vector,

$$\mathbf{e}_3'\mathbf{X} = (0, 0, 1, 0, 0)\begin{bmatrix} 7 & 8 \\ 8 & 10 \\ 0 & 1 \\ 2 & 0 \\ 1 & 1 \end{bmatrix} = (0, 1)$$

to *extract the third row*, or we can postmultiply by $\mathbf{e}_1$,

$$\mathbf{Xe}_1 = \begin{bmatrix} 7 & 8 \\ 8 & 10 \\ 0 & 1 \\ 2 & 0 \\ 1 & 1 \end{bmatrix} \begin{bmatrix} 1 \\ 0 \end{bmatrix} = \begin{bmatrix} 7 \\ 8 \\ 0 \\ 2 \\ 1 \end{bmatrix}$$

to *extract the first column.*

If we *both* pre- and postmultiply $\mathbf{X}$ by these elementary vectors,

$$\mathbf{e}_3'\mathbf{Xe}_1 = (\mathbf{e}_3'\mathbf{X})\mathbf{e}_1 = (0 \quad 1)\begin{bmatrix} 1 \\ 0 \end{bmatrix} = (0)$$

we extract X_{31}, the element at the intersection of the third row and first column.

In general, for any $\mathbf{X}$,

$$\mathbf{e}_i'\mathbf{X} = \mathbf{X}_{i.}'$$

$$\mathbf{Xe}_j = \mathbf{X}_{.j}$$

$$\mathbf{e}_i'\mathbf{Xe}_j = X_{ij}$$

Elementary vectors are used to extract rows, columns, and elements from a matrix.

By using such triple products, a matrix can be multiplied by a unit vector and by an elementary vector. The reader may find it useful to determine the triple products $\mathbf{e}_2'\mathbf{X1}$ and $\mathbf{1}'\mathbf{Xe}_2$.

2.5.4 Matrix Centering

A special matrix multiplication used frequently in multivariate analysis is one that *centers* a matrix. A matrix $\mathbf{W}$ is *centered by columns* if

$$\mathbf{1}'\mathbf{W} = \mathbf{0}'$$

that is, if the sum of the elements in each column is zero; it is *centered by rows* if the sum across each row is zero:

$$\mathbf{W1} = \mathbf{0}$$

A matrix is called *doubly centered* if $\mathbf{1}'\mathbf{W} = \mathbf{0}'$ *and* $\mathbf{W1} = \mathbf{0}$.

Centering a data matrix by columns requires putting each attribute into deviation score form, that is, subtracting the column mean from each observation. This can be accomplished using matrix notation by first defining a matrix called a *centering matrix*. The centering matrix will always be of the form

$$\left[\mathbf{I} - \frac{\mathbf{11}'}{k}\right]$$

where **I** is the identity matrix and **1** is the unit vector, both of order k. The centering matrix for $k = 3$ would look like this:

$$\left[\begin{pmatrix} 1 & 0 & 0 \\ 0 & 1 & 0 \\ 0 & 0 & 1 \end{pmatrix} - \tfrac{1}{3}\begin{pmatrix} 1 & 1 & 1 \\ 1 & 1 & 1 \\ 1 & 1 & 1 \end{pmatrix}\right] = \begin{bmatrix} \frac{2}{3} & -\frac{1}{3} & -\frac{1}{3} \\ -\frac{1}{3} & \frac{2}{3} & -\frac{1}{3} \\ -\frac{1}{3} & -\frac{1}{3} & \frac{2}{3} \end{bmatrix}$$

In practice, we need not calculate the centering matrix. Symbolically premultiplying a data matrix by the centering matrix centers the data matrix by columns as desired. Postmultiplying a data matrix by the centering matrix centers the data matrix by rows. However, the centering matrix differs in the two instances. To be conformable, the column-centering (premultiplying) square matrix must be of the same order as the number of rows in the data matrix, while the row-centering (postmultiplying) matrix, also square, must have rows and columns equal in number to the number of columns of the data matrix. Pre- and postmultiplying a data matrix by the two conformable centering matrices doubly centers the matrix.

Because it is more common to center by columns, we shall examine what happens when we premultiply a data matrix by a centering matrix:

$$\left(\mathbf{I} - \frac{\mathbf{11}'}{n}\right)\mathbf{X} = \mathbf{Z} \tag{2.5.11}$$

where **X** is $n \times m$ and the centering matrix is $n \times n$. This equation provides an opportunity to introduce the *distributive property* of *matrix multiplication*:

$$(\mathbf{A} + \mathbf{B})\mathbf{C} = \mathbf{AC} + \mathbf{BC} \tag{2.5.12}$$

As **A** and **B** are of the same size (they must be, recall, if they are to be added), each can be a premultiplier of **C**. Similarly,

$$\mathbf{C}(\mathbf{F} + \mathbf{G}) = \mathbf{CF} + \mathbf{CG}$$

Expanding the centering equation by the distributive principle gives

$$\left(\mathbf{I} - \frac{\mathbf{11}'}{n}\right)\mathbf{X} = \mathbf{IX} - \mathbf{1}(1/n)\mathbf{1}'\mathbf{X}$$

Here we take advantage of another convention in writing matrix products. If **A**, **B**, and **C** are matrices and k is a scalar,

$$(k\mathbf{A})\mathbf{BC} = \mathbf{A}(k\mathbf{B})\mathbf{C} = \mathbf{AB}(k\mathbf{C}) \tag{2.5.13}$$

That is, we can *permute the scalar* to any position in the multiplicative string without changing the value of the product. We chose to place the scalar $1/n$ between the column vector **1** and the row vector **1′X**. The reason for doing this will be provided shortly.

Before continuing with the development of the centered matrix, you may want to verify that multiplication by the identity matrix, either as a prefactor or a postfactor, has no effect; that is,

$$\mathbf{IX} = \mathbf{XI} = \mathbf{X} \tag{2.5.14}$$

As a result,

$$\left(\mathbf{I} - \frac{\mathbf{11'}}{n}\right)\mathbf{X} = \mathbf{X} - \mathbf{1}(1/n)\mathbf{1'X}$$

Recall that the product $\mathbf{1'X}$ is a row of column sums from the matrix $\mathbf{X}$. Dividing each element in this row vector by n, the number of rows summed over, gives a row vector with elements equal to the mean of each column $\mathbf{X}$. For the matrix

$$\mathbf{X} = \begin{bmatrix} 7 & 8 \\ 8 & 10 \\ 0 & 1 \\ 2 & 0 \\ 1 & 1 \end{bmatrix}$$

we have $\mathbf{1'X} = (18, 20)$ and $(1/n)\mathbf{1'X} = (3.6, 4.0)$. For *any* matrix with n rows, $(1/n)\mathbf{1'X}$ is a row vector of column means.

What is the effect of premultiplying this row vector by the column unit vector? What is $\mathbf{1}[(1/n)\mathbf{1'X}]$? For our example,

$$\begin{bmatrix} 1 \\ 1 \\ 1 \\ 1 \\ 1 \end{bmatrix}(3.6 \quad 4.0) = \begin{bmatrix} 3.6 & 4.0 \\ 3.6 & 4.0 \\ 3.6 & 4.0 \\ 3.6 & 4.0 \\ 3.6 & 4.0 \end{bmatrix}$$

The product $\mathbf{1}(1/n)\mathbf{1'X}$ is a matrix, the same size as $\mathbf{X}$, with elements equal to the mean value of the corresponding column of $\mathbf{X}$.

We are ready to illustrate the last computational step in centering $\mathbf{X}$. If we subtract from $\mathbf{X}$ the matrix just computed, we obtain

$$\begin{bmatrix} 7 & 8 \\ 8 & 10 \\ 0 & 1 \\ 2 & 0 \\ 1 & 1 \end{bmatrix} - \begin{bmatrix} 3.6 & 4.0 \\ 3.6 & 4.0 \\ 3.6 & 4.0 \\ 3.6 & 4.0 \\ 3.6 & 4.0 \end{bmatrix} = \begin{bmatrix} 3.4 & 4.0 \\ 4.4 & 6.0 \\ -3.6 & -3.0 \\ -1.6 & -4.0 \\ -2.6 & -3.0 \end{bmatrix}$$

for

$$\mathbf{X} - [\mathbf{1}(1/n)\mathbf{1'X}] = [\mathbf{I} - \mathbf{1}(1/n)\mathbf{1'}]\mathbf{X}$$

which is a matrix of scores in deviation score form, a matrix centered by columns. This can be verified by showing that the new column means are zero, that is, that all the column sums for this new matrix are zero:

$$(1 \quad 1 \quad 1 \quad 1 \quad 1)\begin{bmatrix} 3.4 & 4.0 \\ 4.4 & 6.0 \\ -3.6 & -3.0 \\ -1.6 & -4.0 \\ -2.6 & -3.0 \end{bmatrix} = (0.0 \quad 0.0)$$

or

$$\mathbf{1}'[\mathbf{I} - \mathbf{1}(1/n)\mathbf{1}']\mathbf{X} = 0'$$

This last point can be demonstrated algebraically. The column sums of the column centered matrix must be zero. We premultiply the centering equation by $\mathbf{1}'$:

$$\mathbf{1}'[\mathbf{I} - \mathbf{1}(1/n)\mathbf{1}']\mathbf{X} = [\mathbf{1}' - \mathbf{1}'\mathbf{1}(1/n)\mathbf{1}']\mathbf{X}$$

Because each of the unit vectors represented here contain n elements, the minor product moment $\mathbf{1}'\mathbf{1}$ on the right must be the scalar n. Substituting n for $\mathbf{1}'\mathbf{1}$ gives

$$\mathbf{1}'[\mathbf{I} - \mathbf{1}(1/n)\mathbf{1}']\mathbf{X} = [\mathbf{1}' - n(1/n)\mathbf{1}']\mathbf{X} = (\mathbf{1}' - \mathbf{1}')\mathbf{X} = \mathbf{0}'\mathbf{X} = \mathbf{0}'$$

which is the desired result.

2.5.5 *Variance–Covariance Matrices*

We have just seen how a matrix of deviation scores may be represented by the product

$$\mathbf{Z} = [\mathbf{I} - \mathbf{1}(1/n)\mathbf{1}']\mathbf{X}$$

If we now find the minor product moment of $\mathbf{Z}$, we have a matrix $\mathbf{Z}'\mathbf{Z}$ with diagonal elements equal to the sums of squares of the columns of $\mathbf{Z}$ and off-diagonal elements equal to sums of products between $\mathbf{Z}$ columns:

$$\begin{bmatrix} 3.4 & 4.4 & -3.6 & -1.6 & -2.6 \\ 4.0 & 6.0 & -3.0 & -4.0 & -3.0 \end{bmatrix}\begin{bmatrix} 3.4 & 4.0 \\ 4.4 & 6.0 \\ -3.6 & -3.0 \\ -1.6 & -4.0 \\ -2.6 & -3.0 \end{bmatrix} = \begin{bmatrix} 53.2 & 65.0 \\ 65.0 & 86.0 \end{bmatrix}$$

In terms of the original data matrix $\mathbf{X}$, however, we have in $\mathbf{Z}'\mathbf{Z}$ sums of squares of deviation scores for each attribute and sums of products of deviation scores between each pair of attributes. To obtain a variance–covariance matrix, divide each element of $\mathbf{Z}'\mathbf{Z}$ by n, the number of rows in $\mathbf{X}$: $\mathbf{S} = (1/n)\mathbf{Z}'\mathbf{Z}$. For our computational example,

$$\mathbf{S} = \tfrac{1}{5}\begin{bmatrix} 53.2 & 65.0 \\ 65.0 & 86.0 \end{bmatrix} = \begin{bmatrix} 10.64 & 13.00 \\ 13.00 & 17.20 \end{bmatrix}$$

This variance–covariance matrix can be expressed in terms of **X** as well as the matrix **Z**. We know that a column-centered matrix is represented as

$$\mathbf{Z} = [\mathbf{I} - \mathbf{1}(1/n)\mathbf{1}']\mathbf{X} \tag{2.5.15}$$

and hence that

$$\mathbf{Z}' = \mathbf{X}'[\mathbf{I} - \mathbf{1}(1/n)\mathbf{1}']$$

The minor product moment of **Z** can then be written

$$\mathbf{Z}'\mathbf{Z} = \mathbf{X}'[\mathbf{I} - \mathbf{1}(1/n)\mathbf{1}'][\mathbf{I} - \mathbf{1}(1/n)\mathbf{1}']\mathbf{X}$$

In matrix algebra, as in ordinary algebra, the product of the sums can be expressed systematically by

$$(\mathbf{A} + \mathbf{B})(\mathbf{C} + \mathbf{D}) = \mathbf{AC} + \mathbf{AD} + \mathbf{BC} + \mathbf{BD} \tag{2.5.16}$$

Applying this rule to the center pair of terms in **Z′Z** gives

$$\begin{aligned}
&[\mathbf{I} - \mathbf{1}(1/n)\mathbf{1}'][\mathbf{I} - \mathbf{1}(1/n)\mathbf{1}'] \\
&\quad = \mathbf{I} - \mathbf{1}(1/n)\mathbf{1}' - \mathbf{1}(1/n)\mathbf{1}' + \mathbf{1}(1/n)\mathbf{1}'\mathbf{1}(1/n)\mathbf{1}' \\
&\quad = \mathbf{I} - (2)\mathbf{1}(1/n)\mathbf{1}' + \mathbf{1}(1/n)(n)(1/n)\mathbf{1}' \\
&\quad = \mathbf{I} - \mathbf{1}(1/n)\mathbf{1}'
\end{aligned}$$

The centering matrix multiplied by itself yields the centering matrix. When multiplied by itself, a matrix that yields itself is called an *idempotent* matrix. If **A** is an idempotent matrix, then

$$\mathbf{AA} = \mathbf{A} \tag{2.5.17}$$

With this result taken into account **Z′Z** becomes

$$\mathbf{Z}'\mathbf{Z} = \mathbf{X}'[\mathbf{I} - \mathbf{1}(1/n)\mathbf{1}']\mathbf{X} \tag{2.5.18}$$

or, on carrying through the multiplications,

$$\mathbf{Z}'\mathbf{Z} = [\mathbf{X}' - \mathbf{X}'\mathbf{1}(1/n)\mathbf{1}']\mathbf{X}$$

and

$$\mathbf{Z}'\mathbf{Z} = \mathbf{X}'\mathbf{X} - (1/n)(\mathbf{X}'\mathbf{1})(\mathbf{1}'\mathbf{X})$$

Multiplying **Z′Z** by $1/n$ gives the variance–covariance matrix, which we can now express

$$\mathbf{S} = (1/n)[\mathbf{X}'\mathbf{X} - (1/n)(\mathbf{X}'\mathbf{1})(\mathbf{1}'\mathbf{X})]$$

or

$$\mathbf{S} = (1/n)\{\mathbf{X}'[\mathbf{I} - \mathbf{1}(1/n)\mathbf{1}']\mathbf{X}\} \tag{2.5.19}$$

Thus the determination of the covariance matrix of a set of attributes from the data matrix can be represented as a series of matrix computations:

a. Form a (row) vector of column (attribute) sums: $\mathbf{1}'\mathbf{X}$.
b. Calculate the (column) vector of attribute means: $(1/n)(\mathbf{X}'\mathbf{1})$.
c. Find the major product of the two vectors: $(1/n)(\mathbf{X}'\mathbf{1})(\mathbf{1}'\mathbf{X})$.

d. Calculate the minor product moment of the data matrix: $\mathbf{X'X}$.
e. Subtract the matrix of step (c) from the minor product moment: $\mathbf{Z'Z} = \mathbf{X'X} - (1/n)(\mathbf{X'1})(\mathbf{1'X})$.
f. Multiply each element of this difference matrix by $1/n$: $\mathbf{S} = (1/n)\mathbf{Z'Z}$.

2.5.6 Diagonal Products

The diagonal elements of $\mathbf{S}$ will be variances for the attributes or columns of $\mathbf{X}$. A diagonal matrix of variances for our example can be written,

$$\mathbf{D_S} = \begin{bmatrix} 10.64 & 0 \\ 0 & 17.20 \end{bmatrix}$$

a diagonal matrix of standard deviations

$$\mathbf{D_S^{1/2}} = \begin{bmatrix} 3.2619 & 0 \\ 0 & 4.1473 \end{bmatrix}$$

and a diagonal matrix of reciprocals of the standard deviations

$$\mathbf{D_S^{-1/2}} = \begin{bmatrix} 0.3066 & 0 \\ 0 & 0.2411 \end{bmatrix}$$

The last diagonal matrix can be used to illustrate the special nature of multiplying a matrix by a diagonal matrix. First, postmultiply $\mathbf{Z}$, the matrix of deviation scores, by $\mathbf{D_S^{-1/2}}$:

$$\mathbf{ZD_S^{-1/2}} = \begin{bmatrix} 3.4 & 4.0 \\ 4.4 & 6.0 \\ -3.6 & -3.0 \\ -1.6 & -4.0 \\ -2.6 & -3.0 \end{bmatrix} \begin{bmatrix} 0.3066 & 0 \\ 0 & 0.2411 \end{bmatrix}$$

$$= \begin{bmatrix} 1.0424 & 0.9644 \\ 1.3490 & 1.4466 \\ -1.1038 & -0.7233 \\ -0.4906 & -0.9644 \\ -0.7972 & -0.7233 \end{bmatrix}$$

Notice that the effect is to multiply each element in the first column of $\mathbf{Z}$ by 0.3066 and to multiply each element in the second column of $\mathbf{Z}$ by 0.2411. *Postmultiplication by a diagonal matrix scales the columns of a matrix*:

$$\mathbf{AD} = \mathbf{B}, \qquad \mathbf{B}_{.j} = \mathbf{A}_{.j} D_{jj} \tag{2.5.20}$$

Since $\mathbf{Z}$ was a matrix of deviation scores, $\mathbf{ZD_S^{-1/2}}$ is a matrix of standard scores. That is, each element in $\mathbf{ZD_S^{-1/2}}$ is of the form

$$\left(\mathbf{ZD_S^{-1/2}}\right)_{ij} = Z_{ij}/S_{\mathbf{X}_j} = \left(X_{ij} - \bar{X}_j\right)/S_{\mathbf{X}_j}$$

where $\overline{X}_j$ denotes the mean and $S_{\mathbf{X}_j}$ the standard deviation of the jth column of $\mathbf{X}$.

As a further example, we shall first postmultiply $\mathbf{S}$ by $\mathbf{D}_\mathbf{S}^{-1/2}$:

$$\mathbf{SD}_\mathbf{S}^{-1/2} = \begin{bmatrix} S^2_{\mathbf{X}_1} & \mathrm{Cov}(\mathbf{X}_1, \mathbf{X}_2) \\ \mathrm{Cov}(\mathbf{X}_2, \mathbf{X}_1) & S^2_{\mathbf{X}_2} \end{bmatrix} \begin{bmatrix} 1/S_{\mathbf{X}_1} & 0 \\ 0 & 1/S_{\mathbf{X}_2} \end{bmatrix}$$

$$= \begin{bmatrix} S_{\mathbf{X}_1} & \mathrm{Cov}(\mathbf{X}_1, \mathbf{X}_2)/S_{\mathbf{X}_2} \\ \mathrm{Cov}(\mathbf{X}_2, \mathbf{X}_1)/S_{\mathbf{X}_1} & S_{\mathbf{X}_2} \end{bmatrix}$$

where $\mathrm{Cov}(\mathbf{X}_1, \mathbf{X}_2)$ denotes the covariance between columns 1 and 2 of the matrix $\mathbf{X}$. Numerically,

$$\mathbf{SD}_\mathbf{S}^{-1/2} = \begin{bmatrix} 10.64 & 13.00 \\ 13.00 & 17.20 \end{bmatrix} \begin{bmatrix} 0.3066 & 0 \\ 0 & 0.2411 \end{bmatrix} = \begin{bmatrix} 3.2622 & 3.1343 \\ 3.9858 & 4.1469 \end{bmatrix}$$

Every element in the first column of $\mathbf{S}$ has been divided by the standard deviation of the first attribute (first column of $\mathbf{X}$), and every element in the second column has been divided by the standard deviation of the second attribute. (The diagonal elements of $\mathbf{SD}_\mathbf{S}^{-1/2}$ are slightly different from the elements of $\mathbf{D}_\mathbf{S}^{1/2}$ previously calculated because of rounding errors associated with reporting only four decimal places following each step.)

Next we premultiply $\mathbf{SD}_\mathbf{S}^{-1/2}$ by $\mathbf{D}_\mathbf{S}^{-1/2}$; numerically,

$$\mathbf{D}_\mathbf{S}^{-1/2}\mathbf{SD}_\mathbf{S}^{-1/2} = \begin{bmatrix} 0.3066 & 0 \\ 0 & 0.2411 \end{bmatrix} \begin{bmatrix} 3.2622 & 3.1343 \\ 3.9858 & 4.1469 \end{bmatrix} = \begin{bmatrix} 1.00 & 0.96 \\ 0.96 & 1.00 \end{bmatrix}$$

and symbolically,

$$\mathbf{D}_\mathbf{S}^{-1/2}\mathbf{SD}_\mathbf{S}^{-1/2} = \begin{bmatrix} 1/S_{\mathbf{X}_1} & 0 \\ 0 & 1/S_{\mathbf{X}_2} \end{bmatrix} \begin{bmatrix} S^2_{\mathbf{X}_1} & \mathrm{Cov}(\mathbf{X}_1, \mathbf{X}_2) \\ \mathrm{Cov}(\mathbf{X}_2, \mathbf{X}_1) & S^2_{\mathbf{X}_2} \end{bmatrix}$$

$$\times \begin{bmatrix} 1/S_{\mathbf{X}_1} & 0 \\ 0 & 1/S_{\mathbf{X}_1} \end{bmatrix}$$

or,

$$\mathbf{D}_\mathbf{S}^{-1/2}\mathbf{SD}_\mathbf{S}^{-1/2} = \begin{bmatrix} 1 & \mathrm{Cov}(\mathbf{X}_1, \mathbf{X}_2)/S_{\mathbf{X}_1}S_{\mathbf{X}_2} \\ \mathrm{Cov}(\mathbf{X}_2, \mathbf{X}_1)/S_{\mathbf{X}_1}S_{\mathbf{X}_2} & 1 \end{bmatrix}$$

$$= \begin{bmatrix} r_{\mathbf{X}_1\mathbf{X}_1} & r_{\mathbf{X}_1\mathbf{X}_2} \\ r_{\mathbf{X}_2\mathbf{X}_1} & r_{\mathbf{X}_2\mathbf{X}_2} \end{bmatrix}$$

Premultiplication by a diagonal matrix scales the rows of a matrix:

$$\mathbf{DA} = \mathbf{C}, \qquad \mathbf{C}_{i.} = D_{ii}\mathbf{A}_{i.} \tag{2.5.21}$$

Premultiplying $\mathbf{SD}_\mathbf{S}^{-1/2}$ by $\mathbf{D}_\mathbf{S}^{-1/2}$ produces a matrix of correlations or, more precisely, an *intercorrelation matrix*. If $\mathbf{S}$ is a variance–covariance matrix, then

$$\mathbf{R} = \mathbf{D}_\mathbf{S}^{-1/2}\mathbf{SD}_\mathbf{S}^{-1/2}; \qquad R_{ii} = 1.0, \quad R_{ij} = r_{\mathbf{x}_i\mathbf{x}_j} \tag{2.5.22}$$

is an intercorrelation matrix.

Consider another example of premultiplication by a diagonal matrix, where the matrix $\mathbf{X}$ consisting of the raw scores of odd- and even-numbered items in a test scored for extrinsic job values is premultiplied by a diagonal matrix:

$$\begin{bmatrix} \frac{1}{15} & & & & \\ & \frac{1}{18} & & 0 & \\ & & \frac{1}{1} & & \\ & 0 & & \frac{1}{2} & \\ & & & & \frac{1}{2} \end{bmatrix} \begin{bmatrix} 7 & 8 \\ 8 & 10 \\ 0 & 1 \\ 2 & 0 \\ 1 & 1 \end{bmatrix} = \begin{bmatrix} 0.47 & 0.53 \\ 0.44 & 0.56 \\ 0.00 & 1.00 \\ 1.00 & 0.00 \\ 0.50 & 0.50 \end{bmatrix}$$

The result is a scaling of the performance of each of the five women applicants (scaling by rows or entities), so that we know the proportion of extrinsic score points earned on odd- and even-numbered items, respectively.

Premultiplication by a diagonal matrix scales (multiplies) *every element in a given row by a constant*, the corresponding diagonal element; *postmultiplication by a diagonal scales every element in a given column by a constant*, the corresponding diagonal element. Earlier we encountered a special case of this when we noted that pre- or postmultiplication by the identity matrix, a diagonal matrix of 1s, results in the multiplication of every matrix element by one.

2.5.7 Permutation of Rows or Columns

We mention one more special matrix product, multiplication by a permutation matrix. As its name implies, a permutation matrix rearranges the rows or columns of a matrix. Like multiplication by a diagonal matrix, premultiplication affects the rows, postmultiplication the columns. Consider again the matrix $\mathbf{X}$. First premultiply by a left-side permutation matrix π_L:

$$\pi_L \mathbf{X} = \begin{bmatrix} 0 & 1 & 0 & 0 & 0 \\ 1 & 0 & 0 & 0 & 0 \\ 0 & 0 & 0 & 1 & 0 \\ 0 & 0 & 0 & 0 & 1 \\ 0 & 0 & 1 & 0 & 0 \end{bmatrix} \begin{bmatrix} 7 & 8 \\ 8 & 10 \\ 0 & 1 \\ 2 & 0 \\ 1 & 1 \end{bmatrix} = \begin{bmatrix} 8 & 10 \\ 7 & 8 \\ 2 & 0 \\ 1 & 1 \\ 0 & 1 \end{bmatrix}$$

The left-side permutation matrix was written to rearrange the rows of $\mathbf{X}$ in descending order relative to the scores in the first column of $\mathbf{X}$. Note how this was accomplished:

a. The single 1 in the first row of π_L is in the position corresponding to the row of the postfactor to be moved to the first row of the product.
b. The single 1 in the second row of π_L is in the position corresponding to the row of the postfactor to be moved to the second row of the product.
c. One proceeds similarly for the third and subsequent rows of π_L. The last row of π_L has 1 in the position corresponding to the row of $\mathbf{X}$ that is to appear as the last row in the rearranged matrix.

If we wish to rearrange the columns of $\mathbf{X}$, we can postmultiply by a right-side permutation matrix π_R:

$$\mathbf{X}\pi_R = \begin{bmatrix} 7 & 8 \\ 8 & 10 \\ 0 & 1 \\ 2 & 0 \\ 1 & 1 \end{bmatrix} \begin{bmatrix} 0 & 1 \\ 1 & 0 \end{bmatrix} = \begin{bmatrix} 8 & 7 \\ 10 & 8 \\ 1 & 0 \\ 0 & 2 \\ 1 & 1 \end{bmatrix}$$

Here are the rules for constructing a right-side or column permutation matrix:

a. Place a single 1 in that position of the *first column* of π_R corresponding to the column of the prefactor to be the first column in the product.
b. Place a 1 in the *second column* of π_R in the position of the column to become the second column.
c. Continue in this way until a 1 is placed in the *last column* of π_R in the position corresponding to the column that is to be moved to the last column of the result.

2.6 Matrix Operations and the Computer

In the preceding section we discussed how the matrix operations of transposition, addition, and multiplication can be used in sequence to move from a data matrix to such derived matrices as the mean vector, the variance–covariance matrix, and the intercorrelation matrix. These same building blocks, together with a few more matrix operations that we introduce in subsequent chapters of Part I, are used to present and to explain the multivariate analyses of Parts II–IV. The ability to manipulate matrices in much the same manner as scalar quantities is very important to understanding what is going on in multivariate analysis.

Such procedures also provide ways of obtaining numerical solutions to multivariate problems. However, the multivariate techniques we present are standard ones for which highly efficient computer programs have been written and made broadly available, and we concentrate on these programs when we discuss numerical solutions. On occasion, we utilize matrix operations programs, notably those contained in the highly developed and widely distributed International Mathematical and Statistical Library (IMSL), to illustrate results *and* to acquaint the reader with the use of such a library of matrix routines. As computational aids, however, such resources are most useful when we need to go beyond the standard analyses. We shall not require that the reader learn matrix programming in order to find principal components or to use discriminant function analysis.

Although we have concentrated on matrix computation in the present chapter, our interest in matrix algebra is almost exclusively conceptual or symbolic. Matrix algebra presents us with the easiest way, perhaps the only way, to represent what we wish to accomplish in multivariate analysis. The hands-on experience with matrices that we offer here is intended to help readers to understand what is being accomplished in different matrix manipulations rather than to transform them into row-by-column multiplying machines.

2.7 Summary

This chapter presents the necessary matrix algebra to prepare the reader for the analysis of multivariate data. The researcher must become accustomed to thinking of data from three perspectives. First, we have *scalars*—data on m attributes for each of the n entities. Second, we can consider the data as *vectors*—column vectors on n entities for each of the m attributes or as row vectors of m attribute scores for each of n entities. Finally, at the *matrix* level we have the multivariate data set of observations on m attributes for n entities. We also have many derived matrices based upon the raw data, such as matrices of variances and covariances or intercorrelations. Conceptualizing the observations as a matrix of data is essential to multivariate analysis.

We defined the matrix operations of transposition, addition, subtraction, and multiplication. Sets of simultaneous linear equations such as are encountered in the analysis of multivariate data are most easily represented and solved using these matrix operations. Special matrices such as square, symmetric, diagonal, identity, scalar, permutation, elementary, null, and idempotent matrices were defined and illustrated. We also examined in detail certain frequently encountered matrix products, such as minor and major products, and the effects of multiplying a matrix by the conformable unit vector and centering matrix. The use of many of these concepts and operations was illustrated by the computation of a matrix of variances and covariances.

We do not intend for the reader to become a machine for the multiplication of matrices. Computer programs are widely available for all operations discussed in this chapter. However, the reader must become adept at using matrix notation and reasoning to follow developments based upon matrix algebra. We assume such skills throughout this book.

Exercises

Consider these five matrices: $\mathbf{A}_{(5\times 2)}$, $\mathbf{B}_{(4\times 1)}$, $\mathbf{C}_{(3\times 3)}$, $\mathbf{D}_{(5\times 3)}$, $\mathbf{E}_{(3\times 1)}$.

1. Describe the dimensions of the major product moments of **A**, **C**, and **E**.

2. Describe the dimensions of the minor product moments of **B** and **D**.

3. Give the dimensions of a matrix by which **A** might be premultiplied and describe the dimensions of the product.

4. Give the dimensions of a matrix by which **B** might be premultiplied and describe the dimensions of the product.

5. Give two examples of permissible multiplications each involving a pair of matrices from among the five given. Describe the dimensions of each product.

6. Form a triple product from three of the matrices and give the dimensions of the product.

7. The matrix **X** is partitioned

$$\mathbf{X} = \begin{bmatrix} \mathbf{U} \\ \mathbf{V} \\ \mathbf{W} \end{bmatrix}$$

Write out the major and minor product moments of **X** in terms of these submatrices. Let **U**, **V**, and **W** be 4×2, 3×2, and 2×2, respectively, and describe the dimensions of the submatrices that contribute to $\mathbf{XX}'$ and $\mathbf{X}'\mathbf{X}$.

8. Center, both by rows and by columns,

$$\mathbf{A} = \begin{bmatrix} 1 & 2 & 3 \\ 0 & 3 & 2 \\ 6 & 3 & 44 \\ 5 & 4 & 22 \end{bmatrix}$$

9. A matrix **S** is a 3×3 variance–covariance matrix:

$$\mathbf{S} = \begin{bmatrix} 9.0 & 10.0 & -12.0 \\ 10.0 & 25.0 & 12.0 \\ -12.0 & 12.0 & 16.0 \end{bmatrix}$$

Find $\mathbf{D}_\mathbf{S}^{-1/2}$ and with it construct the intercorrelation matrix.

10. Find the permutation matrix you would use to arrange the columns of **A** in Exercise 8 in the following order:

$$(\mathbf{A}_{.2} \quad \mathbf{A}_{.3} \quad \mathbf{A}_{.1})$$

11. For the matrix $\mathbf{A}$ of Exercise 8 do the following:
a. Find $\mathbf{A}'\mathbf{A}$ and $\text{tr}(\mathbf{A}'\mathbf{A})$.
b. Find $\mathbf{A}\mathbf{A}'$ and $\text{tr}(\mathbf{A}\mathbf{A}')$.
c. What is the relation between the two traces? Describe why this relation should hold.
d. For $\mathbf{e}_3' = (0, 0, 1, 0)$, find $\mathbf{e}_3'\mathbf{A}$.
e. For $\mathbf{e}_3' = (0, 0, 1)$, find $\mathbf{A}\mathbf{e}_3$.

12. For

$$\mathbf{X} = \begin{bmatrix} 4 & 0 & 0 \\ 0 & 9 & 0 \\ 0 & 0 & 36 \end{bmatrix} \quad \text{and} \quad \mathbf{X}^k = \begin{bmatrix} \frac{1}{2} & 0 & 0 \\ 0 & \frac{1}{3} & 0 \\ 0 & 0 & \frac{1}{6} \end{bmatrix}$$

what is the value of k? Why?

13. Let $\mathbf{X}$ be a square matrix. Let $\mathbf{A} = \mathbf{X11}'$ be the same size as $\mathbf{X}$. Express in words and symbols $\text{tr}(\mathbf{A})$ as a function of the elements of $\mathbf{X}$. How does $\text{tr}(\mathbf{A})$ compare with $\text{tr}(\mathbf{1}'\mathbf{1})$?

14. Let $\mathbf{c}$ be a data vector of order n. What is $(\mathbf{1}'\mathbf{c})/n$? Express the same quantity using the notation $\mathbf{1}'\mathbf{1}$.

15. For the vector $\mathbf{c}$ of Exercise 14 interpret the expression

$$(\mathbf{1}'\mathbf{1})^{-1}\mathbf{c}'\left[\mathbf{I} - \mathbf{1}(\mathbf{1}'\mathbf{1})^{-1}\mathbf{1}'\right]\mathbf{c}$$

16. For

$$\mathbf{X} = \begin{bmatrix} 2 & -\frac{1}{2} \\ 2 & +\frac{1}{2} \end{bmatrix}$$

find $\mathbf{X}'\mathbf{X}$.

17. Based on the results of Exercise 16, find the covariance matrix for the columns of $\mathbf{X}$.

18. Find the major and minor product moments of the matrix

$$\mathbf{A} = \begin{bmatrix} 1 & -3 & 1 & -1 \\ 1 & -1 & -1 & 3 \\ 1 & 1 & -1 & -3 \\ 1 & 3 & 1 & 1 \end{bmatrix}$$

and comment on the results.

19. Based on the results of Exercise 18, find the trace of $\mathbf{A}$ and of $\mathbf{A}'\mathbf{A}$.

Additional Reading

For a parallel development of these basic matrix computations at a more detailed level, the reader may consult

Horst, P. (1963). *Matrix Algebra for Social Scientists*. New York: Holt, Rinehart, Winston.

A more geometric treatment of matrix algebra is provided by

Green, P. E. (1978). *Analyzing Multivariate Data*. New York: Dryden.

Chapter 3
The Information in a Matrix

Introduction

Now that we have some familiarity with matrices, let us return to substantive concerns about the data matrix. In the behavioral sciences, attributes are commonly measured on arbitrary scales (e.g., a seven-point scale may represent levels of occupational aspiration), and often the scales for different attributes are quite disparate—and the values difficult to compare (e.g., yearly income in thousands of dollars and years of formal education). To help understand the relations among attributes, we need to be able to standardize or normalize our observations. In this chapter we consider how such transformations can be computed and some of their side-effects known.

Attributes may be related, and a question facing the multivariate analyst is whether a particular attribute is *redundant*; that is, does it add any new information about the entities sampled? We address this question by defining the rank of a matrix. We then show how to find the rank of any matrix, by a technique called triangular factoring. This discussion of rank also provides an opportunity to define two classes of derived matrices important in multivariate analysis, orthogonal and orthonormal matrices.

In Chapter 2 we introduced the matrix arithmetic operations of addition, subtraction, and multiplication. Here, after developing the notion of matrices of full rank (that is, without any redundancies), we introduce the closest thing to matrix "division." We define the reciprocal, or regular inverse, of a matrix and show, for small matrices, how such inverses may be computed. More important, we use triangular factoring and matrix inversion *symbolically* to develop generalized schemes useful for understanding any particular matrix of observations.

The rank of a matrix and its regular inverse, should one exist, will be familiar to the reader with a background in linear algebra. Such a reader may want to move on fairly directly to Chapter 4.

3.1 Metrics and Transformations

The observations that we collect in the behavioral sciences are frequently recorded in *arbitrary metrics*, and the observations of one attribute may be *incommensurable* with those of a second. What we mean by these terms is best illustrated by example. Consider a fraction of the data matrix shown in Table 1.2 for the College Prediction Study; we have transcribed the scores of ten students from that study on the spatial ability test in the first column and their earned grade point average in university physics courses in the second column:

$$\begin{bmatrix} 38 & 2.80 \\ 43 & 2.50 \\ 48 & 2.35 \\ 38 & 3.53 \\ 57 & 3.75 \\ 43 & 2.16 \\ 55 & 3.60 \\ 65 & 4.00 \\ 55 & 1.90 \\ 45 & 2.43 \end{bmatrix}$$

The numbers in the two columns are of quite different magnitudes. The spatial ability test scores have been scaled (each is a linear transformation of the number of questions answered correctly) so that for a defined population of high school students the mean score is 50 and the standard deviation is 10. The metric is clearly arbitrary; the test scores could have been left in their raw form or standardized to any other choice of mean and standard deviation. There is nothing about the attribute of spatial ability to suggest that the numbers 38, 65, and 55 are values in any "true" metric for that attribute.

In the same fashion, and solely by convention, college and university grades in the United States are often based on a range of 0.0–4.0 (failing to most outstanding), but grades could equally well be expressed (as indeed they are at some institutions), as numbers between 0 and 100. Only because we have some familiarity with these two metrics are we able to "see" larger and smaller values in a fragment of the matrix. Frequently, however, we find it difficult to make comparisons among attributes on which the observations are incommensurable (that is, in different metrics).

Two common approaches are used to translate scores into a comparable form. *Standardization of a closed set of observations* is one approach. We can illustrate this with data from the Higher Education Study. In Table 1.1 we reported, for each of the 50 states, per capita income, per capita high school completions, geographic region, and per capita appropriations of state tax funds for higher education during a recent year. The data compose a *closed set* (or population) because all 50 states are included. No sampling was involved in developing the rows of the data matrix. We can express the three attributes of the states—income, high school completions, and appropriations for higher education—in a common metric by standardizing each separately. A common standardization is illustrated in Section 2.5.4. By subtracting the attribute mean from each observation on the attribute and dividing the difference by the standard deviation of the attribute, a new data matrix is produced in which each attribute has been scaled to have a mean of 0 and a variance of 1.

For the Higher Education Study, per capita income averages 6316.22 (with a standard deviation of 1059.94), per capita high school completions has a mean of 0.0158 (and a standard deviation of 0.0018), and per capita appropriations for higher education has a mean of 66.09 and a standard deviation of 20.78. A portion of the standardized data matrix,

$$\mathbf{Z} = \left[\left(\mathbf{I} - \frac{\mathbf{11}'}{50}\right)\mathbf{X}\right]\mathbf{D}_{\mathbf{S}}^{-1/2}, \qquad \mathbf{S} = \tfrac{1}{50}\left[\mathbf{X}'\left(\mathbf{I} - \frac{\mathbf{11}'}{50}\right)\mathbf{X}\right]$$

(where **X** is the "raw score" data matrix and **S** is the variance–covariance matrix for **X**) is

	per capita income	per capita high school completions	per capita appropriations for higher education
Alaska	5.17	1.61	3.70
California	0.98	−0.63	1.54
Hawaii	0.96	0.49	1.56
Oregon	0.12	−0.22	0.77
Washington	0.72	0.57	1.11

While this standardization may only reinforce the raw score impression that Alaska, relative to the other states, has a very large per capita income and expends appreciable funds in support of higher education, it makes clearer some less obvious aspects of the data. For example, all five Pacific states are above average in per capita income, and three (the exceptions being California and Oregon) in per capita number of students completing high school. Comparisons among the columns (attributes) for the selected entities are also possible. Interestingly, with the exception of Alaska, these states appear to appropriate more tax money per capita for higher education than would be expected on the basis of their relative per capita incomes; that is, the standard scores in the third column are higher (less negative) than those in the first column.

When the selection of entities is a sampling from some larger population of individuals, objects, or situations that might have been observed, incommensurable attributes are best transformed by *norm referencing*. Norm referencing transforms an observation on a given entity relative to the distribution of observations characteristic of some normative group or population. As an example, we might have a portion of the data matrix shown in Table 1.2 for the College Prediction Study in which columns 1–3, respectively, contain raw scores (basically the number of items answered correctly) on three tests—vocabulary, quantitative skills, and spatial ability:

$$\begin{bmatrix} 32 & 7 & 3 \\ 16 & 6 & 6 \\ 14 & 23 & 9 \\ 29 & 14 & 3 \end{bmatrix}$$

The test scores are in different scales. For the normative population of high school juniors completing the battery of tests, the means (and standard deviations) for the three attributes are 20.93 (11.27), 14.62 (6.98), and 9.90 (5.90), respectively. With different means and standard deviations it is difficult to compare a student's scores on the vocabulary and spatial ability tests. We can relate the raw scores back to the performance of this normative group, however, and, by adopting a common metric *in that population* for the three tests, facilitate such comparisons. In this instance, we elected to standardize to a mean of 50 and a standard deviation of 100 by using three transformation equations:

$$\begin{aligned} \mathbf{Z}_{\text{Voc}} &= [(\mathbf{X}_{\text{Voc}} - \mathbf{20.93})/11.27]10.0 + \mathbf{50.0} \\ \mathbf{Z}_{\text{QSk}} &= [(\mathbf{X}_{\text{QSk}} - \mathbf{14.62})/6.98]10.0 + \mathbf{50.0} \\ \mathbf{Z}_{\text{SAb}} &= [(\mathbf{X}_{\text{SAb}} - \mathbf{9.90})/5.90]10.0 + \mathbf{50.0} \end{aligned}$$

Applying these transformations provides a norm-referenced data matrix:

$$\begin{bmatrix} 61 & 39 & 38 \\ 46 & 38 & 43 \\ 44 & 62 & 48 \\ 57 & 49 & 38 \end{bmatrix}$$

We are now in a position to say that the student in the second row has a higher vocabulary score than spatial ability score and that the opposite is true for the student in the third row.

The two transformations have each involved centering by columns—of the data matrix itself in the Higher Education Study and of a reference population matrix for the College Prediction Study. We may also want to center by rows as a data matrix transformation—the principle is the same—and sometimes we obtain data matrices that are already centered by rows, or nearly so. In centering by attributes (over entities) we want to be certain that the entities are a closed set or population of interest. Standardizing over a sample, as we shall discuss further in the following section, may actually destroy important information in the data matrix. For example, if we had standardized this 4×3 College Prediction Study data matrix by computing means and standard deviations over the four observations and using these to center the matrix, all three transformed attributes would have had the same mean and we would have lost the information that, compared with the high school normative group, our sample had higher scores on vocabulary than on spatial ability.

Similarly, in centering by rows the attributes over which we are standardizing should define, or at least be systematically representative of, some population. Let us consider two examples. It has been commonly observed that when an attribute is measured by permitting a judge to rate the extent to which the attribute is present in each of a series of entities, the resulting observations or numerical judgments contain information about not only the rated entities but also the rating behavior of the judge. For example, the Preemployment Study described in Section 1.5 involves, among the criterion measures, supervisory ratings of employee performance. Assume that the following matrix was obtained by having each of five supervisors (corresponding to the rows) rate the same ten employees on a ten-point scale (with 1 indicating low performance and 10 indicating high performance):

$$\mathbf{X}' = \begin{bmatrix} 1 & 2 & 3 & 4 & 5 & 6 & 7 & 8 & 9 & 10 \\ 5 & 5 & 5 & 5 & 5 & 6 & 6 & 6 & 6 & 6 \\ 1 & 1 & 1 & 2 & 2 & 2 & 3 & 3 & 3 & 4 \\ 7 & 8 & 8 & 8 & 9 & 9 & 9 & 10 & 10 & 10 \\ 2 & 3 & 3 & 5 & 5 & 5 & 6 & 6 & 7 & 9 \end{bmatrix}$$

The five judges (supervisors) use the rating scale in quite different ways. The first and fifth judges use the full range of the scale. The second judge uses only the middle of the scale, and the third and fourth typically assign low and high ratings, respectively. Within these limitations, the supervisors tend, however, to agree on which employees are "better."

The data matrix could be centered by rows after computing row means (5.5, 5.5, 2.2, 8.8, and 1.5 from top to bottom) to produce

$$\mathbf{X}'\left[\mathbf{I} - \frac{\mathbf{11}'}{10}\right] = \begin{bmatrix} -4.5 & -3.5 & -2.5 & -1.5 & -0.5 & 0.5 & 1.5 & 2.5 & 3.5 & 4.5 \\ -0.5 & -0.5 & -0.5 & -0.5 & -0.5 & 0.5 & 0.5 & 0.5 & 0.5 & 0.5 \\ -1.2 & -1.2 & -1.2 & -0.2 & -0.2 & -0.2 & 0.8 & 0.8 & 0.8 & 1.8 \\ -1.8 & -0.8 & -0.8 & -0.8 & 0.2 & 0.2 & 0.2 & 1.2 & 1.2 & 1.2 \\ -3.1 & -2.1 & -2.1 & -0.1 & -0.1 & -0.1 & 0.9 & 0.9 & 1.9 & 3.9 \end{bmatrix}$$

This matrix "corrects" for the tendency of the different judges to concentrate their ratings in different ranges of the scale. If we wish also to correct for the differential tendency of judges to "spread out" their ratings, we can multiply each row by the reciprocal of the standard deviation of the ratings in that row (2.87, 0.5, 0.98, 0.98, and 1.97, here); this yields

$$\begin{bmatrix} -1.57 & -1.22 & -0.87 & -0.52 & -0.17 & 0.17 & 0.52 & 0.87 & 1.22 & 1.57 \\ -1.00 & -1.00 & -1.00 & -1.00 & -1.00 & 1.00 & 1.00 & 1.00 & 1.00 & 1.00 \\ -1.22 & -1.22 & -1.22 & -0.20 & -0.20 & -0.20 & 0.82 & 0.82 & 0.82 & 1.84 \\ -1.84 & -0.82 & -0.82 & -0.82 & 0.20 & 0.20 & 0.20 & 1.22 & 1.22 & 1.22 \\ -1.57 & -1.07 & -1.07 & -0.05 & -0.05 & -0.05 & 0.46 & 0.46 & 0.96 & 1.98 \end{bmatrix}$$

where, symbolically,

$$\mathbf{W}' = \mathbf{D}_{\mathbf{S}}^{-1/2}\left[\mathbf{X}'\left(\mathbf{I} - \frac{\mathbf{11}'}{10}\right)\right]$$

and

$$\mathbf{S} = \tfrac{1}{10}\left[\mathbf{X}'\left(\mathbf{I} - \frac{\mathbf{11}'}{10}\right)\mathbf{X}\right]$$

This centering or standardization by rows would be sensible only if we had carefully chosen the employees to be rated: To be certain that we are tapping the judges' behaviors, we would select employees to be representative of the full range of employee performance. Had the ten been selected at random or, worse, were they representative of only a small range of possible employee performances, the results would be quite misleading.

We indicated that a data matrix might be nearly centered by rows. This could be a consequence of how certain attributes are observed. For example, in the Vocational Interest Study described in Section 1.5, high school juniors self-assessed their interest in eight occupational areas by completing the Vocational Interest Inventory (VII). The VII has a forced-choice format with each of eight interest areas paired four separate times with each of the other seven to make 112 items. Scale scores are the number of times an interest area is chosen over the competing areas and range from 0 to 28. However, *for each and every student* completing the test, the sum of the scale scores, over the eight scales, is 112. Three possible scale profiles are

$$\mathbf{V}' = \begin{array}{c} \begin{array}{cccccccc} \text{SER} & \text{BUS} & \text{ORG} & \text{TEC} & \text{OUT} & \text{SCI} & \text{CUL} & \text{ART} \end{array} \\ \begin{bmatrix} 17 & 20 & 17 & 14 & 11 & 8 & 11 & 14 \\ 10 & 14 & 16 & 18 & 16 & 14 & 12 & 12 \\ 13 & 13 & 12 & 14 & 16 & 15 & 15 & 14 \end{bmatrix} \end{array}$$

The first profile scores high on BUS (interest in activities requiring business contacts) and low on SCI (science), the second high on TEC (technical interests) and low on SER (service), and the third high on OUT (outdoors activities) and low on ORG (work in organizations). In each profile CUL (general cultural interest) and ART (arts and entertainment) received scores near the mean. The sum of each row is 112 and the average score in each profile is 14.

A psychological test that provides a fixed-sum set of scores on several scales, so that the relative strengths of several attributes are assessed, is termed an *ipsative test*. For the first student in matrix **V′**, interest in business contact is high *relative to* that student's interest in the other seven vocational areas. For such relative observations to be meaningful, the centering or ipsatizing should be over a complete or closed set of attributes. In the present instance, "relative" means relative to *all* other vocational areas of the same generality. We assume that the eight areas exhaust the interests we want to assess.

3.2 Dilemma of Standardization

In multivariate analysis we frequently must interpret observations on a sampling of entities or attributes that are not commensurable or, worse, for which there are no well-established distributions in the populations of interest. How are we to interpret such "unanchored" and probably arbitrarily scaled observations?

Some research questions are concerned only with comparing one horizontal partition of a data matrix with another, such as one "treatment" against another. If the comparisons of concern are apparently *internal* (that is, within the collected data), problems regarding the meaning of the scale and the differences in scale between attributes can be largely ignored. However, we often need to ensure that the sample, although randomly selected, is plausibly representative of some population to which we would like to generalize our findings. Unless we know how the attributes are distributed in that population, we are left only with our confidence in the sampling scheme.

Other research questions depend on contrasting the results at hand with some characteristics of the distribution of an attribute in some population. We very clearly are in some difficulty if we must both infer the population characteristics and evaluate our hypotheses using the same sample.

To resolve the difficulties inherent in trying to work with attributes that have distinctly different scale characteristics, we frequently are led to *standardize* the attributes on the sample. We sometimes do this explicitly by computing a standard score form of the data matrix, but more frequently we do it implicitly by computing and analyzing sample-based correlations rather than covariances or regression constants that employ a norm-referenced metric. In doing this we throw away information about the sampling variability of our attributes. (In Chapter 14, where we discuss principal components analyses, we shall contrast the information that can be extracted from correlation and covariance matrices.)

Our advice, which we recognize cannot be followed in every instance, is to select measurement techniques or instruments that have established normative data. In the behavioral sciences there is a contrasting tendency to decide that the available devices, about which we know *something*, are not as suitable as those we can construct to fit our research interests. We caution against this inclination.

3.3 Amount of Information in a Matrix

An $n \times m$ data matrix contains observations for n entities (the rows) on m attributes (the columns), but the number of "different" attributes may be smaller than m. In multivariate analysis the redundancy of attributes is well defined. Consider an example introduced in Section 2.3. The first and second columns of the 5×3 data matrix

$$\mathbf{X} = \begin{bmatrix} 7 & 8 & 15 \\ 8 & 10 & 18 \\ 0 & 1 & 1 \\ 2 & 0 & 2 \\ 1 & 1 & 2 \end{bmatrix}$$

contain the extrinsic work values scores for odd- and even-numbered items, respectively, for five woman applicants sampled from those in the Preemployment Study. The third column, whose elements are the sums of the corresponding elements in the first two columns, shows the total extrinsic work values scores.

There are *three* attributes represented in matrix **X**—odd-item scores, even-item scores, and all-item scores—but there are only *two* pieces of information about each of our five job applicants. We say that one of the attributes is *redundant*; that is, it gives no additional information about the entities. For example, if we know the first applicant has an odd item score of 7 and an even item score of 8, then we already know, without further observation or measurement, that she has a total item score of 15. If we know that the second applicant has a total score of 18 and an even item score of 10, then we also know that the odd item score must be 8. *Any one* of the three attributes is redundant with respect to the other two. We can throw it away without losing any information about our applicants. Which one we ignore and which two we retain makes no difference in this example.

3.3.1 Rank

In the algebra of matrices the notion of redundancy of information in a matrix is associated with the concept of the *rank* of a matrix. Every matrix has associated with it a nonnegative integer called the rank of that matrix.

Definition. The matrix **X** has *rank* $r(\mathbf{X})$ if (1) **X** can be expressed as the major product of two matrices whose common order is $r(\mathbf{X})$ and (2) **X**

cannot be expressed as the major product of any two matrices whose common order is less than $r(\mathbf{X})$.

The rank of a matrix is never greater than the smaller of the two dimensions of the matrix. Thus the rank of a vertical matrix can never be greater than the number of columns in that matrix. For a vertical matrix **X** of order $n \times m$, the definition means that (1) we can *always* find a pair of matrices A and B such that

$$\mathbf{A}_{(n\times r(\mathbf{X}))}\mathbf{B}'_{(r(\mathbf{X})\times m)} = \mathbf{X}_{(n\times m)}$$

but that (2) we can *never* find a pair of matrices **U** and **V** such that

$$\mathbf{U}_{(n\times p)}\mathbf{V}'_{(p\times m)} = \mathbf{X}_{(n\times m)}$$

if p is less than $r(\mathbf{X})$.

Another way of saying this is that the rank of a vertical (or square) matrix is the *minimum* number of columns that can be linearly combined to produce the matrix. We shall explore this idea using the extrinsic work values data matrix already presented in this section. We have indicated that the matrix can be expressed in three ways:

$$\begin{bmatrix} 7 & 8 \\ 8 & 10 \\ 0 & 1 \\ 2 & 0 \\ 1 & 1 \end{bmatrix} \begin{bmatrix} 1 & 0 & 1 \\ 0 & 1 & 1 \end{bmatrix} = \begin{bmatrix} 7 & 8 & 15 \\ 8 & 10 & 18 \\ 0 & 1 & 1 \\ 2 & 0 & 2 \\ 1 & 1 & 2 \end{bmatrix}$$

$$\begin{bmatrix} 7 & 15 \\ 8 & 18 \\ 0 & 1 \\ 2 & 2 \\ 1 & 2 \end{bmatrix} \begin{bmatrix} 1 & -1 & 0 \\ 0 & 1 & 1 \end{bmatrix} = \begin{bmatrix} 7 & 8 & 15 \\ 8 & 10 & 18 \\ 0 & 1 & 1 \\ 2 & 0 & 2 \\ 1 & 1 & 2 \end{bmatrix}$$

and

$$\begin{bmatrix} 8 & 15 \\ 10 & 18 \\ 1 & 1 \\ 0 & 2 \\ 1 & 2 \end{bmatrix} \begin{bmatrix} -1 & 1 & 0 \\ 1 & 0 & 1 \end{bmatrix} = \begin{bmatrix} 7 & 8 & 15 \\ 8 & 10 & 18 \\ 0 & 1 & 1 \\ 2 & 0 & 2 \\ 1 & 1 & 2 \end{bmatrix}$$

depending on which two attributes we are using to create the three-attributes matrix. The three columns of **X** can be produced as linear combinations of any two columns or attributes. The postfactors on the left give, in each column, the weights to be assigned the two attributes in forming the corresponding columns of **X**. These examples demonstrate that the rank of

X cannot be 3, since we can express **X** as the major product of matrices whose common order is 2, but not that its rank is 2. We may find that **X** can be expressed as the major product of two vectors, matrices whose common order is 1.

First we give some simple properties of the rank and a formal way of deciding the unknown rank of a matrix. By definition, the *only* matrix of rank 0 is the null matrix. Every nonnull matrix, then, has positive rank. The definition also tells us that the rank of every vector except the null vector is 1, its smaller dimension.

3.3.2 Basic and Nonbasic Matrices

A matrix is called *basic* or of full rank if its rank is equal to its smaller order. A matrix is *nonbasic* if its rank is less than its smaller order. Let us construct some nonbasic matrices. The vector **v** consists of vocabulary scores for four students in the college prediction study:

$$\mathbf{v} = \begin{bmatrix} 60 \\ 46 \\ 44 \\ 57 \end{bmatrix}$$

Consider the matrix **X** that results from postmultiplying **v** by a second vector **w**′:

$$\begin{bmatrix} 60 \\ 46 \\ 44 \\ 57 \end{bmatrix} (1 \quad 0 \quad 2 \quad -1) = \begin{bmatrix} 60 & 0 & 120 & -60 \\ 46 & 0 & 92 & -46 \\ 44 & 0 & 88 & -44 \\ 57 & 0 & 114 & -57 \end{bmatrix}$$

The matrix **X** has rank 1: $r(\mathbf{X}) = 1$: It is not a null matrix and hence must have rank $\geqslant 1$, and since it can be expressed as the major product of two matrices whose common order is 1, **X** must have rank $\leqslant 1$. A characteristic of any rank-1 matrix is that the columns (or rows) consist solely of vectors that are either proportional to each other or null vectors. The first, third, and fourth columns of **X** are all proportional, while the second column is a null vector.

The matrix **X** contains only one piece of information about each of the entities in this study. Everything in a row of **X** is based solely on the vocabulary score for one student.

Let us add the quantitative skills scores for these same students to form a matrix

$$\mathbf{V} = \begin{bmatrix} 60 & 39 \\ 46 & 38 \\ 44 & 62 \\ 57 & 49 \end{bmatrix}$$

and from $\mathbf{V}$ construct another matrix $\mathbf{VW}' = \mathbf{X}$:

$$\begin{bmatrix} 60 & 39 \\ 46 & 38 \\ 44 & 62 \\ 57 & 49 \end{bmatrix} \begin{bmatrix} 1 & 0 & 1 & 1 \\ 0 & 1 & -1 & 1 \end{bmatrix} = \begin{bmatrix} 60 & 39 & 21 & 99 \\ 46 & 38 & 8 & 84 \\ 44 & 62 & -18 & 106 \\ 57 & 49 & 8 & 106 \end{bmatrix}$$

This new matrix $\mathbf{X}$ will have rank 2. The columns (or rows) are not proportional one to another, so $r(\mathbf{X}) > 1$.

The entries in each row of $\mathbf{X}$ are based on only two pieces of information, the vocabulary and quantitative skills scores. The third column in $\mathbf{X}$ is the difference between these two scores, and the fourth column is their sum.

If we were given only the matrix $\mathbf{X}$, it would require some effort to establish that it is nonbasic, that its rank is only 2. We could accomplish this by discovering that we can produce the third and fourth columns by weighting and adding the first two columns. (The third and fourth columns of $\mathbf{W}'$ give the weights.)

Many special matrices can be identified as basic by simple inspection. Their rank will then be equal to the smaller of their dimensions.

All proper diagonal matrices are basic matrices. (A proper diagonal matrix has nonzero elements along the diagonal so that no column is a null vector.) Look at

$$\mathbf{D} = \begin{bmatrix} 12 & 0 & 0 \\ 0 & 8 & 0 \\ 0 & 0 & 10 \end{bmatrix}$$

If $\mathbf{D}$ were nonbasic, at least one of the columns could be obtained by appropriately weighting and adding together the other two. This cannot be true for diagonal matrices. If the first column of $\mathbf{D}$ were weighted by a scalar constant other than 0, the result would be a vector with a nonzero first entry. If the second or third column of $\mathbf{D}$ were weighted by any scalar constant, the result would be a vector with a zero first entry. Adding a weighted first column to a weighted second or third column will always produce a vector with a nonzero first element. Thus the first column *cannot* be weighted to produce either the second or third columns. In the same fashion, it can be shown that the second column cannot be weighted to produce either the first or third columns and that the third column cannot be weighted to produce either the first or second columns. In short, no one of the columns of a diagonal matrix can be obtained by weighting and adding other columns of that diagonal matrix. The diagonal matrix is basic. The diagonal matrix has rank equal to its order.

The rank of the transpose of a matrix is equal to the rank of the matrix: $r(\mathbf{X}) = r(\mathbf{X}')$. We shall not prove this. Intuitively, writing rows as columns and columns as rows cannot change the amount of nonredundant information in a matrix.

Permuting the rows, columns, or both of a matrix does not change its rank. Later we shall offer an algebraic argument for this; intuitively, the order in which we write down the entities or attributes in a matrix should not affect the amount of information about each entity contained in the matrix. Thus for the matrix **X** of rank 2 given earlier, the matrix

$$\boldsymbol{\pi}_L \mathbf{X} \boldsymbol{\pi}_R = \begin{bmatrix} 0 & 0 & 1 & 0 \\ 0 & 0 & 0 & 1 \\ 1 & 0 & 0 & 0 \\ 0 & 1 & 0 & 0 \end{bmatrix} \begin{bmatrix} 60 & 39 & 21 & 99 \\ 46 & 38 & 8 & 84 \\ 44 & 62 & -18 & 106 \\ 57 & 49 & 8 & 106 \end{bmatrix}$$

$$\times \begin{bmatrix} 0 & 1 & 0 & 0 \\ 0 & 0 & 1 & 0 \\ 0 & 0 & 0 & 1 \\ 1 & 0 & 0 & 0 \end{bmatrix}$$

$$= \begin{bmatrix} 106 & 44 & 62 & -18 \\ 106 & 57 & 49 & 8 \\ 99 & 60 & 39 & 21 \\ 84 & 46 & 38 & 8 \end{bmatrix}$$

also has rank 2.

The reader may have noticed that permutation matrices are themselves permutations of the rows (or columns) of a diagonal matrix, the identity matrix. Since diagonal matrices are basic, *permutation matrices are basic matrices.*

The ranks of the minor product and major product moments of a matrix equal the rank of that matrix: $r(\mathbf{X}'\mathbf{X}) = r(\mathbf{X}\mathbf{X}') = r(\mathbf{X})$. We shall not prove this, but we illustrate it for the rank-2 matrix $\mathbf{X} = \mathbf{V}\mathbf{W}'$. The minor product moment of **X** can be written three ways:

$$\mathbf{X}'\mathbf{X} = (\mathbf{V}\mathbf{W}')'\mathbf{X} = \mathbf{W}\mathbf{V}'\mathbf{X} = \mathbf{W}(\mathbf{V}'\mathbf{X})$$

The last form is of interest because $\mathbf{V}'\mathbf{X}$ is a 2×4 matrix:

$$\begin{bmatrix} 60 & 46 & 44 & 57 \\ 39 & 38 & 62 & 49 \end{bmatrix} \begin{bmatrix} 60 & 39 & 21 & 99 \\ 46 & 38 & 8 & 84 \\ 44 & 62 & -18 & 106 \\ 57 & 49 & 8 & 104 \end{bmatrix}$$

$$= \begin{bmatrix} 10{,}901 & 9609 & 1292 & 20{,}510 \\ 9609 & 9210 & 399 & 18{,}819 \end{bmatrix}$$

Thus, $\mathbf{X}'\mathbf{X}$ can be written as the major product of two matrices **W** and $\mathbf{V}'\mathbf{X}$

with common order 2:

$$\begin{bmatrix} 1 & 0 \\ 0 & 1 \\ 1 & -1 \\ 1 & 1 \end{bmatrix} \begin{bmatrix} 10{,}901 & 9609 & 1292 & 20{,}510 \\ 9609 & 9210 & 399 & 18{,}819 \end{bmatrix}$$

$$= \begin{bmatrix} 10{,}901 & 9609 & 1292 & 20{,}510 \\ 9609 & 9210 & 399 & 18{,}819 \\ 1292 & 399 & 893 & 1691 \\ 20{,}510 & 18{,}819 & 1691 & 39{,}329 \end{bmatrix}$$

To be sure, all we have shown is that $\mathbf{X'X}$ cannot have rank greater than that of $\mathbf{X}$. In this instance, we can see that $\mathbf{X'X}$ is not null and does not have proportional rows or columns and hence is not of rank less than the rank of $\mathbf{X}$. Thus $r(\mathbf{X'X}) = r(\mathbf{X}) = 2$.

The major product moment can be studied in the same fashion by expressing $\mathbf{XX'}$ in these two ways:

$$\mathbf{XX'} = (\mathbf{VW'})\mathbf{X'} = \mathbf{V}(\mathbf{W'X'})$$

Again, in this way the 4×4 matrix $\mathbf{XX'}$ is shown to be the major product of two matrices with common order 2 ($\mathbf{V}$ is 4×2 and $\mathbf{W'X'}$ is 2×4).

3.3.3 Orthogonal and Orthonormal Matrices

This rule for the rank of product moment matrices leads to the definition of two classes of matrices important in multivariate analyses. If a vertical or square matrix has a proper diagonal matrix as its minor product moment, that matrix is *orthogonal*. Consider

$$\mathbf{G} = \begin{bmatrix} 60 & -7.42029 \\ 46 & -8.42029 \\ 44 & 15.57910 \\ 57 & 2.57971 \end{bmatrix}$$

The minor product moment of $\mathbf{G}$ is diagonal.

$$\mathbf{G'G} = \begin{bmatrix} 10{,}901 & 0 \\ 0 & 375.34425 \end{bmatrix}$$

and thus $\mathbf{G}$ is an orthogonal matrix.

Since $\mathbf{G'G}$ is diagonal, it is basic, with rank equal to the number of columns of $\mathbf{G}$. But the rank of a matrix is equal to the rank of its minor product moment. Thus the rank of $\mathbf{G}$ itself is just the number of columns of $\mathbf{G}$. This argument is quite general: *All orthogonal matrices are basic matrices.*

If a vertical or square matrix has the identity matrix as its minor product moment, that matrix is termed *orthonormal*. We can develop an orthonormal matrix by first writing an orthogonal matrix, say,

$$\mathbf{H} = \begin{bmatrix} 1 & \frac{2}{3} & 0 \\ 1 & -\frac{1}{3} & \frac{1}{2} \\ 1 & -\frac{1}{3} & -\frac{1}{2} \end{bmatrix}$$

with minor product moment

$$\mathbf{H'H} = \begin{bmatrix} 3 & 0 & 0 \\ 0 & \frac{6}{9} & 0 \\ 0 & 0 & \frac{2}{4} \end{bmatrix}$$

In diagonal matrix notation,

$$\mathbf{D}_{\mathbf{H'H}}^{1/2} = \begin{bmatrix} 3^{1/2} & 0 & 0 \\ 0 & \left(\frac{2}{3}\right)^{1/2} & 0 \\ 0 & 0 & \left(\frac{1}{2}\right)^{1/2} \end{bmatrix}$$

$$\mathbf{D}_{\mathbf{H'H}}^{-1/2} = \begin{bmatrix} \left(\frac{1}{3}\right)^{1/2} & 0 & 0 \\ 0 & \left(\frac{3}{2}\right)^{1/2} & 0 \\ 0 & 0 & 2^{1/2} \end{bmatrix}$$

If we now postmultiply $\mathbf{H}$ by $\mathbf{D}_{\mathbf{H'H}}^{1/2}$, we scale each column of $\mathbf{H}$ by the reciprocal of the square root of the sum of squares of the elements of that column:

$$\mathbf{HD}_{\mathbf{H'H}}^{-1/2} = \begin{bmatrix} 1 & \frac{2}{3} & 0 \\ 1 & -\frac{1}{3} & \frac{1}{2} \\ 1 & -\frac{1}{3} & -\frac{1}{2} \end{bmatrix} \begin{bmatrix} \left(\frac{1}{3}\right)^{1/2} & 0 & 0 \\ 0 & \left(\frac{3}{2}\right)^{1/2} & 0 \\ 0 & 0 & 2^{1/2} \end{bmatrix}$$

$$= \begin{bmatrix} 1/3^{1/2} & \left(\frac{2}{3}\right)^{1/2} & 0 \\ 1/3^{1/2} & -1/6^{1/2} & 1/2^{1/2} \\ 1/3^{1/2} & -1/6^{1/2} & 1/2^{1/2} \end{bmatrix} = \mathbf{Q}$$

with minor product moment

$$\mathbf{Q'Q} = \begin{bmatrix} 1 & 0 & 0 \\ 0 & 1 & 0 \\ 0 & 0 & 1 \end{bmatrix} = \mathbf{I}$$

This result may be easier to see if $\mathbf{Q}$ is rewritten with each element expressed as a fraction with least common denominator:

$$\mathbf{Q} = \begin{bmatrix} 2^{1/2}/6^{1/2} & 2/6^{1/2} & 0 \\ 2^{1/2}/6^{1/2} & -1/6^{1/2} & 3^{1/2}/6^{1/2} \\ 2^{1/2}/6^{1/2} & -1/6^{1/2} & -3^{1/2}/6^{1/2} \end{bmatrix}$$

Q is an orthonormal matrix. We obtained it by *normalizing* each column of an orthogonal matrix. Any vector (column of a matrix) can be normalized by dividing each element by the square root of the sum of squares of the elements in the vector. For any *matrix* **X**, then,

$$\mathbf{Z} = \mathbf{X}\mathbf{D}_{\mathbf{X}'\mathbf{X}}^{-1/2}$$

is such that

$$\mathbf{D}_{\mathbf{Z}'\mathbf{Z}} = \mathbf{I}$$

However, $\mathbf{Z}'\mathbf{Z} = \mathbf{I}$ only if **X** is orthogonal and $\mathbf{X}'\mathbf{X}$ is a diagonal matrix.

Because orthonormal matrices are special forms of orthogonal matrices, *orthonormal matrices are always basic matrices.*

If **P** is a *square orthonormal* matrix, then

$$\mathbf{P}'\mathbf{P} = \mathbf{I} \quad \text{and} \quad \mathbf{P}\mathbf{P}' = \mathbf{I}$$

The reader should check this for **Q**. Both the major and minor product moments of square orthonormal matrices are identity matrices.

3.3.4 Two Rank Inequalities

We now introduce without proof two additional useful rules on the rank of a matrix. Later we shall develop enough matrix algebra to show why they hold. The first concerns the rank of the product of two matrices:

Product Rank Inequality. *The rank of the product of two matrices*

a. *is never less than the sum of the ranks of the two factors less their common order* (CO) *and*
b. *is never greater than the smaller of the ranks of the two factors.*

Symbolically,

$$r(\mathbf{A}) + r(\mathbf{B}) - \mathrm{CO}(\mathbf{A}, \mathbf{B}) \leqslant r(\mathbf{AB}) \leqslant \min[r(\mathbf{A}), r(\mathbf{B})] \quad (3.3.1)$$

One consequence of this rule is that *the rank of a matrix is not altered by pre- or postmultiplying that matrix by a square basic matrix.* A square basic matrix is a square matrix that is basic and thus of rank equal to its order. Square basic matrices encountered so far include diagonal, permutation, square orthonormal, and square orthogonal matrices.

This consequence follows from the product rank rule by taking **X** to be vertical ($n \times m$) and of unknown rank and $\mathbf{S}_{(n\times n)}$ and $\mathbf{T}_{(m\times m)}$ to be two square basic matrices. Consider first the result **Y** of premultiplying **X** by **S**:

$$ {}_{n}\begin{bmatrix} \quad \mathbf{S} \quad \end{bmatrix}^{n} {}_{n}\begin{bmatrix} \mathbf{X} \end{bmatrix}^{m} = {}_{n}\begin{bmatrix} \mathbf{Y} \end{bmatrix}^{m} $$

By the product rank rule,

$$ r(\mathbf{S}) + r(\mathbf{X}) - \mathrm{CO}(\mathbf{S},\mathbf{X}) \leqslant r(\mathbf{Y}) = r(\mathbf{SX}) \leqslant \min[r(\mathbf{S}), r(\mathbf{X})] $$

However,

a. $r(\mathbf{S}) = n$, since **S** is basic;
b. $\mathrm{CO}(\mathbf{S},\mathbf{X}) = n$, their common order; and
c. $\min[r(\mathbf{S}), r(\mathbf{X})] = r(\mathbf{X})$, since $r(\mathbf{X})$ cannot be greater than $r(\mathbf{S}) = n$, one of the dimensions of **X**.

With these substitutions,

$$ n + r(\mathbf{X}) - n \leqslant r(\mathbf{Y}) \leqslant r(\mathbf{X}) $$

or

$$ r(\mathbf{X}) \leqslant r(\mathbf{Y}) \leqslant r(\mathbf{X}) $$

which will be true only if

$$ r(\mathbf{Y}) = r(\mathbf{X}) $$

Postmultiplying **X** by a square basic matrix **T**,

$$ {}_{n}\begin{bmatrix} \mathbf{X} \end{bmatrix}^{m} {}_{m}\begin{bmatrix} \quad \mathbf{T} \quad \end{bmatrix}^{m} = {}_{n}\begin{bmatrix} \mathbf{Z} \end{bmatrix}^{m} $$

gives, by the product rank rule,

$$ r(\mathbf{X}) + r(\mathbf{T}) - \mathrm{CO}(\mathbf{X},\mathbf{T}) \leqslant r(\mathbf{Z}) = r(\mathbf{XT}) \leqslant \min[r(\mathbf{X}), r(\mathbf{T})] $$

Again,

a. $r(\mathbf{T}) = m$, because **T** is basic;
b. $\mathrm{CO}(\mathbf{X},\mathbf{T}) = m$, their common order; and
c. $\min[r(\mathbf{X}), r(\mathbf{T})] = r(\mathbf{X})$, since $r(\mathbf{X})$ cannot be greater than $r(\mathbf{T}) = m$, one of the dimensions of **X**.

On substituting these into the product rank inequality we have

$$ r(\mathbf{X}) \leqslant r(\mathbf{Z}) \leqslant r(\mathbf{X}) $$

again implying that

$$ r(\mathbf{Z}) = r(\mathbf{X}) $$

One important implication of this outcome of the product rank rule is that we may scale the rows or columns of a matrix (pre- or postmultiplying by a diagonal matrix) or rearrange the rows or columns (pre- or postmultiplying by a permutation matrix) without altering the amount of information it contains.

Another implication, to be developed more fully in Chapter 5, is the effect of the *rotation* of a matrix. If a vertical matrix **X** is postmultiplied by a square orthonormal matrix, the result will be a rotation of the columns of **X**. For now, we simply preview certain results of rotation by examining the major and minor product moments of the rotated matrix. In the rotation

$$\mathbf{XQ} = \mathbf{Z}$$

where **Q** is square orthonormal, the major product moment of **Z** is

$$\mathbf{ZZ'} = \mathbf{XQQ'X'}$$

or, since $\mathbf{QQ'} = \mathbf{I}$ for a square **Q**,

$$\mathbf{ZZ'} = \mathbf{XIX'} = \mathbf{XX'}$$

The minor product of Z does not simplify in the same way:

$$\mathbf{Z'Z} = \mathbf{Q'X'XQ}$$

Rotating the columns of a data matrix leaves unchanged certain characteristics of the entities or the relations between the entities. From the equation for **ZZ′**, we see that the sum of squares of scores over the new attributes is the same for any entity as it was for the original attributes. Further, the sum of products between any two entities, an off-diagonal element in the major product moment, is identical whether we compute it over the original or the rotated attributes.

Figure 3.1 Pairs of vocabulary ($V_{.1}$) and quantitative skills ($V_{.2}$) scores.

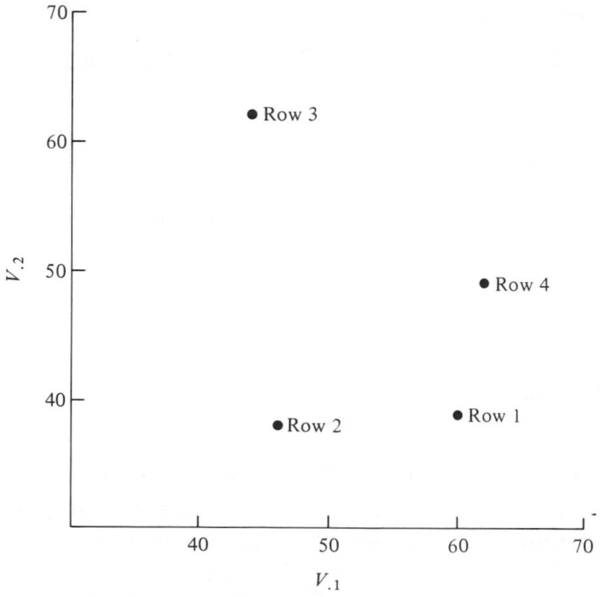

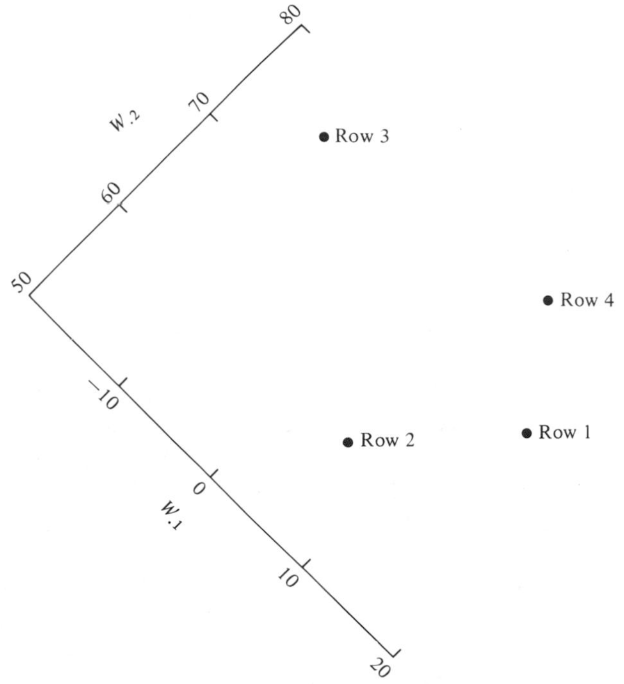

Figure 3.2 Pairs of rotated vocabulary and quantitative skills scores.

Some hint of the origin of the name "rotation" for this matrix procedure is possible by considering an example. We shall rotate **V** (vocabulary and quantitative skills scores) using the 2×2 square orthonormal matrix

$$\mathbf{Q} = \begin{bmatrix} 1/2^{1/2} & 1/2^{1/2} \\ -1/2^{1/2} & 1/2^{1/2} \end{bmatrix}$$

The rotated results are given by $\mathbf{VQ} = \mathbf{W}$;

$$\begin{bmatrix} 60 & 39 \\ 46 & 38 \\ 44 & 62 \\ 62 & 49 \end{bmatrix} \begin{bmatrix} 0.707 & 0.707 \\ -0.707 & 0.707 \end{bmatrix} = \begin{bmatrix} 14.85 & 70.00 \\ 5.66 & 59.40 \\ -12.73 & 74.95 \\ 9.19 & 78.49 \end{bmatrix}$$

In Figures 3.1 and 3.2 the four rows of **V** and **W** have been plotted as four points in a two-dimensional space that have been indexed either by the columns of **V** or the columns of **W**. The coordinate axes of Figure 3.2 have been tipped to reveal that the relative position of the four points has been preserved. It is as if the two axes in Figure 3.1 had been rigidly rotated clockwise, the points remaining fixed, with the $V_{.2}$ axis "becoming" $W_{.1}$, and the $V_{.1}$ axis "moving to" $W_{.2}$. We shall have much more to say about rotations and related transformations of matrices in later chapters.

Just as there is an inequality to help find the rank of a product of two matrices, so there is a

Sum Rank Inequality. *The rank of the sum of two matrices*

a. *is never less than (the absolute value of) the difference in ranks of the two matrices and*
b. *is never greater than the sum of the ranks of the two matrices.*

Symbolically, this rule may be expressed

$$|r(\mathbf{A}) - r(\mathbf{B})| \leqslant r(\mathbf{A} + \mathbf{B}) \leqslant r(\mathbf{A}) + r(\mathbf{B}) \tag{3.3.2}$$

For now only an intuitive justification will be given for this inequality. Consider two data matrices **A** and **B**, each $n \times m$ and each based on the same n entities. Let $r(\mathbf{A}) = p$ ($\leqslant m$) and let $r(\mathbf{B}) = q$ ($\leqslant m$). Matrix **A**, then, contains p nonredundant pieces of information about the n entities, and **B** contains q nonredundant pieces of information about those same entities. How much information about the n entities is there in the new $n \times m$ matrix $\mathbf{A} + \mathbf{B} = \mathbf{C}$?

The right inequality ensures that we cannot create new information about the entities by adding **A** and **B**; at *most* we have $p + q$ nonredundant pieces of information. The left-hand inequality tells us that we cannot mysteriously lose information by adding the two matrices; if $q < p$, then adding **B** to **A** (for the case we are making here, we would more likely be subtracting **B** from **A**) could destroy, in the worst case, only q of the p pieces of information originally in **A**. We would retain in **C** at *least* $p - q$ nonredundant pieces of information. [There is also a second, upper limit to the rank of a sum since the rank of a vertical matrix cannot exceed the column order of that matrix; in this example, $r(\mathbf{C}) \leqslant m$.]

3.4 Finding the Rank of a Matrix

The two inequalities in Section 3.3.4 may help find the rank of a matrix if we know that the matrix was formed as a product or sum of two other matrices whose ranks are known. Even when we know all that, the inequality may tell us only the range in which the rank must lie. Often, of course, we are simply confronted with the matrix, not the factors that gave rise to it. How do we know the rank under such circumstances?

The answer is that we must calculate it. Determining the rank of a matrix is an important aspect of nearly every one of the applied multivariate analysis techniques in this book. If the matrix is at all large—if the smaller dimension is larger than 4 or 5—the computations necessary to find the rank may be time consuming and, because of the number of steps involved, subject to mistakes and round-off errors.

3.4.1 Triangular Factoring

We present in this section a technique for finding the rank of a matrix that has been implemented in computer solutions. We develop it here not to encourage readers to solve such problems by calculations or because we favor its computer implementation. Rather, we offer it so that the reader may understand how it works and how such a solution is related to the properties of rank presented in Section 3.3.

The technique is called *triangular factoring*. It exploits a corollary of the product rank rule: *If a matrix of unknown rank can be expressed as the major product of two basic matrices, then its rank is identical to the common order of those two basic matrices.*

Earlier, in defining rank, we said that a matrix expressible as the major product of two other matrices could not have rank greater than the common order of those two factors. Now we are saying that the rank equals that common order if the two factors are basic. The following diagram illustrates the application of the product rank rule:

$$ {}_{n}\begin{bmatrix} \\ \mathbf{A} \\ \\ \end{bmatrix}^{p} \; {}_{p}[\;\mathbf{B}'\;]^{m} = {}_{n}\begin{bmatrix} \\ \mathbf{X} \\ \\ \end{bmatrix}^{m} $$

For the product to be a major product, $p \leqslant n$ and $p \leqslant m$. Since $\mathbf{A}$ is assumed to be basic, $r(\mathbf{A}) = p$. Since $\mathbf{B}$ is also basic, $r(\mathbf{B}) = p$ as well. Then by substituting into the product rank inequality, we find

$$ p + p - p \leqslant r(\mathbf{AB}') = r(\mathbf{X}) \leqslant \min[p, p] $$

or

$$ p \leqslant r(\mathbf{X}) \leqslant p $$

or finally

$$ r(\mathbf{X}) = p $$

If we can express a matrix $\mathbf{X}$ as the major product of two basic matrices, we know the rank of $\mathbf{X}$. To apply this principle, however, we must know that the two matrices forming the product *are each basic*. Diagonal, permutation, and orthogonal matrices are basic. We now introduce partial triangular matrices, which are also basic. $\mathbf{T}_{\mathrm{L}}$ is a *vertical* or *lower partial triangular* matrix:

$$ \begin{bmatrix} 1 & 0 & 0 \\ 2 & 2 & 0 \\ 2 & 1 & 1 \\ 4 & 2 & 5 \end{bmatrix} $$

The lower partial triangular matrix has zeros everywhere above and to the right of the *main diagonal* of the matrix, which consists of the elements T_{11}, T_{22}, and T_{33}. The main diagonal elements are all nonzero.

The argument that the lower partial triangular matrix is basic is much the same as the argument we gave in the case of the proper diagonal matrix: The pattern of zeros precludes expressing any one of the columns of $\mathbf{T}_L$ as a weighted sum of any of the remaining columns. For example, the first and second columns cannot be weighted and then added together to produce a vector with zeros in both the first and second positions.

The transpose of a lower partial triangular matrix is called an *upper partial triangular matrix*:

$$\mathbf{T}'_U = \begin{bmatrix} 2 & 3 & 4 & 5 \\ 0 & 2 & 4 & 6 \\ 0 & 0 & 1 & 5 \end{bmatrix}$$

An upper partial triangular matrix is horizontal with zeros everywhere to the left and below the main diagonal. An argument that these are basic matrices depends on the principle that $r(\mathbf{T}') = r(\mathbf{T})$.

Square matrices may also be triangular in form, and it is customary to distinguish between a *lower triangular matrix*,

$$\mathbf{T}_L = \begin{bmatrix} 3 & 0 & 0 \\ 6 & 4 & 0 \\ 4 & 7 & 2 \end{bmatrix}$$

and an *upper triangular matrix*,

$$\mathbf{T}_U = \begin{bmatrix} 6 & 5 & 4 & 3 \\ 0 & 3 & 7 & 2 \\ 0 & 0 & 2 & 9 \\ 0 & 0 & 0 & 8 \end{bmatrix}$$

depending on whether the zeros are above or below the diagonal. When the matrix is square, the term "partial" is dropped from the description. Triangular matrices also are basic. Indeed, by definition, triangular matrices are square basic matrices.

Triangular factoring takes its name from the technique's goal, to express an arbitrary matrix as the product of a lower (partial) and an upper (partial) triangular matrix:

$$_n\begin{bmatrix} \mathbf{T}_L \end{bmatrix}^p \; {}_p[\,\mathbf{T}'_U\,]^m = {}_n\begin{bmatrix} \mathbf{X} \end{bmatrix}^m$$

The rank of $\mathbf{X}$ is then shown to be the common order p of the two (partial) triangular, basic factors.

If $\mathbf{X}$ is square and basic, $\mathbf{T}_L$ and $\mathbf{T}_U$ are both square triangular matrices. If $\mathbf{X}$ is vertical and basic, $\mathbf{T}_U$ is a square triangular matrix. If $\mathbf{X}$ is horizontal and basic, $\mathbf{T}_L$ is a square triangular matrix. If $\mathbf{X}$ is nonbasic, $\mathbf{T}_L$ and $\mathbf{T}'_U$ are lower and upper partial triangular matrices, respectively.

Consider this 4×4 matrix, the minor product moment of the design matrix for the Airline Pilot Response Time Study:

$$\mathbf{X'X} = \begin{bmatrix} 48 & 24 & 24 & 12 \\ 24 & 24 & 12 & 12 \\ 24 & 12 & 24 & 12 \\ 12 & 12 & 12 & 12 \end{bmatrix}$$

In a later chapter, we shall see how this particular design matrix was developed. For now, it will suffice that a design matrix carries the predictor information in designed studies and that for the Response Time Study, 48 pilots were randomly divided into four experimental groups of 12. Each group responded to stimuli presented either aurally or visually and either to the left or right sensory field.

We begin the triangular factoring of our $\mathbf{X'X}$ matrix by renaming it ${}_0\mathbf{A}$.

1. Let ${}_1\mathbf{v}$ be the *first* column of ${}_0\mathbf{A}$ and ${}_1\mathbf{w}'$ the *first* row of ${}_0\mathbf{A}$ scaled by $1/({}_0A_{11})$:

$${}_1\mathbf{v} = \begin{bmatrix} 48 \\ 24 \\ 24 \\ 12 \end{bmatrix}, \qquad {}_1\mathbf{w}' = \tfrac{1}{48}(48, 24, 24, 12) = (1, \tfrac{1}{2}, \tfrac{1}{2}, \tfrac{1}{4})$$

2. Form the vector major product ${}_1\mathbf{B} = {}_1\mathbf{v}\,{}_1\mathbf{w}'$:

$$\begin{bmatrix} 48 \\ 24 \\ 24 \\ 12 \end{bmatrix} (1, \tfrac{1}{2}, \tfrac{1}{2}, \tfrac{1}{4}) = \begin{bmatrix} 48 & 24 & 24 & 12 \\ 24 & 12 & 12 & 6 \\ 24 & 12 & 12 & 6 \\ 12 & 6 & 6 & 3 \end{bmatrix}$$

3. Subtract ${}_1\mathbf{B}$ from ${}_0\mathbf{A}$ to produce ${}_1\mathbf{A}$:

$$\begin{bmatrix} 48 & 24 & 24 & 12 \\ 24 & 24 & 12 & 12 \\ 24 & 12 & 24 & 12 \\ 12 & 12 & 12 & 12 \end{bmatrix} - \begin{bmatrix} 48 & 24 & 24 & 12 \\ 24 & 12 & 12 & 6 \\ 24 & 12 & 12 & 6 \\ 12 & 6 & 6 & 3 \end{bmatrix}$$

$$= \begin{bmatrix} 0 & 0 & 0 & 0 \\ 0 & 12 & 0 & 6 \\ 0 & 0 & 12 & 6 \\ 0 & 6 & 6 & 9 \end{bmatrix}$$

4. Let ${}_2\mathbf{v}$ be the *second* column of ${}_1\mathbf{A}$ and ${}_2\mathbf{w}'$ the *second* row of ${}_1\mathbf{A}$ scaled by $1/({}_1A_{22})$:

$$ {}_2\mathbf{v} = \begin{bmatrix} 0 \\ 12 \\ 0 \\ 6 \end{bmatrix}, \qquad {}_2\mathbf{w}' = \tfrac{1}{12}(0, 12, 0, 6) = (0, 1, 0, \tfrac{1}{2}) $$

5. Form the vector major product ${}_2\mathbf{B} = {}_2\mathbf{v}\,{}_2\mathbf{w}'$:

$$ \begin{bmatrix} 0 \\ 12 \\ 0 \\ 6 \end{bmatrix} (0, 1, 0, \tfrac{1}{2}) = \begin{bmatrix} 0 & 0 & 0 & 0 \\ 0 & 12 & 0 & 6 \\ 0 & 0 & 0 & 0 \\ 0 & 6 & 0 & 3 \end{bmatrix} $$

6. Subtract ${}_2\mathbf{B}$ from ${}_1\mathbf{A}$ to form ${}_2\mathbf{A}$:

$$ \begin{bmatrix} 0 & 0 & 0 & 0 \\ 0 & 12 & 0 & 6 \\ 0 & 0 & 12 & 6 \\ 0 & 6 & 6 & 9 \end{bmatrix} - \begin{bmatrix} 0 & 0 & 0 & 0 \\ 0 & 12 & 0 & 6 \\ 0 & 0 & 0 & 0 \\ 0 & 6 & 0 & 3 \end{bmatrix} = \begin{bmatrix} 0 & 0 & 0 & 0 \\ 0 & 0 & 0 & 0 \\ 0 & 0 & 12 & 6 \\ 0 & 0 & 6 & 6 \end{bmatrix} $$

7. Let ${}_3\mathbf{v}$ be the *third* column of ${}_2\mathbf{A}$ and ${}_3\mathbf{w}'$ be the *third* row of ${}_2\mathbf{A}$ scaled by $1/({}_2A_{33})$:

$$ {}_3\mathbf{v} = \begin{bmatrix} 0 \\ 0 \\ 12 \\ 6 \end{bmatrix}, \qquad {}_3\mathbf{w}' = \tfrac{1}{12}(0, 0, 12, 6) = (0, 0, 1, \tfrac{1}{2}) $$

8. Form the vector major product ${}_3\mathbf{B} = {}_3\mathbf{v}\,{}_3\mathbf{w}'$:

$$ \begin{bmatrix} 0 \\ 0 \\ 12 \\ 6 \end{bmatrix} (0, 0, 1, \tfrac{1}{2}) = \begin{bmatrix} 0 & 0 & 0 & 0 \\ 0 & 0 & 0 & 0 \\ 0 & 0 & 12 & 6 \\ 0 & 0 & 6 & 3 \end{bmatrix} $$

9. Subtract ${}_3\mathbf{B}$ from ${}_2\mathbf{A}$ to form ${}_3\mathbf{A}$:

$$ \begin{bmatrix} 0 & 0 & 0 & 0 \\ 0 & 0 & 0 & 0 \\ 0 & 0 & 12 & 6 \\ 0 & 0 & 6 & 6 \end{bmatrix} - \begin{bmatrix} 0 & 0 & 0 & 0 \\ 0 & 0 & 0 & 0 \\ 0 & 0 & 12 & 6 \\ 0 & 0 & 6 & 3 \end{bmatrix} = \begin{bmatrix} 0 & 0 & 0 & 0 \\ 0 & 0 & 0 & 0 \\ 0 & 0 & 0 & 0 \\ 0 & 0 & 0 & 3 \end{bmatrix} $$

10. Let ${}_4\mathbf{v}$ be the *fourth* column of ${}_3\mathbf{A}$ and ${}_4\mathbf{w}'$ be the *fourth* row of ${}_3\mathbf{A}$ scaled by $1/({}_3A_{44})$:

$$ {}_4\mathbf{v} = \begin{bmatrix} 0 \\ 0 \\ 0 \\ 3 \end{bmatrix} \tfrac{1}{3}(0, 0, 0, 3) = (0, 0, 0, 1) $$

11. Form the vector major product ${}_4\mathbf{B} = {}_4\mathbf{v}\,{}_4\mathbf{w}'$:

$$ \begin{bmatrix} 0 \\ 0 \\ 0 \\ 3 \end{bmatrix} (0, 0, 0, 1) = \begin{bmatrix} 0 & 0 & 0 & 0 \\ 0 & 0 & 0 & 0 \\ 0 & 0 & 0 & 0 \\ 0 & 0 & 0 & 3 \end{bmatrix} $$

12. Subtract ${}_4\mathbf{B}$ from ${}_3\mathbf{A}$ to form

$$ \begin{bmatrix} 0 & 0 & 0 & 0 \\ 0 & 0 & 0 & 0 \\ 0 & 0 & 0 & 0 \\ 0 & 0 & 0 & 0 \end{bmatrix} = {}_4\mathbf{A} $$

The resulting difference ${}_4\mathbf{A}$ is the null matrix. This phase of the triangular factoring ceases when a null ${}_i\mathbf{A}$ matrix is reached. If ${}_0\mathbf{A} = \mathbf{X}'\mathbf{X}$ had been of lower rank, fewer cycles would have been required to obtain a null matrix.

From steps 3, 6, 9, and 12 we can write

$$ {}_1\mathbf{A} = {}_0\mathbf{A} - {}_1\mathbf{B} \tag{3.4.1} $$

$$ {}_2\mathbf{A} = {}_1\mathbf{A} - {}_2\mathbf{B} \tag{3.4.2} $$

$$ {}_3\mathbf{A} = {}_2\mathbf{A} - {}_3\mathbf{B} \tag{3.4.3} $$

$$ {}_4\mathbf{A} = {}_3\mathbf{A} - {}_3\mathbf{B} \tag{3.4.4} $$

Substituting for ${}_1\mathbf{A}$ in (3.4.2) from (3.4.1) gives

$$ {}_2\mathbf{A} = {}_0\mathbf{A} - {}_1\mathbf{B} - {}_2\mathbf{B} \tag{3.4.5} $$

while substituting this result into (3.4.3) for ${}_2\mathbf{A}$ gives

$$ {}_3\mathbf{A} = {}_0\mathbf{A} - {}_1\mathbf{B} - {}_2\mathbf{B} - {}_3\mathbf{B} \tag{3.4.6} $$

and substituting this last result for ${}_3\mathbf{A}$ in (3.4.4) yields

$$ \mathbf{0} = {}_0\mathbf{A} - {}_1\mathbf{B} - {}_2\mathbf{B} - {}_3\mathbf{B} - {}_4\mathbf{B} $$

or

$$ \mathbf{0} = {}_0\mathbf{A} - ({}_1\mathbf{B} + {}_2\mathbf{B} + {}_3\mathbf{B} + {}_4\mathbf{B}) $$

or finally

$$ {}_0\mathbf{A} = {}_1\mathbf{B} + {}_2\mathbf{B} + {}_3\mathbf{B} + {}_4\mathbf{B} \tag{3.4.7}$$

In this form we have expressed ${}_0\mathbf{A}$, the matrix with which we started, as the sum of four matrices. Each of these **B** matrices was computed as a major product of vectors, and we can substitute from steps 2, 5, 8, and 11 to write

$$ {}_0\mathbf{A} = {}_1\mathbf{v}\,{}_1\mathbf{w}' + {}_2\mathbf{v}\,{}_2\mathbf{w}' + {}_3\mathbf{v}\,{}_3\mathbf{w}' + {}_4\mathbf{v}\,{}_4\mathbf{w}'$$

we can express this result numerically

$$\begin{bmatrix} 48 & 24 & 24 & 12 \\ 24 & 24 & 12 & 12 \\ 24 & 12 & 24 & 12 \\ 12 & 12 & 12 & 12 \end{bmatrix} = \begin{bmatrix} 48 \\ 24 \\ 24 \\ 12 \end{bmatrix} (1, \tfrac{1}{2}, \tfrac{1}{2}, \tfrac{1}{4}) + \begin{bmatrix} 0 \\ 12 \\ 0 \\ 6 \end{bmatrix} (0, 1, 0, \tfrac{1}{2})$$

$$+ \begin{bmatrix} 0 \\ 0 \\ 12 \\ 6 \end{bmatrix} (0, 0, 1, \tfrac{1}{2}) + \begin{bmatrix} 0 \\ 0 \\ 0 \\ 3 \end{bmatrix} (0, 0, 0, 1)$$

Our reason for expressing the original matrix in the form of a sum of a major product of vectors is that we now may make use of the second rule on the product of partitioned matrices, summarized in Eq. (2.4.5). By symmetrically partitioning the prefactor (vertically) and the postfactor (horizontally), we can represent the product of the two matrices as a sum of matrix products:

$$\mathbf{AB} = (\mathbf{U} \quad \mathbf{V} \quad \mathbf{W}) \begin{bmatrix} \mathbf{X} \\ \mathbf{Y} \\ \mathbf{Z} \end{bmatrix} = \mathbf{UX} + \mathbf{VY} + \mathbf{WZ}$$

Applying this rule in the present situation gives

$$ {}_0\mathbf{A} = {}_1\mathbf{v}\,{}_1\mathbf{w}' + {}_2\mathbf{v}\,{}_2\mathbf{w}' + {}_3\mathbf{v}\,{}_3\mathbf{w}' + {}_4\mathbf{v}\,{}_4\mathbf{w}' = ({}_1\mathbf{v} \quad {}_2\mathbf{v} \quad {}_3\mathbf{v} \quad {}_4\mathbf{v}) \begin{bmatrix} {}_1\mathbf{w}' \\ {}_2\mathbf{w}' \\ {}_3\mathbf{w}' \\ {}_4\mathbf{w}' \end{bmatrix} = \mathbf{VW}'$$

and we see that ${}_0\mathbf{A}$ can be expressed as the product $\mathbf{VW}'$:

$$\begin{bmatrix} 48 & 24 & 24 & 12 \\ 24 & 24 & 12 & 12 \\ 24 & 12 & 24 & 12 \\ 12 & 12 & 12 & 12 \end{bmatrix} = \begin{bmatrix} 48 & 0 & 0 & 0 \\ 24 & 12 & 0 & 0 \\ 24 & 0 & 12 & 0 \\ 12 & 6 & 6 & 3 \end{bmatrix} \begin{bmatrix} 1 & \frac{1}{2} & \frac{1}{2} & \frac{1}{4} \\ 0 & 1 & 0 & \frac{1}{2} \\ 0 & 0 & 1 & \frac{1}{2} \\ 0 & 0 & 0 & 1 \end{bmatrix}$$

In this form, we recognize **V** as a lower triangular matrix and **W**′ as an upper triangular matrix. Both are basic matrices. We see, then, that ${}_0\mathbf{A}$ (or **X**′**X**) has been factored into the major product of two basic matrices and hence has rank equal to their common order, or $r(\mathbf{X}'\mathbf{X}) = 4$.

Our starting matrix ${}_0\mathbf{A}$, you will recall, was a minor product moment matrix **X**′**X**, derived from **X**, a 48×4 design matrix for the Airline Pilots Reaction Time Study. Because the rank of the minor (and major) product moment matrices is the same as the rank of the original matrix, we know that the rank of **X** must also be 4. Because the number of scalar computations required to establish the rank of a matrix increases with the size of that matrix, it would have been a greater calculational effort to determine the rank of the **X** matrix directly than was required to solve for the rank of the 4×4 **X**′**X**. This is a general principle, and the reader will find that in multivariate analysis the smaller minor product moment matrices, derived from vertical matrices, are used in rank determination solutions rather than the vertical matrices themselves.

3.4.2 Nonbasic Example

If we were interested in using this triangular factoring algorithm only to find the rank of a matrix, we would not need to assemble the triangular matrices. It would suffice to determine how many **B** matrices, or, equivalently, how many **V** rows (and **W**′ columns), are necessary to account for the starting matrix. ${}_0\mathbf{A}$ of the previous section proved to be basic. Let us try the triangular factoring algorithm to determine the rank of a matrix we know to be nonbasic.

In Section 3.3 we developed a matrix **X** as the major product of two other matrices **A** and **B**′:

$$\begin{bmatrix} 60 & 39 \\ 46 & 38 \\ 44 & 62 \\ 57 & 49 \end{bmatrix} \begin{bmatrix} 1 & 0 & 1 & 1 \\ 0 & 1 & -1 & 1 \end{bmatrix} = \begin{bmatrix} 60 & 39 & 21 & 99 \\ 46 & 38 & 8 & 84 \\ 44 & 62 & -18 & 106 \\ 57 & 49 & 8 & 106 \end{bmatrix}$$

For four students in the College Prediction Study, matrix **A** carried vocabulary scores in the first column and quantitative skills scores in the second column. The third and fourth columns of **X** carried differences and sums, respectively, between these two test performances.

In the triangular factoring of **X** we find first ${}_1\mathbf{v}' = (60, 46, 44, 57)$ and ${}_1\mathbf{w}' = (1, 39/60, 21/60, 99/60)$ and thus ${}_1\mathbf{B} = {}_1\mathbf{v}\,{}_1\mathbf{w}'$, so that

$${}_1\mathbf{B} = \begin{bmatrix} 60 & 39 & 21 & 99 \\ 46 & (46)(39)/60 & (46)(21)/60 & (44)(99)/60 \\ 44 & (44)(39)/60 & (44)(21)/60 & (44)(99)/60 \\ 57 & (57)(39)/60 & (57)(21)/60 & (57)(99)/60 \end{bmatrix}$$

Subtracting ${}_1\mathbf{B}$ from ${}_0\mathbf{A}$ to produce ${}_1\mathbf{A}$ gives

$${}_1\mathbf{A} = \begin{bmatrix} 0 & 0 & 0 & 0 \\ 0 & 486/60 & -486/60 & 486/60 \\ 0 & 2004/60 & -2004/60 & 2004/60 \\ 0 & 717/60 & -717/60 & 717/60 \end{bmatrix} = {}_0\mathbf{A} - {}_1\mathbf{B}$$

(Note that to avoid decimals we have retained the common denominator of 60.)

From ${}_1\mathbf{A}$ we now obtain ${}_2\mathbf{v}' = (0, 486/60, 2004/60, 717/60)$ and ${}_2\mathbf{w}' = (0, 1, -1, 1)$. If we calculate ${}_2\mathbf{B} = {}_2\mathbf{v}\,{}_2\mathbf{w}'$, we obtain

$${}_2\mathbf{B} = \begin{bmatrix} 0 & 0 & 0 & 0 \\ 0 & 486/60 & -486/60 & 486/60 \\ 0 & 2004/60 & -2004/60 & 2004/60 \\ 0 & 717/60 & -717/60 & 717/60 \end{bmatrix}$$

and subtracting this from ${}_1\mathbf{A}$ produces the null matrix:

$${}_1\mathbf{A} - {}_2\mathbf{B} = {}_2\mathbf{A} = 0$$

For this second example, then, ${}_0\mathbf{A}$ can be written

$${}_0\mathbf{A} = {}_1\mathbf{B} + {}_2\mathbf{B} = {}_1\mathbf{v}\,{}_1\mathbf{w}' + {}_2\mathbf{v}\,{}_2\mathbf{w}' = ({}_1\mathbf{v} \quad {}_2\mathbf{v})\begin{bmatrix} {}_1\mathbf{w}' \\ {}_2\mathbf{w}' \end{bmatrix}$$

or, numerically, $\mathbf{X} = \mathbf{T}_L\mathbf{T}'_U$:

$$\begin{bmatrix} 60 & 39 & 21 & 99 \\ 46 & 38 & 8 & 84 \\ 44 & 62 & -18 & 106 \\ 57 & 49 & 8 & 106 \end{bmatrix} = \begin{bmatrix} 60 & 0 \\ 46 & 486/60 \\ 44 & 2004/60 \\ 57 & 717/60 \end{bmatrix} \times \begin{bmatrix} 1 & 39/60 & 21/60 & 99/60 \\ 0 & 1 & -1 & 1 \end{bmatrix}$$

The rank of $\mathbf{X}$ is 2, the common order of the basic matrices (partial triangular matrices) whose major product recreates $\mathbf{X}$.

In computing an ${}_i\mathbf{w}$ vector we had to divide each element of ${}_{(i-1)}\mathbf{A}_{i.}$ by ${}_{(i-1)}A_{ii}$. We have no trouble following this rule so long as ${}_{(i-1)}A_{ii}$ is not zero. When one of these "diagonals" is zero, we are stymied. To avoid this problem, most triangular factoring routines "pivot" not on the diagonals of the ${}_i\mathbf{A}$ matrices but on the largest (absolute value) element in that matrix. The $\mathbf{v}$ vector is the column containing that largest element, and the $\mathbf{w}$ vector is the row containing that largest element, after each element has been divided by that largest element.

Following this modified algorithm, we might first pivot on $X_{34} = {}_0A_{34} = 106$, in which case

$${}_1\mathbf{v}' = (99, 84, 106, 106) \quad \text{and} \quad {}_1\mathbf{w}' = (44/106, 62/106, -18/106, 1)$$

which gives ${}_1\mathbf{B} = {}_1\mathbf{v}\,{}_1\mathbf{w}'$, so that

$$ {}_1\mathbf{B} = \begin{bmatrix} (99)(44)/106 & (99)(62)/106 & (99)(-18)/106 & 99 \\ (84)(44)/106 & (84)(62)/106 & (84)(-18)/106 & 84 \\ 44 & 62 & -18 & 106 \\ 44 & 62 & -18 & 106 \end{bmatrix} $$

which, when subtracted from ${}_0A$, gives

$$ {}_1\mathbf{A} = \begin{bmatrix} 2004/106 & -2004/106 & 4008/106 & 0 \\ 1180/106 & -1180/106 & 2360/106 & 0 \\ 0 & 0 & 0 & 0 \\ 1378/106 & -1378/106 & 2756/106 & 0 \end{bmatrix} = {}_0\mathbf{A} - {}_1\mathbf{B} $$

In the second cycle, we want to pivot on ${}_1A_{13} = 4008/106$, the largest element. The second set of factors, then, are

$$ {}_2\mathbf{v}' = (4008/106, 2360/106, 0, 2756/106) \quad \text{and} \quad {}_2\mathbf{w}' = (\tfrac{1}{2}, -\tfrac{1}{2}, 1, 0) $$

Their major product, ${}_2\mathbf{B} = {}_2\mathbf{v}\,{}_2\mathbf{w}'$, will be equal to ${}_1\mathbf{A}$, so that, again, $0 = {}_1\mathbf{B} + {}_2\mathbf{B} = {}_1\mathbf{v}\,{}_1\mathbf{w}' + {}_2\mathbf{v}\,{}_2\mathbf{w}' = ({}_1\mathbf{v}, {}_2\mathbf{v})_1\mathbf{w}' = \mathbf{V}\mathbf{X}'$.

By making the numerical substitution, however, we obtain for $\mathbf{X} = \mathbf{V}\mathbf{W}'$

$$ \begin{bmatrix} 60 & 39 & 21 & 99 \\ 46 & 38 & 8 & 84 \\ 44 & 62 & -18 & 106 \\ 57 & 49 & 8 & 106 \end{bmatrix} = \begin{bmatrix} 99 & 4008/106 \\ 84 & 2360/106 \\ 106 & 0 \\ 106 & 2756/106 \end{bmatrix} \times \begin{bmatrix} 44/106 & 62/106 & -18/106 & 1 \\ \tfrac{1}{2} & -\tfrac{1}{2} & 1 & 0 \end{bmatrix} $$

We note that the factors $\mathbf{V}$ and $\mathbf{W}'$ are not partial triangular matrices. Hence we do not have immediate assurance that $\mathbf{V}$ and $\mathbf{W}'$ are basic matrices.

Each *is* basic, however. Whenever this modified triangular factoring algorithm is followed, with pivots on the largest element in the matrix, the rows of the prefactor $\mathbf{V}$ may always be rearranged to form a lower (partial) triangular matrix,

$$ \boldsymbol{\pi}_L \mathbf{V} = \mathbf{T}_L $$

and the columns of the postfactor $\mathbf{W}'$ may always be rearranged to form an upper (partial) triangular matrix,

$$ \mathbf{W}'\boldsymbol{\pi}_R = \mathbf{T}_U' $$

For the present example, the appropriate permutation matrices are

$$\pi_L = \begin{bmatrix} 0 & 0 & 1 & 0 \\ 1 & 0 & 0 & 0 \\ 0 & 1 & 0 & 0 \\ 0 & 0 & 0 & 1 \end{bmatrix} \quad \text{and} \quad \pi_R = \begin{bmatrix} 0 & 1 & 0 & 0 \\ 0 & 0 & 1 & 0 \\ 0 & 0 & 0 & 1 \\ 1 & 0 & 0 & 0 \end{bmatrix}$$

The argument showing that $\mathbf{V}$ and $\mathbf{W}'$ are basic is completed as follows: Because π_L and π_R are *square basic* matrices, $\mathbf{T}_L$ must have the same rank as $\mathbf{V}$ and $\mathbf{T}'_U$ must have the same rank as $\mathbf{W}'$ (multiplying by a square basic matrix does not alter rank). Since $\mathbf{T}_L$ and $\mathbf{T}'_U$ *are known to be basic*, $\mathbf{V}$ and $\mathbf{W}'$ must also be basic. The modified triangular factoring rule expresses the initial matrix as the major product of two basic matrices; these basic factors may be permutations of (partial) triangular matrices rather than (partial) triangular matrices.

3.5 Regular Inverse

In Chapter 2 we presented rules for adding (and subtracting) matrices and for multiplying one matrix by another. To this point, however, we have not offered a matrix equivalent for division. While we shall not speak of dividing one matrix by another, a matrix operation rather analogous to scalar division does exist, and we develop that here.

First let us review. The rank of a matrix (including vectors and scalars) is a nonnegative integer no larger than the smaller of the two dimensions of the matrix (vector, or scalar). The only matrices with rank of zero are, by definition, null matrices (including null vectors and the scalar zero). All other matrices have positive rank. In particular, any nonzero scalar (viewed as a 1×1 matrix) must have rank 1. In fact, any nonzero scalar is a square basic matrix of order 1.

In scalar algebra, for every nonzero scalar quantity, say, $x = 2.5$, there is a second quantity, termed the *reciprocal of* x, written $1/x$ ($1/2.5 = 0.4$). The product of a nonzero scalar and its reciprocal is always the scalar 1. Further, multiplying a second scalar, say, y, by the reciprocal of x, is the same as dividing y by x:

$$y(1/x) = y/x$$

Theorem. *Every square basic matrix, not just those of order* 1, *has associated with it a reciprocal. This reciprocal is a second square basic matrix, of the same order as the original matrix. If* $\mathbf{X}$ *is a square basic matrix, its reciprocal is written* $\mathbf{X}^{-1}$ *and*

$$\mathbf{X}\mathbf{X}^{-1} = \mathbf{X}^{-1}\mathbf{X} = \mathbf{I} \qquad (3.5.1)$$

That is, if we pre- or postmultiply a square basic matrix by its reciprocal, the result is the identity matrix.

Equation (3.5.1) certainly holds for the scalar and its reciprocal, with the identity matrix of order 1 being simply the scalar 1. Now we consider the problem of finding the reciprocal of larger square basic matrices. First, however, a change in nomenclature is desirable. It has become customary to refer to the matrix $\mathbf{X}^{-1}$ that satisfies the conditions of (3.5.1) as the *regular inverse* of $\mathbf{X}$. Every square basic matrix has a regular inverse, which will be a square basic matrix of the same order. Pre- or postmultiplying a square basic matrix by its regular inverse yields the identity matrix.

Theorem. *Only square basic matrices have regular inverses.*

If $\mathbf{X}$ is square but nonbasic or if $\mathbf{X}$ is nonsquare, *there is no matrix* $\mathbf{X}^{-1}$ to satisfy (3.5.1).

For certain square basic matrices, the regular inverse may be computed rather easily. For others we are well advised to employ carefully designed and evaluated computer routines (as in the IMSL package). In this section we shall first point out the easily inverted matrices and then provide a way of extending the triangular factoring introduced in the last section to find the regular inverse in more general cases.

The easiest square basic matrix to invert is the permutation matrix. *The regular inverse of a permutation matrix is its transpose*:

$$\pi'\pi = \pi\pi' = \mathbf{I} \tag{3.5.2}$$

The two examples, π_L and π_R, introduced in the preceding section, illustrate this point:

$$\begin{bmatrix} 0 & 0 & 1 & 0 \\ 1 & 0 & 0 & 0 \\ 0 & 1 & 0 & 0 \\ 0 & 0 & 0 & 1 \end{bmatrix} \begin{bmatrix} 0 & 1 & 0 & 0 \\ 0 & 0 & 1 & 0 \\ 1 & 0 & 0 & 0 \\ 0 & 0 & 0 & 1 \end{bmatrix} = \begin{bmatrix} 1 & & & \\ & 1 & & 0 \\ & & 1 & \\ 0 & & & 1 \end{bmatrix}$$

and

$$\begin{bmatrix} 0 & 0 & 0 & 1 \\ 1 & 0 & 0 & 0 \\ 0 & 1 & 0 & 0 \\ 0 & 0 & 1 & 0 \end{bmatrix} \begin{bmatrix} 0 & 1 & 0 & 0 \\ 0 & 0 & 1 & 0 \\ 0 & 0 & 0 & 1 \\ 1 & 0 & 0 & 0 \end{bmatrix} = \begin{bmatrix} 1 & & & \\ & 1 & & 0 \\ & & 1 & \\ 0 & & & 1 \end{bmatrix}$$

The permutation matrix, then, is a square orthonormal matrix. Any square orthonormal matrix has a regular inverse in its transposed form.

A second class of easily inverted square basic matrices is the proper diagonal matrix. The notation introduced earlier for diagonal matrices hints at the process. *The regular inverse of a proper diagonal matrix is a second proper diagonal matrix whose diagonal elements equal the reciprocals of the diagonal elements of the original matrix.* In Section 3.1 we developed an example based on the supervisory ratings of five different supervisors. When rating the same ten employees, the five produced ratings with standard deviations given in the diagonal of the matrix

$$\mathbf{D_S} = \begin{bmatrix} 2.87 & & & & \\ & 0.50 & & 0 & \\ & & 0.98 & & \\ & 0 & & 0.98 & \\ & & & & 1.97 \end{bmatrix}$$

The regular inverse of $\mathbf{D_S}$, written $\mathbf{D_S^{-1}}$, will be a second diagonal matrix:

$$\mathbf{D_S^{-1}} = \begin{bmatrix} 0.3484 & & & & \\ & 2.0000 & & 0 & \\ & & 1.0204 & & \\ & 0 & & 1.0204 & \\ & & & & 0.5076 \end{bmatrix}$$

Each diagonal element in $\mathbf{D_S^{-1}}$ is the reciprocal of the corresponding diagonal element in $\mathbf{D_S}$. (We have again used a shorter notation describing only the nonzero—diagonal—elements.)

A third class of square basic matrices for which inverses are rather easily determined is those that are 2×2. Here we can do more than announce the result; we can apply our definition of the regular inverse to develop the answer. As an example consider the 2×2 upper-left partition of the minor product moment of the design matrix for the Airline Pilot Reaction Time Study, which gives us the matrix

$$\mathbf{S} = \begin{bmatrix} 48 & 24 \\ 24 & 24 \end{bmatrix}$$

for which we wish to find the regular inverse. By the defining equation (3.5.1) we can write the matrix equation $\mathbf{SS^{-1}} = \mathbf{I}$:

$$\begin{bmatrix} 48 & 24 \\ 24 & 24 \end{bmatrix} \begin{bmatrix} (S^{-1})_{11} & (S^{-1})_{12} \\ (S^{-1})_{21} & (S^{-1})_{22} \end{bmatrix} = \begin{bmatrix} 1 & 0 \\ 0 & 1 \end{bmatrix}$$

and from it write down the four scalar equations (each a row-by-column multiplication) that link the left- and right-hand sides:

$$48(S^{-1})_{11} + 24(S^{-1})_{21} = 1 \quad (3.5.3)$$

$$48(S^{-1})_{12} + 24(S^{-1})_{22} = 0 \quad (3.5.4)$$

$$24(S^{-1})_{11} + 24(S^{-1})_{21} = 0 \quad (3.5.5)$$

$$24(S^{-1})_{12} + 24(S^{-1})_{22} = 1 \quad (3.5.6)$$

Subtracting (3.5.5) from (3.5.3) yields

$$24(S^{-1})_{11} = 1$$

or

$$(S^{-1})_{11} = \tfrac{1}{24} \tag{3.5.7}$$

Similarly, subtracting (3.5.6) from (3.5.4) yields

$$24(S^{-1})_{12} = -1$$

or

$$(S^{-1})_{12} = -\tfrac{1}{24} \tag{3.5.8}$$

Substituting the result of (3.5.7) into (3.5.3) gives

$$2 + 24(S^{-1})_{21} = 1$$

or

$$(S^{-1})_{21} = -\tfrac{1}{24} \tag{3.5.9}$$

Finally, substituting from (3.5.8) into (3.5.4) gives

$$-2 + 24(S^{-1})_{22} = 0$$

or

$$(S^{-1})_{22} = \tfrac{2}{24} \tag{3.5.10}$$

We can write the complete regular inverse from (3.5.7)–(3.5.10). Using the result as a premultiplier of **S** results in the product

$$\begin{bmatrix} \tfrac{1}{24} & -\tfrac{1}{24} \\ -\tfrac{1}{24} & \tfrac{2}{24} \end{bmatrix} \begin{bmatrix} 48 & 24 \\ 24 & 24 \end{bmatrix} = \begin{bmatrix} 1 & 0 \\ 0 & 1 \end{bmatrix}$$

for $\mathbf{S}^{-1}\mathbf{S} = \mathbf{I}$.

The steps in computing the 2 × 2 regular inverse in this example may be presented in a closed form. If **A** is a 2 × 2 square basic matrix with elements,

$$\mathbf{A} = \begin{bmatrix} A_{11} & A_{12} \\ A_{21} & A_{22} \end{bmatrix}$$

then the regular inverse of **A** can be written

$$\mathbf{A}^{-1} = \frac{1}{A_{11}A_{22} - A_{12}A_{21}} \begin{bmatrix} A_{22} & -A_{12} \\ -A_{21} & A_{11} \end{bmatrix} \tag{3.5.11}$$

For the example, this computation yields

$$\mathbf{A}^{-1} = \frac{1}{(48)(24) - (24)(24)} \begin{bmatrix} 24 & -24 \\ -24 & 48 \end{bmatrix}$$

or

$$\mathbf{A}^{-1} = \frac{1}{(24)(24)}\begin{bmatrix} 24 & -24 \\ -24 & 48 \end{bmatrix}$$

which is $\mathbf{S}^{-1}$. [Note that a nonbasic square matrix of order 2×2 will be detected by Eq. (3.5.11). The difference $A_{11}A_{22} - A_{12}A_{21}$ is zero for a nonbasic matrix, precluding the computation of a regular inverse.]

The regular inverse follows two rules similar to those for transposes. First, if the square matrix $\mathbf{A}$ has an inverse $\mathbf{A}^{-1}$, then

$$(\mathbf{A}^{-1})^{-1} = \mathbf{A} \tag{3.5.12}$$

The regular inverse of a regular inverse is the original matrix. The reader may want to apply Eq. (3.5.12) to $\mathbf{S}^{-1}$ to confirm that its inverse is the matrix $\mathbf{S}$.

The second rule relates to the regular inverse of a product. If $\mathbf{A}$, $\mathbf{B}$, and $\mathbf{C}$ are square basic matrices of the same order, then the three regular inverses $\mathbf{A}^{-1}, \mathbf{B}^{-1}, \mathbf{C}^{-1}$ exist. Also, if $\mathbf{A}$, $\mathbf{B}$, and $\mathbf{C}$ are square basic matrices of the same order, then the product $\mathbf{ABC} = \mathbf{G}$ exists and is also a square basic matrix. This follows from the rule developed earlier that multiplication by a square basic matrix does not alter the rank of a matrix: $r(\mathbf{AB}) = r(\mathbf{A})$; $r[(\mathbf{AB})\mathbf{C}] = r(\mathbf{AB}) = r(\mathbf{A})$. Since $\mathbf{G}$ is square and basic, it too will have a regular inverse. The regular inverse of $\mathbf{G}$ will be related to those of $\mathbf{A}, \mathbf{B}, \mathbf{C}$ as follows:

$$(\mathbf{ABC})^{-1} = \mathbf{C}^{-1}\mathbf{B}^{-1}\mathbf{A}^{-1} \tag{3.5.13}$$

The inverse of a product is the product of the inverses taken in reverse order. As an example, consider the product $\mathbf{SD} = \mathbf{W}$:

$$\begin{bmatrix} 48 & 24 \\ 24 & 24 \end{bmatrix}\begin{bmatrix} \frac{1}{24} & 0 \\ 0 & \frac{1}{12} \end{bmatrix} = \begin{bmatrix} 2 & 2 \\ 1 & 2 \end{bmatrix}$$

According to (3.5.13), $\mathbf{W}^{-1} = \mathbf{D}^{-1}\mathbf{S}^{-1}$, or

$$\begin{bmatrix} 24 & 0 \\ 0 & 12 \end{bmatrix}\begin{bmatrix} \frac{1}{24} & -\frac{1}{24} \\ -\frac{1}{24} & \frac{2}{24} \end{bmatrix} = \begin{bmatrix} 1 & -1 \\ -\frac{1}{2} & 1 \end{bmatrix}$$

The correctness of this solution may be confirmed by (a) solving for $\mathbf{W}^{-1}$ directly using Eq. (3.5.11) and (b) computing $\mathbf{WW}^{-1}$ or $\mathbf{W}^{-1}\mathbf{W}$ to make certain the identity matrix results.

3.5.1 Finding Inverses Using Triangular Factoring

Diagonal, permutation, and 2×2 matrices are only a few of the square basic matrices whose inverses are sought in multivariate analyses. How can we find the regular inverse of a more general (and larger) square basic $n \times n$ matrix? For $n > 4$, the computations become lengthy and susceptible to error. Such solutions are best left to well-prepared computer programs. However, to show that such solutions are not "magic," we outline here one approach to finding the regular inverse of a square basic 4 × 4 matrix.

The technique that we shall illustrate is in two steps. The first step involves the triangular factoring of the square matrix, as developed in the preceding section. The reason for this approach is that the inversion of a triangular matrix, though not trivial, is more easily accomplished than the inversion of a general square basic matrix. In the preceding section we showed that triangular factoring of a square basic matrix leads to an expression of that matrix as the product of lower and upper triangular matrices, each of which is a square basic matrix:

$$\mathbf{S} = \mathbf{T}_{\mathrm{L}}\mathbf{T}_{\mathrm{U}}$$

By the rule on the inverse of a product, the regular inverse of the matrix **S** can then be written in terms of its triangular factors:

$$\mathbf{S}^{-1} = (\mathbf{T}_{\mathrm{L}}\mathbf{T}_{\mathrm{U}})^{-1} = \mathbf{T}_{\mathrm{U}}^{-1}\mathbf{T}_{\mathrm{L}}^{-1}$$

In Section 3.4, using the minor product moment of the Reaction Time Study design matrix, we obtained this specific triangular factoring $\mathbf{S} = \mathbf{T}_{\mathrm{L}}\mathbf{T}_{\mathrm{U}}$:

$$\begin{bmatrix} 48 & 24 & 24 & 12 \\ 24 & 24 & 12 & 12 \\ 24 & 12 & 24 & 12 \\ 12 & 12 & 12 & 12 \end{bmatrix} = \begin{bmatrix} 48 & 0 & 0 & 0 \\ 24 & 12 & 0 & 0 \\ 24 & 0 & 12 & 0 \\ 12 & 6 & 6 & 3 \end{bmatrix} \begin{bmatrix} 1 & \frac{1}{2} & \frac{1}{2} & \frac{1}{4} \\ 0 & 1 & 0 & \frac{1}{2} \\ 0 & 0 & 1 & \frac{1}{2} \\ 0 & 0 & 0 & 1 \end{bmatrix}$$

The regular inverse of **S** can now be found from the inverses of $\mathbf{T}_{\mathrm{L}}$ and $\mathbf{T}_{\mathrm{U}}$. Before computing that, however, we point out a simplification that occurs when the matrix **S** is *symmetric* as well as square and basic.

If **S** is a basic symmetric matrix, then the right triangular factor $\mathbf{T}_{\mathrm{U}}$ has rows proportional to the columns of the left triangular factor $\mathbf{T}_{\mathrm{L}}$. In fact, $\mathbf{T}_{\mathrm{U}}$ can be written

$$\mathbf{T}_{\mathrm{U}} = \mathbf{D}_{\mathbf{T}_{\mathrm{L}}}^{-1}\mathbf{T}_{\mathrm{L}}'$$

In the present example,

$$\begin{bmatrix} 1 & \frac{1}{2} & \frac{1}{2} & \frac{1}{4} \\ 0 & 1 & 1 & \frac{1}{2} \\ 0 & 0 & 1 & \frac{1}{2} \\ 0 & 0 & 0 & 1 \end{bmatrix} = \begin{bmatrix} \frac{1}{48} & & & 0 \\ & \frac{1}{12} & & \\ & & \frac{1}{12} & \\ 0 & & & \frac{1}{3} \end{bmatrix} \begin{bmatrix} 48 & 24 & 24 & 12 \\ 0 & 12 & 0 & 6 \\ 0 & 0 & 12 & 6 \\ 0 & 0 & 0 & 3 \end{bmatrix}$$

By substituting $\mathbf{D}_{\mathbf{T}_{\mathrm{L}}}^{-1}\mathbf{T}_{\mathrm{L}}'$ for $\mathbf{T}_{\mathrm{U}}$, we can write

$$\mathbf{S} = \mathbf{T}_{\mathrm{L}}\mathbf{D}_{\mathbf{T}_{\mathrm{L}}}^{-1}\mathbf{T}_{\mathrm{L}}'$$

and

$$\mathbf{S}^{-1} = (\mathbf{T}'_{\mathrm{L}})^{-1}\mathbf{D}_{\mathbf{T}_{\mathrm{L}}}\mathbf{T}_{\mathrm{L}}^{-1} \tag{3.5.14}$$

by the inverse product rule.

We begin the inversion of $\mathbf{S}$ by showing how a triangular matrix is inverted. For notational simplicity, we rename $\mathbf{T}_{\mathrm{L}} = \mathbf{T}$ and $\mathbf{T}_{\mathrm{L}}^{-1} = \mathbf{a}$, where it is required that the product of the known $\mathbf{T}$ and unknown $\mathbf{a}$ be $\mathbf{I}$:

$$\begin{bmatrix} 48 & 0 & 0 & 0 \\ 24 & 12 & 0 & 0 \\ 24 & 0 & 12 & 0 \\ 12 & 6 & 6 & 3 \end{bmatrix} \begin{bmatrix} a_{11} & a_{12} & a_{13} & a_{14} \\ a_{21} & a_{22} & a_{23} & a_{24} \\ a_{31} & a_{32} & a_{33} & a_{34} \\ a_{41} & a_{42} & a_{43} & a_{44} \end{bmatrix} = \begin{bmatrix} 1 & & & 0 \\ & 1 & & \\ & & 1 & \\ 0 & & & 1 \end{bmatrix}$$

1. From this we can see that

$$\mathbf{T}'_{1.}\mathbf{a}_{.1} = (48, 0, 0, 0)\begin{bmatrix} a_{11} \\ a_{21} \\ a_{31} \\ a_{41} \end{bmatrix} = 1 \qquad \text{only if} \quad a_{11} = \tfrac{1}{48}$$

2. $\mathbf{T}'_{1.}\mathbf{a}_{.2} = 0$, $\mathbf{T}'_{1.}\mathbf{a}_{.3} = 0$, and $\mathbf{T}'_{1.}\mathbf{a}_{.4} = 0$ only if $a_{12} = a_{13} = a_{14} = 0$.

We now have

$$\begin{bmatrix} 48 & 0 & 0 & 0 \\ 24 & 12 & 0 & 0 \\ 24 & 0 & 12 & 0 \\ 12 & 6 & 6 & 3 \end{bmatrix} \begin{bmatrix} \frac{1}{48} & 0 & 0 & 0 \\ a_{21} & a_{22} & a_{23} & a_{24} \\ a_{31} & a_{32} & a_{33} & a_{34} \\ a_{41} & a_{42} & a_{43} & a_{44} \end{bmatrix} = \begin{bmatrix} 1 & & & 0 \\ & 1 & & \\ & & 1 & \\ 0 & & & 1 \end{bmatrix}$$

3. From this we can conclude that

$$\mathbf{T}'_{2.}\mathbf{a}_{.1} = (24, 12, 0, 0)\begin{bmatrix} \frac{1}{48} \\ a_{21} \\ a_{31} \\ a_{41} \end{bmatrix} = 0 \qquad \text{only if} \quad a_{21} = -\tfrac{1}{24}$$

4. $\mathbf{T}'_{2.}\mathbf{a}_{.2} = 1$ only if $12a_{22} = 1$, or $a_{22} = \frac{1}{12}$.

5. $\mathbf{T}'_{2.}\mathbf{a}_{.3} = 0$ and $\mathbf{T}'_{2.}\mathbf{a}_{.4} = 0$ only if $a_{23} = a_{24} = 0$.

This now gives us

$$\begin{bmatrix} 48 & 0 & 0 & 0 \\ 24 & 12 & 0 & 0 \\ 24 & 0 & 12 & 0 \\ 12 & 6 & 6 & 3 \end{bmatrix} \begin{bmatrix} \frac{1}{48} & 0 & 0 & 0 \\ -\frac{1}{24} & \frac{1}{12} & 0 & 0 \\ a_{31} & a_{32} & a_{33} & a_{34} \\ a_{41} & a_{42} & a_{43} & a_{44} \end{bmatrix} = \begin{bmatrix} 1 & & & 0 \\ & 1 & & \\ & & 1 & \\ 0 & & & 1 \end{bmatrix}$$

6. Continuing, we see that

$$\mathbf{T}'_{3.}\mathbf{a}_{.1} = (24, 0, 12, 0)\begin{bmatrix} \frac{1}{48} \\ -\frac{1}{24} \\ a_{31} \\ a_{42} \end{bmatrix} = 0 \qquad \text{only if} \quad a_{31} = -\tfrac{1}{24}$$

7. $\mathbf{T}'_{3.}\mathbf{a}_{.2} = 0$ only if $a_{32} = 0$.

8. $\mathbf{T}'_{3.}\mathbf{a}_{.3} = 1$ only if $a_{33} = \frac{1}{12}$.

9. $\mathbf{T}'_{3.}\mathbf{a}_{.4} = 0$ only if $a_{34} = 0$.

We now have three rows of **a**:

$$\begin{bmatrix} 48 & 0 & 0 & 0 \\ 24 & 12 & 0 & 0 \\ 24 & 0 & 12 & 0 \\ 12 & 6 & 6 & 3 \end{bmatrix}\begin{bmatrix} \frac{1}{48} & 0 & 0 & 0 \\ -\frac{1}{24} & \frac{1}{24} & 0 & 0 \\ -\frac{1}{24} & 0 & \frac{1}{12} & 0 \\ a_{41} & a_{42} & a_{43} & a_{44} \end{bmatrix} = \begin{bmatrix} 1 & & & 0 \\ & 1 & & \\ & & 1 & \\ 0 & & & 1 \end{bmatrix}$$

10. The inverse is completed by noting that

$$\mathbf{T}_{4.}\mathbf{a}_{.1} = (12, 6, 6, 3)\begin{bmatrix} \frac{1}{48} \\ -\frac{1}{24} \\ -\frac{1}{24} \\ a_{41} \end{bmatrix} = 0 \qquad \text{only if} \quad a_{41} = \tfrac{1}{12}$$

11. $\mathbf{T}'_{4.}\mathbf{a}_{.2} = 0$ only if $3(a_{42}) = -\frac{1}{2}$, or $a_{42} = -\frac{1}{6}$.

12. $\mathbf{T}'_{4.}\mathbf{a}_{.3} = 0$ only if $a_{43} = -\frac{1}{6}$.

13. $\mathbf{T}'_{4.}\mathbf{a}_{.4} = 1$ only if $a_{44} = \frac{1}{3}$.

The regular inverse of **T**, then, is completely specified:

$$\begin{bmatrix} 48 & 0 & 0 & 0 \\ 24 & 12 & 0 & 0 \\ 24 & 0 & 12 & 0 \\ 12 & 6 & 6 & 3 \end{bmatrix}\begin{bmatrix} \frac{1}{48} & 0 & 0 & 0 \\ -\frac{1}{24} & \frac{1}{12} & 0 & 0 \\ -\frac{1}{24} & 0 & \frac{1}{12} & 0 \\ \frac{1}{12} & -\frac{1}{6} & -\frac{1}{6} & \frac{1}{3} \end{bmatrix} = \begin{bmatrix} 1 & & & 0 \\ & 1 & & \\ & & 1 & \\ 0 & & & 1 \end{bmatrix}$$

We have obtained one of the three inverses, $\mathbf{T}_L^{-1}$, required in (3.5.14). The product $\mathbf{D}_{\mathbf{T}_L}\mathbf{T}_L^{-1}$ is computed:

$$\begin{bmatrix} 48 & & & 0 \\ & 12 & & \\ & & 12 & \\ 0 & & & 3 \end{bmatrix}\mathbf{T}_L^{-1} = \begin{bmatrix} 1 & 0 & 0 & 0 \\ -\frac{1}{2} & 1 & 0 & 0 \\ -\frac{1}{2} & 0 & 1 & 0 \\ \frac{1}{4} & -\frac{1}{2} & -\frac{1}{2} & 1 \end{bmatrix}$$

leaving only the task of finding $(\mathbf{T}'_L)^{-1}$. This is easily accomplished by invoking the following principle: *If a square basic matrix* $\mathbf{S}$ *has an inverse* $\mathbf{S}^{-1}$, *then its transpose* $\mathbf{S}'$ *has an inverse that is the transpose of the inverse of* $\mathbf{S}$, *that is,*

$$(\mathbf{S}')^{-1} = (\mathbf{S}^{-1})' \tag{3.5.15}$$

In the present example, then,

$$(\mathbf{T}'_L)^{-1} = (\mathbf{T}_L^{-1})' = \begin{bmatrix} \frac{1}{48} & -\frac{1}{24} & -\frac{1}{24} & \frac{1}{12} \\ 0 & \frac{1}{12} & 0 & -\frac{1}{6} \\ 0 & 0 & \frac{1}{12} & -\frac{1}{6} \\ 0 & 0 & 0 & \frac{1}{3} \end{bmatrix}$$

and by (3.5.14) $[(\mathbf{T}'_L)^{-1}(\mathbf{D}_{\mathbf{T}_L}\mathbf{T}_L^{-1}) = \mathbf{S}^{-1}]$

$$\begin{bmatrix} \frac{1}{48} & -\frac{1}{24} & -\frac{1}{24} & \frac{1}{12} \\ 0 & \frac{1}{12} & 0 & -\frac{1}{6} \\ 0 & 0 & \frac{1}{12} & -\frac{1}{6} \\ 0 & 0 & 0 & \frac{1}{3} \end{bmatrix} \begin{bmatrix} 1 & 0 & 0 & 0 \\ -\frac{1}{2} & 1 & 0 & 0 \\ -\frac{1}{2} & 0 & 1 & 0 \\ \frac{1}{4} & -\frac{1}{2} & -\frac{1}{2} & 1 \end{bmatrix}$$

$$= \begin{bmatrix} \frac{1}{12} & -\frac{1}{12} & -\frac{1}{12} & \frac{1}{12} \\ -\frac{1}{12} & \frac{1}{6} & \frac{1}{12} & -\frac{1}{6} \\ -\frac{1}{12} & \frac{1}{12} & \frac{1}{6} & -\frac{1}{6} \\ \frac{1}{12} & -\frac{1}{6} & -\frac{1}{6} & \frac{1}{3} \end{bmatrix}$$

Note that the result satisfies the principle that *the inverse of a symmetric matrix is a symmetric matrix.*

We have shown how the triangular factoring method is used to find, relatively easily, the inverse of a matrix, particularly a symmetric matrix. The computational labor is still nontrivial, however, and in practice the computation of inverses is left to the computer.

3.5.2 Inversion of Symmetrically Partitioned Matrices

One final approach, which may be used in the computation of the inverse of a larger square matrix but is more important for its value in symbolizing certain matrix results, is the *partition rule for inversion*. This rule can be stated in different forms. We present it here in terms of finding the inverse of a larger square basic matrix, given the inverse of a certain partition of the matrix. We assume that a square matrix A is *symmetrically* partitioned so that

$$\mathbf{A} = \begin{bmatrix} {}_1\mathbf{A} & {}_2\mathbf{A} \\ {}_3\mathbf{A} & {}_4\mathbf{A} \end{bmatrix}$$

The symmetric partitioning ensures that ${}_1\mathbf{A}$ and ${}_4\mathbf{A}$ are both square matrices. Assume we know or can compute ${}_1\mathbf{A}^{-1}$. Then the regular inverse of the larger matrix $\mathbf{A}$ is

$$\mathbf{A}^{-1} = \left[\begin{array}{c|c} {}_1\mathbf{A}^{-1} + {}_1\mathbf{A}^{-1}{}_2\mathbf{A}\mathbf{B}^{-1}{}_3\mathbf{A}'{}_1\mathbf{A}^{-1} & -{}_1\mathbf{A}^{-1}{}_2\mathbf{A}\mathbf{B}^{-1} \\ \hline -\mathbf{B}^{-1}{}_3\mathbf{A}'{}_1\mathbf{A}^{-1} & \mathbf{B}^{-1} \end{array}\right] \tag{3.5.16}$$

where

$$\mathbf{B} = {}_4\mathbf{A} - {}_3\mathbf{A}'{}_1\mathbf{A}^{-1}{}_2\mathbf{A} \tag{3.5.17}$$

This partitioning rule can be illustrated by finding, again, the inverse of

$$\mathbf{S} = \left[\begin{array}{cc|cc} 48 & 24 & 24 & 12 \\ 24 & 24 & 12 & 12 \\ \hline 24 & 12 & 24 & 12 \\ 12 & 12 & 12 & 12 \end{array}\right] = \begin{bmatrix} {}_1\mathbf{A} & {}_2\mathbf{A} \\ {}_3\mathbf{A} & {}_4\mathbf{A} \end{bmatrix}$$

Earlier, Eq. (3.5.11) was applied to find the inverse of ${}_1\mathbf{A}$:

$${}_1\mathbf{A}^{-1} = \begin{bmatrix} \frac{1}{24} & -\frac{1}{24} \\ -\frac{1}{24} & \frac{2}{24} \end{bmatrix}$$

From the result, we can now make the following computations:

$${}_3\mathbf{A}'{}_1\mathbf{A}^{-1} = \begin{bmatrix} 24 & 12 \\ 12 & 12 \end{bmatrix}\begin{bmatrix} \frac{1}{24} & -\frac{1}{24} \\ -\frac{1}{24} & \frac{2}{24} \end{bmatrix} = \begin{bmatrix} \frac{1}{2} & 0 \\ 0 & \frac{1}{2} \end{bmatrix} \tag{3.5.18}$$

$${}_1\mathbf{A}^{-1}{}_2\mathbf{A} = \begin{bmatrix} \frac{1}{24} & -\frac{1}{24} \\ -\frac{1}{24} & \frac{2}{24} \end{bmatrix}\begin{bmatrix} 24 & 12 \\ 12 & 12 \end{bmatrix} = \begin{bmatrix} \frac{1}{2} & 0 \\ 0 & \frac{1}{2} \end{bmatrix} \tag{3.5.19}$$

and from (3.5.18),

$$({}_3\mathbf{A}'{}_1\mathbf{A}^{-1}){}_2\mathbf{A} = \begin{bmatrix} \frac{1}{2} & 0 \\ 0 & \frac{1}{2} \end{bmatrix}\begin{bmatrix} 24 & 12 \\ 12 & 12 \end{bmatrix} = \begin{bmatrix} 12 & 6 \\ 6 & 6 \end{bmatrix} \tag{3.5.20}$$

By (3.5.17) and (3.5.20), the matrix $\mathbf{B} = {}_4\mathbf{A} - {}_3\mathbf{A}'{}_1\mathbf{A}^{-1}{}_2\mathbf{A}$ is given by

$$\begin{bmatrix} 24 & 12 \\ 12 & 12 \end{bmatrix} - \begin{bmatrix} 12 & 6 \\ 6 & 6 \end{bmatrix} = \begin{bmatrix} 12 & 6 \\ 6 & 6 \end{bmatrix}$$

and $\mathbf{B}^{-1}$ by

$$\frac{1}{(12)(6)-(6)(6)}\begin{bmatrix} 6 & -6 \\ -6 & 12 \end{bmatrix} = \begin{bmatrix} \frac{1}{6} & -\frac{1}{6} \\ -\frac{1}{6} & \frac{2}{6} \end{bmatrix} = \mathbf{B}^{-1} \tag{3.5.21}$$

From (3.5.18) and (3.5.21),

$$-\mathbf{B}^{-1}({}_3\mathbf{A}'{}_1\mathbf{A}^{-1}) = \begin{bmatrix} -\frac{1}{6} & \frac{1}{6} \\ \frac{1}{6} & -\frac{2}{6} \end{bmatrix}\begin{bmatrix} \frac{1}{2} & 0 \\ 0 & \frac{1}{2} \end{bmatrix} = \begin{bmatrix} -\frac{1}{12} & \frac{1}{12} \\ \frac{1}{12} & -\frac{1}{6} \end{bmatrix} \tag{3.5.22}$$

from (3.5.19) and (3.5.21),

$$(-{}_1\mathbf{A}^{-1}{}_2\mathbf{A})\mathbf{B}^{-1} = \begin{bmatrix} -\frac{1}{2} & 0 \\ 0 & -\frac{1}{2} \end{bmatrix} \begin{bmatrix} \frac{1}{6} & -\frac{1}{6} \\ -\frac{1}{6} & \frac{2}{6} \end{bmatrix} = \begin{bmatrix} -\frac{1}{12} & \frac{1}{12} \\ \frac{1}{12} & -\frac{1}{6} \end{bmatrix} \tag{3.5.23}$$

and from (3.5.19) and (3.5.22),

$$({}_1\mathbf{A}^{-1}{}_2\mathbf{A})\mathbf{B}^{-1}{}_3\mathbf{A}'\,{}_1\mathbf{A}^{-1} = \begin{bmatrix} \frac{1}{2} & 0 \\ 0 & \frac{1}{2} \end{bmatrix} \begin{bmatrix} \frac{1}{12} & -\frac{1}{12} \\ -\frac{1}{12} & \frac{1}{6} \end{bmatrix} = \begin{bmatrix} \frac{1}{24} & -\frac{1}{24} \\ -\frac{1}{24} & \frac{1}{12} \end{bmatrix} \tag{3.5.24}$$

Adding the result in (3.5.24) to ${}_1\mathbf{A}^{-1}$ yields

$${}_1\mathbf{A}^{-1} + {}_1\mathbf{A}^{-1}{}_2\mathbf{A}\mathbf{B}^{-1}{}_3\mathbf{A}'\,{}_1\mathbf{A}^{-1} = \begin{bmatrix} \frac{1}{12} & -\frac{1}{12} \\ -\frac{1}{12} & \frac{1}{6} \end{bmatrix} \tag{3.5.25}$$

All the partitions needed to form $\mathbf{A}^{-1}$ according to (3.5.16) are given by (3.5.21)–(3.5.23) and (3.5.25). Collecting these results gives

$$\mathbf{A}^{-1} = \begin{bmatrix} \frac{1}{12} & -\frac{1}{12} & -\frac{1}{12} & \frac{1}{12} \\ -\frac{1}{12} & \frac{1}{6} & \frac{1}{12} & -\frac{1}{6} \\ -\frac{1}{12} & \frac{1}{12} & \frac{1}{6} & -\frac{1}{6} \\ \frac{1}{12} & -\frac{1}{6} & -\frac{1}{6} & \frac{1}{3} \end{bmatrix}$$

The inverse, found here by the partitioning rule, is identical to the solution obtained using the triangular factoring method. The importance, we emphasize again, of both approaches is that they will allow us in subsequent chapters to represent important matrix operations.

3.6 Summary

Since few ratio scale measurements are used in the behavioral sciences, we often must transform the raw data so that information from different attributes may be commensurable. In such transformations as centering the data matrix by columns, some information may be lost. This issue of the amount of information in a matrix is addressed by defining the rank of a matrix $\mathbf{X}$, $r(\mathbf{X})$. A matrix whose rank is equal to its smaller dimension is basic. A matrix is nonbasic if its rank is less than its smaller order. Several basic and nonbasic matrices were discussed. Two rules for finding the rank of products of matrices with known rank were defined: the product rank inequality and sum rank inequality. These rules were applied to investigate the effects of centering, rotating, and orthogonalizing matrices. We illustrated the triangular factoring procedure for finding the rank of basic and nonbasic matrices.

Finding the triangular factors of a square basic matrix can be used as an intermediate step in finding the regular inverse of the square basic matrix. The regular inverse plays a vital role in the matrix algebra of multivariate analysis. As we noted in introducing the inverse, such matrices provide the matrix analog to algebraic division. One purpose in introducing the inverse at this point, however, is to call attention to the very large number of ways in which a matrix may be factored into a product of basic matrices or factors.

In defining the rank of a matrix, we noted that any matrix could be expressed as the major product of two basic matrices with common order equal to the rank of the matrix **X**. That is, a vertical $n \times m$ matrix **X** with $r(\mathbf{X}) = p \leqslant m$ may be expressed as the product

$$\mathbf{X}_{(n\times m)} = \mathbf{A}_{(n\times p)}\mathbf{B}'_{(p\times m)}$$

of the two basic matrices **A** and **B**. When **X** is a data matrix, carrying m scores for each of n entities, it is natural to interpret **A** as containing $p \leqslant m$ scores for each of the entities from which, using the weights that make up the columns of **B**′, all m scores (columns) in **X** can be recreated. While there may be redundancies among the columns of **X**, there will be none among the columns of the basic matrix **A**.

Factoring a matrix into basic factors, then, would seem to be a step toward finding out something of central importance about the original matrix. If p is less than m, the p nonredundant scores or dimensions ought to be more basic than the m columns with their redundancies that comprise **X**. While intuitively that is true, our ability to use the matrix **A** to help us to understand **X** is drastically limited when we know that **X** can be represented as the product of infinitely many pairs of basic matrices.

Section 3.4 presented one way of factoring a matrix into the major product of two basic matrices. By this technique, the resulting factors will be either (partial) triangular matrices or permutations of (partial) triangular matrices. Symbolically, a largest-element pivoting factoring expresses **X** as

$$\mathbf{X}_{(n\times m)} = \mathbf{L}_{(n\times p)}\mathbf{R}'_{(p\times m)} \tag{3.6.1}$$

where $r(\mathbf{X}) = p$; **L** and **R**′ are both basic with rank p. If **S** is *any* square basic $p \times p$ matrix, then **X** may also be represented as the major product of the two basic matrices **A** and **B**′, where

$$\mathbf{A} = \mathbf{LS} \qquad \text{and} \qquad \mathbf{B}' = \mathbf{S}^{-1}\mathbf{R}'$$

inasmuch as

$$\mathbf{AB}' = \mathbf{LSS}^{-1}\mathbf{R}' = \mathbf{LIR}' = \mathbf{LR}' = \mathbf{X}$$

The matrix **S** may be a diagonal matrix, in which case the columns of **L** are scaled to produce the columns of **A** and the rows of **R**′ are scaled (by $\mathbf{S}^{-1}$) to produce the rows of **B**′. Alternatively, **S** may be a permutation matrix that rearranges the columns of **L**, while its inverse (the transpose of

S if it is a permutation matrix) rearranges the rows of **R**′. Or **S** may be any one of a number of general square basic matrices of order p, in which case we cannot easily predict the relation of **A** to **L** or of **B**′ to **R**′.

Which pair of basic matrices provides the best representation of **X**? We chose the original, triangular factors only because they were recognizable as basic matrices. They teach us little more about **X** than its rank. Transforming the two factors by multiplying by a square basic matrix and its inverse does not change the rank of **X**. We thus have another representation of **X**—however, another arbitrary one. In the next chapter we consider a nonarbitrary factoring of a matrix into what we call its *basic structure*. This basic structure representation will not only give the rank of the matrix but also serves to clarify the goals of most of our multivariate analyses.

Exercises

Use the following two matrices for Exercises 1–9:

$$\mathbf{A} = \begin{bmatrix} 1 & 1 & 1 & 1 \\ 1 & 2 & 4 & 8 \\ 1 & 3 & 9 & 27 \\ 1 & 4 & 16 & 64 \end{bmatrix}, \qquad \mathbf{X} = \begin{bmatrix} 1 & 4 & 1 \\ 1 & 5 & 10 \\ 2 & 6 & 12 \end{bmatrix}$$

1. Standardize **X** by columns.

2. Standardize **A** by rows.

3. Find the rank of **X**.

4. Find the rank of **X**, pivoting on ${}_0X_{11}$, ${}_1X_{22}$, and ${}_2X_{33}$ in order.

5. Find the regular inverse of **X** using the triangular factors found in Exercise 3.

6. Find triangular factors of **X**, pivoting on the *largest* elements. How do these results compare with those of Exercise 3?

7. Find the regular inverse of **X** using the factors found in Exercise 6. How do the inverses of Exercises 5 and 8 (below) compare?

8. Find **A′A**, the rank of **A′A** and, if defined, the regular inverse of **A′A** partitioned into 2 × 2 submatrices.

9. Find tr($\mathbf{X}^{-1}$). How does this compare with tr(**X**)?

In the remaining exercises,

$$\mathbf{Y} = \begin{bmatrix} \frac{2}{3} & -\frac{1}{3} & -\frac{1}{3} \\ -\frac{1}{3} & \frac{2}{3} & -\frac{1}{3} \\ -\frac{1}{3} & -\frac{1}{3} & \frac{2}{3} \end{bmatrix}$$

10. Find the triangular factors of **Y**. What is the rank of **Y**?

11. Compute **Y′Y**. What do you conclude about **Y**?

Additional Reading

The following reference provides a good discussion of problems of standardizing scores on behavioral science measures:

Angoff, W. H. (1971). Scales, norms and equivalent scores, in *Educational Measurement*, 2nd ed. (R. L. Thorndike, ed). Washington, DC: American Council on Education.

For an extended discussion on the vector basis for the rank of a matrix and the regular inverse a good source is

Green, P. E. (1976). *Mathematical Tools for Applied Multivariate Analysis*. New York: Academic.

Chapter 4
The Basic Structure of a Matrix

Introduction

Chapter 3 emphasized the importance to multivariate analysis of determining the rank of a matrix and showed that such decomposition techniques as triangular factoring yield only arbitrary factors of matrices. In this chapter we define a nonarbitrary decomposition of a matrix with real-valued elements, its *basic structure*. Such a unique representation of a matrix as the product of its basic orthonormals and a basic diagonal plays an important role in developing and understanding techniques of multivariate analysis.

We define basic structure from an algebraic and geometric point of view. We then apply the definition to find the basic structure of a number of special matrices—scalars, vectors, diagonals, and product moments—and to develop the *generalized inverse* of any real matrix and the *power* or *root* of a product moment matrix. These specialized matrices recur in the analysis of data matrices. A technique for finding the basic structure of a matrix is then developed and illustrated.

Students of linear algebra will see certain similarities between the basic structure and the notion of characteristic equations. There are some differences, however, particularly in our extensive use of functions of the basic structure of Gramian or product moment matrices, and we urge the reader to become comfortable with our notation.

4.1 Some Factorings of a Data Matrix

In Chapter 3 we presented a 4×2 matrix showing the deviation scores of four students on the verbal and quantitative tests in the College Prediction Study. This matrix, centered by columns, and its minor product moment are

$$\mathbf{X} = (\mathbf{V} \quad \mathbf{Q}) = \begin{bmatrix} 9.25 & -5.50 \\ -6.75 & -6.50 \\ -8.75 & 10.50 \\ 6.25 & 1.50 \end{bmatrix}$$

$$\mathbf{X'X} = \begin{bmatrix} 246.75 & -89.50 \\ -89.50 & 185.00 \end{bmatrix}$$

Because **X** is column centered, the diagonal and off-diagonal elements of **X′X** are proportional, respectively, to variances and covariances.

We know that this data matrix can be represented as the major product of two basic matrices,

$$\mathbf{X} = \mathbf{AB'}$$

where the common order of **A** and **B′** is the rank of **X**. The prefactor must have the same number of rows as there are entities in the data matrix and a number of columns equal to the rank of the data matrix. Similarly, the number of rows in the postfactor must equal the rank of **X** and the number of columns must equal the number of attributes in the data matrix. (In this example we know that **X** is basic—the columns are not mutually proportional—and the common order of **A** and **B′** is 2.)

If we think of **X** as being *produced from* **A** and **B′**, it is natural to interpret the row-by-column matrix multiplication definition,

$$X_{ij} = \mathbf{A}'_{i.}\mathbf{B}_j$$

as saying that the score earned by the ith entity on the jth attribute is a weighted sum of the "factor" scores for the ith entity (the elements of the ith row of **A**). The weights, of course, are the elements of the jth column of **B′** (or, equivalently, the jth row of **B**). If the data matrix were not basic, an entity would have fewer contributing factor scores than observed attribute scores.

Thinking of the prefactor **A** as containing, row by row, sets of factor scores for each of the entities, the columns of the postfactor **B′** provide the weights to be used in combining these factor scores to produce the attribute scores of the data matrix. What characteristics would we like these contributing factor scores to have? In Section 3.6 we saw that any factoring was arbitrary in the sense that, through multiplication by any conformable square basic matrix and its regular inverse, it may be replaced with another factoring that reproduces the data matrix. How do we choose among alternative factorings?

As a starting point, let us consider a particular factoring of the data matrix $\mathbf{X} = (\mathbf{V}, \mathbf{Q}) = \mathbf{AB}'$:

$$\begin{bmatrix} 9.25 & -5.50 \\ -6.75 & -6.50 \\ -8.75 & 10.50 \\ 6.25 & 1.50 \end{bmatrix} = \begin{bmatrix} 9.25 & -14.75 \\ -6.75 & 0.25 \\ -8.75 & 19.25 \\ 6.25 & -4.75 \end{bmatrix} \begin{bmatrix} 1.0 & 1.0 \\ 0.0 & 1.0 \end{bmatrix}$$

Here we have chosen to take the "factors" in **A** to be the verbal (deviation) score **V** and the difference **Q** − **V** between the quantitative and verbal (deviation) scores. The minor product moment of this prefactor,

$$\mathbf{A}'\mathbf{A} = \begin{bmatrix} 246.75 & -336.25 \\ -336.25 & 610.75 \end{bmatrix}$$

is again proportional to a variance–covariance matrix. Both columns of **A** are centered. Here we observe that the two factors **V** and **Q** − **V** covary together; they are not linearly independent.

We may, instead, choose to represent **V** and **Q**, themselves correlated (see **X′X**), in terms of a pair of uncorrelated factors. One such pair of factors consists of **V** itself and a second vector made up of those parts of the **Q** scores that are orthogonal to **V**.

This set of factors may be obtained as a transformation of **A** by matrices **F** and **C**, where **F** = **AC**:

$$\begin{bmatrix} 9.25 & -2.1449 \\ -6.75 & -8.9483 \\ -8.75 & 7.3262 \\ 6.25 & 3.7670 \end{bmatrix} = \begin{bmatrix} 9.25 & -14.75 \\ -6.75 & 0.25 \\ -8.75 & 19.25 \\ 6.25 & -4.75 \end{bmatrix} \begin{bmatrix} 1.0 & -\left(\dfrac{-336.25}{246.75}\right) \\ 0 & 1.0 \end{bmatrix}$$

(A later chapter will show how the desired **C** matrix is formed from elements of **A′A**.) The columns of **F** are orthogonal:

$$\mathbf{F}'\mathbf{F} = \begin{bmatrix} 246.75 & 0 \\ 0 & 152.5362 \end{bmatrix}$$

Since we now have

$$\mathbf{X} = \mathbf{AB}' \qquad \text{and} \qquad \mathbf{F} = \mathbf{AC}$$

we can reexpress the data matrix **X** as the product

$$\mathbf{X} = \mathbf{FG}', \qquad \text{where} \quad \mathbf{G}' = \mathbf{C}^{-1}\mathbf{B}'$$

That is,

$$\mathbf{G}' = \begin{bmatrix} 1.0 & -\left(\dfrac{-336.25}{246.75}\right) \\ 0 & 1.0 \end{bmatrix}^{-1} \begin{bmatrix} 1.0 & 1.0 \\ 0 & 1.0 \end{bmatrix}$$

$$= \begin{bmatrix} 1.0 & -\left(\dfrac{336.25}{246.75}\right) \\ 0 & 1.0 \end{bmatrix} \begin{bmatrix} 1.0 & 1.0 \\ 0 & 1.0 \end{bmatrix}$$

$$= \begin{bmatrix} 1.0 & -0.3627 \\ 0 & 1.0 \end{bmatrix}$$

and $\mathbf{X} = \mathbf{FG}'$:

$$\begin{bmatrix} 9.25 & -5.50 \\ -6.75 & -6.50 \\ -8.75 & 10.50 \\ 6.25 & 1.50 \end{bmatrix} = \begin{bmatrix} 9.25 & -2.1449 \\ -6.75 & -8.9483 \\ -8.75 & 7.3262 \\ 6.25 & 3.7670 \end{bmatrix} \begin{bmatrix} 1.0 & -0.3627 \\ 0 & 1.0 \end{bmatrix}$$

We now have a second factoring of **X**. The two factors (the columns of **F**) are orthogonal—they provide an orthogonal basis for **X**. However, in **F′F**, the two factors are not of the same length or, since **F** is centered by columns, do not have the same variance.

It is often the case that we wish to express a matrix in terms of a pair of factors that have equal variance as well as zero covariance. One way would be to scale each column of **F** to have unit sum of squares. This *normalization* is accomplished by postmultiplying **F** by a diagonal matrix:

$$\mathbf{H} = \begin{bmatrix} 9.25 & -2.1449 \\ -6.75 & -8.9483 \\ -8.75 & 7.3262 \\ 6.25 & 3.7670 \end{bmatrix} \begin{bmatrix} \dfrac{1}{(246.75)^{1/2}} & 0 \\ 0 & \dfrac{1}{(152.5362)^{1/2}} \end{bmatrix}$$

$$= \begin{bmatrix} 0.5889 & -0.1737 \\ 0.4297 & -0.7245 \\ -0.5570 & 0.5932 \\ 0.3979 & 0.3050 \end{bmatrix} = \mathbf{FD}_{\mathbf{F'F}}^{-1/2}$$

The matrix **H** is orthonormal: $\mathbf{H}'\mathbf{H} = \mathbf{I}$. The data matrix **X** is given by the product $\mathbf{X} = \mathbf{HJ}'$, where

$$\mathbf{J}' = \left(\mathbf{D}_{\mathbf{F'F}}^{-1/2}\right)^{-1}\mathbf{G}' = \mathbf{D}_{\mathbf{F'F}}^{1/2}\mathbf{G}'$$

That is,

$$\mathbf{J}' = \begin{bmatrix} (246.75)^{1/2} & 0 \\ 0 & (152.5362)^{1/2} \end{bmatrix} \begin{bmatrix} 1.0 & -0.3627 \\ 0 & 1.0 \end{bmatrix}$$

$$= \begin{bmatrix} (246.75)^{1/2} & (-0.3627)(246.75)^{1/2} \\ 0 & (152.5362)^{1/2} \end{bmatrix}$$

$\mathbf{X} = \mathbf{HJ}'$ now becomes

$$\begin{bmatrix} 9.25 & -5.50 \\ -6.75 & -6.50 \\ -8.75 & 10.50 \\ 6.25 & 1.50 \end{bmatrix} = \begin{bmatrix} 0.5889 & -0.1737 \\ -0.4297 & -0.7245 \\ -0.5570 & 0.5932 \\ 0.3979 & 0.3050 \end{bmatrix} \begin{bmatrix} 15.708 & -5.697 \\ 0 & 12.351 \end{bmatrix}$$

This third factoring of the data matrix $\mathbf{X}$ expresses the observed attributes (columns of $\mathbf{X}$) as functions of two orthonormal factors (the columns of $\mathbf{H}$). These factors are said to provide an *orthonormal basis* for the data matrix $\mathbf{X}$. That is, we can construct the columns of $\mathbf{X}$ by weighting and combining the columns of the orthonormal $\mathbf{H}$. (Indeed, the matrix $\mathbf{J}'$ gives us the weights we need.) It will be convenient to refer to $\mathbf{H}$ as an orthonormal basis. Where the data matrix is centered, these factors will be uncorrelated and of equal variance.

It would be even more convenient if the orthonormal basis were unique. The matrix $\mathbf{H}$, as we have defined it, is not unique. Consider this algebraic example. Let $\mathbf{H}$ be an orthonormal basis for $\mathbf{X}$. That is,

$$\mathbf{X} = \mathbf{HJ}' \qquad \text{and} \qquad \mathbf{H}'\mathbf{H} = \mathbf{I}$$

Let $\mathbf{Q}$ be any square orthonormal with order equal to the number of columns of $\mathbf{H}$. Then $\mathbf{K} = \mathbf{HQ}$ is also an orthonormal basis for $\mathbf{X}$ because

$$\mathbf{K}'\mathbf{K} = \mathbf{Q}'\mathbf{H}'\mathbf{HQ} = \mathbf{Q}'\mathbf{IQ} = \mathbf{Q}'\mathbf{Q} = \mathbf{I}$$

and $\mathbf{X}$ can be expressed in terms of $\mathbf{K}$:

$$\mathbf{X} = \mathbf{K}(\mathbf{Q}'\mathbf{J}') = \mathbf{HQQ}'\mathbf{J}' = \mathbf{HJ}'$$

For the computational example, with which we have been working, the reader may be interested in developing a second orthonormal basis, $\mathbf{K} = \mathbf{HQ}$, where

$$\mathbf{Q} = \begin{bmatrix} 1/\sqrt{2} & 1/\sqrt{2} \\ 1/\sqrt{2} & -1/\sqrt{2} \end{bmatrix}$$

Either symbolically or computationally we have no reason to prefer $\mathbf{H}$ to $\mathbf{K}$ as an orthonormal basis for $\mathbf{X}$. Further, $\mathbf{H}$ and $\mathbf{K}$ were the result of starting from a factoring of $\mathbf{X}$ into $\mathbf{V}$ and $\mathbf{Q} - \mathbf{V}$. If we had chosen to begin, say, with $\mathbf{Q}$ and $\mathbf{V} - \mathbf{Q}$ or with $\mathbf{V} + \mathbf{Q}$ and $\mathbf{V} - \mathbf{Q}$, we would have been led to yet different orthonormal bases.

Fortunately, for the typical data matrix a *unique* orthonormal basis can be found.

4.2 Definition of the Basic Structure of a Matrix

Nearly every real-valued matrix may be uniquely factored into what is called its *basic structure*. In the sections that follow we discuss some of the exceptions and demonstrate how the basic structure is found. First we give its definition.

If $\mathbf{X}$ is a real-valued, nonhorizontal, and non-null matrix, of order $n \times m$ with rank r ($\leqslant m$), then $\mathbf{X}$ may be written as the triple product

$$\mathbf{X} = \mathbf{P\Delta Q}' \tag{4.2.1}$$

where $\mathbf{P}$ is $n \times r$, $\mathbf{\Delta}$ is $r \times r$, and $\mathbf{Q}'$ is $r \times m$. $\mathbf{P}$ and $\mathbf{Q}$ are orthonormal matrices ($\mathbf{P'P} = \mathbf{I}$ and $\mathbf{Q'Q} = \mathbf{I}$) referred to, respectively, as the *left* and *right orthonormals* to (the basic structure of) **X**. $\boldsymbol{\Delta}$ is a diagonal matrix, called the *basic diagonal* to **X**. The diagonal elements of $\boldsymbol{\Delta}$ are all positive and, further, they are always arranged in nonincreasing order:

$$\Delta_{11} \geqslant \Delta_{22} \geqslant \Delta_{33} \geqslant \cdots \geqslant \Delta_{rr} \tag{4.2.2}$$

The basic structure of a matrix is a factoring of the matrix into a product of three basic matrices. If we rewrite (4.2.1) in the form

$$\mathbf{X} = \mathbf{PA'}, \qquad \text{where} \quad \mathbf{A'} = \boldsymbol{\Delta}\mathbf{Q'}$$

we see that the basic structure does give an orthonormal basis for **X**. In Section 4.4 we examine the conditions under which the orthonormal basis is unique.

The basic structure $\mathbf{X} = \mathbf{P}\boldsymbol{\Delta}\mathbf{Q'}$ of Section 4.1 will be found in Section 4.6. However, we give the results here:

$$\begin{bmatrix} 9.25 & -5.50 \\ -6.75 & -6.50 \\ -8.75 & 10.50 \\ 6.25 & 1.50 \end{bmatrix} = \begin{bmatrix} 0.608 & 0.081 \\ -0.131 & -0.837 \\ -0.750 & 0.316 \\ 0.239 & 0.440 \end{bmatrix} \times \begin{bmatrix} 17.62 & 0 \\ 0 & 11.01 \end{bmatrix} \begin{bmatrix} 0.8143 & -0.5803 \\ 0.5803 & 0.8143 \end{bmatrix}$$

The elements of the basic diagonal are positive and descending in magnitude, and the reader may verify that **P** and **Q** are both orthonormal matrices. (As this particular **X** is itself basic, having rank equal to its number of columns, its right orthonormal **Q** will be a *square orthonormal.* Thus, not only $\mathbf{Q'Q} = \mathbf{I}$ but $\mathbf{QQ'} = \mathbf{I}$ as well.)

The three parts of Figure 4.1 depict geometrically how the rows of **X** can be built up from the basic structure of **X**. In Figure 4.1a, scores in the first column of the left orthonormal **P** have been plotted, horizontally against scores in the second column of **P**. Each of the four entities contributing to **X**, each of the rows of **P**, appears as a numbered point in this two-dimensional space. Although there are but four points, their positions may convey to the reader that (1) the columns of **P** are centered (the points are distributed about the origin), (2) the columns of **P** are mutually orthogonal (the plot is of two uncorrelated attributes), and (3) the columns of **P** have equal sums of squares (or variances as the columns are centered). Thus Figure 4.1a displays an orthonormal basis.

We now postmultiply **P** by the basic diagonal to give

$$\mathbf{P}\boldsymbol{\Delta} = \begin{bmatrix} 10.72 & 0.89 \\ -2.30 & -9.21 \\ -13.23 & 3.47 \\ 4.22 & 4.85 \end{bmatrix}$$

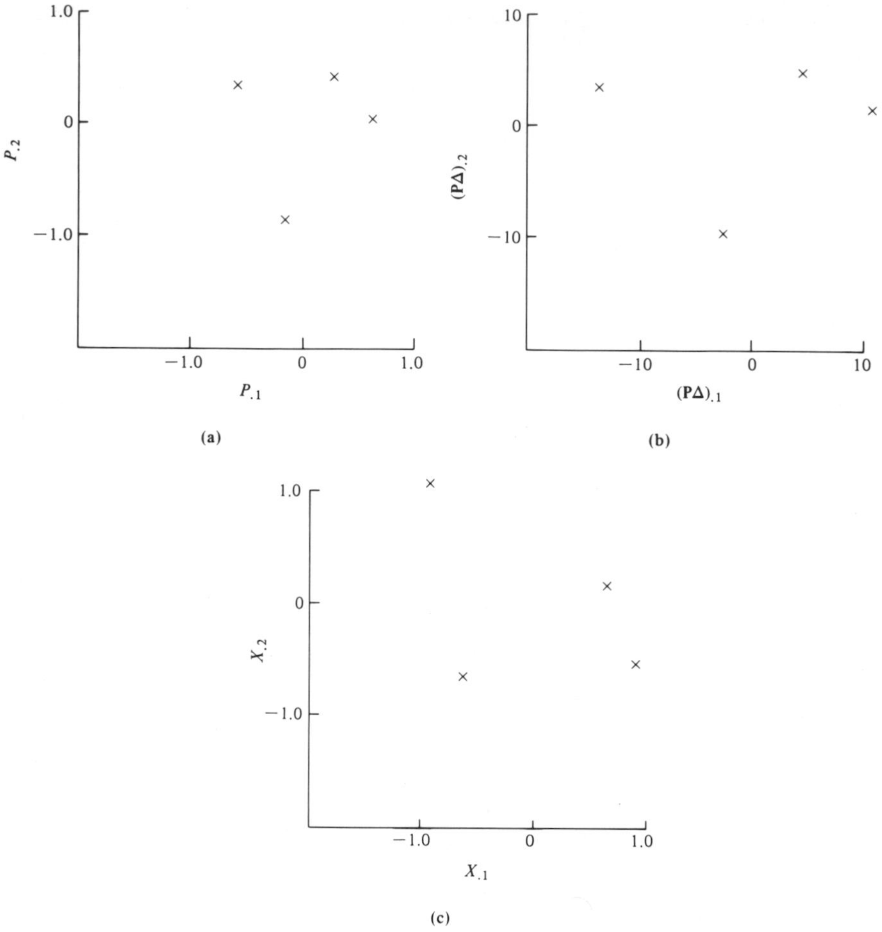

Figure 4.1 **(a)** A data matrix **X** and **(b)** its basic orthogonal **PΔ** and **(c)** basic orthonormal **P**.

Postmultiplying by a diagonal matrix, we saw earlier, scales the columns of the prefactor. Here every element in the first column of **P** has been multiplied by 17.62, the first basic diagonal, and every element in the second column has been scaled by 11.01, or Δ_{22}. The rows of the product matrix **PΔ** are plotted in Figure 4.1b.

The effect of this scaling by the basic diagonal is to stretch the two axes, but not uniformly. The four observations are now more widely dispersed on the first than on the second dimension. The reader may note that the two dimensions remain uncorrelated, however. That is, the columns of **PΔ** are orthogonal, though no longer orthonormal:

$$(\mathbf{P\Delta})'(\mathbf{P\Delta}) = \mathbf{\Delta P'P\Delta} = \mathbf{\Delta I \Delta} = \mathbf{\Delta}^2$$

Finally, we complete the formation of **X** by postmultiplying **PΔ** by the right orthonormal **Q′**:

$$\begin{bmatrix} 10.72 & 0.89 \\ -2.30 & -9.21 \\ -13.23 & 3.47 \\ 4.22 & 4.85 \end{bmatrix} \begin{bmatrix} 0.8143 & -0.5803 \\ 0.5803 & 0.8143 \end{bmatrix} = \begin{bmatrix} 9.25 & -5.50 \\ -6.75 & -6.50 \\ -8.75 & 10.50 \\ 6.25 & 1.50 \end{bmatrix}$$

The rows of **X** are plotted in Figure 4.1c. Postmultiplying **PΔ** by the square orthonormal **Q′** is said to "rotate" the columns. We shall have more to say about this type of transformation in Chapter 6. For now, note that the four data points have exactly the same positions, relative to one another, in the **X** plot as in the **PΔ** plot. The columns of **X**, however, are no longer orthogonal.

4.3 Basic Structure as a Transformation of the Data Matrix

So far we have stressed finding a basis for a data matrix. That is, we have thought about how **X** might be produced. In terms of basic structure, we represented **X** in Eq. (4.2.1) by

$$\mathbf{X} = \mathbf{P\Delta Q'}$$

indicating that **X** could be created from the orthonormal columns of **P** by postmultiplying by the weighting matrix **ΔQ′**.

Now we would like to turn things around a bit. Some insights into the nature of the basic structure of **X** will be gained if we ask not how **X** can be formed but how it can be *transformed*. In particular, consider the result of postmultiplying the matrix in basic structure form by the transpose of its right orthonormal:

$$\mathbf{XQ} = \mathbf{P\Delta Q'Q} = \mathbf{P\Delta} \tag{4.3.1}$$

As in Section 4.2, the postmultiplication by a square orthonormal "rotates" the columns of a matrix. Here, however, it is the columns of **X** that have been rotated, producing **PΔ**. From what we saw in the earlier section, the effect of this rotation is to provide orthogonal dimensions; the columns of **PΔ** are orthogonal. Figure 4.1 shows that these particular orthogonal dimensions preserve the relative positions of the entities. *Postmultiplying a vertical matrix by the transpose of its right orthonormal produces a matrix orthogonal by columns.*

We may now take this transformation a step further, effectively working back from Figure 4.1c to Figure 4.1a by postmultiplying the product **XQ** by the inverse of the basic diagonal of **X** to yield (4.3.1):

$$\mathbf{XQ\Delta^{-1}} = \mathbf{P\Delta\Delta^{-1}} = \mathbf{P} \tag{4.3.2}$$

The matrix **P** is orthonormal by columns.

By first postmultiplying the vertical matrix **X** by the transpose to its right orthonormal and then postmultiplying again by the inverse of the basic diagonal, an orthonormal matrix, the left orthonormal to **X**, is produced.

We can, and shall, look on the basic structure of a matrix in two ways: as providing both a (frequently) unique orthogonal (and orthonormal) basis for that matrix *and* a (frequently) unique orthogonal (and orthonormal) matrix to which that starting matrix may be transformed.

4.4 Uniqueness of the Basic Structure

It is beyond the scope of this book to prove that every real-valued matrix has a (possibly unique) basic structure of the form defined in Section 4.2. In Section 4.6 we shall outline a technique for finding the basic structure when it is unique. Here we demonstrate the conditions under which the basic structure to a matrix is not unique.

The basic diagonal $\boldsymbol{\Delta}$ has been defined [see (4.2.2)] to have nonincreasing positive diagonal elements:

$$\Delta_{11} \geqslant \Delta_{22} \geqslant \Delta_{33} \geqslant \cdots \geqslant \Delta_{rr}$$

If the basic diagonal elements are *strictly decreasing*,

$$\Delta_{11} > \Delta_{22} > \Delta_{33} > \cdots > \Delta_{rr}$$

the basic structure is unique. If, however, two (or more) of the basic diagonal elements are equal, the basic structure is not unique.

The nonuniqueness is easily illustrated using permutation matrices. Let $\mathbf{X} = \mathbf{P}\boldsymbol{\Delta}\mathbf{Q}'$ satisfy the basic structure conditions and let $\Delta_{ii} = \Delta_{(i+1),(i+1)}$. We define the permutation matrix

$$\boldsymbol{\pi} = \left[\begin{array}{c|cc|c} I & 0 & 0 & 0 \\ \hline 0' & 0 & 1 & 0' \\ 0' & 1 & 0 & 0' \\ \hline 0 & 0 & 0 & I \end{array}\right] \begin{array}{c} \\ i \\ i+1 \\ \\ \end{array}$$
$$\qquad\qquad i \quad i+1$$

Premultiplying $\boldsymbol{\Delta}$ by $\boldsymbol{\pi}$ interchanges the ith and the $(i+1)$th rows of $\boldsymbol{\Delta}$. Postmultiplying the product $\boldsymbol{\pi}\boldsymbol{\Delta}$ by $\boldsymbol{\pi}'$ interchanges the columns of $\boldsymbol{\pi}\boldsymbol{\Delta}$. The triple product $\boldsymbol{\pi}\boldsymbol{\Delta}\boldsymbol{\pi}'$ is a diagonal matrix, equal to $\boldsymbol{\Delta}$, except that the ith and $(i+1)$th diagonal elements have been interchanged. Since $\Delta_{ii} = \Delta_{(i+1),(i+1)}$, $\boldsymbol{\Delta}\boldsymbol{\pi}\boldsymbol{\Delta}'$ also has nonincreasing diagonal elements.

Remembering now that the permutation matrix has its transpose as its inverse, we can write

$$\mathbf{X} = (\mathbf{P}\boldsymbol{\pi}')(\boldsymbol{\pi}\boldsymbol{\Delta}\boldsymbol{\pi}')(\boldsymbol{\pi}\mathbf{Q}') = \mathbf{P}\boldsymbol{\Delta}\mathbf{Q}'$$

Not only does $\boldsymbol{\pi}\boldsymbol{\Delta}\boldsymbol{\pi}'$ satisfy the conditions of a basic diagonal, but $\mathbf{P}\boldsymbol{\pi}'$ and $\boldsymbol{\pi}\mathbf{Q}'$ are similarly left and right orthonormals:

$$\boldsymbol{\pi}\mathbf{P}'\mathbf{P}\boldsymbol{\pi}' = \boldsymbol{\pi}\boldsymbol{\pi}' = \mathbf{I} \qquad \text{and} \qquad \boldsymbol{\pi}\mathbf{Q}'\mathbf{Q}\boldsymbol{\pi}' = \boldsymbol{\pi}\boldsymbol{\pi}' = \mathbf{I}$$

As a result the basic structure of $\mathbf{X}$ may be given either as $\mathbf{X} = \mathbf{P\Delta Q'}$ or as $\mathbf{X} = (\mathbf{P}\pi')\pi\Delta\pi'(\pi\mathbf{Q}')$ and is not unique.

With few exceptions, data matrices will have unique basic structures. In Section 4.5 we shall remark on some derived or operating matrices that may not have unique basic structures.

4.5 Basic Structures of Special Matrices

First, we consider the basic structures of matrices related to a matrix $\mathbf{X}$ with known basic structure. By Eq. (4.2.1),

$$\mathbf{X} = \mathbf{P\Delta Q'}$$

The transpose of $\mathbf{X}$ is written directly, of course, as

$$\mathbf{X'} = \mathbf{Q\Delta P'} \tag{4.5.1}$$

The transpose of $\mathbf{X}$ has the same basic diagonal as does $\mathbf{X}$, while the left and right orthonormals to $\mathbf{X}'$ are, respectively, the transposes of the right and left orthonormals to $\mathbf{X}$.

4.5.1 Basic Structures of Product Moments

Using (4.2.1) and (4.5.1) to represent the minor and major product moments of $\mathbf{X}$ yields

$$\mathbf{X'X} = \mathbf{Q\Delta P'P\Delta Q'} = \mathbf{Q\Delta^2 Q'} \tag{4.5.2}$$

and

$$\mathbf{XX'} = \mathbf{P\Delta Q'Q\Delta P'} = \mathbf{P\Delta^2 P'} \tag{4.5.3}$$

The expressions on the right of (4.5.2) and (4.5.3) satisfy the basic structure conditions, $\mathbf{Q}$ and $\mathbf{P}$ remain orthonormal matrices, and $\mathbf{\Delta}^2$ is a basic diagonal matrix since $\Delta_{ii} \geqslant \Delta_{jj}$ implies $\Delta_{ii}^2 \geqslant \Delta_{jj}^2$ for all positive Δ_{ii} and Δ_{jj}.

The left and right orthonormals of product moment matrices are mutual transposes, and the basic diagonal of a product moment matrix is the "square" of the basic diagonal of the original real matrix.

From (4.5.2) or (4.5.3) and the rule that $\mathrm{tr}(\mathbf{A'A}) = \mathrm{tr}(\mathbf{AA'})$, we establish that

$$\mathrm{tr}(\mathbf{X'X}) = \mathrm{tr}(\mathbf{XX'}) = \mathrm{tr}[(\mathbf{P\Delta})(\mathbf{\Delta P'})] = \mathrm{tr}[(\mathbf{\Delta P'})(\mathbf{P\Delta})] = \mathrm{tr}(\mathbf{\Delta}^2)$$

The sum of the diagonals of a product moment matrix is identical to the sum of the basic diagonals for that matrix.

The minor product moment matrix for the $\mathbf{X}$ of Sections 4.1 and 4.2 written in basic form is $\mathbf{X'X} = \mathbf{Q\Delta^2Q'}$:

$$\begin{bmatrix} 246.75 & -89.50 \\ -89.50 & 185.00 \end{bmatrix} = \begin{bmatrix} 0.8143 & 0.5803 \\ -0.5803 & 0.8143 \end{bmatrix} \begin{bmatrix} 310.5 & 0 \\ 0 & 121.2 \end{bmatrix} \times \begin{bmatrix} 0.8143 & -0.5803 \\ 0.5803 & 0.8143 \end{bmatrix}$$

4.5.2 Basic Structure and the Regular Inverse

If a matrix $\mathbf{W}$ is square and basic, then it has a regular inverse, $\mathbf{W}^{-1}$ such that $\mathbf{WW}^{-1} = \mathbf{W}^{-1}\mathbf{W} = \mathbf{I}$. Such a matrix $\mathbf{W}$ has basic structure

$$\mathbf{W} = \mathbf{pdq'} \tag{4.5.4}$$

where $\mathbf{p}$ and $\mathbf{q}$ are both square orthonormals. Thus all three terms on the right of (4.5.4) are square basic matrices, each with its own inverse.

Applying the rule on the inverse of a product of square basic matrices yields

$$\mathbf{W}^{-1} = (\mathbf{pdq'})^{-1} = (\mathbf{q'})^{-1}\mathbf{d}^{-1}\mathbf{p}^{-1} = \mathbf{qd}^{-1}\mathbf{p'} \tag{4.5.5}$$

since the inverse of a square orthonormal is its transpose. [The reader may wish to pre- or postmultiply the right-hand side of (4.5.4) by the right-hand side of (4.5.5) to verify that an inverse has been identified.]

If we know the basic structure of a square basic matrix, then we can find the regular inverse of that matrix directly. For example, since data matrix $\mathbf{X}$ is basic, its minor product moment is a square basic matrix. In basic structure notation, the regular inverse of this minor product moment is given, from (4.5.2), by

$$(\mathbf{X'X})^{-1} = \mathbf{Q\Delta^{-2}Q'} \tag{4.5.6}$$

By components we compute

$$\begin{bmatrix} 0.8143 & 0.5803 \\ -0.5803 & 0.8143 \end{bmatrix} \begin{bmatrix} 1/310.5 & 0 \\ 0 & 1/121.2 \end{bmatrix} \begin{bmatrix} 0.8143 & -0.5803 \\ 0.5803 & 0.8143 \end{bmatrix}$$

or

$$\begin{bmatrix} 0.8143 & 0.5803 \\ -0.5803 & 0.8143 \end{bmatrix} \begin{bmatrix} 0.00322 & 0 \\ 0 & 0.00825 \end{bmatrix} \begin{bmatrix} 0.8143 & -0.5803 \\ 0.5803 & 0.8143 \end{bmatrix}$$

Completion of the example and verification that the resultant is a regular inverse are left as an exercise.

This illustrates a point that needs to be kept in mind. Although the regular inverse of a square basic matrix can be expressed in terms of the basic structure of the original matrix $\mathbf{W}^{-1} = \mathbf{qd}^{-1}\mathbf{p'}$, where $\mathbf{W}$ has basic structure $\mathbf{W} = \mathbf{pdq'}$, this does not give directly the basic structure of $\mathbf{W}^{-1}$.

Although $\mathbf{q}$ and $\mathbf{p}'$ are orthonormals, $\mathbf{d}^{-1}$ in general cannot be a basic diagonal. Since the elements of $\mathbf{d}$ must be nonincreasing, the elements of $\mathbf{d}^{-1}$ must be nondecreasing: if $d_{11} \geqslant d_{22} \geqslant \cdots \geqslant d_{rr}$, then $1/d_{11} \leqslant 1/d_{22} \leqslant \cdots \leqslant 1/d_{rr}$. For the numerical example,

$$\boldsymbol{\Delta}^2 = \begin{bmatrix} 310.5 & 0 \\ 0 & 121.2 \end{bmatrix} \quad \text{and} \quad \boldsymbol{\Delta}^{-2} = \begin{bmatrix} 0.00322 & 0 \\ 0 & 0.00825 \end{bmatrix}$$

The permutation argument employed earlier can be used to find the basic structure for $\mathbf{W}^{-1}$. For this we define the special permutation matrix

$$\boldsymbol{\pi}_{\mathrm{R}} = \begin{bmatrix} 0 & 0 & 0 & \cdots & 1 \\ \vdots & \vdots & \vdots & & \vdots \\ 0 & 0 & 1 & \cdots & 0 \\ 0 & 1 & 0 & \cdots & 0 \\ 1 & 0 & 0 & \cdots & 0 \end{bmatrix}$$

We call this the *reversing* permutation matrix. Premultiplying a matrix by $\boldsymbol{\pi}_{\mathrm{R}}$ reverses the order of the rows in that matrix, and postmultiplying a matrix by $\boldsymbol{\pi}_{\mathrm{R}}$ reverses the order of the columns. If we both pre- and postmultiply a diagonal matrix by $\boldsymbol{\pi}_{\mathrm{R}}$, we reverse the order of the diagonal elements.

Thus $\boldsymbol{\pi}_{\mathrm{R}}\mathbf{d}^{-1}\boldsymbol{\pi}_{\mathrm{R}} = \mathbf{D}$ is a diagonal matrix for which $D_{11} \geqslant D_{22} \geqslant \cdots \geqslant D_{rr}$. Keeping in mind that the inverse of an (orthonormal) permutation matrix is its transpose and that this particular permutation matrix $\boldsymbol{\pi}_{\mathrm{R}}$ is symmetric (having itself as its transpose), we can rewrite (4.5.5):

$$\mathbf{W}^{-1} = (\mathbf{q}\boldsymbol{\pi}_{\mathrm{R}})(\boldsymbol{\pi}_{\mathrm{R}}\mathbf{d}^{-1}\boldsymbol{\pi}_{\mathrm{R}})(\boldsymbol{\pi}_{\mathrm{R}}\mathbf{p}') \tag{4.5.7}$$

The right-hand side now meets the criteria for a basic structure representation of $\mathbf{W}^{-1}$: $\mathbf{q}\boldsymbol{\pi}_{\mathrm{R}}$ is a columnar rearrangement of $\mathbf{q}$ and hence remains a left orthonormal (by columns); $\boldsymbol{\pi}_{\mathrm{R}}\mathbf{p}'$, similarly, is a row rearrangement of $\mathbf{p}'$ and is a right orthonormal (by rows) and, as just noted, $\boldsymbol{\pi}_{\mathrm{R}}\mathbf{d}^{-1}\boldsymbol{\pi}_{\mathrm{R}}$ is a basic diagonal with positive nonincreasing elements. The right-hand side of (4.5.7), then, gives the basic structure of the regular inverse of the matrix whose basic structure is identified in (4.5.4).

Thus the basic structure of the regular inverse of our minor product moment matrix $(\mathbf{X}'\mathbf{X})^{-1}$ is a rearrangement of the elements of the matrices

$$(\mathbf{Q}\boldsymbol{\pi}_{\mathrm{R}})(\boldsymbol{\pi}_{\mathrm{R}}\boldsymbol{\Delta}^{-2}\boldsymbol{\pi}_{\mathrm{R}})(\boldsymbol{\pi}_{\mathrm{R}}\mathbf{Q}') = \begin{bmatrix} 0.5803 & 0.8143 \\ 0.8143 & -0.5803 \end{bmatrix} \begin{bmatrix} 0.00825 & 0 \\ 0 & 0.00322 \end{bmatrix} \times \begin{bmatrix} 0.5803 & 0.8143 \\ 0.8143 & -0.5803 \end{bmatrix}$$

4.5.3 Basic Structure and a Generalized Inverse

In Chapter 3 we developed the concept of a regular inverse. For any square and basic matrix $\mathbf{S}$, there exists a second square and basic matrix $\mathbf{S}^{-1}$ such that $\mathbf{SS}^{-1} = \mathbf{S}^{-1}\mathbf{S} = \mathbf{I}$. This regular inverse is unique and is defined only for square basic matrices.

Now we introduce a second kind of matrix inverse, a *generalized inverse*. The generalized inverse is at once broader and more limited in scope than the regular inverse. It is broader in the sense that we can find a generalized inverse for every real matrix, not just for square basic ones. It is more limited in that the generalized inverse frequently is not unique and, for multivariate analyses at the level at which we shall be studying them, will be of less importance.

For a real matrix $\mathbf{A}$, a generalized inverse $\mathbf{A}^+$ must satisfy the two equations

$$\mathbf{A}(\mathbf{A}^+\mathbf{A}) = (\mathbf{AA}^+)\mathbf{A} = \mathbf{A} \tag{4.5.8}$$

and

$$\mathbf{A}^+(\mathbf{AA}^+) = (\mathbf{A}^+\mathbf{A})\mathbf{A}^+ = \mathbf{A}^+ \tag{4.5.9}$$

The reader may note that if $\mathbf{A}$ is square and basic, taking $\mathbf{A}^+ = \mathbf{A}^{-1}$ will satisfy both conditions. In general, however, neither $\mathbf{AA}^+$ nor $\mathbf{A}^+\mathbf{A}$ need be the identity matrix.

If we take $\mathbf{X}$ to be a nonhorizontal matrix of order $n \times m$ with basic structure given by (4.2.1),

$$\mathbf{X} = \mathbf{P\Delta Q}'$$

then one way of finding a generalized inverse is in terms of these basic structure matrices:

$$\mathbf{X}^+ = \mathbf{Q\Delta}^{-1}\mathbf{P}' \tag{4.5.10}$$

For $\mathbf{X}^+$ so defined,

$$\mathbf{X}^+\mathbf{X} = \mathbf{Q\Delta}^{-1}\mathbf{P}'\mathbf{P\Delta Q}' = \mathbf{QQ}' \tag{4.5.11}$$

$$\mathbf{XX}^+ = \mathbf{P\Delta Q}'\mathbf{Q\Delta}^{-1}\mathbf{P}' = \mathbf{PP}' \tag{4.5.12}$$

and, thus for (4.5.8) and (4.5.9),

$$\mathbf{X}(\mathbf{X}^+\mathbf{X}) = \mathbf{P\Delta Q}'\mathbf{QQ}' = \mathbf{P\Delta Q}' = \mathbf{X}$$

$$\mathbf{X}^+(\mathbf{XX}^+) = \mathbf{Q\Delta}^{-1}\mathbf{P}'\mathbf{PP}' = \mathbf{Q\Delta}^{-1}\mathbf{P}' = \mathbf{X}^+$$

If the nonhorizontal matrix $\mathbf{X}$ is basic, having rank equal to its column order, then its right orthonormal $\mathbf{Q}'$ is square and in (4.5.11) $\mathbf{X}^+\mathbf{X} = \mathbf{QQ}' = \mathbf{I}$. *Premultiplying a vertical basic matrix by its generalized inverse yields the identity matrix*. This is of interest, because most data matrices are vertical basic matrices.

For the nonhorizontal basic matrix, the generalized inverse is often obtained not from the basic structure, but from the regular inverse of the product moment matrix:

$$\mathbf{X}^+ = (\mathbf{X}'\mathbf{X})^{-1}\mathbf{X}' = (\mathbf{Q\Delta}^{-2}\mathbf{Q}')(\mathbf{Q\Delta P}') = \mathbf{Q\Delta}^{-1}\mathbf{P}'$$

If the nonhorizontal matrix **S** is not only basic but square, then both its left and right orthonormal matrices are square and in (4.5.11) and (4.5.12) $\mathbf{X}^+\mathbf{X} = \mathbf{I}$ and $\mathbf{X}\mathbf{X}^+ = \mathbf{I}$. *For a square basic matrix, the generalized and regular inverses are identical*: $\mathbf{X}^{-1} = \mathbf{X}^+$.

4.5.4 Basic Structures of Types of Matrices

To provide some additional familiarity with basic structure notions, we examine the basic structures of some simple real matrices.

Scalars. Scalars are 1×1 matrices. Except for the scalar 0, which has rank 0 by definition, every scalar is a square basic matrix. In terms of basic structure, we can express a scalar

$$\begin{aligned} 2 &= (1)(2)(1) \\ s &= pdq' \end{aligned}$$

if the scalar is positive, or

$$\begin{aligned} -0.5 &= (-1)(0.5)(1) \\ s &= pdq' \end{aligned}$$

if the scalar is negative. The scalars p and q' are orthonormal matrices and d is a basic diagonal.

A rule for determining these outcomes is that for the nonzero real scalar s, the basic structure has elements

$$\begin{aligned} p &= \left[s(s^2)^{-1/2}\right] &&\text{(the left orthonormal)} \\ d &= (s^2)^{1/2} &&\text{(the basic diagonal)} \\ q' &= 1 &&\text{(the right orthonormal)} \end{aligned}$$

where $(s^2)^{1/2}$ is the positive square root of s^2 and $(s^2)^{-1/2}$ is the reciprocal of that square root.

Vectors. Each column vector, other than the null vector, is a basic matrix with rank 1. The basic structure of the nonnull column vector $\mathbf{v}$, will have elements

$$\begin{aligned} \mathbf{p} &= \mathbf{v}(\mathbf{v}'\mathbf{v})^{-1/2} \\ d &= (\mathbf{v}'\mathbf{v})^{1/2} \\ q' &= 1 \qquad \text{(the scalar)} \end{aligned}$$

The rule here is a simple extension of the scalar rule.

Below are three examples of basic structure for vectors:

$$\mathbf{e}_2 = \begin{bmatrix} 0 \\ 1 \\ 0 \\ 0 \end{bmatrix} = \begin{bmatrix} 0 \\ 1 \\ 0 \\ 0 \end{bmatrix}(1)(1)$$

$$\mathbf{1} = \begin{bmatrix} 1 \\ 1 \\ 1 \end{bmatrix} = \begin{bmatrix} 1/\sqrt{3} \\ 1/\sqrt{3} \\ 1/\sqrt{3} \end{bmatrix}(\sqrt{3})(1)$$

$$\mathbf{w} = \begin{bmatrix} 3 \\ -4 \end{bmatrix} = \begin{bmatrix} \frac{3}{5} \\ -\frac{4}{5} \end{bmatrix}(5)(1)$$

Rank-1 Matrices. The basic structure of a major product of nonnull vectors, a rank-1 matrix, is easily obtained as an application of the vector rule. The basic structure of $\mathbf{1w}'$, for example, is

$$\mathbf{1w}' = \begin{bmatrix} 1 \\ 1 \\ 1 \end{bmatrix}(3 \quad -4) = \begin{bmatrix} 1/\sqrt{3} \\ 1/\sqrt{3} \\ 1/\sqrt{3} \end{bmatrix}(5\sqrt{3})(\tfrac{3}{5} \quad -\tfrac{4}{5})$$

Diagonal Matrices. Finding the basic structure of a diagonal matrix follows on transforming that diagonal into a basic diagonal. Let us consider the diagonal matrix

$$\mathbf{D} = \begin{bmatrix} 1 & 0 & 0 \\ 0 & -4 & 0 \\ 0 & 0 & 2 \end{bmatrix}$$

$\mathbf{D}$ is not a basic diagonal; not only are the elements not in descending order, but one of them is negative.

As a first step toward transforming $\mathbf{D}$, we can obtain a diagonal matrix with all positive elements by premultiplying by a *sign matrix*. A sign matrix is a diagonal matrix whose diagonal elements are 1 and -1. An appropriate sign matrix here is

$$\mathbf{J} = \begin{bmatrix} 1 & 0 & 0 \\ 0 & -1 & 0 \\ 0 & 0 & 1 \end{bmatrix}$$

so that

$$\mathbf{JD} = \begin{bmatrix} 1 & 0 & 0 \\ 0 & 4 & 0 \\ 0 & 0 & 2 \end{bmatrix}$$

with all positive elements.

To rearrange the diagonal elements into a descending order, we pre- and postmultiply **JD** by a permutation matrix and its transpose. Taking

$$\boldsymbol{\pi} = \begin{bmatrix} 0 & 1 & 0 \\ 0 & 0 & 1 \\ 1 & 0 & 0 \end{bmatrix}$$

we obtain

$$\boldsymbol{\pi}\mathbf{JD} = \begin{bmatrix} 0 & 4 & 0 \\ 0 & 0 & 2 \\ 1 & 0 & 0 \end{bmatrix}$$

and

$$\boldsymbol{\pi}\mathbf{JD}\boldsymbol{\pi}' = \begin{bmatrix} 4 & 0 & 0 \\ 0 & 2 & 0 \\ 0 & 0 & 1 \end{bmatrix}$$

This last matrix is the basic diagonal of **D**. The right orthonormal is simply the permutation matrix $\boldsymbol{\pi}$, and the left orthonormal is the product

$$\mathbf{J}\boldsymbol{\pi}' = \begin{bmatrix} 0 & 0 & 1 \\ 1 & 0 & 0 \\ 0 & 1 & 0 \end{bmatrix}$$

Putting these results together gives

$$\mathbf{D} = (\mathbf{J}\boldsymbol{\pi}')(\boldsymbol{\pi}\mathbf{JD}\boldsymbol{\pi}')(\boldsymbol{\pi}) = \mathbf{JJD} = \mathbf{D}$$

since $\mathbf{JJ} = \mathbf{I}$. Numerically, the basic structure of **D** is

$$\begin{bmatrix} 0 & 0 & 1 \\ -1 & 0 & 0 \\ 0 & 1 & 0 \end{bmatrix}\begin{bmatrix} 4 & 0 & 0 \\ 0 & 2 & 0 \\ 0 & 0 & 1 \end{bmatrix}\begin{bmatrix} 0 & 1 & 0 \\ 0 & 0 & 1 \\ 1 & 0 & 0 \end{bmatrix} = \begin{bmatrix} 1 & 0 & 0 \\ 0 & -4 & 0 \\ 0 & 0 & 2 \end{bmatrix} = \mathbf{P\Delta Q}'$$

Orthogonal and Orthonormal Matrices. If the nonhorizontal matrix **M** is orthonormal by columns ($\mathbf{M}'\mathbf{M} = \mathbf{I}$), then the left orthonormal can be **M** itself and both the basic diagonal and the right orthonormal identity matrices: $\mathbf{M} = \mathbf{MII}$. Note that the orthonormal has equal basic diagonal elements. Hence the basic structure is not unique.

If the nonhorizontal matrix **K** is *orthogonal* by columns ($\mathbf{K}'\mathbf{K} = \mathbf{D}$), then **K** has basic structure

$$\mathbf{K} = (\mathbf{KD}^{-1/2}\boldsymbol{\pi}')(\boldsymbol{\pi}\mathbf{D}^{1/2}\boldsymbol{\pi}')(\boldsymbol{\pi})$$

where the permutation matrices are needed to ensure the appropriate order of the diagonal elements. Thus for

$$\mathbf{K} = \begin{bmatrix} 1 & 1 & 1 \\ 1 & 1 & -1 \\ 1 & -1 & 1 \\ 1 & -1 & -1 \end{bmatrix}$$

it follows that

$$\mathbf{K'K} = \begin{bmatrix} 4 & 0 & 0 \\ 0 & 4 & 0 \\ 0 & 0 & 4 \end{bmatrix}$$

and *one* of the basic structures of **K** is given as

$$\begin{bmatrix} \frac{1}{2} & \frac{1}{2} & \frac{1}{2} \\ \frac{1}{2} & \frac{1}{2} & -\frac{1}{2} \\ \frac{1}{2} & -\frac{1}{2} & \frac{1}{2} \\ \frac{1}{2} & -\frac{1}{2} & -\frac{1}{2} \end{bmatrix} \begin{bmatrix} 2 & 0 & 0 \\ 0 & 2 & 0 \\ 0 & 0 & 2 \end{bmatrix} \begin{bmatrix} 1 & 0 & 0 \\ 0 & 1 & 0 \\ 0 & 0 & 1 \end{bmatrix} = \mathbf{P\Delta Q'}$$

With equal diagonal elements the structure, again, is not unique.

Symmetric Idempotent Matrices. The symmetric matrix **A** is *idempotent* if $\mathbf{AA} = \mathbf{A}$. Since **A** is symmetric, we can also write the identifying equation in the form $\mathbf{A'A} = \mathbf{A}$. In this form we can see that **A** is a product moment matrix and, as such, has a basic structure of the form

$$\mathbf{A} = \mathbf{Q\Delta Q'} \tag{4.5.13}$$

That is, based on (4.5.2), the left and right orthonormals of the symmetric idempotent matrix are mutual transposes. Writing the idempotency equation from (4.5.13) gives

$$\mathbf{AA} = \mathbf{Q\Delta Q'Q\Delta Q'} = \mathbf{Q\Delta^2 Q'} = \mathbf{A}$$

This result, that the basic diagonal of **A** may be expressed either as $\mathbf{\Delta}$ or as $\mathbf{\Delta}^2$, is equivalent to asserting that $\mathbf{\Delta} = \mathbf{\Delta}^2$. For what diagonal matrix will $\mathbf{D} = \mathbf{D}^2$? The answer, which we shall not prove, is that $\mathbf{\Delta}$ must be the identity matrix. That is, the symmetric idempotent matrix has basic structure $\mathbf{A} = \mathbf{QIQ'}$; equivalently, the symmetric idempotent matrix is the major product moment of an orthonormal matrix,

$$\mathbf{A} = \mathbf{QQ'} \tag{4.5.14}$$

Since the general symmetric idempotent matrix is not the identity matrix, the **Q** of (4.5.14) is not square. In turn, this tells us that every symmetric idempotent matrix (other than the identity matrix itself) is *rank deficient*, that is, has rank less than its order, or is nonbasic. [We see that for the idempotent matrix the basic diagonal elements are again equal. Thus **Q** is not unique. If **A** is $m \times m$ with rank r, then **Q** will be $m \times r$. For any $r \times r$

square orthonormal $\mathbf{p}$, the product $\mathbf{P} = \mathbf{Qp}$ will also satisfy (4.5.14): $\mathbf{A} = \mathbf{PP}' = \mathbf{Qpp}'\mathbf{Q}' = \mathbf{QQ}'$.] The reader should also consider the conditions under which the products $\mathbf{XX}^+$ and $\mathbf{X}^+\mathbf{X}$ will be idempotent matrices.

4.5.5 Powers of Product Moments

We can now consider the result of multiplying a general product moment matrix by itself. For example, the minor product moment $\mathbf{X}'\mathbf{X}$ has basic structure given by Eq. (4.5.2):

$$\mathbf{X}'\mathbf{X} = \mathbf{Q}\mathbf{\Delta}^2\mathbf{Q}'$$

and multiplying $\mathbf{X}'\mathbf{X}$ by itself gives

$$(\mathbf{X}'\mathbf{X})(\mathbf{X}'\mathbf{X}) = (\mathbf{X}'\mathbf{X})^2 = \mathbf{Q}\mathbf{\Delta}^4\mathbf{Q}'$$

The orthonormals of $(\mathbf{X}'\mathbf{X})^2$ are the same as the orthonormals of $\mathbf{X}'\mathbf{X}$, and the basic diagonal of $(\mathbf{X}'\mathbf{X})^2$ is the square of the basic diagonal of $\mathbf{X}'\mathbf{X}$.

These properties generalize to higher powers of $\mathbf{X}'\mathbf{X}$. That is, we can continue multiplying $\mathbf{X}'\mathbf{X}$ by itself until we have obtained $(\mathbf{X}'\mathbf{X})^k$. This matrix, $\mathbf{X}'\mathbf{X}$ raised to the kth power, has basic structure

$$(\mathbf{X}'\mathbf{X})^k = \mathbf{Q}\mathbf{\Delta}^{2k}\mathbf{Q}' \tag{4.5.15}$$

where the basic structure of $\mathbf{X}$ is given by Eq. (4.2.1): $\mathbf{X} = \mathbf{P}\mathbf{\Delta}\mathbf{Q}'$. In Section 4.6 we use this property of the *powered* matrix to illustrate one technique for finding the basic structure of an arbitrary data matrix.

We can also use the powering property of product moment matrices to *define* matrices that we would not actually compute by powering. For example, we can define the matrix $(\mathbf{X}'\mathbf{X})^{1/2} = \mathbf{Q}\mathbf{\Delta}\mathbf{Q}'$. This matrix is a square root of the matrix $\mathbf{X}'\mathbf{X}$ in the sense that

$$(\mathbf{X}'\mathbf{X})^{1/2}(\mathbf{X}'\mathbf{X})^{1/2} = \mathbf{Q}\mathbf{\Delta}\mathbf{Q}'\mathbf{Q}\mathbf{\Delta}\mathbf{Q}' = \mathbf{Q}\mathbf{\Delta}^2\mathbf{Q}' = \mathbf{X}'\mathbf{X}$$

4.6 Finding the Basic Structure of a Matrix

In Section 4.2 we stated but did not derive the basic structure matrices for the verbal and quantitative deviation scores in the College Prediction Study. In this section, we introduce Hotelling's powering method and use it to find the basic diagonal and left and right orthonormals of a matrix.

The basic structure of a data matrix $\mathbf{X}$ is most easily found by first taking the minor product moment of $\mathbf{X}$ and from this smaller matrix determining the right orthonormal $\mathbf{Q}$. We can then find $\mathbf{\Delta}$ by recalling from Eq. (4.5.2) that

$$\mathbf{X}'\mathbf{X} = \mathbf{Q}\mathbf{\Delta}^2\mathbf{Q}'$$

Premultiplying by $\mathbf{Q}'$ and postmultiplying by $\mathbf{Q}$ gives

$$\mathbf{Q}'\mathbf{X}'\mathbf{X}\mathbf{Q} = \mathbf{Q}'\mathbf{Q}\boldsymbol{\Delta}^2\mathbf{Q}'\mathbf{Q} = \boldsymbol{\Delta}^2$$

or

$$\boldsymbol{\Delta} = (\mathbf{Q}'\mathbf{X}'\mathbf{X}\mathbf{Q})^{1/2} \tag{4.6.1}$$

The left orthonormal of matrix $\mathbf{X}$ may be obtained by working with the known $\mathbf{X}$, $\mathbf{Q}$, and $\boldsymbol{\Delta}$. Starting with the basic structure representation of Eq. (4.2.1), $\mathbf{X} = \mathbf{P}\boldsymbol{\Delta}\mathbf{Q}'$, we postmultiply by $\mathbf{Q}\boldsymbol{\Delta}^{-1}$, which gives

$$\mathbf{X}\mathbf{Q}\boldsymbol{\Delta}^{-1} = \mathbf{P}\boldsymbol{\Delta}\mathbf{Q}'\mathbf{Q}\boldsymbol{\Delta}^{-1} = \mathbf{P} \tag{4.6.2}$$

In summary, finding the elements of $\mathbf{Q}$ allows us to determine the basic diagonal $\boldsymbol{\Delta}$, and the left orthonormal $\mathbf{P}$.

4.6.1 Matrix Powering and Basic Structure

To use Hotelling's method, first recall from Eq. (4.5.15) that

$$(\mathbf{X}'\mathbf{X})^k = \mathbf{Q}\boldsymbol{\Delta}^{2k}\mathbf{Q}'$$

When a product moment is raised to any power, the basic orthonormals of the minor product moment $\mathbf{Q}$ are unchanged.

Recall that the elements of $\boldsymbol{\Delta}$ are in descending order and assume that $\mathbf{X}'\mathbf{X}$ is raised to a very high power k. Now divide both sides of (4.5.15) by the first element in $\boldsymbol{\Delta}^{2k}$ (which we shall now write with a single subscript):

$$\left(\frac{1}{\Delta_1^{2k}}\right)(\mathbf{X}'\mathbf{X})^k = \mathbf{Q}\left[\left(\frac{1}{\Delta_1^{2k}}\right)\boldsymbol{\Delta}^{2k}\right]\mathbf{Q}' \tag{4.6.3}$$

In expanded form, the middle term on the right can be written

$$\left(\frac{1}{\Delta_1^{2k}}\right)\boldsymbol{\Delta}^{2k} = \begin{bmatrix} 1 & 0 & \cdots & 0 \\ 0 & \left(\dfrac{\Delta_2}{\Delta_1}\right)^{2k} & \cdots & 0 \\ \vdots & \vdots & & \vdots \\ 0 & 0 & \cdots & \left(\dfrac{\Delta_r}{\Delta_1}\right)^{2k} \end{bmatrix}$$

If $\Delta_1 > \Delta_2$, we may select the power k large enough so that the ratio

$$\left(\frac{\Delta_2}{\Delta_1}\right)^{2k}$$

is 0 to some specified number of decimal places. If we do this, then all elements in the diagonal matrix

$$(1/\Delta_1^{2k})\boldsymbol{\Delta}^{2k}$$

except the first element are 0. The result is an *elementary* matrix $\mathbf{E}_{11} = \mathbf{e}_1\mathbf{e}_1'$, with a 1 in the upper diagonal position and 0s elsewhere.

With the appropriate choice of k, (4.6.3) can be rewritten

$$(1/\Delta_1)^{2k}(\mathbf{X}'\mathbf{X})^k = \mathbf{Q}\mathbf{e}_1\mathbf{e}_1'\mathbf{Q}'$$

or, on carrying through the postmultiplication of $\mathbf{Q}$ by $\mathbf{e}_1$,

$$(1/\Delta_1)^{2k}(\mathbf{X}'\mathbf{X})^k = \mathbf{Q}_{.1}\mathbf{Q}_{.1}'$$

which is equivalent to

$$(\mathbf{X}'\mathbf{X})^k = c\mathbf{Q}_{.1}\mathbf{Q}_{.1}' \tag{4.6.4}$$

Postmultiplying this result by any vector ${}_0\mathbf{V}$ not orthogonal to $\mathbf{Q}_{.1}$ gives

$$(\mathbf{X}'\mathbf{X})^k\,{}_0\mathbf{V} = s\mathbf{Q}_{.1} \tag{4.6.5}$$

where s is a scalar.

This indicates that the product of an arbitrary vector and a sufficiently high power of $\mathbf{X}'\mathbf{X}$ is, to a specified number of decimal places, proportional to the first column of the right orthonormal of the basic structure of $\mathbf{X}$. Since $\mathbf{Q}$ is, by definition orthonormal, $\mathbf{Q}_{.1}$ can then be found by normalizing the vector obtained in (4.6.5).

The process of raising $\mathbf{X}'\mathbf{X}$ to a power can be simplified computationally. Suppose we postmultiply $\mathbf{X}'\mathbf{X}$ by any vector ${}_0\mathbf{V}$ not orthogonal to $\mathbf{Q}_{.1}$. The resulting vector is $(\mathbf{X}'\mathbf{X})\,{}_0\mathbf{V} = {}_1\mathbf{V}$. Now postmultiply $\mathbf{X}'\mathbf{X}$ by ${}_1\mathbf{V}$, and the resulting ${}_2\mathbf{V}$, etc.:

$$(\mathbf{X}'\mathbf{X})\,{}_1\mathbf{V} = {}_2\mathbf{V}$$
$$(\mathbf{X}'\mathbf{X})\,{}_2\mathbf{V} = {}_3\mathbf{V}$$
$$\vdots$$

If we substitute back, $(\mathbf{X}'\mathbf{X})^3\,{}_0\mathbf{V} = {}_3\mathbf{V}$. In general, then,

$$(\mathbf{X}'\mathbf{X})^k\,{}_0\mathbf{V} = {}_k\mathbf{V} \tag{4.6.6}$$

This suggests that postmultiplying the minor product moment of $\mathbf{X}$ by an arbitrary vector and continuing to postmultiply $\mathbf{X}'\mathbf{X}$ by the resultant product vector in successive iterations corresponds to raising $\mathbf{X}'\mathbf{X}$ to a power and then postmultiplying by the initial arbitrary vector. In this approach many fewer computations are required: If $\mathbf{X}'\mathbf{X}$ is an $m \times m$ matrix, direct powering requires m times as many scalar computations at each iteration.

It is common to take ${}_0\mathbf{V}$ as the conformable unit vector. Furthermore, dividing the product $(\mathbf{X}'\mathbf{X})\,{}_i\mathbf{V}$ by the largest element in the resulting product vector keeps the elements in subsequent multiplying vectors within reasonable limits of magnitude. Eventually, the ${}_i\mathbf{V}$ must converge, so within the limits of decimal accuracy, ${}_k\mathbf{V}$ and ${}_{k+1}\mathbf{V}$ are equal, with elements proportional to $\mathbf{Q}_{.1}$, indicated in (4.6.5). Since $\mathbf{Q}$ is orthonormal, we can find $\mathbf{Q}_{.1}$ from ${}_k\mathbf{V}$ by normalizing ${}_k\mathbf{V}$:

$$\mathbf{Q}_{.1} = {}_k\mathbf{V}({}_k\mathbf{V}'\,{}_k\mathbf{V})^{-1/2} \tag{4.6.7}$$

We now have the elements of the first column of the right orthonormal of $\mathbf{X}$. The first element in the basic diagonal can now be found from Eq. (4.6.1):

$$\Delta_1^2 = \mathbf{Q}'_{.1}(\mathbf{X}'\mathbf{X})\mathbf{Q}_{.1} \tag{4.6.8}$$

4.6.2 Example

Let us now illustrate this procedure with the matrix of verbal and quantitative deviation scores from the College Prediction Study.

$$\mathbf{X} = \begin{bmatrix} 9.25 & -5.50 \\ -6.75 & -6.50 \\ -8.75 & 10.50 \\ 6.25 & 1.50 \end{bmatrix}$$

First, we find the minor product moment:

$$\mathbf{X}'\mathbf{X} = \begin{bmatrix} 246.75 & -89.50 \\ -89.50 & 185.00 \end{bmatrix}$$

Then we multiply $\mathbf{X}'\mathbf{X}$ by a succession of powering vectors, beginning with the conformable unit vector as ${}_0\mathbf{V}$. After each multiplication of $\mathbf{X}'\mathbf{X}$ by ${}_i\mathbf{V}$, we divide the resultant vector ${}_{i+1}\mathbf{U}$ by its largest element and begin another iteration. This scaling prevents the sequence of vectors from becoming either very large or very small. As each element in a given vector is multiplied by the same scaling constant, the scaled vector itself approaches a vector proportional to $\mathbf{Q}_{.1}$. When ${}_k\mathbf{V}$ and ${}_{k+1}\mathbf{V}$ are equal within the limits we have set, the resultant vector is proportional to $\mathbf{Q}_{.1}$. Normalizing this vector gives us $\mathbf{Q}_{.1}$.

1. $(\mathbf{X}'\mathbf{X})\,{}_0\mathbf{V} = {}_1\mathbf{U}$:

$$\begin{bmatrix} 246.75 & -89.50 \\ -89.50 & 185.00 \end{bmatrix}\begin{bmatrix} 1 \\ 1 \end{bmatrix} = \begin{bmatrix} 157.25 \\ 95.50 \end{bmatrix}, \qquad {}_1\mathbf{V} = \begin{bmatrix} 1.00 \\ 0.6073 \end{bmatrix}$$

2. $(\mathbf{X}'\mathbf{X})\,{}_1\mathbf{V} = {}_2\mathbf{U}$:

$$\mathbf{X}\mathbf{X}'\begin{bmatrix} 1 \\ 0.6073 \end{bmatrix} = \begin{bmatrix} 192.397 \\ 22.8505 \end{bmatrix}, \qquad {}_2\mathbf{V} = \begin{bmatrix} 1.000 \\ 0.1188 \end{bmatrix}$$

3. $(\mathbf{X}'\mathbf{X})\,{}_2\mathbf{V} = {}_3\mathbf{U}$:

$$\mathbf{X}\mathbf{X}'\begin{bmatrix} 1.000 \\ 0.1188 \end{bmatrix} = \begin{bmatrix} 236.1174 \\ -67.522 \end{bmatrix}, \qquad {}_3\mathbf{V} = \begin{bmatrix} 1.000 \\ -0.2859 \end{bmatrix}$$

4. $(\mathbf{X}'\mathbf{X})\,{}_3\mathbf{V} = {}_4\mathbf{U}$:

$$\mathbf{X}'\mathbf{X}\begin{bmatrix} 1 \\ -0.2859 \end{bmatrix} = \begin{bmatrix} 257.383 \\ -142.395 \end{bmatrix}, \qquad {}_4\mathbf{V} = \begin{bmatrix} 1 \\ -0.5532 \end{bmatrix}$$

5. $(\mathbf{X}'\mathbf{X})\,{}_4\mathbf{V} = {}_5\mathbf{U}$:

$$\mathbf{X}'\mathbf{X}\begin{bmatrix} 1 \\ -0.5532 \end{bmatrix} = \begin{bmatrix} 296.261 \\ -191.842 \end{bmatrix}, \qquad {}_5\mathbf{V} = \begin{bmatrix} 1 \\ -0.6475 \end{bmatrix}$$

6. $(\mathbf{X}'\mathbf{X})\,{}_5\mathbf{V} = {}_6\mathbf{U}$:

$$\mathbf{X}'\mathbf{X}\begin{bmatrix} 1 \\ -0.6475 \end{bmatrix} = \begin{bmatrix} 304.701 \\ -209.288 \end{bmatrix}, \qquad {}_6\mathbf{V} = \begin{bmatrix} 1 \\ -0.6869 \end{bmatrix}$$

7. $(\mathbf{X}'\mathbf{X})\,{}_6\mathbf{V} = {}_7\mathbf{U}$:

$$\mathbf{X}'\mathbf{X}\begin{bmatrix} 1 \\ -0.6869 \end{bmatrix} = \begin{bmatrix} 308.228 \\ -217.577 \end{bmatrix}, \qquad {}_7\mathbf{V} = \begin{bmatrix} 1 \\ -0.7059 \end{bmatrix}$$

8. $(\mathbf{X}'\mathbf{X})\,{}_7\mathbf{V} = {}_8\mathbf{U}$:

$$\mathbf{X}'\mathbf{X}\begin{bmatrix} 1 \\ -0.7059 \end{bmatrix} = \begin{bmatrix} 309.928 \\ -220.092 \end{bmatrix}, \qquad {}_8\mathbf{V} = \begin{bmatrix} 1 \\ -0.7101 \end{bmatrix}$$

9. $(\mathbf{X}'\mathbf{X})\,{}_8\mathbf{V} = {}_9\mathbf{U}$:

$$\mathbf{X}'\mathbf{X}\begin{bmatrix} 1 \\ -0.7101 \end{bmatrix} = \begin{bmatrix} 310.304 \\ -220.869 \end{bmatrix}, \qquad {}_9\mathbf{V} = \begin{bmatrix} 1 \\ -0.7118 \end{bmatrix}$$

10. $(\mathbf{X}'\mathbf{X})\,{}_9\mathbf{V} = {}_{10}\mathbf{U}$:

$$\mathbf{X}'\mathbf{X}\begin{bmatrix} 1 \\ -0.7118 \end{bmatrix} = \begin{bmatrix} 310.456 \\ -221.183 \end{bmatrix}, \qquad {}_{10}\mathbf{V} = \begin{bmatrix} 1 \\ -0.7124 \end{bmatrix}$$

11. $(\mathbf{X}'\mathbf{X})\,{}_{10}\mathbf{V} = {}_{11}\mathbf{U}$:

$$\mathbf{X}'\mathbf{X}\begin{bmatrix} 1 \\ -0.7124 \end{bmatrix} = \begin{bmatrix} 310.510 \\ 221.294 \end{bmatrix}, \qquad {}_{11}\mathbf{V} = \begin{bmatrix} 1 \\ -0.7127 \end{bmatrix}$$

12. $(\mathbf{X}'\mathbf{X})\,{}_{11}\mathbf{V} = {}_{12}\mathbf{U}$:

$$\mathbf{X}'\mathbf{X}\begin{bmatrix} 1 \\ -0.7127 \end{bmatrix} = \begin{bmatrix} 310.537 \\ 221.350 \end{bmatrix}, \qquad {}_{12}\mathbf{V} = \begin{bmatrix} 1 \\ -0.7128 \end{bmatrix}$$

13. $(\mathbf{X}'\mathbf{X})\,{}_{12}\mathbf{V} = {}_{13}\mathbf{U}$:

$$\mathbf{X}'\mathbf{X}\begin{bmatrix} 1 \\ -0.7128 \end{bmatrix} = \begin{bmatrix} 310.546 \\ 221.368 \end{bmatrix}, \qquad {}_{13}\mathbf{V} = \begin{bmatrix} 1 \\ -0.7128 \end{bmatrix}$$

The difference between ${}_{12}\mathbf{V}$ and ${}_{13}\mathbf{V}$ is less than 0.0001, and we thus consider ${}_{13}\mathbf{V}$ proportional to $\mathbf{Q}_{.1}$. Normalizing ${}_{13}\mathbf{V}$ thus produces

$$\mathbf{Q}_{.1} = \begin{bmatrix} 1 \\ -0.7128 \end{bmatrix} \left[(1 \quad -0.7128) \begin{pmatrix} 1 \\ -0.7128 \end{pmatrix} \right]^{-1/2} = \begin{bmatrix} 0.8143 \\ -0.5803 \end{bmatrix}$$

To find Δ_1 we use the relation (4.6.7),

$$\Delta_1 = [\mathbf{Q}'_{.1}(\mathbf{X}'\mathbf{X})\mathbf{Q}_{.1}]^{1/2}$$

to give

$$\begin{aligned} \Delta_1 &= \left[(0.8143 \quad -0.5803) \begin{pmatrix} 246.75 & -89.50 \\ -89.50 & 185.00 \end{pmatrix} \begin{pmatrix} 0.8143 \\ -0.5803 \end{pmatrix} \right]^{1/2} \\ &= (310.571)^{1/2} = 17.62 \end{aligned}$$

To find $\mathbf{Q}_{.2}$ and Δ_2, note that $\mathbf{X}'\mathbf{X}$ can be written

$$\mathbf{X}'\mathbf{X} = (\mathbf{Q}_{.1} \quad \mathbf{Q}_{.2}) \begin{bmatrix} \Delta_1^2 & 0 \\ 0 & \Delta_2^2 \end{bmatrix} \begin{bmatrix} \mathbf{Q}'_{.1} \\ \mathbf{Q}'_{.2} \end{bmatrix} = (\mathbf{Q}_{.1}\Delta_1^2 \quad \mathbf{Q}_{.2}\Delta_2^2) \begin{bmatrix} \mathbf{Q}'_{.1} \\ \mathbf{Q}'_{.2} \end{bmatrix}$$

or, by representing the matrix product as a sum of major products of vectors,

$$\mathbf{X}'\mathbf{X} = \mathbf{Q}_{.1}\Delta_1^2\mathbf{Q}'_{.1} + \mathbf{Q}_{.2}\Delta_2^2\mathbf{Q}'_{.2}$$

We can now subtract $\mathbf{Q}_{.1}\Delta_1^2\mathbf{Q}'_{.1}$ from $\mathbf{X}'\mathbf{X}$ to form a *residual matrix*:

$${}_1(\mathbf{X}'\mathbf{X}) = \mathbf{X}'\mathbf{X} - \mathbf{Q}_{.1}\Delta_1^2\mathbf{Q}'_{.1} = \mathbf{Q}_{.2}\Delta_2^2\mathbf{Q}'_{.2}$$

from which $\mathbf{Q}_{.2}$ and Δ_2 can be found by the same powering argument used to develop $\mathbf{Q}_{.1}$ and Δ_1. Let us illustrate by continuing the example:

$$\begin{aligned} {}_1(\mathbf{X}'\mathbf{X}) &= \begin{bmatrix} 246.75 & -89.50 \\ -89.50 & 185 \end{bmatrix} - \begin{bmatrix} 205.89 & -146.72 \\ -146.72 & 104.56 \end{bmatrix} \\ &= \begin{bmatrix} 40.86 & 57.22 \\ 57.22 & 80.44 \end{bmatrix} \end{aligned}$$

1. ${}_1(\mathbf{X}'\mathbf{X})\,{}_0\mathbf{V} = {}_1\mathbf{U}$:

$$\begin{bmatrix} 40.86 & 57.22 \\ 57.22 & 80.44 \end{bmatrix} \begin{bmatrix} 1 \\ 1 \end{bmatrix} = \begin{bmatrix} 98.08 \\ 137.66 \end{bmatrix}, \qquad {}_1\mathbf{V} = \begin{bmatrix} 0.71248 \\ 1 \end{bmatrix}$$

2. ${}_1(\mathbf{X}'\mathbf{X})\,{}_1\mathbf{V} = {}_2\mathbf{U}$:

$${}_1(\mathbf{X}'\mathbf{X}) \begin{bmatrix} 0.71248 \\ 1 \end{bmatrix} = \begin{bmatrix} 86.33 \\ 121.209 \end{bmatrix}, \qquad {}_2\mathbf{V} = \begin{bmatrix} 0.71225 \\ 1 \end{bmatrix}$$

3. ${}_1(\mathbf{X}'\mathbf{X})\,{}_2\mathbf{V} = {}_3\mathbf{U}$:

$${}_1(\mathbf{X}'\mathbf{X}) \begin{bmatrix} 0.71225 \\ 1 \end{bmatrix} = \begin{bmatrix} 86.32 \\ 121.192 \end{bmatrix}, \qquad {}_3\mathbf{V} = \begin{bmatrix} 0.71224 \\ 1 \end{bmatrix}$$

4. ${}_1(\mathbf{X'X})\,{}_3\mathbf{V} = {}_4\mathbf{U}$:

$$ {}_1(\mathbf{X'X})\begin{bmatrix} 0.71224 \\ 1 \end{bmatrix} = \begin{bmatrix} 86.32 \\ 121.198 \end{bmatrix}, \qquad {}_4\mathbf{V} = \begin{bmatrix} 0.71224 \\ 1 \end{bmatrix} $$

Thus

$$ \mathbf{Q}_{.2} = \frac{{}_4\mathbf{V}}{({}_4\mathbf{V}'\,{}_4\mathbf{V})^{1/2}} = \begin{bmatrix} 0.5803 \\ 0.8143 \end{bmatrix} $$

$$ \Delta_2 = [\mathbf{Q}'_{.2}(\mathbf{X'X})\mathbf{Q}_{.2}]^{1/2} = (121.18)^{1/2} = 11.01 $$

$$ \mathbf{Q} = \begin{bmatrix} 0.8143 & 0.5803 \\ -0.5803 & 0.8143 \end{bmatrix} $$

$$ \mathbf{\Delta} = \begin{bmatrix} 17.62 & 0 \\ 0 & 11.01 \end{bmatrix}, \qquad \mathbf{\Delta}^2 = \begin{bmatrix} 310.57 & 0 \\ 0 & 121.18 \end{bmatrix} $$

The product moment $\mathbf{X'X} = \mathbf{Q\Delta}^2\mathbf{Q}'$ is now available:

$$ \begin{bmatrix} 246.75 & -89.50 \\ -89.50 & 185.00 \end{bmatrix} = \begin{bmatrix} 0.8143 & 0.5803 \\ -0.5803 & 0.8143 \end{bmatrix}\begin{bmatrix} 310.57 & 0 \\ 0 & 121.18 \end{bmatrix} \times \begin{bmatrix} 0.8143 & -0.5803 \\ 0.5803 & 0.8143 \end{bmatrix} $$

However, finding the complete basic structure of $\mathbf{X}$ requires that we compute $\mathbf{P}$, the left orthonormal, as well. From Eq. (4.6.2), $\mathbf{P} = \mathbf{XQ\Delta}^{-1}$:

$$ \mathbf{P} = \begin{bmatrix} 9.25 & -5.50 \\ -6.75 & -6.50 \\ -8.75 & 10.50 \\ 6.25 & 1.50 \end{bmatrix}\begin{bmatrix} 0.8143 & 0.5803 \\ -0.5803 & 0.8143 \end{bmatrix}\begin{bmatrix} 17.62 & 0 \\ 0 & 11.01 \end{bmatrix}^{-1} = \begin{bmatrix} 0.608 & 0.081 \\ -0.131 & -0.837 \\ -0.750 & 0.315 \\ 0.239 & 0.440 \end{bmatrix} $$

The solutions for $\mathbf{Q}'$, $\mathbf{\Delta}$, and $\mathbf{P}$ are the results stated in Section 4.2:

$$ \begin{bmatrix} 9.25 & -5.50 \\ -6.75 & -6.50 \\ -8.75 & 10.50 \\ 6.25 & 1.50 \end{bmatrix} = \begin{bmatrix} 0.608 & 0.081 \\ -0.131 & -0.837 \\ -0.750 & 0.315 \\ 0.239 & 0.440 \end{bmatrix}\begin{bmatrix} 17.62 & 0 \\ 0 & 11.01 \end{bmatrix} \times \begin{bmatrix} 0.8143 & -0.5803 \\ 0.5803 & 0.8143 \end{bmatrix} $$

$$ \mathbf{X} = \mathbf{P\Delta Q}' $$

Had the rank of $\mathbf{X'X}$ been greater than 2, the iterative determination of successive vectors $\mathbf{Q}_{.i}$ and elements Δ_i would result from "powering" successive residual matrices,

$$_2(\mathbf{X'X}) = {}_1(\mathbf{X'X}) - \mathbf{Q}_{.2}\Delta_2^2\mathbf{Q'}_{.2}$$

$$_3(\mathbf{X'X}) = {}_2(\mathbf{X'X}) - \mathbf{Q}_{.3}\Delta_3^2\mathbf{Q'}_{.3}$$

4.7 Some Notes on Computation

The preceding section illustrated the computation of the basic structure of a data matrix by the powering method. We chose this method because its logic makes it relatively easy to develop. As a computational technique, however, it has certain limitations. First, it depends on the separation in value of the elements of the basic diagonal; if two adjacent diagonal elements are close in size, many iterations are required before the ratio of the powers of the elements approaches zero. In the limit, the procedure will fail to find a solution if two diagonal elements are identical. Second, the powering method we described extracts one orthonormal vector and an associated diagonal element at a time, cycling through successive residual matrices to complete the basic structure solution.

Basic structure algorithms for digital computers are considerably more efficient. Computer algorithms provide solutions for diagonal elements close in size and solve simultaneously for the complete orthonormal and basic diagonal matrices. The International Mathematical and Statistical Library (IMSL) computer routine EIGRS, for example, employs a Householder reduction that transforms the real product moment matrix into what is known as its tridiagonal form before solving for what are termed the matrix's eigenvalues and eigenvectors. For a product moment matrix, the eigenvalues and eigenvectors are the same as the elements of the basic diagonal and columns of the orthonormal matrix, respectively. Program P4M in the Biomedical Computer Programs P Series (BMDP) uses the Householder reduction to define the principal components. The SPSS routine FACTOR uses a different approach, termed a Cholesky decomposition, for its principal components solution. The principal components, as we shall show in Chapter 14, refer to the basic structure of a covariance or correlation matrix.

4.8 Summary

In this chapter we discussed various factorings of the data matrix. By pointing to the arbitrary characteristics of triangular and other approaches to factoring a matrix, we emphasized the uniqueness of the basic structure decomposition $\mathbf{X} = \mathbf{P\Delta Q'}$. The basic structures of many special matrices were derived. Consideration of the basic structure of powers of product moment matrices led to an illustration of the method of powered vectors for finding the columns of **Q**, from which the elements of **Δ** and **P** can be calculated.

The basic structure decomposition of minor product moment matrices is important as an end in itself, as when we find the principal components of an intercorrelation matrix. Basic structure orthonormalizations for data and derived matrices are also important in the analysis of multivariate data. The reader must understand the use of basic structure representations in both contexts.

Exercises

Consider the following matrices:

$$\mathbf{A} = \begin{bmatrix} 1 & 1 & 1 & 1 \\ 1 & 2 & 4 & 8 \\ 1 & 3 & 9 & 27 \\ 1 & 4 & 16 & 64 \end{bmatrix}$$

$$\mathbf{y}' = (6, 7, 3), \qquad \mathbf{1}' = (1, 1, 1, 1, 1)$$

$$\mathbf{G} = \begin{bmatrix} 2 & -2 & -4 \\ -1 & 3 & 4 \\ 1 & -2 & -3 \end{bmatrix}, \qquad \mathbf{X} = \begin{bmatrix} 1 & 7 & 5 \\ 2 & 4 & 9 \\ 3 & 6 & 2 \\ 6 & 5 & 0 \\ 8 & 3 & 9 \end{bmatrix}$$

1. Find the basic structure of $\mathbf{1}'\mathbf{1}$.

2. Find the basic structure of $\mathbf{y}$.

3. Demonstrate that $\mathbf{G}$ is idempotent.

4. Find the product $\mathbf{XG}$. What is the rank of $\mathbf{XG}$?

5. Find the regular inverse of $\mathbf{A}$ by the partitioned matrix rule.

6. Why can we not find a regular inverse for $\mathbf{G}$?

7. Find the basic structure of $\mathbf{X}$.

8. Denote the basic structure of $\mathbf{X}$ by $\mathbf{X} = \mathbf{PDQ}'$. Compute $\mathbf{XQ}$ and $(\mathbf{XQ})'(\mathbf{XQ}) = \mathbf{Q}'\mathbf{X}'\mathbf{XQ}$. How does this result relate to $\mathrm{tr}(\mathbf{X}'\mathbf{X})$? Interpret $\mathbf{XQ}$?

9. Find the variance–covariance matrix for the columns of $\mathbf{X}$. Find the inverse of this variance–covariance matrix.

10. If a matrix $\mathbf{B}$ is a general symmetric idempotent matrix, show that $\text{tr}(\mathbf{B}) = \text{rank}(\mathbf{B})$.

11. If a matrix $\mathbf{B}$ is a general symmetric idempotent matrix, show that $\mathbf{I} - \mathbf{B}$ is also an idempotent matrix.

12. What is the rank of the $\mathbf{I} - \mathbf{B}$ in Exercise 11?

13. Show that the only full rank idempotent matrix is the identity matrix.

Additional Reading

The following provides additional discussion of the basic structure of product moment matrices and their powers as well as of rank reduction processes:

Horst, P. (1963). *Matrix Algebra for Social Scientists*. New York: Holt, Rinehart, Winston.

The reader will find a discussion of the relation of the basic structure to the characteristic equation of a matrix in

Green, P. E. (1976). *Mathematical Tools for Applied Multivariate Analysis*. New York: Academic.

For the reader interested in a catalog of mathematical results relating to basic structure (singular value decomposition for mathematicians) and idempotent matrices, the following is excellent:

Graybill, F. A. (1969). *Introduction to Matrices with Applications in Statistics*. Belmont, CA: Wadsworth.

Hotelling first proposed his powering solution in the following journal article:

Hotelling, H. (1933). Analysis of a complex of statistical variables, *Journal of Educational Psychology* 24:417–441.

Chapter 5
Transforming Data Matrices

Introduction

Every question about multivariate observations requires that we *transform* those observations in some fashion. In this chapter we introduce the classes of transformations that are most useful.

By far the most commonly applied are *linear transformations*, by which we form new attributes as a weighted sum of our original measures. What are some characteristics of these transformed attributes? Have we lost any information in forming the composite measures? We address these questions making use of the basic structure of the transformation.

A sum of squares of deviations is frequently used in data analysis as a measure of error or of the goodness of fit of some theoretical model to a set of data. That sum of squares is a function or transformation of the data matrix, often a *quadratic* transformation. In this chapter we introduce the reader to the interpretation and algebra of quadratic (and related bilinear) transformations to help us understand how hypotheses about the structure of multivariate measurements can be evaluated.

The important third class of multivariate transformations comprises the *spherizing* transformations. The name derives from the goal of transforming a set of attributes into a new set of mutually independent attributes, each possessing the same amount of variability. If such a transformed data set were plotted, the observations would form a "sphere" in three or more dimensions. More prosaically, we want transformations that rid the data of the annoyances of correlated attributes and of attributes based on different metrics and having quite different variances. Not surprisingly, basic structure plays an important role in finding these transformations.

This is the next to last of the chapters that present the tools of multivariate analysis. Although we continue our interest in manipulating matrices, even the reader with a fairly strong background in linear algebra probably will find new material here. Quadratic and spherizing transformations are strongly motivated by statistical and data analytic goals and may not be widely known.

5.1 Multivariate Analysis: Directed Data Transformations

All multivariate research begins with the formation of a data matrix by systematically recording the observations of m attributes on n entities. The data matrix may be partitioned between attributes (as between predictor and criterion measures, for example) or between entities (as between groups differentially treated).

The analysis of these multivariate data nearly always takes the form of asking whether, or how well, the data matrix or a partition of the data matrix may be *transformed* so as to have certain desirable properties. The differences among the multivariate analyses reflect *differences among the goals* of multivariate researches. These, in turn, manifest themselves in differences in the transformations of the partitions of the data matrix.

For example, the analysis of variance approach (whether there is one or more than one dependent variable) asks whether the assignment of entities to treatment conditions "explains" or "accounts for" the criterion data (the dependent variable scores). The predictor part of the data matrix reflects the treatment history of the experimental entities, and the analytic task is to transform these predictor data linearly so as to minimize, under some hypothesis, the discrepancies between the transformations (estimated criterion scores) and the observed criterion data.

In another approach, that of principal components analysis, the data matrix is transformed so as to produce, for a specified rank, the best-fitting approximation to that data matrix. For example, in the Grade Prediction Study, we might ask how well a matrix of rank 2 can reproduce the set of three test scores and three university grade point averages earned by participants in the College Prediction Study.

The transformational target in both these instances is an empirical matrix of observations. By contrast, in a rotational problem we want to transform an empirical matrix to match best a hypothetically structured matrix and to measure the accuracy attained.

In these and other analyses, multivariate research questions are answered by transforming the data matrix and then measuring the closeness of fit of the transformed matrix to a goal. Multivariate analyses provide ways of phrasing the goals, implementing the transformations, and assessing the closeness of attainment. Before considering different multivariate analyses, we shall review some of the general properties of matrix transformations.

5.2 Linear Transformations of Attributes

The most frequently encountered transformations are linear. You probably already know some properties of these transformations, which we now place in a matrix context.

If $\mathbf{X}$ is an $n \times m$ data matrix and $\mathbf{w}$ is an m-element vector, then

$$\mathbf{Xw} = \mathbf{y} \tag{5.2.1}$$

produces an n-element vector. Each of the n entities in $\mathbf{X}$ has a score on this new attribute $\mathbf{y}$, and this score is a *linear function* of an entity's scores on the m attributes described by the columns of $\mathbf{X}$. The vector $\mathbf{y}$ is a linear transformation (or weighted combination) of the columns of $\mathbf{X}$. Each column of $\mathbf{X}$ has been multiplied by a scaling constant, the corresponding element of the vector $\mathbf{w}$, and the scaled columns added together to produce $\mathbf{y}$. If the scaling constants or weights are arrayed as a diagonal matrix rather than as a vector, this might appear clearer: $\mathbf{XD_w1} = \mathbf{y}$. Since $\mathbf{D_w1} = \mathbf{w}$, the two equations are interchangeable.

If $\boldsymbol{\mu}_\mathbf{X}$ is an m-element vector of means (sample or population) of the attributes making up the columns of $\mathbf{X}$ and if $\boldsymbol{\Sigma}_\mathbf{X}$ is a symmetric $m \times m$ variance–covariance matrix (again, either sample or population) for those same attributes, then

$$\mathbf{w}'\boldsymbol{\mu}_\mathbf{X} = \boldsymbol{\mu}_\mathbf{y} \tag{5.2.2}$$

and

$$\mathbf{w}'\boldsymbol{\Sigma}_\mathbf{X}\mathbf{w} = \boldsymbol{\Sigma}_\mathbf{y} \tag{5.2.3}$$

are, respectively, the mean and variance of the attribute $\mathbf{y}$, the result of the linear transformation.

If you are unfamiliar with this result, consider further the effect of adding the odd and even extrinsic work values scores discussed in Section 3.3. Then (5.2.1) becomes

$$\begin{bmatrix} 7 & 8 \\ 8 & 10 \\ 0 & 1 \\ 2 & 0 \\ 1 & 1 \end{bmatrix} \begin{bmatrix} 1 \\ 1 \end{bmatrix} = \begin{bmatrix} 15 \\ 18 \\ 1 \\ 2 \\ 2 \end{bmatrix}$$

The odd and even item scores, for five job applicants, make up columns 1 and 2, respectively, of the matrix $\mathbf{X}$. The sample mean vector for $\mathbf{X}$ is $\boldsymbol{\mu}'_\mathbf{X} = (3.6, 4.0)$ and the sample variance–covariance matrix is

$$\boldsymbol{\Sigma}_\mathbf{X} = \begin{bmatrix} 10.64 & 13.00 \\ 13.00 & 17.20 \end{bmatrix}$$

The mean of the **y** observations, then, ought to be

$$\mu_y = (1 \quad 1)\begin{bmatrix} 3.6 \\ 4.0 \end{bmatrix} = 7.6$$

and the variance of the **y**, similarly, should be given by

$$\begin{aligned}\Sigma_y &= (1 \quad 1)\begin{bmatrix} 10.64 & 13.00 \\ 13.00 & 17.20 \end{bmatrix}\begin{bmatrix} 1 \\ 1 \end{bmatrix} \\ &= (1 \quad 1)\begin{bmatrix} 23.64 \\ 30.20 \end{bmatrix} = 53.84\end{aligned}$$

These results may be checked by directly calculating the mean and variance of the set of five scores making up the vector **y**.

Two or more linear transformations of the same set of attributes can be obtained simultaneously if the vector **w** is replaced by a matrix **W**. Extending the example just given yields **XW** = **Y**:

$$\begin{bmatrix} 7 & 8 \\ 8 & 10 \\ 0 & 1 \\ 2 & 0 \\ 1 & 1 \end{bmatrix}\begin{bmatrix} 1 & 1 \\ 1 & -1 \end{bmatrix} = \begin{bmatrix} 15 & -1 \\ 18 & -2 \\ 1 & -1 \\ 2 & 2 \\ 2 & 0 \end{bmatrix}$$

Equations (5.2.2) and (5.2.3) generalize to these new situations:

$$\mathbf{W}'\mu_X = \mu_Y \tag{5.2.4}$$

$$\mathbf{W}'\Sigma_X\mathbf{W} = \Sigma_Y \tag{5.2.5}$$

However, μ_Y is a *vector* of means for the two transformed columns,

$$\begin{bmatrix} 1 & 1 \\ 1 & -1 \end{bmatrix}\begin{bmatrix} 3.6 \\ 4.0 \end{bmatrix} = \begin{bmatrix} 7.6 \\ -0.4 \end{bmatrix}$$

and Σ_Y becomes a variance–covariance matrix,

$$\begin{aligned}\begin{bmatrix} 1 & 1 \\ 1 & -1 \end{bmatrix}\begin{bmatrix} 10.64 & 13.00 \\ 13.00 & 17.20 \end{bmatrix}\begin{bmatrix} 1 & 1 \\ 1 & -1 \end{bmatrix} &= \begin{bmatrix} 23.64 & 30.20 \\ -2.36 & -4.20 \end{bmatrix}\begin{bmatrix} 1 & 1 \\ 1 & -1 \end{bmatrix} \\ &= \begin{bmatrix} 53.84 & -6.56 \\ -6.56 & 1.84 \end{bmatrix}\end{aligned}$$

The first and second columns of **Y** have variances 53.84 and 1.84, respectively, and the covariance between the two sets of observations is -6.56.

5.3 Rank of a Set of Linear Transformations

How much of the information about the entities presented in the original data matrix **X** is preserved in the transformation? How does the rank of **Y** compare with the rank of **X**? The answer depends, as might be expected, on the rank of **W**, the weight or *transformation* matrix.

Let **X** be an $n \times m$ vertical matrix and **W** be an $m \times p$ matrix. The product **Y** is $n \times p$, and by the *product rank rule* the rank of **Y** is constrained by relation (3.3.1):

$$r(\mathbf{X}) + r(\mathbf{W}) - m \leqslant r(\mathbf{Y}) \leqslant \min[r(\mathbf{X}), r(\mathbf{W})]$$

The right inequality reminds us that we cannot increase the amount of information by postmultiplying **X** by a transformation matrix; the rank of **Y** cannot be greater than the rank of **X**. We shall consider the implications of the product rank rule for different cases.

I. *If* **X** *is a basic matrix*, then (a) $r(\mathbf{X}) = m$ (the number of columns in the vertical matrix), and (b) $r(\mathbf{W}) \leqslant r(\mathbf{X})$ since the number of rows of **W** must be m. Making these two substitutions into (3.3.1) permits the following generalization. *If* **X** *is a basic vertical matrix, then the transformed matrix* **XW** = **Y** *has rank equal to that of the transformation matrix*: $r(\mathbf{Y}) = r(\mathbf{W})$.

II. *If the rank of the transformation matrix is equal to the number of columns in* **X**, i.e., if $r(\mathbf{W}) = m$, then the rank inequality simplifies to

$$r(\mathbf{X}) \leqslant r(\mathbf{Y}) \leqslant r(\mathbf{X})$$

or, more simply, $r(\mathbf{Y}) = r(\mathbf{X})$. *If the rank of the transformation matrix* **W** *is the same as the number of columns in the vertical matrix* **X**, *then the transformed matrix* **XW** = **Y**, *has the same rank as* **X**: $r(\mathbf{Y}) = r(\mathbf{X})$. The most common application of this second principle involves square basic transformation matrices. As shown earlier, multiplication by a square basic matrix does not affect the rank of a matrix. The rule for case II holds as well, though, for *horizontal* basic **W** matrices. These transformation matrices will also be of rank m, the number of columns of **X**.

III. *If the ranks of both the data matrix* **X** *and the transformation matrix* **W** *are less than the number of columns of* **X**, *the rank of the transformed matrix* **XW** = **Y** *is not known*. What we do know is that $r(\mathbf{Y})$ *cannot* exceed $r(\mathbf{X})$ or $r(\mathbf{W})$, whichever is smaller.

Confronted with a set of linear transformations of the form **XW** = **Y**, the first thing to note is whether the number of columns to **Y**, the number of "new" attributes, is less than, equal to, or greater than the number of "old" attributes, or columns, of **X**. Of course, new information has not been created—the rank of **Y** cannot possibly be greater than m, the number of columns in the vertical **X** matrix—but the rank of **Y** cannot be determined until the ranks of **X** and **W** are considered. Even then there are instances of case III where the rank of **Y** may have to be numerically evaluated, by triangular factoring for example.

5.4 Basic Structure of Transforming Matrices

We can apply some of the basic structure concepts introduced in Chapter 4 to the properties of transformation matrices. Note that the transformation matrix itself has a basic structure. That is, the transformation matrix may be represented as

$$\mathbf{W} = \mathbf{P\Delta Q'} \tag{5.4.1}$$

where **P** and **Q** are orthonormal matrices and Δ is the basic diagonal. The transformation matrix also has a *generalized* inverse,

$$\mathbf{W}^{+} = \mathbf{Q\Delta}^{-1}\mathbf{P'} \tag{5.4.2}$$

Since **W** is $m \times p$, $\mathbf{W}^+$ is $p \times m$ and it is possible to write the product

$$\mathbf{Y}_{(n\times p)}\mathbf{W}^{+}_{(p\times m)} = \mathbf{U}_{(n\times m)}$$

or, on substituting $\mathbf{Y} = \mathbf{XW}$ and Eqs. (5.4.1) and (5.4.2),

$$\mathbf{U} = \mathbf{YW}^{+} = \mathbf{XWW}^{+} = \mathbf{XP\Delta Q'Q\Delta}^{-1}\mathbf{P'} = \mathbf{XPP'} \tag{5.4.3}$$

If **P** is *square* as well as orthonormal, $\mathbf{PP'} = \mathbf{I}$ and thus $\mathbf{YW}^{+} = \mathbf{X}$. The matrix **X** can be produced by a set of linear transformations of **Y**. The transformation of **X** to **Y** is then said to be *invertible*; that is, we can get back to **X** from **Y**.

The left orthonormal matrix **P** is square whenever $\mathbf{W} = \mathbf{P\Delta Q'}$, an $m \times p$ matrix, has rank m, that is, when **W** is a square or horizontal *basic* matrix. To put it another way, the conditions for a transformation to be invertible are just those of case II of Section 5.3. For case II, recall, $r(\mathbf{Y}) = r(\mathbf{X})$, so that all of the information in **X** was carried over to **Y**. Intuitively, it is possible to get back to **X** from **Y** because all of the information is still there. If **W** is a square basic transformation matrix, the generalized inverse $\mathbf{W}^+$ is also the regular inverse $\mathbf{W}^{-1}$.

In Section 5.2 we considered a transformation $\mathbf{XW} = \mathbf{Y}$ given by

$$\begin{bmatrix} 7 & 8 \\ 8 & 10 \\ 0 & 1 \\ 2 & 0 \\ 1 & 1 \end{bmatrix} \begin{bmatrix} 1 & 1 \\ 1 & -1 \end{bmatrix} = \begin{bmatrix} 15 & -1 \\ 18 & -2 \\ 1 & -1 \\ 2 & 2 \\ 2 & 0 \end{bmatrix}$$

If **W** is a basic matrix, this transformation is invertible. Because **W** does have a regular inverse,

$$\mathbf{W}^{-1} = \frac{1}{(1)(-1) - (1)(1)} \begin{bmatrix} -1 & -1 \\ -1 & 1 \end{bmatrix} = \begin{bmatrix} \frac{1}{2} & \frac{1}{2} \\ \frac{1}{2} & -\frac{1}{2} \end{bmatrix}$$

the transformation matrix is basic and of rank equal to the number of attributes in **X**. The matrix **X** may be recovered from **Y** by the inverse transformation $\mathbf{YW}^{-1} = \mathbf{X}$:

$$\begin{bmatrix} 15 & -1 \\ 18 & -2 \\ 1 & -1 \\ 2 & 2 \\ 2 & 0 \end{bmatrix} \begin{bmatrix} \frac{1}{2} & \frac{1}{2} \\ \frac{1}{2} & -\frac{1}{2} \end{bmatrix} = \begin{bmatrix} 7 & 8 \\ 8 & 10 \\ 0 & 1 \\ 2 & 0 \\ 1 & 1 \end{bmatrix}$$

This same example may be extended to show an invertible nonsquare transformation. Two columns may be added to $\mathbf{W}$ to produce the transformation $\mathbf{XW}' = \mathbf{Y}$:

$$\begin{bmatrix} 7 & 8 \\ 8 & 10 \\ 0 & 1 \\ 2 & 0 \\ 1 & 1 \end{bmatrix} \begin{bmatrix} 1 & 0 & 1 & 1 \\ 0 & 1 & 1 & -1 \end{bmatrix} = \begin{bmatrix} 7 & 8 & 15 & -1 \\ 8 & 10 & 18 & -2 \\ 0 & 1 & 1 & -1 \\ 2 & 0 & 2 & 2 \\ 1 & 1 & 2 & 0 \end{bmatrix}$$

The matrix $\mathbf{W}'$ is no longer square. It does, however, have rank 2, the number of columns of $\mathbf{X}$ (as is quickly verified—the rows of $\mathbf{W}'$ would be proportional if the rank were 1).

The basic structure of $\mathbf{W}$ may be represented as in Eq. (5.4.1), $\mathbf{W} = \mathbf{P\Delta Q}'$, and the basic structure of $\mathbf{W}'$ as $\mathbf{W}' = \mathbf{Q\Delta P}'$. The $\mathbf{W}'$ matrix thus has a generalized inverse given by

$$(\mathbf{W}')^+ = \mathbf{P\Delta}^{-1}\mathbf{Q}'$$

If the transformation $\mathbf{XW}' = \mathbf{Y}$ is invertible, then the matrix $\mathbf{X}$ ought to be recoverable from $\mathbf{Y}$ by the transformation $\mathbf{Y(W')}^+ = \mathbf{X}$.

In assessing the basic structure of $\mathbf{W}$ (or $\mathbf{W}'$) we shall make use of the shortcut of determining first the basic structure of the minor product moment $\mathbf{W'W}$. Recall from Chapter 4 and (5.4.1) that the basic structure of $\mathbf{W'W}$ is $\mathbf{W'W} = \mathbf{Q\Delta^2 Q}'$. Numerically (but not uniquely), this is

$$\begin{bmatrix} 3 & 0 \\ 0 & 3 \end{bmatrix} = \begin{bmatrix} 1 & 0 \\ 0 & 1 \end{bmatrix} \begin{bmatrix} 3 & 0 \\ 0 & 3 \end{bmatrix} \begin{bmatrix} 1 & 0 \\ 0 & 1 \end{bmatrix}$$

The generalized inverse can now be constructed. First, recall that the second (left) orthonormal of $\mathbf{W}$ can be obtained from (5.4.1): $\mathbf{P} = \mathbf{WQ\Delta}^{-1}$. This result substituted into the expression for the generalized inverse gives

$$(\mathbf{W}')^+ = (\mathbf{WQ\Delta}^{-1})\mathbf{\Delta}^{-1}\mathbf{Q}' = \mathbf{WQ\Delta}^{-2}\mathbf{Q}'$$

From the numerical results,

$$\mathbf{Q\Delta}^{-2}\mathbf{Q}' = \begin{bmatrix} 1 & 0 \\ 0 & 1 \end{bmatrix} \begin{bmatrix} \frac{1}{3} & 0 \\ 0 & \frac{1}{3} \end{bmatrix} \begin{bmatrix} 1 & 0 \\ 0 & 1 \end{bmatrix} = \begin{bmatrix} \frac{1}{3} & 0 \\ 0 & \frac{1}{3} \end{bmatrix}$$

and the generalized inverse is given by

$$\begin{bmatrix} 1 & 0 \\ 0 & 1 \\ 1 & 1 \\ 1 & -1 \end{bmatrix} \begin{bmatrix} \frac{1}{3} & 0 \\ 0 & \frac{1}{3} \end{bmatrix} = \begin{bmatrix} \frac{1}{3} & 0 \\ 0 & \frac{1}{3} \\ \frac{1}{3} & \frac{1}{3} \\ \frac{1}{3} & -\frac{1}{3} \end{bmatrix}$$

The invertibility of the transformation $\mathbf{XW} = \mathbf{Y}$ is shown, then, by computing $\mathbf{Y}(\mathbf{W}')^{+} = \mathbf{X}$,

$$\begin{bmatrix} 7 & 8 & 15 & -1 \\ 8 & 10 & 18 & -2 \\ 0 & 1 & 1 & -1 \\ 2 & 0 & 2 & 2 \\ 1 & 1 & 2 & 0 \end{bmatrix} \begin{bmatrix} \frac{1}{3} & 0 \\ 0 & \frac{1}{3} \\ \frac{1}{3} & \frac{1}{3} \\ \frac{1}{3} & -\frac{1}{3} \end{bmatrix} = \begin{bmatrix} 7 & 8 \\ 8 & 10 \\ 0 & 1 \\ 2 & 0 \\ 1 & 1 \end{bmatrix}$$

We shall have more to say about the basic structure of transformation matrices when we consider in detail specific linear transformations.

5.5 Rank Reduction Transformation

In this section we develop, in general form, a particularly important linear transformation. It is commonly used in multivariate analysis when a data or derived matrix is to have some influence removed. As it typically results in a residual matrix with rank smaller than that of the original matrix, this transformation is referred to as the rank reduction transformation. The square transformation matrix involved has a distinctive basic structure and normally is itself nonbasic.

Three examples of the rank reduction transformation have been presented in earlier chapters. We may use these by way of introduction.

In Chapter 2 we introduced the centering matrix $\mathbf{I} - \mathbf{1}(\mathbf{1}'\mathbf{1})^{-1}\mathbf{1}'$ and noted that after postmultiplication of the $n \times m$ data matrix $\mathbf{X}$ by a suitably dimensioned centering matrix,

$$\mathbf{Z} = \mathbf{X}\left[\mathbf{I} - \mathbf{1}(\mathbf{1}'\mathbf{1})^{-1}\mathbf{1}'\right] \tag{5.5.1}$$

one obtains a matrix centered by rows. That is, each row of $\mathbf{Z}$ has zero sum:

$$\mathbf{Z1} = \mathbf{0} \tag{5.5.2}$$

In Chapter 3 we introduced a technique for determining the rank of a matrix, called triangular factoring. Beginning with some $n \times m$ matrix ${}_0\mathbf{A}$, we found a residual matrix

$${}_1\mathbf{A} = {}_0\mathbf{A} - {}_1\mathbf{v}\,{}_1\mathbf{w}' \tag{5.5.3}$$

by subtracting from ${}_0\mathbf{A}$ a major product of two vectors determined from ${}_0\mathbf{A}$: If ${}_0A_{ij}$ denotes the largest element of ${}_0\mathbf{A}$, then

$${}_1\mathbf{v} = {}_0\mathbf{A}_{.j} \tag{5.5.4}$$

and

$$ {}_1\mathbf{w}' = (1/{}_0A_{ij})\,{}_0\mathbf{A}'_{i.} \tag{5.5.5}$$

Substituting from (5.5.4) and (5.5.5) into (5.5.3), we can write

$$ {}_1\mathbf{A} = {}_0\mathbf{A} - {}_0\mathbf{A}_{.j}(1/{}_0A_{ij})\,{}_0\mathbf{A}'_{i.} \tag{5.5.6}$$

Now, since by (5.5.6)

$$ {}_0\mathbf{A}\mathbf{e}_j = {}_0\mathbf{A}_{.j} \qquad \text{and} \qquad {}_0\mathbf{A}'_{i.}\mathbf{e}_j = {}_0A_{ij}$$

we can further rewrite ${}_1\mathbf{A}$ in a form comparable to the row-centering example:

$$ {}_1\mathbf{A} = {}_0\mathbf{A}\left[\mathbf{I} - \mathbf{e}_j\left({}_0\mathbf{A}'_{i.}\mathbf{e}_j\right)^{-1}{}_0\mathbf{A}'_{i.}\right] \tag{5.5.7}$$

Finally, in Section 4.6, as part of our demonstration of the powering method for finding the basic structure of a matrix, we found a residual matrix ${}_1(\mathbf{X}'\mathbf{X})$ by subtracting from the $m \times m$ minor product moment matrix $\mathbf{X}'\mathbf{X}$ a product:

$$ {}_1(\mathbf{X}'\mathbf{X}) = (\mathbf{X}'\mathbf{X}) - \mathbf{Q}_{.1}\Delta_1^2\mathbf{Q}'_{.1} \tag{5.5.8}$$

We shall transform this expression into a form similar to (5.5.1) or (5.5.7). First, note the result of postmultiplying the transpose of an orthonormal matrix by any one of its columns: $\mathbf{Q}'\mathbf{Q}_{.1} = \mathbf{e}_1$. The column $\mathbf{Q}_{.1}$ is orthogonal to all rows of $\mathbf{Q}'$ except the first. Now we can premultiply the result by the basic diagonal matrix for $\mathbf{X}'\mathbf{X}$:

$$\Delta^2\mathbf{Q}'\mathbf{Q}_{.1} = \Delta^2\mathbf{e}_1 = \mathbf{e}_1\Delta_1^2$$

The result is a vector all of whose elements are zero save the first, which equals the first diagonal element of Δ^2.

Next we premultiply again, this time by the orthonormal matrix $\mathbf{Q}$:

$$\mathbf{Q}\Delta^2\mathbf{Q}'\mathbf{Q}_{.1} = \mathbf{Q}\mathbf{e}_1\Delta_1^2 = \mathbf{Q}_{.1}\Delta_1^2$$

(Postmultiplying a matrix by an $\mathbf{e}_j$ vector pulls out the jth column of that matrix.) Finally, we postmultiply this last result by $\mathbf{Q}'_{.1}$:

$$(\mathbf{Q}\Delta^2\mathbf{Q}')\mathbf{Q}_{.1}\mathbf{Q}'_{.1} = \mathbf{Q}_{.1}\Delta_1^2\mathbf{Q}'_{.1}$$

Recalling the basic structure of $\mathbf{X}'\mathbf{X}$, so that $\mathbf{X}'\mathbf{X} = \mathbf{Q}\Delta^2\mathbf{Q}'$, we may write

$$\mathbf{X}'\mathbf{X}\mathbf{Q}_{.1}\mathbf{Q}'_{.1} = \mathbf{Q}_{.1}\Delta_1^2\mathbf{Q}'_{.1} \tag{5.5.9}$$

We can now substitute the left side of this equation for the equivalent term on the right in (5.5.8):

$$ {}_1(\mathbf{X}'\mathbf{X}) = (\mathbf{X}'\mathbf{X}) - (\mathbf{X}'\mathbf{X})\mathbf{Q}_{.1}\mathbf{Q}'_{.1} = \mathbf{X}'\mathbf{X}(\mathbf{I} - \mathbf{Q}_{.1}\mathbf{Q}'_{.1})$$

This can then be brought to the desired form,

$$ {}_1(\mathbf{X'X}) = \mathbf{X'X}\left[\mathbf{I} - \mathbf{Q}_{.1}(\mathbf{Q}'_{.1}\mathbf{Q}_{.1})^{-1}\mathbf{Q}'_{.1}\right] \tag{5.5.10} $$

if we recall that $\mathbf{Q}'_{.1}\mathbf{Q}_{.1} = 1$.

The three equations (5.5.1), (5.5.7), and (5.5.10) are now all of the form

$$ {}_{i+1}\mathbf{X} = {}_i\mathbf{X}\left[\mathbf{I} - \mathbf{u}(\mathbf{v'u})^{-1}\mathbf{v'}\right] \tag{5.5.11} $$

where $\mathbf{u}$ and $\mathbf{v}$ are two conformable vectors such that $\mathbf{v'u} \neq 0$.

The postmultiplier in this expression,

$$ \mathbf{A} = \mathbf{I} - \mathbf{u}(\mathbf{v'u})^{-1}\mathbf{v'} \tag{5.5.12} $$

deserves examination. First, note that

$$ \mathbf{Au} = \left[\mathbf{I} - \mathbf{u}(\mathbf{v'u})^{-1}\mathbf{v'}\right]\mathbf{u} = \mathbf{u} - \mathbf{u}(\mathbf{v'u})^{-1}\mathbf{v'u} = \mathbf{u} - \mathbf{u} = \mathbf{0} \tag{5.5.13} $$

The rows of $\mathbf{A}$ are orthogonal to the vector $\mathbf{u}$. Also, since

$$ {}_{i+1}\mathbf{X} = {}_i\mathbf{XA} $$

we know that

$$ {}_{i+1}\mathbf{Xu} = {}_i\mathbf{XAu} = {}_i\mathbf{X0} = \mathbf{0} \tag{5.5.14} $$

The rows of ${}_{i+1}\mathbf{X}$ are also orthogonal to the vector $\mathbf{u}$.

We exploited Eq. (5.5.14) in our centering and triangular factoring examples. In centering by rows, we have $\mathbf{u} = \mathbf{1}$ and, from (5.5.14), $\mathbf{Z1} = \mathbf{0}$, so the row sums are zero. In the triangular factoring example of (5.5.7), we have $\mathbf{u} = \mathbf{e}_j$, so that ${}_1\mathbf{Ae}_j = {}_1\mathbf{A}_{.j} = \mathbf{0}$. That is, we have "zeroed out" a selected column of the residual matrix.

We now apply (5.5.14) with (5.5.10) to the basic structure by powering. Now, $\mathbf{u} = \mathbf{Q}_{.1}$, gives ${}_1(\mathbf{X'X})\mathbf{Q}_{.1} = 0$. The residual matrix is orthogonal to the first column of the orthonormal matrix.

Just as $\mathbf{Au} = \mathbf{0}$, so too we find that

$$ \mathbf{v'A} = \mathbf{v'}\left[\mathbf{I} - \mathbf{u}(\mathbf{v'u})^{-1}\mathbf{v'}\right] = \mathbf{v'} - \mathbf{v'u}(\mathbf{v'u})^{-1}\mathbf{v'} = \mathbf{v'} - \mathbf{v'} = \mathbf{0} \tag{5.5.15} $$

The columns of $\mathbf{A}$ are orthogonal to the vector $\mathbf{v}$. [If we were considering examples in which a matrix was being premultiplied by a matrix of the form of $\mathbf{A}$, we could develop an equivalent to (5.5.14); the columns of the reduced or residual matrix would be orthogonal to $\mathbf{v}$ as well.]

We also note that the matrix $\mathbf{A}$ is idempotent.

$$ \begin{aligned} \mathbf{AA} &= \left[\mathbf{I} - \mathbf{u}(\mathbf{v'u})^{-1}\mathbf{v'}\right]\left[\mathbf{I} - \mathbf{u}(\mathbf{v'u})^{-1}\mathbf{v'}\right] \\ &= \left[\mathbf{I} - \mathbf{u}(\mathbf{v'u})^{-1}\mathbf{v'} - \mathbf{u}(\mathbf{v'u})^{-1}\mathbf{v'} + \mathbf{u}(\mathbf{v'u})^{-1}\mathbf{v'u}(\mathbf{v'u})^{-1}\mathbf{v'}\right] \\ &= \left[\mathbf{I} - 2\mathbf{u}(\mathbf{v'u})^{-1}\mathbf{v'} + \mathbf{u}(\mathbf{v'u})^{-1}\mathbf{v'}\right] \\ &= \left[\mathbf{I} - \mathbf{u}(\mathbf{v'u})^{-1}\mathbf{v'}\right] = \mathbf{A} \end{aligned} \tag{5.5.16} $$

In Section 4.5 we observed that idempotent matrices, at least symmetric idempotent matrices, are not basic. What can we say about the rank of **A** here? To answer this, we apply first the sum rank rule and then the product rank rule. **A** is a sum of two matrices, so by (5.5.12) we can apply the inequality

$$\left|r(\mathbf{I}) - r\left[-\mathbf{u}(\mathbf{v}'\mathbf{u})^{-1}\mathbf{v}'\right]\right| \leqslant r(\mathbf{A}) \leqslant r(\mathbf{I}) + r\left[-\mathbf{u}(\mathbf{v}'\mathbf{u})^{-1}\mathbf{v}'\right]$$

The identity matrix is of full rank, that is, $r(\mathbf{I}) = m$, while the $m \times m$ product $-\mathbf{u}(\mathbf{v}'\mathbf{u})^{-1}\mathbf{v}'$ is a major product of vectors and hence has rank 1. Substituting in the rank inequality gives

$$m - 1 \leqslant r(\mathbf{A}) \leqslant m + 1$$

or, since **A** is only $m \times m$,

$$m - 1 \leqslant r(\mathbf{A}) \leqslant m \tag{5.5.17}$$

We next apply the product rank rule to the relation $\mathbf{Au} = \mathbf{0}$. We know at the outset that the rank of this product (the null vector) is 0. The product rank rule requires that

$$r(\mathbf{A}) + r(\mathbf{u}) - \mathrm{CO}(\mathbf{A}, \mathbf{u}) \leqslant r(\mathbf{Au}) = 0$$

The vector **u** has rank 1, while the common order of **A** and **U** is m, so

$$r(\mathbf{A}) + 1 - m = r(\mathbf{A}) - (m - 1) \leqslant 0$$

or

$$r(\mathbf{A}) \leqslant m - 1 \tag{5.5.18}$$

From (5.5.17) and (5.5.18) we have $m - 1 \leqslant r(\mathbf{A}) \leqslant m - 1$, which requires that

$$r(\mathbf{A}) = m - 1 \tag{5.5.19}$$

The $m \times m$ idempotent matrix **A** in (5.5.12) has rank $m - 1$.

We can extend what we have learned about **A** to a more general case. Let us define **U** and **V** to be $m \times p$ basic matrices, each with rank p. Further, let us define **V** and **U** so that the $p \times p$ minor product $\mathbf{V}'\mathbf{U}$ is *basic*. Then $\mathbf{V}'\mathbf{U}$ has a regular inverse, and we can define a new matrix

$$\mathbf{A} = \mathbf{I} - \mathbf{U}(\mathbf{V}'\mathbf{U})^{-1}\mathbf{V}' \tag{5.5.20}$$

(If $\mathbf{U} = \mathbf{V}$, then **A** is symmetric.) The properties of this **A** parallel those of the **A** of (5.5.12):

$$\mathbf{AU} = \mathbf{0} \tag{5.5.21}$$

$${}_0\mathbf{XAU} = {}_1\mathbf{XU} = \mathbf{0} \tag{5.5.22}$$

$$\mathbf{V}'\mathbf{A} = \mathbf{0} \tag{5.5.23}$$

$$\mathbf{V}'\mathbf{A}\,{}_0\mathbf{X} = \mathbf{V}'\,{}_1\mathbf{X} = \mathbf{0} \tag{5.5.24}$$

$$\mathbf{AA} = \mathbf{A} \tag{5.5.25}$$

$$r(\mathbf{A}) = m - p \tag{5.5.26}$$

These properties are demonstrated exactly as for **v** and **u**.

We can now reduce the rank of a basic matrix. If ${}_0\mathbf{X}$ is $n \times m$ with $r({}_0\mathbf{X}) = m$ and we define

$${}_1\mathbf{X} = {}_0\mathbf{XA} = {}_0\mathbf{X}\left[\mathbf{I} - \mathbf{U}(\mathbf{V}'\mathbf{U})^{-1}\mathbf{V}'\right] \tag{5.5.27}$$

then we can find the rank of ${}_1\mathbf{X}$. By the product rank rule,

$$r({}_0\mathbf{X}) + r(\mathbf{A}) - \mathrm{CO}({}_0\mathbf{X}, \mathbf{A}) \leqslant r({}_0\mathbf{XA}) = r({}_1\mathbf{X}) \leqslant \min\left[r(\mathbf{A}), r({}_0\mathbf{X})\right]$$

Now, $m + (m - p) - m \leqslant r({}_1\mathbf{X}) \leqslant m - p$, so that

$$r({}_1\mathbf{X}) = m - p \tag{5.5.28}$$

Postmultiplying the vertical and basic matrix ${}_0\mathbf{X}$ by the idempotent **A** of (5.5.20) reduces the rank of that matrix by p from m to $m - p$, where p is the rank of the basic **U** and **V**. We shall encounter many applications of this rank reduction principle.

The effect of postmultiplying a nonbasic ${}_0\mathbf{X}$ by a general **A** of the form of (5.5.20) cannot be determined exactly. If we take ${}_0\mathbf{X}$ to have rank $q < m$, the product rank rule ensures only that for ${}_1\mathbf{X}$ of (5.5.27)

$$q - p \leqslant r({}_1\mathbf{X}) \leqslant q \tag{5.5.29}$$

We may not alter the rank at all—or we may reduce it by as much as p (reducing the matrix to a null matrix if $p \geqslant q$). The exact result depends on the proper choice of matrices **U** and **V**. The principle can be illustrated by considering further some examples that we have already treated. First, centering a matrix by rows does not reduce the rank of that matrix by 1 if the row sums are already 0. Similarly, in triangular factoring, zeroing out the jth column cannot reduce the rank of a matrix if the jth column is already filled with 0s.

Transformations by rank reducing (and often symmetric) idempotent matrices play an important role in our development in Chapter 8 of the general linear model underlying multiple regression and fixed effects analysis of variance. Similar transformations are also central to our extension of this model to the multivariate criterion in Chapters 10 and 11. Finally, rank reduction transformations underlie the analysis in Chapter 14 of interdependence techniques—principal components and factor analysis—that seek interpretable lower rank approximations to a data or derived matrix.

5.6 Quadratic and Bilinear Transformations

We shall have much more to say about the applications of linear transformations in subsequent chapters. There are, however, other types of transformations important to multivariate analysis.

In multivariate analysis, as in more elementary data analysis, sums of squares and sums of cross products play an important role. In understanding their function, it will help to think of them as transformations of the data matrix. Indeed, optimizing certain properties of these transformations is the key to many analyses.

We begin with a single vector. We let a vector **x** represent a set of eight scores obtained on the Vocational Interest Inventory for one participant in the Vocational Interest Study. In Section 3.1 it was noted that the sum of eight scores for this test is always 112.

	SER	BUS	ORG	TEC	OUT	SCI	CUL	ART
$\mathbf{x}' = ($	17	20	17	14	11	8	11	14 $)$

In Chapter 2 we introduced a centering matrix that could be used, at least symbolically, to convert a column of scores into deviation form. For the vector of Vocational Interest Inventory scores, $[\mathbf{I} - \mathbf{1}(\frac{1}{8})\mathbf{1}']\mathbf{x} = \mathbf{z}$ is in deviation score form. The centering matrix is a symmetric 8×8 matrix. For the present example, $\mathbf{1}'\mathbf{x} = 112$, the mean of the eight scores is 14, and the deviation vector is

$$\mathbf{z}' = (3, 6, 3, 0, -3, -6, -3, 0)$$

The *sum of squared deviations* of the eight scores about the mean of the eight is given by the minor product moment of this vector, $\mathbf{z}'\mathbf{z}$. We may write this quantity

$$\mathbf{z}'\mathbf{z} = \mathbf{x}'[\mathbf{I} - \mathbf{1}(\tfrac{1}{8})\mathbf{1}'][\mathbf{I} - \mathbf{1}(\tfrac{1}{8})\mathbf{1}']\mathbf{x}$$

based on the symmetry of the centering matrix, or

$$\mathbf{z}'\mathbf{z} = \mathbf{x}'[\mathbf{I} - \mathbf{1}(\tfrac{1}{8})\mathbf{1}']\mathbf{x} = 108 \qquad (5.6.1)$$

based on the idempotency of the centering matrix.

The sum of squared deviations is expressed in (5.6.1) as a quadratic form (or transformation) of the vector **x**.

Definition. If **y** is a vector of observations and **S** is a symmetric matrix, then $\mathbf{v} = \mathbf{y}'\mathbf{S}\mathbf{y}$ is termed a *quadratic form* in **y**, and the matrix **S** is termed the *matrix of the form*.

As we noted in Chapter 4 [Eq. (4.5.2)], some, but not all, symmetric matrices have basic structures that can be written $\mathbf{S} = \mathbf{Q}\boldsymbol{\Delta}^2\mathbf{Q}'$. These symmetric matrices are called *nonnegative* matrices, inasmuch as each can be expressed as the major product moment of a matrix,

$$\mathbf{S} = \mathbf{A}\mathbf{A}' \qquad \text{where} \quad \mathbf{A} = \mathbf{Q}\boldsymbol{\Delta} \qquad (5.6.2)$$

and, in consequence, each of the diagonal elements of **S** is nonnegative.

When $\mathbf{S}$ is nonnegative (and symmetric) the quadratic form

$$\mathbf{v} = \mathbf{y}'\mathbf{S}\mathbf{y} \tag{5.6.3}$$

can be interpreted as a sum of squares of linearly transformed scores.

Definition. A *nonnegative quadratic form* is a sum of squares of k linearly transformed scores, where k is the rank of $\mathbf{S}$, the symmetric, nonnegative matrix of the form.

This definition of the quadratic form will become clear if we follow up on Eq. (5.6.2) to write the nonnegative quadratic form $\mathbf{v} = \mathbf{y}'\mathbf{A}\mathbf{A}'\mathbf{y}$.

This result may be represented pictorially:

$${}_1[\mathbf{v}]^1 = {}_1[\ \mathbf{y}'\ ]^n\ {}_n\begin{bmatrix}\mathbf{A}\end{bmatrix}^k\ {}_k\begin{bmatrix}\mathbf{A}'\end{bmatrix}^n\ {}_n\begin{bmatrix}\mathbf{y}\end{bmatrix}^1,$$

It will be noted that the product $\mathbf{y}'\mathbf{A}$ is a k-element row vector, with transpose $\mathbf{A}'\mathbf{y}$, and the quadratic form $\mathbf{v}$ may be expressed as

$${}_1[\mathbf{v}]^1 = {}_1[\ \mathbf{y}'\mathbf{A}\ ]^k\ {}_k\begin{bmatrix}\mathbf{A}'\mathbf{y}\end{bmatrix}^1$$

or

$$\mathbf{v} = (\mathbf{y}'\mathbf{A})(\mathbf{A}'\mathbf{y})$$

In this last form, it is clear that the nonnegative quadratic form is a sum of squares of k elements. Each of the k elements is a linear transformation of the n elements in $\mathbf{y}$. Further, the k sets of weights employed in these linear transformations are mutually orthogonal. That is, $\mathbf{A}'\mathbf{A} = \mathbf{\Delta}\mathbf{Q}'\mathbf{Q}\mathbf{\Delta} = \mathbf{\Delta}^2$, the basic diagonal of $\mathbf{S} = \mathbf{Q}\mathbf{\Delta}^2\mathbf{Q}' = \mathbf{A}\mathbf{A}'$. All of these properties will be important in the use of quadratic forms, in the multivariate analysis of variance for example.

For now, however, a few representative quadratic forms may be examined. The sum of squares about the mean, as a quadratic form, employs the centering matrix as the matrix of the form. Because the centering matrix is symmetric and idempotent ($\mathbf{S} = \mathbf{S}'$ and $\mathbf{SS} = \mathbf{S}$), it can be expressed (see Chapter 4) as a major product moment (of itself!): $\mathbf{SS}' = \mathbf{S}$. The centering matrix is therefore a nonnegative matrix. Indeed, we shall make considerable use of the fact that all symmetric idempotent matrices are nonnegative.

In discussing rank reduction in Chapter 4, we noted that the centering matrix is not a basic matrix; rather its rank is one less than its order. Thus the 8×8 matrix of the form in Eq. (5.6.1) is of rank 7, and the sum of squares of the eight vocational interest scores, taken about their mean, is in fact a sum of squares of seven linear transformations of the eight scores. (In Chapter 8, in discussing the general linear model, this sum of squares of

$n - 1$ transformations of n scores will be related to the loss of a degree of freedom brought about by computing the sample mean.)

A quadratic form employing a diagonal matrix $\mathbf{D}$ with all positive elements is nonnegative. $\mathbf{D} = (\mathbf{D}^{1/2})(\mathbf{D}^{1/2})'$. Such a quadratic form results if each attribute represented in the original vector is *scaled* before the sum of squares is computed. If $\mathbf{D}$ is $n \times n$, the quadratic form $\mathbf{w} = \mathbf{y}'\mathbf{D}\mathbf{y}$ is a sum of squares of n transformations of the n variables. (A special case of this is the sum of squares $\mathbf{y}'\mathbf{y} = \mathbf{y}'\mathbf{I}\mathbf{y}$.)

What if the n-element vector $\mathbf{y}$ in Eq. (5.6.3) is replaced by an $n \times k$ matrix? Consider a set of Vocational Interest Inventory scores for three high school students:

$$\mathbf{Y}' = \begin{bmatrix} 17 & 20 & 17 & 14 & 11 & 8 & 11 & 14 \\ 10 & 14 & 16 & 18 & 16 & 14 & 12 & 12 \\ 13 & 13 & 12 & 14 & 16 & 15 & 15 & 14 \end{bmatrix}$$
$$\quad \text{SER} \quad \text{BUS} \quad \text{ORG} \quad \text{TEC} \quad \text{OUT} \quad \text{SCI} \quad \text{CUL} \quad \text{ART}$$

The product $[\mathbf{I} - \mathbf{1}(\frac{1}{8})\mathbf{1}']\mathbf{Y} = \mathbf{Z}$ is an 8×3 matrix of deviation scores within each of the three student's profiles:

$$\mathbf{Z}' = \begin{bmatrix} 3 & 6 & 3 & 0 & -3 & -6 & -3 & 0 \\ -4 & 0 & 2 & 4 & 2 & 0 & -2 & -2 \\ -1 & -1 & -2 & 0 & 2 & 1 & 1 & 0 \end{bmatrix}$$

The minor product moment of $\mathbf{Z}$ is a 3×3 symmetric matrix,

$$\mathbf{Z}'\mathbf{Z} = \begin{bmatrix} 108 & -6 & -30 \\ -6 & 48 & 2 \\ -30 & 2 & 12 \end{bmatrix}$$

with diagonal elements equal to the sums of squares of the three columns of $\mathbf{Z}$ and with off-diagonal elements equal to the sums of products of pairs of columns of $\mathbf{Z}$. We can write

$$\begin{aligned} \mathbf{Z}'\mathbf{Z} &= \mathbf{Y}'[\mathbf{I} - \mathbf{1}(\tfrac{1}{8})\mathbf{1}'][\mathbf{I} - \mathbf{1}(\tfrac{1}{8})\mathbf{1}']\mathbf{Y} \\ &= \mathbf{Y}'[\mathbf{I} - \mathbf{1}(\tfrac{1}{8})\mathbf{1}']\mathbf{Y} \end{aligned}$$

a result very similar to (5.6.1). Again because the centering matrix is nonnegative and symmetric, expressions of this form provide a useful generalization of quadratic forms.

Theorem. *If* $\mathbf{Y}$ *is an* $n \times m$ *matrix of observations and if* $\mathbf{S}$ *is an* $n \times n$ *symmetric matrix, then* $\mathbf{V} = \mathbf{Y}'\mathbf{S}\mathbf{Y}$ *is an* $m \times m$ *symmetric matrix with the following properties*:

a. V_{ii} *is a quadratic form in the ith column of* $\mathbf{Y}$ $(i = 1, 2, \ldots, m)$, *and*
b. V_{ij} *is a bilinear form in the ith and jth columns of* $\mathbf{Y}$ $(i, j = 1, 2, \ldots m,\ i \neq j)$.

We again focus on the nonnegative symmetric matrix. When $\mathbf{S}$ is nonnegative, it can be replaced by $\mathbf{S} = \mathbf{Q\Delta^2 Q'}$:

$$\mathbf{V} = \mathbf{Y'Q\Delta^2Q'Y} = \mathbf{Y'Q\Delta\Delta Q'Y}$$

permitting us to take $\mathbf{A} = \mathbf{Q\Delta}$, as we did for the quadratic form. Then the $m \times m$ matrix $\mathbf{V}$ can be expressed $\mathbf{V} = \mathbf{Y'AA'Y}$.

Keeping in mind the dimensions of the factors,

$$_m\Big[\ \mathbf{V}\ \Big]^m = {_m}[\ \mathbf{Y'}\ ]^n\ {_n}\Big[\ \mathbf{A}\ \Big]^k\ {_k}[\ \mathbf{A'}\ ]^n\ {_n}\Big[\ \mathbf{Y}\ \Big]^m$$

we can see that the product $\mathbf{Y'A}$ is an $m \times k$ matrix:

$${_m}[\ \mathbf{W'}\ ]^k = {_m}[\ \mathbf{Y'}\ ]^n\ {_n}\Big[\ \mathbf{A}\ \Big]^k$$

The rank of the symmetric matrix $\mathbf{S}$ is $k \leqslant n$. Each row of $\mathbf{W'}$ consists of k linear transformations of the n elements making up a particular column of $\mathbf{Y}$. The weights for the k transformations, the columns of $\mathbf{A}$, are orthogonal to each other, $\mathbf{A'A} = \mathbf{\Delta}^2$, the basic diagonal of $\mathbf{S}$.

Expressing the matrix $\mathbf{V}$ in terms of $\mathbf{W}$,

$$_m\Big[\ \mathbf{V}\ \Big]^m = {_m}[\ \mathbf{W'}\ ]^k\ {_k}\Big[\ \mathbf{W}\ \Big]^m$$

we can see that each diagonal element of $\mathbf{V}$ is a sum of squares of k linear transformations of the elements in a column of $\mathbf{Y}$ and that each off-diagonal element of $\mathbf{V}$ is a sum of products of these k transformations applied to two different columns of $\mathbf{Y}$. This may be made clearer by writing two typical elements of $\mathbf{V}$:

$${_1}[V_{ii}]^1 = {_1}[\ \mathbf{Y}'_{.i}\ ]^n\ {_n}\Big[\ \mathbf{A}\ \Big]^k\ {_k}[\ \mathbf{A'}\ ]^n\ {_n}\Big[\ \mathbf{Y}_{.i}\ \Big]^1$$

and

$${_1}\Big[V_{ij}\Big]^1 = {_1}[\ \mathbf{Y}'_{.i}\ ]^n\ {_n}\Big[\ \mathbf{A}\ \Big]^k\ {_k}[\ \mathbf{A'}\ ]^n\ {_n}\Big[\ \mathbf{Y}_{.j}\ \Big]^1$$

The product $\mathbf{A'Y}_{.i}$ is a k-element vector, each element a linear transformation of the n elements in $\mathbf{Y}_{.i}$. The sum of squares of these k elements is V_{ii}. The product $\mathbf{A'Y}_{.j}$ is, similarly, a k-element vector of linear transformations of the n elements in $\mathbf{Y}_{.j}$. The minor product of $\mathbf{A'Y}_{.i}$ and $\mathbf{A'Y}_{.j}$ gives V_{ij}.

In the Vocational Interest Inventory examples, the diagonal elements of $\mathbf{Z'Z}$ give sums of squares of deviations of the three students' profiles from a "flat" profile, a score of 14 on each of the eight scales. The three profiles were selected to be flatter, moving from the first to the second to the third student. The off-diagonal elements of $\mathbf{Z'Z}$ reflect the extent to which *pairs* of the interest profiles vary together. A high positive off-diagonal element would indicate a coincidence of peaks and valleys for two profiles, whereas a large negative off-diagonal element would signal that two profiles are out of phase—one being high on the scales where the other is low and vice versa. There is some indication here that students 1 and 3 are out of phase.

The nonnegative form plays a vital role in the general linear model developed in Chapter 8 for a single dependent or criterion attribute. Moreover, the extension to a symmetric matrix of quadratic and bilinear forms is necessary in those multivariate analyses that grow out of the multivariate general linear model, appropriate when there are several criterion attributes—the multivariate analysis of variance, for example.

5.7 Spherizing Transformations

Vector normalization, described in Section 3.3, can also be thought of as a transformation. The goal of normalization is to scale a vector, multiplying each element by a constant, so that the resulting vector will have a sum of squares equal to 1.0. Let us again consider the following vector of vocabulary scores for four students from the College Prediction Study; the scores are in deviation score form, so that $\mathbf{1}'\mathbf{x} = 0$:

$$\mathbf{x}' = (9.25, \ -6.75, \ -8.75, 6.25)$$

This vector is normalized by dividing each element by the square root of the sum of squares of its elements. That is, $\mathbf{z} = \mathbf{x}(\mathbf{x}'\mathbf{x})^{-1/2}$ and we may verify that

$$\mathbf{z}'\mathbf{z} = (\mathbf{x}'\mathbf{x})^{-1/2}\mathbf{x}'\mathbf{x}(\mathbf{x}'\mathbf{x})^{-1/2} = \mathbf{x}'\mathbf{x}(\mathbf{x}'\mathbf{x})^{-1} = 1.0$$

For the numerical example, $\mathbf{x}'\mathbf{x} = 246.75$, $(\mathbf{x}'\mathbf{x})^{1/2} = 15.708278$, $\mathbf{z}' = (0.58886, -0.42971, -0.55703, 0.39708)$, and $\mathbf{z}'\mathbf{z} = 1.0$.

It was also noted in Section 3.3 that an orthogonal matrix $\mathbf{X}$ for which $\mathbf{X}'\mathbf{X} = \mathbf{D}$, a diagonal, could be transformed into an orthonormal matrix by forming the product $\mathbf{Z} = \mathbf{X}\mathbf{D}^{-1/2}$. The minor product moment of $\mathbf{Z}$ is indeed $\mathbf{Z}'\mathbf{Z} = (\mathbf{D}^{-1/2}\mathbf{X}')(\mathbf{X}\mathbf{D}^{-1/2}) = \mathbf{D}^{-1/2}\mathbf{D}\mathbf{D}^{-1/2} = \mathbf{I}$.

Each column vector in $\mathbf{Z}$ is normalized, and each column is orthogonal to all other columns. Orthonormalizing a deviation form matrix $(\mathbf{1}'\mathbf{X} = \mathbf{0}')$, would mean that the columns (attributes if $\mathbf{X}$ is a data matrix) had been transformed so that (1) each transformed column has a zero mean and (2) each transformed column has the same variance. (If $\mathbf{1}'\mathbf{Z}_{.i} = 0$ and $\mathbf{Z}'_{.i}\mathbf{Z}_{.i} = 1.0$, then $\text{Var}(\mathbf{Z}_{.i}) = 1/n$.) Moreover, (3) each transformed column is linearly independent of all others; the covariance between any two columns is zero. (If $\mathbf{1}'\mathbf{Z}_{.i} = \mathbf{1}'\mathbf{Z}_{.j} = 0$ and if $\mathbf{Z}'_{.i}\mathbf{Z}_{.j} = 0$, then $\text{Cov}(\mathbf{Z}_{.i}, \mathbf{Z}_{.j}) = (1/n)0 = 0$.)

Such a transformation will be very helpful in multivariate analysis. It permits control for varying scales for the different columns of a matrix (different column variances) and for correlations or partial dependences among the columns.

Care must be taken, however, that no information is lost in the transformation. To ensure this, it is usual to require that the transformation be reversible; that is, the original matrix **X** should be expressible as a linear transformation of the orthonormal matrix obtained. Because of this condition, transforming to an orthonormal matrix is frequently referred to as finding an orthonormal basis for the original matrix. That is, every column of **X** can be represented as a linear combination of the columns of the orthonormal matrix.

The following algebraic argument indicates how an orthonormal basis may be found for an arbitrary matrix. Let **X** be a vertical $n \times m$ matrix. We require a transformation

$$\mathbf{X}_{(n\times m)}\mathbf{W}_{(m\times p)} = \mathbf{Y}_{(n\times p)} \qquad \text{such that} \quad \mathbf{Y'Y} = \mathbf{I} \tag{5.7.1}$$

It is also required that **W** be chosen such that

$$\mathbf{YW}^{+} = \mathbf{X} \tag{5.7.2}$$

where $\mathbf{W}^+$ is the generalized inverse of **W**, to ensure that **X** can be recreated from the orthonormal **Y**.

In Section 4.2 we suggested that the basic structure $\mathbf{X} = \mathbf{P\Delta Q'}$, given by Eq. (4.2.1), holds an answer. The left orthonormal **P** is the required basis. Not only do we have $\mathbf{P'P} = \mathbf{I} = \mathbf{Y'Y}$, satisfying (5.6.1), but (4.2.1) tells us how to recreate **X** from **P** and satisfy (5.6.2): $\mathbf{PW}^+ = \mathbf{X} = \mathbf{P\Delta Q'}$ is satisfied by taking

$$\mathbf{W}^{+} = \mathbf{\Delta Q'} = \mathbf{I\Delta Q'} \tag{5.7.3}$$

Of what matrix can this be a generalized inverse? Our solution for a generalized inverse suggests

$$\mathbf{W} = \mathbf{Q\Delta}^{-1} = \mathbf{Q\Delta}^{-1}\mathbf{I} \tag{5.7.4}$$

and we may check that the two triple products

$$\mathbf{W}^{+}\mathbf{WW}^{+} = \mathbf{\Delta Q'Q\Delta}^{-1}\mathbf{\Delta Q'} = \mathbf{\Delta Q'} = \mathbf{W}^{+}$$
$$\mathbf{WW}^{+}\mathbf{W} = \mathbf{Q\Delta}^{-1}\mathbf{\Delta Q'Q\Delta}^{-1} = \mathbf{Q\Delta}^{-1} = \mathbf{W}$$

satisfy the generalized inverse conditions.

Thus the orthonormal basis of **X** is, from (4.2.3), the left orthonormal of its basic structure, $\mathbf{X}(\mathbf{Q\Delta}^{-1}) = \mathbf{P}$, and **X** is represented in this orthonormal basis simply by completing the basic structure equation (4.2.1),

$$\mathbf{P}(\mathbf{\Delta Q'}) = \mathbf{X}$$

In practice **Q** and **Δ** are obtained by finding the basic structure of the minor product moment $\mathbf{X'X} = \mathbf{Q\Delta}^2\mathbf{Q'}$. To indicate that a specific orthonormal basis is being sought, $\mathbf{X} = \mathbf{YW'} = \mathbf{P}(\mathbf{\Delta Q'})$ will be referred to as a *spherizing* transformation. The name suggests the result of equating the variances of the columns and eliminating any correlation between columns. If rows of the transformed matrix were plotted as points in space the resulting cluster of points would be a circle in two dimensions, a sphere in three-dimensional space, and a k-dimensional hypersphere in k-space.

The spherizing transformation can be illustrated by considering a matrix containing two scores, vocabulary and quantitative skills, for a sample of four students in the College Prediction Study:

$$\mathbf{X} = \begin{bmatrix} 9.25 & -5.5 \\ -6.75 & -6.5 \\ -8.75 & 10.5 \\ 6.25 & 1.5 \end{bmatrix}$$

Both sets of scores are in deviation form, $\mathbf{1}'\mathbf{X} = \mathbf{0}'$. Vocabulary scores are in the first column, quantitative skills scores in the second.

The minor product moment matrix is

$$\mathbf{X}'\mathbf{X} = \begin{bmatrix} 246.75 & -89.50 \\ -89.50 & 185.00 \end{bmatrix}$$

The minor product moment has the basic structure

$$\mathbf{X}'\mathbf{X} = \mathbf{Q}\mathbf{\Delta}^2\mathbf{Q}' = \begin{bmatrix} 0.8143 & 0.5803 \\ -0.5803 & 0.8143 \end{bmatrix} \begin{bmatrix} 310.55 & 0 \\ 0 & 121.20 \end{bmatrix}$$
$$\times \begin{bmatrix} 0.8143 & -0.5803 \\ 0.5803 & 0.8143 \end{bmatrix}$$

From the basic diagonal one can compute

$$\mathbf{\Delta}^{-1} = \begin{bmatrix} 0.0567 & 0 \\ 0 & 0.0908 \end{bmatrix}$$

and from this result, in turn,

$$\mathbf{Q}\mathbf{\Delta}^{-1} = \begin{bmatrix} 0.0462 & 0.0527 \\ -0.0329 & 0.0740 \end{bmatrix}$$

Finally, an orthonormal basis for $\mathbf{X}$ is given by $\mathbf{X}(\mathbf{Q}\mathbf{\Delta}^{-1}) = \mathbf{P}$:

$$\begin{bmatrix} 9.25 & -5.5 \\ -6.75 & -6.5 \\ -8.75 & 10.5 \\ 6.25 & 1.5 \end{bmatrix} \begin{bmatrix} 0.0462 & 0.0527 \\ -0.0329 & 0.0740 \end{bmatrix} = \begin{bmatrix} 0.6083 & 0.0805 \\ -0.1313 & -0.8367 \\ -0.7497 & 0.3159 \\ 0.2394 & 0.4404 \end{bmatrix}$$

the left orthonormal in the basic structure of $\mathbf{X}$.

The four observations are plotted in Figure 5.1, using both $\mathbf{X}_{.i}$ and $\mathbf{P}_{.i}$ coordinates. The movement toward a circular arrangement is fairly clear.

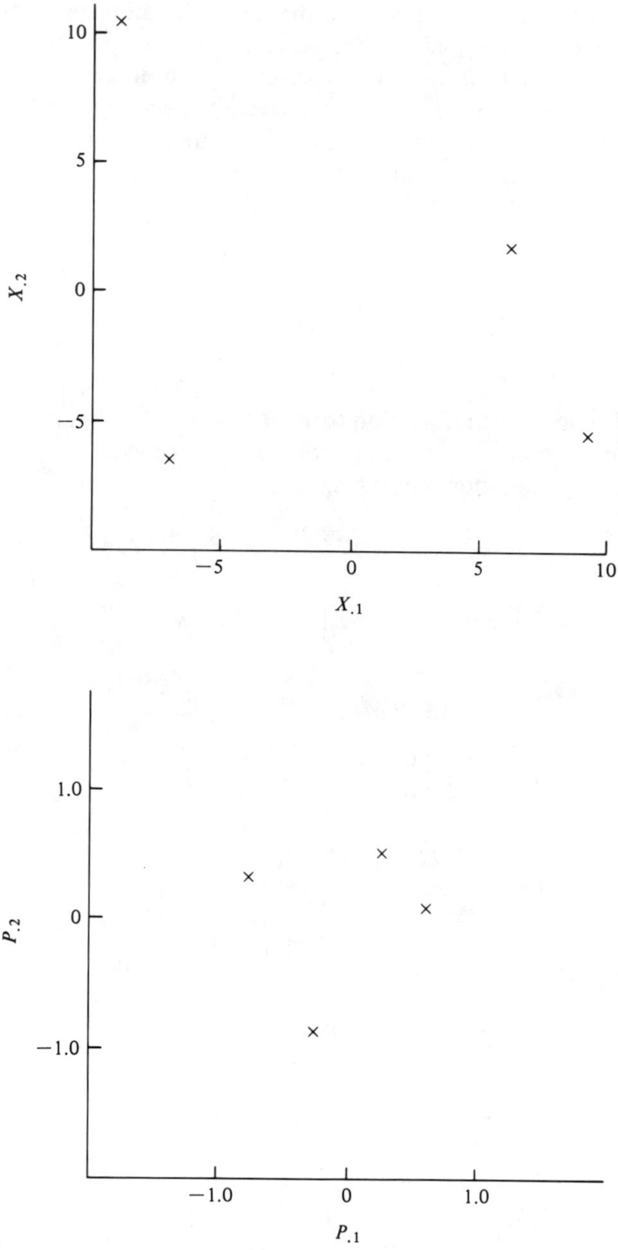

Figure 5.1 Spherizing transformation of the data matrix **X**.

5.8 Summary

The goals of multivariate analysis usually require transforming the data matrix. This chapter discussed the most significant types: linear, rank reduction, quadratic, bilinear, and spherizing transformations.

The transformation matrix **W** transforms the data matrix **X** linearly if $\mathbf{XW} = \mathbf{Y}$. Linear transformations affect the means of the attributes as $\boldsymbol{\mu}_{\mathbf{Y}} = \mathbf{W}'\boldsymbol{\mu}_{\mathbf{X}}$ and the variance–covariance matrix of the attributes as $\boldsymbol{\Sigma}_{\mathbf{Y}} = \mathbf{W}'\boldsymbol{\Sigma}_{\mathbf{X}}\mathbf{W}$. We distinguished three cases, according to the rank of **Y**. If **X** is basic, then $\mathbf{Y} = \mathbf{XW}$ has rank equal to the rank of the transformation matrix **W**. If the rank of **W** is the same as the number of columns in the vertical matrix **X**, then **Y** has the same rank as **X**. If the ranks of **X** and **W** are less than the number of columns of **X**, then the rank of **Y** is no greater than that of **X** or **W**, whichever is smaller. Postmultiplying **X** by its conformable centering matrix was explored as an example of a transformation in which the rank of the transformation matrix is less than the number of columns of **X**.

A quadratic transformation of the columns gives a sum of squares of the attributes, and a bilinear transformation gives the cross products of the columns. Each diagonal element of $\mathbf{V} = \mathbf{W}'\mathbf{XW}$ is a sum of squares of the linear transformations of the columns in **X**, and each off-diagonal element of **V** is the sum of products of the transformations applied to two different columns of **X**.

If the vector **x** is multiplied by $(\mathbf{x}'\mathbf{x})^{-1/2}$ or if the basic matrix **X** is postmultiplied by $(\mathbf{X}'\mathbf{X})^{-1/2} = (\mathbf{Q}\boldsymbol{\Delta}^{-1}\mathbf{Q}')$, the transformed columns are mutually independent and the sum of squares is the same for each column. Such a spherizing transformation plays a major role in multivariate analysis. One of the most common spherizing transformations is derived from the basic structure of **X**. If $\mathbf{X} = \mathbf{P}\boldsymbol{\Delta}\mathbf{Q}'$, where **P** is orthonormal, then $\mathbf{XQ}\boldsymbol{\Delta}^{-1} = \mathbf{P}$ and $\mathbf{W} = \mathbf{Q}\boldsymbol{\Delta}^{-1}$ is a transformation matrix that spherizes **X**. Such a transformation matrix plays an important role since it represents the data in terms of an equal-variance independent basis.

Exercises

Consider the following two matrices:

$$\mathbf{A} = \begin{bmatrix} 1 & -3 & 1 & -1 \\ 1 & -1 & -1 & 3 \\ 1 & 1 & -1 & -3 \\ 1 & 3 & 1 & 1 \end{bmatrix}, \qquad \mathbf{X} = \begin{bmatrix} 1 & 3 & 4 & 4 \\ 2 & 6 & 7 & 8 \\ 3 & 8 & 9 & 9 \\ 5 & 6 & 9 & 10 \\ 6 & 10 & 9 & 9 \end{bmatrix}$$

1. Find the rank of the transformation matrix **A**.

2. Find the inverse of **A**.

3. What can you say in advance about the rank of **XA**?

4. Compute **XA**. What is the rank of **XA**?

5. Express **X′X** as a set of quadratic and bilinear forms in the columns of **X**.

6. Center **X** by columns. Express this result in terms of a centering matrix.

7. Find the mean vector and variance–covariance matrix for **X**.

8. Compute from the results of Exercise 7 and the matrix **A** the mean vector and variance–covariance matrix for the columns of **XA** = **Y**. How are they related to those for **X**?

9. Find an orthonormal basis for **A**.

10. Plot the first two columns of **X** one against the other.

11. Find the basic structure of the first two columns of $\mathbf{X}_{(2)} = \mathbf{pdq}'$.

12. Plot the two columns of $\mathbf{X}_{(2)}\mathbf{q}$ one against the other. How do these results compare with the plot of Exercise 10?

Additional Reading

For another look at the role of the basic structure of the transforming matrix in determining the effect of a linear transformation, the reader can consult

Green, P. E. (1976). *Mathematical Tools for Applied Multivariate Analysis*. New York: Academic.

Quadratic forms are taken up in many first level mathematical statistics texts. For a good mathematical introduction the following can be recommended:

Hogg, R. V., and Craig, A. T. (1970). *Introduction to Mathematical Statistics*, 3rd ed. New York: Macmillan.

Chapter 6
Optimizing Transformations

Introduction

In multivariate analysis we seek not just to transform the data matrix but to transform it optimally. What we mean by "optimal" will change with the goal of the analysis, but in all cases we can use the techniques of the differential calculus to help us locate that best transformation. In this chapter we review the elementary principles and results of differentiation and offer two extensions useful in developing techniques for analyzing the data matrix.

For the reader who has had an introductory calculus course, the first part of this chapter offers an opportunity for a quick brushup. We hope we have also provided enough detail that the reader with no calculus background will be able to understand how optimizing works, at least in the simplest cases.

A multivariate transformation often must satisfy more than one condition. For example, we might want a linear transformation of a set of predictors that would not only correlate as strongly as possible with a criterion but would, at the same time, be uncorrelated with some control attribute. We show how to build such a *side condition* into a function to be optimized through the use of Lagrangian multipliers. Although easily understood, the use of Lagrangian multipliers is not commonly taught in the first year of calculus.

The functions to be optimized in multivariate analysis are usually functions of several variables, the complete set of predictor attributes, for example. To facilitate the solution of such optimization problems, we present a set of rules for expressing symbolically the result of simultaneously differentiating with respect to all of the relevant variables. Use of these rules is a decided improvement over working out the partial derivative for each variable and then aggregating these results.

With this chapter we complete the presentation of the necessary matrix manipulation techniques, and in Chapter 7 we turn to using these techniques to answer specific research questions.

6.1 Multivariate Analyses and Optimizing Transformations

In the analysis of multivariate observations we seek not just to transform the data matrix but to find the *best* of a certain class of transformations. For example, in the Preemployment Study we might be concerned with how well job satisfaction can be predicted from two attributes, extent of work experience and amount of formal education. We could determine a linear transformation of the two predictors.

$$\text{Predicted(job satisfaction)}_i = a + b_1(\text{work experience})_i + b_2(\text{education})_i$$

choosing the constants a, b_1, and b_2 in the transformation so that the predicted values are close to the observed values. One commonly used criterion is to choose constants so as to minimize the sum of squares of discrepancies between the observed and predicted criterion values:

$$\sum_{i=1}^{N} [\text{Observed(job satisfaction)}_i - \text{Predicted(job satisfaction)}_i]^2$$

In Chapter 7 we show how constants can be selected to meet this least squares criterion.

Similarly, in principal components analysis a linear transformation of a set of attributes is sought such that the set of transformed observations has maximum variability. In the multivariate analysis of variance, estimators of population means and variances are sought that maximize the likelihood that the sample would have been obtained under different hypothesized conditions. Finally, in discriminant function analysis the linear combination of a set of attributes is sought that best discriminates among members of different groups.

In each instance, as will be clear in succeeding chapters, achievement of the best transformation can be understood in terms of some aspects of elementary differential calculus. In particular, we shall want to find the minimum or maximum value of a function, and we shall do so by evaluating and setting equal to zero derivatives or partial derivatives.

6.2 Elementary Derivatives, Maxima, and Minima

When we say, or write, that *y is a function of x*, we are expressing the notion that a variable y may take on different values, depending on the value of a variable x. The value of the *function* y is related to the value of the argument

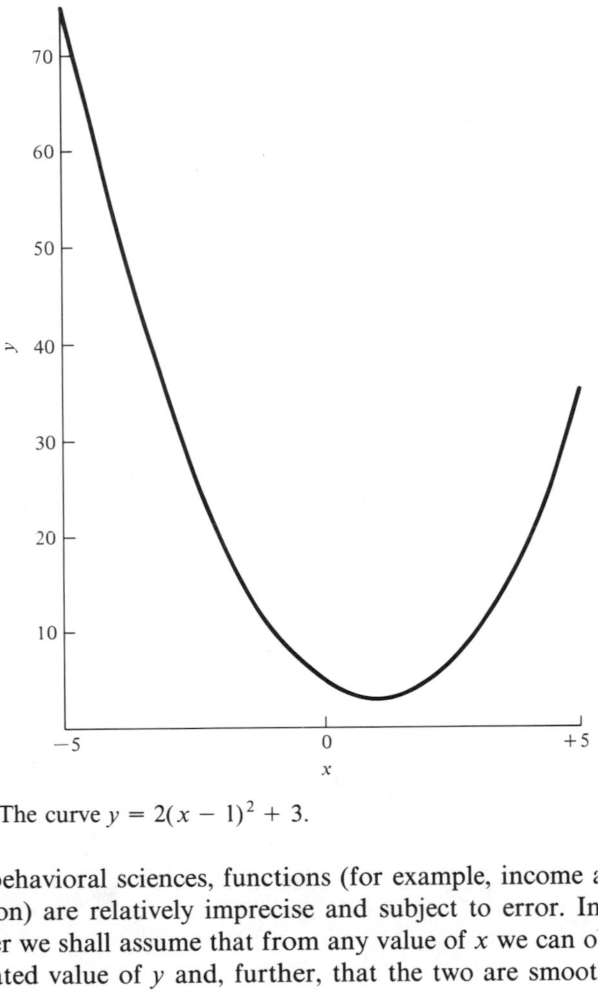

Figure 6.1 The curve $y = 2(x - 1)^2 + 3$.

x. In the behavioral sciences, functions (for example, income as a function of education) are relatively imprecise and subject to error. In contrast, in this chapter we shall assume that from any value of x we can obtain exactly the associated value of y and, further, that the two are smoothly, continuously related.

As an example, Figure 6.1 plots values of the function

$$y = f(x) = 2(x - 1)^2 + 3$$

against values of x; a few of the sample points are given in Table 6.1. The value of the function clearly changes over the range of arguments displayed. From the left, the function dips down until $x = 1$ and then rises. By inspection of the plot we can tell that, at least over the range of x from -5 to 5, this function is a minimum when $x = 1$. Differential calculus will permit us to reach that conclusion without having to plot the function against its argument.

Differential calculus is concerned with the *rate* at which a function y changes as its argument x changes. Consider some arbitrary value in

Table 6.1[a] $y = 2(x-1)^2 + 3$

x	-5	-4	-3	-2	-1	0	1	2	3	4	5
y	75	53	35	21	11	5	3	5	11	21	35

[a]See Figure 6.1.

the range of the argument and call it x_0. The value of the function when $x = x_0$ is

$$\begin{aligned} y_0 &= f(x_0) = 2(x_0 - 1)^2 + 3 \\ &= 2(x_0^2 - 2x_0 + 1) + 3 \\ &= 2x_0^2 - 4x_0 + 5 \end{aligned}$$

Consider the effect of taking x to be just *slightly* larger (more positive than x_0) by a small positive quantity Δx. We shall call this second value x_1. The value of y when $x_1 = x_0 + \Delta x$ is

$$\begin{aligned} y_1 &= 2(x_0 + \Delta x)^2 - 4(x_0 + \Delta x) + 5 \\ &= 2x_0^2 - 4x_0 + 5 + 2\,\Delta x^2 + 4x_0\Delta x - 4\,\Delta x \end{aligned}$$

The *change* in the value of the function as x changes from x_0 to $x_0 + \Delta x$ is given by

$$\Delta y = y_1 - y_0 = 2\,\Delta x^2 + 4x_0\Delta x - 4\,\Delta x$$

The *rate of change* of y is simply the ratio of the change in y to the change in x:

$$\begin{aligned} \frac{\Delta y}{\Delta x} &= \frac{2\,\Delta x^2 + 4x_0\Delta x - 4\,\Delta x}{\Delta x} \\ &= 2\,\Delta x + 4x_0 - 4 \end{aligned}$$

The *derivative* of $y = f(x)$, symbolized dy/dx and usually read "the derivative of y with respect to x," is defined as the value assumed by $\Delta y/\Delta x$ *in the limit as* Δx *becomes smaller and smaller*, approaching zero. That is, the derivative of a function $f(x)$ is the *instantaneous rate of change* in $f(x)$—the rate at which the function changes when the argument x changes only infinitesimally.

For the function described in Figure 6.1, $y = 2(x-1)^2 + 3$, we write

$$\begin{aligned} dy/dx &= \lim_{\Delta x \to 0} (\Delta y/\Delta x) \\ &= \lim_{\Delta x \to 0} (2\,\Delta x + 4x - 4) \\ &= 2(0) + 4x - 4 \\ &= 4x - 4 \end{aligned}$$

Frequently, as here, the derivative of a function of x is a second function of x. That is, dy/dx may itself assume different values, depending upon the value of x. The derivative dy/dx was defined as an instantaneous rate of change in $f(x)$. The value it assumes for any selected value of x can be interpreted as the slope of the straight line drawn tangent to the curve $y = f(x)$ at that selected value of x. For example, at the point where $x = 3$, the slope of the tangent to the curve $y = 2(x - 1)^2 + 3$ is 8—dy/dx evaluated for $x = 3$.

Notice that as y reaches its minimum value in the example the slope of the tangent to the curve—moving from left to right—changes from negative to positive. At the minimum, where $x = 1$, the slope of the tangent is actually zero; the tangent is a horizontal line. The instantaneous rate of change in y as a function of x is zero. If y reached a unique maximum value in the plotted range of the argument x; the slope of the tangent at the maximum would also be zero. The change in tangent slope for a maximum, however, would be from positive to negative as values of x increase.

We use the fact that the derivative is zero at an extremum to find the value of an argument that will optimize a function. If we wish, without examining a plot, to find the value of x that minimizes the function

$$y = f(x) = 2(x - 1)^2 + 3$$

we obtain the derivative, set the derivative equal to zero, and solve for x. That is, we ask for the value of x for which $dy/dx = 0$.

We find

$$dy/dx = 4x - 4 = 0$$

i.e., $4x = 4$, or $x = 1$. The result is consistent with our inspection of the plotted function, but how would we know without looking at a plot that the value of x satisfying the equation $dy/dx = 0$ provides a minimum value rather than a maximum value (or neither) for y?

A mathematical technique for distinguishing maxima and minima may be illustrated with a second algebraic function. In Figure 6.2 the function

$$y = x^3 - 3x^2 - 9x$$

has been plotted for a range of values of x displayed in Table 6.2. The function was chosen because it possesses both a local maximum and a local minimum. (We call them *local* extrema because the function reaches a maximum or minimum—or, as here, both—for values of x in the *interior* of a closed interval. Maxima or minima that are approached only at the limits of an interval—often as x approaches ∞ or $-\infty$—are seldom useful in optimizing functions in applied multivariate analysis.)

To find dy/dx we first compute

$$\begin{aligned} y + \Delta y = f(x + \Delta x) &= (x + \Delta x)^3 - 3(x + \Delta x)^2 - 9(x + \Delta x) \\ &= x^3 + 3x^2\Delta x + 3x\Delta x^2 + \Delta x^3 \\ &\quad - 3x^2 - 6x\Delta x - 3\Delta x^2 - 9x - 9\Delta x \end{aligned}$$

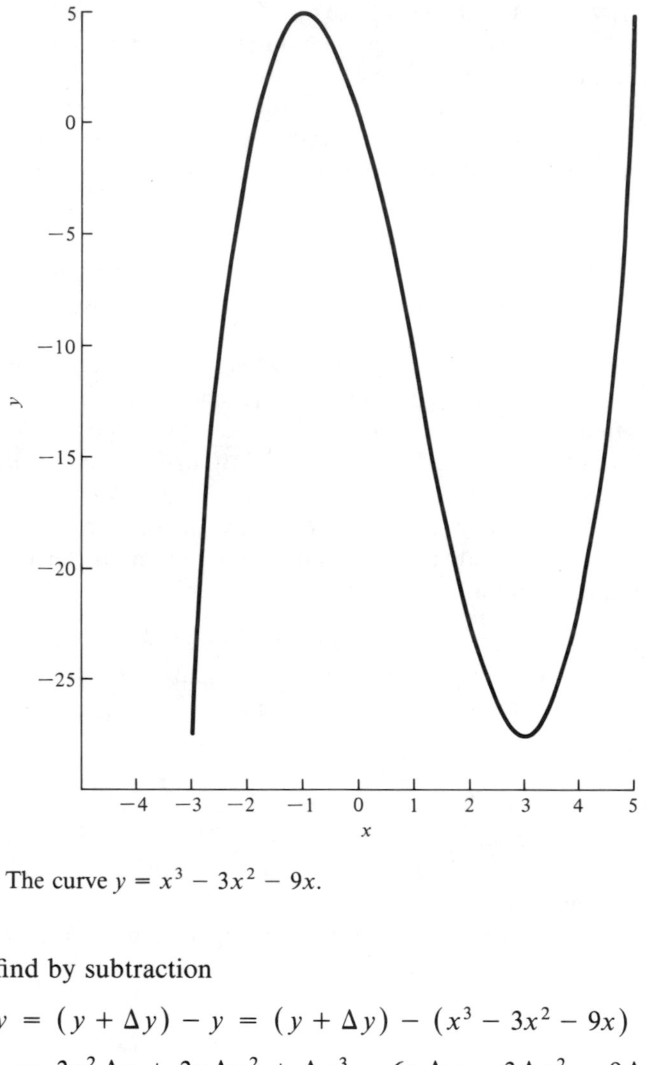

Figure 6.2 The curve $y = x^3 - 3x^2 - 9x$.

and then find by subtraction

$$\begin{aligned}\Delta y &= (y + \Delta y) - y = (y + \Delta y) - (x^3 - 3x^2 - 9x) \\ &= 3x^2 \Delta x + 3x \Delta x^2 + \Delta x^3 - 6x \Delta x - 3 \Delta x^2 - 9 \Delta x\end{aligned}$$

Dividing Δy by Δx yields

$$\Delta y / \Delta x = 3x^2 + 3x \Delta x + \Delta x^2 - 6x - 3 \Delta x - 9$$

The derivative dy/dx is the limit of $\Delta y / \Delta x$ as Δx becomes smaller and smaller, approaching zero:

$$dy/dx = 3x^2 - 6x - 9$$

the other terms becoming zero. Setting the derivative equal to zero gives $3x^2 - 6x - 9 = 0$ or $x^2 - 2x - 3 = 0$ or $(x - 3)(x + 1) = 0$ which has two solutions, $x = 3$ and $x = -1$.

SENIOR YEAR

Fall Semester	Credit Hrs.
Major Area	3
Electives	12
	15

Students must complete the Arts and S
level courses.

Three to six credits will be granted
if practicable.

*For B.A. in Law Enforcement, substit
Humanities in accordance with Plan II
The B.A. with Spanish as a language i
who intend to work in Law Enforcement
speaking people.

Nelson Denny
ACT

SS reading
NS reading

ultimate goal
is to use ACT
instead of ND

Table 6.2[a] $y = x^3 - 3x^2 - 9x$

x	-3	-2	-1	0	1	2	3	4	5
y	-27	-2	5	0	-11	-22	-27	-20	5
dy/dx	36	15	0	-9	-12	-9	0	15	36

[a]See Figure 6.2.

We see again that the values of x that satisfy $dy/dx = 0$ are those that in a plot yield a local maximum (when $x = -1$) or local minimum (when $x = 3$) for y.

Values of dy/dx are included in Table 6.2. Note that dy/dx, the slope of the tangent to the function, changes *from positive to negative* as a *maximum* value of y is achieved and *from negative to positive* as a *minimum* value of y is reached. The shift in sign of dy/dx can be used as a test of whether a minimum or a maximum value to y has been identified. This requires solving for the value of dy/dx for values of x adjacent to those that satisfy $dy/dx = 0$.

This interpretation of the *change* in dy/dx suggests an alternative test, based on the second derivative. The expression dy/dx is the *first derivative* of y with respect to x. But dy/dx is itself a function of x. If we name this latter function

$$u = dy/dx = 3x^2 - 6x - 9$$

we can ask questions just as of the function $y = x^3 - 3x^2 - 9x$. In particular, we can investigate the rate of change in u as a function of x.

Following the usual strategy,

$$\begin{aligned} u + \Delta u &= 3(x + \Delta x)^2 - 6(x + \Delta x) - 9 \\ &= 3x^2 + 6x\,\Delta x + 3\,\Delta x^2 - 6x - 6\,\Delta x - 9 \\ \Delta u &= (u + \Delta u) - u = 6x\,\Delta x + 3\,\Delta x^2 - 6\,\Delta x \\ \Delta u/\Delta x &= 6x + 3\,\Delta x - 6 \end{aligned}$$

The derivative of u with respect to x, du/dx, is the limit of $\Delta u/\Delta x$ as Δx approaches zero, or

$$du/dx = 6x - 6$$

Since u is itself a derivative, $u = dy/dx$, we have taken the derivative of a derivative: $du/dx = d(dy/dx)/dx$. This is referred to as a *second derivative*. We have differentiated y twice (successively) with respect to x,

$$du/dx = d(dy/dx)/dx = d^2y/dx^2$$

and we have on the far right the customary notation.

For the function $y = x^3 - 3x^2 - 9x$ the first derivative (with respect to x) is $dy/dx = 3x^2 - 6x - 9$ and the second derivative (also with respect to

x) is $d^2y/dx^2 = 6x - 6$. Similarly, for function Figure 6.1, $y = 2(x - 1)^2 + 3$, the first derivative is $dy/dx = 4x - 4$ and the *second* derivative

$$d^2y/dx^2 = d(4x - 4)/dx = 4$$

(The reader may verify this by starting with the function $u = 4x - 4$, then computing $u + \Delta u$, Δu, $\Delta u/\Delta x$, and finally obtaining du/dx or d^2y/dx^2 as the limit of this ratio as Δx approaches zero.)

The importance of the second derivative in determining extrema x_e of a function is made clear by the following:

Theorem. *Let $dy/dx = 0$ when $x = x_e$.*

a. *If d^2y/dx^2 is positive when $x = x_e$, then y reaches a local minimum when $x = x_e$.*
b. *If d^2y/dx^2 is negative when $x = x_e$, then y reaches a local maximum when $x = x_e$.*
c. *If d^2y/dx^2 is zero when $x = x_e$, then y reaches what is called a stationary point, neither a maximum nor a minimum when $x = x_e$.*

For the first of our illustrations, $y = 2(x - 1)^2 + 3$, $dy/dx = 0$ for $x = 1$. The value of the second derivative, $d^2y/dx^2 = 4$, is *positive* for $x = 1$. As a result,

$$y = 2(x - 1)^2 + 3$$

is a *minimum* when $x = 1$.

For the second illustration, consider $y = x^3 - 3x^2 - 9x$; $dy/dx = 0$ for $x = -1$ and for $x = 3$. The value of the second derivative, $d^2y/dx^2 = 6(x - 1)$, is *negative* for $x = -1$ and *positive* for $x = 3$. Thus

$$y = x^3 - 3x^2 - 9x$$

is a *maximum* when $x = -1$ and a *minimum* when $x = 3$.

6.3 Rules of Differentiation

Having to find $y + \Delta y$, Δy, $\Delta y/\Delta x$, and the limit of this as Δx goes to zero for each $y = f(x)$ encountered would be quite a chore. Fortunately, there are rules that permit us to write down, more or less directly, dy/dx for most functions of x. Although we shall not attempt to prove any of these rules, we summarize the more useful of them here.

6.3.1 Algebraic Functions

If y is an algebraic function of the form

$$y = ax^k$$

for constants a and k, then dy/dx can be written

$$dy/dx = kax^{k-1}$$

Examples are

$$u = 3x^2, \qquad du/dx = (2)(3)x^{2-1} = 6x$$

$$v = 1/x = x^{-1}, \qquad dv/dx = (-1)x^{-1-1} = -x^{-2} = -1/x^2$$

$$w = \sqrt{x} = x^{1/2}, \qquad dw/dx = \tfrac{1}{2}x^{(1/2)-1} = \tfrac{1}{2}x^{-1/2} = \tfrac{1}{2}(1/\sqrt{x})$$

$$y = 12 = (12)(1) = 12x^0, \qquad dy/dx - 0x^{0-1} = 0/x - 0$$

The last example justifies the statement that the *derivative of a constant is zero.*

6.3.2 Exponential and Logarithmic Functions

In a second major class of functions encountered in multivariate analyses x appears in an *exponent*:

$$y = ab^u \qquad \text{where} \quad u = f(x)$$

or in the argument of a *logarithm*,

$$w = c \log v, \qquad \text{where} \quad v = f(x)$$

Here we take log to be the *natural logarithm*: If $g = \log h$, then $h = e^g$, where e is the Napierian constant, 2.71828. . . .

The derivative of an exponential function $y = ab^u$ is written

$$\frac{dy}{dx} = a \log(b)\, b^u \left(\frac{du}{dx}\right)$$

Examples are

$$u = e^x, \qquad du/dx = 1 \log(e)\, e^x(1) = e^x$$

since $\log e = 1$;

$$v = (3)2^{x^2}, \qquad \frac{dv}{dx} = 3 \log(2)\, 2^{x^2} \frac{d(x^2)}{dx}$$

$$= 3 \log(2)\, 2^{x^2} 2x$$

$$= (6x)2^{x^2} \log 2$$

$$w = ke^{-(1/2)x}, \qquad dw/dx = ke^{-(1/2)x}\left(-\tfrac{1}{2}\right)$$

$$= -\tfrac{1}{2}ke^{-(1/2)x}$$

The derivative of a logarithmic function $y = c \log v$ is

$$\frac{dy}{dx} = c\frac{1}{v}\frac{dv}{dx}$$

Examples are

$$u = \log x, \qquad du/dx = 1/x$$

$$v = \log x^2, \qquad dv/dx = (1/x^2)(2x) = 2/x$$

$$w = \log\sqrt{x} = \log x^{1/2}, \qquad dw/dx = (1/x^{1/2})\tfrac{1}{2}x^{-1/2}$$

$$= \tfrac{1}{2}x^{-1/2}x^{-1/2}$$

$$= \tfrac{1}{2}x^{-1} = 1/2x$$

6.3.3 *Functions of Functions*

Most often the functions requiring differentiation are made up of more than one term involving the variable of differentiation x. We shall present here four rules useful in differentiating complex functions.

If the function to be differentiated is the *sum* of two (or more) functions, $y = u + v$, u and v each a function of x, then the derivative is a sum of derivatives:

$$dy/dx = du/dx + dv/dx$$

Thus for Figure 6.2, $y = x^3 - 3x^2 - 9x$, the first derivative is

$$dy/dx = d(x^3)/dx - d(3x^2)/dx - d(9x)/dx = 3x^2 - 6x - 9$$

and the second derivative is

$$\begin{aligned} d^2y/dx^2 &= d(3x^2)/dx - d(6x)/dx - d(9)/dx = 6x - 6 - 0 \\ &= 6x - 6 \end{aligned}$$

If the function to be differentiated is the *product* of two functions, $y = uv$ with u and v each a function of x, then the derivative is

$$dy/dx = v(du/dx) + u(dv/dx)$$

To illustrate, if $y = (x^2 - 2)(x + 3)$ then

$$\begin{aligned} dy/dx &= (x + 3)(2x) + (x^2 - 2)(1) = 2x^2 + 6x + x^2 - 2 \\ &= 3x^2 + 6x - 2 \end{aligned}$$

The reader may verify this outcome by multiplying the two terms and differentiating the result using the sum rule.

If the function to be differentiated is the *ratio* of two functions, $y = u/v$, u and v each a function of x, then the derivative is

$$\frac{dy}{dx} = \frac{1}{v^2}\left(v\frac{du}{dx} - u\frac{dv}{dx}\right)$$

As a simple example, consider $y = x^4/x^2$;

$$dy/dx = (1/x^4)[x^2(4x^3) - x^4(2x)] = (1/x^4)(2x^5) = 2x$$

As a check, we could express $y = x^4/x^2 = x^2$ and apply the algebraic differentiation rule.

If the function to be differentiated is a *composite* function, $y = f(u)$, where u is a function of x, the derivative is

$$\frac{dy}{dx} = \frac{dy}{du}\frac{du}{dx}$$

The function of Figure 6.1 provides an example. If, in $y = 2(x - 1)^2 + 3$ we let $u = (x - 1)$, y may be rewritten as a function of u:

$$y = 2u^2 + 3$$

The derivative of y with respect to u is $dy/du = 4u$, while the derivative of u with respect to x is $du/dx = 1$. Putting the two results together gives

$$dy/dx = (dy/du)(du/dx) = (4u)(1) = 4(x - 1) = 4x - 4$$

the result already obtained.

6.4 Partial Differentiation

In multivariate analyses the functions to be optimized are usually functions of more than one variable. We must therefore find the maximum or the minimum of a function with respect to two or more arguments.

As an example, consider a goal in least squares prediction, a topic more fully developed in Chapter 7. For each of n entities we have both a predictor observation x_i and a criterion observation y_i, and we wish to develop a linear prediction equation,

$$\hat{y}_i = a + bx_i$$

choosing the values for *both* a and b so that the sum of the squares of the discrepancies between the predicted and observed values of the criterion,

$$\sum_{i=1}^{n} e_i^2 = \sum_{i=1}^{n} (y_i - \hat{y}_i)^2 = \sum_{i=1}^{n} (y_i - a - bx_i)^2$$

is a minimum.

By expanding the term on the far right the function to be minimized can be written

$$\sum_i e_i^2 = \sum_i y_i^2 - 2a\sum_i y_i - 2b\sum_i x_i y_i + na^2 + 2ab\sum_i x_i + b^2\sum_i x_i^2 \tag{6.4.1}$$

For a *given* set of x and y observations $\Sigma_i e_i^2$ can take different values depending on the values of a and b. In other words, $\Sigma_i e_i^2$ is a function of a and of b.

Optimizing a function with two or more arguments is accomplished in essentially the same fashion as optimizing a function with a single argument. The difference is that we must differentiate the function separately and successively with respect to each of the arguments, set each of the resulting expressions equal to zero, and then solve that set of simultaneous equations for the several unknowns. The derivative of a function y with respect to one of several arguments, say, x, is called a *partial derivative* of y with respect to x, symbolized $\partial y/\partial x$.

For the least squares example the two partial derivatives are

$$\partial\left(\sum_i e_i^2\right)\Big/\partial a = -2\sum_i y_i + 2na + 2b\sum_i x_i \tag{6.4.2}$$

and

$$\partial\left(\sum_i e_i^2\right) \Big/ \partial b = -2\sum_i x_i y_i + 2a\sum_i x_i + 2b\sum_i x_i^2 \qquad (6.4.3)$$

Setting (6.4.2) equal to zero gives

$$0 = -\sum_i y_i + na + b\sum_i x_i$$

a can be expressed as a function of b:

$$a = \frac{1}{n}\left(\sum_i y_i - b\sum_i x_i\right) \qquad (6.4.4)$$

Substituting the solution for a into (6.4.3) and setting the resulting expression for $\partial(\Sigma_i e_i^2)/\partial b$ equal to zero yields

$$0 = -\sum_i x_i y_i + \frac{1}{n}\left(\sum_i y_i - b\sum_i x_i\right)\sum_i x_i + b\sum_i x_i^2$$

or

$$b\sum_i x_i^2 - \frac{1}{n}b\left(\sum_i x_i\right)^2 = \sum_i x_i y_i - \frac{1}{n}\sum_i x_i \sum_i y_i$$

so that

$$b = \left(\sum_i x_i y_i - \frac{1}{n}\sum_i x_i \sum y_i\right) \Big/ \left[\sum_i x_i^2 - \frac{1}{n}\left(\sum_i x_i\right)^2\right] \qquad (6.4.5)$$

Although we shall not interpret the optimizing solutions here, a and b from (6.4.4) and (6.4.5) may more succinctly be expressed

$$a = \bar{y} - b\bar{x} \qquad (6.4.6)$$

where

$$b = \mathrm{Cov}(x, y)/\mathrm{Var}(x) \qquad (6.4.7)$$

Whether a minimum, maximum, or neither has been reached in the simultaneous solution of equations in which each partial derivative has been set equal to zero depends, by extension of simple differentiation, on the second-order partial derivatives. Because these include *mixed* derivatives—i.e., terms of the form $d(dy/dx)/dz = d^2y/dx\,dz$—as well as the results of differentiating twice with respect to the same argument (for example, $d(dy/dx)/dx = d^2y/dx^2$), the rules for deciding between minima, maxima, and other stationary points are complex. We shall not attempt to include them in this review.

The reader usually does not need to apply the rules about second derivatives to understand the optimization problems in the remainder of this text. In each instance, the desired extremum is the only one locally defined. For instance, in the least squares problem just presented, it is fairly obvious that a maximum for $\Sigma_i e_i^2$ is approached only as a and b become very large.

6.5 Side Conditions in Optimizing Transformations

As noted in Section 6.4, some optimizing transformations may not be definable because they are unbounded. This is frequently the case when we want some function to take the largest possible value. By imposing one or more additional restrictions on the solution, however, a local maximum may be found.

As an example, we consider an optimizing problem closely associated with principal components, a topic to be developed in Chapter 14. We have observations on two attributes, say, verbal aptitude X and quantitative aptitude Y. A new variable Z is formed as a linear combination of the two scores,

$$Z_i = b_X X_i + b_Y Y_i$$

and we want to select the weights b_X and b_Y, so that this new variable has maximum variability.

We know that the variance of this linear transformation can be expressed in terms of the variances of X and Y and the covariance between the two:

$$\mathrm{Var}(Z) = b_X^2 \mathrm{Var}(X) + b_Y^2 \mathrm{Var}(Y) + 2 b_X b_Y \mathrm{Cov}(X, Y)$$

If $\mathrm{Cov}(X, Y)$ is nonnegative, $\mathrm{Var}(Z)$ becomes larger and larger as we allow b_X and b_Y to assume larger and larger positive values. There is no local maximum for $\mathrm{Var}(Z)$.

We can find a local maximum, however, for a function very closely related to $\mathrm{Var}(Z)$. First, notice that for any set of positive weights $Z = b_X X + b_Y Y$, multiplying each by a positive constant k defines a second new variable

$$W = (k b_X) X + (k b_Y) Y$$

such that

$$\mathrm{Var}(W) = k^2 \mathrm{Var}(Z)$$

In other words, for any choice of b_X and b_Y, we can make $\mathrm{Var}(W)$ as large as we like by the appropriate choice of k. That is not a very interesting result. More interesting is whether there is a unique choice of b_X and b_Y that is better than all others.

How can we restrict the candidates for this initial set? We have seen already it will be of little value to consider all possible initial sets. For any set of weights we can define a positive constant c such that

$$(c b_X)^2 + (c b_Y)^2 = 1$$

All we need to do is to define c as

$$c = \sqrt{b_X^2 + b_Y^2}$$

Conversely, any set of weights can be expressed as a scaling of weights with sum of squares equal to unity:

$$\begin{bmatrix} b_X \\ b_Y \end{bmatrix} = \frac{1}{c}\begin{bmatrix} a_X \\ a_Y \end{bmatrix}, \qquad a_X^2 + a_Y^2 = 1$$

We can, therefore, restrict ourselves to sets of weights with sum of squares equal to 1. We then try to maximize

$$\operatorname{Var}(Z) = b_X^2 \operatorname{Var}(X) + b_Y^2 \operatorname{Var}(Y) + 2b_X b_Y \operatorname{Cov}(X, Y) \tag{6.5.1}$$

subject to the restriction

$$b_X^2 + b_Y^2 = 1 \tag{6.5.2}$$

We seek an extremum of a function under a *side condition*. To solve this problem by differential calculus we use *Lagrangian* or *undefined multipliers*. In essence, the function to be optimized is replaced with one that incorporates all side conditions. This new function consists of the original one plus an additional term for each side condition of the form

$$-\lambda G$$

where λ is the Lagrangian multiplier and the side condition has been written $G = 0$. In the present example,

$$G = b_X^2 + b_Y^2 - 1 \tag{6.5.3}$$

since the desired side condition is $b_X^2 + b_Y^2 = 1$.

The function to be maximized is written

$$\begin{aligned} F(b_X, b_Y) &= \operatorname{Var}(Z) - \lambda G \\ &= b_X^2 \operatorname{Var}(X) + b_Y^2 \operatorname{Var}(Y) + 2b_X b_Y \operatorname{Cov}(X, Y) \\ &\quad -\lambda\left(b_X^2 + b_Y^2 - 1\right) \end{aligned} \tag{6.5.4}$$

One obtains a local extremum of this function by differentiating with respect to each of the unknowns, *including* the Lagrangian multiplier, and setting the resulting partial derivatives equal to zero:

$$\partial F/\partial b_X = 2b_X \operatorname{Var}(X) + 2b_Y \operatorname{Cov}(X, Y) - 2b_X\lambda = 0 \tag{6.5.5}$$

$$\partial F/\partial b_Y = 2b_Y \operatorname{Var}(Y) + 2b_X \operatorname{Cov}(X, Y) - 2b_Y\lambda = 0 \tag{6.5.6}$$

$$\partial F/\partial \lambda = 1 - \left(b_X^2 + b_Y^2\right) = 0 \tag{6.5.7}$$

We now see why the side condition is included in the composite function: Equations (6.5.5)–(6.5.7) ensure that $b_X^2 + b_Y^2 = 1$ when F is at its maximum.

Solving the set of three simultaneous equations in three unknowns is somewhat taxing. It is easier to approach the problem through matrix notation. Equations (6.5.5) and (6.5.6) can be written together

$$\begin{bmatrix} \text{Var}(X) & \text{Cov}(X, Y) \\ \text{Cov}(X, Y) & \text{Var}(Y) \end{bmatrix} \begin{bmatrix} b_X \\ b_Y \end{bmatrix} = \lambda \begin{bmatrix} b_X \\ b_Y \end{bmatrix} \quad \text{or} \quad \mathbf{Sb} = \lambda\mathbf{b} \tag{6.5.8}$$

while (6.5.7) becomes

$$\mathbf{b'b} = 1 \tag{6.5.9}$$

Premultiplying (6.5.8) by $\mathbf{b}'$ gives

$$\mathbf{b'Sb} = \lambda\mathbf{b'b} \tag{6.5.10}$$

Since $\mathbf{b'b} = 1$ and, by scalar equation (6.5.1), $\mathbf{b'Sb} = \text{Var}(Z)$, the Lagrangian multiplier is just

$$\lambda = \text{Var}(Z) \tag{6.5.11}$$

the quantity we seek to maximize.

Equation (6.5.8) is of a form encountered throughout the second half of this text. A symmetric matrix (a product moment matrix) is postmultiplied by a vector and the resulting vector is *proportional* to the postfactor. Equations of this form are satisfied if we represent the product moment in basic structure form and take the postmultiplying vector to be a column of the (left) orthonormal to that basic structure.

The basic structure of $\mathbf{S}$ can be expressed

$$\mathbf{S} = \mathbf{Q\Delta}^2\mathbf{Q}' \tag{6.5.12}$$

With $\mathbf{b} = \mathbf{Q}_{\cdot i}$ the product $\mathbf{Sb}$ becomes $\mathbf{Sb} = \mathbf{Q\Delta}^2\mathbf{Q'Q}_{\cdot i}$. The vector $\mathbf{Q}_{\cdot i}$ is orthogonal to all columns of $\mathbf{Q}$ (other than the ith), $\mathbf{Q'Q}_{\cdot i} = \mathbf{e}_i$, and we have

$$\mathbf{Sb} = \mathbf{Q\Delta}^2\mathbf{e}_i$$

The product $\mathbf{\Delta}^2\mathbf{e}_i$ is a vector that is null except in the ith position and can be written $\mathbf{\Delta}^2\mathbf{e}_i = \Delta_i^2\mathbf{e}_i$, so $\mathbf{Sb}$ can be expressed $\mathbf{Sb} = \Delta_i^2\mathbf{Qe}_i$, but postmultiplying $\mathbf{Q}$ by $\mathbf{e}_i$ yields the ith column of $\mathbf{Q}$: $\mathbf{Sb} = \Delta_i^2\mathbf{Q}_{\cdot i}$. Remembering that $\mathbf{b} = \mathbf{Q}_{\cdot i}$, we obtain

$$\mathbf{SQ}_{\cdot i} = \Delta_i^2\mathbf{Q}_{\cdot i} \tag{6.5.13}$$

the desired form.

Comparing (6.5.13) with (6.5.8), we see that Δ_i^2 is the constant of proportionality—and thus the variance λ of the linear transformation. Equation (6.5.13) does not provide a unique solution. There are as many solutions as there are columns in $\mathbf{Q}$. (The present example assumed $\mathbf{S}$ to be 2×2 and thus to have rank no greater than 2. The method of analysis outlined here generalizes, however, to the linear transformation of k attributes.) However, the elements of the basic diagonal are arranged in *descending* order of magnitude, so that $\Delta_1^2 \geqslant \Delta_2^2$. One of the solutions, therefore, provides the required maximum: $\text{Var}(Z) = \lambda = \Delta_1^2$. For $\mathbf{b} = \mathbf{Q}_{\cdot 1}$, then, the transformed variable has maximum variability, namely, Δ_1^2.

In this example a side condition was introduced to provide for a closed solution to an optimizing problem. In Chapters 7 and 8 a second use of side conditions will be critical. While unconstrained or full model maximum likelihood estimators of parameters of the linear regression model are realizable and play a role in applications of the model, it is also important to estimate these same parameters subject to one or more linear constraints. This will be done by incorporating the linear constraints as side conditions to the expression to be optimized.

6.6 Matrix Differentiation and Optimization

In Section 6.5 a set of partial derivatives was collected in a matrix (vector) as an aid to finding a solution to the equations that resulted from differentiation. The several partial derivatives, however, were obtained one at a time. In this section we show how to differentiate an entire matrix, a collection of functions, and how to form a derivative with respect to an *array* (a vector or matrix) of arguments. Matrix differentiation, sometimes called symbolic matrix differentiation, is simply a set of rules for writing out in an orderly fashion a series of ordinary single-function single-argument derivatives. This can both simplify the development of multivariate analysis and make the results easier to interpret.

6.6.1 *Differentiation of a Matrix with Respect to a Scalar*

Consider a matrix, say,

$$\mathbf{X} = \begin{bmatrix} y^2 & my + c & 1/y \\ 5 & 6y & y^2 + 7 \end{bmatrix}$$

each of whose elements can be regarded as a function of the scalar y.

Definition. The *derivative of a matrix with respect to a scalar* is a second matrix, of the same order as the first, with elements obtained by differentiating each of the corresponding elements of the original matrix.

Thus in our example,

$$\begin{aligned} d\mathbf{X}/dy &= \begin{bmatrix} d(y^2)/dy & d(my + c)/dy & d(1/y)/dy \\ d5/dy & d(6y)/dy & d(y^2 + 7)/dy \end{bmatrix} \\ &= \begin{bmatrix} 2y & m & -1/y^2 \\ 0 & 6 & 2y \end{bmatrix} \end{aligned}$$

Differentiation might be employed to minimize or maximize simultaneously several different functions of the same unknown parameter y.

6.6.2 Differentiation of a Scalar with Respect to a Matrix

Now we consider a scalar quantity y as a function of the several elements of a matrix $\mathbf{X}$. For example, let

$$y = u^2 v + w/t = f(\mathbf{X}) \qquad \text{where} \quad \mathbf{X} = \begin{bmatrix} u & v \\ w & t \end{bmatrix}$$

Definition. The *derivative of a scalar with respect to a matrix* is a second matrix, of the same order as the first, with elements obtained by differentiating the scalar with respect to each of the elements of the original matrix.

For the example,

$$dy/d\mathbf{X} = \begin{bmatrix} dy/dX_{11} & dy/dX_{12} \\ dy/dX_{21} & dy/dX_{22} \end{bmatrix} = \begin{bmatrix} 2uv & u^2 \\ 1/t & -w/t^2 \end{bmatrix}$$

This is the kind of matrix differentiation that we encounter most frequently —optimizing a single function of several unknown variables.

6.6.3 Differentiation of a Matrix with Respect to a Matrix

Suppose

$$\mathbf{X} = \begin{bmatrix} X_{11} & X_{12} \\ X_{21} & X_{22} \end{bmatrix}$$

is related to a second matrix $\mathbf{Y}$ as $\mathbf{AX} = \mathbf{Y}$, say,

$$\begin{bmatrix} 1 & 2 \\ 3 & 0 \\ 5 & -1 \end{bmatrix} \begin{bmatrix} X_{11} & X_{12} \\ X_{21} & X_{11} \end{bmatrix} = \begin{bmatrix} X_{11} + 2X_{21} & X_{12} + 2X_{22} \\ 3X_{11} & 3X_{12} \\ 5X_{11} - X_{21} & 5X_{12} - X_{22} \end{bmatrix}$$

The matrix $\mathbf{Y}$ is a function of the matrix $\mathbf{X}$. We may also think of each element of $\mathbf{Y}$ as a function of the elements of $\mathbf{X}$; the matrix $\mathbf{Y}$, then, represents *several* functions, each of *several* variables. Differentiating each of the $2 \times 3 = 6$ functions in $\mathbf{Y}$ with respect to each of the $2 \times 2 = 4$ variables in $\mathbf{X}$ would produce 24 ordinary derivatives.

The result of differentiating a matrix with respect to another matrix is not a simple matrix; there is no natural way to arrange the 24 derivatives in a rectangular array. One gives instead a series of matrices. There are two ways in which this series of matrices may be developed, based, respectively, on Sections 6.6.1 and 6.6.2.

1. The matrix **Y** may be differentiated with respect to each of the scalar elements in **X**, producing four matrices each the size of **Y**:

$$d\mathbf{Y}/dX_{11} = \begin{bmatrix} 1 & 0 \\ 3 & 0 \\ 5 & 0 \end{bmatrix}, \qquad d\mathbf{Y}/dX_{12} = \begin{bmatrix} 0 & 1 \\ 0 & 3 \\ 0 & 5 \end{bmatrix}$$

$$d\mathbf{Y}/dX_{21} = \begin{bmatrix} 2 & 0 \\ 0 & 0 \\ -1 & 0 \end{bmatrix}, \qquad d\mathbf{Y}/dX_{22} = \begin{bmatrix} 0 & 2 \\ 0 & 0 \\ 0 & -1 \end{bmatrix}$$

2. Each of the scalar elements in **Y** may be differentiated with respect to the matrix **X**, producing six matrices each the size of **X**:

$$dY_{11}/d\mathbf{X} = \begin{bmatrix} 1 & 0 \\ 2 & 0 \end{bmatrix}, \qquad dY_{12}/d\mathbf{X} = \begin{bmatrix} 0 & 1 \\ 0 & 2 \end{bmatrix}$$

$$dY_{21}/d\mathbf{X} = \begin{bmatrix} 3 & 0 \\ 0 & 0 \end{bmatrix}, \qquad dY_{22}/d\mathbf{X} = \begin{bmatrix} 0 & 3 \\ 0 & 0 \end{bmatrix}$$

$$dY_{31}/d\mathbf{X} = \begin{bmatrix} 5 & 0 \\ -1 & 0 \end{bmatrix}, \qquad dY_{32}/d\mathbf{X} = \begin{bmatrix} 0 & 5 \\ 0 & -1 \end{bmatrix}$$

The same 24 scalar elements are distributed over the two sets of matrices. We shall need to distinguish, then, in differentiating one matrix with respect to another, which arrangement of the functions is required. In multivariate analysis the context will make it clear whether we need to take the function or the argument to be a *scalar*.

6.6.4 Symbolic Derivatives

So far all of the derivatives in this chapter, including matrix derivatives, have been produced by brute force. We have applied the rules of scalar differentiation element by element, until all the elements of one, or perhaps two, matrices were exhausted. We have simply presented some rules for organizing the resulting scalar derivatives. Matrix algebra can aid further, permitting the functional relation between **X** and **Y** to be written as a matrix equation. The reader may have noted in the example involving differentiating $\mathbf{Y} = \mathbf{AX}$ that the result consisted of matrices made up of rows or columns of the matrix **A** augmented by null vectors. This result will now be expressed systematically.

Recall that the elementary matrix $\mathbf{E}_{ij}$ is everywhere null except at the intersection of the ith row and jth column. The ijth element has the value 1. The importance of the elementary matrix is that it operates to "pull out" a row or column from the matrix by which it is multiplied and to augment that row or column with null vectors.

With elementary matrices the matrix by matrix derivatives of $\mathbf{Y} = \mathbf{AX}$ can be written

$$dY/dX_{ij} = \mathbf{AE}_{ij}$$

for the matrix by scalar form and

$$dY_{ij}/d\mathbf{X} = \mathbf{A'E}_{ij}$$

for the scalar by matrix form. We give an example of each:

$$d\mathbf{Y}/dX_{21} = \mathbf{AE}_{21} = \begin{bmatrix} 1 & 2 \\ 3 & 0 \\ 5 & -1 \end{bmatrix} \begin{bmatrix} 0 & 0 \\ 1 & 0 \end{bmatrix} = \begin{bmatrix} 2 & 0 \\ 0 & 0 \\ -1 & 0 \end{bmatrix}$$

$$dY_{12}/d\mathbf{X} = \mathbf{A'E}_{12} = \begin{bmatrix} 1 & 3 & 5 \\ 2 & 0 & -1 \end{bmatrix} \begin{bmatrix} 0 & 1 \\ 0 & 0 \\ 0 & 0 \end{bmatrix} = \begin{bmatrix} 0 & 1 \\ 0 & 2 \end{bmatrix}$$

which agree with direct scalar differentiation. We offer no proof of these rules, which we have adapted, together with the other differentiation rules presented in Table 6.3, from the pioneering work of Dwyer and MacPhail cited at the end of this chapter.

Suppose, however, that $\mathbf{Y}$ is not of the simple form $\mathbf{AX}$. For example let the function $\mathbf{Y}$ be the minor product moment of $\mathbf{X}_{(3\times 2)}$. The elements of $\mathbf{Y}$ are

$$\begin{bmatrix} X_{11} & X_{21} & X_{31} \\ X_{12} & X_{22} & X_{32} \end{bmatrix} \begin{bmatrix} X_{11} & X_{12} \\ X_{21} & X_{22} \\ X_{31} & X_{32} \end{bmatrix}$$

$$= \begin{bmatrix} X_{11}^2 + X_{21}^2 + X_{31}^2 & X_{11}X_{12} + X_{21}X_{22} + X_{31}X_{32} \\ X_{12}X_{11} + X_{22}X_{21} + X_{32}X_{31} & X_{12}^2 + X_{22}^2 + X_{32}^2 \end{bmatrix}$$

As an example of matrix-by-scalar differentiation we can write

$$d\mathbf{Y}/dX_{31} = \begin{bmatrix} 2X_{31} & X_{32} \\ X_{32} & 0 \end{bmatrix}$$

Note that it is the same order as $\mathbf{Y}$. Similarly, $dY_{12}/d\mathbf{X}$ is one of the scalar-by-matrix derivatives:

$$dY_{12}/d\mathbf{X} = \begin{bmatrix} X_{12} & X_{11} \\ X_{22} & X_{21} \\ X_{32} & X_{31} \end{bmatrix}$$

The derivative with respect to $\mathbf{X}$ is the same order as $\mathbf{X}$.

From Table 6.3, the symbolic rule for matrix-by-scalar differentiation in this instance is

$$d(\mathbf{X'X})/dX_{ij} = (\mathbf{E}_{ij})'\mathbf{X} + \mathbf{X'E}_{ij}$$

Table 6.3 Table of Matrix Derivatives[a]

$\mathbf{Y}$	$d\mathbf{Y}/dX_{ij}$	$dY_{ij}/d\mathbf{X}$	$dy/d\mathbf{X}$	$d(\mathrm{tr}\,\mathbf{Y})/d\mathbf{X}$
$\mathbf{X}$	$\mathbf{E}_{ij}$	$\mathbf{E}_{ij}$	$\mathbf{1}$	$\mathbf{I}$
$\mathbf{AB}$	$\mathbf{0}$	$\mathbf{0}$	$\mathbf{0}$	$\mathbf{0}$
$\mathbf{AX}$	$\mathbf{AE}_{ij}$	$\mathbf{A'E}_{ij}$	$\mathbf{A'}$	$\mathbf{A'}$
$\mathbf{XB}$	$\mathbf{E}_{ij}\mathbf{B}$	$\mathbf{E}_{ij}\mathbf{B'}$	$\mathbf{B'}$	$\mathbf{B'}$
$\mathbf{AX'}$	$\mathbf{A}(\mathbf{E}_{ij})'$	$(\mathbf{E}_{ij})'\mathbf{A}$	$\mathbf{A}$	$\mathbf{A}$
$\mathbf{X'B}$	$(\mathbf{E}_{ij})'\mathbf{B}$	$\mathbf{B}(\mathbf{E}_{ij})'$	$\mathbf{B}$	$\mathbf{B}$
$\mathbf{X'X}$	$(\mathbf{E}_{ij})'\mathbf{X} + \mathbf{X'E}_{ij}$	$\mathbf{X}(\mathbf{E}_{ij})' + \mathbf{XE}_{ij}$	$2\mathbf{X}$	$2\mathbf{X}$
$\mathbf{XX'}$	$\mathbf{E}_{ij}\mathbf{X'} + \mathbf{X}(\mathbf{E}_{ij})'$	$\mathbf{E}_{ij}\mathbf{X} + (\mathbf{E}_{ij})'\mathbf{X}$	$2\mathbf{X}$	$2\mathbf{X}$
$\mathbf{AXB}$	$\mathbf{AE}_{ij}\mathbf{B}$	$\mathbf{A'E}_{ij}\mathbf{B'}$	$\mathbf{A'B'}$	$\mathbf{A'B'}$
$\mathbf{A'XA}$[b]	—	$\mathbf{AE}_{ij}\mathbf{A'} + \mathbf{A}(\mathbf{E}_{ij})'\mathbf{A'} - \mathbf{D}_{[\mathbf{AE}_{ij}\mathbf{A'}]}$	$2\mathbf{AA'} - \mathbf{D}_{\mathbf{AA'}}$	$2\mathbf{AA'} - \mathbf{D}_{\mathbf{AA'}}$
$\mathbf{AX'X}$	$\mathbf{A}(\mathbf{E}_{ij})'\mathbf{X} + \mathbf{AX'E}_{ij}$	$\mathbf{X}(\mathbf{E}_{ij})'\mathbf{A} + \mathbf{XA'E}_{ij}$	$2\mathbf{AX}$	$\mathbf{X}(\mathbf{A} + \mathbf{A'})$
$\mathbf{X'BX}$	$(\mathbf{E}_{ij})'\mathbf{BX} + \mathbf{X'BE}_{ij}$	$\mathbf{BX}(\mathbf{E}_{ij})' + \mathbf{B'XE}_{ij}$	$(\mathbf{B} + \mathbf{B'})\mathbf{X}$	$(\mathbf{B} + \mathbf{B'})\mathbf{X}$
$\mathbf{X'XB}$	$(\mathbf{E}_{ij})'\mathbf{XB} + \mathbf{X'E}_{ij}\mathbf{B}$	$\mathbf{XB}(\mathbf{E}_{ij})' + \mathbf{XE}_{ij}\mathbf{B'}$	$2\mathbf{BX}$	$\mathbf{X}(\mathbf{B} + \mathbf{B'})$
$\mathbf{AXX'}$	$\mathbf{AE}_{ij}\mathbf{X'} + \mathbf{AX}(\mathbf{E}_{ij})'$	$\mathbf{A'XE}_{ij} + (\mathbf{E}_{ij})'\mathbf{AX}$	$2\mathbf{AX}$	$(\mathbf{A'} + \mathbf{A})\mathbf{X}$
$\mathbf{XBX'}$	$\mathbf{E}_{ij}\mathbf{BX'} + \mathbf{XB}(\mathbf{E}_{ij})'$	$\mathbf{E}_{ij}\mathbf{XB'} + (\mathbf{E}_{ij})'\mathbf{XB}$	$\mathbf{X}(\mathbf{B} + \mathbf{B'})$	$\mathbf{X}(\mathbf{B} + \mathbf{B'})$
$\mathbf{XX'B}$	$\mathbf{E}_{ij}\mathbf{X'B} + \mathbf{X}(\mathbf{E}_{ij})'\mathbf{B}$	$\mathbf{E}_{ij}\mathbf{B'X} + \mathbf{B}(\mathbf{E}_{ij})'\mathbf{X}$	$2\mathbf{BX}$	$(\mathbf{B'} + \mathbf{B})\mathbf{X}$
$\mathbf{XX'X}$	$\mathbf{E}_{ij}\mathbf{X'X} + \mathbf{X}(\mathbf{E}_{ij})'\mathbf{X} + \mathbf{XX'E}_{ij}$	$\mathbf{E}_{ij}\mathbf{X'X} + \mathbf{X}(\mathbf{E}_{ij})'\mathbf{X} + \mathbf{XX'E}_{ij}$	$3\mathbf{X}^2$	$\mathbf{X'X} + \mathbf{XX} + \mathbf{XX'}$
$\mathbf{X'XX'}$	$(\mathbf{E}_{ij})'\mathbf{XX'} + \mathbf{XE}_{ij}\mathbf{X'} + \mathbf{X'X}(\mathbf{E}_{ij})'$	$\mathbf{XX'}(\mathbf{E}_{ij})' + \mathbf{XE}_{ij}\mathbf{X} + (\mathbf{E}_{ij})'\mathbf{X'X}$	$3\mathbf{X}^2$	$\mathbf{XX'} + \mathbf{XX} + \mathbf{X'X}$

[a]Adapted from Dwyer and MacPhail (1948).
[b]$\mathbf{X}$ symmetric.

which we verify for the example $d(\mathbf{X}'\mathbf{X})/dX_{31}$:

$$\begin{bmatrix} 0 & 0 & 1 \\ 0 & 0 & 0 \end{bmatrix} \begin{bmatrix} X_{11} & X_{12} \\ X_{12} & X_{22} \\ X_{31} & X_{32} \end{bmatrix} + \begin{bmatrix} X_{11} & X_{21} & X_{31} \\ X_{12} & X_{22} & X_{32} \end{bmatrix} \begin{bmatrix} 0 & 0 \\ 0 & 0 \\ 1 & 0 \end{bmatrix}$$

$$= \begin{bmatrix} X_{31} & X_{32} \\ 0 & 0 \end{bmatrix} + \begin{bmatrix} X_{31} & 0 \\ X_{32} & 0 \end{bmatrix} = \begin{bmatrix} 2X_{31} & X_{32} \\ X_{32} & 0 \end{bmatrix}$$

The symbolic rule for the scalar-by-matrix derivative is

$$d\left[(\mathbf{X}'\mathbf{X})_{ij}\right]/d\mathbf{X} = \mathbf{X}(\mathbf{E}_{ij})' + \mathbf{X}\mathbf{E}_{ij}$$

Applying this rule to $d[(\mathbf{X}'\mathbf{X})_{12}]/d\mathbf{X}$ yields

$$\begin{bmatrix} X_{11} & X_{12} \\ X_{21} & X_{22} \\ X_{31} & X_{32} \end{bmatrix} \begin{bmatrix} 0 & 0 \\ 1 & 0 \end{bmatrix} + \begin{bmatrix} X_{11} & X_{12} \\ X_{21} & X_{22} \\ X_{31} & X_{32} \end{bmatrix} \begin{bmatrix} 0 & 1 \\ 0 & 0 \end{bmatrix}$$

$$= \begin{bmatrix} X_{12} & 0 \\ X_{22} & 0 \\ X_{32} & 0 \end{bmatrix} + \begin{bmatrix} 0 & X_{11} \\ 0 & X_{21} \\ 0 & X_{31} \end{bmatrix} = \begin{bmatrix} X_{12} & X_{11} \\ X_{22} & X_{21} \\ X_{32} & X_{31} \end{bmatrix}$$

again identical to the result obtained by direct scalar differentiation.

Table 6.3 includes derivatives of matrix functions encountered in the development of multivariate analysis. The first column identifies the matrix function, while the second and third columns give, respectively, the symbolic rules for the matrix-by-scalar and scalar-by-matrix derivatives. The elementary matrices required in the second column are those of the same order as $\mathbf{X}$, while the elementary matrices of the third column are of the same order as $\mathbf{Y}$. Because many of the matrix derivatives encountered in optimizing multivariate transformations are derivatives of scalar functions, a fourth column has been included for $\mathbf{Y} = (y)$, a scalar; it is obtained from the third column by taking $\mathbf{E}_{ij} = 1$, the scalar 1. The reader should note that taking a scalar Y imposes restrictions on the order of one or more of the factors in the expression. For example, $\mathbf{AX}$ is a scalar only if $\mathbf{A}$ and $\mathbf{X}$ are each scalar or if $\mathbf{A}$ and $\mathbf{X}$ are, respectively, conformable row and column vectors.

When several criterion attributes are present in a multivariate design it is common, as we shall see, to optimize the trace of a square (frequently symmetric) matrix. The final column contains $d(\operatorname{tr}\mathbf{Y})/d\mathbf{X}$, easily obtained from $d(Y_{ij})/d\mathbf{X}$ by noting that

$$d(\operatorname{tr}\mathbf{Y})/d\mathbf{X} = \sum_{i=1}^{k} d(Y_{ii})/d\mathbf{X}$$

This is readily obtained from the third column of Table 6.3 by taking

$\mathbf{E}_{ij} = (\mathbf{E}_{ij})' = \mathbf{I}$. Throughout Table 6.3 **A** and **B** are taken to be *constant* matrices independent of the elements of **X**.

One special case included is a symmetric matrix pre- and postmultiplied by a constant matrix and its transpose. The resulting derivatives with respect to **X** are somewhat different from those based on an arbitrary **X** pre- and postmultiplied by different constant matrices. The results are given only for scalar **Y** and tr(**Y**) because it is rare to require the derivative with respect to a specific element of **X**.

We can apply the symbolic differentiation rules directly to the maximization problem from principal components. Given an $n \times m$ data matrix **X**, we seek an m-element transformation vector **b** such that $\mathbf{Xb} = \mathbf{Z}$ is a set of n observations with maximum variability. If **S** is the $m \times m$ (symmetric) variance–covariance matrix based on **X**, then the variance of the transformed observations is given by $\text{Var}(\mathbf{Z}) = \mathbf{b'Sb}$.

We seek to maximize Var(**Z**). We saw earlier, however, that to do so we must impose a further restriction on the vector **b**. We take this to be $\mathbf{b'b} = 1$, or $\mathbf{b'b} - 1 = 0$. Incorporating this side condition, with associated undefined multipliers, into the function to be maximized gives

$$F = \mathbf{b'Sb} - \lambda(\mathbf{b'b} - 1)$$

We can now differentiate F with respect to all m elements of **b** simultaneously:

$$dF/d\mathbf{b} = d(\mathbf{b'Sb})/d\mathbf{b} - d(\lambda\mathbf{b'b})/d\mathbf{b}$$

We use column 4 of Table 6.3 because F is a scalar. The first term is of the form **X′BX**, so its derivative is

$$d(\mathbf{b'Sb})/d\mathbf{b} = (\mathbf{S} + \mathbf{S'})\mathbf{b} = 2\mathbf{Sb}$$

since **S** is symmetric. Similarly, for $\lambda\mathbf{b'b}$ we adopt the result for **AX′X**,

$$d(\lambda\mathbf{b'b})/d\mathbf{b} = 2\lambda\mathbf{b}$$

Putting the two results together gives $dF/d\mathbf{b} = 2(\mathbf{Sb} - \lambda\mathbf{b})$. It was this expression, set equal to zero, that we used earlier to solve for **b**.

To differentiate F with respect to λ, we can use the fourth row of Table 6.3,

$$\begin{aligned} dF/d\lambda &= d[-\lambda(\mathbf{b'b} - 1)]/d\lambda = d[\lambda(1 - \mathbf{b'b})]/d\lambda \\ &= (1 - \mathbf{b'b})' = 1 - \mathbf{b'b} \end{aligned}$$

and then set the derivative equal to zero to obtain $1 - \mathbf{b'b} = 0$, the side condition introduced with the multiplier λ.

6.7 Summary

Not all transformations of a data matrix, of course, are of equal interest. We may seek to obtain a particular derived matrix, such as the variance–covariance matrix. At other times, however, we are interested in transforming the data optimally—so that some function of the data has its optimal value. For this purpose, later chapters will depend on differential calculus, and this chapter summarized the rules the reader must recognize in order to understand our work.

We presented the use of first and second derivatives of functions to find maximum and minimum points. Exponential and logarithmic functions were discussed, and the treatment was extended to partial differentiation.

Often we obtain an optimal solution that is unique only up to an unknown constant of proportionality. By building in other constraints on the optimization process through the use of Lagrangian multipliers, we can obtain a unique solution. For example, in solving for the principal components of an intercorrelation matrix, we assume that the weighted composite must account for the maximum variability and that the sum of squares of the set of weights is unity.

The rules for differentiation can be extended to matrices, but rather than differentiating matrix sums and products by "brute force," it is far easier to adopt a symbolic matrix calculus. Table 6.3 shows how many matrix derivatives may be expressed; we shall refer to it often throughout this book.

Exercises

1. Show that the sum of squares of a set of values is a minimum if taken about the mean of that set. That is, use the rules of differentiation to show that $\Sigma_i(X_i - a)^2$ is a minimum for $a = (1/n)\Sigma_i X_i$.

2. Find a minimum for $y = 3x^2 + 6x - 2$. Graph the function in the region of that minimum.

3. Find the derivative of $\log x^2$.

4. Consider the matrix equation

$$s = \mathbf{b}'\begin{bmatrix} 1 & 0 & 1 \\ 0 & 2 & 3 \\ 1 & 1 & 4 \end{bmatrix}\mathbf{b}$$

where **b** is a vector. Use Table 6.3 to express $ds/d\mathbf{b}$ as a matrix, and evaluate the result for $\mathbf{b}' = (1, 1, 1)$.

Additional Reading

Another discussion of optimizing under side conditions as well as a review of elementary principles of differentiation is given by

Green, P. E. (1976). *Mathematical Tools for Applied Multivariate Analysis*. New York: Academic.

The table of matrix derivatives was adapted from the following original work:

Dwyer, P. S., and MacPhail, M. S. (1948). Symbolic matrix derivatives, *Annals of Mathematical Statistics* 19:517–534.

Other approaches to matrix calculus are provided by the following two references:

Bentler, P. M., and Lee, S. Y. (1975). Some extensions of matrix calculus, *General Systems* 20:145–150.

McDonald, R. P., and Swaminathan, H. (1973). A simple matrix calculus with applications to multivariate analysis, *General Systems* 18:37–54.

PART II
BASIC STRUCTURE AND THE UNIVARIATE CRITERION

In Part II we look at applications of the basic tools to the analysis of research designs involving one measured criterion attribute but multiple predictors. The approach we adopt is based on the general linear model (GLM), which permits us not only to present some very useful approaches to the univariate criterion—e.g., multiple regression and fixed effects analysis of variance and covariance—but also to lay the groundwork for problems with multivariate criteria to be discussed in Part III.

Many research questions require that, to the extent permitted by the data, we explain the variability in the criterion attribute as a function of one or many predictor attributes. We address this question from two perspectives. In Chapter 7 we develop a data bound model, appropriate when our interest is in explaining the data at hand and not in generalizing relations to other observations or to any superordinate source of the observed data. Such a model permits us to address only a limited class of research problems, but certain of the concepts—e.g., partial correlation and linear constraints—serve as a link to a more general approach.

More often than explaining the data at hand, we want to treat these as sampled from some populations of potential observations and to draw inferences about these populations. In Chapter 8 we provide an introduction to the very extensively used linear regression model—the GLM. We present both the sampling and structural aspects to the model and, by extending the discussion in Chapter 7, develop broadly applicable techniques for testing hypotheses and estimating the magnitude of effects of interest.

Chapter 9 develops specific forms of the GLM. We derive the fixed effects analysis of variance and the analysis of covariance from a GLM perspective. We also present a number of types of applications of multiple linear regression.

Part II describes the classical linear regression model. We use basic structure concepts to more clearly link popular techniques of analysis to an underlying statistical model and to lay the foundation for our consideration of the analysis of multivariate criteria.

Chapter 7

Least Squares: Data Fitting

Introduction

How accurately can we express a criterion attribute as a linear transformation of one or more predictor attributes? How much of the variability in a criterion can be accounted for by the variability in one or more predictors? In this chapter, we provide a way of answering these and related questions when the span of concern is limited to a particular data matrix. In this case, the *least squares* approach is descriptive; it speaks only to the data at hand. In the chapters that follow we shall present models that, when their assumptions are honored, provide ways of drawing inferences about, not just the data at hand but the *sources* of those data.

However, in discussing least squares fitting we introduce concepts that not only make it possible to work with data bound models but have sufficient generality that they are useful with other models as well. One of these concepts is the *constraint matrix*, which is used to capture in systematic fashion the differences among several equations to be fitted to a set of observations. In least squares, we can use the constraint matrix to develop the result of requiring, for example, that two predictors be given equal weight in predicting a criterion or that a particular attribute be given zero weight when developing a prediction equation.

We also develop the concepts of *multiple* and *partial correlations* as useful indices in the least squares approach. The squared multiple correlation provides a measure of the proportion of criterion variability accounted for by a set of predictors, while the partial correlation speaks to the unique contribution of one predictor to the prediction of the criterion.

7.1 Least Squares Fitting Procedure

As a first step toward developing the least squares approach, consider the data from the Higher Education Study. Recall from Chapter 1 that in this study four observations were made on each of the 50 states in the United States: per capita income, per capita high school completions, geographic region, and annual per capita spending in support of higher education. The data include information on each state, and we do not consider the states as a sample from some larger population. A portion of the data matrix is

$$\begin{bmatrix} 5227.84 & 0.01568 & 6 & 76.26 \\ 11{,}800.00 & 0.01869 & 8 & 143.03 \\ 6130.98 & 0.01366 & 7 & 90.90 \end{bmatrix}$$

The per capita amount spent by each state on higher education could be related to the other characteristics. For example, some theorists would expect states with larger per capita incomes or greater per capita high school completions to appropriate more funds for the support of colleges and universities than less prosperous states or states with fewer high school graduates.

One approach to investigating this possibility is by least squares. Here least squares fitting is applied to determine the optimal linear weighting of the two predictors, where optimality is defined in the following manner: Given a linear prediction of the spending of the jth state,

$$\begin{aligned}(\text{predicted spending})_j &= b_1 + b_2(\text{income})_j \\ &\quad + b_3(\text{high school completions})_j\end{aligned}$$

the difference between predicted and actual spending is calculated. This discrepancy, called the *error of prediction*, is squared, and the squared errors are then summed over the 50 states. The sum of squares of errors will be a function of the weights b_1, b_2, b_3 chosen for the prediction equation. The optimal, *least squares*, weights are those that minimize this sum of squares of errors.

The prediction equation, the errors of prediction, and the sum of squares of errors of prediction can be represented in matrix form. We note first that our measures of income, education, and expenditures are such that the observations have greatly different magnitudes. To avoid computational problems, we may restate income in thousands of dollars and high school graduations as the number per thousand. With these changes, the predictor matrix $\mathbf{X}_{(50\times 3)}$ is written

$$\mathbf{X} = \begin{bmatrix} 1 & 5.227 & 15.68 \\ 1 & 11.800 & 18.69 \\ 1 & 6.131 & 13.66 \\ \vdots & \vdots & \vdots \\ 1 & 7.072 & 16.99 \end{bmatrix}$$

The first column is a unit vector and the second and third columns contain the income and high school completion data. Then a conformable three-element vector is defined: $\mathbf{b}' = (b_1, b_2, b_3)$. The product $\mathbf{Xb}$ is a 50-element column vector with typical element of the form

$$(\mathbf{Xb})_j = b_1 + b_2 X_{j2} + b_3 X_{j3} \tag{7.1.1}$$

The elements of vector $\mathbf{Xb}$ are linear weightings of the predictor measures or attributes, and the elements of $\mathbf{b}$ are to be chosen such that $\mathbf{Xb}$ is a vector of *predicted* criterion observations. The prediction equation is now expressed

$$\hat{\mathbf{y}} = \mathbf{Xb} \tag{7.1.2}$$

The vector $\hat{\mathbf{y}}$ contains *estimates* or predictions of higher education appropriations, state by state. A second vector $\mathbf{y}$, the *criterion* vector contains the *actual* appropriations. The difference between these two,

$$\mathbf{e} = \mathbf{y} - \hat{\mathbf{y}} \tag{7.1.3}$$

is the vector of discrepancies: e_j is the difference between the actual amount allocated for higher education in the jth state and the amount predicted. The vector $\mathbf{e}$ is called the *error vector*.

The sum of squares of errors, then, is given by the minor product moment of $\mathbf{e}$:

$$\mathbf{e}'\mathbf{e} = (\mathbf{y} - \hat{\mathbf{y}})'(\mathbf{y} - \hat{\mathbf{y}}) \tag{7.1.4}$$

If we substitute $\mathbf{Xb}$ for $\hat{\mathbf{y}}$ in this equation, we can express the sum of squares of errors in terms of the predictor matrix $\mathbf{X}$, the criterion vector $\mathbf{y}$, and a weighting vector $\mathbf{b}$:

$$\mathbf{e}'\mathbf{e} = (\mathbf{y} - \mathbf{Xb})'(\mathbf{y} - \mathbf{Xb}) \tag{7.1.5}$$

As we are given $\mathbf{X}$ and $\mathbf{y}$, the sum of squares of errors depends only on the choice of weighting vector. In *least squares* we choose $\mathbf{b}$ so as to minimize $\mathbf{e}'\mathbf{e}$.

Before applying some of the optimizing principles developed in Chapter 6, Eq. (7.1.5) will be expanded,

$$\begin{aligned} \mathbf{e}'\mathbf{e} &= (\mathbf{y}' - \mathbf{b}'\mathbf{X}')(\mathbf{y} - \mathbf{Xb}) \\ &= \mathbf{y}'\mathbf{y} - \mathbf{y}'\mathbf{Xb} - \mathbf{b}'\mathbf{X}'\mathbf{y} + \mathbf{b}'\mathbf{X}'\mathbf{Xb} \end{aligned}$$

Because $\mathbf{y}'\mathbf{Xb}$ is the transpose of $\mathbf{b}'\mathbf{X}'\mathbf{y}$, *a scalar quantity*, the two must be equal, $\mathbf{y}'\mathbf{Xb} = \mathbf{b}'\mathbf{X}'\mathbf{y}$, and (7.1.5) can be written

$$\mathbf{e}'\mathbf{e} = \mathbf{y}'\mathbf{y} - 2\mathbf{b}'\mathbf{X}'\mathbf{y} + \mathbf{b}'\mathbf{X}'\mathbf{Xb} \tag{7.1.6}$$

The values of $\mathbf{b}$ that minimize $\mathbf{e}'\mathbf{e}$, as given in (7.1.6), may now be sought by differentiating the scalar $\mathbf{e}'\mathbf{e}$ with respect to the vector $\mathbf{b}$ and then setting the resulting vector equal to the null vector. That is, $\mathbf{e}'\mathbf{e}$ will be minimized if $\mathbf{b}$ is the vector that satisfies $\partial\mathbf{e}'\mathbf{e}/\partial\mathbf{b} = 0$, or

$$\partial\mathbf{e}'\mathbf{e}/\partial\mathbf{b} = \partial(\mathbf{y}'\mathbf{y})/\partial\mathbf{b} + \partial(-2\mathbf{b}'\mathbf{X}'\mathbf{y})/\partial\mathbf{b} + \partial(\mathbf{b}'\mathbf{X}'\mathbf{X}\mathbf{b})/\partial\mathbf{b} = \mathbf{0} \tag{7.1.7}$$

Each of the terms on the right may be evaluated using the matrix differentiation rules of Table 6.3.

First,

$$\partial(\mathbf{y}'\mathbf{y})/\partial\mathbf{b} = \mathbf{0} \tag{7.1.8}$$

inasmuch as the vector **b** does not contribute to the term under differentiation. Now

$$\partial(-2\mathbf{b}'\mathbf{X}'\mathbf{y})/\partial\mathbf{b} = -2\mathbf{X}'\mathbf{y}, \qquad \partial(\mathbf{b}'\mathbf{X}'\mathbf{X}\mathbf{b})/\partial\mathbf{b} = 2\mathbf{X}'\mathbf{X}\mathbf{b} \tag{7.1.9}$$

Substituting this into (7.1.7) gives

$$\partial(\mathbf{e}'\mathbf{e})/\partial\mathbf{b} = -2\mathbf{X}'\mathbf{y} + 2\mathbf{X}'\mathbf{X}\mathbf{b} = \mathbf{0} \tag{7.1.10}$$

or

$$\mathbf{X}'\mathbf{X}\mathbf{b} = \mathbf{X}'\mathbf{y} \tag{7.1.11}$$

If the predictor matrix **X** is *basic*, then the minor product moment $\mathbf{X}'\mathbf{X}$ is a square basic matrix and possesses a regular inverse, $(\mathbf{X}'\mathbf{X})^{-1}$. Premultiplying the two sides of Eq. (7.1.11) by $(\mathbf{X}'\mathbf{X})^{-1}$ leads to a solution for **b**:

$$\mathbf{b} = (\mathbf{X}'\mathbf{X})^{-1}\mathbf{X}'\mathbf{y} \tag{7.1.12}$$

Based on these optimal values for the elements of **b**, the vector of predicted criterion values, from (7.1.2), is given by

$$\hat{\mathbf{y}} = \mathbf{X}\mathbf{b} = \mathbf{X}(\mathbf{X}'\mathbf{X})^{-1}\mathbf{X}'\mathbf{y} \tag{7.1.13}$$

the error vector by

$$\mathbf{e} = \mathbf{y} - \hat{\mathbf{y}} = \mathbf{y} - \mathbf{X}(\mathbf{X}'\mathbf{X})^{-1}\mathbf{X}'\mathbf{y} = \left[\mathbf{I} - \mathbf{X}(\mathbf{X}'\mathbf{X})^{-1}\mathbf{X}'\right]\mathbf{y} \tag{7.1.14}$$

and the sum of squares of errors by

$$\mathbf{e}'\mathbf{e} = \mathbf{y}'\left[\mathbf{I} - \mathbf{X}(\mathbf{X}'\mathbf{X})^{-1}\mathbf{X}'\right]\left[\mathbf{I} - \mathbf{X}(\mathbf{X}'\mathbf{X})^{-1}\mathbf{X}'\right]\mathbf{y} = \mathbf{y}'\left[\mathbf{I} - \mathbf{X}(\mathbf{X}'\mathbf{X})^{-1}\mathbf{X}'\right]\mathbf{y} \tag{7.1.15}$$

as $\mathbf{I} - \mathbf{X}(\mathbf{X}'\mathbf{X})^{-1}\mathbf{X}'$ is in the class of idempotent matrices discussed in Section 5.5.

For computational purposes $\mathbf{e}'\mathbf{e}$ may be expressed in different forms:

$$\mathbf{e}'\mathbf{e} = \mathbf{y}'\mathbf{y} - \mathbf{y}'\mathbf{X}(\mathbf{X}'\mathbf{X})^{-1}\mathbf{X}'\mathbf{y} \tag{7.1.16}$$

produced by multiplying through on the right in (7.1.15);

$$\mathbf{e}'\mathbf{e} = \mathbf{y}'\mathbf{y} - \mathbf{y}'\mathbf{X}\mathbf{b} \tag{7.1.17}$$

produced by substituting for **b** from (7.1.12);

$$\mathbf{e}'\mathbf{e} = \mathbf{y}'\mathbf{y} - \mathbf{y}'\hat{\mathbf{y}} \tag{7.1.18}$$

if **Xb** is replaced by $\hat{\mathbf{y}}$; or, by rewriting (7.1.18),

$$\mathbf{e}'\mathbf{e} = \mathbf{y}'(\mathbf{y} - \hat{\mathbf{y}}) = \mathbf{y}'\mathbf{e} \tag{7.1.19}$$

Equations (7.1.15)–(7.1.19) provide not only alternative means of computating $\mathbf{e}'\mathbf{e}$ but alternative symbolic representations. These will assist the reader as additional aspccts of the general linear model are developed.

A typical computation leading to the vector **b** and the sum of squares of errors or prediction can now be illustrated for the Higher Education Study. With the data in Table 1.5,

$$\mathbf{X}'\mathbf{X} = \begin{bmatrix} 50.000 & 315.811 & 788.990 \\ 315.811 & 2049.785 & 4993.658 \\ 788.990 & 4993.658 & 12{,}605.919 \end{bmatrix}$$

$$(\mathbf{X}'\mathbf{X})^{-1} = \begin{bmatrix} 2.131 & -0.097 & -0.095 \\ -0.097 & 0.018 & -0.001 \\ -0.095 & -0.001 & 0.0065 \end{bmatrix}$$

$\mathbf{y}'\mathbf{X} = (3304.291, 21{,}469.137, 52{,}645.091)$, $\mathbf{y}'\mathbf{y} = 239{,}534.085$,

$$\mathbf{b} = \begin{bmatrix} -39.875 \\ 10.398 \\ 2.553 \end{bmatrix}$$

$\mathbf{y}'\mathbf{X}\mathbf{b} = 225{,}876.622$, and $\mathbf{e}'\mathbf{e} = 13{,}657.463$.

This sum of squares of errors based on optimally weighting the two predictors, income and completions, might be compared with the sum of squares of errors that would result if predictions of state higher education spending were to be made on the basis of no predictive information. With no information to distinguish among states, the same prediction should be made for each of the $n = 50$ states; i.e., $\hat{\mathbf{y}} = k\mathbf{1}$. The sum of squares of errors becomes, in this instance,

$$\begin{aligned} \mathbf{e}'\mathbf{e} &= (\mathbf{y} - k\mathbf{1})'(\mathbf{y} - k\mathbf{1}) = \mathbf{y}'\mathbf{y} - 2k\mathbf{1}'\mathbf{y} + k^2\mathbf{1}'\mathbf{1} \\ &= \mathbf{y}'\mathbf{y} - 2k\mathbf{1}'\mathbf{y} + k^2 n \end{aligned}$$

Finding a value of k to minimize $\mathbf{e}'\mathbf{e}$ requires setting the derivative of this equal to zero:

$$0 = \partial(\mathbf{e}'\mathbf{e})/\partial k = 0 - 2(\mathbf{1}'\mathbf{y}) + 2kn$$

or $k = (1/n)\mathbf{1}'\mathbf{y}$.

Not surprisingly, in the case of no information for distinguishing among the 50 states, the constant prediction that minimizes the sum of squares of errors is the *criterion mean*; the sum of squares about the mean is smaller than the sum of squares about any other single value. Substituting this solution for k into the equation for $\mathbf{e'e}$ gives

$$\begin{aligned}\mathbf{e'e} &= \mathbf{y'y} - 2(1/n)\mathbf{1'y1'_2y} + (1/n)^2\mathbf{1'y1'y}n \\ &= \mathbf{y'y} - (1/n)(\mathbf{1'y})^2\end{aligned}$$

In the higher education study, the mean education expenditure of the states is $\mathbf{1'y} = 3304.291$, $(1/n)\mathbf{1'y} = 66.086$, and $\mathbf{e'e} = 21{,}167.437$. The sum of squares of errors of prediction drops from 21,167.4 to 13,657.5 with the introduction of the two predictors, and the averaged squared error $\mathbf{e'e}/n$ drops from 423.35 (which may be recognized as the criterion variance) to 273.15, sometimes called the variance of the errors or, more simply, the *error variance*.

7.1.1 Least Squares with a Nonbasic Predictor Matrix

To solve for the least squares weighting vector $\mathbf{b}$ we have required so far that $\mathbf{X'X}$ be a square basic matrix, possessing a regular inverse, in which case [see (7.1.12)] $\mathbf{b} = (\mathbf{X'X})^{-1}\mathbf{X'y}$. Can a vector $\mathbf{b}$ be found to minimize $\mathbf{e'e}$ when the predictor matrix $\mathbf{X}$, and hence its minor product moment, is not basic?

We address this question only briefly, showing one approach to the problem. Most of the readily available computer programs require that $\mathbf{X}$ be a basic matrix. The reader is advised that, although solutions for nonbasic predictor sets are available, it is better to select predictor attributes so as to eliminate linear dependencies among them.

An approach to the solution with a nonbasic $\mathbf{X}$ begins with Eq. (7.1.11): $\mathbf{X'Xb} = \mathbf{X'y}$. A vector $\mathbf{b}$ that satisfies this equation must also minimize $\mathbf{e'e}$. While $\mathbf{X'X}$ no longer possesses a regular inverse, it does have a *generalized* inverse. We define

$$(\mathbf{X'X})^+ = \mathbf{Q}\Delta^{-2}\mathbf{Q'} \tag{7.1.20}$$

to be the generalized inverse of a product moment matrix with basic structure $\mathbf{X'X} = \mathbf{Q}\Delta^2\mathbf{Q'}$. Premultiplying both sides of (7.1.11) by $(\mathbf{X'X})^+$ gives

$$(\mathbf{X'X})^+(\mathbf{X'X})\mathbf{b} = \mathbf{Q}\Delta^{-2}\mathbf{Q'Q}\Delta^2\mathbf{Q'b} = \mathbf{QQ'b} = (\mathbf{X'X})^+\mathbf{X'y} \tag{7.1.21}$$

The matrix $\mathbf{X}$ has itself a basic structure with elements common to those of $\mathbf{X'X}$, namely, $\mathbf{X} = \mathbf{P}\Delta\mathbf{Q'}$. Premultiplying both sides of (7.1.21) by $\mathbf{X}$ yields

$$\mathbf{XQQ'b} = \mathbf{P}\Delta\mathbf{Q'QQ'b} = \mathbf{P}\Delta\mathbf{Q'b} = \mathbf{Xb} = \mathbf{X}\left[(\mathbf{X'X})^+\mathbf{X'y}\right]$$

or

$$\hat{\mathbf{y}} = \mathbf{Xb} = \mathbf{X}\left[(\mathbf{X'X})^+\mathbf{X'y}\right] \tag{7.1.22}$$

Equation (7.1.22) indicates that least squares predictions of the criterion values can be obtained even when the predictor matrix is nonbasic. Further, this equation shows that one way of obtaining the vector of predicted criterion scores is to compute

$$\hat{\mathbf{y}} = \mathbf{Xb}^+ \quad \text{where} \quad \mathbf{b}^+ = (\mathbf{X'X})^+ \mathbf{X'y} \tag{7.1.23}$$

is a weighting vector. However, this solution for the least squares weighting vector **b** is not unique. If $\mathbf{b}^+$ is a solution vector, then so too is the vector

$$\mathbf{b}^v = \mathbf{b}^+ + (\mathbf{I} - \mathbf{QQ}')\mathbf{v} \tag{7.1.24}$$

where **v** is any nonnull vector.

This multiplicity of solutions is illustrated by premultiplying (7.1.24) by **X**:

$$\begin{aligned} \mathbf{Xb}^v &= \mathbf{Xb}^+ + \mathbf{X}(\mathbf{I} - \mathbf{QQ}')\mathbf{v} = \mathbf{Xb}^+ + \mathbf{P\Delta Q}'(\mathbf{I} - \mathbf{QQ}')\mathbf{v} \\ &= \mathbf{Xb}^+ + (\mathbf{P\Delta Q}' - \mathbf{P\Delta Q}')\mathbf{v} = \mathbf{Xb}^+ \end{aligned}$$

Both $\mathbf{Xb}^+$ and $\mathbf{Xb}^v$ give identical estimates of **y**. Hence $\mathbf{b}^+$ and $\mathbf{b}^v$ yield the same sum of squares of errors.

The sum of squares of errors is given in the nonbasic case, from (7.1.22), by

$$\begin{aligned} \mathbf{e'e} &= (\mathbf{y} - \hat{\mathbf{y}})'(\mathbf{y} - \hat{\mathbf{y}}) = \mathbf{y'y} - \mathbf{y'X}(\mathbf{X'X})^+ \mathbf{X'y} \\ &= \mathbf{y'y} - \mathbf{y'Xb}^+ = \mathbf{y'y} - \mathbf{y'}\hat{\mathbf{y}} \end{aligned} \tag{7.1.25}$$

The form is identical to (7.1.16)–(7.1.19) for basic **X′X**, with the generalized inverse substituted for the regular inverse. [The curious reader may note that deriving (7.1.25) makes use of the property of the generalized inverse that $(\mathbf{X'X})^+(\mathbf{X'X})(\mathbf{X'X})^+ = (\mathbf{X'X})^+$.]

For nonbasic **X**, the least squares approach may be illustrated with the following small data set:

$$\mathbf{X} = \begin{bmatrix} 1 & 1 & 0 \\ 1 & 1 & 0 \\ 1 & 1 & 0 \\ 1 & 1 & 0 \\ 1 & 0 & 1 \\ 1 & 0 & 1 \\ 1 & 0 & 1 \\ 1 & 0 & 1 \end{bmatrix}, \qquad \mathbf{y} = \begin{bmatrix} 15 \\ 12 \\ 11 \\ 8 \\ 13 \\ 16 \\ 9 \\ 12 \end{bmatrix}$$

The predictor matrix **X** is clearly nonbasic; for example, the third column can be produced by subtracting the second from the first. The two predictors (we shall assume) correspond to the two attributes *received an influenza vaccine* (yes = 1, no = 0) in the second column and *did not receive an influenza vaccine* (yes = 1, no = 0) in the third column. The criterion vector contains the number of days lost from work owing to illness during the winter months of 1981–1982. Data are for eight elementary school teachers.

Carrying through the computations,

$$\mathbf{X}'\mathbf{X} = \begin{bmatrix} 8 & 4 & 4 \\ 4 & 4 & 0 \\ 4 & 0 & 4 \end{bmatrix}, \qquad \mathbf{X}'\mathbf{y} = \begin{bmatrix} 96 \\ 46 \\ 50 \end{bmatrix}$$

$\mathbf{y}'\mathbf{y} = 1204$, and $\mathbf{1}'\mathbf{y} = 96$. The basic structure of $\mathbf{X}'\mathbf{X}$ is given by

$$\mathbf{X}'\mathbf{X} = \begin{bmatrix} 0.816 & 0.000 \\ -0.408 & -0.707 \\ -0.408 & 0.707 \end{bmatrix} \begin{bmatrix} 12 & 0 \\ 0 & 4 \end{bmatrix} \begin{bmatrix} 0.816 & -0.408 & -0.408 \\ 0.000 & -0.707 & 0.707 \end{bmatrix}$$
$$= \mathbf{Q}\boldsymbol{\Delta}^2\mathbf{Q}'$$

Our definition of basic structure is such that $\mathbf{Q}$ should be 3×2 and $\boldsymbol{\Delta}^2$ should be 2×2. As a result, the generalized inverse $(\mathbf{X}'\mathbf{X})^+ = \mathbf{Q}\boldsymbol{\Delta}^{-2}\mathbf{Q}'$ is

$$(\mathbf{X}'\mathbf{X})^+ = \begin{bmatrix} 0.0556 & -0.0278 & -0.0278 \\ -0.0278 & 0.1389 & -0.1111 \\ -0.0278 & -0.1111 & 0.1389 \end{bmatrix}$$

The solution vector $\mathbf{b}^+$, from (7.1.23), is given by

$$(\mathbf{b}^+)' = \left[(\mathbf{X}'\mathbf{X})^+\mathbf{X}'\mathbf{y}\right]' = (8.0, 3.5, 4.5)$$

and the predicted values of $\mathbf{y}$ are provided by

$$\hat{\mathbf{y}}' = (\mathbf{X}\mathbf{b}^+)' = (11.5, 11.5, 11.5, 11.5, 12.5, 12.5, 12.5, 12.5)$$

The minor product $\mathbf{y}'\hat{\mathbf{y}} = 1154$ and, by (7.1.25), $\mathbf{e}'\mathbf{e} = \mathbf{y}'\mathbf{y} - \mathbf{y}'\hat{\mathbf{y}} = 50$.

To illustrate the nonuniqueness of $\mathbf{b}^+$, consider the arbitrary vector $\mathbf{v}' = (1.0, 2.0, 1.0)$. A new set of least squares weights can be obtained from (7.1.24):

$$\mathbf{b}^{\mathrm{v}} = \mathbf{b}^+ + (\mathbf{I} + \mathbf{Q}\mathbf{Q}')\mathbf{v} = \mathbf{b}^+ + \mathbf{v} - \mathbf{Q}(\mathbf{Q}'\mathbf{v})$$

Using these new weights, one finds that the elements of $\mathbf{X}\mathbf{b}^{\mathrm{v}}$ are the same as those of $\mathbf{X}\mathbf{b}^+$; i.e., $\mathbf{X}\mathbf{b}^{\mathrm{v}} = \mathbf{X}\mathbf{b}^+ = \hat{\mathbf{y}}$. Whether $\mathbf{b}^+$ or $\mathbf{b}^{\mathrm{v}}$ (or any one of a very large set of other three-element vectors) is used to weight the three columns of $\mathbf{X}$, the same set of least squares predictions of $\mathbf{y}$ results.

Following the advice we offered earlier, we shall find another least squares solution to the present problem by first replacing $\mathbf{X}$ with a basic matrix. It was noted that the third column in $\mathbf{X}$ contains no new information about the entities. We shall simply drop the third column, making $\mathbf{X}$ a basic matrix of rank 2:

$$\mathbf{X}' = \begin{bmatrix} 1 & 1 & 1 & 1 & 1 & 1 & 1 & 1 \\ 1 & 1 & 1 & 1 & 0 & 0 & 0 & 0 \end{bmatrix}$$

For this new $\mathbf{X}$, the computations yield

$$\mathbf{X}'\mathbf{X} = \begin{bmatrix} 8 & 4 \\ 4 & 4 \end{bmatrix}, \qquad \mathbf{X}'\mathbf{y} = \begin{bmatrix} 96 \\ 46 \end{bmatrix}, \qquad \mathbf{y}'\mathbf{y} = 1204, \qquad \mathbf{1}'\mathbf{y} = 96$$

From these,

$$(\mathbf{X}'\mathbf{X})^{-1} = \begin{bmatrix} \frac{1}{4} & -\frac{1}{4} \\ -\frac{1}{4} & \frac{1}{2} \end{bmatrix}, \qquad (\mathbf{X}'\mathbf{X})^{-1}\mathbf{X}'\mathbf{y} = \begin{bmatrix} 12.5 \\ -1.0 \end{bmatrix} = \mathbf{b}$$

$$\hat{\mathbf{y}}' = (11.5, 11.5, 11.5, 11.5, 12.5, 12.5, 12.5, 12.5)$$

Notice that the same set of least squares estimates is obtained here as from the nonbasic three-column predictor matrix. In consequence, $\mathbf{e}'\mathbf{e}$ must also be identical. Clearly no predictive information was lost by omitting the final column.

The reader will note that, whether $\mathbf{X}$ is basic or not, the average of the predicted values of $\mathbf{y}$ equals the average of the $\mathbf{y}$ values. This property is a direct extension of Eq. (7.1.11): $\mathbf{X}'(\mathbf{Xb}) = \mathbf{X}'\mathbf{y}$. As the first column of $\mathbf{X}$ is the unit vector, we can write $\mathbf{1}'(\mathbf{Xb}) = \mathbf{1}'\mathbf{y}$ or $\mathbf{1}'\hat{\mathbf{y}} = \mathbf{1}'\mathbf{y}$, so that

$$(1/n)\mathbf{1}'\hat{\mathbf{y}} = (1/n)\mathbf{1}'\mathbf{y}$$

or

$$\bar{\hat{y}} = \bar{y}$$

7.2 Constraining the Least Squares Weights

In Section 7.1 techniques were illustrated for finding a set of weights for m predictor attributes such that the sum of those weighted predictors would most closely match, in the least squares sense, a vector of criterion observations for the same entities. If the m predictor attributes contain no linear dependencies, then the set of weights is unique, but $\mathbf{e}'\mathbf{e}$, the sum of squares of errors of prediction, can always be found.

Frequently research questions require that we determine (if we can) not only the optimal least squares weights and the associated sum of squares of errors *but* also the weights and squared errors with additional specified restrictions imposed on the solution.

The least squares solution of Section 7.1 is an *unconstrained* solution. Here we shall develop a systematic way of imposing restrictions, or *constraints*, on the least squares weights and observing the consequences. As an example of a constraint, we shall consider anew the Higher Education Study data. In the unconstrained least squares solution, weights were found for income and high school completions that would minimize the errors associated with predicting state spending on higher education. The resulting sum of squares of errors was found to be $\mathbf{e}'\mathbf{e} = 13{,}657.5$ and (dividing by 50) the average squared error of prediction (across states) 273.15. What would the errors be if spending were predicted solely from income and if differences in per capita high school completions among the states were ignored, or equivalently, if we constrained b_3, the weight given per capita high school completions, to be zero?

An intuitive, and numerically correct, answer can be found using the results of Section 7.1 simply by employing a predictor matrix from which completion data have been dropped. Now we have

$$\mathbf{X'X} = \begin{bmatrix} 50.000 & 315.811 \\ 315.811 & 2049.785 \end{bmatrix}, \qquad (\mathbf{X'X})^{-1} = \begin{bmatrix} 0.745 & -0.115 \\ -0.115 & 0.018 \end{bmatrix}$$

$$\mathbf{y'X} = (3304.30, 21{,}469.13), \qquad \mathbf{y'y} = 239{,}534.085$$

$$\mathbf{b'} = (-2.58, 10.87), \qquad \mathbf{e'e} = 14{,}660.66, \qquad \mathbf{e'e}/n = 293.21$$

As the reader may have guessed, the errors of prediction are greater based on income alone than on income and high school completions together, 14,660.66 compared with 13,657.46.

Dropping vectors (columns) of predictor data is one way of imposing constraints on the least squares solution. Dropping a predictor attribute in the least squares solution is conceptually identical to stipulating that the weight for that attribute is constrained exactly to zero. This, however, is only one possible type of constraint.

Consider *increasing* the number of predictors in the Higher Education Study by including the 0–1 columns that carry information about the geographic region in which each state is located. One form of the predictor matrix would now contain ten columns $\mathbf{X}_{.1}$ = state income, $\mathbf{X}_{.2}$ = state high school completions, $\mathbf{X}_{.3}$ = New England state (1 if true, 0, if not), $\mathbf{X}_{.4}$ = Middle Atlantic state (1 if true, 0 if not), ..., $\mathbf{X}_{.10}$ = Pacific state (1 if true, 0 if not).

Note that the unit vector no longer appears as the first column of the predictor matrix. Because each state is in exactly one of the eight geographic regions, the sum of the last eight columns is the unit vector. Including the unit vector would have made the predictor matrix nonbasic.

Associated with the predictor matrix in the least squares model is a vector of weights $\mathbf{b}$ such that predictions of the criterion are of the form $\hat{\mathbf{y}} = \mathbf{Xb}$. Again, because for each state the scores on the final eight attributes consist of seven zeros and one unity, the form of the prediction for each state is

$$\begin{aligned}(\hat{y}_j, \text{predicted spending}) = {} & b_1(X_{j1}, \text{state per capita income}) \\ & + b_2(X_{j2}, \text{state per capita high school} \\ & \qquad \text{completions}) \\ & + b_k\end{aligned}$$

The additive constant b_k in the linear prediction equation varies from state to state and gives the least squares weight associated with the regional attribute corresponding to the region in which the state is located.

An interesting constraint to consider in this case, then, is one stipulating that the least squares weights for the regions are all the same: $b_3 = b_4 = b_5 = b_6 = b_7 = b_8 = b_9 = b_{10}$. How would the sum of squares of errors change between permitting each region to enjoy a separate additive constant and requiring the additive constant to be the same for all states? Before attempting to answer this particular question, a conceptual and computational scheme for imposing constraints will be developed.

7.2.1 Constraint Matrix

The scheme to be developed will generalize to all linear constraints on **b** that can be expressed in the form

$$\mathbf{C}'\mathbf{b} = \mathbf{0} \tag{7.2.1}$$

If **b** is an m-element vector, then **C** must be $m \times k$. We make two assumptions about **C**: $k \leqslant m$ and **C** is basic. **C** is then of rank k. We say that $\mathbf{C}'\mathbf{b} = \mathbf{0}$ imposes k constraints on **b**. Equation (7.2.1) provides a standard way of phrasing these constraints.

For the Higher Education Study, the desired constraint $b_3 = 0$—no weight should be given to high school completions in predicting state spending—could be expressed

$$\mathbf{C}'\mathbf{b} = (0 \quad 0 \quad 1)\begin{bmatrix} b_1 \\ b_2 \\ b_3 \end{bmatrix} = 0$$

The rank of **C** is 1, and one constraint is imposed on the three-element vector **b**.

For the second example, $\mathbf{C}'\mathbf{b} = \mathbf{0}$ could be

$$\begin{bmatrix} 0 & 0 & 1 & -1 & 0 & 0 & 0 & 0 & 0 & 0 \\ 0 & 0 & 1 & 0 & -1 & 0 & 0 & 0 & 0 & 0 \\ 0 & 0 & 1 & 0 & 0 & -1 & 0 & 0 & 0 & 0 \\ 0 & 0 & 1 & 0 & 0 & 0 & -1 & 0 & 0 & 0 \\ 0 & 0 & 1 & 0 & 0 & 0 & 0 & -1 & 0 & 0 \\ 0 & 0 & 1 & 0 & 0 & 0 & 0 & 0 & -1 & 0 \\ 0 & 0 & 1 & 0 & 0 & 0 & 0 & 0 & 0 & -1 \end{bmatrix}\begin{bmatrix} b_1 \\ b_2 \\ b_3 \\ b_4 \\ b_5 \\ b_6 \\ b_7 \\ b_8 \\ b_9 \\ b_{10} \end{bmatrix} = \begin{bmatrix} 0 \\ 0 \\ 0 \\ 0 \\ 0 \\ 0 \\ 0 \end{bmatrix}$$

The seven constraints imposed are $b_3 - b_4 = 0$, $b_3 - b_5 = 0$, $b_3 - b_6 = 0$, $b_3 - b_7 = 0$, $b_3 - b_8 = 0$, $b_3 - b_9 = 0$, and $b_3 - b_{10} = 0$, *or* equivalently, $b_3 = b_4$, $b_3 = b_5$, $b_3 = b_6$, $b_3 = b_7$, $b_3 = b_8$, $b_3 = b_9$, and $b_3 = b_{10}$. It is not necessary to impose additional equivalences, say, $b_4 = b_5$, as the row (0, 0, 0, 1, -1, 0, 0, 0, 0, 0) carrying that restriction would be linearly dependent in this case on the first and second rows of $\mathbf{C}'$; the second row minus the first gives the proposed additional row.

The seven constraints on the geographic region could have been expressed by a different matrix $\mathbf{C}'$, for example,

$$\begin{bmatrix} 0 & 0 & 1 & -1 & 0 & 0 & 0 & 0 & 0 & 0 \\ 0 & 0 & 0 & 1 & -1 & 0 & 0 & 0 & 0 & 0 \\ 0 & 0 & 0 & 0 & 1 & -1 & 0 & 0 & 0 & 0 \\ 0 & 0 & 0 & 0 & 0 & 1 & -1 & 0 & 0 & 0 \\ 0 & 0 & 0 & 0 & 0 & 0 & 1 & -1 & 0 & 0 \\ 0 & 0 & 0 & 0 & 0 & 0 & 0 & 1 & -1 & 0 \\ 0 & 0 & 0 & 0 & 0 & 0 & 0 & 0 & 1 & -1 \end{bmatrix}$$

This time the equality of the eight regions is expressed by $b_3 = b_4$, $b_4 = b_5$, $b_5 = b_6$, $b_6 = b_7$, $b_7 = b_8$, $b_8 = b_9$, and $b_9 = b_{10}$. When more than one constraint is imposed, it is frequently possible to express those constraints in different forms.

7.2.2 Least Squares Solution with Constraints

With the constraints phrased in the form $\mathbf{C'b} = \mathbf{0}$ a general approach to the least squares solution can be used. Now, we seek a weighting vector, ${}_c\mathbf{b}$, that possesses two properties. First, the vector must satisfy the condition $\mathbf{C'b} = \mathbf{0}$; it must ensure that the constraints are satisfied. Second, the vector must minimize the sum of squares of the differences between the elements of the predicted criterion vector based upon the constrained vectors ${}_c\mathbf{b}$,

$$\hat{\mathbf{y}}_c = \mathbf{X}\,{}_c\mathbf{b}$$

and the elements of the observed criterion vector $\mathbf{y}$.

That is, if we first define a constrained error vector $\mathbf{f} = \mathbf{y} - \mathbf{x}\,{}_c\mathbf{b}$, then what is required is a vector ${}_c\mathbf{b}$ to minimize

$$\mathbf{f'f} = (\mathbf{y} - \mathbf{X}\,{}_c\mathbf{b})'(\mathbf{y} - \mathbf{X}\,{}_c\mathbf{b}) \tag{7.2.2}$$

subject to the side condition

$$\mathbf{C}'\,{}_c\mathbf{b} = \mathbf{0} \tag{7.2.3}$$

Optimization under side conditions can be accomplished through the use of Lagrangian multipliers (Section 6.3). The composite function to be minimized can be written by (7.2.2) and (7.2.3) as

$$\phi_c = \mathbf{f'f} - \boldsymbol{\lambda}'\mathbf{C}'\,{}_c\mathbf{b} = \mathbf{y'y} - 2\,{}_c\mathbf{b}'\mathbf{X}'\mathbf{y} + {}_c\mathbf{b}'\mathbf{X}'\mathbf{X}\,{}_c\mathbf{b} - {}_c\mathbf{b}'\mathbf{C}\boldsymbol{\lambda} \tag{7.2.4}$$

The desired ${}_c\mathbf{b}$ is found by setting each of the two derivatives $\partial\phi_c/\partial\,{}_c\mathbf{b}$ and $\partial\phi_c/\partial\boldsymbol{\lambda}$ equal to m- and k-element null vectors, respectively, and solving the resulting equations. The Lagrangian multiplier $\boldsymbol{\lambda}$ will be a k-element vector where $\mathbf{C}'$ is $k \times m$.

.090

~~.056~~

.101

9 H. S. GP

10 category of classes

Differentiate (7.2.4) first with respect to ${}_{c}\mathbf{b}$:

$$\partial\phi_c/\partial\,{}_{c}\mathbf{b} = -2\mathbf{X}'\mathbf{y} + 2\mathbf{X}'\mathbf{X}\,{}_{c}\mathbf{b} - \mathbf{C}\lambda \tag{7.2.5}$$

and then with respect to λ:

$$\partial\phi_c/\partial\lambda = \mathbf{C}'\,{}_{c}\mathbf{b} \tag{7.2.6}$$

Setting (7.2.5) equal to the null vector and rearranging terms gives $(\mathbf{X}'\mathbf{X})\,{}_{c}\mathbf{b} - \mathbf{X}'\mathbf{y} = \mathbf{C}(\lambda/2)$, which can be premultiplied by $(\mathbf{X}'\mathbf{X})^{-1}$ to yield

$${}_{c}\mathbf{b} - (\mathbf{X}'\mathbf{X})^{-1}\mathbf{X}'\mathbf{y} = (\mathbf{X}'\mathbf{X})^{-1}\mathbf{C}(\lambda/2) \tag{7.2.7}$$

We assume $\mathbf{X}$ is a basic predictor matrix.

If (7.2.7) is next premultiplied by $\mathbf{C}'$, the result is

$$\mathbf{C}'\,{}_{c}\mathbf{b} - \mathbf{C}'(\mathbf{X}'\mathbf{X})^{-1}\mathbf{X}'\mathbf{y} = \mathbf{C}'(\mathbf{X}'\mathbf{X})^{-1}\mathbf{C}(\lambda/2) \tag{7.2.8}$$

However, setting the derivative in (7.2.6) equal to the null vector, $\mathbf{C}'\,{}_{c}\mathbf{b} = 0$, returns the desired side condition, which may be substituted into the left side of (7.2.8) to give

$$-\mathbf{C}'(\mathbf{X}'\mathbf{X})^{-1}\mathbf{X}'\mathbf{y} = \mathbf{C}'(\mathbf{X}'\mathbf{X})^{-1}\mathbf{C}(\lambda/2) \tag{7.2.9}$$

Because $\mathbf{C}$ is basic and of rank no greater than the rank of $\mathbf{X}$, the product rank rule could be used to show that $\mathbf{C}'(\mathbf{X}'\mathbf{X})^{-1}\mathbf{C}$ is a square basic matrix. The exercise is left to the reader. The two sides of (7.2.9) may be premultiplied by the regular inverse, $[\mathbf{C}'(\mathbf{X}'\mathbf{X})^{-1}\mathbf{C}]^{-1}$, to define $\lambda/2$:

$$\lambda/2 = -\left[\mathbf{C}'(\mathbf{X}'\mathbf{X})^{-1}\mathbf{C}\right]^{-1}\mathbf{C}'(\mathbf{X}'\mathbf{X})^{-1}\mathbf{X}'\mathbf{y} \tag{7.2.10}$$

If Eq. (7.2.7) is now rewritten in terms of ${}_{c}\mathbf{b}$,

$${}_{c}\mathbf{b} = (\mathbf{X}'\mathbf{X})^{-1}\mathbf{X}'\mathbf{y} + (\mathbf{X}'\mathbf{X})^{-1}\mathbf{C}(\lambda/2)$$

and the right side of (7.2.10) is substituted for $\lambda/2$, we have a solution for ${}_{c}\mathbf{b}$:

$${}_{c}\mathbf{b} = (\mathbf{X}'\mathbf{X})^{-1}\mathbf{X}'\mathbf{y} - (\mathbf{X}'\mathbf{X})^{-1}\mathbf{C}\left[\mathbf{C}'(\mathbf{X}'\mathbf{X})^{-1}\mathbf{C}\right]^{-1}\mathbf{C}'(\mathbf{X}'\mathbf{X})^{-1}\mathbf{X}'\mathbf{y} \tag{7.2.11}$$

This equation appears quite involved. Obtaining ${}_{c}\mathbf{b}$ is not trivial, and the reader will almost certainly want to solve for ${}_{c}\mathbf{b}$ by using a computer program. In practice, the solution is not as complex as (7.2.11) would suggest. First, it is most likely the *unconstrained* least squares weights from Eq. (7.1.12),

$$\mathbf{b} = (\mathbf{X}'\mathbf{X})^{-1}\mathbf{X}'\mathbf{y}$$

would have already been obtained, and the constrained least squares weights could be computed as

$${}_{c}\mathbf{b} = \mathbf{b} - (\mathbf{X}'\mathbf{X})^{-1}\mathbf{C}\left[\mathbf{C}'(\mathbf{X}'\mathbf{X})^{-1}\mathbf{C}\right]^{-1}\mathbf{C}'\mathbf{b} \tag{7.2.12}$$

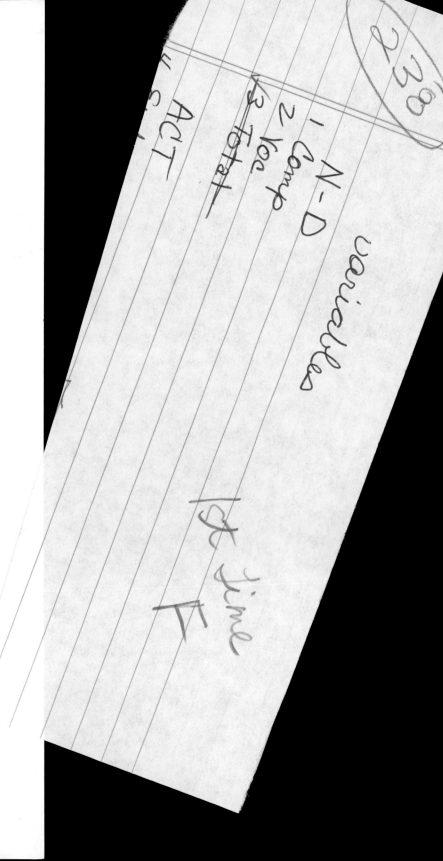

by substituting **b** into (7.2.11). Second, very often only one constraint at a time is imposed on the vector **b**. That is, $\mathbf{C}'$ is frequently a row vector and $\mathbf{C}'(\mathbf{X}'\mathbf{X})^{-1}\mathbf{C}$, in consequence, is a scalar quantity, very easily inverted.

How is the sum of squares of errors of prediction to be expressed when the predictions are constrained? In (7.2.2) this quantity was defined as

$$\mathbf{f}'\mathbf{f} = (\mathbf{y} - \mathbf{X}\,{}_c\mathbf{b})'(\mathbf{y} - \mathbf{X}\,{}_c\mathbf{b}) = \mathbf{y}'\mathbf{y} - 2\,{}_c\mathbf{b}'\mathbf{X}'\mathbf{y} + {}_c\mathbf{b}'\mathbf{X}'\mathbf{X}\,{}_c\mathbf{b}$$

Although the algebra is not obvious, we can rewrite ${}_c\mathbf{b}$ from (7.2.11) as

$${}_c\mathbf{b} = \left\{\mathbf{I} - (\mathbf{X}'\mathbf{X})^{-1}\mathbf{C}\left[\mathbf{C}'(\mathbf{X}'\mathbf{X})^{-1}\mathbf{C}\right]^{-1}\mathbf{C}'\right\}\mathbf{b}$$

and then make use of the fact that the premultiplier of **b** is another example of the rank-reducing idempotent matrix of the form $\mathbf{I} - \mathbf{U}(\mathbf{V}'\mathbf{U})^{-1}\mathbf{V}'$. Using this idempotency and substituting $(\mathbf{X}'\mathbf{X})^{-1}\mathbf{X}'\mathbf{y}$ for **b**, we can establish that ${}_c\mathbf{b}'(\mathbf{X}'\mathbf{X})\,{}_c\mathbf{b} = {}_c\mathbf{b}'\mathbf{X}'\mathbf{y}$. As a result, (7.2.2) simplifies to $\mathbf{f}'\mathbf{f} = \mathbf{y}'\mathbf{y} - {}_c\mathbf{b}'\mathbf{X}'\mathbf{y}$. Or, substituting for ${}_c\mathbf{b}$ from (7.2.11), we have the constrained sum of squares of errors:

$$\mathbf{f}'\mathbf{f} = \mathbf{y}'\mathbf{y} - \mathbf{y}'\mathbf{X}(\mathbf{X}'\mathbf{X})^{-1}\mathbf{X}'\mathbf{y} + \mathbf{y}'\mathbf{X}(\mathbf{X}'\mathbf{X})^{-1}\mathbf{C}\left[\mathbf{C}'(\mathbf{X}'\mathbf{X})^{-1}\mathbf{C}\right]^{-1}\mathbf{C}'(\mathbf{X}'\mathbf{X})^{-1}\mathbf{X}'\mathbf{y} \tag{7.2.13}$$

This equation is not, certainly, a handy way of computing $\mathbf{f}'\mathbf{f}$, but it allows us to make two points about the form of this quantity. First, note that the first two terms on the right of (7.2.13) equal

$$\mathbf{e}'\mathbf{e} = \mathbf{y}'\mathbf{y} - \mathbf{y}'\mathbf{X}(\mathbf{X}'\mathbf{X})^{-1}\mathbf{X}'\mathbf{y}$$

the *unconstrained* error sum of squares, as given in (7.1.16). As a result, (7.2.13) allows us to express the differences between the constrained and unconstrained sums of squares of errors as

$$\mathbf{f}'\mathbf{f} - \mathbf{e}'\mathbf{e} = \mathbf{y}'\mathbf{X}(\mathbf{X}'\mathbf{X})^{-1}\mathbf{C}\left[\mathbf{C}'(\mathbf{X}'\mathbf{X})^{-1}\mathbf{C}\right]^{-1}\mathbf{C}'(\mathbf{X}'\mathbf{X})^{-1}\mathbf{X}'\mathbf{y} \tag{7.2.14}$$

or

$$\mathbf{f}'\mathbf{f} - \mathbf{e}'\mathbf{e} = \mathbf{b}'\mathbf{C}\left[\mathbf{C}'(\mathbf{X}'\mathbf{X})^{-1}\mathbf{C}\right]^{-1}\mathbf{C}'\mathbf{b} \tag{7.2.15}$$

Frequently **b** and $\mathbf{e}'\mathbf{e}$ are known before ${}_c\mathbf{b}$ and $\mathbf{f}'\mathbf{f}$, so it is common to obtain ${}_c\mathbf{b}$ from (7.2.12) and $\mathbf{f}'\mathbf{f}$ from (7.2.15).

The second point has to do with the form not only of $\mathbf{f}'\mathbf{f}$, but of $\mathbf{e}'\mathbf{e}$ and of $\mathbf{f}'\mathbf{f} - \mathbf{e}'\mathbf{e}$ as well. From (7.1.15), (7.2.15), and (7.2.13), respectively, these are

$$\mathbf{e}'\mathbf{e} = \mathbf{y}'\left[\mathbf{I} - \mathbf{X}(\mathbf{X}'\mathbf{X})^{-1}\mathbf{X}'\right]\mathbf{y}$$

$$\mathbf{f}'\mathbf{f} - \mathbf{e}'\mathbf{e} = \mathbf{y}'\left\{\mathbf{X}(\mathbf{X}'\mathbf{X})^{-1}\mathbf{C}\left[\mathbf{C}'(\mathbf{X}'\mathbf{X})^{-1}\mathbf{C}\right]^{-1}\mathbf{C}'(\mathbf{X}'\mathbf{X})^{-1}\mathbf{X}'\right\}\mathbf{y}$$

and

$$\mathbf{f}'\mathbf{f} = \mathbf{y}'\left\{\mathbf{I} - \mathbf{X}(\mathbf{X}'\mathbf{X})^{-1}\mathbf{X}' + \mathbf{X}(\mathbf{X}'\mathbf{X})^{-1}\mathbf{C}\left[\mathbf{C}'(\mathbf{X}'\mathbf{X})^{-1}\mathbf{C}\right]^{-1}\mathbf{C}(\mathbf{X}'\mathbf{X})^{-1}\mathbf{X}'\right\}\mathbf{y}$$

Each of the three terms is a *quadratic* form in $\mathbf{y}$:

$$\begin{aligned} \mathbf{e'e} &= \mathbf{y'Ay} \\ \mathbf{f'f} - \mathbf{e'e} &= \mathbf{y'By} \\ \mathbf{f'f} &= \mathbf{y'(A + B)y} \end{aligned}$$

This is because $\mathbf{A}$, $\mathbf{B}$, and $\mathbf{A + B}$ are all symmetric matrices. (If $\mathbf{S}$ is a symmetric matrix, then $\mathbf{USU'}$ is a symmetric matrix. If $\mathbf{A}$ and $\mathbf{B}$ are both symmetric matrices, then $\mathbf{A + B}$ and $\mathbf{A - B}$ are also symmetric.)

We also know that $\mathbf{e'e}$ is a nonnegative quadratic form in $\mathbf{y}$. This follows because $\mathbf{I - X(X'X)^{-1}X'}$ is idempotent as well as symmetric.

The quadratic form $\mathbf{f'f - e'e}$ can also be shown to be a nonnegative form. First we define a matrix $\mathbf{U = X(X'X)^{-1}C}$, in terms of which we can express the matrix $\mathbf{C'(X'X)^{-1}C}$ as a product moment matrix

$$\mathbf{U'U} = \mathbf{C'(X'X)^{-1}X'X(X'X)^{-1}C} = \mathbf{C'(X'X)^{-1}C}$$

and, in turn, the $\mathbf{f'f - e'e}$ of (7.2.14) as $\mathbf{f'f - e'e = y'[U(U'U)^{-1}U']y}$. Because $\mathbf{U(U'U)^{-1}U'}$ is idempotent and symmetric, it is also a nonnegative matrix.

Both $\mathbf{e'e}$ and $\mathbf{f'f - e'e}$ are sums of squares of linear transformations of the criterion vector $\mathbf{y}$. Chapter 5 established that the idempotent matrix reducing form $\mathbf{I - U(V'U)^{-1}V'}$ has rank $n - m$ if both $\mathbf{U}$ and $\mathbf{V}$ are vertical basic matrices of size $n \times m$.

For the quadratic form $\mathbf{e'e = y'Ay}$, the matrix of the form $\mathbf{A = I - X(X'X)^{-1}X'}$ will have rank $n - m$, where the predictor matrix $\mathbf{X}$ has n rows and m columns. As a result, $\mathbf{e'e}$ can be interpreted as the sum of squares of $n - m$ linear transformations of the n criterion observations.

In the case of $\mathbf{f'f - e'e = y'By}$, the matrix of the form

$$\mathbf{B} = \mathbf{X(X'X)^{-1}C\left[C'(X'X)^{-1}C\right]^{-1}C'(X'X)^{-1}X'}$$

can be shown, through repeated application of the product rank rule, to have rank $k \leqslant m$, the number of constraints imposed on $\mathbf{b}$, or equivalently, the number of columns in the vertical basic matrix $\mathbf{C}$. The difference $\mathbf{f'f - e'e}$, then, can be thought of as a sum of squares of k linear transformations of the criterion observations.

The finding that $\mathbf{f'f - e'e}$ is a sum of squares suggests that $\mathbf{f'f}$ cannot be less than $\mathbf{e'e}$; i.e., $\mathbf{f'f - e'e}$ cannot be negative. While this can be shown to be true in other ways, we shall simply remind the reader that $\mathbf{e'e}$ is the *minimum sum of squares* of errors possible given free choice of the weighting vector $\mathbf{b}$, whereas $\mathbf{f'f}$ is the minimum when we restrict our search for $\mathbf{b}$ to those vectors that also satisfy the equation $\mathbf{C'b = 0}$. The restricted solution cannot provide any better fit than the unrestricted one.

In fact, $\mathbf{f'f - e'e}$ is usually interpreted as the *increase in the sum of squares of errors* associated with imposing on the least squares solution a particular constraint or set of constraints.

7.3 Data-Bound Multiple and Partial Correlations

The weighting of several predictor attributes so that their sum best approximates a criterion measure, in the least squares sense, is closely related to the concept of the *multiple correlation*.

The multiple correlation problem can be stated in the following way. We have a set of observations for some n entities on m predictors and on one criterion and can represent them as a predictor matrix $\mathbf{X}_{(n\times m)}$ and a criterion vector $\mathbf{y}_{(n\times 1)}$. What linear transformation of the m predictor attributes is maximally correlated with the criterion and what is the magnitude of that correlation? In other words, what is the maximum correlation attainable between the criterion and some linear transformation of the m predictor attributes? The reader should note that $\mathbf{X}$ has a predictor attribute in each column (the first column of $\mathbf{X}$ is *not* the unit vector). That is, we seek an m-element weighting vector $\mathbf{a}$ such that the vector of transformed predictor scores

$$\hat{\mathbf{y}} = \mathbf{Xa}$$

is as highly correlated with $\mathbf{y}$ as possible.

We shall assume that we have already computed from the data the following summary matrices: $\mathbf{r}_{\mathbf{XX}}$, an $m \times m$ symmetric matrix of correlations among the m predictor attributes; $\mathbf{r}_{\mathbf{Xy}}$, an m-element vector of correlations of the m predictors with the criterion; $\mathbf{D}_{\mathbf{S_X}}$, a diagonal matrix of standard deviations for the m predictors; and $S_{\mathbf{y}}$, the criterion standard deviation.

The variance–covariance matrix for the m predictors is given by

$$\mathbf{D}_{\mathbf{S_X}}\mathbf{r}_{\mathbf{XX}}\mathbf{D}_{\mathbf{S_X}} = \mathbf{Cov}(\mathbf{X}) \tag{7.3.1}$$

and the covariances of the predictors with the criterion form a vector

$$S_{\mathbf{y}}\mathbf{D}_{\mathbf{S_X}}\mathbf{r}_{\mathbf{Xy}} = \mathbf{Cov}(\mathbf{X}, \mathbf{y}) \tag{7.3.2}$$

Based on the properties of linear transformations developed in Chapter 5, $\mathrm{Var}(\hat{\mathbf{y}}) = \mathrm{Var}(\mathbf{Xa}) = \mathbf{a}'\,\mathbf{Cov}(\mathbf{X})\,\mathbf{a}$, so that

$$\mathrm{Var}(\hat{\mathbf{y}}) = \mathbf{a}'\mathbf{D}_{\mathbf{S_X}}\mathbf{r}_{\mathbf{XX}}\mathbf{D}_{\mathbf{S_X}}\mathbf{a} \tag{7.3.3}$$

and the covariance of $\hat{\mathbf{y}}$ with $\mathbf{y}$ is

$$\begin{aligned}\mathrm{Cov}(\hat{\mathbf{y}}, \mathbf{y}) &= \mathrm{Cov}(\mathbf{Xa}, \mathbf{y}) = \mathbf{a}'\,\mathbf{Cov}(\mathbf{X}, \mathbf{y}) \\ &= \mathbf{a}'S_{\mathbf{y}}\mathbf{D}_{\mathbf{S_X}}\mathbf{r}_{\mathbf{Xy}} = S_{\mathbf{y}}\mathbf{a}'\mathbf{D}_{\mathbf{S_X}}\mathbf{r}_{\mathbf{Xy}}\end{aligned} \tag{7.3.4}$$

The correlation between $\hat{\mathbf{y}}$ and $\mathbf{y}$ is written from (7.3.3) and (7.3.4) based on the definition of a product moment correlation:

$$\begin{aligned}r_{\hat{\mathbf{y}}\mathbf{y}} &= \mathrm{Cov}(\hat{\mathbf{y}}, \mathbf{y})/S_{\hat{\mathbf{y}}}S_{\mathbf{y}} = S_{\mathbf{y}}\mathbf{a}'\mathbf{D}_{\mathbf{S_X}}\mathbf{r}_{\mathbf{Xy}}/S_{\mathbf{y}}\left(\mathbf{a}'\mathbf{D}_{\mathbf{S_X}}\mathbf{r}_{\mathbf{XX}}\mathbf{D}_{\mathbf{S_X}}\mathbf{a}\right)^{1/2} \\ &= \mathbf{a}'\mathbf{D}_{\mathbf{S_X}}\mathbf{r}_{\mathbf{Xy}}/\left(\mathbf{a}'\mathbf{D}_{\mathbf{S_X}}\mathbf{r}_{\mathbf{XX}}\mathbf{D}_{\mathbf{S_X}}\mathbf{a}\right)^{1/2}\end{aligned} \tag{7.3.5}$$

It is easy to note from (7.3.5) that any vector of weights proportional to the vector **a** produces a set of transformed scores $\hat{\mathbf{y}}$ that correlates with **y** to exactly the same degree as does **Xa**. For example, if $\mathbf{b} = 2\mathbf{a}$, then the correlation between **Xb** and **y** is

$$r_{\mathbf{Xb},\mathbf{y}} = \mathbf{b}'\mathbf{D}_{S_\mathbf{X}}\mathbf{r}_{\mathbf{Xy}}/(\mathbf{b}'\mathbf{D}_{S_\mathbf{X}}\mathbf{r}_{\mathbf{XX}}\mathbf{D}_{S_\mathbf{X}}\mathbf{b})^{1/2} = 2\mathbf{a}'\mathbf{D}_{S_\mathbf{X}}\mathbf{r}_{\mathbf{Xy}}/(2\mathbf{a}'\mathbf{D}_{S_\mathbf{X}}\mathbf{r}_{\mathbf{XX}}\mathbf{D}_{S_\mathbf{X}}2\mathbf{a})^{1/2}$$

$$= \mathbf{a}'\mathbf{D}_{S_\mathbf{X}}\mathbf{r}_{\mathbf{Xy}}/(\mathbf{a}'\mathbf{D}_{S_\mathbf{X}}\mathbf{r}_{\mathbf{XX}}\mathbf{D}_{S_\mathbf{X}}\mathbf{a})^{1/2} = r_{\mathbf{Xa},\mathbf{y}}$$

To find a unique solution for the transformation that maximizes the correlation, it is necessary to restrict in some way the scaling of the elements of **a**. One way is to state as a side condition that the variance of the transformed scores, the elements of $\hat{\mathbf{y}}$, is equal to some specified value. With $S_\mathbf{y}^2$ known and $S_{\hat{\mathbf{y}}}^2$ specified, the correlation between **y** and $\hat{\mathbf{y}}$ is maximized by maximizing the covariance between the two.

That is, we can maximize $\mathrm{Cov}(\mathbf{y}, \hat{\mathbf{y}})$ subject to the side condition $\mathrm{Var}(\hat{\mathbf{y}}) = k$, a constant. The composite function with Lagrangian multiplier is then written

$$\phi = \mathrm{Cov}(\mathbf{y}, \hat{\mathbf{y}}) - \lambda[\mathrm{Var}(\hat{\mathbf{y}}) - k]$$

or, from (7.3.3) and (7.3.4),

$$\phi = S_\mathbf{y}\mathbf{a}'\mathbf{D}_{S_\mathbf{X}}\mathbf{r}_{\mathbf{Xy}} - \lambda(\mathbf{a}'\mathbf{D}_{S_\mathbf{X}}\mathbf{r}_{\mathbf{XX}}\mathbf{D}_{S_\mathbf{X}}\mathbf{a} - k)$$

Differentiating ϕ with respect to **a** and setting the resulting equation equal to zero gives

$$S_\mathbf{y}\mathbf{D}_{S_\mathbf{X}}\mathbf{r}_{\mathbf{Xy}} = \mathbf{D}_{S_\mathbf{X}}\mathbf{r}_{\mathbf{XX}}\mathbf{D}_{S_\mathbf{X}}\mathbf{a}(2\lambda) \tag{7.3.6}$$

If the predictor data matrix is basic, **Cov(X)** is a square basic matrix and both sides of (7.3.6) may then be premultiplied by $(\mathbf{D}_{S_\mathbf{X}}\mathbf{r}_{\mathbf{XX}}\mathbf{D}_{S_\mathbf{X}})^{-1} = \mathbf{D}_{S_\mathbf{X}}^{-1}\mathbf{r}_{\mathbf{XX}}^{-1}\mathbf{D}_{S_\mathbf{X}}^{-1}$:

$$S_\mathbf{y}\mathbf{D}_{S_\mathbf{X}}^{-1}\mathbf{r}_{\mathbf{XX}}\mathbf{r}_{\mathbf{Xy}} = \mathbf{a}(2\lambda) \tag{7.3.7}$$

The vector **a** is proportional to the product on the left. We can find the correlation between **y** and $\hat{\mathbf{y}}$ even though we do not know the proportionality constant 2λ. This is because the correlation between **y** and $\hat{\mathbf{y}}$ is identical to that between **y** and any constant scaling of $\hat{\mathbf{y}}$, say, $(2\lambda)\hat{\mathbf{y}} = \mathbf{Xa}(2\lambda)$.

From (7.3.3) and (7.3.7),

$$\mathrm{Var}[\mathbf{Xa}(2\lambda)] = (S_\mathbf{y}\mathbf{D}_{S_\mathbf{X}}^{-1}\mathbf{r}_{\mathbf{XX}}^{-1}\mathbf{r}_{\mathbf{Xy}})'(\mathbf{D}_{S_\mathbf{X}}\mathbf{r}_{\mathbf{XX}}\mathbf{D}_{S_\mathbf{X}})(S_\mathbf{y}\mathbf{D}_{S_\mathbf{X}}^{-1}\mathbf{r}_{\mathbf{XX}}^{-1}\mathbf{r}_{\mathbf{Xy}})$$

$$= S_\mathbf{y}^2\mathbf{r}'_{\mathbf{Xy}}\mathbf{r}_{\mathbf{XX}}^{-1}\mathbf{r}_{\mathbf{Xy}} \tag{7.3.8}$$

and, from (7.3.4) and (7.3.7),

$$\mathrm{Cov}[\mathbf{Xa}(2\lambda), \mathbf{y}] = (S_\mathbf{y}\mathbf{D}_{S_\mathbf{X}}^{-1}\mathbf{r}_{\mathbf{XX}}^{-1}\mathbf{r}_{\mathbf{Xy}})'(S_\mathbf{y}\mathbf{D}_{S_\mathbf{X}}\mathbf{r}_{\mathbf{Xy}}) = S_\mathbf{y}^2\mathbf{r}'_{\mathbf{Xy}}\mathbf{r}_{\mathbf{XX}}^{-1}\mathbf{r}_{\mathbf{Xy}} \tag{7.3.9}$$

7.3.1 Squared Multiple Correlation

We now can define the multiple correlation, symbolized $R_{\mathbf{y}.\mathbf{X}}$ and read as the correlation between the single criterion measure **y** and the (most optimal weighting of) several (multiple) predictors $\mathbf{X}_{.1}, \mathbf{X}_{.2}, \ldots, \mathbf{X}_{.k}$ in terms of the ratio

$$R_{\mathbf{y}.\mathbf{X}} = \mathrm{Cov}[\mathbf{Xa}(2\lambda), \mathbf{y}]/S_{\mathbf{y}}\{\mathrm{Var}[\mathbf{Xa}(2\lambda)]\}^{1/2}$$

It will be easier, however, first to obtain the square of the multiple correlation $R^2_{\mathbf{y}.\mathbf{X}}$ from (7.3.8) and (7.3.9):

$$R^2_{\mathbf{y}.\mathbf{X}} = \frac{S_{\mathbf{y}}^2 S_{\mathbf{y}}^2 (\mathbf{r}'_{\mathbf{X}\mathbf{y}} \mathbf{r}^{-1}_{\mathbf{X}\mathbf{X}} \mathbf{r}_{\mathbf{X}\mathbf{y}})(\mathbf{r}'_{\mathbf{X}\mathbf{y}} \mathbf{r}^{-1}_{\mathbf{X}\mathbf{X}} \mathbf{r}_{\mathbf{X}\mathbf{y}})}{S_{\mathbf{y}}^2 S_{\mathbf{y}}^2 (\mathbf{r}'_{\mathbf{X}\mathbf{y}} \mathbf{r}^{-1}_{\mathbf{X}\mathbf{X}} \mathbf{r}_{\mathbf{X}\mathbf{y}})} = \mathbf{r}'_{\mathbf{X}\mathbf{y}} \mathbf{r}^{-1}_{\mathbf{X}\mathbf{X}} \mathbf{r}_{\mathbf{X}\mathbf{y}} \qquad (7.3.10)$$

Equation (7.3.10) gives a standard computational determination of the squared multiple correlation for a set of observations. If the regular inverse of the matrix of predictor intercorrelations is both pre- and postmultiplied by the vector of predictor–criterion correlations, the product is the squared multiple correlation, the square of the maximum attainable correlation in the data between the criterion and *any* linear combination of the predictors.

We have yet to deal with the constant of proportionality problem in calculating the weights **a**. In setting up the composite ϕ to be differentiated we constrained Var(**Xa**) to be equal to a constant k. The usual choice is

$$\mathrm{Var}(\hat{\mathbf{y}}) = \mathrm{Var}(\mathbf{Xa}) = R^2_{\mathbf{y}.\mathbf{X}} S^2_{\mathbf{y}}$$

As the squared multiple correlation approaches its upper limit of 1.0, the variance of $\hat{\mathbf{y}}$ approaches the variance of the criterion and, as the squared multiple correlation approaches its lower limit of 0.0, the variance of $\hat{\mathbf{y}}$ approaches zero as well.

If we note now that the weighting of (7.3.7), $\mathbf{a}(2\lambda) = S_{\mathbf{y}} \mathbf{D}^{-1}_{S_{\mathbf{X}}} \mathbf{r}^{-1}_{\mathbf{X}\mathbf{X}} \mathbf{r}_{\mathbf{X}\mathbf{y}}$ produces a transformation $\mathbf{Xa}(2\lambda)$ that, according to (7.3.8) and (7.3.10), has variance $S^2_{\mathbf{y}} \mathbf{R}^2_{\mathbf{y}.\mathbf{X}}$, then the appropriate solution for the unknown is $\lambda = \frac{1}{2}$ and the weighting vector is seen to be

$$\mathbf{a} = S_{\mathbf{y}} \mathbf{D}^{-1}_{S_{\mathbf{X}}} \mathbf{r}^{-1}_{\mathbf{X}\mathbf{X}} \mathbf{r}_{\mathbf{X}\mathbf{y}} \qquad (7.3.11)$$

The composite predictor based on these weights, $\hat{\mathbf{y}} = \mathbf{Xa}$ is maximally correlated with **y**.

The square of this maximum correlation, known as the *squared multiple correlation*, is given by (7.3.10), while the transformed predictor scores have variance

$$S^2_{\hat{\mathbf{y}}} = R^2_{\mathbf{y}.\mathbf{X}} S^2_{\mathbf{y}} \qquad (7.3.12)$$

and identical covariance between the criterion and the linear combination of predictors:

$$\mathrm{Cov}(\mathbf{y}, \hat{\mathbf{y}}) = R^2_{\mathbf{y}.\mathbf{X}} S^2_{\mathbf{y}} \qquad (7.3.13)$$

Note from (7.3.11) that when the criterion and all k predictors have the same variance over the n observations, when $\mathbf{D}_{S_\mathbf{x}} = S_\mathbf{y}\mathbf{I}$, the appropriate weights are given by

$$\boldsymbol{\beta} = \mathbf{r}_{\mathbf{XX}}^{-1}\mathbf{r}_{\mathbf{Xy}} \tag{7.3.14}$$

These *beta weights* are the weights that would be given the several predictors if all attributes had been standardized on the data.

A common multiple correlation (and least squares, as we shall see directly) computational procedure is first to find beta weights and then to compute the squared correlation

$$R_{\mathbf{y}.\mathbf{X}}^2 = \boldsymbol{\beta}'\mathbf{r}_{\mathbf{Xy}} \tag{7.3.15}$$

The raw, unstandardized, or **b** weights are

$$\mathbf{b} = S_\mathbf{y}\mathbf{D}_{S_\mathbf{x}}^{-1}\boldsymbol{\beta} \tag{7.3.16}$$

There is a direct link between the multiple correlation and least squares fitting problems. This can be understood most directly by considering the least squares solution when both predictors and criterion are in *deviation form*, so that $\mathbf{1}'\mathbf{X} = \mathbf{0}'$ and $\mathbf{1}'\mathbf{y} = 0$. When these conditions are met,

$$\mathbf{X}'\mathbf{X} = n\,\mathbf{Cov}(\mathbf{X}) = n\mathbf{D}_{S_\mathbf{x}}\mathbf{r}_{\mathbf{XX}}\mathbf{D}_{S_\mathbf{x}}, \qquad \mathbf{X}'\mathbf{y} = n\,\mathbf{Cov}(\mathbf{X},\mathbf{y}) = n\mathbf{D}_{S_\mathbf{x}}\mathbf{r}_{\mathbf{Xy}}S_\mathbf{y}$$

with the result that the *least squares weights*,

$$\begin{aligned}\mathbf{b} &= (\mathbf{X}'\mathbf{X})^{-1}\mathbf{X}'\mathbf{y} = \left[\mathbf{D}_{S_\mathbf{x}}^{-1}\mathbf{r}_{\mathbf{XX}}^{-1}\mathbf{D}_{S_\mathbf{x}}^{-1}(1/n)\right]\left[n\mathbf{D}_{S_\mathbf{x}}\mathbf{r}_{\mathbf{Xy}}S_\mathbf{y}\right] \\ &= S_\mathbf{y}\mathbf{D}_{S_\mathbf{x}}^{-1}\mathbf{r}_{\mathbf{XX}}^{-1}\mathbf{r}_{\mathbf{Xy}}\end{aligned}$$

are identically the multiple correlation **b** weights.

A numerical example will make the point even more clearly. For the Higher Education Study we have, in the least squares model, a predictor matrix with three columns (the first column consisting of the unit vector) and a criterion vector. From these data,

$$\mathbf{X}'\mathbf{X} = \begin{bmatrix} 50 & 315.811 & 788.990 \\ 315.811 & 2049.7852 & 4993.658 \\ 788.990 & 4993.658 & 12{,}605.919 \end{bmatrix}, \qquad \mathbf{X}'\mathbf{y} = \begin{bmatrix} 3304.290 \\ 21{,}469.137 \\ 52{,}645.091 \end{bmatrix}$$

$$\mathbf{y}'\mathbf{y} = 239{,}534.85 \qquad \mathbf{1}'\mathbf{X} = \begin{bmatrix} 50 \\ 315.811 \\ 788.990 \end{bmatrix}, \qquad \mathbf{1}'\mathbf{y} = 3304.290$$

From the inverse of $\mathbf{X}'\mathbf{X}$ and $\mathbf{X}'\mathbf{y}$, the least squares weights are

$$\mathbf{b} = \begin{bmatrix} -39.87 \\ 10.40 \\ 2.55 \end{bmatrix}$$

The calculation of the sum of squares of the errors of prediction gave $\mathbf{e}'\mathbf{e} = \mathbf{y}'\mathbf{y} - \mathbf{b}'(\mathbf{X}'\mathbf{y}) = 13{,}657.463$.

For the multiple correlation solution, we first calculate a variance–covariance matrix for the two predictors and the covariance of the two predictors with the criterion.

The variance of the criterion is

$$S_{\mathbf{y}}^2 = \tfrac{1}{50}[\mathbf{y}'\mathbf{y} - (1/50)(\mathbf{y}'\mathbf{1})(\mathbf{1}'\mathbf{y})] = 423.35$$

The variance–covariance matrix for the two predictors is given by

$$(1/n)[\mathbf{X}'\mathbf{X} - (1/n)\mathbf{X}'\mathbf{1}\mathbf{1}'\mathbf{X}] = \begin{bmatrix} 1.101 & 0.204 \\ 0.204 & 3.116 \end{bmatrix}$$

and the covariance of the two predictors with the criterion is

$$(1/n)[\mathbf{X}'\mathbf{y} - (1/n)\mathbf{X}'\mathbf{1}\mathbf{1}'\mathbf{y}] = \begin{bmatrix} 11.961 \\ 10.098 \end{bmatrix}$$

From these results,

$$\mathbf{D}_{S_{\mathbf{X}}} = \begin{bmatrix} 1.049 & 0 \\ 0 & 1.765 \end{bmatrix}, \qquad S_{\mathbf{y}} = 20.575$$

$$\mathbf{r}_{\mathbf{XX}} = \begin{bmatrix} 1 & 0.110 \\ 0.110 & 1.000 \end{bmatrix}, \qquad \mathbf{r}_{\mathbf{Xy}} = \begin{bmatrix} 0.554 \\ 0.278 \end{bmatrix}$$

$$\boldsymbol{\beta} = \mathbf{r}_{\mathbf{XX}}^{-1}\mathbf{r}_{\mathbf{Xy}} = \begin{bmatrix} 0.530 \\ 0.220 \end{bmatrix}, \qquad \mathbf{b} = S_{\mathbf{y}}\mathbf{D}_{S_{\mathbf{X}}}^{-1}\boldsymbol{\beta} = \begin{bmatrix} 10.398 \\ 2.553 \end{bmatrix}$$

$$R_{\mathbf{y}.\mathbf{X}}^2 = \mathbf{r}_{\mathbf{Xy}}'\boldsymbol{\beta} = \mathbf{r}_{\mathbf{Xy}}'\mathbf{r}_{\mathbf{XX}}^{-1}\mathbf{r}_{\mathbf{Xy}} = 0.355, \qquad \mathrm{Var}(\hat{\mathbf{y}}) = R_{\mathbf{y}.\mathbf{X}}^2 S_{\mathbf{y}}^2 = 148.17$$

From these results we see that the last two least squares **b** weights are identical to the multiple correlation **b** weights. Both are to be applied to the two predictor attributes, state per capita income and state per capita high school completions.

How can the first least squares weight be fitted into the multiple correlation model? Up to this point, in presenting the multiple correlation solution, we have decided upon a scaling of the multiplicative **b** weights but have said nothing about an additive constant b_0, for a linear equation of the form $\hat{y}_i = b_0 + b_1 X_{i1} + b_2 X_{i2}$. Clearly, $R_{\mathbf{y}.\mathbf{X}}^2$ is unaffected by the choice. One requirement that may be desirable—one that establishes even more closely the link between least squares and multiple correlation—is that the average **y** value be equal to the average $\hat{\mathbf{y}}$ value:

$$\bar{y} = \bar{\hat{y}} = (1/n)(\mathbf{1}'\mathbf{X})\mathbf{b} + b_0$$

or

$$b_0 = \bar{y} - (1/n)(\mathbf{1}'\mathbf{X})\mathbf{b} \tag{7.3.17}$$

For the example we see that $b_0 = -39.875$ is identical to the first least squares weight.

Finally, the link between least squares and multiple correlation that is probably of greatest practical importance to the reader is between $\mathbf{e'e}$, the sum of squares of errors of prediction in the least squares model, and $R^2_{\mathbf{y.X}}$, the squared multiple correlation coefficient. Intuitively, if there is a strong correlation between the criterion and some linear transformation of the predictors, then one ought to find relatively little error in the prediction of the criterion as a linear function of the predictor attributes. Large R^2 values should be associated with small $\mathbf{e'e}$ values.

The form of the relation between $\mathbf{e'e}$ and $R^2_{\mathbf{y.X}}$ is again most easily shown by considering the computation of $\mathbf{e'e}$ for $\mathbf{X}$ and $\mathbf{y}$ in deviation score form. We can then substitute into (7.1.16), $\mathbf{e'e} = \mathbf{y'y} - \mathbf{y'X(X'X)^{-1}X'y}$, to get

$$\begin{aligned} \mathbf{e'e} &= nS_{\mathbf{y}}^2 - nS_{\mathbf{y}}\mathbf{r}'_{\mathbf{Xy}}\mathbf{D}_{S_{\mathbf{X}}}\left(n\mathbf{D}_{S_{\mathbf{X}}}\mathbf{r}_{\mathbf{XX}}\mathbf{D}_{S_{\mathbf{X}}}\right)^{-1}nS_{\mathbf{y}}\mathbf{D}_{S_{\mathbf{X}}}\mathbf{r}_{\mathbf{Xy}} \\ &= nS_{\mathbf{y}}^2 - nS_{\mathbf{y}}\mathbf{r}'_{\mathbf{Xy}}\mathbf{D}_{S_{\mathbf{X}}}\mathbf{D}_{S_{\mathbf{X}}}^{-1}\mathbf{r}_{\mathbf{XX}}^{-1}\mathbf{D}_{S_{\mathbf{X}}}(1/n)\left(nS_{\mathbf{y}}\mathbf{D}_{S_{\mathbf{X}}}\mathbf{r}_{\mathbf{Xy}}\right) \\ &= nS_{\mathbf{y}}^2 - nS_{\mathbf{y}}^2\mathbf{r}'_{\mathbf{Xy}}\mathbf{r}_{\mathbf{XX}}^{-1}\mathbf{r}_{\mathbf{Xy}} \end{aligned}$$

or

$$\mathbf{e'e}/n = S_{\mathbf{y}}^2\left(1 - R^2_{\mathbf{y.X}}\right) \tag{7.3.18}$$

The reader may want to compare $\mathbf{e'e}$ and $R^2_{\mathbf{y.X}}$ for the Higher Education Study data given earlier to verify this relation.

Equation (7.3.18) is of value because it permits us to go back and forth between R^2 and $\mathbf{e'e}$, provided we know the criterion variance $S_{\mathbf{y}}^2$. The average squared error of prediction in least squares is equal to the criterion variance minus R^2 times the criterion variance.

If $R^2 = 0$, as is the case if the predictors are uncorrelated to the criterion, the least squares errors are those that result if each criterion observation were given the same prediction, that of the criterion mean. If $R^2 = 1$, there exists a linear transformation of the predictors that is perfectly correlated with the criterion and the least squares errors go to zero.

If in the Higher Education Study we had used but one predictor, then the multiple correlation would be simply the zero order product moment correlation. For example, correlating state income with state spending gives a $\boldsymbol{\beta}$ weight of the form

$$\beta = r_{\mathbf{XX}}^{-1}r_{\mathbf{Xy}} = (1)r_{\mathbf{Xy}} = r_{\mathbf{Xy}} = 0.554$$

which is the correlation itself (now $r_{\mathbf{XX}}$ is the scalar 1.0 and $r_{\mathbf{Xy}}$ is also a scalar). Similarly,

$$b = S_{\mathbf{y}}\mathbf{D}_{S_{\mathbf{X}}}^{-1}\beta = r_{\mathbf{Xy}}(S_{\mathbf{y}}/S_{\mathbf{X}}) = 0.0109$$

and

$$R^2_{\mathbf{y.X}} = \beta' r_{\mathbf{Xy}} = r^2_{\mathbf{Xy}} = 0.307$$

If we consider that we have dropped high school completions as a predictor by constraining its least squares weight to zero, we can write a constrained form of (7.3.18):

$$\mathbf{f'f}/n = S_{\mathbf{y}}^2\left(1 - r_{\mathbf{Xy}}^2\right)$$

Together these two results express the difference in the sum of squares of errors associated with dropping one or more predictions:

$$\mathbf{f'f}/n - \mathbf{e'e}/n = S_{\mathbf{y}}^2\left(1 - R_{\mathbf{y.x}}^2\right) - S_{\mathbf{y}}^2\left(1 - R_{\mathbf{y.x,z}}^2\right) = S_{\mathbf{y}}^2\left(R_{\mathbf{y.x,z}}^2 - R_{\mathbf{y.x}}^2\right) \tag{7.3.19}$$

The notation $R_{\mathbf{y.x,z}}^2$ indicates that two predictors $\mathbf{x}$ and $\mathbf{z}$ are weighted to predict $\mathbf{y}$ (in the least squares sense). The errors associated with two predictors, $\mathbf{e'e}$, are compared to those that result when only the $\mathbf{x}$ predictor is employed. *The difference in the squared multiple correlations is proportional to the difference in the average squared errors of prediction.*

For the prediction of state spending based on one and two predictors, the numerical results are

$$\mathbf{f'f} = 14{,}660.66, \qquad \mathbf{e'e} = 13{,}657.46, \qquad R_{\mathbf{y.X_1X_2}}^2 = 0.355$$

$$R_{\mathbf{y.X_1}}^2 = r_{\mathbf{yX_1}}^2 = 0.307, \qquad S_{\mathbf{y}}^2 = 423.35$$

7.3.2 Partial Correlation

Equation (7.3.18) and other equivalent expressions are frequently used to interpret the errors of prediction as reflecting the amount or proportion of criterion variability that has not been accounted for. That is, $R_{\mathbf{y.x}}^2$ is loosely spoken of as the proportion of criterion variance accounted for by the predictors in the set $\mathbf{x}$ and $1 - R_{\mathbf{y.x}}^2$ as the proportion of criterion variance not accounted for. The meaning of "accounted for" or "not accounted for" seems to us to be strictly limited to what is expressly stated in Eq. (7.3.18).

This "proportion of variance accounted for" interpretation extends to Eq. (7.3.19). The *incremental* or *additional* amount of criterion variance accounted for by $\mathbf{x}$ and $\mathbf{z}$ together, above that accounted for by $\mathbf{x}$ alone, is indicated by the increase in the squared multiple correlation—from $R_{\mathbf{y.x}}^2$ to $R_{\mathbf{y.x,z}}^2$.

In this section we introduce the concept of *partial variables* and their correlations as a way of explaining incremental or additive contributions to the reduction of errors in least squares prediction. We begin with the simplest case of predictions based on one or two predictor attributes. For the Higher Education Study we observed the following zero order and multiple correlations:

$$R_{\text{spending.income,completions}}^2 = 0.355, \qquad R_{\text{s.i,c}} = 0.596$$

$$r_{\text{spending.income}}^2 = 0.307, \qquad r_{\text{si}} = 0.554$$

$$r_{\text{spending.completions}}^2 = 0.077, \qquad r_{\text{sc}} = 0.278$$

$$r_{\text{income.completions}}^2 = 0.012, \qquad r_{\text{ic}} = 0.110$$

Income variability from state to state accounts for about 31% of the variability in higher education spending; $r_{si}^2 = 0.307$. Income and high school completions *together* account for about 36% of the spending variance. The *difference* of 5% is the incremental contribution of completions as a predictor.

The reader will note, however, that the *incremental contribution* is *less than* the proportion of criterion variability that could have been accounted for by high school completions alone. If that attribute was used as the only predictor of spending, $r_{sc}^2 = 0.08$, completion variability from state to state would account for nearly 8% of the variability in spending. Why is the *incremental* contribution less than this basal contribution? The "answer" lies in the fact that completion and *income* are themselves correlated. Here there is a correlation of $r_{ic} = 0.110$ between the two. If completions and income were uncorrelated, then we would observe an equality between the incremental and basal contributions and $R_{s.c,i}^2 = r_{sc}^2 + r_{si}^2$ or

$$R_{s.c,i}^2 - r_{sc}^2 = r_{si}^2$$

Except in controlled laboratory studies, uncorrelated predictors are a rarity; we need to know how to deal with correlated predictors.

In introducing least square prediction, we defined an error variable [see (7.1.3)] $\mathbf{e} = \mathbf{y} - \mathbf{Xb}$, where the transformation vector $\mathbf{b}$ was selected so as to minimize the sum of squares of the discrepancies between $\hat{\mathbf{y}} = \mathbf{Xb}$ and $\mathbf{y}$.

We shall, however, broaden the use of the least squares procedure beyond estimating the criterion in a particular study and prefer therefore to also call $\mathbf{e}$ a *partial variable*. That is, a partial variable, written $\mathbf{y.X}$, is defined by

$$\mathbf{y.X} = \mathbf{y} - \mathbf{Xb} = \mathbf{y} - \mathbf{X}(\mathbf{X'X})^{-1}\mathbf{X'y} = \left[\mathbf{I} - \mathbf{X}(\mathbf{X'X})^{-1}\mathbf{X'}\right]\mathbf{y} \tag{7.3.20}$$

where $\mathbf{b}$ (a vector if there is more than one column in the matrix $\mathbf{X}$) has been selected to minimize the difference, in the least square sense, between $\mathbf{y}$ and $\mathbf{Xb}$. In forming $\mathbf{y.X}$, we say that $\mathbf{y}$ has had (the effects of) $\mathbf{X}$ "partialed" from it.

The partial variable has some interesting properties. The vector equation (7.1.11), $\mathbf{X'y} = \mathbf{X'Xb}$, holds in particular for $\mathbf{X}_{.1} = \mathbf{1}$; that is, $\mathbf{1'y} = \mathbf{1'Xb}$ or $\mathbf{1'(y - Xb)} = \mathbf{1'y.X} = 0$. This tells us that the average value of the partial variable (the average error in least squares prediction) is zero:

$$\overline{\mathbf{y.X}} = 0 \tag{7.3.21}$$

Since $\mathbf{y.X}$ is centered about zero, its variance can be written $S_{\mathbf{y.X}}^2 = (1/n)(\mathbf{y.X})'(\mathbf{y.X})$ or, from (7.3.20),

$$S_{\mathbf{y.X}}^2 = (1/n)\mathbf{y'}\left[\mathbf{I} - \mathbf{X}(\mathbf{X'X})^{-1}\mathbf{X'}\right]\mathbf{y}$$

From (7.1.16) and (7.3.18), however, we recognize that the right side of this equation is

$$S_{\mathbf{y.X}}^2 = S_{\mathbf{y}}^2\left(1 - R_{\mathbf{y.X}}^2\right) \tag{7.3.22}$$

The partial variable has a mean of zero and a variance usually *smaller* than that of the original or zero order variable. (**y** here is the original variable, from which the effect of **X** is being partialed.) The amount by which the variance is reduced, we note again, depends on the (multiple) correlation between **y** and **X**.

Other important characteristics of **y.X** are its covariances with other attributes. Consider first the covariance between **y.X** and **X**, which can be expressed as the matrix product

$$\begin{aligned}\mathbf{Cov}(\mathbf{X}, \mathbf{y.X}) &= (1/n)\{\mathbf{X}'[\mathbf{I} - \mathbf{1}(\mathbf{1}'\mathbf{1})^{-1}\mathbf{1}']\}[\mathbf{y} - \mathbf{Xb}] \\ &= (1/n)[(\mathbf{X}'\mathbf{y} - \mathbf{X}'\mathbf{Xb}) - (1/n)(\mathbf{X}'\mathbf{1})(\mathbf{1}'\mathbf{y} - \mathbf{1}'\mathbf{Xb})]\end{aligned}$$

Again, by (7.1.11) $\mathbf{X}'\mathbf{Xb} = \mathbf{X}'\mathbf{y}$ and $\mathbf{1}'\mathbf{Xb} = \mathbf{1}'\mathbf{y}$, so that the covariance between **X** and **y.X** must be zero. Consequently, the correlation between **X** and **y.X** is also zero:

$$\mathbf{Cov}(\mathbf{X}, \mathbf{y.X}) = \mathbf{r}(\mathbf{X}, \mathbf{y.X}) = \mathbf{0} \tag{7.3.23}$$

After the effects of **X** have been partialed from **y**, the residual or partial variable is uncorrelated with **X**. It is important to note that Eq. (7.3.23) is in the form of a vector. The *partial variable* **y.X** has *covariance of zero with each column of* **X**.

A direct consequence of this result is that **y.X** has zero covariance (and correlation) with any linear transformation of the columns of **X**:

$$\mathrm{Cov}(\mathbf{Xa}, \mathbf{y.X}) = \mathbf{a}'\,\mathbf{Cov}(\mathbf{X}, \mathbf{y.X}) = \mathbf{a}'\mathbf{0} = 0 = r_{\mathbf{Xa},\mathbf{y.X}} \tag{7.3.24}$$

For the particular linear transformation $\hat{\mathbf{y}} = \mathbf{Xb}$ this is

$$\mathrm{Cov}(\hat{\mathbf{y}}, \mathbf{y.X}) = 0 = r_{\hat{\mathbf{y}},\mathbf{y.X}} \tag{7.3.25}$$

The covariance between **y.X** and **y** can be obtained from

$$\begin{aligned}\mathrm{Cov}(\mathbf{y}, \mathbf{y.X}) &= (1/n)\{\mathbf{y}'[\mathbf{I} - \mathbf{1}(\mathbf{1}'\mathbf{1})^{-1}\mathbf{1}']\}(\mathbf{y} - \mathbf{Xb}) \\ &= (1/n)[(\mathbf{y}'\mathbf{y} - \mathbf{y}'\mathbf{Xb}) - (1/n)\mathbf{y}'1(\mathbf{1}'\mathbf{y} - \mathbf{1}'\mathbf{Xb})] \\ &= (1/n)(\mathbf{y}'\mathbf{y} - \mathbf{y}'\mathbf{Xb})\end{aligned}$$

Again from (7.1.16) and (7.3.18), this result gives

$$\mathrm{Cov}(\mathbf{y}, \mathbf{y.X}) = S_{\mathbf{y}}^2\left(1 - R_{\mathbf{y.X}}^2\right) \tag{7.3.26}$$

From (7.3.22) and (7.3.26), the correlation between **y** and **y.X** is

$$r_{\mathbf{y},\mathbf{y.X}} = S_{\mathbf{y}}^2\left(1 - R_{\mathbf{y.X}}^2\right)/S_{\mathbf{y}}\left[S_{\mathbf{y}}^2\left(1 - R_{\mathbf{y.X}}^2\right)\right]^{1/2} = \left(1 - R_{\mathbf{y.X}}^2\right)^{1/2} \tag{7.3.27}$$

If relatively little is partialed from **y** (if $R^2_{\mathbf{y}.\mathbf{X}}$ is quite small), then the partial variable and the original **y** are highly correlated.

Finally, let us see what we can say about the covariance and correlation between **y**.**X** and an outside variable **z**, one that is neither **y** nor one of the attributes included among the columns of **X**. The covariance is

$$\begin{aligned}\mathrm{Cov}(\mathbf{z},\mathbf{y}.\mathbf{X}) &= (1/n)\{\mathbf{z}'[\mathbf{I} - \mathbf{1}(\mathbf{1}'\mathbf{1})^{-1}\mathbf{1}']\}(\mathbf{y} - \mathbf{Xb}) \\ &= (1/n)\{\mathbf{z}'[\mathbf{I} - \mathbf{1}(\mathbf{1}'\mathbf{1})^{-1}\mathbf{1}']\mathbf{y} - \mathbf{z}'[\mathbf{I} - \mathbf{1}(\mathbf{1}'\mathbf{1})^{-1}\mathbf{1}']\mathbf{Xb}\} \\ &= \mathrm{Cov}(\mathbf{z},\mathbf{y}) - \mathrm{Cov}(\mathbf{z},\mathbf{Xb}) = \mathrm{Cov}(\mathbf{z},\mathbf{y}) - \mathbf{Cov}(\mathbf{z},\mathbf{X})'\mathbf{b}\end{aligned}$$

Converting this covariance into a correlation gives

$$r_{\mathbf{z}(\mathbf{y}.\mathbf{X})} = (r_{\mathbf{yz}}S_{\mathbf{y}}S_{\mathbf{z}} - S_{\mathbf{z}}\mathbf{b}'\mathbf{D}_{\mathbf{S_X}}\mathbf{r_{Xz}})/S_{\mathbf{z}}S_{\mathbf{y}}(1 - R^2_{\mathbf{y}.\mathbf{X}})^{1/2}$$

or, on substituting for **b** and $R^2_{\mathbf{y}.\mathbf{X}}$ from (7.3.14) into (7.3.16),

$$r_{\mathbf{z}(\mathbf{y}.\mathbf{X})} = (r_{\mathbf{yz}} - \mathbf{r}'_{\mathbf{Xy}}\mathbf{r}^{-1}_{\mathbf{XX}}\mathbf{r}_{\mathbf{Xz}})/(1 - \mathbf{r}'_{\mathbf{Xy}}\mathbf{r}^{-1}_{\mathbf{XX}}\mathbf{r}_{\mathbf{Xy}})^{1/2} \tag{7.3.28}$$

This is termed a *semipartial* (or *part*) correlation. The name derives from the idea that a partial variable is being correlated with a raw or nonpartial variable. (We shall shortly contrast this correlation with a partial correlation, a correlation involving *two* partial variables.)

If **x** is a vector—if the effect of only one other attribute is being partialed from **y**—the *semipartial* correlation is said to be *first-order* and (7.3.28) simplifies to

$$r_{\mathbf{z}(\mathbf{y}.\mathbf{x})} = (r_{\mathbf{yz}} - r_{\mathbf{xy}}r_{\mathbf{xz}})/(1 - r^2_{\mathbf{xy}})^{1/2} \tag{7.3.29}$$

Let us rename the three measures in the Higher Education Study: **x**, state per capita high school completions; **y**, state per capita income; and **z**, higher education spending. We can relate the semipartial correlation to the incremental contribution of a predictor: From $r_{\mathbf{xy}} = 0.110$, $r_{\mathbf{xz}} = 0.278$, and $r_{\mathbf{yz}} = 0.554$, we compute $r_{\mathbf{z}(\mathbf{y}.\mathbf{x})} = 0.527$.

Earlier we determined that the squared multiple correlation between spending and the best linear combination of high school completions and income is $R^2_{\mathbf{z}.\mathbf{xy}} = 0.355$, and we described the incremental contribution of income as the *difference* between this squared multiple correlation and the square of the zero order correlation between spending and completions: $R^2_{\mathbf{z}.\mathbf{yx}} - r^2_{\mathbf{zx}} = 0.355 - 0.077 = 0.278$. This incremental contribution is identical to the square of the semipartial correlation between appropriations and income with completions partialed from income:

$$R^2_{\mathbf{z}.\mathbf{x},\mathbf{y}} - R^2_{\mathbf{zx}} = r^2_{\mathbf{z}(\mathbf{y}.\mathbf{x})} = 0.355 - 0.077 = (0.527)^2 \tag{7.3.30}$$

Intuitively, this can be interpreted as the result of having "replaced" the predictor *income* (which was correlated with completions) with a predictor *income* × *completions* that is uncorrelated with completions. Equation (7.3.30) states that the sum of the contributions of the two uncorrelated predictors is identically their joint contribution to the prediction of the criterion.

There is more than one way, of course, to create two uncorrelated variables from two correlated ones. We could have partialed income from completions and broken down the joint contribution of the two predictors into $R^2_{\mathbf{z.x,y}} = r^2_{\mathbf{zy}} + r^2_{\mathbf{z(x.y)}}$, thus showing the *incremental* contribution of per capita high school completions.

Equation (7.3.30) generalizes to more than two predictors, although we shall not provide numerical examples here. The joint contribution of three predictor measures might be broken down as

$$R^2_{\mathbf{z.w,x,y}} = r^2_{\mathbf{zw}} + r^2_{\mathbf{z(x.w)}} + r^2_{\mathbf{z(y.x,w)}} \tag{7.3.31}$$

Here **w** has been partialed from **x** and both **x** and **w** partialed from **y**. The three variables **w**, **x.w**, and **y.x, w** will be mutually uncorrelated. Note that the first two terms on the right define a squared multiple correlation: $R^2_{\mathbf{z.w,x}} = r^2_{\mathbf{zw}} + r^2_{\mathbf{z(x.w)}}$. Equation (7.3.31) might therefore be rewritten

$$R^2_{\mathbf{z.w,x,y}} = R^2_{\mathbf{z.w,x}} + r^2_{\mathbf{z(y.w,x)}}$$

or, by rearranging terms,

$$r^2_{\mathbf{z(y.x,w)}} = R^2_{\mathbf{z.w,x,y}} - R^2_{\mathbf{z.w,x}} \tag{7.3.32}$$

In this form a squared semipartial correlation again gives the *incremental* contribution of a third predictor.

Now we shall develop the related concept of *partial correlation*. We begin by defining two partial variables **y.X** and **z.X**. Note that it is the effect of the same (set of) variables that we partial from both **y** and **z**. That is,

$$\mathbf{y.X} = \mathbf{y} - \mathbf{Xb} = \mathbf{y} - \mathbf{X}(\mathbf{X'X})^{-1}\mathbf{X'y} = \left[\mathbf{I} - \mathbf{X}(\mathbf{X'X})^{-1}\mathbf{X'}\right]\mathbf{y}$$

and

$$\mathbf{z.X} = \mathbf{Z} - \mathbf{Xa} = \mathbf{z} - \mathbf{X}(\mathbf{X'X})^{-1}\mathbf{X'z} = \left[\mathbf{I} - \mathbf{X}(\mathbf{X'X})^{-1}\mathbf{X'}\right]\mathbf{z}$$

The properties derived for **y.X** hold as well for **z.X**:

1. $\overline{\mathbf{z.X}} = 0$.
2. $S^2_{\mathbf{z.X}} = S^2_{\mathbf{z}}(1 - R^2_{\mathbf{z.X}})$.
3. $\mathbf{r}_{(\mathbf{X,z.X})} = \mathbf{0}$.
4. $\mathrm{Cov}(\mathbf{Xb}, \mathbf{z.X}) = 0$.
5. $r_{\mathbf{z,z.X}} = (1 - R^2_{\mathbf{z.X}})^{1/2}$.
6. $r_{\mathbf{y(z.X)}} = (r_{\mathbf{yz}} - \mathbf{r}'_{\mathbf{Xz}}\mathbf{r}^{-1}_{\mathbf{XX}}\mathbf{r}_{\mathbf{Xy}})/(1 - \mathbf{r}'_{\mathbf{Xz}}\mathbf{r}^{-1}_{\mathbf{XX}}\mathbf{r}_{\mathbf{Xz}})^{1/2}$.

The new relation we wish to develop is the correlation $r_{\mathbf{(y.X)(z.X)}}$ between the two partial variables. First, the covariance between the two can be expressed

$$\begin{aligned}\mathrm{Cov}(\mathbf{y.X}, \mathbf{z.X}) &= (1/n)\{(\mathbf{y} - \mathbf{Xb})'[\mathbf{I} - (1/n)\mathbf{11}'](\mathbf{z} - \mathbf{Xa})\} \\ &= \mathrm{Cov}(\mathbf{y}, \mathbf{z}) - \mathbf{b}'\,\mathrm{Cov}(\mathbf{X}'\mathbf{z}) - \mathbf{a}'\,\mathrm{Cov}(\mathbf{X}'\mathbf{y}) + \mathbf{b}'\,\mathrm{Cov}(\mathbf{X})\,\mathbf{a}\end{aligned}$$

Again, by (7.3.14)–(7.3.16), the vectors **a** and **b** may be expressed

$$\mathbf{a} = S_{\mathbf{z}}\mathbf{D}_{S_{\mathbf{X}}}^{-1}\mathbf{r}_{\mathbf{XX}}^{-1}\mathbf{r}_{\mathbf{Xz}} \qquad \text{and} \qquad \mathbf{b} = S_{\mathbf{y}}\mathbf{D}_{S_{\mathbf{X}}}^{-1}\mathbf{r}_{\mathbf{XX}}^{-1}\mathbf{r}_{\mathbf{Xy}}$$

When these are substituted into the covariance and the summations completed, we find

$$\mathrm{Cov}(\mathbf{y.X}, \mathbf{z.X}) = S_{\mathbf{y}}S_{\mathbf{z}}\left(r_{\mathbf{yz}} - \mathbf{r}'_{\mathbf{Xy}}\mathbf{r}_{\mathbf{XX}}^{-1}\mathbf{r}_{\mathbf{Xz}}\right)$$

The partial correlation is then found by dividing this covariance by the product of the standard deviations for the two partial variables:

$$\begin{aligned}r_{(\mathbf{y.X})(\mathbf{z.X})} &= \frac{S_{\mathbf{y}}S_{\mathbf{z}}\left(r_{\mathbf{yz}} - \mathbf{r}'_{\mathbf{Xy}}\mathbf{r}_{\mathbf{XX}}^{-1}\mathbf{r}_{\mathbf{Xz}}\right)}{\left[S_{\mathbf{y}}^2\left(1 - R_{\mathbf{y.X}}^2\right)\right]^{1/2}\left[S_{\mathbf{z}}^2\left(1 - R_{\mathbf{z.X}}^2\right)\right]^{1/2}} \\ &= \frac{r_{\mathbf{yz}} - \mathbf{r}'_{\mathbf{Xy}}\mathbf{r}_{\mathbf{XX}}^{-1}\mathbf{r}_{\mathbf{Xz}}}{\left(1 - R_{\mathbf{y.X}}^2\right)^{1/2}\left(1 - R_{\mathbf{z.X}}^2\right)^{1/2}}\end{aligned} \qquad (7.3.33)$$

The reader may note that this is quite similar to the semipartial $r_{\mathbf{z}(\mathbf{y.X})}$ given by (7.3.28). The partial correlation has the same sign as either this semipartial or

$$r_{\mathbf{y}(\mathbf{z.X})} = \left(r_{\mathbf{yz}} - \mathbf{r}'_{\mathbf{Xz}}\mathbf{r}_{\mathbf{XX}}^{-1}\mathbf{r}_{\mathbf{Xy}}\right)/\left(1 - R_{\mathbf{z.X}}^2\right)^{1/2}$$

but is usually larger in magnitude than either:

$$r_{(\mathbf{y.X})(\mathbf{z.X})} = r_{\mathbf{z}(\mathbf{y.X})}/\left(1 - R_{\mathbf{z.X}}^2\right)^{1/2} = r_{\mathbf{y}(\mathbf{z.X})}/\left(1 - R_{\mathbf{y.X}}^2\right)^{1/2}$$

Apart from its logical interpretation, the partial correlation can also be related to the multiple correlation and to the incremental contribution of predictors in the least squares model. Recalling that $r_{\mathbf{z}(\mathbf{y.X})}^2$ can be interpreted as the incremental contribution of **y** to the prediction of **z**—the content of (7.3.32)—and noting that the squared partial correlation can be written in terms of a squared semipartial,

$$r_{(\mathbf{z.X})(\mathbf{y.X})}^2 = r_{\mathbf{y}(\mathbf{z.X})}^2/\left(1 - R_{\mathbf{y.X}}^2\right)$$

we may write the squared partial correlation as the ratio

$$r_{(\mathbf{z.X})(\mathbf{y.X})}^2 = \left(R_{\mathbf{y.X,z}}^2 - R_{\mathbf{y.X}}^2\right)/\left(1 - R_{\mathbf{y.X}}^2\right) \qquad (7.3.34)$$

In this ratio the incremental contribution of **z** to the prediction of **y** is compared to the unaccounted for criterion variability $1 - R_{\mathbf{y.X}}^2$. On the right-hand side of (7.3.34), then, we have the proportion of the remaining variability (remaining after $\mathbf{y}$ has been predicted by $\mathbf{X}$) that can be accounted for by the additional attribute $\mathbf{z}$.

This introduction to partial variables, their variances, and their correlations will be reexamined and extended in later chapters. Multivariate analysis is often concerned with determining the effect of partialing one measure, or set of measures, from another. Sometimes, as here, the partialing is of observed attributes. At other times, we shall want to partial a *composite* or *hypothetical* variate from certain observations. In either event, we are led to the further study of partial variables.

7.4 Summary

In this chapter we examined the least squares approach to transforming predictor attributes to optimally account for variance in a set of criterion observations. We defined the least squares function of the errors, which was to be minimized. Then, using the tools of matrix algebra and optimization techniques, we solved for the elements of the vector $\mathbf{b}$ in the linear transformation $\mathbf{Xb} = \hat{\mathbf{y}}$.

We next examined the effects of redundant information in the predictor attributes on the estimation of the criterion. Two degrees of redundancy were discussed. In the case of total redundancy (a nonbasic matrix of predictor attributes), we found that although for a given set of predictor information the sum of squares of the errors of prediction is unique, many weighting vectors $\mathbf{b}$ lead to the value of $\mathbf{e}'\mathbf{e}$. We suggested two solutions, one based on basic structure considerations and the other consisting simply in eliminating the redundant column of $\mathbf{X}$. Recognizing that predictors contribute differentially to the prediction of the criterion, we also compared the fit of weighted composites of the predictors that incorporate no predictor information (the case of $\mathbf{b}$ null), and we discussed predicting from some of the predictor information. We gave equations incorporating all, some, or none of the predictor attributes as constraints. This approach is very general and will be used in later chapters to express various relations among predictors and criteria.

Further consideration of the errors of prediction leads to the more general formulation of errors as partial variables. Expressions are then obtained for semipartial correlations between partial variables and raw scores and partial correlations representing the relation between two partial variables.

The data-bound descriptive model we explored in this chapter provides an introduction to more general concepts such as constraint matrices, partial correlations, semipartial correlations, and multiple correlations. The descriptive model is also useful in its own regard when we are interested in investigating relations between a criterion and predictors within a set of observations.

Exercises

1. You are given the following values for two predictors $\mathbf{X}_1$ and $\mathbf{X}_2$ and for a criterion $\mathbf{y}$:

$\mathbf{X}_1$	$\mathbf{X}_2$	y
0	0	10
10	0	30
0	10	35
10	10	55
20	10	80
10	20	95
5	0	20
0	5	25
5	5	35

a. Find the means and variances of the three attributes.
b. Find the best fitting (least squares) equation for predicting **y** from $\mathbf{X}_1$ and $\mathbf{X}_2$.
c. Find the correlations among the three attributes. How do you interpret these within the least squares model?
d. Find the partial correlation between $\mathbf{X}_2$ and **y** (partialing out $\mathbf{X}_1$).

2. If 65% of the variance in **z** is associated with variation in **x** and 70% of the variation in **z** is associated with variation in **y**, what can be said about the magnitude of the correlation between **x** and **y**?

3. Why is it possible for some authors of statistics books to discuss prediction equations without any reference to the correlation coefficient?

4. If the correlation between **x** and **y** is 0.60 and between **x** and **w** is 0.80, what are the maximum and minimum values r_{yw} can achieve?

5. Define a set of three attributes of interest to you. Then decide upon a particular partial correlation involving the three. Describe this partial correlation as a zero order correlation between two attributes. What are the two attributes?

6. Many clinical psychologists administer the Rorschach (ink blot) test. Becausc the score for any particular category (say, C, the use of color) is influenced by the total number of responses (a score R), some attempt to control for R by using a percentage score: $100(C/R)$ = percent color. Suggest and justify another procedure that would take care of individual differences in R, in the sense of yielding a score for color independent of R.

7. On the basis of the least squares model, interpret this equation:

$$\sum_{i=1}^{n} (Y_i - \bar{Y})^2 = \sum_{i=1}^{n} (Y_i - \hat{Y}_i)^2 + \sum_{i=1}^{n} (\hat{Y}_i - \bar{Y})^2$$

8. Show that $\mathbf{I} - (\mathbf{X}'\mathbf{X})^{-1}\mathbf{C}[\mathbf{C}'(\mathbf{X}'\mathbf{X})^{-1}\mathbf{C}]^{-1}\mathbf{C}'$ is an idempotent matrix.

9. Based on the discussion of Section 7.2.2, show that $\mathbf{C}'(\mathbf{X}'\mathbf{X})^{-1}\mathbf{C}$ is a square basic matrix.

Additional Reading

The least squares approach is presented in considerable detail in the following book:

Daniel, C., and Wood, F. S. (1971). *Fitting Equations to Data*. New York: Wiley.

For a further discussion of the sample multiple and partial correlation coefficients, the reader may find this source of interest:

Thorndike, R. M. (1978). *Correlational Procedures for Research*. New York: Gardner.

To see how the least squares model can be exploited in the behavioral sciences, the following offers some interesting examples:

Ward, J. H., and Jennings, E. (1973). *Introduction to Linear Models*. Englewood Cliffs, NJ: Prentice-Hall.

Chapter 8

Linear Regression: Some Theory

Introduction

No data analytic approach is as extensively applied as is *linear regression*. Its basic premise is that the value we *expect* our criterion attribute to assume is a linear transformation of the values of our predictors. That assumption is attractive, whether we have strong theoretical reasons to believe in a linear relation or simply no reason to believe that any more complicated transformation should be necessary.

To make the regression approach into something with which we can test hypotheses, contrast models, and draw inferences about different populations of potential observations, we need to make some further assumptions, including some about how our data were sampled. In this chapter we develop the classical normal regression model; that is, we assume that our criterion observations are sampled independently from populations all of which are normal in distributional form and have the same variance. The population means, which are assumed to be functions of the different values taken on by one or more of our predictors, may differ. Our interest tends to center on testing the degree to which our data fit hypotheses about those population means.

We address this interest through the development of techniques, employing the maximum likelihood principle, for estimating the unknown parameters of these normal populations. We then are able to consider a procedure for testing hypotheses about population means, building upon the notion of linear constraints introduced in Chapter 7.

The resulting general linear model (GLM) is an extremely useful one. It underlies the analysis of variance (ANOVA) and analysis of covariance as well as multiple regression. An understanding of its statistical and structural basis is absolutely vital to its effective use. The reader who already knows about ANOVA and multiple regression will still benefit by studying this chapter: it provides a preparation for Part III, and the GLM gives a new perspective on these standard techniques.

8.1 Linear Regression Problem

The least squares approach and the concepts of multiple and partial correlation developed in Chapter 7 are limited in their analytic usefulness because they are essentially descriptive and data bound. For a specific set of n observations on m predictors and one criterion, the least squares weights for those predictors provide us with a linear composite that most closely matches the criterion (in the least squares sense). The multiple correlation, or its square, gives the closeness of the match, and the partial and semipartial correlations describe relations that exist *in that set of data* after we have made one (or more) attributes independent of others.

As long as we have before us the complete collection of observations, as we did in the Higher Education Study, these techniques meet our needs; more often, however, we have only a sample from the collection of potentially relevant observations. Still, we would like our study of the data at hand to permit us to make some claims, however guarded, about what we would expect to find were we to look at a new or different sample or to have revealed to us the characteristics of the larger population from which we have sampled. In short, we shall frequently want to make *inferences* from observations.

Making inferences requires some assumptions about the origin of the data. In this section we develop some of the more basic aspects of the sampling, inference, and hypothesis-testing model so widely used in behavioral science research. The reader is probably already familiar with some aspects to it, if not with the general model itself. We call this the *linear regression* or *general linear model* (GLM).

The basic concept underlying regression is that of a systematic relation (regression) of the means of populations of criterion observations to selected predetermined values of one or more predictor attributes. The regression is said to be linear if the criterion means can be assumed to be linear functions of these predictor values.

An example may make these ideas clearer. Assume that the same reading comprehension test, in French, is administered to every student enrolled in a French language class in a large metropolitan public school district. The resulting test scores are then organized into eight sets, depending on whether the score was earned by a student enrolled in the first, second, third,..., or eighth semester of study of French. The results might look like Figure 8.1, where the eight distributions of scores have been arrayed side by side. First-semester students earned scores between 0 and 40, for example,

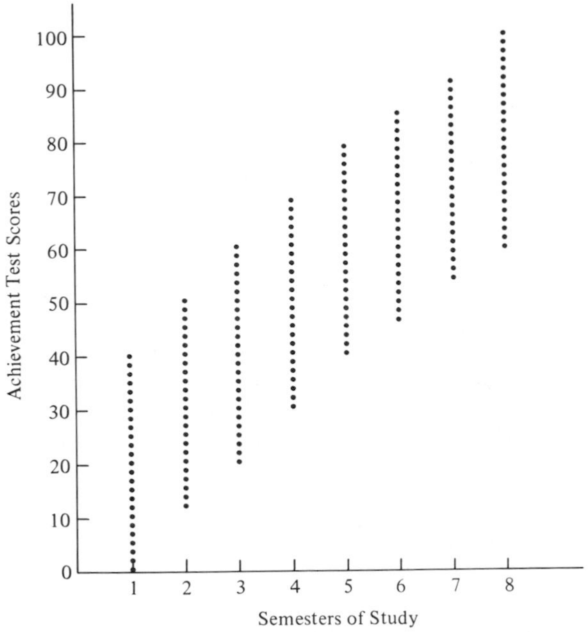

Figure 8.1 Distribution of French reading comprehension scores.

whereas those in their eighth semester of study obtained scores between 58 and 98.

Although not every student in a given semester of study earned the same score (there is *variability* in each of the score distributions), the average of the score distribution increases as the number of semesters of study increases. In technical terms, we say there is a *regression* of reading comprehension scores on number of semesters of study. Equivalently, the average or *expected value* of the comprehension test score distribution is a function of the number of semesters of study. Here we have carefully constructed the example so that the function is linear: the average of each of the eight distributions increases linearly with the number of semesters of study. The equation in this instance is

$$E(\text{comprehension score} \mid \text{semesters of study}) = 12 + 8(\text{semesters of study})$$

for semester values of $1, 2, \ldots, 8$. (The term on the left is read "The expected value of the comprehension score, given that we specify the number of semesters of study is....")

In the example we would know that a comprehension test score for a student in the third semester course is an observation from a distribution with a mean (expected value) of 36, whereas a test score for a student in the fifth semester course is a selection from a distribution with a mean of 52. To illustrate how linear regression can be represented in matrix form, assume that a set of six comprehension test scores, making up a *criterion vector*

$$\mathbf{y} = \begin{bmatrix} 30 \\ 15 \\ 24 \\ 42 \\ 61 \\ 42 \end{bmatrix}$$

has been observed, where the first two scores were obtained by first-semester students, the next two by a pair of third-semester enrollees, and the last two by students in their fifth semester of French study. With this information on the source of the scores, we can write a vector of *expected values* for the six observations:

$$\mathbf{E}(\mathbf{y}) = \begin{bmatrix} 20 \\ 20 \\ 36 \\ 36 \\ 52 \\ 52 \end{bmatrix}$$

The vector $\mathbf{E}(\mathbf{y})$ has the same number of elements as $\mathbf{y}$, and each element of $\mathbf{E}(\mathbf{y})$ is the mean of the population from which the corresponding element of $\mathbf{y}$ was selected, drawn, or chosen. For example, the first two elements of $\mathbf{y}$ are scores for first semester students, observations from a population of scores with a mean of 20.

Since we assume these expected test scores regress linearly on the number of semesters of study, we should be able to write the vector of expected values $\mathbf{E}(\mathbf{y})$ as a linear function of a set of predictor scores arrayed as a *predictor matrix*. A common way of expressing this linear function is

$$\mathbf{E}(\mathbf{y}) = \mathbf{X}\boldsymbol{\beta} \tag{8.1.1}$$

where $\mathbf{X}$ is an $n \times m$ predictor matrix, $\boldsymbol{\beta}$ is an m-element vector of *regression weights*, and $\mathbf{E}(\mathbf{y})$ is the n-element vector of population means. On the right of Eq. (8.1.1) we have a linear transformation of the columns of $\mathbf{X}$.

For our six criterion observations and an appropriate choice of $\mathbf{X}$ and $\boldsymbol{\beta}$, Eq. (8.1.1) provides

$$\begin{bmatrix} 20 \\ 20 \\ 36 \\ 36 \\ 52 \\ 52 \end{bmatrix} = \begin{bmatrix} 1 & 1 \\ 1 & 1 \\ 1 & 3 \\ 1 & 3 \\ 1 & 5 \\ 1 & 5 \end{bmatrix} \begin{bmatrix} 12 \\ 8 \end{bmatrix}$$

The two elements of $\boldsymbol{\beta}$ are the additive and multiplicative constants, respectively, of the linear equation linking the number of semesters of study (given in the second column of $\mathbf{X}$) to the expected comprehension test score (given on the left of the equation).

8.2 Maximum Likelihood Estimation of Parameters

Equation (8.1.1) expresses the fundamental relation in the linear regression model. It allows us to state the criterion value that is expected (on the average) given a vector of predictor scores. As a practical matter, however, Eq. (8.1.1) by itself is limited in value, as it requires that the elements of $\boldsymbol{\beta}$ be known. Since these, in turn, give information about the means of populations, they could be known only if we could observe and inspect the populations directly. Normally we cannot accumulate information about entire populations; rather, as mentioned earlier, we would like to infer something about these populations—in particular, how criterion population means are related to predictor values—from studying only a sample of observations.

Although the elements of $\boldsymbol{\beta}$ usually are unknown, the linear regression model permits us to estimate and test hypotheses about $\boldsymbol{\beta}$ on the basis of a sample of observations. To accomplish this, the GLM requires that we satisfy three assumptions about the populations sampled and the nature of the sampling:

1. Each criterion population sampled consists of *normally distributed* observations.
2. Each of these normal distributions has a *common variance.*
3. The n criterion observations are obtained *independently.*

That is, we shall assume the vector of criterion observations consists of elements chosen independently from normal distributions having the same variance but with potentially different means given by $\mathbf{E(y)} = \mathbf{X}\boldsymbol{\beta}$. The matrix $\mathbf{X}$ consists of known predictor scores and the vector $\boldsymbol{\beta}$ contains unknown population parameters.

The predictor matrix $\mathbf{X}$ is sometimes referred to as a *design matrix* since it indicates how the sample was chosen; that is, which normal distributions were sampled and how frequently. The design matrix for the six French comprehension scores,

$$\mathbf{X} = \begin{bmatrix} 1 & 1 \\ 1 & 1 \\ 1 & 3 \\ 1 & 3 \\ 1 & 5 \\ 1 & 5 \end{bmatrix}$$

indicates that we independently tested the French reading comprehension of two first-semester, two third-semester, and two fifth-semester students. Each row of $\mathbf{X}$ refers to a normal distribution from which an observation was sampled. If two (or more) rows of $\mathbf{X}$ are identical, then two (or more) of the corresponding elements of $\mathbf{y}$ (criterion observations) were drawn from the same normal distribution.

To estimate $\boldsymbol{\beta}$ requires that we take into account the properties of normal distributions. The distribution of test scores for third-semester students is pictured in Figure 8.2. The normal distribution is *symmetric* about its mean with values close to the mean more likely than values further removed. These distributional properties are summarized, mathematically, in the normal *probability density*

$$p(y_i) = f(y_i, \mu, \sigma^2) = (1/2\pi\sigma^2)^{1/2} e^{(-1/2)[(y_i-\mu)/\sigma]^2} \tag{8.2.1}$$

It is this equation that generates the bell-shaped curve: $p(y_i)$ is the height of the curve for $y = y_i$. The curve depends upon two parameters: μ is the mean of the distribution and σ^2 is its variance. $p(y_i)$ is a maximum for $y_i = \mu$ and falls off symmetrically as the value of y_i increases or decreases.

As a probability density, Eq. (8.2.1) is a function of varying values of y_i; a value close to μ is more probable than a value considerably larger or smaller. The equation has a second aspect to it: it is also a *likelihood function*; that is, it expresses the relative likelihood of observing a given value y_i for *different* values of the parameters μ and σ^2. This second interpretation of the function is important when y_i is known but μ and σ^2 are unknown.

For the linear regression model, Eq. (8.2.1) may be rewritten, taking note of the fact that the population mean μ is a linear function of the design matrix:

$$L(y_i|\boldsymbol{\beta}, \sigma^2) = (1/2\pi\sigma^2)^{1/2} \exp\{(-\tfrac{1}{2})[(y_i - \mathbf{X}'_{i.}\boldsymbol{\beta})/\sigma]^2\} \tag{8.2.2}$$

This equation gives the likelihood of the ith criterion observation as a function of the ith row of the design matrix, which is known, and the unknown vector $\boldsymbol{\beta}$ and σ^2. (An expression of the form $\exp k$ is the same as e^k.) If we have observed a sample of n criterion values, we could write an equation of this form for each of those observations.

Figure 8.2 Distribution of French reading comprehension scores, third semester.

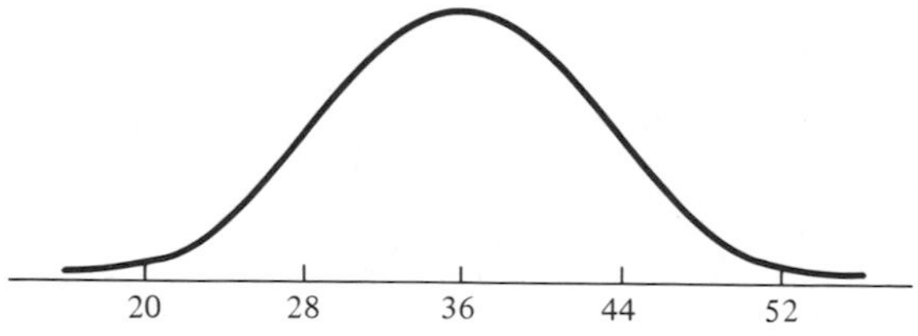

Just as the probabilities of independent events may be multiplied to produce the probability of their joint occurrence, the product of the likelihoods of each of a set of independent observations will give us the *likelihood of the entire sample*. If we can assume our n criterion observations were independently sampled (i.e., how the ith observation was selected was uninfluenced by how any of the other observations were obtained), the likelihood of the complete set may be written as the product of n terms, each of the form of the right side of Eq. (8.2.2):

$$\begin{aligned} L(\mathbf{y}|\boldsymbol{\beta}, \sigma^2) &= \prod_{i=1}^{n} L(y_i|\boldsymbol{\beta}, \sigma^2) \\ &= (1/2\pi\sigma^2)^{n/2} \prod_{i=1}^{n} \exp\{(-\tfrac{1}{2})[(y_i - \mathbf{X}'_{i.}\boldsymbol{\beta})/\sigma]^2\} \\ &= (1/2\pi\sigma^2)^{n/2} \exp\left\{(-\tfrac{1}{2}) \sum_{i=1}^{n} [(y_i - \mathbf{X}'_{i.}\boldsymbol{\beta})/\sigma]^2\right\} \\ &= (2\pi\sigma^2)^{-n/2} \exp\left[(-1/2\sigma^2) \sum_{i=1}^{n} (y_i - \mathbf{X}'_{i.}\boldsymbol{\beta})^2\right] \\ &= (2\pi\sigma^2)^{-n/2} \exp[(-1/2\sigma^2)(\mathbf{y} - \mathbf{X}\boldsymbol{\beta})'(\mathbf{y} - \mathbf{X}\boldsymbol{\beta})] \end{aligned} \tag{8.2.3}$$

Note that in the final equation the minor product moment of the vector $\mathbf{y} - \mathbf{X}\boldsymbol{\beta}$ replaces the equivalent scalar sum of squares expression

$$\sum_{i=1}^{n} (y_i - \mathbf{X}'_{i.}\boldsymbol{\beta})^2$$

(In writing the likelihood of the vector of criterion observations we noted that $(1/2\pi\sigma^2)^{1/2}$ was an unchanging multiplier of each term and then employed the rule on a product of exponential terms: $e^a e^b = e^{a+b}$.)

Eq. (8.2.3) gives the likelihood of observing a particular $\mathbf{y}$ *given* a design matrix $\mathbf{X}$, a regression vector $\boldsymbol{\beta}$, and a common variance σ^2 for the populations sampled. For a particular set of observations, what influences the value of $L(\mathbf{y})$? The matrix $\mathbf{X}$ is fixed: We decided how the criterion observations were to be sampled before they were observed. The design matrix will not vary within a linear regression problem. The parameters $\boldsymbol{\beta}$ and σ^2, however, are unknown. Since their values will influence the likelihood of the sample, we can use the property that different choices of $\boldsymbol{\beta}$ and σ^2 yield different likelihoods for an observed sample to estimate these unknown parameters.

We have already assumed that the distributions from which our criterion observations were sampled were normal in form, with a common variance σ^2, and with means that are linear functions of $\mathbf{X}$ [Eq. (8.1.1)]. Once we choose or estimate $\boldsymbol{\beta}$ and σ^2, we have completely specified the distributions sampled; that is, with values for $\boldsymbol{\beta}$ and σ^2 we can compute the likelihood for the sample. Here we shall use the *maximum likelihood* principle to estimate the unknown parameters. Briefly, this principle states that a reasonable estimator of an unknown parameter, based on a particular set of data, is the value of that parameter in the population (or populations) which is most likely to have provided the data. That is, the maximum likelihood estimators of $\boldsymbol{\beta}$ and σ^2 are the values for those parameters which give the sample likelihood, Eq. (8.2.3), its largest possible value. No other choice of estimators for $\boldsymbol{\beta}$ and σ^2 will specify a set of normal populations from which it is more likely that the sample of criterion observations was selected.

The matrix differentiation rules of Chapter 6 can be used to find values of $\boldsymbol{\beta}$ and σ^2 that maximize Eq. (8.2.3). First, though, we note that the sample likelihood is always nonnegative, and therefore the logarithm of the likelihood is maximized for the same values of $\boldsymbol{\beta}$ and σ^2 as maximize the likelihood itself. This is important because it is easier to show how $\log[L(\mathbf{y})]$ is maximized. [For convenience we again write $\log_e$ as log and $L(\mathbf{y}|\boldsymbol{\beta}, \sigma^2)$ as $L(\mathbf{y})$.] The log likelihood of $\mathbf{y}$ is

$$\log[L(\mathbf{y})] = \log\left[(2\pi)^{(-n/2)}\right] + \log\left[(\sigma^2)^{(-n/2)}\right] + (-1/2\sigma^2)(\mathbf{y} - \mathbf{X}\boldsymbol{\beta})'(\mathbf{y} - \mathbf{X}\boldsymbol{\beta}) \qquad (8.2.4)$$

We consider first the role of $\boldsymbol{\beta}$ in maximizing this function. $\boldsymbol{\beta}$ enters only into the third term on the right. As both σ^2 and the minor product moment are positive, the entire third term is negative. To maximize the (positive) log likelihood, then, we need to minimize the contribution of this third term; with respect to $\boldsymbol{\beta}$ we need to minimize $(\mathbf{y} - \mathbf{X}\boldsymbol{\beta})'(\mathbf{y} - \mathbf{X}\boldsymbol{\beta})$. We can accomplish this by recalling how we minimized a similar form in Chapter 7. There we sought a vector $\mathbf{b}$ that would satisfy the least squares criterion and minimize $\mathbf{e}'\mathbf{e} = (\mathbf{y} - \mathbf{X}\mathbf{b})'(\mathbf{y} - \mathbf{X}\mathbf{b})$. The maximum likelihood estimator of $\boldsymbol{\beta}$, then, must be

$$\hat{\boldsymbol{\beta}} = (\mathbf{X}'\mathbf{X})^{-1}\mathbf{X}'\mathbf{y} \qquad (8.2.5)$$

if the design matrix is basic. Recall that the nonunique vector

$$\boldsymbol{\beta}^{+} = (\mathbf{X}'\mathbf{X})^{+}\mathbf{X}'\mathbf{y} \qquad (8.2.6)$$

minimizes $(\mathbf{y} - \mathbf{X}\boldsymbol{\beta})'(\mathbf{y} - \mathbf{X}\boldsymbol{\beta})$ if $\mathbf{X}$ is nonbasic. Equation (8.2.5) gives a unique maximum likelihood estimator for $\boldsymbol{\beta}$ so long as $\mathbf{X}$ is of full rank. Rather than further develop the estimation problem for the rank-deficient design matrix, we suggest eliminating redundancies or linear dependences before pursuing a solution.

If $\hat{\boldsymbol{\beta}}$ is substituted for $\boldsymbol{\beta}$ in Eq. (8.1.1), a vector of *estimated population means* results:

$$\hat{\mathbf{E}}(\mathbf{y}) = \mathbf{X}\hat{\boldsymbol{\beta}} = \mathbf{X}(\mathbf{X}'\mathbf{X})^{-1}\mathbf{X}'\mathbf{y} \qquad (8.2.7)$$

The difference vector $\mathbf{y} - \hat{\mathbf{E}}(\mathbf{y}) = \mathbf{y} - \mathbf{X}\hat{\boldsymbol{\beta}}$ has as its elements discrepancies between the criterion observations and the maximum likelihood estimators of the means of the populations from which the observations were sampled. The minor product moment of this difference vector is a sum of squares of deviations of criterion observations about their estimated means:

$$[\mathbf{y} - \hat{\mathbf{E}}(\mathbf{y})]'[\mathbf{y} - \hat{\mathbf{E}}(\mathbf{y})] = (\mathbf{y} - \mathbf{X}\hat{\boldsymbol{\beta}})'(\mathbf{y} - \mathbf{X}\hat{\boldsymbol{\beta}})$$

We can exploit the least squares development again, particularly Eqs. (7.1.15)–(7.1.18), to note that this sum of squared deviations is computationally equivalent to $\mathbf{e}'\mathbf{e}$:

$$\begin{aligned}[\mathbf{y} - \hat{\mathbf{E}}(\mathbf{y})]'[\mathbf{y} - \hat{\mathbf{E}}(\mathbf{y})] &= \mathbf{y}'\mathbf{y} - \mathbf{y}'\mathbf{X}(\mathbf{X}'\mathbf{X})^{-1}\mathbf{X}'\mathbf{y} = \mathbf{y}'\mathbf{y} - \mathbf{y}'\mathbf{X}\hat{\boldsymbol{\beta}} \\ &= \mathbf{y}'\left[\mathbf{I} - \mathbf{X}(\mathbf{X}'\mathbf{X})^{-1}\mathbf{X}'\right]\mathbf{y} = \mathbf{e}'\mathbf{e} \qquad (8.2.8)\end{aligned}$$

This result, in turn, allows us to express the sum of squared deviations in the linear regression model in terms of the squared sample multiple correlation, using Eq. (7.3.18):

$$[\mathbf{y} - \hat{\mathbf{E}}(\mathbf{y})]'[\mathbf{y} - \hat{\mathbf{E}}(\mathbf{y})] = nS_{\mathbf{y}}^2\left(1 - R_{\mathbf{y}.\mathbf{X}}^2\right) \qquad (8.2.9)$$

One purpose in developing this sum of squared deviations is to relate it to the maximum likelihood estimator of the second unknown parameter σ^2. Differentiating the log likelihood expression of Eq. (8.2.4) with respect to σ^2 gives

$$\begin{aligned}\frac{\partial[\log L(\mathbf{y})]}{\partial\sigma^2} &= \frac{\partial[(-n/2)\log\sigma^2]}{\partial\sigma^2} + \frac{\partial[(-1/2\sigma^2)(\mathbf{y} - \mathbf{X}\boldsymbol{\beta})'(\mathbf{y} - \mathbf{X}\boldsymbol{\beta})]}{\partial\sigma^2} \\ &= (-n/2)(1/\sigma^2) + \tfrac{1}{2}(1/\sigma^2)^2(\mathbf{y} - \mathbf{X}\boldsymbol{\beta})'(\mathbf{y} - \mathbf{X}\boldsymbol{\beta})\end{aligned}$$

Setting this equal to zero we can solve, first, for n:

$$n = (1/\hat{\sigma}^2)(\mathbf{y} - \mathbf{X}\hat{\boldsymbol{\beta}})'(\mathbf{y} - \mathbf{X}\hat{\boldsymbol{\beta}})$$

and then for $\hat{\sigma}^2$:

$$\hat{\sigma}^2 = (1/n)(\mathbf{y} - \mathbf{X}\hat{\boldsymbol{\beta}})'(\mathbf{y} - \mathbf{X}\hat{\boldsymbol{\beta}}) \qquad (8.2.10)$$

The maximum likelihood estimator of σ^2 is the average squared deviation of the criterion observations from their estimated means. It is related to the least squares $\mathbf{e}'\mathbf{e}$ and the squared sample multiple correlation, using the results of Eqs. (8.2.8) and (8.2.9), by

$$\hat{\sigma}^2 = (1/n)\mathbf{e}'\mathbf{e} = (1/n)\mathbf{y}'\left[\mathbf{I} - \mathbf{X}(\mathbf{X}'\mathbf{X})^{-1}\mathbf{X}'\right]\mathbf{y} = (1/n)(\mathbf{y}'\mathbf{y} - \mathbf{y}'\mathbf{X}\hat{\boldsymbol{\beta}}) \qquad (8.2.11)$$

and

$$\hat{\sigma}^2 = S_{\mathbf{y}}^2\left(1 - R_{\mathbf{y}.\mathbf{X}}^2\right) \qquad (8.2.12)$$

The reader should keep in mind that although there are computational similarities between the linear regression estimation equations and the least squares and multiple correlation solutions, the *interpretation* of $\mathbf{X}\hat{\boldsymbol{\beta}}$ and $\hat{\sigma}^2$ as estimators of population means and a population variance depends on the reasonableness of the assumptions made in the linear regression model *and* on our adherence to the sample selection paradigm—adoption of a specific $\mathbf{X}$ or design matrix and independent selection of the observations $\mathbf{y}$. Least squares and multiple correlation computer programs may be used to obtain linear regression estimates but do not justify their interpretation.

If we assume the vector $\mathbf{y}$ of six French comprehension scores given in Section 8.1 represents an independent sampling that follows the specification of the design matrix, we can use those data to illustrate the maximum likelihood estimation computations.

$$\mathbf{y}' = (30, 15, 24, 42, 61, 42), \qquad \mathbf{X} = \begin{bmatrix} 1 & 1 \\ 1 & 1 \\ 1 & 3 \\ 1 & 3 \\ 1 & 5 \\ 1 & 5 \end{bmatrix}$$

$$\mathbf{y}'\mathbf{y} = 8950, \qquad \mathbf{y}'\mathbf{X} = (214, 758)$$

$$\mathbf{X}'\mathbf{X} = \begin{bmatrix} 6 & 18 \\ 18 & 70 \end{bmatrix}, \qquad (\mathbf{X}'\mathbf{X})^{-1} = \tfrac{1}{96}\begin{bmatrix} 70 & -18 \\ -18 & 6 \end{bmatrix}$$

$$\hat{\boldsymbol{\beta}}' = \mathbf{y}'\mathbf{X}(\mathbf{X}'\mathbf{X})^{-1} = \tfrac{1}{96}(1336, 696) = (13.9167, 7.2500)$$

$$\hat{\boldsymbol{\beta}}'\mathbf{X}' = \hat{\mathbf{E}}(\mathbf{y})' = (21.17, 21.17, 35.67, 35.67, 50.17, 50.17)$$

$$\hat{\boldsymbol{\beta}}'\mathbf{X}'\mathbf{y} = \mathbf{y}'\mathbf{X}(\mathbf{X}'\mathbf{X})^{-1}\mathbf{X}'\mathbf{y} = 8473.67$$

$$\hat{\sigma}^2 = (1/n)(\mathbf{y}'\mathbf{y} - \hat{\boldsymbol{\beta}}'\mathbf{X}'\mathbf{y}) = \tfrac{1}{6}(8950 - 8473.6) = 79.39$$

We also have, for $\mathbf{x} = \mathbf{X}_{.2}$

$$S_{\mathbf{y}}^2 = (1/n)\left[\mathbf{y}'\mathbf{y} - (1/n)(\mathbf{1}'\mathbf{y})^2\right] = \tfrac{1}{6}\left[8950 - \tfrac{1}{6}(214)^2\right] = 219.556$$

$$S_{\mathbf{x}}^2 = \tfrac{1}{6}\left[70 - \tfrac{1}{6}(18)^2\right] = 2.667$$

$$\operatorname{Cov}(\mathbf{x}, \mathbf{y}) = \tfrac{1}{6}\left[758 - \tfrac{1}{6}(18)(214)\right] = 19.333$$

$$R_{\mathbf{y}.\mathbf{x}}^2 = (19.333)^2/[(219.556)(2.667)] = 0.638$$

$$\hat{\sigma}^2 = S_{\mathbf{y}}^2\left(1 - R_{\mathbf{y}.\mathbf{x}}^2\right) = (219.556)(0.362) = 79.39$$

The estimated criterion variance—the variance within each of the semesters of study distributions of test scores—was computed here in two ways.

8.3 Sampling Distribution of Maximum Likelihood Estimators

In the previous computational example, $\boldsymbol{\beta}$ was estimated by a vector whose elements differed from the population values (12 and 8), values that we were privileged to know in advance. That is, sampling six observations did not unerringly reveal the linear structure to the population means. How close were we, though, and what should we expect from a study like the one described?

Normally, of course, we do not know the values of the population means and are unable to compare our sample estimates with those values. Rather, in statistical estimation we usually base our evaluation of a particular estimator on a comparison with the estimators that might have resulted from similar experiments. For example, if we were once again independently to choose and test two first-semester, two third-semester, and two fifth-semester students of French, we would anticipate those six students to give a somewhat different set of comprehension scores. If the vector $\mathbf{y}$ of comprehension scores were different, the computations summarized at the end of the last section could well yield different estimates of $\boldsymbol{\beta}$ or σ^2. Repeating the experiment over and over, each time independently selecting another set of six students of French for testing (each time, of course, according to the same design), a distribution of $\hat{\boldsymbol{\beta}}$ vectors could be built up, as could a distribution of estimators of $\hat{\sigma}^2$.

The distribution of values of an estimator that is approached as the repetition of an experiment is continued, essentially without end, is the estimator's *sampling distribution*. If we are able to learn something about that distribution—its mean, its variability, whether it follows some known statistical form—we often can make some judgment about the "true" value of the parameter under estimation or about how well we are likely to have estimated it.

Consider first the estimator $\hat{\boldsymbol{\beta}} = [(\mathbf{X}'\mathbf{X})^{-1}\mathbf{X}']\mathbf{y}$ of $\boldsymbol{\beta}$. Brackets have been inserted to emphasize that the design matrix, and hence $(\mathbf{X}'\mathbf{X})^{-1}\mathbf{X}'$, is fixed by the design of the study and would not change in the hypothetical repetitions of the experiment. From this equation $\hat{\boldsymbol{\beta}}$ can in any trial of the experiment therefore be viewed as a linear transformation of $\mathbf{y}$. $\mathbf{y}$ may change from trial to trial, but transformation weights $(\mathbf{X}'\mathbf{X})^{-1}\mathbf{X}'$ do not change.

From Chapter 5, then, we can express the mean and variance of the sampling distribution of $\hat{\boldsymbol{\beta}}$ in terms of the mean and variance of $\mathbf{y}$ and the transformation weights. Because $\mathbf{y}$ and $\hat{\boldsymbol{\beta}}$ are both vectors, the means $\mathbf{E}(\mathbf{y})$ and $\mathbf{E}(\hat{\boldsymbol{\beta}})$ are vectors, and the variances $\mathbf{Var}(\mathbf{y})$ and $\mathbf{Var}(\hat{\boldsymbol{\beta}})$ are symmetric variance–covariance matrices. Applying the rules on linear transformations gives the mean vector of the sampling distribution of $\hat{\boldsymbol{\beta}}$:

$$\mathbf{E}(\hat{\boldsymbol{\beta}}) = \left[(\mathbf{X}'\mathbf{X})^{-1}\mathbf{X}'\right]\mathbf{E}(\mathbf{y})$$

The basic assumption of the linear regression model, however, is that of Eq. (8.1.1), $\mathbf{E}(\mathbf{y}) = \mathbf{X}\boldsymbol{\beta}$, and if this substitution is made we find

$$\mathbf{E}(\hat{\boldsymbol{\beta}}) = \left[(\mathbf{X}'\mathbf{X})^{-1}\mathbf{X}'\right]\mathbf{X}\boldsymbol{\beta} = \boldsymbol{\beta} \tag{8.3.1}$$

On average, this equation tells us, $\hat{\boldsymbol{\beta}} = \boldsymbol{\beta}$. Because $\boldsymbol{\beta}$ and $\hat{\boldsymbol{\beta}}$ are vectors this equality must be true for each of the elements of $\hat{\boldsymbol{\beta}}$. An estimator whose sampling distribution has as its mean the value of the parameter being estimated is said to be an *unbiased estimator*. The sampling distribution of $\hat{\beta}_i$ is clustered about β_i, the true but unknown value of the parameter.

How closely, however, are the values of $\hat{\beta}_i$ distributed around β_i? Answering this question requires finding the variance of the sampling distribution of $\hat{\boldsymbol{\beta}}$. Applying the rule for the variance of linear transformations to (8.2.5) results in

$$\mathbf{Var}(\hat{\boldsymbol{\beta}}) = \left[(\mathbf{X}'\mathbf{X})^{-1}\mathbf{X}'\right]\mathbf{Var}(\mathbf{y})\left[\mathbf{X}(\mathbf{X}'\mathbf{X})^{-1}\right]$$

$\mathbf{Var}(\mathbf{y})$ is a symmetric $n \times n$ matrix whose diagonal elements are variances. In the French comprehension experiment (where $n = 6$), $[\mathrm{Var}(y)]_{11}$, for example, is the variance that would be observed in the test scores of first-semester students if one student were selected and tested for each trial of the experiment. (Here we are concerned only with the scores that would appear as y_1 on repeated trials of the experiment.) The variance of these scores is the variance σ^2 of the population of scores from which we sample on each trial. In general, because the linear regression model assumes that the variance is the same in each of the populations from which sampling occurs, we know that each diagonal element of $\mathbf{Var}(\mathbf{y})$ is equal to σ^2, a constant but unknown variance.

Another assumption made in the linear regression model, and one researchers need to honor in the conduct of studies, is that the n criterion observations are selected independently. Criterion observation y_k, for example, is selected in a way that has nothing at all to do with the values of any of the other criterion scores; knowing that the first selected first-semester student scored high on the comprehension test should say nothing about the scores earned by any of the other five students sampled.

The independence condition can be translated into the following statistical statement:

$$\mathrm{Cov}(y_i, y_j) = 0$$

Over a long sequence of trials of the experiment, the distributions of scores for any two criterion vector positions will be uncorrelated. For example, the scores of the first and second selected third-semester students would be uncorrelated if the two were selected independently on each trial of the experiment. The covariance just described is a typical off-diagonal element of the matrix $\mathbf{Var}(\mathbf{y})$.

The variance–covariance matrix **Var(y)** is thus a diagonal matrix, with all diagonal elements equal to σ^2:

$$\mathbf{Var}(\mathbf{y}) = \sigma^2\mathbf{I} \tag{8.3.2}$$

The variance–covariance matrix for the sampling distribution of $\hat{\boldsymbol{\beta}}$ becomes

$$\begin{aligned}\mathbf{Var}(\hat{\boldsymbol{\beta}}) &= \left[(\mathbf{X}'\mathbf{X})^{-1}\mathbf{X}'\right]\mathbf{Var}(\mathbf{y})\left[\mathbf{X}(\mathbf{X}'\mathbf{X})^{-1}\right] \\ &= \left[(\mathbf{X}'\mathbf{X})^{-1}\mathbf{X}'\right]\sigma^2\mathbf{I}\left[\mathbf{X}(\mathbf{X}'\mathbf{X})^{-1}\right] \\ &= \sigma^2\left[(\mathbf{X}'\mathbf{X})^{-1}\mathbf{X}'\right]\left[\mathbf{X}(\mathbf{X}'\mathbf{X})^{-1}\right] \\ &= \sigma^2(\mathbf{X}'\mathbf{X})^{-1} \end{aligned} \tag{8.3.3}$$

The diagonal elements of the $(m \times m)$ $\mathbf{Var}(\hat{\boldsymbol{\beta}})$ are variances for the sampling distributions of each of the m elements of $\hat{\boldsymbol{\beta}}$. The off-diagonal elements are covariances between, for example, $\hat{\beta}_i$ and $\hat{\beta}_j$. The elements of $\mathbf{Var}(\hat{\boldsymbol{\beta}})$ are a function of the unknown σ^2 and the matrix $(\mathbf{X}'\mathbf{X})^{-1}$. Equation (8.3.3) suggests that we can exert appreciable control over the sampling variance of elements of $\hat{\boldsymbol{\beta}}$ and of the covariance between $\hat{\beta}_i$ and $\hat{\beta}_j$ through the selection of a design matrix—through the design of the study. We shall have more to say on this in later sections. For now, however, note that (a) the covariance between estimators of the elements of $\hat{\boldsymbol{\beta}}$ is zero if $(\mathbf{X}'\mathbf{X})^{-1}$ is diagonal, (b) $(\mathbf{X}'\mathbf{X})^{-1}$ is diagonal if $\mathbf{X}'\mathbf{X}$ is diagonal, and (c) $\mathbf{X}'\mathbf{X}$ is diagonal if $\mathbf{X}$ is an orthogonal matrix.

A third important aspect to the sampling distribution of $\hat{\boldsymbol{\beta}}$ is derived from (8.2.5),

$$\hat{\boldsymbol{\beta}} = \left[(\mathbf{X}'\mathbf{X})^{-1}\mathbf{X}'\right]\mathbf{y}$$

based again on the fact that each element of $\hat{\boldsymbol{\beta}}$ is a linear transformation of the n criterion observations. In particular, each element of $\hat{\boldsymbol{\beta}}$ is a linear transformation of n independent and normally distributed observations. We make use of the statistical theorem that a weighted sum of independent normal observations is itself an observation from a normal distribution.

Summarizing what we know about the sampling distribution of $\hat{\boldsymbol{\beta}}$, $\hat{\beta}_i$ has a normal sampling distribution with mean equal to the estimated parameter, $E(\hat{\beta}_i) = \beta_i$ and with variance equal to the product of the common criterion population variance and the ith diagonal element of the inverse of the minor product moment of the design matrix $\mathrm{Var}(\hat{\beta}_i) = \sigma^2[(\mathbf{X}'\mathbf{X})^{-1}]_{ii}$.

The assumptions and sampling requirements of the linear regression model allowed the expression of the sampling distribution of the estimates of $\boldsymbol{\beta}$. If we knew the value of the parameter σ^2, we would now be able to test hypotheses about the value of any particular β_i. Since $\hat{\beta}_i$ has a normal distribution with mean β_i and a variance completely determined by σ^2 and the design matrix, hypothesis-testing logic could be employed to test whether it is likely that an observed value of $\hat{\beta}_i$ could have been sampled from a distribution with a mean β_i, the hypothesized value.

We do not, of course, know σ^2, so we must learn something too of the sampling distribution of its estimator. From (8.2.11) the maximum likelihood estimator

$$\hat{\sigma}^2 = \mathbf{y}'\left[(1/n)\left(\mathbf{I} - \mathbf{X}(\mathbf{X}'\mathbf{X})^{-1}\mathbf{X}'\right)\right]\mathbf{y}$$

is a quadratic form in $\mathbf{y}$. Quadratic forms in a vector that consists of normally distributed observations have some well-known properties. We shall use these, first, to develop the sampling distribution of $\hat{\sigma}^2$ and, then, in Section 8.4, to show how hypotheses are tested in the general linear model.

One relevant theorem on quadratic forms of normally distributed observations is the following.

Theorem. *If* $\mathbf{y}$ *is a vector of normally distributed observations with mean vector* $\mathbf{E}(\mathbf{y})$ *and variance–covariance matrix* $\mathbf{Var}(\mathbf{y})$, *then the quadratic form* $\mathbf{Q}(\mathbf{A}) = [\mathbf{y} - \mathbf{E}(\mathbf{y})]'\mathbf{A}[\mathbf{y} - \mathbf{E}(\mathbf{y})]$ *is distributed as a (central) chi-squared variable with* $r(\mathbf{A})$ *degrees of freedom, where* $r(\mathbf{A})$ *is the rank of* $\mathbf{A}$ *the symmetric matrix of the form, provided that* $\mathbf{A}$ *is such that the product* $\mathbf{A}\,\mathbf{Var}(\mathbf{y})\,\mathbf{A} = \mathbf{A}$.

The (central) chi-squared variables have well-known sampling distributions (see Appendix A), so referring an observation to a chi-squared distribution will be useful. Perhaps the best known chi-squared observation is given by $\mathbf{Q}(\mathbf{A})$ when $\mathbf{y}$ is sampled such that $\mathbf{E}(\mathbf{y}) = \mathbf{0}$, $\mathbf{Var}(\mathbf{y}) = \mathbf{I}$, and $\mathbf{A}$ is taken to be $\mathbf{I}$; a sum of squares of n independently distributed normal observations each with mean zero and variance one is a chi-squared observation with n degrees of freedom.

To express the sampling distribution of $\hat{\sigma}^2$, we apply this theorem to three quadratic forms which arise in linear regression. If the true value of $\boldsymbol{\beta}$ were known, we could write a vector of deviations from the population mean as

$$\mathbf{y} - \mathbf{E}(\mathbf{y}) = \mathbf{y} - \mathbf{X}\boldsymbol{\beta} = \left(\mathbf{y} - \mathbf{X}\hat{\boldsymbol{\beta}}\right) + \left(\mathbf{X}\hat{\boldsymbol{\beta}} - \mathbf{X}\boldsymbol{\beta}\right) \qquad (8.3.4)$$

That is, we could partition the deviation between each criterion observation and its population mean into the sum of the difference between that observation and the maximum likelihood estimate of the mean and the discrepancy between this estimate and the true mean.

One reason for representing $\mathbf{y} - \mathbf{X}\boldsymbol{\beta}$ in this form is that the two vectors on the right of Eq. (8.3.4) are orthogonal:

$$\begin{aligned}(\mathbf{y} - \mathbf{X}\hat{\boldsymbol{\beta}})'(\mathbf{X}\hat{\boldsymbol{\beta}} - \mathbf{X}\boldsymbol{\beta}) &= \mathbf{y}'\mathbf{X}(\mathbf{X}'\mathbf{X})^{-1}\mathbf{X}'\mathbf{y} - \mathbf{y}'\mathbf{X}\boldsymbol{\beta} \\ &\quad - \mathbf{y}'\mathbf{X}(\mathbf{X}'\mathbf{X})^{-1}\mathbf{X}'\mathbf{X}(\mathbf{X}'\mathbf{X})^{-1}\mathbf{X}'\mathbf{y} + \mathbf{y}'\mathbf{X}(\mathbf{X}'\mathbf{X})^{-1}\mathbf{X}'\mathbf{X}\boldsymbol{\beta} \\ &= 0\end{aligned}$$

[As in the discussion of least squares, the errors of prediction over the sample of n observations are uncorrelated with any linear transformation of the predictors. $\mathbf{X}\hat{\boldsymbol{\beta}}$ and $\mathbf{X}\boldsymbol{\beta}$ are each linear transformations of $\mathbf{X}$.]

A consequence of this orthogonality is that the sum of squares of the deviations of the criterion observations about their population means, the minor product moment of the left-hand side of (8.3.4), can be written as the sum of two other minor product moments:

$$(\mathbf{y} - \mathbf{X}\boldsymbol{\beta})'(\mathbf{y} - \mathbf{X}\boldsymbol{\beta}) = (\mathbf{y} - \mathbf{X}\hat{\boldsymbol{\beta}})'(\mathbf{y} - \mathbf{X}\hat{\boldsymbol{\beta}}) + (\mathbf{X}\hat{\boldsymbol{\beta}} - \mathbf{X}\boldsymbol{\beta})'(\mathbf{X}\hat{\boldsymbol{\beta}} - \mathbf{X}\boldsymbol{\beta}) \tag{8.3.5}$$

The sum of the squares about the population means equals the sum of squares about the estimated means plus the sum of squares of the errors of estimation of the means. This *decomposition* of the sum of squares plays an important role in the general linear model.

The left side of (8.3.5) is a quadratic form that has some of the hallmarks of a chi-squared observation,

$$[\mathbf{y} - \mathbf{E}(\mathbf{y})]'\mathbf{I}[\mathbf{y} - \mathbf{E}(\mathbf{y})]$$

although the triple product involving $\mathbf{I}$, the *matrix of the quadratic form* and $\mathbf{Var}(\mathbf{y})$ fails to return the matrix of the form:

$$\mathbf{I}\,\mathbf{Var}(\mathbf{y})\,\mathbf{I} = \mathbf{I}\mathbf{I}(\sigma)^2\mathbf{I} = \mathbf{I}(\sigma)^2 \neq \mathbf{I}$$

If the identity matrix is replaced as the matrix of the quadratic form by the symmetric matrix $\mathbf{A} = [\mathbf{Var}(\mathbf{y})]^{-1} - 1/\sigma^2\mathbf{I}$, a quadratic form is produced that does have the chi-squared properties. That is,

$$(1/\sigma^2)(\mathbf{y} - \mathbf{X}\boldsymbol{\beta})'(\mathbf{y} - \mathbf{X}\boldsymbol{\beta}) = [\mathbf{y} - \mathbf{E}(\mathbf{y})]'\big[(1/\sigma^2)\mathbf{I}\big][\mathbf{y} - \mathbf{E}(\mathbf{y})]$$

is a quadratic form such that

$$\mathbf{A}\,\mathrm{Var}(\mathbf{y})\,\mathbf{A} = \big[(1/\sigma^2)\mathbf{I}\big](\sigma^2\mathbf{I})\big[(1/\sigma^2)\mathbf{I}\big] = \big[(1/\sigma^2)\mathbf{I}\big] = \mathbf{A}$$

The matrix $\mathbf{A}$ is $n \times n$ diagonal and $r(\mathbf{A}) = n$. Thus we find that the sampling distribution of $(1/\sigma^2)(\mathbf{y} - \mathbf{X}\boldsymbol{\beta})'(\mathbf{y} - \mathbf{X}\boldsymbol{\beta})$ is that of the chi-squared variable with n degrees of freedom. Because $\boldsymbol{\beta}$ is not known to us, we cannot use this result directly. We can use the result, however, to multiply through on both sides of (8.3.5) by $1/\sigma^2$ and then consider the two terms on the right.

We shall first develop $(1/\sigma^2)(\mathbf{X}\hat{\boldsymbol{\beta}} - \mathbf{X}\boldsymbol{\beta})'(\mathbf{X}\hat{\boldsymbol{\beta}} - \mathbf{X}\boldsymbol{\beta})$. To get this in the form of a quadratic in $\mathbf{y} - \mathbf{E}(\mathbf{y})$ we can first rewrite $\mathbf{X}\hat{\boldsymbol{\beta}} - \mathbf{X}\boldsymbol{\beta}$, substituting for $\hat{\boldsymbol{\beta}}$, as $[\mathbf{X}(\mathbf{X}'\mathbf{X})^{-1}\mathbf{X}']\mathbf{y} = \mathbf{X}\hat{\boldsymbol{\beta}}$. Next, noting that $[\mathbf{X}(\mathbf{X}'\mathbf{X})^{-1}\mathbf{X}']\mathbf{X}\boldsymbol{\beta} = \mathbf{X}\boldsymbol{\beta}$, $\mathbf{X}\hat{\boldsymbol{\beta}} - \mathbf{X}\boldsymbol{\beta}$ can also be written

$$\mathbf{X}(\mathbf{X}'\mathbf{X})^{-1}\mathbf{X}'(\mathbf{y} - \mathbf{X}\boldsymbol{\beta}) = \mathbf{X}(\mathbf{X}'\mathbf{X})^{-1}\mathbf{X}'[\mathbf{y} - \mathbf{E}(\mathbf{y})]$$

From this,

$$(1/\sigma^2)(\mathbf{X}\hat{\boldsymbol\beta} - \mathbf{X}\boldsymbol\beta)'(\mathbf{X}\hat{\boldsymbol\beta} - \mathbf{X}\boldsymbol\beta) = [\mathbf{y} - \mathbf{E}(\mathbf{y})]'\left[(1/\sigma^2)\mathbf{X}(\mathbf{X}'\mathbf{X})^{-1}\mathbf{X}'\right] \times [\mathbf{y} - \mathbf{E}(\mathbf{y})] \qquad (8.3.6)$$

The matrix of this quadratic form is $\mathbf{B} = [(1/\sigma^2)\mathbf{X}(\mathbf{X}'\mathbf{X})^{-1}\mathbf{X}']$, and the product $\mathbf{B}\,\mathrm{Var}(\mathbf{y})\,\mathbf{B} = \mathbf{B}(\sigma^2\mathbf{I})\mathbf{B}$ yields $\mathbf{B}$. The rank of $\mathbf{B}$ is the rank of $\mathbf{X}$, the number of columns in $\mathbf{X}$ or the number of elements m in β.

Thus the sampling distribution of $(1/\sigma^2)(\mathbf{X}\hat{\boldsymbol\beta} - \mathbf{X}\boldsymbol\beta)'(\mathbf{X}\hat{\boldsymbol\beta} - \mathbf{X}\boldsymbol\beta)$ is that of the chi-squared variable with m degrees of freedom, where m is the number of elements in the unknown vector $\boldsymbol\beta$, the number of $\boldsymbol\beta$ parameters that must be estimated.

Finally, $(1/\sigma^2)(\mathbf{y} - \mathbf{X}\hat{\boldsymbol\beta})'(\mathbf{y} - \mathbf{X}\hat{\boldsymbol\beta})$ may also be developed as a chi-squared observation. Substituting for $\hat{\boldsymbol\beta}$,

$$\mathbf{y} - \mathbf{X}\hat{\boldsymbol\beta} = \mathbf{y} - \mathbf{X}(\mathbf{X}'\mathbf{X})^{-1}\mathbf{X}'\mathbf{y} = \left[\mathbf{I} - \mathbf{X}(\mathbf{X}'\mathbf{X})^{-1}\mathbf{X}'\right]\mathbf{y}$$

Then, using the result $[\mathbf{I} - \mathbf{X}(\mathbf{X}'\mathbf{X})^{-1}\mathbf{X}']\mathbf{X}\boldsymbol\beta = \mathbf{0}$,

$$\mathbf{y} - \mathbf{X}\hat{\boldsymbol\beta} = \left[\mathbf{I} - \mathbf{X}(\mathbf{X}'\mathbf{X})^{-1}\mathbf{X}'\right](\mathbf{y} - \mathbf{X}\boldsymbol\beta)$$

Substituting this and recognizing the idempotency of $\mathbf{I} - \mathbf{X}(\mathbf{X}'\mathbf{X})^{-1}\mathbf{X}'$ shows that

$$(1/\sigma^2)(\mathbf{y} - \mathbf{X}\hat{\boldsymbol\beta})'(\mathbf{y} - \mathbf{X}\hat{\boldsymbol\beta}) = (\mathbf{y} - \mathbf{X}\boldsymbol\beta)'(1/\sigma^2)\left[\mathbf{I} - \mathbf{X}(\mathbf{X}'\mathbf{X})^{-1}\mathbf{X}'\right] \times (\mathbf{y} - \mathbf{X}\boldsymbol\beta) \qquad (8.3.7)$$

is in the desired form. The triple product $\mathbf{C}\,\mathbf{Var}(\mathbf{y})\,\mathbf{C}$ for $\mathbf{C} = (1/\sigma^2)[\mathbf{I} - \mathbf{X}(\mathbf{X}'\mathbf{X})^{-1}\mathbf{X}']$ satisfies the theorem condition, $\mathbf{C}\,\mathbf{Var}(\mathbf{y})\,\mathbf{C} = \mathbf{C}$. The term on the left of Eq. (8.3.7), then, has a chi-squared sampling distribution with $r(\mathbf{C})$ degrees of freedom.

What is $r(\mathbf{C})$, the rank of $\mathbf{I} - \mathbf{X}(\mathbf{X}'\mathbf{X})^{-1}\mathbf{X}'$? By the results of Section 5.5 and knowing that $r[\mathbf{X}(\mathbf{X}'\mathbf{X})^{-1}\mathbf{X}'] = m$, we are assured that $r(\mathbf{C}) = n - m$. We know, then, that

$$(1/\sigma^2)(\mathbf{y} - \mathbf{X}\hat{\boldsymbol\beta})(\mathbf{y} - \mathbf{X}\hat{\boldsymbol\beta}) = (1/\sigma^2)\mathbf{e}'\mathbf{e} = (1/\sigma^2)n(\hat\sigma^2) = n(\hat\sigma^2/\sigma^2)$$

has the sampling distribution of a chi-squared variable with $n - m$ degrees of freedom.

One of the properties of the general central chi-squared distribution is an expected value identical to the number of degrees of freedom:

$$E(\chi_p^2) = p$$

Thus for the maximum likelihood estimator of σ^2, the sampling distribution mean may be found from $E[n(\hat\sigma^2/\sigma^2] = n - m$ to be

$$E(\hat\sigma^2) = [(n - m)/n]\sigma^2 \qquad (8.3.8)$$

The maximum likelihood estimator of σ^2 is not an unbiased estimator of the parameter: $E(\hat{\sigma}^2) \neq (\sigma^2)$. Equation (8.3.8) tells how to create an unbiased estimator from $\mathbf{e}'\mathbf{e}$. Defining a second estimator

$$\hat{\hat{\sigma}}^2 = [1/(n - m)]\mathbf{e}'\mathbf{e} \tag{8.3.9}$$

we see that $E(\hat{\hat{\sigma}}^2) = [1/(n - m)]E(\mathbf{e}'\mathbf{e}) = [\sigma^2/(n - m)]E(\mathbf{e}'\mathbf{e}/\sigma^2)$, or, since $E(\mathbf{e}'\mathbf{e}/\sigma^2) = n - m$, $E(\hat{\hat{\sigma}}^2) = \sigma^2$.

8.4 Hypothesis Testing

The primary purpose of developing the sampling distribution for $\hat{\boldsymbol{\beta}}$ and $\mathbf{e}'\mathbf{e}$ (or one of its relatives, $\hat{\sigma}^2$ or $\hat{\hat{\sigma}}^2$) is to facilitate testing hypotheses about the population values of $\boldsymbol{\beta}$ or σ^2 and to permit us to establish some *confidence bounds* for sample-based estimators.

8.4.1 Testing Hypotheses About σ^2

Because $\boldsymbol{\beta}$ carries information about *differences* between populations (namely, in their mean values), we are typically more interested in its estimator then we are in our estimate of the variance, which the linear regression model assumes to be constant across populations. Nevertheless, we can and do use the chi-squared property of the sampling distribution of $\mathbf{e}'\mathbf{e}$ to inquire into σ^2.

As an example, let us go back to the French comprehension test. We sampled two observations from each of three populations, estimated the two β parameters required for the assumed linear regression of $\mathbf{y}$ on $\mathbf{X}$, $\mathbf{E}(\mathbf{y}) = \mathbf{X}\boldsymbol{\beta}$, and computed $(\mathbf{y} - \mathbf{X}\hat{\boldsymbol{\beta}})'(\mathbf{y} - \mathbf{X}\hat{\boldsymbol{\beta}}) = 476.33$ and $\hat{\sigma}^2 = 79.39$. An unbiased estimator of σ^2 is given by $\hat{\hat{\sigma}}^2 = \mathbf{e}'\mathbf{e}/(n - m) = 476.33/(6 - 2) = 119.08$.

On the basis of prior work with this French comprehension test, perhaps on the basis of the standardization data collected when the test was constructed, we believe the standard deviation within any semester-of-study population ought to be eight score points, so that the variance σ^2 ought to be 64. If that is the correct hypothesis, then the sampling distribution of any of

$$\mathbf{e}'\mathbf{e}/64 = \mathbf{e}'\mathbf{e}/\sigma^2$$

$$(6)\hat{\sigma}^2/64 = n\hat{\sigma}^2/\sigma^2$$

$$(6 - 2)\hat{\hat{\sigma}}^2/64 = (n - m)\hat{\hat{\sigma}}^2/\sigma^2$$

or

$$(6)S_{\mathbf{y}}^2\left(1 - R_{\mathbf{y}.\mathbf{X}}^2\right)/64 = (n)S_{\mathbf{y}}^2\left(1 - R_{\mathbf{y}.\mathbf{X}}^2\right)/\sigma^2$$

is that of a chi-squared variable with $n - m = 4$ degrees of freedom. (Remember that the S_y^2 of the multiple and partial correlation section is a measure of the variance of the criterion measures across as well as within populations from which observations have been sampled. It is the total variance of the sample of n criterion scores.)

The argument we develop for the distribution for $\hat{\sigma}^2$ could be equally well constructed for any of the other relatives of $\mathbf{e}'\mathbf{e}$. We choose $\hat{\sigma}^2$ because, as an unbiased estimator of σ^2, it may seem more natural to relate it to hypotheses about σ^2. Since the product $(n - m)\hat{\sigma}^2/\sigma^2$ follows the chi-squared distribution with $n - m$ degrees of freedom, we can use the distribution tabulated in the appendix to find the frequency with which we should expect large and small values of $(n - m)\hat{\sigma}^2/\sigma^2$. For $n - m = 6 - 2 = 4$, values smaller than 0.297, 0.711, and 1.064 will occur on 1%, 5%, and 10% of repetitions of the experiment, and values greater than 13.276, 9.488, and 7.780 will also occur 1%, 5%, and 10% of the times the study is replicated.

Since $(n - m)(\hat{\sigma})^2/\sigma^2 \sim \chi^2_{n-m}$ ($\sim$ is read "is distributed as"), we can describe the sampling distribution for $\hat{\sigma}^2$ directly by rewriting this distributional statement $\hat{\sigma}^2 \sim [\sigma^2/(n - m)]\chi^2_{n-m}$. Large and small values in the sampling distribution of $\hat{\sigma}^2$ are found by multiplying the corresponding large and small values of χ^2_{n-m} by the ratio $\sigma^2/(n - m)$. For $n - m = 6 - 2 = 4$ and the hypothesized σ^2 of 64, then, the unbiased estimator $\hat{\sigma}^2$ has a sampling distribution in which values smaller than 4.75, 11.38, and 17.02 make up 1%, 5%, and 10% of the distribution and values greater than 212.42, 151.81, and 124.48 also form 1%, 5%, and 10% of the population of $\hat{\sigma}^2$. The average value in the sampling distribution is then 64, as $\hat{\sigma}^2$ unbiasedly estimates σ^2.

We could have decided, knowing in advance that $\hat{\sigma}^2$ would have this sampling distribution, that we would reject the hypothesis that $\sigma^2 = 64$ if we obtained a value of $\hat{\sigma}^2$ less than 4.75 or greater than 212.42. In the present instance the calculated value of $\hat{\sigma}^2 = 119.08$ is well within these limits and the hypothesis would not be rejected. In this example, we have set the α level of our test—that is, the *probability of rejecting the hypothesis when it is correct*—at 0.02. There would be little likelihood, then, of making this *type I error*.

The small size of the sample, however, gives this test of the hypothesis that $\sigma^2 = 64$ little power against alternative values of σ^2. That is, there is an appreciable probability that we would fail to reject the hypothesis that $\sigma = 8$ ($\sigma^2 = 64$) when in fact $\sigma = 6$ ($\sigma^2 = 36$) or $\sigma = 10$ ($\sigma^2 = 100$). The reader is encouraged to consider this issue by using the chi-squared table to approximate the frequency with which values between 4.75 and 212.42 occur when $\sigma = 6$ or when $\sigma = 10$. These are instances in which a *type II error* would result: the hypothesized value of $\sigma = 8$ could not be rejected although $\sigma = 6$ or $\sigma = 10$. To understand the effect of increasing the sample size, the reader may wish to repeat this exercise assuming ten students were sampled from each of the three levels of study.

The estimator $\hat{\hat{\sigma}}^2$ is a *point estimator*. It represents an attempt to estimate from the sample the unknown σ^2 with a single value, a point on the continuum of possible values. An alternative to point estimation is to use the sample data to provide an *interval estimate* of the parameter. The point estimator resulting from any trial of the experiment quite often does not correspond exactly to the parameter being estimated. Interval estimation involves describing a range or interval of possible values for the parameter in such a way that we can make some statement about our confidence that intervals constructed in that fashion will include the unknown parameter.

For example, a 90% confidence interval for σ^2 is constructed about $\hat{\hat{\sigma}}^2$ in such a way that if intervals were constructed in the same fashion for each value in the sampling distribution of $\hat{\hat{\sigma}}^2$, 90% of those intervals would overlap or contain the parameter σ^2. Figure 8.3 describes the sampling distribution for $\hat{\hat{\sigma}}^2$ where $n - m = 4$. By reference to the tabulated values for the χ^2 distribution with 4 degrees of freedom, we see that 5% of the $\hat{\hat{\sigma}}^2$ values will be smaller than $0.178\sigma^2$ (i.e., $0.178 = 0.711/4$) and another 5% will be larger than $2.372\sigma^2$ (where $2.372 = 9.488/4$). The middle 90% of the sampling distribution for $\hat{\hat{\sigma}}^2$ lies between these two values with $E(\hat{\hat{\sigma}}^2) = \sigma^2$.

Figure 8.3 The sampling distribution for $\hat{\hat{\sigma}}^2$ with $n - m = 4$.

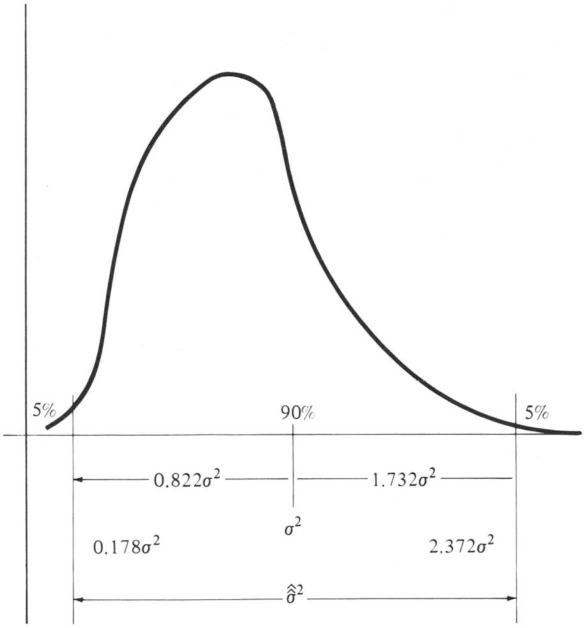

The middle 90% of the sampling distribution ranges from a smallest value of

$$\hat{\sigma}_S^2 = 0.178\sigma^2 = \sigma^2 - 0.822\sigma^2 \tag{8.4.1}$$

to a largest value of

$$\hat{\sigma}_L^2 = 2.372\sigma^2 = \sigma^2 + 1.372\sigma^2 \tag{8.4.2}$$

If an interval constructed about $\hat{\sigma}^2$ is to cover or reach σ^2 when $\hat{\sigma}^2 = \hat{\sigma}_S^2$, it must reach upward $0.822\sigma^2$ from $\hat{\sigma}^2$, based on Eq. (8.4.1). However, based on this same equation, $\sigma^2 = \hat{\sigma}^2/0.178$ when $\hat{\sigma}^2 = \hat{\sigma}_S^2$. The upper limit of our confidence interval is

$$\hat{\sigma}^2 + \hat{\sigma}^2(0.822/0.178) = \hat{\sigma}^2/0.178 \tag{8.4.3}$$

By a similar argument, the confidence interval will reach σ^2 when $\hat{\sigma}^2 = \hat{\sigma}_L^2$, based on Eq. (8.4.2), if the interval extends downward $1.372\sigma^2$ or $1.372(\hat{\sigma}^2/2.372)$ from $\hat{\sigma}^2$. The lower limit of the 90% confidence interval for σ^2 when $n - m = 4$ is

$$\hat{\sigma}^2 - \hat{\sigma}^2(1.372/2.372) = \hat{\sigma}^2/2.372 \tag{8.4.4}$$

Intervals constructed in this fashion cover σ^2 whenever

$$\hat{\sigma}_S^2 \leqslant \hat{\sigma}^2 \leqslant \hat{\sigma}_L^2$$

The interval is not wide enough to reach σ^2 when $\hat{\sigma}^2$ is outside this range.

Equations (8.4.1)–(8.4.4) describe the 90% confidence interval for σ^2 when $n - m = 4$. We can state these results more generally. The $(100 - 2\alpha)$% confidence interval for σ^2 has a lower limit of

$$\mathrm{CI}(L)_{\alpha,(n-m)} = \hat{\sigma}^2/U_{\alpha,(n-m)} \tag{8.4.5}$$

and an upper limit of

$$\mathrm{CI}(U)_{\alpha,(n-m)} = \hat{\sigma}^2/L_{\alpha,(n-m)} \tag{8.4.6}$$

where $U_{\alpha,(n-m)}$ and $L_{\alpha,(n-m)}$ are values, given in multiples of σ^2, which separate the highest and lowest α% of the sampling distribution of $\hat{\sigma}^2$ ($U_{5,4} = 2.372$ and $L_{5,4} = 0.178$.)

These limits are more usefully phrased in terms of percentage points of the appropriate chi-squared distributions:

$$U_{\alpha,(n-m)} = \chi^2_{(100-\alpha),(n-m)}/(n - m)$$

and

$$L_{\alpha,(n-m)} = \chi^2_{\alpha,(n-m)}/(n - m)$$

giving

$$\mathrm{CI}(L)_{\alpha,(n-m)} = (n - m)\hat{\sigma}^2/\chi^2_{(100-\alpha),(n-m)} \tag{8.4.7}$$

and

$$\mathrm{CI}(U)_{\alpha,(n-m)} = (n-m)\hat{\sigma}^2/\chi^2_{\alpha,(n-m)} \tag{8.4.8}$$

To illustrate these procedures for the six observations on the French comprehension test, we obtained a point estimate, $\hat{\sigma}^2 = 119.08$. For the interval estimate, then, we have a lower limit of $119.08/2.372 = 50.20$ and an upper limit of $119.08/0.178 = 669$.

The confidence interval or interval estimate of σ^2 is as wide as it is in this instance because of the very small sample size.

We have seen how to evaluate an estimate of the common but unknown criterion variance σ^2 either by testing hypotheses or by forming a confidence interval. Greater interest usually centers on the distribution of $\hat{\boldsymbol{\beta}}$. That different populations from which we are sampling may have different means is frequently of more theoretical concern than that the populations may all have a particular value of the common variance.

8.4.2 Likelihood Ratio Test of Hypotheses About $\boldsymbol{\beta}$

We have chosen to discuss the sampling distribution of $\hat{\sigma}^2$ first because it presents somewhat fewer problems and, having solved those, we are in a better position to discuss hypotheses about $\boldsymbol{\beta}$. The sampling distribution of $\hat{\sigma}^2$ is dependent, of course, on the value of σ^2. Having specified or hypothesized the value of σ^2, however, the sampling distribution of $\hat{\sigma}^2$ is completely known and hypothesis testing is easily and directly accomplished.

With $\hat{\boldsymbol{\beta}}$ it is not quite as easy. In Section 8.3 we showed that the sampling distribution of $\hat{\beta}_i$, or the ith element of $\hat{\boldsymbol{\beta}}$, was normal with a mean value of β_i and a variance of $\sigma^2[(\mathbf{X}'\mathbf{X})^{-1}]_{ii}$. If we hypothesize a value for β_i, we specify the mean of the sampling distribution of $\hat{\beta}_i$. The variance of the sampling distribution, however, remains unknown so long as σ^2 is not known. Thus we cannot immediately evaluate the hypothesis that, for example, $\beta_i = 0$ even if we know that $\hat{\beta}_i$ would be then distributed normally about that mean.

In linear regression a major difficulty is coping with the unknown σ^2, frequently not of interest itself but a determiner of the sampling distribution of statistics that are of concern. The strategy that we summarize here, makes use of a general approach to developing a test statistic termed the *likelihood ratio*. This approach is associated with the use, illustrated in Section 8.2, of maximum likelihood estimators of unknown parameters.

Equation (8.2.3) gave the likelihood of a sample of n criterion observations collected in accordance with the linear regression model:

$$L(\mathbf{y}|\boldsymbol{\beta},\sigma^2) = (2\pi\sigma^2)^{-n/2}\exp[(-1/2\sigma^2)(\mathbf{y}-\mathbf{X}\boldsymbol{\beta})'(\mathbf{y}-\mathbf{X}\boldsymbol{\beta})]$$

The numerical value of this likelihood, given a design matrix $\mathbf{X}$ and a sample vector of criterion observations $\mathbf{y}$, depends on the (unknown) values of the parameters $\boldsymbol{\beta}$ and σ^2. We exploited this principle in obtaining the maximum likelihood estimators of $\boldsymbol{\beta}$ and σ^2 [see (8.2.5) and (8.2.10)] $\hat{\boldsymbol{\beta}} = (\mathbf{X}'\mathbf{X})^{-1}\mathbf{X}'\mathbf{y}$ and $\hat{\sigma}^2 = (1/n)(\mathbf{y} - \mathbf{X}\hat{\boldsymbol{\beta}})'(\mathbf{y} - \mathbf{X}\hat{\boldsymbol{\beta}})$.

If $\hat{\boldsymbol{\beta}}$ and $\hat{\sigma}^2$ are substituted for $\boldsymbol{\beta}$ and σ^2 in Eq. (8.2.3), then the likelihood of the sample becomes as large as possible. No other values substituted for the unknown $\boldsymbol{\beta}$ and σ^2 lead to a larger value of the sample likelihood. An (unrestricted) estimate of the likelihood of the sample is given by

$$L(\mathbf{y}|\hat{\boldsymbol{\beta}}, \hat{\sigma}^2) = (2\pi\hat{\sigma}^2)^{-n/2} \exp[(-1/2\hat{\sigma}^2)(\mathbf{y} - \mathbf{X}\hat{\boldsymbol{\beta}})'(\mathbf{y} - \mathbf{X}\hat{\boldsymbol{\beta}})] \tag{8.4.9}$$

where the maximum likelihood estimators have been substituted in the likelihood equation.

Assume that we know (or hypothesize) the value of $\boldsymbol{\beta}$. Taking $\boldsymbol{\beta}$ as known in Eq. (8.2.3) gives a second estimate of the likelihood of the sample. Again σ^2 remains unknown. We could, however, find the maximum likelihood estimator of σ^2 subject to the restriction that $\boldsymbol{\beta}$ is known. The reader may repeat the differentiation that led to finding a maximum likelihood estimator of σ^2 [see the development of (8.2.10)] but treating $\boldsymbol{\beta}$ as known; the result is

$$\tilde{\sigma}^2 = (1/n)(\mathbf{y} - \mathbf{X}\boldsymbol{\beta})'(\mathbf{y} - \mathbf{X}\boldsymbol{\beta}) \tag{8.4.10}$$

Equation (8.2.3) takes on a numerical value when $\boldsymbol{\beta}$ is known and σ^2 is replaced by $\tilde{\sigma}^2$:

$$L(\mathbf{y}|\boldsymbol{\beta}, \tilde{\sigma}^2) = (2\pi\tilde{\sigma}^2)^{-n/2} \exp[(-1/2\tilde{\sigma}^2)(\mathbf{y} - \mathbf{X}\boldsymbol{\beta})'(\mathbf{y} - \mathbf{X}\boldsymbol{\beta})] \tag{8.4.11}$$

This estimate of the likelihood of the sample, by comparison with (8.4.9), is restricted by our stipulation of the vector $\boldsymbol{\beta}$.

Before continuing, it will be convenient to rewrite (8.4.9) by replacing $\hat{\sigma}^2$ with the right side of Eq. (8.2.10):

$$\begin{aligned} L(\mathbf{y}|\hat{\boldsymbol{\beta}}) = {} & (2\pi)^{-n/2}[(1/n)(\mathbf{y} - \mathbf{X}\hat{\boldsymbol{\beta}})'(\mathbf{y} - \mathbf{X}\hat{\boldsymbol{\beta}})]^{-n/2} \\ & \times \exp\{-\tfrac{1}{2}(1/n)^{-1}[(\mathbf{y} - \mathbf{X}\hat{\boldsymbol{\beta}})'(\mathbf{y} - \mathbf{X}\hat{\boldsymbol{\beta}})]^{-1} \\ & \times [(\mathbf{y} - \mathbf{X}\hat{\boldsymbol{\beta}})'(\mathbf{y} - \mathbf{X}\hat{\boldsymbol{\beta}})]\} \end{aligned}$$

or, after simplification,

$$L(\mathbf{y}|\hat{\boldsymbol{\beta}}) = (2\pi e)^{-n/2}[(\mathbf{y} - \mathbf{X}\hat{\boldsymbol{\beta}})'(\mathbf{y} - \mathbf{X}\hat{\boldsymbol{\beta}})]^{-n/2} \tag{8.4.12}$$

We similarly rewrite Eq. (8.4.11), replacing $\tilde{\sigma}^2$ with (8.4.10):

$$L(\mathbf{y}|\boldsymbol{\beta}) = (2\pi e)^{-n/2}[(\mathbf{y} - \mathbf{X}\boldsymbol{\beta})'(\mathbf{y} - \mathbf{X}\boldsymbol{\beta})]^{-n/2} \tag{8.4.13}$$

Because $\hat{\beta}$ and $\hat{\sigma}^2$ were chosen to maximize (8.2.3), we can be sure that

$$L(\mathbf{y}|\boldsymbol{\beta}) \leqslant L(\mathbf{y}|\hat{\boldsymbol{\beta}})$$

or, equivalently, that the *likelihood ratio*

$$\lambda = L(\mathbf{y}|\boldsymbol{\beta})/L(\mathbf{y}|\hat{\boldsymbol{\beta}}) \leqslant 1.0 \tag{8.4.14}$$

is the ratio of two estimates of the likelihood, the estimate in the numerator being more restricted (by hypothesis) than the estimate in the denominator. If our hypothesis is "correct"—if it is consistent with the sample observations—the two likelihood estimates should be close in value. Imposing the hypothesis on our likelihood estimate should have little effect on the numerical solution and λ should be "close to 1.0" or "large." On the other hand, if our hypothesis is "incorrect" (not consistent with the data), then the restricted likelihood estimate should be reduced in magnitude and λ should be "small."

This logic of the likelihood ratio suggests, then, that we should reject the hypothesis when λ is small and fail to reject it when λ is large. To give meaning to large and small, however, we need to know something about the sampling distribution of λ or of a statistic closely related to λ. A test statistic developed in this fashion is called a *likelihood ratio test*.

We can continue the definition of a likelihood ratio test for hypotheses about $\boldsymbol{\beta}$ by substituting from Eqs. (8.4.12) and (8.4.13) into (8.4.14):

$$\lambda = \frac{[(\mathbf{y} - \mathbf{X}\boldsymbol{\beta})'(\mathbf{y} - \mathbf{X}\boldsymbol{\beta})]^{-n/2}}{[(\mathbf{y} - \mathbf{X}\hat{\boldsymbol{\beta}})'(\mathbf{y}'\mathbf{X}\hat{\boldsymbol{\beta}})]^{-n/2}} = \left[\frac{(\mathbf{y} - \mathbf{X}\boldsymbol{\beta})'(\mathbf{y} - \mathbf{X}\boldsymbol{\beta})}{(\mathbf{y} - \mathbf{X}\hat{\boldsymbol{\beta}})'(\mathbf{y} - \mathbf{X}\hat{\boldsymbol{\beta}})}\right]^{-n/2}$$

$$= \left[\frac{(\mathbf{y} - \mathbf{X}\hat{\boldsymbol{\beta}})'(\mathbf{y} - \mathbf{X}\hat{\boldsymbol{\beta}})}{(\mathbf{y} - \mathbf{X}\boldsymbol{\beta})'(\mathbf{y} - \mathbf{X}\boldsymbol{\beta})}\right]^{n/2}$$

Rather than continue with λ, we define a closely related but simpler statistic,

$$\Lambda = \lambda^{2/n} = (\mathbf{y} - \mathbf{X}\hat{\boldsymbol{\beta}})'(\mathbf{y} - \mathbf{X}\hat{\boldsymbol{\beta}})/(\mathbf{y} - \mathbf{X}\boldsymbol{\beta})'(\mathbf{y} - \mathbf{X}\boldsymbol{\beta}) \tag{8.4.15}$$

This new statistic, although not itself a likelihood ratio, obeys the same logic: Λ is large or small as λ itself is large or small. Again, we would reject or not reject the hypothesis depending upon whether Λ is small or large.

We can begin to learn something about the sampling distribution of Λ if we first substitute for $(\mathbf{y} - \mathbf{X}\boldsymbol{\beta})'(\mathbf{y} - \mathbf{X}\boldsymbol{\beta})$ using the sum of squares decomposition of Eq. (8.3.5),

$$\Lambda = \frac{(\mathbf{y} - \mathbf{X}\hat{\boldsymbol{\beta}})'(\mathbf{y} - \mathbf{X}\hat{\boldsymbol{\beta}})}{(\mathbf{y} - \mathbf{X}\hat{\boldsymbol{\beta}})'(\mathbf{y} - \mathbf{X}\hat{\boldsymbol{\beta}}) + (\mathbf{X}\hat{\boldsymbol{\beta}} - \mathbf{X}\boldsymbol{\beta})'(\mathbf{X}\hat{\boldsymbol{\beta}} - \mathbf{X}\boldsymbol{\beta})}$$

and then multiply numerator and denominator through by $1/\sigma^2$,

$$\Lambda = \frac{(1/\sigma^2)(\mathbf{y} - \mathbf{X}\hat{\boldsymbol{\beta}})'(\mathbf{y} - \mathbf{X}\hat{\boldsymbol{\beta}})}{[(1/\sigma^2)(\mathbf{y} - \mathbf{X}\hat{\boldsymbol{\beta}})'(\mathbf{y} - \mathbf{X}\hat{\boldsymbol{\beta}})] + [(1/\sigma^2)(\mathbf{X}\hat{\boldsymbol{\beta}} - \mathbf{X}\boldsymbol{\beta})'(\mathbf{X}\hat{\boldsymbol{\beta}} - \mathbf{X}\boldsymbol{\beta})]} \tag{8.4.16}$$

The importance of these substitutions is that Λ now is made up of terms with known sampling distributions. Previously, we demonstrated that

$$(1/\sigma^2)(\mathbf{y} - \mathbf{X}\hat{\boldsymbol{\beta}})'(\mathbf{y} - \mathbf{X}\hat{\boldsymbol{\beta}}) \sim \chi^2_{n-m}$$

and that

$$(1/\sigma^2)(\mathbf{X}\hat{\boldsymbol{\beta}} - \mathbf{X}\boldsymbol{\beta})'(\mathbf{X}\hat{\boldsymbol{\beta}} - \mathbf{X}\boldsymbol{\beta}) \sim \chi^2_{m}$$

Both statistics, at least under the hypothesis, have χ^2 distributions, the first with $n - m$ degrees of freedom, the second with m.

That Λ is made up of terms that separately have known sampling distributions is not sufficient to tell us the sampling distribution of Λ. To reach that point we need to exploit two additional properties of chi-squared observations. The first is an extension of the statistical theorem (introduced in Section 8.3) on quadratic forms of normally distributed observations.

Theorem. *If* **y** *is a vector of normally distributed observations drawn from populations with means given by the vector* **E(y)** *and with variances and covariances given by the matrix* **Var(y)**, *then*

$$\mathbf{Q(A)} = [\mathbf{y} - \mathbf{E(y)}]'\mathbf{A}[\mathbf{y} - \mathbf{E(y)}], \qquad \mathbf{Q(B)} = [\mathbf{y} - \mathbf{E(y)}]'\mathbf{B}[\mathbf{y} - \mathbf{E(y)}]$$

have independent chi-squared sampling distributions with $r(\mathbf{A})$ *and* $r(\mathbf{B})$ *degrees of freedom, respectively, provided that*

$$\mathbf{A}\,\mathbf{Var(y)}\,\mathbf{A} = \mathbf{A}, \qquad \mathbf{B}\,\mathbf{Var(y)}\,\mathbf{B} = \mathbf{B}, \qquad \mathbf{A}\,\mathbf{Var(y)}\,\mathbf{B} = \mathbf{0}$$

This last condition, the relation between the two matrices **A** and **B**, indicates whether two χ^2 variables have independent sampling distributions.

This theorem is important as it leads to several approaches to finding a sampling distribution for Λ or for some simple transformation of Λ. The most frequently employed approach, the one we shall develop, is dependent on a second property:

Theorem. *The ratio of two independently distributed* χ^2 *observations, each divided by its degrees of freedom, is an observation following the F distribution with degrees of freedom N and D, where N and D are, respectively, the degrees of freedom of the numerator and denominator* χ^2 *observations.*

Applying, first, the theorem on independence to the two χ^2 observations involved in Λ, we obtain from Eqs. (8.3.6) and (8.3.7) the two quadratic forms

$$(1/\sigma^2)(\mathbf{X}\hat{\boldsymbol{\beta}} - \mathbf{X}\boldsymbol{\beta})'(\mathbf{X}\hat{\boldsymbol{\beta}} - \mathbf{X}\boldsymbol{\beta}) = [\mathbf{y} - \mathbf{E(y)}]'\left[(1/\sigma^2)\mathbf{X}(\mathbf{X}'\mathbf{X})^{-1}\mathbf{X}'\right] \times [\mathbf{y} - \mathbf{E(y)}]$$

and

$$(1/\sigma^2)(\mathbf{y} - \mathbf{X}\hat{\boldsymbol{\beta}})'(\mathbf{y} - \mathbf{X}\hat{\boldsymbol{\beta}}) = [\mathbf{y} - \mathbf{E}(\mathbf{y})]'(1/\sigma^2)\left[\mathbf{I} - \mathbf{X}(\mathbf{X}'\mathbf{X})^{-1}\mathbf{X}'\right] \times [\mathbf{y} - \mathbf{E}(\mathbf{y})]$$

The triple product prescribed by the theorem takes the form

$$\begin{aligned}\mathbf{A}\,\mathbf{Var}(\mathbf{y})\,\mathbf{B} &= (1/\sigma^2)\mathbf{X}(\mathbf{X}'\mathbf{X})^{-1}\mathbf{X}'\sigma^2\mathbf{I}(1/\sigma^2)\left[\mathbf{I} - \mathbf{X}(\mathbf{X}'\mathbf{X})^{-1}\mathbf{X}'\right] \\ &= (1/\sigma^2)\mathbf{X}(\mathbf{X}'\mathbf{X})^{-1}\mathbf{X}'\left[\mathbf{I} - \mathbf{X}(\mathbf{X}'\mathbf{X})^{-1}\mathbf{X}'\right] = \mathbf{0}\end{aligned}$$

Thus

$$(1/\sigma^2)(\mathbf{y} - \mathbf{X}\hat{\boldsymbol{\beta}})'(\mathbf{y} - \mathbf{X}\hat{\boldsymbol{\beta}}) \qquad \text{and} \qquad (1/\sigma^2)(\mathbf{X}\hat{\boldsymbol{\beta}} - \mathbf{X}\boldsymbol{\beta})'(\mathbf{X}\hat{\boldsymbol{\beta}} - \mathbf{X}\boldsymbol{\beta})$$

are independently distributed χ^2 observations under the conditions of the linear regression model and under the assumption that $\boldsymbol{\beta}$ is known or correctly hypothesized.

Looking back at Eq. (8.4.16), it may not be clear how to apply the second statistical finding. If, however, we divide numerator and denominator of (8.4.16) by the numerator, Λ becomes

$$\Lambda = \left[1 + \frac{(1/\sigma^2)(\mathbf{X}\hat{\boldsymbol{\beta}} - \mathbf{X}\boldsymbol{\beta})'(\mathbf{X}\hat{\boldsymbol{\beta}} - \mathbf{X}\boldsymbol{\beta})}{(1/\sigma^2)(\mathbf{y} - \mathbf{X}\hat{\boldsymbol{\beta}})'(\mathbf{y} - \mathbf{X}\hat{\boldsymbol{\beta}})}\right]^{-1}$$

In this form it is clear that Λ is large if the ratio is small and that Λ is small if the ratio is large. Rather than Λ, we evaluate the ratio

$$G = \frac{(1/\sigma^2)(\mathbf{X}\hat{\boldsymbol{\beta}} - \mathbf{X}\boldsymbol{\beta})'(\mathbf{X}\hat{\boldsymbol{\beta}} - \mathbf{X}\boldsymbol{\beta})}{(1/\sigma^2)(\mathbf{y} - \mathbf{X}\hat{\boldsymbol{\beta}})'(\mathbf{y} - \mathbf{X}\hat{\boldsymbol{\beta}})}$$

(However, now we would reject the hypothesis about $\boldsymbol{\beta}$ if G were large and not reject that hypothesis if G were small, unlike Λ or λ.)

The statistic G is a ratio of independently distributed χ^2 observations. It comes close to satisfying the F-distribution requirement. Multiplying G by $(n - m)/m$ provides a final statistic:

$$\begin{aligned}F = [(n - m)/m]G &= \frac{[(1/\sigma^2)(\mathbf{X}\hat{\boldsymbol{\beta}} - \mathbf{X}\boldsymbol{\beta})'(\mathbf{X}\hat{\boldsymbol{\beta}} - \mathbf{X}\boldsymbol{\beta})]/(m)}{[(1/\sigma^2)(\mathbf{y} - \mathbf{X}\hat{\boldsymbol{\beta}})'(\mathbf{y} - \mathbf{X}\hat{\boldsymbol{\beta}})]/(n - m)} \\ &= \frac{[(\mathbf{X}\hat{\boldsymbol{\beta}} - \mathbf{X}\boldsymbol{\beta})'(\mathbf{X}\hat{\boldsymbol{\beta}} - \mathbf{X}\boldsymbol{\beta})]/m}{[(\mathbf{y} - \mathbf{X}\hat{\boldsymbol{\beta}})'(\mathbf{y} - \mathbf{X}\hat{\boldsymbol{\beta}})]/(n - m)} = \frac{(\hat{\boldsymbol{\beta}} - \boldsymbol{\beta})'\mathbf{X}'\mathbf{X}(\hat{\boldsymbol{\beta}} - \boldsymbol{\beta})/m}{\mathbf{e}'\mathbf{e}/(n - m)}\end{aligned} \tag{8.4.17}$$

The F statistic is the ratio of two independently distributed χ^2 observations each divided by its degrees of freedom. Hence, it is distributed as the F statistical variate. Since the multiplier $(n - m)/m$ is positive, F obeys the same logic as G. Small values of F are to be expected when the hypothesis about $\boldsymbol{\beta}$ is true, large values of F when the hypothesis is not correct. Thus we have a statistic with a known distribution and a rule for relating that distribution to hypothesis testing: Reject the hypothesis if F is large. Finally, and most important, in forming the ratio of the two χ^2 variates, the unknown parameter σ^2, which was a part of each of the χ^2 variates, cancels out. The final ratio in (8.4.17) is based only on known arrays—$\mathbf{X}$, $\mathbf{y}$, and $\hat{\boldsymbol{\beta}}$—and a hypothesized array $\boldsymbol{\beta}$.

Under the hypothesis, the statistic F has as its sampling distribution the F distribution with m and $n - m$ degrees of freedom.

Equation (8.4.17) provides a way, then, of testing the hypothesis that β is a vector of known elements. Drawing once again on the example of the French comprehension test, $\hat{\boldsymbol{\beta}}' = (13.9167, 7.25)$, $\mathbf{e}'\mathbf{e} = 476.33$, $n = 6$, $m = 2$, and

$$\mathbf{X}'\mathbf{X} = \begin{bmatrix} 6 & 18 \\ 18 & 70 \end{bmatrix}$$

The two parameters β_1 and β_2, the reader may recall, are involved in the linear equation

$$E(\text{comprehension semesters of study}) = \beta_1 + \beta_2(\text{semesters of study})$$

The particular model for a linear increase in comprehension scores with greater semesters of study hypothesized that the values of β_1 and β_2 are 12 and 8, respectively.

The vector $(\hat{\boldsymbol{\beta}} - \boldsymbol{\beta})' = (1.9167, -0.75)$, and pre- and postmultiplying $\mathbf{X}'\mathbf{X}$ by this vector yields the scalar 11.0417. The F ratio, then, is

$$F = \frac{11.0417/2}{476.33/(6-2)} = \frac{5.5208}{119.08} = 0.046$$

The very small value of F does not permit us to reject the hypothesis that $\boldsymbol{\beta}' = (12, 8)$. From the tabulated values for $F_{(2,4)}$ in the Appendix, $F \geqslant 6.94$ is required to reject the hypothesis at the $\alpha = 0.05$ level.

8.4.3 *Testing Constraints on* $\boldsymbol{\beta}$

The F ratio of Eq. (8.4.17) *simultaneously* tests hypotheses about *all* values of $\boldsymbol{\beta}$. A rejection of the hypothesized $\boldsymbol{\beta}$ vector tells us that the model used to generate the vector does not fit the data at hand. Sometimes this outcome (or the negation of it) is sufficient for the research question of interest. More often, however, when $\boldsymbol{\beta}$ consists of several elements, we want to determine systematically which parts of the model fit and which do not. One way of testing hypotheses about elements of $\boldsymbol{\beta}$ employs the use of constraints on the elements of $\boldsymbol{\beta}$.

Recall that we imposed one or more such constraints on the least squares weighting vector $\mathbf{b}$ by including the equation $\mathbf{C}'\mathbf{b} = \mathbf{0}$ in the expression that was to be minimized. The number of constraints imposed was equal to the rank of the matrix $\mathbf{C}$, usually the number of rows in the basic matrix $\mathbf{C}'$.

We can do something similar in the linear regression model, phrasing specific hypotheses about the elements of $\boldsymbol{\beta}$ in the form

$$\mathbf{C}'\boldsymbol{\beta} = \boldsymbol{\Gamma} \tag{8.4.18}$$

This equation is a generalized form of the procedure introduced for least squares. The linear weightings of the elements of $\boldsymbol{\beta}$ imposed by the elements in a row of $\mathbf{C}'$ need not sum to zero. Here some other value may be specified by hypothesis. For example, separating the two parts of the complete hypothesis for the French comprehension data could be accomplished by the two constraints

$$(1 \quad 0)\begin{bmatrix}\beta_1\\ \beta_2\end{bmatrix} = 12, \qquad (0 \quad 1)\begin{bmatrix}\beta_1\\ \beta_2\end{bmatrix} = 8$$

Nonzero constraints could have been developed in the least squares problem just as well. (We introduce them here to illustrate both possibilities.) In practice, as later discussions will demonstrate, most hypotheses of interest can be tested employing zero sum constraints, i.e., taking $\boldsymbol{\Gamma} = \mathbf{0}$. Additionally, computer implementations of the general linear model that accommodate nonzero constraints are fewer, less readily available, and somewhat more complicated to use than those programs that permit either a full range of zero constraints or are restricted to zero constraints that drop one or more columns from the design matrix by setting selected elements of β equal to zero.

The constrained maximum likelihood solution proceeds very much as did the constrained least squares solution. The log likelihood equation (8.2.4) is amended to include the side condition by introducing a vector of Lagrangian multipliers $\boldsymbol{\lambda}$:

$$\begin{aligned}\phi &= \log(2\pi)^{-n/2} + \log(\sigma^2)^{-n/2}\\ &\quad - (1/2\sigma^2)(\mathbf{y} - \mathbf{X}\boldsymbol{\beta})'(\mathbf{y} - \mathbf{X}\boldsymbol{\beta}) - \boldsymbol{\lambda}'(\mathbf{C}'\boldsymbol{\beta} - \boldsymbol{\Gamma})\end{aligned} \tag{8.4.19}$$

By differentiating ϕ successively with respect to $\boldsymbol{\beta}$, σ^2, and $\boldsymbol{\lambda}$, setting the results equal to zero, and then solving that set of simultaneous equations for $\boldsymbol{\beta}$ and σ^2, *constrained maximum likelihood* estimators ${}_c\hat{\boldsymbol{\beta}}$ and ${}_c\hat{\sigma}^2$ are obtained. This process results in

$$\begin{aligned}{}_c\hat{\boldsymbol{\beta}} &= (\mathbf{X}'\mathbf{X})^{-1}\mathbf{X}'\mathbf{y} - (\mathbf{X}'\mathbf{X})^{-1}\mathbf{C}\left[\mathbf{C}'(\mathbf{X}'\mathbf{X})^{-1}\mathbf{C}\right]^{-1}\mathbf{C}'(\mathbf{X}'\mathbf{X})^{-1}\mathbf{X}'\mathbf{y}\\ &\quad + (\mathbf{X}'\mathbf{X})^{-1}\mathbf{C}\left[\mathbf{C}'(\mathbf{X}'\mathbf{X})^{-1}\mathbf{C}\right]^{-1}\boldsymbol{\Gamma}\\ &= \hat{\boldsymbol{\beta}} - (\mathbf{X}'\mathbf{X})^{-1}\mathbf{C}\left[\mathbf{C}'(\mathbf{X}'\mathbf{X})^{-1}\mathbf{C}\right]^{-1}(\mathbf{C}'\hat{\boldsymbol{\beta}} - \boldsymbol{\Gamma})\end{aligned} \tag{8.4.20}$$

Note that for $\boldsymbol{\Gamma} = \mathbf{0}$ Eq. (8.4.20) is identical to the solution developed in (7.2.12) for the least square weights ${}_c\mathbf{b}$ constrained by $\mathbf{C}'\mathbf{b} = \mathbf{0}$. The constrained maximum likelihood estimator of σ^2 is defined analogously to the unconstrained estimator

$$_{c}\hat{\sigma}^2 = (1/n)(\mathbf{y} - \mathbf{X}_c\hat{\boldsymbol{\beta}})'(\mathbf{y} - \mathbf{X}_c\hat{\boldsymbol{\beta}}) = (1/n)\mathbf{f}'\mathbf{f} \tag{8.4.21}$$

and is, like $\mathbf{e}'\mathbf{e}/n$, a biased estimator of σ^2.

The logic of the likelihood ratio approach to constructing a test statistic carries over to the hypotheses associated with constraints on β. In parallel with (8.4.14), the likelihood ratio is given as

$$\lambda = L(\mathbf{y}|\mathbf{C}'\hat{\boldsymbol{\beta}} = \boldsymbol{\Gamma})/L(\mathbf{y}|\hat{\boldsymbol{\beta}}) < 1.0 \tag{8.4.22}$$

Likelihood estimates can be found for the two conditions (constrained and unconstrained maximum likelihood estimation of the unknown $\boldsymbol{\beta}$ and σ^2), and the resulting likelihood ratio is

$$\lambda = (\mathbf{e}'\mathbf{e}/\mathbf{f}'\mathbf{f})^{n/2}$$

The sums of squares $\mathbf{e}'\mathbf{e}$ and $\mathbf{f}'\mathbf{f}$ are given by Eqs. (8.2.11) and (8.4.21). Working from λ as before, we first write

$$\Lambda = (\lambda)^{2/n} = \mathbf{e}'\mathbf{e}/\mathbf{f}'\mathbf{f}$$

Next we expand the denominator of this ratio,

$$\Lambda = \frac{\mathbf{e}'\mathbf{e}}{(\mathbf{f}'\mathbf{f} - \mathbf{e}'\mathbf{e}) + \mathbf{e}'\mathbf{e}}$$

and, finally, rewrite it as

$$\Lambda = [1 + (\mathbf{f}'\mathbf{f} - \mathbf{e}'\mathbf{e})/\mathbf{e}'\mathbf{e}]^{-1}$$

This form leads to a hypothesis testing logic: Rather than reject the hypothesis that $\mathbf{C}'\boldsymbol{\beta} = \boldsymbol{\Gamma}$ when λ is small, reject when

$$G = (\mathbf{f}'\mathbf{f} - \mathbf{e}'\mathbf{e})/\mathbf{e}'\mathbf{e}$$

is large. An important result here is the analogous representation of $\mathbf{f}'\mathbf{f} - \mathbf{e}'\mathbf{e}$, the quantity described in Chapter 7 as the increase in the sum of squares of errors of prediction:

$$\mathbf{f}'\mathbf{f} - \mathbf{e}'\mathbf{e} = (\mathbf{y} - \mathbf{X}_c\hat{\boldsymbol{\beta}})'(\mathbf{y} - \mathbf{X}_c\hat{\boldsymbol{\beta}}) - (\mathbf{y} - \mathbf{X}\hat{\boldsymbol{\beta}})'(\mathbf{y} - \mathbf{X}\hat{\boldsymbol{\beta}}) \tag{8.4.23}$$

Without tracing the algebraic development, this difference can be expressed

$$\begin{aligned} \mathbf{f}'\mathbf{f} - \mathbf{e}'\mathbf{e} &= (\boldsymbol{\Gamma} - \mathbf{C}'\hat{\boldsymbol{\beta}})'[\mathbf{C}'(\mathbf{X}'\mathbf{X})^{-1}\mathbf{C}]^{-1}(\boldsymbol{\Gamma} - \mathbf{C}'\hat{\boldsymbol{\beta}}) \\ &= [\boldsymbol{\beta} - (\mathbf{X}'\mathbf{X})^{-1}\mathbf{X}'\mathbf{y}]'\mathbf{C}[\mathbf{C}'(\mathbf{X}'\mathbf{X})^{-1}\mathbf{C}]^{-1}\mathbf{C}'[\boldsymbol{\beta} - (\mathbf{X}'\mathbf{X})^{-1}\mathbf{X}'\mathbf{y}] \end{aligned} \tag{8.4.24}$$

The first expression in Eq. (8.4.24) is of computational value, as $\boldsymbol{\Gamma}$ and $\mathbf{C}$ are specified by the hypothesis while $\mathbf{X}'\mathbf{X}$ and $\hat{\boldsymbol{\beta}} = (\mathbf{X}'\mathbf{X})^{-1}\mathbf{X}'\mathbf{y}$ are given by the data. We use the last form of (8.4.24) to develop another version of the test of the hypothesis. In the least squares development we showed that $\mathbf{f}'\mathbf{f} - \mathbf{e}'\mathbf{e}$ is a quadratic form for $\boldsymbol{\Gamma} = \mathbf{0}$. We now consider nonzero $\boldsymbol{\Gamma}$; moreover, we shall establish that $(\mathbf{f}'\mathbf{f} - \mathbf{e}'\mathbf{e})/\sigma^2$ has a chi-squared sampling distribution with $r(\mathbf{C})$ degrees of freedom.

The demonstration is straightforward once one notes that the right side of (8.4.24) can also be expressed

$$\begin{aligned}\mathbf{f}'\mathbf{f} - \mathbf{e}'\mathbf{e} &= [(\mathbf{X}'\mathbf{X})\boldsymbol{\beta} - \mathbf{X}'\mathbf{y}]'(\mathbf{X}'\mathbf{X})^{-1}\mathbf{C}\left[\mathbf{C}'(\mathbf{X}'\mathbf{X})^{-1}\mathbf{C}\right]^{-1} \\ &\quad \times \mathbf{C}'(\mathbf{X}'\mathbf{X})^{-1}[(\mathbf{X}'\mathbf{X})\boldsymbol{\beta} - \mathbf{X}'\mathbf{y}] \\ &= (\mathbf{X}\boldsymbol{\beta} - \mathbf{y})'\mathbf{X}(\mathbf{X}'\mathbf{X})^{-1}\mathbf{C}\left[\mathbf{C}'(\mathbf{X}'\mathbf{X})^{-1}\mathbf{C}\right]^{-1}\mathbf{C}'(\mathbf{X}'\mathbf{X})^{-1}\mathbf{X}'(\mathbf{X}\boldsymbol{\beta} - \mathbf{y})\end{aligned}$$

and

$$(\mathbf{f}'\mathbf{f} - \mathbf{e}'\mathbf{e})/\sigma^2 = (\mathbf{y} - \mathbf{X}\boldsymbol{\beta})'\mathbf{A}(\mathbf{y} - \mathbf{X}\boldsymbol{\beta}) = [\mathbf{y} - \mathbf{E}(\mathbf{y})]'\mathbf{A}[\mathbf{y} - \mathbf{E}(\mathbf{y})]$$

where

$$\mathbf{A} = (1/\sigma^2)\left[\mathbf{X}(\mathbf{X}'\mathbf{X})^{-1}\mathbf{C}\left[\mathbf{C}'(\mathbf{X}'\mathbf{X})^{-1}\mathbf{C}\right]^{-1}\mathbf{C}'(\mathbf{X}'\mathbf{X})^{-1}\mathbf{X}'\right]$$

The matrix $\mathbf{A}$ is symmetric and has rank equal to the rank of $\mathbf{C}$, the number of constraints imposed on the vector $\boldsymbol{\beta}$. Moreover, $\mathbf{A}\,\mathbf{Var}(\mathbf{y})\,\mathbf{A} = \mathbf{A}$ and $(\mathbf{f}'\mathbf{f} - \mathbf{e}'\mathbf{e})/\sigma^2 \sim \chi^2_{r(\mathbf{C})}$.

Furthermore, $(\mathbf{f}'\mathbf{f} - \mathbf{e}'\mathbf{e})/\sigma^2$ is distributed independently of $\mathbf{e}'\mathbf{e}/\sigma^2$. For this to be true, it is necessary that $\mathbf{A}\,\mathbf{Var}(\mathbf{y})\,\mathbf{B} = 0$, where

$$\mathbf{B} = (1/\sigma^2)\left[\mathbf{I} - \mathbf{X}(\mathbf{X}'\mathbf{X})^{-1}\mathbf{X}'\right]$$

The reader may carry through the multiplication to establish the independence.

As indicated in Section 8.4.2, the ratio of the two independently distributed chi-squared variables, each divided by its degrees of freedom, is distributed as an F variate:

$$F = \frac{(\mathbf{f}'\mathbf{f} - \mathbf{e}'\mathbf{e})/r(\mathbf{C})}{\mathbf{e}'\mathbf{e}/(n - m)} \sim F_{r(\mathbf{C}),(n-m)} \qquad (8.4.25)$$

This F ratio is large when the likelihood-ratio-based G is large, and it is therefore widely used for testing hypotheses in the linear regression model. The numerator, based on a hypothesis expressed as a set of one or more linear constraints on the elements of $\boldsymbol{\beta}$, $\mathbf{C}'\boldsymbol{\beta} = \boldsymbol{\Gamma}$, is obtained from (8.4.24) and the denominator from (8.2.11).

There are, as we shall see, special forms of (8.4.24) depending on the design matrix $\mathbf{X}$ and the constraint matrix $\mathbf{C}$. For instance, if $\boldsymbol{\Gamma} = \boldsymbol{\beta}$ and $\mathbf{C} = \mathbf{I}$, (8.4.24) becomes the numerator in the simultaneous test of all elements of $\boldsymbol{\beta}$:

$$\mathbf{f}'\mathbf{f} - \mathbf{e}'\mathbf{e} = (\boldsymbol{\beta} - \hat{\boldsymbol{\beta}})'(\mathbf{X}'\mathbf{X})(\boldsymbol{\beta} - \hat{\boldsymbol{\beta}})$$

Similarly, when $\boldsymbol{\Gamma} = \mathbf{0}$ and the rows of $\mathbf{C}'$ are $\mathbf{e}_i'$ vectors, i.e., when the hypothesis involves dropping predictors by setting one or more elements of $\boldsymbol{\beta}$ equal to zero, we can use the multiple correlation results analogous to (7.3.18) and (7.3.19) to write

$$F = \frac{(\mathbf{f}'\mathbf{f} - \mathbf{e}'\mathbf{e})/r(\mathbf{C})}{\mathbf{e}'\mathbf{e}/(n-m)} = \frac{\left(R^2_{\mathbf{y}.\mathbf{X},\mathbf{Z}} - R^2_{\mathbf{y}.\mathbf{X}}\right)/r(\mathbf{C})}{\left(1 - R^2_{\mathbf{y}.\mathbf{X},\mathbf{Z}}\right)/(n-m)}$$

8.4.4 *Multiple Hypotheses on* $\boldsymbol{\beta}$

The F ratio of (8.4.25) provides a test of any hypothesis about the elements of $\boldsymbol{\beta}$ that can be phrased in the form $\mathbf{C}'\boldsymbol{\beta} = \boldsymbol{\Gamma}$. Almost without exception we shall test more than one hypothesis in a study, impose more than one set of constraints on the $\boldsymbol{\beta}$ parameters, and evaluate for each either the decrease in the likelihood of $\mathbf{y}$ or the increase in the sum of squares of the errors over the *unconstrained* solution.

There may be some uncertainty, however, in how to interpret these presumably different hypotheses, particularly when the design is such that the elements of $\boldsymbol{\beta}$ have correlated or nonindependent sampling distributions. Each constraint is a linear transformation of the elements of $\boldsymbol{\beta}$ and, as such, has a sampling distribution with mean and variance dependent on the mean vector and variance–covariance matrix for the sampling distributions of the elements of $\hat{\boldsymbol{\beta}}$. Based on (8.3.1) and (8.3.3) the constraint $\mathbf{C}'\hat{\boldsymbol{\beta}}$ has a sampling distribution with mean

$$\mathbf{E}(\mathbf{C}'\hat{\boldsymbol{\beta}}) = \mathbf{C}'\mathbf{E}(\hat{\boldsymbol{\beta}}) = \mathbf{C}'\boldsymbol{\beta} \tag{8.4.26}$$

and variance

$$\mathbf{Var}(\mathbf{C}'\hat{\boldsymbol{\beta}}) = \mathbf{C}'\,\mathbf{Var}(\hat{\boldsymbol{\beta}})\,\mathbf{C} = \sigma^2\mathbf{C}'(\mathbf{X}'\mathbf{X})^{-1}\mathbf{C} \tag{8.4.27}$$

If there are two hypotheses—two separately evaluated sets of constraints—each a linear transformation of $\boldsymbol{\beta}$, the two sampling distributions must also have a covariance. For the two constraints ${}_1\mathbf{C}'\boldsymbol{\beta} = {}_1\boldsymbol{\Gamma}$ and ${}_2\mathbf{C}'\boldsymbol{\beta} = {}_2\boldsymbol{\Gamma}$,

$$\mathbf{Cov}\left({}_1\mathbf{C}'\hat{\boldsymbol{\beta}}, {}_2\mathbf{C}'\hat{\boldsymbol{\beta}}\right) = {}_1\mathbf{C}'\,\mathbf{Var}(\hat{\boldsymbol{\beta}})\,{}_2\mathbf{C} = \sigma^2\,{}_1\mathbf{C}'(\mathbf{X}'\mathbf{X})^{-1}\,{}_2\mathbf{C} \tag{8.4.28}$$

This covariance may be nonzero: the constraints and the hypotheses they represent need not be mutually independent.

Consider a simple example. The two hypotheses are $\beta_1 = 0$ and $\beta_2 = 0$. As linear constraints, these hypotheses may be written

$$\mathbf{e}_1'\boldsymbol{\beta} = 0, \qquad \mathbf{e}_2'\boldsymbol{\beta} = 0$$

Thus the sampling covariance between the two constraints is

$$\mathrm{Cov}\left(\mathbf{e}_1'\hat{\boldsymbol{\beta}}, \mathbf{e}_2'\hat{\boldsymbol{\beta}}\right) = \sigma^2\mathbf{e}_1'(\mathbf{X}'\mathbf{X})^{-1}\mathbf{e}_2 = \sigma^2\left[(\mathbf{X}'\mathbf{X})^{-1}\right]_{12}$$

and is likely to be nonzero if the columns of the design matrix are not mutually orthogonal. If the covariance between the constraints is not zero, the tests of the two hypotheses are not independent. One may be troubled, then, at the prospect of rejecting the hypothesis that $\beta_2 = 0$ simply because $\hat{\beta}_2$ is correlated with $\hat{\beta}_1$.

We shall outline a general way of expressing constraints that may be tested at least *quasi-independently*. We call this the method of *hierarchical constraints*. The independence (we shall explain in due course the use of the modifier "quasi") has, in our opinion, distinct advantages, although other writers find the interpretation of hierarchical hypotheses cumbersome, if not unnatural. We shall review these arguments in the next chapter, where we discuss the analysis of variance as an application of the GLM.

Consider, again, the two hypotheses $\beta_1 = 0$ and $\beta_2 = 0$. This time, however, we take the two constraints ${}_1\mathbf{C}'\boldsymbol{\beta} = {}_1\boldsymbol{\Gamma}$,

$$(1 \quad 0)\begin{bmatrix} \beta_1 \\ \beta_2 \end{bmatrix} = 0$$

and ${}_2\mathbf{C}'\boldsymbol{\beta} = {}_2\boldsymbol{\Gamma}$,

$$\begin{bmatrix} 1 & 0 \\ 0 & 1 \end{bmatrix}\begin{bmatrix} \beta_1 \\ \beta_2 \end{bmatrix} = \begin{bmatrix} 0 \\ 0 \end{bmatrix}$$

We say these two constraints are *hierarchical* or *hierarchically ordered* because one is contained within the other:

$${}_2\mathbf{C}' = \begin{bmatrix} {}_1\mathbf{C}' \\ 0 \quad 1 \end{bmatrix} \qquad \text{and} \qquad {}_2\boldsymbol{\Gamma} = \begin{bmatrix} {}_1\boldsymbol{\Gamma} \\ 0 \end{bmatrix}$$

Before going any further, let us take the more general case of ${}_2\mathbf{C}$ and ${}_2\boldsymbol{\Gamma}$ of the form

$${}_2\mathbf{C} = ({}_1\mathbf{C} \quad {}_2\mathbf{K}), \qquad {}_2\boldsymbol{\Gamma} = \begin{bmatrix} {}_1\boldsymbol{\Gamma} \\ {}_2\mathbf{G} \end{bmatrix}$$

Here ${}_1\mathbf{C}'\boldsymbol{\beta} = {}_1\boldsymbol{\Gamma}$ imposes p constraints on the m elements of $\boldsymbol{\beta}$, and ${}_2\mathbf{C}'\boldsymbol{\beta} = {}_2\boldsymbol{\Gamma}$ imposes $p + q\ (\leqslant m)$ constraints. If we find maximum likelihood estimators of $\boldsymbol{\beta}$, constrained first by ${}_1\mathbf{C}'\boldsymbol{\beta} = {}_1\boldsymbol{\Gamma}$ and then by ${}_2\mathbf{C}'\boldsymbol{\beta} = {}_2\boldsymbol{\Gamma}$, we obtain, using (8.4.23), the two quantities

$$\mathbf{f}'\mathbf{f} - \mathbf{e}'\mathbf{e} = ({}_1\boldsymbol{\Gamma} - {}_1\mathbf{C}'\hat{\boldsymbol{\beta}})'\left[{}_1\mathbf{C}'(\mathbf{X}'\mathbf{X})^{-1}{}_1\mathbf{C}\right]^{-1}({}_1\boldsymbol{\Gamma} - {}_1\mathbf{C}'\hat{\boldsymbol{\beta}})$$

and

$$\mathbf{g}'\mathbf{g} - \mathbf{e}'\mathbf{e} = ({}_2\boldsymbol{\Gamma} - {}_2\mathbf{C}'\hat{\boldsymbol{\beta}})'\left[{}_2\mathbf{C}'(\mathbf{X}'\mathbf{X})^{-1}{}_2\mathbf{C}\right]^{-1}({}_2\boldsymbol{\Gamma} - {}_2\mathbf{C}'\hat{\boldsymbol{\beta}})$$

Since $(\mathbf{f}'\mathbf{f} - \mathbf{e}'\mathbf{e})/\sigma^2$ and $(\mathbf{g}'\mathbf{g} - \mathbf{e}'\mathbf{e})/\sigma^2$ have chi-squared sampling distributions with p and $p + q$ degrees of freedom, respectively, and each is distributed *independently* of $\mathbf{e}'\mathbf{e}/\sigma^2$, we can compute and evaluate the two F

ratios

$$\frac{(\mathbf{f}'\mathbf{f} - \mathbf{e}'\mathbf{e})/p}{(\mathbf{e}'\mathbf{e})/(n-m)} \sim F_{p,(n-m)} \quad \text{and} \quad \frac{(\mathbf{g}'\mathbf{g} - \mathbf{e}'\mathbf{e})/(p+q)}{(\mathbf{e}'\mathbf{e})/(n-m)} \sim F_{(p+q),(n-m)}$$

The first tests the hypothesis that ${}_1\mathbf{C}'\boldsymbol{\beta} = {}_1\boldsymbol{\Gamma}$, the second whether ${}_1\mathbf{C}'\boldsymbol{\beta} = {}_1\boldsymbol{\Gamma}$ and ${}_2\mathbf{K}'\boldsymbol{\beta} = {}_2\mathbf{G}$. The two tests of hypotheses are clearly not independent; the second includes the first.

Now consider the sum of squares of prediction errors (SSE):

1. $\mathbf{e}'\mathbf{e}$ is the unconstrained SSE.
2. $\mathbf{f}'\mathbf{f}$ is the SSE when we require that ${}_1\mathbf{C}'\boldsymbol{\beta} = {}_1\boldsymbol{\Gamma}$.
3. $\mathbf{g}'\mathbf{g}$ is the SSE when we require both ${}_1\mathbf{C}'\boldsymbol{\beta} = {}_1\boldsymbol{\Gamma}$ and ${}_2\mathbf{K}'\boldsymbol{\beta} = {}_2\mathbf{G}$.

Because $\mathbf{g}'\mathbf{g}$ is the result of imposing constraints in addition to those that yielded $\mathbf{f}'\mathbf{f}$, $\mathbf{f}'\mathbf{f} \leqslant \mathbf{g}'\mathbf{g}$. Indeed, the difference $\mathbf{g}'\mathbf{g} - \mathbf{f}'\mathbf{f}$ is interpretable as an increase in SSE occasioned by requiring that ${}_2\mathbf{K}'\boldsymbol{\beta} = {}_2\mathbf{G}$ as well as ${}_1\mathbf{C}'\boldsymbol{\beta} = {}_1\boldsymbol{\Gamma}$.

Hierarchical constraints, in effect, allow us to break up the overall increase in SSE, $\mathbf{g}'\mathbf{g} - \mathbf{e}'\mathbf{e}$, into two parts, $\mathbf{f}'\mathbf{f} - \mathbf{e}'\mathbf{e}$ and $\mathbf{g}'\mathbf{g} - \mathbf{f}'\mathbf{f}$. More important, we can then test these two parts, again, quasi-independently. By establishing that $(\mathbf{g}'\mathbf{g} - \mathbf{f}'\mathbf{f})/\sigma^2$ has a sampling distribution that is

1. a chi-squared variable with q degrees of freedom,
2. independent of $(\mathbf{f}'\mathbf{f} - \mathbf{e}'\mathbf{e})/\sigma^2$, and
3. independent of $\mathbf{e}'\mathbf{e}/\sigma^2$.

The two F ratios

$$\frac{(\mathbf{f}'\mathbf{f} - \mathbf{e}'\mathbf{e})/p}{\mathbf{e}'\mathbf{e}/(n-m)} \sim F_{p,(n-m)} \qquad \text{and} \qquad \frac{(\mathbf{g}'\mathbf{g} - \mathbf{f}'\mathbf{f})/q}{\mathbf{e}'\mathbf{e}/(n-m)} \sim F_{q,(n-m)}$$

provide quasi-independent tests of the hypotheses ${}_1\mathbf{C}'\boldsymbol{\beta} = {}_1\boldsymbol{\Gamma}$ and ${}_2\mathbf{K}'\boldsymbol{\beta} = {}_2\mathbf{G}$ (given that ${}_1\mathbf{C}'\boldsymbol{\beta} = {}_1\boldsymbol{\Gamma}$). The F ratios are only quasi-independent rather than fully independent because of their common denominator, $\mathbf{e}'\mathbf{e}/(n-m)$.

The reader should note the conditional form of the second hypothesis. We are asking, "Does imposing the restriction ${}_2\mathbf{K}'\boldsymbol{\beta} = {}_2\mathbf{G}$ on the sample data affect the fit of the model, *given that* we have already imposed the restriction that ${}_1\mathbf{C}'\boldsymbol{\beta} = {}_1\boldsymbol{\Gamma}$?" The interpretation will be easier to understand if we examine the special case in which the design matrix consists only of the unit vector and two predictor attributes $\mathbf{x}$ and $\mathbf{z}$ and the two constraints are

$${}_1\mathbf{C}'\boldsymbol{\beta} = (0 \quad 1 \quad 0)\begin{bmatrix} \beta_1 \\ \beta_{\mathbf{x}} \\ \beta_{\mathbf{z}} \end{bmatrix} = 0 = {}_1\Gamma$$

and

$$\begin{bmatrix} {}_1\mathbf{C}' \\ {}_2\mathbf{K}' \end{bmatrix}\boldsymbol{\beta} = \begin{bmatrix} 0 & 1 & 0 \\ 0 & 0 & 1 \end{bmatrix}\begin{bmatrix} \beta_1 \\ \beta_{\mathbf{x}} \\ \beta_{\mathbf{z}} \end{bmatrix} = \begin{bmatrix} 0 \\ 0 \end{bmatrix} = \begin{bmatrix} {}_1\Gamma \\ {}_2G \end{bmatrix}$$

The first constraint requires that we set the predictive weight for $\mathbf{x}$ equal to zero, dropping $\mathbf{x}$ as a predictor. The second constraint matrix requires that we drop both $\mathbf{x}$ and $\mathbf{z}$. Parallel to the development in Section 7.3 on partial and multiple correlation, we can write for this example

$$\begin{aligned} \mathbf{e}'\mathbf{e} &= nS_{\mathbf{y}}^2\left(1 - R_{\mathbf{y}.\mathbf{x},\mathbf{z}}^2\right) \\ \mathbf{f}'\mathbf{f} - \mathbf{e}'\mathbf{e} &= nS_{\mathbf{y}}^2\left(R_{\mathbf{y}.\mathbf{x},\mathbf{z}}^2 - r_{\mathbf{yz}}^2\right) \\ \mathbf{g}'\mathbf{g} - \mathbf{e}'\mathbf{e} &= nS_{\mathbf{y}}^2\left(R_{\mathbf{y}.\mathbf{x},\mathbf{z}}^2 - 0\right) = nS_{\mathbf{y}}^2\left(R_{\mathbf{y}.\mathbf{x},\mathbf{z}}^2\right) \end{aligned}$$

and then obtain by subtraction $\mathbf{g}'\mathbf{g} - \mathbf{f}'\mathbf{f} = nS_{\mathbf{y}}^2(r_{\mathbf{yz}}^2)$.

The two F ratios,

$$\frac{\mathbf{f}'\mathbf{f} - \mathbf{e}'\mathbf{e}}{\mathbf{e}'\mathbf{e}/(n-3)} = \frac{R_{\mathbf{y}.\mathbf{x},\mathbf{z}}^2 - r_{\mathbf{yz}}^2}{1 - R_{\mathbf{y}.\mathbf{x},\mathbf{z}}^2} = \frac{r_{\mathbf{y}(\mathbf{x}.\mathbf{z})}^2}{1 - R_{\mathbf{y}.\mathbf{x},\mathbf{z}}^2} \sim F_{1,(n-3)}$$

and

$$\frac{\mathbf{g}'\mathbf{g} - \mathbf{f}'\mathbf{f}}{\mathbf{e}'\mathbf{e}/(n-3)} = \frac{r_{\mathbf{yz}}^2}{1 - R_{\mathbf{y}.\mathbf{x},\mathbf{z}}^2} \sim F_{1,(n-3)}$$

test the *incremental* contribution of $\mathbf{x}$ ("Does dropping $\mathbf{x}$ make a difference, given that we retain $\mathbf{z}$ the other predictor?") and the *marginal* contribution of $\mathbf{z}$ ("Does dropping $\mathbf{z}$ make a difference, given that we have already dropped the other predictor?").

In gaining independence for the two tests of hypotheses, we have had to make the tests asymmetric. This may pose problems for the investigator unless there is an a priori order in which effects should be evaluated. That is, to take advantage of this approach (indeed, to interpret correlated effects), it is necessary that some effects be conditional on others that, in turn, can be evaluated for their direct contribution. Note, for example, that the order in which $\mathbf{x}$ and $\mathbf{z}$ are removed could make a difference; $r_{\mathbf{yx}}$ might be significant but not $r_{\mathbf{y}(\mathbf{z}.\mathbf{x})}$ (or vice versa).

A better practice where the investigator has control is to design studies so that the design attributes—or important contrasts among the attributes—are independent. Though we shall not prove the point, if

$${}_1\mathbf{C}'(\mathbf{X}'\mathbf{X})^{-1}\,{}_2\mathbf{K} = \mathbf{0}$$

then the tests

$$\frac{(\mathbf{g}'\mathbf{g} - \mathbf{f}'\mathbf{f})/q}{\mathbf{e}'\mathbf{e}/(n-m)} \sim F_{q,(n-m)} \qquad \text{and} \qquad \frac{(\mathbf{h}'\mathbf{h} - \mathbf{e}'\mathbf{e})/q}{\mathbf{e}'\mathbf{e}/(n-m)} \sim F_{q,(n-m)}$$

are identical, where $\mathbf{h}'\mathbf{h}$ is the SSE associated with imposing *only* the q constraints ${}_2\mathbf{K}'\boldsymbol{\beta} = {}_2\mathbf{G}$. As the hierarchical procedure provides independent, symmetric tests when the constraints are in fact independent and provides independent though nonsymmetric tests when the constraints are correlated, it is a technique worth mastering in designing studies.

We turn now to a demonstration that $(\mathbf{g}'\mathbf{g} - \mathbf{f}'\mathbf{f})/\sigma^2$ has the desired distribution properties. We know that $(\mathbf{f}'\mathbf{f} - \mathbf{e}'\mathbf{e})/\sigma^2$ and $(\mathbf{g}'\mathbf{g} - \mathbf{e}'\mathbf{e})/\sigma^2$ are each chi-squared observations and can be written as the quadratic forms

$$(\mathbf{f}'\mathbf{f} - \mathbf{e}'\mathbf{e})/\sigma^2 = [\mathbf{y} - \mathbf{E}(\mathbf{y})]'(1/\sigma^2)\left\{\mathbf{X}(\mathbf{X}'\mathbf{X})^{-1}\,{}_1\mathbf{C}\left[{}_1\mathbf{C}'(\mathbf{X}'\mathbf{X})^{-1}\,{}_1\mathbf{C}\right]^{-1} \times {}_1\mathbf{C}'(\mathbf{X}'\mathbf{X})^{-1}\mathbf{X}'\right\}[\mathbf{y} - \mathbf{E}(\mathbf{y})]$$

and

$$(\mathbf{g}'\mathbf{g} - \mathbf{e}'\mathbf{e})/\sigma^2 = [\mathbf{y} - \mathbf{E}(\mathbf{y})]'(1/\sigma^2)\left\{\mathbf{X}(\mathbf{X}'\mathbf{X})^{-1}\,{}_2\mathbf{C}\left[{}_2\mathbf{C}'(\mathbf{X}'\mathbf{X})^{-1}\,{}_2\mathbf{C}\right]^{-1} \times {}_2\mathbf{C}'(\mathbf{X}'\mathbf{X})^{-1}\mathbf{X}'\right\}[\mathbf{y} - \mathbf{E}(\mathbf{y})]$$

For notational simplicity let $\mathbf{u} = \mathbf{y} - \mathbf{E}(\mathbf{y})$ and $\mathbf{V} = \mathbf{X}(\mathbf{X}'\mathbf{X})^{-1}$. Then

$$(\mathbf{g}'\mathbf{g} - \mathbf{f}'\mathbf{f})/\sigma^2 = (\mathbf{u}'\mathbf{V})(1/\sigma)^2\left\{{}_2\mathbf{C}\left[{}_2\mathbf{C}'(\mathbf{X}'\mathbf{X})^{-1}\,{}_2\mathbf{C}\right]^{-1}\,{}_2\mathbf{C}' - {}_1\mathbf{C}\left[{}_1\mathbf{C}'(\mathbf{X}'\mathbf{X})^{-1}\,{}_1\mathbf{C}\right]^{-1}\right\}{}_1\mathbf{C}'(\mathbf{V}'\mathbf{u}) \qquad (8.4.29)$$

Again for simplicity, let $\mathbf{C} = {}_1\mathbf{C}$ and $\mathbf{K} = {}_2\mathbf{K}$; ${}_2\mathbf{C}'(\mathbf{X}'\mathbf{X})^{-1}\,{}_2\mathbf{C}$ can now be written

$$\begin{bmatrix}\mathbf{C}'\\ \mathbf{K}'\end{bmatrix}(\mathbf{X}'\mathbf{X})^{-1}(\mathbf{C}\quad\mathbf{K}) = \begin{bmatrix}\mathbf{C}'(\mathbf{X}'\mathbf{X})^{-1}\mathbf{C} & \mathbf{C}'(\mathbf{X}'\mathbf{X})^{-1}\mathbf{K}\\ \mathbf{K}'(\mathbf{X}'\mathbf{X})^{-1}\mathbf{C} & \mathbf{K}'(\mathbf{X}'\mathbf{X})^{-1}\mathbf{K}\end{bmatrix}$$

Let the partitioning be resymbolized

$$\begin{bmatrix}\mathbf{S}_{11} & \mathbf{S}_{12}\\ \mathbf{S}'_{12} & \mathbf{S}_{22}\end{bmatrix} = \begin{bmatrix}\mathbf{C}'(\mathbf{X}'\mathbf{X})^{-1}\mathbf{C} & \mathbf{C}'(\mathbf{X}'\mathbf{X})^{-1}\mathbf{K}\\ \mathbf{K}'(\mathbf{X}'\mathbf{X})^{-1}\mathbf{C} & \mathbf{K}'(\mathbf{X}'\mathbf{X})^{-1}\mathbf{K}\end{bmatrix}$$

We can now use our rule on the inverse of a symmetrically partitioned matrix to write

$$\left[{}_2\mathbf{C}'(\mathbf{X}'\mathbf{X})^{-1}\,{}_2\mathbf{C}\right]^{-1} = \begin{bmatrix}\mathbf{S}_{11}^{-1} + \mathbf{S}_{11}^{-1}\mathbf{S}_{12}\mathbf{T}_2^{-1}\mathbf{S}'_{12}\mathbf{S}_{11}^{-1} & -\mathbf{S}_{11}^{-1}\mathbf{S}_{12}\mathbf{T}_2^{-1}\\ -\mathbf{T}_2^{-1}\mathbf{S}'_{12}\mathbf{S}_{11}^{-1} & \mathbf{T}_2^{-1}\end{bmatrix}$$

where $\mathbf{T}_2 = \mathbf{S}_{22} - \mathbf{S}'_{12}\mathbf{S}_{11}^{-1}\mathbf{S}_{12}$. Premultiplying $[{}_2\mathbf{C}'(\mathbf{X}'\mathbf{X})^{-1}\,{}_2\mathbf{C}]^{-1}$ by ${}_2\mathbf{C} = (\mathbf{C}, \mathbf{K})$ gives

$$\left(\mathbf{C}\mathbf{S}_{11}^{-1} + \mathbf{C}\mathbf{S}_{11}^{-1}\mathbf{S}_{12}\mathbf{T}_2^{-1}\mathbf{S}'_{12}\mathbf{S}_{11}^{-1} - \mathbf{K}\mathbf{T}_2^{-1}\mathbf{S}'_{12}\mathbf{S}_{11}^{-1},\ -\mathbf{C}\mathbf{S}_{11}^{-1}\mathbf{S}_{12}\mathbf{T}_2^{-1} + \mathbf{K}\mathbf{T}_2^{-1}\right)$$

and then postmultiplying this result by

$${}_2\mathbf{C}' = \begin{bmatrix}\mathbf{C}'\\ \mathbf{K}'\end{bmatrix}$$

gives

$${}_2\mathbf{C}\left[{}_2\mathbf{C}'(\mathbf{X}'\mathbf{X})^{-1}\,{}_2\mathbf{C}\right]{}_2\mathbf{C}' = \mathbf{C}\mathbf{S}_{11}^{-1}\mathbf{C}' + \mathbf{C}\mathbf{S}_{11}^{-1}\mathbf{S}_{12}\mathbf{T}_2^{-1}\mathbf{S}'_{12}\mathbf{S}_{11}^{-1}\mathbf{C}' - \mathbf{K}\mathbf{T}_2^{-1}\mathbf{S}'_{12}\mathbf{S}_{11}^{-1}\mathbf{C}' - \mathbf{C}\mathbf{S}_{11}^{-1}\mathbf{S}_{12}\mathbf{T}_2^{-1}\mathbf{K}' + \mathbf{K}\mathbf{T}_2^{-1}\mathbf{K}'$$

The terms on the right can be regrouped as

$$ {}_2\mathbf{C}\left[{}_2\mathbf{C}'(\mathbf{X}'\mathbf{X})^{-1}{}_2\mathbf{C}\right]^{-1}{}_2\mathbf{C}' = \mathbf{C}\mathbf{S}_{11}^{-1}\mathbf{C}' + (\mathbf{K} - \mathbf{C}\mathbf{S}_{11}^{-1}\mathbf{S}_{12})\mathbf{T}_2^{-1}(\mathbf{K} - \mathbf{C}\mathbf{S}_{11}^{-1}\mathbf{S}_{12})' $$

Recalling that $\mathbf{C} = {}_1\mathbf{C}$ and that $\mathbf{S}_{11}^{-1} = [{}_1\mathbf{C}'(\mathbf{X}'\mathbf{X})^{-1}{}_1\mathbf{C}]^{-1}$, we have

$$ \mathbf{C}\mathbf{S}_{11}^{-1}\mathbf{C}' = {}_1\mathbf{C}\left[{}_1\mathbf{C}'(\mathbf{X}'\mathbf{X})^{-1}{}_1\mathbf{C}\right]^{-1}{}_1\mathbf{C}' $$

and the difference

$$ \begin{aligned} &{}_2\mathbf{C}\left[{}_2\mathbf{C}'(\mathbf{X}'\mathbf{X})^{-1}{}_2\mathbf{C}\right]^{-1}{}_2\mathbf{C}' - {}_1\mathbf{C}\left[{}_1\mathbf{C}'(\mathbf{X}'\mathbf{X})^{-1}{}_1\mathbf{C}\right]^{-1}{}_1\mathbf{C}' \\ &= (\mathbf{K} - \mathbf{C}\mathbf{S}_{11}^{-1}\mathbf{S}_{12})\mathbf{T}_2^{-1}(\mathbf{K} - \mathbf{C}\mathbf{S}_{11}^{-1}\mathbf{S}_{12})' \end{aligned} $$

Letting

$$ \mathbf{W} = \mathbf{K} - \mathbf{C}\mathbf{S}_{11}^{-1}\mathbf{S}_{12} $$

we are ready to rewrite $(\mathbf{g}'\mathbf{g} - \mathbf{f}'\mathbf{f})/\sigma^2$ from (8.4.29):

$$ [\mathbf{y} - \mathbf{E}(\mathbf{y})]'(1/\sigma^2)\mathbf{V}\mathbf{W}\mathbf{T}_2^{-1}\mathbf{W}'\mathbf{V}'[\mathbf{y} - \mathbf{E}(\mathbf{y})] $$

The matrix of the quadratic form

$$ \mathbf{B} = (1/\sigma^2)(\mathbf{V}\mathbf{W}\mathbf{T}_2^{-1}\mathbf{W}'\mathbf{V}') $$

is symmetric since $\mathbf{T}_2$, and hence $\mathbf{T}_2^{-1}$, is symmetric. The rank of $\mathbf{B}$ is equal to the rank of $\mathbf{T}_2$, that is, to the rank of $\mathbf{K}$, the number of additional constraints imposed by ${}_2\mathbf{K}'\boldsymbol{\beta} = {}_2\mathbf{G}$.

The final pieces of evidence needed to substantiate that $(\mathbf{g}'\mathbf{g} - \mathbf{f}'\mathbf{f})/\sigma^2$ is a chi-squared observation are that $\mathbf{B}$ can be represented as a product moment matrix and that $\mathbf{B}\,\mathbf{Var}(\mathbf{y})\,\mathbf{B} = \mathbf{B}$. Explicitly,

$$ \mathbf{B}\,\mathbf{Var}(\mathbf{y})\,\mathbf{B} = (1/\sigma^2)(\sigma^2)(1/\sigma^2)(\mathbf{V}\mathbf{W}\mathbf{T}_2^{-1}\mathbf{W}'\mathbf{V}')(\mathbf{V}\mathbf{W}\mathbf{T}_2^{-1}\mathbf{W}'\mathbf{V}') $$

Now, if it can be shown that $\mathbf{W}'\mathbf{V}'\mathbf{V}\mathbf{W} = \mathbf{T}_2$, then

$$ \mathbf{B}\,\mathbf{Var}(\mathbf{y})\,\mathbf{B} = (1/\sigma^2)(\mathbf{V}\mathbf{W}\mathbf{T}_2^{-1}\mathbf{W}'\mathbf{V}') = \mathbf{B} $$

We begin with

$$ \mathbf{V}'\mathbf{V} = (\mathbf{X}'\mathbf{X})^{-1}\mathbf{X}'\mathbf{X}(\mathbf{X}'\mathbf{X})^{-1} = (\mathbf{X}'\mathbf{X})^{-1} $$

and then pre- and postmultiply by $\mathbf{W}'$ and $\mathbf{W}$:

$$ \begin{aligned} \mathbf{W}'\mathbf{V}'\mathbf{V}\mathbf{W} &= (\mathbf{K} - \mathbf{C}\mathbf{S}_{11}^{-1}\mathbf{S}_{12})'(\mathbf{X}'\mathbf{X})^{-1}(\mathbf{K} - \mathbf{C}\mathbf{S}_{11}^{-1}\mathbf{S}_{12}) \\ &= \mathbf{K}'(\mathbf{X}'\mathbf{X})^{-1}\mathbf{K} - \mathbf{K}'(\mathbf{X}'\mathbf{X})^{-1}\mathbf{C}\mathbf{S}_{11}^{-1}\mathbf{S}_{12} \\ &\quad - \mathbf{S}_{12}'\mathbf{S}_{11}^{-1}\mathbf{C}'(\mathbf{X}'\mathbf{X})^{-1}\mathbf{K} + \mathbf{S}_{12}'\mathbf{S}_{11}^{-1}\mathbf{C}'(\mathbf{X}'\mathbf{X})^{-1}\mathbf{C}\mathbf{S}_{11}^{-1}\mathbf{S}_{12} \end{aligned} $$

Recalling that in the partitioning notation

$$ \begin{aligned} \mathbf{K}'(\mathbf{X}'\mathbf{X})^{-1}\mathbf{K} &= \mathbf{S}_{22}, \qquad & \mathbf{K}'(\mathbf{X}'\mathbf{X})^{-1}\mathbf{C} &= \mathbf{S}_{12}' \\ \mathbf{C}'(\mathbf{X}'\mathbf{X})^{-1}\mathbf{K} &= \mathbf{S}_{12}, \qquad & \mathbf{C}'(\mathbf{X}'\mathbf{X})^{-1}\mathbf{C} &= \mathbf{S}_{11} \end{aligned} $$

we can write

$$\begin{aligned}\mathbf{W'V'VW} &= \mathbf{S}_{22} - \mathbf{S}'_{12}\mathbf{S}_{11}^{-1}\mathbf{S}_{12} - \mathbf{S}'_{12}\mathbf{S}_{11}^{-1}\mathbf{S}_{12} + \mathbf{S}_{12}^{-1}\mathbf{S}_{11}^{-1}\mathbf{S}_{11}\mathbf{S}_{11}^{-1}\mathbf{S}_{12}\\ &= \mathbf{S}_{22} - \mathbf{S}'_{12}\mathbf{S}_{11}^{-1}\mathbf{S}_{12} = \mathbf{T}_2\end{aligned}$$

Thus, $(\mathbf{g'g} - \mathbf{f'f})/\sigma^2$ is distributed as a chi-squared observation with q degrees of freedom.

To show directly that $(\mathbf{g'g} - \mathbf{f'f})/\sigma^2$ is independent of $(\mathbf{f'f} - \mathbf{e'e})/\sigma^2$ and $\mathbf{e'e}/\sigma^2$, we need to demonstrate that $\mathbf{B}\,\mathbf{Var(y)}\,\mathbf{H} = \mathbf{0}$ and $\mathbf{B}\,\mathbf{Var(y)}\,\mathbf{M} = \mathbf{0}$, where $\mathbf{H}$ and $\mathbf{M}$ are the matrices of the quadratic forms $(\mathbf{f'f} - \mathbf{e'e})/\sigma^2$ and $\mathbf{e'e}/\sigma^2$, respectively:

$$\begin{aligned}\mathbf{H} &= (1/\sigma^2)\left\{\mathbf{X}(\mathbf{X'X})^{-1}\,{}_1\mathbf{C}\left[{}_1\mathbf{C}'(\mathbf{X'X})^{-1}\,{}_1\mathbf{C}\right]^{-1}\,{}_1\mathbf{C}'(\mathbf{X'X})^{-1}\mathbf{X}'\right\}\\ &= (1/\sigma^2)\mathbf{VCS}_{11}^{-1}\mathbf{C'V'}\end{aligned}$$

and

$$\mathbf{M} = (1/\sigma^2)\left[\mathbf{I} - \mathbf{X}(\mathbf{X'X})^{-1}\mathbf{X}'\right] = (1/\sigma^2)(\mathbf{I} - \mathbf{VX}')$$

For

$$\mathbf{B}\,\mathbf{Var(y)}\,\mathbf{H} = (1/\sigma^2)(\sigma^2)(1/\sigma^2)\mathbf{VWT}_2^{-1}\mathbf{W'V'VCS}_{11}^{-1}\mathbf{C'V}$$

we start with $\mathbf{V'V} = (\mathbf{X'X})^{-1}$ premultiply by $\mathbf{W}'$ and postmultiply by $\mathbf{C}$ to give

$$\begin{aligned}\mathbf{W'V'VC} &= \left(\mathbf{K} - \mathbf{CS}_{11}^{-1}\mathbf{S}_{12}\right)'(\mathbf{X'X})^{-1}\mathbf{C}\\ &= \mathbf{K}'(\mathbf{X'X})\mathbf{C} - \mathbf{S}'_{12}\mathbf{S}_{11}^{-1}\mathbf{C}'(\mathbf{X'X})^{-1}\mathbf{C}\\ &= \mathbf{S}_{12} - \mathbf{S}_{12}\mathbf{S}_{11}^{-1}\mathbf{S}_{11} = \mathbf{0}\end{aligned}$$

Thus, $\mathbf{B}\,\mathbf{Var(y)}\,\mathbf{H} = \mathbf{0}$, and the sampling distributions of the chi-squared observations $(\mathbf{g'g} - \mathbf{f'f})/\sigma^2$ and $(\mathbf{f'f} - \mathbf{e'e})/\sigma^2$ are independent.

Similarly, for

$$\mathbf{B}\,\mathbf{Var(y)}\,\mathbf{M} = (1/\sigma^2)(\sigma^2)(1/\sigma^2)\mathbf{VWT}_2^{-1}\mathbf{W'V'}(\mathbf{I} - \mathbf{VX}')$$

the product $\mathbf{V}'(\mathbf{I} - \mathbf{VX}') = \mathbf{V}' - \mathbf{V'VX}' = \mathbf{V}' - (\mathbf{X'X})^{-1}\mathbf{X}' = (\mathbf{X'X})^{-1}\mathbf{X}' - (\mathbf{X'X})^{-1}\mathbf{X}' = \mathbf{0}$, so $\mathbf{B}\,\mathbf{Var(y)}\,\mathbf{M} = \mathbf{0}$, and $(\mathbf{g'g} - \mathbf{f'f})/\sigma^2$ is distributed independently of $\mathbf{e'e}/\sigma^2$ as well.

Although we developed both the proof and example of hierarchical constraints as if there were only two hypotheses of interest, given by the sets of constraints ${}_1\mathbf{C}'\boldsymbol{\beta} = {}_1\boldsymbol{\Gamma}$ and ${}_2\mathbf{K}'\boldsymbol{\beta} = {}_2\mathbf{G}$, it should be emphasized that both the principle and the practice extend to three, four, and more separate hypotheses (and constraint sets). Clearly, for example, we could have begun with the constraint ${}_2\mathbf{C}'\boldsymbol{\beta} = {}_2\boldsymbol{\Gamma}$ and then introduced the additional constraints ${}_3\mathbf{K}'\boldsymbol{\beta} = {}_3\mathbf{G}$, finding the chi-squared distributed SSE associated with the latter could be tested independently of the former constraints *provided* we were willing to test ${}_3\mathbf{K}'\boldsymbol{\beta} = {}_3\mathbf{G}$ conditional upon ${}_2\mathbf{C}'\boldsymbol{\beta} = {}_2\boldsymbol{\Gamma}$.

In practice, the only requirement is that the constraints be ordered so that each successive set can include all of those that went before. In the aggregate, however, the total number of constraints cannot exceed the number of elements of $\boldsymbol{\beta}$. That is, we can never impose more than m constraints on the m $\boldsymbol{\beta}$ parameter estimates. Although the sets of constraints may be of any size, within the above limits, we shall point out later that it is frequently easier to interpret hypotheses that require adding only one constraint at a time. Multiple degrees of freedom tests, those involving more than one constraint on $\boldsymbol{\beta}$, are often ambiguous in meaning when the data result in the rejection of the hypothesis.

8.4.5 Interval Estimates of $\boldsymbol{\beta}$

A confidence interval estimate of a β_i can be developed by recalling that the sampling distribution of $\hat{\beta}_i$ is normal with mean β_i and variance $\sigma^2[(\mathbf{X}'\mathbf{X})^{-1}]_{ii}$. A 95% confidence interval for β_i, then, should include the symmetrically distributed middle 95% of the $\hat{\beta}_i$ distribution. The intervals need to be wide enough so that 95% of them reach β_i from $\hat{\beta}_i$. The problem of course, is that the desired width is a function of the unknown σ^2. The larger σ^2, the wider the confidence intervals need to be.

This difficulty may be overcome with quite general results. We begin by considering the single constraint $\mathbf{e}_i'\boldsymbol{\beta} = g$. By the development in Section 8.3 we know that the estimator of this constraint has sampling distribution

$$\mathbf{e}_i'\hat{\boldsymbol{\beta}} \sim N\left(g, \sigma^2\mathbf{e}_i'(\mathbf{X}'\mathbf{X})^{-1}\mathbf{e}_i\right)$$

Equivalently, the statistic

$$Z_i = \frac{\mathbf{e}_i'\hat{\boldsymbol{\beta}} - g}{\sigma\left[\mathbf{e}_i'(\mathbf{X}'\mathbf{X})^{-1}\mathbf{e}_i\right]^{1/2}} = \frac{\hat{\beta}_i - g}{\sigma\left[(\mathbf{X}'\mathbf{X})^{-1}\right]_{ii}^{1/2}} = \frac{\hat{\beta}_i - \beta_i}{\sigma\left[(\mathbf{X}'\mathbf{X})^{-1}\right]_{ii}^{1/2}} \tag{8.4.30}$$

is an observation from the normal distribution with mean 0 and variance 1.

Just as we also observed in Section 8.3 that

$$(\mathbf{y} - \mathbf{X}\hat{\boldsymbol{\beta}})'(\mathbf{y} - \mathbf{X}\hat{\boldsymbol{\beta}}) = \mathbf{e}'\mathbf{e}$$

is distributed independently of

$$(\mathbf{X}\hat{\boldsymbol{\beta}} - \boldsymbol{\beta}\mathbf{X})'(\mathbf{X}\hat{\boldsymbol{\beta}} - \mathbf{X}\boldsymbol{\beta}) = (\hat{\boldsymbol{\beta}} - \boldsymbol{\beta})'\mathbf{X}'\mathbf{X}(\hat{\boldsymbol{\beta}} - \boldsymbol{\beta})$$

it is also possible to show that $\mathbf{e}'\mathbf{e}$ is distributed independently of $\hat{\beta}_i - \beta_i$ or of

$$Z_i = (\hat{\beta}_i - \beta_i)/\sigma\left[(\mathbf{X}'\mathbf{X})^{-1}\right]_{ii}^{1/2}$$

We shall not provide the proof but point out simply that in both instances what is asserted is that the variability in the estimator is uncorrelated with the variability of observations about the estimator.

This sampling independence of $\mathbf{e}'\mathbf{e}$ and Z_i is used in much the same way as was the independence of $\mathbf{e}'\mathbf{e}$ and $\mathbf{f}'\mathbf{f} - \mathbf{e}'\mathbf{e}$: The ratio of a normally distributed observation (with mean 0 and variance 1) to the square root of a chi-squared observation (the latter divided by its degrees of freedom) is distributed as Student's t (with the same number of degrees of freedom as the chi-squared observation) provided that the normal and chi-squared observations have independent sampling distributions. That is, from (8.4.30),

$$t = (\hat{\beta}_i - \beta_i)\sqrt{\frac{n - m}{\mathbf{e}'\mathbf{e}\left[(\mathbf{X}'\mathbf{X})^{-1}\right]_{ii}}} \sim t_{n-m} \qquad (8.4.31)$$

(recall that $\mathbf{e}'\mathbf{e}/\sigma^2 \sim \chi^2_{n-m}$).

This equation may be rewritten isolating β_i:

$$\beta_i = \hat{\beta}_i - \sqrt{\frac{\mathbf{e}'\mathbf{e}[(\mathbf{X}'\mathbf{X})]^{-1}_{ii}}{(n - m)}}\, t \qquad (8.4.32)$$

A confidence interval is now established by substituting appropriate values for t. For example, a 95% confidence interval about $\hat{\beta}_i$ is obtained by substituting the values from the random variable t_{n-m} that cuts off the smallest and largest 2.5% of the observations.

If we note the direct relation between the t_{n-m} and the $F_{1,n-m}$ distributions—sampling observations from t_{n-m} and then squaring each produces $F_{1,n-m}$—it will do as well to find that value of $F_{1,n-m}$ that cuts off the largest 5% of the observations. The positive and negative square roots of this value may then be substituted for t in (8.4.32) to give end points for the 95% confidence interval about $\hat{\beta}_i$:

$$L_{\hat{\beta}_i} = \hat{\beta}_i \pm \sqrt{\frac{\mathbf{e}'\mathbf{e}\left[(\mathbf{X}'\mathbf{X})^{-1}\right]_{ii}}{n - m} F_{0.95,1,n-m}} \qquad (8.4.33)$$

For an application let us turn once again to the French comprehension test. We had previously calculated for this example $\hat{\boldsymbol{\beta}}' = (13.9167,\ 7.25)$, $\mathbf{e}'\mathbf{e} = 476.33$, $n = 6$, $m = 2$, and

$$(\mathbf{X}'\mathbf{X})^{-1} = \tfrac{1}{96}\begin{bmatrix} 70 & -18 \\ -18 & 6 \end{bmatrix}$$

For $F_{1,4}$, 95% of the observations lies below 7.71. Thus, for $\hat{\beta}_1$, we have

THESIS MODEL

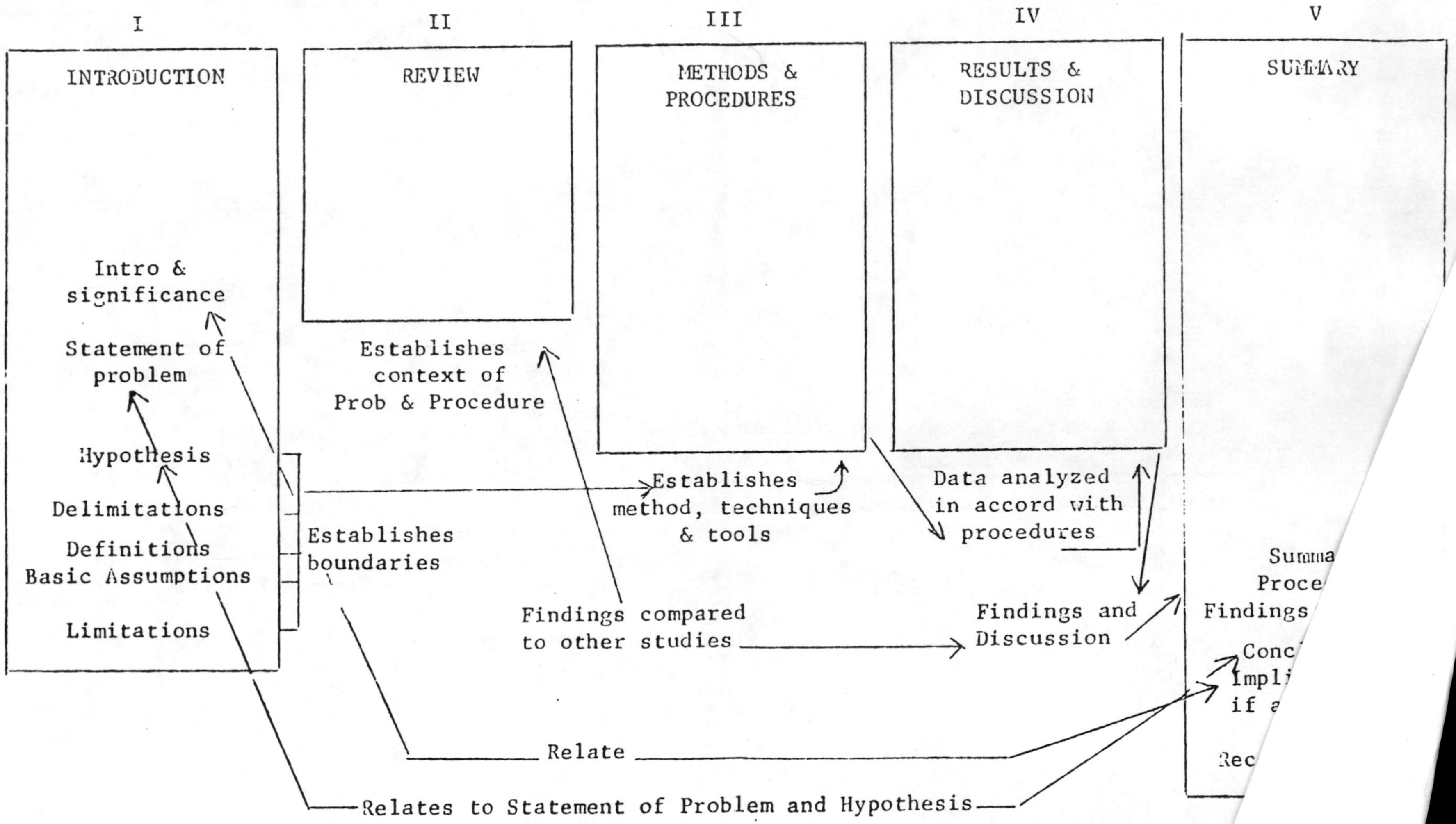

NOTE: Size of box does not represent importance of length of chapter.

limits given by

$$L_{\beta_1} = 13.9167 \pm \sqrt{\frac{(476.33)(70/96)}{4}}\sqrt{7.71} = 13.917 \pm 25.874$$

$$= -11.96 \quad \text{and} \quad 39.49$$

Equations (8.4.30) and (8.4.33) generalize to any single constraint on $\boldsymbol{\beta}$. That is, for any $\mathbf{C}'\boldsymbol{\beta} = \Gamma$ where $\mathbf{C}'$ is a row vector, and not just for $\mathbf{C}' = \mathbf{e}_i'$,

$$Z = (\mathbf{C}'\hat{\boldsymbol{\beta}} - \Gamma)/\sigma\left[\mathbf{C}'(\mathbf{X}'\mathbf{X})^{-1}\mathbf{C}\right]^{1/2} \tag{8.4.34}$$

is an observation from the normal distribution with mean 0 and variance 1 and

$$L_{\mathbf{C}'\hat{\boldsymbol{\beta}}} = \mathbf{C}'\hat{\boldsymbol{\beta}} \pm \sqrt{\frac{\mathbf{e}'\mathbf{e}\left[\mathbf{C}'(\mathbf{X}'\mathbf{X})^{-1}\mathbf{C}\right]F_{0.95,1,n-m}}{n-m}} \tag{8.4.35}$$

gives the limits for a 95% confidence interval about $\mathbf{C}'\hat{\boldsymbol{\beta}}$. Narrower or wider intervals are constructed by taking alternative percentage points of the $F_{1,n-m}$ distribution.

8.5 Summary

This chapter presented the theoretical development for the general linear regression model (GLM), which is widely used in the estimation of parameters and testing of hypotheses. The GLM assumes that the criterion data are independent observations drawn from distributions with means which are linear transformations of the predictor attributes. Furthermore, the maximum likelihood method of estimating the means and variance of these criterion distributions assumes that each is normally distributed with equal variance.

Based on these assumptions, we derived the estimates of $\boldsymbol{\beta}$ and σ^2 and explored the shape and characteristics of the sampling distributions of $\hat{\boldsymbol{\beta}}$ and $\hat{\sigma}^2$, the maximum likelihood estimators. Following a brief review of the logic of hypothesis testing and the discussion of theorems relating to the distribution of quadratic forms and ratios of independent chi-squared distributions, we defined interval estimates of σ^2 and the elements of $\boldsymbol{\beta}$.

Finally, since most research questions are translated into hypotheses about the elements of $\boldsymbol{\beta}$, we generalized our use in Chapter 7 of constraint matrices to express such hypotheses in the GLM. Combined with the theorems specifying the appropriate sampling distributions, we developed a procedure for comparing the likelihood of $\mathbf{y}$ under various constraints on the elements of $\boldsymbol{\beta}$.

Chapter 9 extends this discussion of the GLM by illustrating the commonly used techniques of analysis of variance and covariance and multiple regression as specific cases of the GLM. In addition, many of the concepts and theorems provide a basis for the discussion of the analysis of multivariate criteria in Part III.

Exercises

1. For each of the following multiple regression equations, sketch and describe the surface of $E(y)$. (Assume that X_1 and X_2, the predictors, have the same means and variances in the design; say, 50 and 100, respectively.)
   a. $E(y) = 50 + 0X_1 + 0X_2$.
   b. $E(y) = 6 + 2X_1 + 0X_2$.
   c. $E(y) = 6 + 0X_1 + 3X_2$.
   d. $E(y) = 6 + 2X_1 + 3X_2$.
   e. $E(y) = 6 + 3X_1 + 0X_2 + 3(X_2)^2$.
   f. $E(y) = 6 + 3X_1 + 2X_2 - 2(X_1 - 50)(X_2 - 50)$.

2. Distinguish the least squares and linear regression approaches. What is the difference in interpretation given to the correlation coefficient in the two approaches?

3. Distinguish part and partial correlations. How is the squared part correlation related to the addition of predictor variables in multiple regression?

4. Derive Eq. (8.4.20) from (8.4.19).

5. Assume that we sampled ten students at each of the three levels of study in the French comprehension example.
   a. For what values of $\hat{\sigma}^2$ would the hypothesis that $\sigma^2 = 64$ not be rejected for $\alpha = 0.02$?
   b. For what values of $\hat{\sigma}^2$ would the hypothesis that $\sigma^2 = 64$ not be rejected for $\alpha = 0.10$?
   c. If $\hat{\sigma}^2 = 119.08$, construct a 90% confidence interval.
   d. If $\hat{\sigma}^2 = 119.08$, construct a 98% confidence interval.

Additional Reading

For more advanced topics in linear regression the following is a good reference:

Draper, N., and Smith, H. (1981). *Applied Regression Analysis*, 2nd ed. New York: Wiley.

This book is a bit more technical but provides very good coverage of the relation of ANOVA and analysis of covariance to the general linear model:

Searle, S. R. (1971). *Linear Models*. New York: Wiley.

Chapter 9

Linear Regression: Some Applications

Introduction

In this chapter we present some illustrations of how the general linear model is adapted to specific research questions. We consider instances in which the predictors are measured, categorical, and a mixture of the two.

Section 9.1 develops the fixed effects analysis of variance, including the factorial design and other randomized groups designs, as an example of the GLM. *Design templates* based on mutually exclusive categorical variables, dummy variables, or effect coding are presented. We outline the use of a hierarchy of submodels to permit the quasi-independent testing of effects when the k populations have been sampled in an unbalanced fashion.

We extend our discussion of the categorical predictors employed for the ANOVA design with a development of designs additionally incorporating one or more measured predictors, as in the analysis of covariance or the closely related concomitant variable design. These applications of the GLM are discussed in Section 9.2.

Four variations of multiple regression prediction are presented in Section 9.3. These include examples of curve fitting, polynomial regression, testing for the linearity of regression, and evaluating the contribution of a moderator variable.

The applications presented in this chapter do not exhaust the range of the GLM. They were chosen to stimulate the reader's curiosity about that range.

9.1 Fixed Effects Analysis of Variance Designs

The linear regression model we have been developing is frequently employed as the inferential or statistical underpinning to hypothesis testing in the analysis of variance of fixed effects designs. We shall briefly develop this application, beginning with a simple ANOVA design, that of k randomized groups.

As our example we employ part of the data shown in Table 1.4 for the Airline Pilots Response Time Study design. Forty-eight commercial airline pilots were recruited for participation in this hypothetical study by randomly selecting names from the captains' roster of a major airline. After recruitment the 48 were randomly divided into four groups of 12. The times taken by the pilots to respond by depressing a switch to a series of stimuli were then recorded. For one group of pilots the stimuli were light flashes presented to the right visual field; light flashes were presented to the left visual field of a second group. The other two groups received aural stimulation, directed for the one group to the left ear and for the other to the right. Our interest is in finding whether there are differences in the speed with which pilots respond to the different conditions of stimulation. Although not all pilots respond equally quickly (there is variability among them) do they, *on the average*, respond more quickly to visual or to aural stimuli, and to stimulation on the left or to sounds or lights on the right?

The picture of the world from which we are sampling, then, might be something like this. If we could observe the switch-closing time of *all* pilots in response to left visual stimulation, we would observe one distribution of scores. Similarly, if we could have them all respond to right visual, left aural, and right aural signals, we would observe a second, third, and fourth distribution of response times. The means of these four distributions might be different. We cannot observe the four distributions of scores directly, as we cannot measure the performance of all pilots under all stimulus conditions. Rather, we want to make some trustworthy inferences about the means of the four distributions and about differences among them on the basis of sampling only a few observations from each. If we can justify the assumptions of the linear regression model, we have a way of doing just that.

Recall from Section 8.2 that we need to assume the following:

1. We have selected observations independently from the four distributions.
2. Each of the distributions is normal with a common variance.
3. The means of those distributions are linearly related to a set of predictor attributes.

Our techniques for randomly selecting a group of pilots and then randomly assigning each to one of the four stimulus conditions satisfies the independent selection criterion. Subject to some statistical verification, we can provisionally accept the assumption of normal distributions with equal variance. What, however, about the linear assumption itself?

At the heart of the linear regression model is the assumption that for a vector of 48 criterion observations (48 response times, one for each of our tested pilots) there is a corresponding vector of expected values $\mathbf{E}(\mathbf{y})$ that is a linear function or transformation of the predictor or design matrix. This is symbolized in Eq. (8.1.1), $\mathbf{E}(\mathbf{y}) = \mathbf{X}\boldsymbol{\beta}$.

Let us assume that in recording the vector $\mathbf{y}$ that we have grouped the criterion observations so that the first 12 observations are for those pilots responding to left visual (LV) stimuli, the next 12 from the right visual (RV) group, followed by the left aural (LA) and right aural (RA) respondents. The first 12 entries in $\mathbf{E}(\mathbf{y})$ should all be identically equal to the mean of the population of all pilot response times to LV stimulation, the second 12 should all equal the RV mean, and the third and fourth blocks of $\mathbf{E}(\mathbf{y})$ values each consist of 12 repetitions, respectively, of the LA and RA population means. Only four possibly distinct values can occur in $\mathbf{E}(\mathbf{y})$, as we are sampling from only four distributions. What predictor or design matrix would be linearly related to such an expectation vector?

9.1.1 Design Template

The rows of $\mathbf{X}$ must be arranged in the same manner as the rows of $\mathbf{y}$ and $\mathbf{E}(\mathbf{y})$. The first 12 rows contain predictor or design information about the 12 pilots assigned to the LV treatment condition. The only information about those pilots that is relevant to the present design, however, is that they were all assigned to this condition. Thus, the first 12 rows in $\mathbf{X}$ should all be identical. Similar arguments hold for the second, third, and fourth sets of 12 rows in $\mathbf{X}$. We can distinguish no more than four different rows in $\mathbf{X}$. We can choose a design such that $\mathbf{X}_{1.}$, $\mathbf{X}_{13.}$, $\mathbf{X}_{25.}$, and $\mathbf{X}_{37.}$ represent the four types.

We consider next its column order. The rank of $\mathbf{X}$ cannot be greater than 4 inasmuch as all 48 of the rows can be expressed as linear combinations of four vectors, the rows $\mathbf{X}_{1.}$, $\mathbf{X}_{13.}$, $\mathbf{X}_{25.}$, and $\mathbf{X}_{37.}$ If we stipulate that $\mathbf{X}$ is basic, the design matrix cannot have more than four columns. The number of columns in $\mathbf{X}$ corresponds to the number of elements in the vector $\boldsymbol{\beta}$.

In the rest of the discussion we introduce the notion of a *design template matrix*, which we designate ${}_{\mathrm{T}}\mathbf{X}$ and which consists of the four unique rows of $\mathbf{X}$:

$$
{}_{\mathrm{T}}\mathbf{X} = \begin{bmatrix} \mathbf{X}'_{1.} \\ \mathbf{X}'_{13.} \\ \mathbf{X}'_{25.} \\ \mathbf{X}'_{37.} \end{bmatrix}
$$

The four population means make up an *expectation template*, given by the product

$$
{}_{\mathrm{T}}\mathbf{E}(\mathbf{y}) = \begin{bmatrix} \mathbf{E}(\mathbf{y}\,|\,\mathrm{LV}) \\ \mathbf{E}(\mathbf{y}\,|\,\mathrm{RV}) \\ \mathbf{E}(\mathbf{y}\,|\,\mathrm{LA}) \\ \mathbf{E}(\mathbf{y}\,|\,\mathrm{LV}) \end{bmatrix} = \begin{bmatrix} \mathbf{X}'_{1.} \\ \mathbf{X}'_{13.} \\ \mathbf{X}'_{25.} \\ \mathbf{X}'_{37.} \end{bmatrix} \boldsymbol{\beta} = {}_{\mathrm{T}}\mathbf{X}\boldsymbol{\beta} \tag{9.1.1}
$$

Each of the four population means is a linear transformation of the elements of $\boldsymbol{\beta}$. For the randomized groups design we place no restrictions, *at the outset*, on how the four means are related among themselves. Posing and testing such restrictions is done with linear constraints on $\boldsymbol{\beta}$. However, to ensure that the elements of the expectation template are linearly independent, the design template ${}_{\mathrm{T}}\mathbf{X}$ must satisfy certain conditions. It is important that there can be no linear dependencies among the rows of ${}_{\mathrm{T}}\mathbf{X}$. (If the third row of ${}_{\mathrm{T}}\mathbf{X}$, for example, could be expressed as a linear combination of rows 1, 2, and 4, then the third mean in the expectation template would be that same linear combination of the first, second, and fourth means.) This tells us that ${}_{\mathrm{T}}\mathbf{X}$ must be a basic matrix and the rank of ${}_{\mathrm{T}}\mathbf{X}$ must be 4. In turn, this means ${}_{\mathrm{T}}\mathbf{X}$ must have no fewer than four columns. Earlier, however, we saw that ${}_{\mathrm{T}}\mathbf{X}$ could have no more than four columns. The conclusion, then, is that ${}_{\mathrm{T}}\mathbf{X}$ must be a *square basic matrix* of order equal to the number of randomized groups in the design.

If ${}_{\mathrm{T}}\mathbf{X}$ is a square basic matrix, then $\mathbf{X}$ itself is also basic. This may be demonstrated by expressing the design matrix as the product

$$\mathbf{X} = \mathbf{L}\,{}_{\mathrm{T}}\mathbf{X} \tag{9.1.2}$$

where $\mathbf{L}$ is a matrix made up of null and unit vectors arranged in the following way (1 and 0 are 12-element column vectors for the four-group Response Time Study):

$$\mathbf{L} = \left[\begin{array}{cccc} 1 & 0 & 0 & 0 \\ \hdashline 0 & 1 & 0 & 0 \\ \hdashline 0 & 0 & 1 & 0 \\ \hdashline 0 & 0 & 0 & 1 \\ \hdashline \end{array}\right] \begin{array}{l} 12 \\ 24 \\ 36 \\ 48 \end{array}$$

To verify that $\mathbf{X} = \mathbf{L}\,{}_{\mathrm{T}}\mathbf{X}$, $\mathbf{X}$ may be constructed as a sum of major products of vectors: $\mathbf{X} = \mathbf{L}_{.1}\mathbf{X}'_{1.} + \mathbf{L}_{.2}\mathbf{X}'_{13.} + \mathbf{L}_{.3}\mathbf{X}'_{25.} + \mathbf{L}_{.4}\mathbf{X}'_{37.}$. Since $\mathbf{L}$ is a basic matrix (*proof*: $\mathbf{L}$ is an orthogonal matrix, as $\mathbf{L}'\mathbf{L}$ is diagonal), and ${}_{\mathrm{T}}\mathbf{X}$ is a square basic matrix, their product is basic.

Aside from ensuring that ${}_{\mathrm{T}}\mathbf{X}$ is a square basic matrix, how should it be constructed to be appropriate to the k-randomized-groups design? An answer from the point of view of the linear regression model, is that *it makes no difference which kth-order square basic matrix is used as* ${}_{\mathrm{T}}\mathbf{X}$. Different writers follow different traditions in constructing ${}_{\mathrm{T}}\mathbf{X}$ or $\mathbf{X}$, and we shall review some of these. *In estimating* $\mathbf{E}(\mathbf{y})$ or in testing hypotheses about the elements of ${}_{\mathrm{T}}\mathbf{E}(\mathbf{y})$, however, *the results will be indistinguishable whatever the choice of* ${}_{\mathrm{T}}\mathbf{X}$.

In satisfying the reader that this is true, we can adopt an even more general definition of how the design matrix is written. We need not insist that all observations drawn from one population be grouped together or that the same number of observations be drawn from each population. The design matrix can then be written as the triple product

$$\mathbf{X} = \boldsymbol{\pi}\mathbf{L}\,{}_{\mathrm{T}}\mathbf{X} \tag{9.1.3}$$

where the permutation matrix $\boldsymbol{\pi}$ allows the rows of $\mathbf{X}$ to be "mixed up" with respect to group origin; $\mathbf{L}$ remains a partitioned matrix of null and unit vectors such that $\mathbf{L}'\mathbf{L} = \mathbf{D}$ is a diagonal matrix with D_i equal to the number of observations drawn from the ith population; and ${}_{\mathrm{T}}\mathbf{X}$ is an arbitrary kth-order square basic matrix. (If the jth observation in the vector $\mathbf{y}$ is from the ith population, then the jth row of $\mathbf{X}$ is identical to the ith row of ${}_{\mathrm{T}}\mathbf{X}$.)

The vector of expected values for $\mathbf{y}$, the vector $\mathbf{E}(\mathbf{y})$, is estimated by Eq. (8.2.7):

$$\hat{\mathbf{E}}(\mathbf{y}) = \mathbf{X}\hat{\boldsymbol{\beta}} = \mathbf{X}\left[(\mathbf{X}'\mathbf{X})^{-1}\mathbf{X}'\mathbf{y}\right]$$

Substituting for $\mathbf{X}$ in this equation,

$$\begin{aligned}\hat{\mathbf{E}}(\mathbf{y}) &= \boldsymbol{\pi}\mathbf{L}\,{}_{\mathrm{T}}\mathbf{X}[{}_{\mathrm{T}}\mathbf{X}'\mathbf{L}'\boldsymbol{\pi}'\boldsymbol{\pi}\mathbf{L}\,{}_{\mathrm{T}}\mathbf{X}]^{-1}\,{}_{\mathrm{T}}\mathbf{X}'\mathbf{L}'\boldsymbol{\pi}'\mathbf{y}\\ &= \boldsymbol{\pi}\mathbf{L}\,{}_{\mathrm{T}}\mathbf{X}[{}_{\mathrm{T}}\mathbf{X}'\mathbf{L}'\mathbf{L}\,{}_{\mathrm{T}}\mathbf{X}]^{-1}\,{}_{\mathrm{T}}\mathbf{X}'\mathbf{L}'\boldsymbol{\pi}'\mathbf{y}\end{aligned}$$

and recalling that $\boldsymbol{\pi}$ is a square orthonormal matrix and that all square basic matrices, including ${}_{\mathrm{T}}\mathbf{X}$, have regular inverses, we see that $\hat{\mathbf{E}}(\mathbf{y})$ is independent of choice of ${}_{\mathrm{T}}\mathbf{X}$:

$$\begin{aligned}\hat{\mathbf{E}}(\mathbf{y}) &= \boldsymbol{\pi}\mathbf{L}\,{}_{\mathrm{T}}\mathbf{X}\,{}_{\mathrm{T}}\mathbf{X}^{-1}\mathbf{D}^{-1}({}_{\mathrm{T}}\mathbf{X}')^{-1}\,{}_{\mathrm{T}}\mathbf{X}'\mathbf{L}'\boldsymbol{\pi}'\mathbf{y}\\ &= \boldsymbol{\pi}\mathbf{L}\mathbf{D}^{-1}\mathbf{L}'\boldsymbol{\pi}\mathbf{y} \qquad \left(\mathbf{D}^{-1} = (\mathbf{L}'\mathbf{L})^{-1}\right)\end{aligned} \tag{9.1.4}$$

What is the nature of $\hat{\mathbf{E}}(\mathbf{y})$ for the randomized groups design? First, the triple product $\mathbf{L}'\boldsymbol{\pi}'\mathbf{y}$ is a k-element column vector. The k elements are the sums of the criterion observations drawn, respectively, from each of the k populations. $\mathbf{D}$ is a diagonal matrix, with the number of observations from the first to the kth populations arrayed along the diagonal. The product $\mathbf{D}^{-1}\mathbf{L}'\boldsymbol{\pi}'\mathbf{y}$, then, is a k-element vector of sample criterion means. Premultiplying this mean vector by $\boldsymbol{\pi}\mathbf{L}$ gives an n-element vector of these k sample means arranged, with respect to group membership, in the same order as $\mathbf{y}$. Thus $\hat{\mathbf{E}}(\mathbf{y})$ contains the sample criterion mean for the group to which each $\mathbf{y}$ observation belongs.

If $\hat{\mathbf{E}}(\mathbf{y})$ does not depend on the choice of ${}_{\mathrm{T}}\mathbf{X}$, then neither does $\mathbf{e}'\mathbf{e} = [\mathbf{y} - \hat{\mathbf{E}}(\mathbf{y})]'[\mathbf{y} - \hat{\mathbf{E}}(\mathbf{y})]$.

Also, from the relation [(9.1.1)] between the expectation template and the design template ${}_{\mathrm{T}}\mathbf{E}(\mathbf{y}) = {}_{\mathrm{T}}\mathbf{X}\boldsymbol{\beta}$, any hypothesized linear relations among the population means,

$$\mathbf{H}'\,{}_{\mathrm{T}}\mathbf{E}(\mathbf{y}) = \boldsymbol{\Gamma} \tag{9.1.5}$$

can be translated into linear constraints on $\boldsymbol{\beta}$ by taking the constraint matrix to be

$$\mathbf{C}' = \mathbf{H}'{}_{\mathrm{T}}\mathbf{X} \tag{9.1.6}$$

That is, $\mathbf{C}'\boldsymbol{\beta} = \mathbf{H}'{}_{\mathrm{T}}\mathbf{X}\boldsymbol{\beta} = \mathbf{H}'{}_{\mathrm{T}}\mathbf{E}(\mathbf{y}) = \boldsymbol{\Gamma}$. Thus, whatever our choice of ${}_{\mathrm{T}}\mathbf{X}$, we can test any hypothesis of the form $\mathbf{H}'{}_{\mathrm{T}}\mathbf{E}(\mathbf{y}) = \boldsymbol{\Gamma}$.

When we phrase a hypothesis as a linear function of the population means, we shall term this a *contrast* among means. When we phrase the hypothesis as a linear function of $\boldsymbol{\beta}$, making it specific to the choice of ${}_{\mathrm{T}}\mathbf{X}$ in the case of randomized groups designs, we shall term it a *constraint* on $\boldsymbol{\beta}$. Rows of $\mathbf{H}'$ are *contrasts*; rows of $\mathbf{C}'$ are *constraints*.

We shall deal briefly with three of the practical alternative choices for ${}_{\mathrm{T}}\mathbf{X}$. The first, sometimes termed the use of *categorical variables*, or mutually exclusive categorical variables, defines ${}_{\mathrm{T}}\mathbf{X} = \mathbf{I}$. Each of the columns of the design matrix is a 0–1 variable and each column is associated with one of the randomized groups. Each row of $\mathbf{X}$ contains exactly $k - 1$ zeros and a single 1, the one occurring in the column associated with the group to which the corresponding $\mathbf{y}$ observation belonged.

With categorical variable columns making up the design matrix and given that ${}_{\mathrm{T}}\mathbf{E}(\mathbf{y}) = {}_{\mathrm{T}}\mathbf{X}\boldsymbol{\beta}$, the elements of $\boldsymbol{\beta}$ will be the population means: ${}_{\mathrm{T}}\mathbf{E}(\mathbf{y}) = \boldsymbol{\beta}$. Further, the design matrix is orthogonal, $\mathbf{X}'\mathbf{X} = \mathbf{D}$, with $D_i = n_i$, the number of observations in the ith group, and $\mathbf{X}'\mathbf{y}$ is a vector of criterion sums for the k groups.

Using data shown in Table 1.4 for the Response Time Study, where y_i is the average response time of the ith pilot over a block of trials, we have for categorical coding

$$\mathbf{y}'\mathbf{y} = 30{,}582{,}878.51$$

$$\mathbf{X}'\mathbf{X} = \begin{bmatrix} 12 & & & 0 \\ & 12 & & \\ & & 12 & \\ 0 & & & 12 \end{bmatrix}$$

$$(\mathbf{X}'\mathbf{y})' = [9855, 8711, 9924, 9174]$$

$$(\mathbf{X}'\mathbf{X})^{-1} = \begin{bmatrix} \frac{1}{12} & & & 0 \\ & \frac{1}{12} & & \\ & & \frac{1}{12} & \\ 0 & & & \frac{1}{12} \end{bmatrix}$$

$$(\mathbf{X}'\mathbf{X})^{-1}\mathbf{X}'\mathbf{y} = \begin{bmatrix} 821.25 \\ 725.92 \\ 827.00 \\ 764.50 \end{bmatrix}$$

$${}_{\mathrm{T}}\hat{\mathbf{E}}(\mathbf{y})' = ({}_{\mathrm{T}}\mathbf{X}\hat{\boldsymbol{\beta}})' = \hat{\boldsymbol{\beta}}' = (821.25, 725.92, 827.00, 764.50)$$

$$\mathbf{e}'\mathbf{e} = \mathbf{y}'\mathbf{y} - \mathbf{y}'\mathbf{X}\hat{\boldsymbol{\beta}} = 945{,}326.17$$

A second common choice of ${}_{\mathrm{T}}\mathbf{X}$ is the use of *dummy variables*, as they are called. Dummy variable coding is quite similar to categorical variable coding with two changes. First, the design matrix includes a first column that is the *unit* vector. This represents a direct extension of the more general linear regression tradition, in which the first entry in $\boldsymbol{\beta}$ is the additive constant in the regression equation

$$[E(\mathbf{y})]_i = \beta_1 + \beta_2 X_{i2} + \beta_3 X_{i3} + \beta_m X_{im}$$

For each row of the matrix equation $\mathbf{E(y)} = \mathbf{X}\boldsymbol{\beta}$, $\mathbf{X}_{i1} = 1$.

Under dummy variable coding the remaining columns of the design matrix are the same as for categorical variable coding. Indeed, the name derives from the notion that dummy variables are used to code group membership. Since ${}_{\mathrm{T}}\mathbf{X}$ must be basic, one of the categorical variables must be dropped to make room for the unit vector. Traditionally, the categorical variable coding the *last* of the randomized groups is the one dropped, in the sense that the observations making up $\mathbf{y}$ are clustered frequently by groups with the final n_k entries being those sampled from the kth or last population.

The dummy variable coding design template for the four group study is

$${}_{\mathrm{T}}\mathbf{X} = \begin{bmatrix} 1 & 1 & 0 & 0 \\ 1 & 0 & 1 & 0 \\ 1 & 0 & 0 & 1 \\ 1 & 0 & 0 & 0 \end{bmatrix}$$

Because of the relation between the design template and the expectation template, the population means are given in terms of the dummy variable elements:

$$[{}_{\mathrm{T}}E(\mathbf{y})]_i = \begin{cases} \beta_1 + \beta_{i+1} & \text{for } i = 1, 2, \ldots, k-1 \\ \beta_1 & \text{for } i = k \end{cases}$$

For the data of the Response Time Study, the dummy coding gives

$$\mathbf{y}'\mathbf{y} = 30{,}582{,}878.51$$

$$(\mathbf{X}'\mathbf{y})' = (37{,}644,\ 9{,}855,\ 8{,}711,\ 9{,}924)$$

$$\mathbf{X}'\mathbf{X} = \begin{bmatrix} 48 & 12 & 12 & 12 \\ 12 & 12 & 0 & 0 \\ 12 & 0 & 12 & 0 \\ 12 & 0 & 0 & 12 \end{bmatrix}$$

$$(\mathbf{X}'\mathbf{X})^{-1} = \begin{bmatrix} 0.083 & -0.083 & -0.083 & -0.083 \\ -0.083 & 0.167 & 0.083 & 0.083 \\ -0.083 & 0.083 & 0.167 & 0.083 \\ -0.083 & 0.083 & 0.083 & 0.167 \end{bmatrix}$$

$$\hat{\boldsymbol{\beta}} = (\mathbf{X}'\mathbf{X})^{-1}\mathbf{X}'\mathbf{y} = \begin{bmatrix} 764.50 \\ 56.75 \\ -38.58 \\ 62.50 \end{bmatrix}$$

$$_{\mathrm{T}}\hat{\mathbf{E}}(\mathbf{y})' = ({}_{\mathrm{T}}\mathbf{X}\hat{\boldsymbol{\beta}})' = (821.25, 725.92, 827.00, 764.50)$$

and $\mathbf{e}'\mathbf{e} = 945{,}326.17$. As indicated earlier, $_{\mathrm{T}}\hat{\mathbf{E}}(\mathbf{y})$ and $\mathbf{e}'\mathbf{e}$ are the same as under categorical variable coding.

A third common form of the design matrix for the k randomized groups design uses *effect coding*. The intent of effect coding is to mirror in the elements of $\boldsymbol{\beta}$ certain analysis of variance conventions. There is an overall effect common to the k treatments; there are k treatment or group effects; and these treatment effects are deviations from the overall effect, summing to zero across the k groups. The design template for effect coding of randomized groups is identical to that for dummy variable coding except for the final row. In this final row, the template for the kth group, the zeros in the second through kth columns are replaced with entries of -1.

The effect coding design template for the four group study is

$$_{\mathrm{T}}\mathbf{X} = \begin{bmatrix} 1 & 1 & 0 & 0 \\ 1 & 0 & 1 & 0 \\ 1 & 0 & 0 & 1 \\ 1 & -1 & -1 & -1 \end{bmatrix}$$

In this design template, the k population means are expressed in terms of $\boldsymbol{\beta}$ as

$$[{}_{\mathrm{T}}E(\mathbf{y})]_i = \begin{cases} \beta_1 + \beta_{i+1} & \text{for} \quad i = 1, 2, \ldots, (k-1) \\ \beta_1 - \sum\limits_{j=2}^{k} \beta_j & \text{for} \quad i = k \end{cases}$$

The overall effect is given by β_1 while β_2 through β_k give the treatment effects for groups 1 through $k - 1$. As the group effects must sum to zero, the effect for the kth group is the negative of the sum of the group effects for the first $k - 1$ groups.

Using the Response Time Study data to illustrate the effect coding, we obtain

$$\mathbf{y}'\mathbf{y} = 30{,}582{,}878.51$$

$$\mathbf{y}'\mathbf{X} = (37{,}644,\ 1{,}144,\ -69,\ 681)$$

$$\mathbf{X}'\mathbf{X} = \begin{bmatrix} 48 & 0 & 0 & 0 \\ 0 & 24 & 12 & 12 \\ 0 & 12 & 24 & 12 \\ 0 & 12 & 12 & 24 \end{bmatrix}$$

$$(\mathbf{X}'\mathbf{X})^{-1} = \begin{bmatrix} 0.021 & 0.000 & 0.000 & 0.000 \\ 0.000 & 0.063 & -0.021 & -0.021 \\ 0.000 & -0.021 & 0.063 & -0.021 \\ 0.000 & -0.021 & -0.021 & 0.063 \end{bmatrix}$$

$$\hat{\boldsymbol{\beta}}' = (784.67, 36.58, -58.75, 42.33)$$

$${}_{\mathrm{T}}\hat{\mathbf{E}}(\mathbf{y})' = (821.25, 725.92, 827.00, 764.50)$$

$$\mathbf{e}'\mathbf{e} = 945{,}326.17$$

Again the estimates of the population means are identical to the earlier solutions as is the value of $\mathbf{e}'\mathbf{e}$. This latter quantity is used in any estimation of the unknown variance common to the four populations, σ^2.

The interpretation of $\boldsymbol{\beta}$ changes, however, with a different selection of ${}_{\mathrm{T}}\mathbf{X}$. As a result, the *constraints* necessary to test a particular hypothesis will also be different. As an example, consider the following hypothesis for the Response Time Study: $\mathbf{E}(\mathbf{y}|\mathrm{LA}) = \mathbf{E}(\mathbf{y}|\mathrm{RA})$. If the stimulus is aural, the response time will be the same whether the stimulus is directed to the right or to the left ear; that is, the means for two of the populations are hypothesized to be the same. In terms of the expectation template, this hypothesis can be put into algebraic form as the *contrast* $\mathbf{H}'\,{}_{\mathrm{T}}\mathbf{E}(\mathbf{y}) = \Gamma$:

$$(0 \quad 0 \quad 1 \quad -1)\begin{bmatrix} \mathbf{E}(\mathbf{y}|\mathrm{LV}) \\ \mathbf{E}(\mathbf{y}|\mathrm{RV}) \\ \mathbf{E}(\mathbf{y}|\mathrm{LA}) \\ \mathbf{E}(\mathbf{y}|\mathrm{RA}) \end{bmatrix} = 0$$

This hypothesis can be converted to constraints on $\boldsymbol{\beta}$ for each of the different design matrices by using the relation $\mathbf{C}' = \mathbf{H}'\,{}_{\mathrm{T}}\mathbf{X}$.

For *categorical variable coding*, ${}_{\mathrm{T}}\mathbf{X} = \mathbf{I}$ and $\mathbf{C}' = \mathbf{H}' = (0, 0, 1, -1)$. In terms of $\boldsymbol{\beta}$, then, the hypothesis is that $\beta_3 - \beta_4 = 0$, or that $\beta_3 = \beta_4$. For *dummy variable coding*, $\mathbf{C}'$ is

$$(0 \quad 0 \quad 1 \quad -1)\begin{bmatrix} 1 & 1 & 0 & 0 \\ 1 & 0 & 1 & 0 \\ 1 & 0 & 0 & 1 \\ 1 & 0 & 0 & 0 \end{bmatrix} = (0 \quad 0 \quad 0 \quad 1)$$

The hypothesis is that $\beta_4 = 0$. For *effect variable coding*, the relation is

$$(0 \quad 0 \quad 1 \quad -1)\begin{bmatrix} 1 & 1 & 0 & 0 \\ 1 & 0 & 1 & 0 \\ 1 & 0 & 0 & 1 \\ 1 & -1 & -1 & -1 \end{bmatrix} = (0 \quad 1 \quad 1 \quad 2)$$

The hypothesis is that $\beta_2 + \beta_3 + 2\beta_4 = 0$ or, restated in terms of the group effects, that the third and fourth group effects are the same:

$$\beta_4 = -(\beta_2 + \beta_3 + \beta_4)$$

Each of the constraints makes the same sense, keeping in mind how $\boldsymbol{\beta}$ and ${}_{\mathrm{T}}\mathbf{E}(\mathbf{y})$ are related under the different design codings. Further, the value computed for the increase in the SSE, $\mathbf{f}'\mathbf{f} - \mathbf{e}'\mathbf{e}$, is the same under all three codings. The reader is invited to verify this for the three constraints using the data in Table 1.4.

9.1.2 *ANOVA Sums of Squares*

In Section 9.1.1 we noted that, for the unconstrained maximum likelihood solution in the randomized groups design, $[\hat{E}(\mathbf{y})]_i$ is the sample mean computed over the group of criterion observations that includes $\mathbf{y}_i$. Thus $\mathbf{e}'\mathbf{e}$ is a sum of squares of discrepancies of the elements of $\mathbf{y}$ taken about the several group sample means. In analysis of variance terms, it is the *within groups sum of squares*. Divided by $n_1 + n_2 \cdots + n_k - k$, it provides the *mean square within groups*, the error term for testing hypotheses about differences in group (population) means.

The *between groups sum of squares* in the analysis of variance of randomized groups corresponds to the increase in SSE associated with imposing the condition that all k population means are equal to the same value. Intuitively, if each of the populations has the same mean, there is no between group variability, other than that occasioned by sampling. Increases in SSE in the analysis of variance, as throughout linear regression, indicate an increasing lack of fit of the data at hand to the model as the model becomes more restricted through the imposition of constraints, directly on $\boldsymbol{\beta}$ or indirectly through contrasts on ${}_{\mathrm{T}}\mathbf{E}(\mathbf{y})$.

To compute the between groups sum of squares for the k-randomized-groups design, we compute $\mathbf{f}'\mathbf{f} - \mathbf{e}'\mathbf{e}$ associated with the $k - 1$ constraints necessary to set the k populations means equal to one another. There are several ways in which these constraints might be phrased. We develop them here as a set of $k - 1$ *contrasts*, linear functions of the expectation template ${}_{\mathrm{T}}\mathbf{E}(\mathbf{y})$.

The k population means will all be equal if the following $k - 1$ relations are satisfied:

$$[{}_{\mathrm{T}}E(\mathbf{y})]_i = [{}_{\mathrm{T}}E(\mathbf{y})]_{i+1} \qquad \text{for} \quad i = 1, 2, 3, \ldots, k - 1$$

Quantities equal to a common quantity are equal to each other, of course, and it is not necessary to require explicitly that, for example, $[{}_{\mathrm{T}}E(\mathbf{y})]_2 = [{}_{\mathrm{T}}E(\mathbf{y})]_4$. The contrast or $\mathbf{H}'$ matrix that conveys these restrictions for a four group design is

$$\mathbf{H}' = \begin{bmatrix} 1 & -1 & 0 & 0 \\ 0 & 1 & -1 & 0 \\ 0 & 0 & 1 & -1 \end{bmatrix}$$

for $\mathbf{H}'\,{}_{\mathrm{T}}\mathbf{E}(\mathbf{y}) = 0$.

If the design matrix followed the categorical variable coding convention, the associated *constraint* matrix is

$$\begin{bmatrix} 1 & -1 & 0 & 0 \\ 0 & 1 & -1 & 0 \\ 0 & 0 & 1 & -1 \end{bmatrix} \begin{bmatrix} 1 & 0 & 0 & 0 \\ 0 & 1 & 0 & 0 \\ 0 & 0 & 1 & 0 \\ 0 & 0 & 0 & 1 \end{bmatrix} = \begin{bmatrix} 1 & -1 & 0 & 0 \\ 0 & 1 & -1 & 0 \\ 0 & 0 & 1 & -1 \end{bmatrix}$$

The four elements of $\boldsymbol{\beta}$ are constrained all to be equal.

Not surprisingly, under these three constraints, and because for categorical variable coding the elements of $\boldsymbol{\beta}$ are population means, the common value estimated for these elements is the overall sample $\mathbf{y}$ mean, and $\mathbf{f'f}$ is correspondingly the sum of squares of the criterion observations about this overall sample mean. It should be noted that this same interpretation of $\mathbf{f'f}$ would prevail had we developed and imposed constraints for the dummy variable or effect coding variable design matrix or for any other design matrix based on a square basic ${}_{\mathrm{T}}\mathbf{X}$.

Using the categorical variable coding $\mathbf{C}'$ to develop the between groups sum of squares for the Response Time Study data set based on (8.4.24),

$$\mathbf{f'f} - \mathbf{e'e} = (\boldsymbol{\Gamma} - \mathbf{C}'\hat{\boldsymbol{\beta}})'\left[\mathbf{C}'(\mathbf{X'X})^{-1}\mathbf{C}\right]^{-1}(\boldsymbol{\Gamma} - \mathbf{C}'\hat{\boldsymbol{\beta}})$$

we compute

$$\mathbf{C}'\hat{\boldsymbol{\beta}} = \begin{bmatrix} \hat{\beta}_1 - \hat{\beta}_2 \\ \hat{\beta}_2 - \hat{\beta}_3 \\ \hat{\beta}_3 - \hat{\beta}_4 \end{bmatrix} = \begin{bmatrix} 821.25 - 725.92 \\ 725.92 - 827.00 \\ 827.00 - 764.50 \end{bmatrix}$$

$$\mathbf{C}'(\mathbf{X'X})^{-1}\mathbf{C} = \mathbf{C}'\left(\tfrac{1}{12}\mathbf{I}\right)\mathbf{C} = \tfrac{1}{12}\mathbf{C'C} = \tfrac{1}{12}\begin{bmatrix} 2 & -1 & 0 \\ -1 & 2 & -1 \\ 0 & -1 & 2 \end{bmatrix}$$

$$\left[\mathbf{C}'(\mathbf{X'X})^{-1}\mathbf{C}\right]^{-1} = (12)\begin{bmatrix} 2 & -1 & 0 \\ 1 & 2 & -1 \\ 0 & -1 & 2 \end{bmatrix}^{-1} = \tfrac{12}{4}\begin{bmatrix} 3 & 2 & 1 \\ 2 & 4 & 2 \\ 1 & 2 & 3 \end{bmatrix}$$

and, because $\boldsymbol{\Gamma}' = (0, 0, 0)$, $\mathbf{f'f} - \mathbf{e'e} = 83{,}864.50$.

As a check on the interpretation of $\mathbf{f'f}$ as the sum of squared deviations about the overall sample mean, $\mathbf{f'f}$ ought to be determined by

$$\mathbf{y}'\left[\mathbf{I} - \mathbf{1}(\mathbf{1'1})^{-1}\mathbf{1}'\right]\mathbf{y} = \mathbf{y'y} - \mathbf{y'1}(1/n)\mathbf{1'y}$$

$\mathbf{y'y} = 30{,}582{,}878.51$, $\mathbf{y'1} = 37{,}664.0016$, $1/n = \frac{1}{48}$, $\mathbf{f'f} = 1{,}029{,}190.67$. In Section 9.1.1 $\mathbf{e'e}$ was found to be 945,326.17 and $\mathbf{f'f} - \mathbf{e'e}$ has just been computed as 83,864.50. Together these give a value for $\mathbf{f'f}$ identical to this sum of squares about the overall sample mean.

The hypothesis of no group differences would be evaluated by computing from Eq. (8.4.25) the ratio

$$\frac{(\mathbf{f}'\mathbf{f} - \mathbf{e}'\mathbf{e})/(k-1)}{\mathbf{e}'\mathbf{e}/(N-k)}$$

which, by hypothesis, is an observation from $F_{(k-1),(N-k)}$, where $N = n_1 + n_2 + \cdots + n_k$ is the total number of observations. For the Response Time Study data,

$$\frac{83{,}864.50/3}{945{,}326.17/44} = 1.30$$

and the resulting value, not significant at the 0.05 level for $F_{3,44}$, leads us not to reject the hypothesis of no group differences.

9.1.3 Hypothesis Testing in Randomized Groups ANOVA

Usually, however, the simple hypothesis of no group differences will not answer the research questions for which the k-group study was designed. Rather, two or more separate questions are to be put to the data. In the Response Time Study, for example, we may be interested in two questions:

Do pilots respond as quickly to stimuli on the right as they do to stimuli on the left?

Do pilots respond with the same speed to visual as to aural stimuli?

These are rather imprecise, but we shall see how we might interpret them within the design of the study.

Consider two alternative ways of converting these two questions into contrasts among the population means, the elements of ${}_{\mathrm{T}}\mathbf{E}(\mathbf{y})$. The first set of contrasts is

$${}_1\mathbf{H}' = (1, 1, -1, -1), \qquad {}_2\mathbf{H}' = (1, -1, 1, -1)$$

The first contrast translates the visual–aural hypothesis into the form

$$\mathbf{E}(\mathbf{y}\,|\,\mathrm{LV}) + \mathbf{E}(\mathbf{y}\,|\,\mathrm{RV}) = \mathbf{E}(\mathbf{y}\,|\,\mathrm{LA}) + \mathbf{E}(\mathbf{y}\,|\,\mathrm{RA})$$

If we sum over the left and right presentation conditions, the contrast states, no differences appear in the average response time to visual and to aural stimuli.

Similarly, ${}_2\mathbf{H}$ describes a relation among the four population means,

$$\mathbf{E}(\mathbf{y}\,|\,\mathrm{LV}) + \mathbf{E}(\mathbf{y}\,|\,\mathrm{LA}) = \mathbf{E}(\mathbf{y}\,|\,\mathrm{RV}) + \mathbf{E}(\mathbf{y}\,|\,\mathrm{RA})$$

which stipulates that if we sum over the visual and aural conditions, no difference appears in the average response time of pilots to stimuli presented in the left and in the right sensory fields.

Summing over the second facet of the study, however, may not be the reader's idea of how one or both of the hypotheses should be translated. It might, instead, be useful to test a model requiring that there be no visual–aural response time differences for stimuli presented to the left and for stimuli presented to the right. The two contrasts

$$ {}_3\mathbf{H}' = \begin{bmatrix} 1 & 0 & -1 & 0 \\ 0 & 1 & 0 & -1 \end{bmatrix} $$

restrict the linear regression model so that

$$ \mathbf{E}(\mathbf{y}|\mathrm{LV}) = \mathbf{E}(\mathbf{y}|\mathrm{LA}), \qquad \mathbf{E}(\mathbf{y}|\mathrm{RV}) = \mathbf{E}(\mathbf{y}|\mathrm{RA}) $$

In the same way, the hypothesis of no difference in response time between left and right presentations can be required to be true for both visual and aural presentations through the contrasts

$$ {}_4\mathbf{H}' = \begin{bmatrix} 1 & -1 & 0 & 0 \\ 0 & 0 & 1 & -1 \end{bmatrix} $$

The choice among these contrasts—and among others we have yet to present—depends, among other things, on a more precise statement of the researcher's substantive interest in the study. We introduce the sets of contrasts ${}_1\mathbf{H},\ldots,{}_4\mathbf{H}$, at this point to illustrate one consideration in the selection or generation of hypotheses.

For clarity of interpretation it is frequently desirable to have two (sets of) contrasts that are statistically independent. We saw in Section 8.4.4 that two sets of *constraints* on $\boldsymbol{\beta}$ of the form ${}_i\mathbf{C}'\boldsymbol{\beta} = {}_i\boldsymbol{\Gamma}$ and ${}_j\mathbf{C}'\boldsymbol{\beta} = {}_j\boldsymbol{\Gamma}$ have a sampling covariance, given by Eq. (8.4.28), of

$$ \begin{aligned} \mathbf{Cov}\left({}_i\mathbf{C}'\hat{\boldsymbol{\beta}}, {}_j\mathbf{C}'\hat{\boldsymbol{\beta}}\right) &= {}_i\mathbf{C}'\left[\mathbf{Var}(\hat{\boldsymbol{\beta}})\right]{}_j\mathbf{C} \\ &= \sigma^2 {}_i\mathbf{C}'(\mathbf{X}'\mathbf{X})^{-1}{}_j\mathbf{C} \end{aligned} $$

The two constraints are linearly independent if this covariance is $\mathbf{0}$ (a null vector or matrix if ${}_i\mathbf{C}$, ${}_j\mathbf{C}$, or both impose more than one constraint on $\boldsymbol{\beta}$).

We can rephrase this covariance in terms of contrasts among population means rather than in terms of constraints by using the design template to express the relation between the two as

$$ {}_i\mathbf{C}' = {}_i\mathbf{H}'{}_\mathrm{T}\mathbf{X} $$

and writing

$$ (\mathbf{X}'\mathbf{X})^{-1} = ({}_\mathrm{T}\mathbf{X})^{-1}\mathbf{D}^{-1}({}_\mathrm{T}\mathbf{X}')^{-1} $$

With these two results substituted, the covariance becomes

$$ \mathbf{Cov}\left[{}_i\mathbf{H}'{}_\mathrm{T}\hat{\mathbf{E}}(\mathbf{y}),\ {}_j\mathbf{H}'{}_\mathrm{T}\hat{\mathbf{E}}(\mathbf{y})\right] = \sigma^2 {}_i\mathbf{H}'\mathbf{D}^{-1}{}_j\mathbf{H} \qquad (9.1.7) $$

where $\mathbf{D}$ is diagonal with elements $D_i = n_i$, the number of observations selected from the ith population.

If the same number of observations is drawn from each of the k populations, $\mathbf{D} = (N/k)\mathbf{I}$, $\mathbf{D}^{-1} = (k/N)\mathbf{I}$, and

$$\mathbf{Cov}\big({}_i\mathbf{H'}_{\mathrm{T}}\mathbf{E(y)},\ {}_j\mathbf{H'}_{\mathrm{T}}\mathbf{E(y)}\big) = (k/N)\sigma^2\,{}_i\mathbf{H'}_j\mathbf{H} \tag{9.1.8}$$

A randomized group design with the same number of observations or replications in each of the groups is called a *balanced design*. *Orthogonal* contrasts among population means in a balanced design are linearly independent and can be tested quasi-independently:

$$\mathbf{Cov}\big({}_i\mathbf{H'}_{\mathrm{T}}\mathbf{E(y)},\ {}_j\mathbf{H'}_{\mathrm{T}}\mathbf{E(y)}\big) = \mathbf{0} \qquad \text{if} \quad {}_i\mathbf{H'}_j\mathbf{H} = \mathbf{0}$$

Applying this result to the Response Time Study, we find

$${}_1\mathbf{H'}_2\mathbf{H} = 0, \qquad {}_3\mathbf{H'}_4\mathbf{H} = \begin{bmatrix} 1 & -1 \\ -1 & 1 \end{bmatrix}$$

$${}_1\mathbf{H'}_4\mathbf{H} = (0, 0), \qquad {}_2\mathbf{H'}_3\mathbf{H} = (0, 0)$$

Of interest here is that while the two overall hypotheses (summing or averaging over levels of the second facet) yield independent contrasts, ${}_1\mathbf{H}$ and ${}_2\mathbf{H}$, the hypotheses by levels define contrasts ${}_3\mathbf{H}$ and ${}_4\mathbf{H}$ that have correlated sampling distributions and are not independently testable. We do find, however, that a hypothesis by levels regarding sensory field, for example, can be tested independently of an overall hypothesis regarding sense modality.

9.1.4 Factorial Design

In fact, the Response Time Study was designed as a 2×2 *factorial* study, and the contrasts or constraints of usual interest have yet to be developed. A factorial design is a *fully crossed* design among two or more factors or *effects*. Each level of the first factor (sense modality) has been paired with each level of the second factor (sensory field) in determining the populations from which we sampled theoretically or, equivalently, the treatment groups to which participant airline pilots were assigned. A 2×2 factorial design is a four-group design, a 2×3 factorial design requires six groups, and a $2 \times 3 \times 2$ factorial design, one with three factors, calls for 12 randomized groups.

The essential virtue of the factorial design is that it permits the evaluation of not only the impact of each factor on the criterion variable but also any *interaction* among those factors. Is the effect of factor A different if we move from one level of factor B to another or is the influence of A the same at each level of B? If A and B do not interact, if the effect of varying levels of A is constant for all levels of B, then the effects of A and B are said to be *additive*. In the analysis of the factorial design, the first analyses considered are those relating to interactions among the factors. Indeed, the analysis depends on the outcome of those tests for interactions.

In Chapter 12, on multivariate analysis of variance, we shall consider designs of higher order than the 2×2 example, building on the analysis developed here. In the Response Time Study, the hypothesis of no interaction between the two factors, modality and field, can be expressed in terms of the means of the four populations from which we sampled, either as

$$\mathbf{E}(\mathbf{y}|\mathrm{LV}) - \mathbf{E}(\mathbf{y}|\mathrm{LA}) = \mathbf{E}(\mathbf{y}|\mathrm{RV}) - \mathbf{E}(\mathbf{y}|\mathrm{RA})$$

or as

$$\mathbf{E}(\mathbf{y}|\mathrm{LV}) - \mathbf{E}(\mathbf{y}|\mathrm{RV}) = \mathbf{E}(\mathbf{y}|\mathrm{LA}) - \mathbf{E}(\mathbf{y}|\mathrm{RA})$$

The difference (in expected response time) between visual and auditory stimulation is the same for the left and right sensory field; equivalently, the difference between stimulation to the left and to the right sensory fields is the same for visual and auditory signals. In each instance the difference in expected response time is the criterion measure.

The hypothesis of additivity (no interaction) expressed in either of these forms yields the following contrast among population means:

$$(1 \quad -1 \quad -1 \quad 1)\begin{bmatrix} \mathbf{E}(\mathbf{y}|\mathrm{LV}) \\ \mathbf{E}(\mathbf{y}|\mathrm{RV}) \\ \mathbf{E}(\mathbf{y}|\mathrm{LA}) \\ \mathbf{E}(\mathbf{y}|\mathrm{RA}) \end{bmatrix} = 0$$

This contrast—or the associated constraint, given a particular choice of design template—can then be tested, using the F ratio test developed earlier in the chapter.

Assuming this hypothesis cannot be rejected, that is, assuming the sample data are consistent with an additive model, it is then usual to test for the overall effect of each of the two factors. The contrast ${}_2\mathbf{H}' = (1, -1, 1, -1)$ stipulates that there is no sensory field effect, and ${}_3\mathbf{H}' = (1, 1, -1, -1)$ evaluates the hypothesis of no difference in response time between stimuli presented aurally and visually. (The subscripts to the $\mathbf{H}$ vectors in this discussion of the analysis of the factorial design should not be confused with the similarly subscripted $\mathbf{H}$ vectors and matrices used in Section 9.1.3.)

If the hypothesis of additivity is rejected, however, there is little reason to test directly for the overall effect of each factor. Having concluded, for example, that the sensory modality effect is different for the right than for the left sensory fields, it is preferable to follow up by asking if there is a modality effect in the right field and, separately, if there is a modality effect in the left field than to ask if, summing over the two fields, there is a modality effect. The contrasts by levels, then, are ${}_4\mathbf{H}' = (1, 0, -1, 0)$ and ${}_5\mathbf{H}' = (0, 1, 0, -1)$.

Consider again the issue of independence of contrasts for the balanced design:

$$\begin{bmatrix} 1 & -1 & -1 & 1 \\ 1 & -1 & 1 & -1 \\ 1 & 1 & -1 & -1 \end{bmatrix} \begin{bmatrix} 1 & 1 & 1 \\ -1 & -1 & 1 \\ -1 & 1 & -1 \\ 1 & -1 & -1 \end{bmatrix} = \begin{bmatrix} 4 & 0 & 0 \\ 0 & 4 & 0 \\ 0 & 0 & 4 \end{bmatrix}$$

and the interaction and two overall effects are independently testable.

If the hypothesis of additivity is rejected, a choice should then be made as to which factor is to be evaluated by levels of the second factor. As noted in Section 9.1.3, the two hypotheses by levels are not independent:

$$\begin{bmatrix} {}_{A4}\mathbf{H}' \\ {}_{A5}\mathbf{H}' \end{bmatrix} = \begin{bmatrix} 1 & 0 & -1 & 0 \\ 0 & 1 & 0 & -1 \end{bmatrix}$$

or

$$\begin{bmatrix} {}_{B4}\mathbf{H}' \\ {}_{B5}\mathbf{H}' \end{bmatrix} = \begin{bmatrix} 1 & -1 & 0 & 0 \\ 0 & 0 & 1 & -1 \end{bmatrix}$$

For the balanced 2×2 factorial design, then, three orthogonal contrasts serve to account for the among populations mean differences as reflected in the sample. If the additive model is not rejected, the contrasts are ${}_1\mathbf{H}$, ${}_2\mathbf{H}$, and ${}_3\mathbf{H}$, and if the hypothesis of no interaction is rejected, the contrasts are (one choice of the sets) ${}_4\mathbf{H}$, ${}_5\mathbf{H}$, and the overall test for the remaining effect.

What do we mean when we say that either of these sets of contrasts accounts for the between groups mean differences? Recall that (8.4.24) expressed the between groups sum of squares in the context of the constrained maximum likelihood model:

$$\mathbf{f}'\mathbf{f} - \mathbf{e}'\mathbf{e} = (\boldsymbol{\Gamma} - \mathbf{C}'\hat{\boldsymbol{\beta}})'\left[\mathbf{C}'(\mathbf{X}'\mathbf{X})^{-1}\mathbf{C}\right]^{-1}(\boldsymbol{\Gamma} - \mathbf{C}'\hat{\boldsymbol{\beta}})$$

with $\boldsymbol{\Gamma} = \mathbf{0}$, $\mathbf{X}'\mathbf{X} = \mathbf{D}$, and

$$\mathbf{C}' = \begin{bmatrix} 1 & -1 & 0 & 0 \\ 0 & 1 & -1 & 0 \\ 0 & 0 & 1 & -1 \end{bmatrix}$$

The constraints necessary to set the four population means equal are given by $\mathbf{C}'\boldsymbol{\beta} = \mathbf{0}$ for the vector $\boldsymbol{\beta}$ associated with the categorical variable design matrix for the randomized groups design.

Restricting discussion to constraints of the form $\mathbf{C}'\boldsymbol{\beta} = \mathbf{0}$, then, Eq. (8.4.24) may be rewritten

$$\mathbf{f}'\mathbf{f} - \mathbf{e}'\mathbf{e} = \hat{\boldsymbol{\beta}}'\mathbf{C}\left[\mathbf{C}'(\mathbf{X}'\mathbf{X})^{-1}\mathbf{C}\right]^{-1}\mathbf{C}'\hat{\boldsymbol{\beta}}$$

Consider a second set of constraints $\mathbf{K}'\boldsymbol{\beta} = \mathbf{0}$, where $\mathbf{K}' = \mathbf{S}\mathbf{C}'$ and $\mathbf{S}$ is any square basic matrix. The increase in SSE associated with this set of constraints is given by

$$\mathbf{g}'\mathbf{g} - \mathbf{e}'\mathbf{e} = \hat{\boldsymbol{\beta}}'\mathbf{K}\left[\mathbf{K}'(\mathbf{X}'\mathbf{X})^{-1}\mathbf{K}\right]^{-1}\mathbf{K}'\hat{\boldsymbol{\beta}}$$
$$= \hat{\boldsymbol{\beta}}'\mathbf{C}\mathbf{S}'\left[\mathbf{S}\mathbf{C}(\mathbf{X}'\mathbf{X})^{-1}\mathbf{C}\mathbf{S}'\right]^{-1}\mathbf{S}\mathbf{C}'\hat{\boldsymbol{\beta}}$$

The square basic **S** has a regular inverse, so we can also write

$$\mathbf{g}'\mathbf{g} - \mathbf{e}'\mathbf{e} = \hat{\boldsymbol{\beta}}'\mathbf{C}\mathbf{S}'(\mathbf{S}')^{-1}\left[\mathbf{C}'(\mathbf{X}'\mathbf{X})^{-1}\mathbf{C}\right]^{-1}\mathbf{S}^{-1}\mathbf{S}\mathbf{C}'\hat{\boldsymbol{\beta}}$$
$$= \hat{\boldsymbol{\beta}}'\mathbf{C}\left[\mathbf{C}'(\mathbf{X} - \mathbf{X})^{-1}\mathbf{C}\right]^{-1}\mathbf{C}'\hat{\boldsymbol{\beta}}$$

identical to $\mathbf{f}'\mathbf{f} - \mathbf{e}'\mathbf{e}$.

We have just demonstrated that the two sets of constraints $\mathbf{C}'\boldsymbol{\beta} = \mathbf{0}$ and $\mathbf{K}'\boldsymbol{\beta} = \mathbf{0}$ lead to the same reduction in SSE, where $\mathbf{K}' = \mathbf{S}\mathbf{C}'$ and **S** is a square basic matrix. We can also apply this result to contrasts of the form $\mathbf{H}'{}_{\mathrm{T}}\mathbf{E}(\mathbf{y}) = \mathbf{0}$ for the randomized groups design simply by taking $\mathbf{X}'\mathbf{X} = \mathbf{D}$ and rewriting $\mathbf{f}'\mathbf{f} - \mathbf{e}'\mathbf{e} = [\hat{\mathbf{E}}(\mathbf{y})]'\mathbf{H}(\mathbf{H}'\mathbf{D}^{-1}\mathbf{H})^{-1}\mathbf{H}'[\hat{\mathbf{E}}(\mathbf{y})]$.

One set of contrasts that accounts for the between groups differences is

$$\mathbf{H}' = \begin{bmatrix} 1 & -1 & 0 & 0 \\ 0 & 1 & -1 & 0 \\ 0 & 0 & 1 & -1 \end{bmatrix}$$

Three of the possible square basic transformations of $\mathbf{H}'$ are

$$_1\mathbf{S}\mathbf{H}' = \begin{bmatrix} 1 & 0 & -1 \\ 1 & 0 & 0 \\ 1 & 2 & 1 \end{bmatrix}\begin{bmatrix} 1 & -1 & 0 & 0 \\ 0 & 1 & -1 & 0 \\ 0 & 0 & 1 & -1 \end{bmatrix}$$
$$= \begin{bmatrix} 1 & -1 & -1 & 1 \\ 1 & -1 & 0 & 0 \\ 1 & 1 & -1 & -1 \end{bmatrix} = \begin{bmatrix} {}_1\mathbf{H}' \\ {}_2\mathbf{H}' \\ {}_3\mathbf{H}' \end{bmatrix}$$

$$_2\mathbf{S}\mathbf{H}' = \begin{bmatrix} 1 & 1 & 0 \\ 0 & 1 & 1 \\ 1 & 0 & 1 \end{bmatrix}\begin{bmatrix} 1 & -1 & 0 & 0 \\ 0 & 1 & -1 & 0 \\ 0 & 0 & 1 & -1 \end{bmatrix}$$
$$= \begin{bmatrix} 1 & 0 & -1 & 0 \\ 0 & 1 & 0 & -1 \\ 1 & -1 & 1 & -1 \end{bmatrix} = \begin{bmatrix} {}_{A4}\mathbf{H}' \\ {}_{A5}\mathbf{H}' \\ {}_{A6}\mathbf{H}' \end{bmatrix}$$

$$_3\mathbf{S}\mathbf{H}' = \begin{bmatrix} 1 & 0 & 0 \\ 0 & 0 & 1 \\ 1 & 2 & 1 \end{bmatrix}\begin{bmatrix} 1 & -1 & 0 & 0 \\ 0 & 1 & -1 & 0 \\ 0 & 0 & 1 & -1 \end{bmatrix}$$
$$= \begin{bmatrix} 1 & -1 & 0 & 0 \\ 0 & 0 & 1 & -1 \\ 1 & 1 & -1 & -1 \end{bmatrix} = \begin{bmatrix} {}_{B4}\mathbf{H}' \\ {}_{B5}\mathbf{H}' \\ {}_{B6}\mathbf{H}' \end{bmatrix}$$

The three matrices on the right are the three sets of contrasts developed for the balanced 2×2 factorial design. The reader will find it fairly easy to establish that ${}_1\mathbf{S}$, ${}_2\mathbf{S}$, and ${}_3\mathbf{S}$ are each basic and, hence, that each of the three sets of contrasts accounts for the between groups differences in that design. Furthermore, as the contrasts in each set are now mutually orthogonal, there is no overlap in the between groups differences accounted for by the contrasts in a particular set. As the following analyses for the Response Time Study show, the between group sum of squares is the sum of the reductions in SSE associated with ${}_1\mathbf{H}$, ${}_2\mathbf{H}$, and ${}_3\mathbf{H}$ or with ${}_4\mathbf{H}$, ${}_5\mathbf{H}$, and ${}_6\mathbf{H}$.

The *unbalanced k*-groups design is one in which the N criterion observations are not drawn equally from the k populations. Rather, some groups have more observations than others.

Table 9.1 summarizes one way of adjusting the flow of hypothesis testing for the $A \times B$ factorial design depending on whether the design is balanced or unbalanced and on whether the hypothesis of an *additive* model is rejected. Because we think the contrasts ${}_1\mathbf{H}, \ldots, {}_5\mathbf{H}$ provide the natural hypotheses to be tested in this design, we prefer to evaluate them as independently as possible. In the case of the unbalanced design, we turn to the hierarchical approach developed in Section 8.4.4.

Use of the hierarchical approach requires that the two main effects be tested somewhat differently when the additive model is not rejected. As presented in Table 9.1, the factor B is tested in the presence of A (does it make an *incremental* contribution to determining group criterion differences?), whereas the factor A is tested as an isolated source of the between groups differences on the criterion measure. As noted earlier, the

Table 9.1 Use of Contrasts in the 2×2 Factorial Design[a]

Effect	Balanced design		Unbalanced design	
$A \times B$	${}_1\mathbf{H}$: $\mathbf{f'f} - \mathbf{e'e}$ F num: $\mathbf{f'f} - \mathbf{e'e}$		${}_1\mathbf{H}$: $\mathbf{f'f} - \mathbf{e'e}$ F num: $\mathbf{f'f} - \mathbf{e'e}$	
	Additive model	Interactive model	Additive model	Interactive model
B	${}_2\mathbf{H}$: $\mathbf{g'g} - \mathbf{e'e}$ F num: $\mathbf{g'g} - \mathbf{e'e}$		${}_1\mathbf{H}, {}_2\mathbf{H}$: $\mathbf{q'q} - \mathbf{e'e}$ F num: $\mathbf{q'q} - \mathbf{f'f}$	
A	${}_3\mathbf{H}$: $\mathbf{h'h} - \mathbf{e'e}$ F num: $\mathbf{h'h} - \mathbf{e'e}$	${}_4\mathbf{H}$: $\mathbf{j'j} - \mathbf{e'e}$ F num: $\mathbf{j'j} - \mathbf{e'e}$ ${}_5\mathbf{H}$: $\mathbf{k'k} - \mathbf{e'e}$ F num: $\mathbf{k'k} - \mathbf{e'e}$	${}_1\mathbf{H}, {}_2\mathbf{H}, {}_3\mathbf{H}$: $\mathbf{r'r} - \mathbf{e'e}$ F num: $\mathbf{r'r} - \mathbf{q'q}$	${}_1\mathbf{H}, {}_4\mathbf{H}$: $\mathbf{s's} - \mathbf{e'e}$ F num: $\mathbf{s's} - \mathbf{f'f}$ ${}_1\mathbf{H}, {}_5\mathbf{H}$: $\mathbf{t't} - \mathbf{e'e}$ F num: $\mathbf{t't} - \mathbf{f'f}$

[a] Each contrast has a single degree of freedom. The denominator for the F ratio in each instance is $\mathbf{e'e}/(N - k)$ where N observations are distributed over the k groups. Contrasts ${}_4\mathbf{H}$ and ${}_5\mathbf{H}$ refer here to the first set in the text. That is, A is the modality factor and B the field factor in the Response Time Study. Depending upon the researcher's interest, the two factors could be reversed. F num indicates the sum of squares in the numerator of the F ratio.

difference is akin to the difference between evaluating the zero order correlation r_{yA} and the semipartial correlation $r_{y(B.A)}$. In designing an analysis for the unbalanced study, the researcher must decide which effect is to be evaluated conditionally.

Notice that the contrasts ${}_4\mathbf{H}$ and ${}_5\mathbf{H}$ are independent even for the unbalanced design, so they need not be hierarchically arranged in testing the effect of A by levels of B.

An alternative to the hierarchical approach to the analysis of the unbalanced design is to require the individual contrasts to be independent. We do not favor this approach, as it frequently leads to testing hypotheses that are not the ones of experimental interest. We shall illustrate this point by tracing through one procedure sometimes used to form orthogonal contrasts.

The contrasts ${}_1\mathbf{H}$, ${}_2\mathbf{H}$, and ${}_3\mathbf{H}$ are mutually independent in the balanced design. That is, ${}_1\mathbf{H}'\mathbf{D}^{-1}{}_2\mathbf{H} = \mathbf{0}$, ${}_1\mathbf{H}'\mathbf{D}^{-1}{}_3\mathbf{H} = \mathbf{0}$, and ${}_2\mathbf{H}'\mathbf{D}^{-1}{}_3\mathbf{H} = \mathbf{0}$ when $\mathbf{D} = (N/k)\mathbf{I}$. For the general unbalanced design, $D_i = n_i$, and these triple products do not yield zero. However, the contrasts can be adjusted so the three triple products are each zero.

Consider the following three equations, in which the 1s in the three contrasts have been replaced by unknowns a, b, c, and d:

$$(a \quad -b \quad -c \quad d)\begin{bmatrix} 1/n_1 & & & 0 \\ & 1/n_2 & & \\ & & 1/n_3 & \\ 0 & & & 1/n_4 \end{bmatrix}\begin{bmatrix} a \\ -b \\ c \\ -d \end{bmatrix} = 0$$

$$(a \quad -b \quad -c \quad d)\begin{bmatrix} 1/n_1 & & & 0 \\ & 1/n_2 & & \\ & & 1/n_3 & \\ 0 & & & 1/n_4 \end{bmatrix}\begin{bmatrix} a \\ b \\ -c \\ -d \end{bmatrix} = 0$$

$$(a \quad -b \quad c \quad -d)\begin{bmatrix} 1/n_1 & & & 0 \\ & 1/n_2 & & \\ & & 1/n_3 & \\ 0 & & & 1/n_4 \end{bmatrix}\begin{bmatrix} a \\ b \\ -c \\ -d \end{bmatrix} = 0$$

Carrying through the multiplication yields

$$\begin{aligned} a^2/n_1 + b^2/n_2 - c^2/n_3 - d^2/n_4 &= 0 \\ a^2/n_1 - b^2/n_2 + c^2/n_3 - d^2/n_4 &= 0 \\ a^2/n_1 - b^2/n_2 - c^2/n_3 + d^2/n_4 &= 0 \end{aligned}$$

Three equations are not sufficient to specify completely the values of the four unknowns. However, the three equations may be evaluated simultaneously to establish the following relation among a, b, c, and d:

$$a^2/n_1 = b^2/n_2 = c^2/n_3 = d^2/n_4$$

The squares of the unknowns must be constant multiples of the group sizes:

$$a^2 = gn_1, \qquad b^2 = gn_2, \qquad c^2 = gn_3, \qquad d^2 = gn_4$$

The choice of proportionality constant g is arbitrary and does not affect any outcome; the scalar constant can be thought of as a 1×1 square basic matrix and as a multiplier of a contrast has no effect. One choice is $g = 1/n_1$, which makes $a^2 = 1$, $b^2 = n_2/n_1$, $c^2 = n_3/n_1$, and $d^2 = n_4/n_1$. With these values, the three contrasts

$${}_{\mathrm{I}}\mathbf{H}' = \left((n_1/n_1)^{1/2}, -(n_2/n_1)^{1/2}, -(n_3/n_1)^{1/2}, (n_4/n_1)^{1/2}\right)$$

$${}_{\mathrm{II}}\mathbf{H}' = \left((n_1/n_1)^{1/2}, -(n_2/n_1)^{1/2}, (n_3/n_1)^{1/2}, -(n_4/n_1)^{1/2}\right)$$

$${}_{\mathrm{III}}\mathbf{H}' = \left((n_1/n_1)^{1/2}, (n_2/n_1)^{1/2}, -(n_3/n_1)^{1/2}, -(n_4/n_1)^{1/2}\right)$$

are such that ${}_{\mathrm{I}}\mathbf{H}'\mathbf{D}^{-1}\,{}_{\mathrm{II}}\mathbf{H}' = \mathbf{0}$, ${}_{\mathrm{I}}\mathbf{H}'\mathbf{D}^{-1}\,{}_{\mathrm{III}}\mathbf{H}' = \mathbf{0}$, and ${}_{\mathrm{II}}\mathbf{H}'\mathbf{D}^{-1}\,{}_{\mathrm{III}}\mathbf{H} = \mathbf{0}$.

The three contrasts are independent. The problem is that they represent hypotheses that lack obvious appeal. For example, ${}_{\mathrm{I}}\mathbf{H}'\,{}_{\mathrm{T}}\mathbf{E}(\mathbf{y})$ translates to

$$\mathbf{E}(\mathbf{y}|\mathrm{LV}) - (n_2/n_1)^{1/2}\mathbf{E}(\mathbf{y}|\mathrm{RV}) = (n_3/n_1)^{1/2}\mathbf{E}(\mathbf{y}|\mathrm{LA}) - (n_4/n_1)^{1/2}\mathbf{E}(\mathbf{y}|\mathrm{RA})$$

This is not the same as the additivity hypothesis

$$\mathbf{E}(\mathbf{y}|\mathrm{LV}) - \mathbf{E}(\mathbf{y}|\mathrm{RV}) = \mathbf{E}(\mathbf{y}|\mathrm{LA}) - \mathbf{E}(\mathbf{y}|\mathrm{RA})$$

Incidentally, two additional design templates are common in factorial designs. They are extensions of the dummy variable and effect codings for the k-randomized-groups design. We can illustrate the two for a 3×2 design, one in which factor A is present at $a = 3$ levels and factor B at $b = 2$ levels, giving rise to six randomized groups.

Dummy variable coding. The first column of ${}_{\mathrm{T}}\mathbf{X}$ always contains a 1. This is followed by $a - 1$ columns, where A is represented at a levels. Each of these columns is a 0–1 variable, 1 if that row is at that level of the factor A and 0 otherwise. There is no column for the last level of A. These columns are followed in turn by $b - 1$ 0–1 variables coding membership (1) or nonmembership (0) in each of the first $b - 1$ levels of B. Again, there is no column for the last, the bth, level of B. Finally, the design matrix contains an additional $(a - 1)(b - 1)$ columns, each the direct product of one of the columns for factor A and one of the columns for factor B. A *direct product* of two vectors, defined for two vectors of the same length, is a third vector of that length with elements equal to the scalar products of the two elements in the corresponding positions of the original vectors. We represent the direct product $\mathbf{a} * \mathbf{b} = \mathbf{c}$ where $c_i = a_i b_i$.

For the 3×2 factorial, then, the dummy variable design template is

randomized group	design template row					
A_1B_1	1	1	0	1	1	0
A_1B_2	1	1	0	0	0	0
A_2B_1	1	0	1	1	0	1
A_2B_2	1	0	1	0	0	0
A_3B_1	1	0	0	1	0	0
A_3B_2	1	0	0	0	0	0
		A_1	A_2	B_1	A_1B_1	A_2B_1

There are six columns to the basic design matrix for the 3×2 study. Columns 2 and 3 are the A columns, column 4 is the single B column, and columns 5 and 6 are the direct products of columns 2 and 3, respectively, with column 4.

Effect coding. Effect coding for the factorial design is much like dummy variable coding. The only differences are that the last level of the factor A is coded -1 rather than 0 in all the columns for A and that the last level of B is similarly coded -1 in all the columns for B. The final $(a - 1)(b - 1)$ columns are, again, direct products of the resulting columns for A and B. For the 3×2 design the template is

randomized group	design template row					
A_1B_1	1	1	0	1	1	0
A_1B_2	1	1	0	−1	−1	0
A_2B_1	1	0	1	1	0	1
A_2B_2	1	0	1	−1	0	−1
A_3B_1	1	−1	−1	1	−1	−1
A_3B_2	1	−1	−1	−1	1	1
		A_1	A_2	B_1	A_1B_1	A_2B_1

We shall now use factorial effect coding for the 2×2 design to rephrase contrasts ${}_1\mathbf{H}$ through ${}_5\mathbf{H}$ as constraints and then use those to evaluate numerically the results of the Response Time Study. The 2×2 design template,

$$
{}_{\mathrm{T}}\mathbf{X} = \begin{bmatrix} 1 & 1 & 1 & 1 \\ 1 & 1 & -1 & -1 \\ 1 & -1 & 1 & -1 \\ 1 & -1 & -1 & 1 \end{bmatrix}
$$

yields constraints $\mathbf{H}'\,{}_{\mathrm{T}}\mathbf{X} = \mathbf{C}'$:

$$
\begin{bmatrix} 1 & -1 & -1 & 1 \\ 1 & -1 & 1 & -1 \\ 1 & 1 & -1 & -1 \\ 1 & 0 & -1 & 0 \\ 0 & 1 & 0 & -1 \end{bmatrix}
\begin{bmatrix} 1 & 1 & 1 & 1 \\ 1 & 1 & -1 & -1 \\ 1 & -1 & 1 & -1 \\ 1 & -1 & -1 & 1 \end{bmatrix}
=
\begin{bmatrix} 0 & 0 & 0 & 4 \\ 0 & 0 & 4 & 0 \\ 0 & 4 & 0 & 0 \\ 0 & 2 & 0 & 2 \\ 0 & 2 & 0 & -2 \end{bmatrix}
$$

$$
\begin{bmatrix} {}_1\mathbf{H}' \\ {}_2\mathbf{H}' \\ {}_3\mathbf{H}' \\ {}_4\mathbf{H}' \\ {}_5\mathbf{H}' \end{bmatrix} {}_{\mathrm{T}}\mathbf{X} = \begin{bmatrix} {}_1\mathbf{C}' \\ {}_2\mathbf{C}' \\ {}_3\mathbf{C}' \\ {}_4\mathbf{C}' \\ {}_5\mathbf{C}' \end{bmatrix}
$$

Since row scaling of $\mathbf{C}'$ does not affect the outcome, we can as well take the contrasts to be

$$
\begin{bmatrix} 0 & 0 & 0 & 1 \\ 0 & 0 & 1 & 0 \\ 0 & 1 & 0 & 0 \\ 0 & 1 & 0 & 1 \\ 0 & 1 & 0 & -1 \end{bmatrix} = \begin{bmatrix} {}_1\mathbf{C}' \\ {}_2\mathbf{C}' \\ {}_3\mathbf{C}' \\ {}_4\mathbf{C}' \\ {}_5\mathbf{C}' \end{bmatrix}
$$

Use of these contrasts for the response times of the 48 pilots at time 1 yields the results given in Table 9.2. The only comparison resulting in a decision to

Table 9.2 Factorial Contrasts for the Response Time Study

Source of variation	Sum of squares	df	Mean square	F
Modality	5896.33	1	5896.33	0.274
Direction	74,734.08	1	74,734.08	3.478
Modality × direction	3234.08	1	3234.08	0.151
$\mathbf{e}'\mathbf{e}$	945,326.17	44	27,954.83	

reject the hypothesis of no differences in reaction time is that for the direction of stimulus input. The mean reaction time for input to the right ear or eye, 745.21 msec, is significantly less [$F(1, 44) = 3.478$, $p < 0.05$] than the mean reaction time to input to the left ear or eye, 824.13 msec.

This analysis has partitioned $\mathbf{f}'\mathbf{f} - \mathbf{e}'\mathbf{e} = 83{,}864.50$ into the three sources of modality, direction, and modality–direction interaction.

9.2 Concomitant Variable and Covariance Analysis

Fixed effects analysis of variance, as developed in Section 9.1, represents a version of the linear regression model in which the design matrix consists entirely of categorical variables. In Sections 8.2–8.9 we presented the general linear regression model which can also be directly applied to designs with measured (and likely correlated) predictor attributes. Section 9.3 will restate this, giving emphasis to some special applications. In the present section we discuss two common designs in which both categorical and measured design variables are present.

The usual motivation for *concomitant* or *covariance* designs is that the dependent or criterion attribute is known to be linearly influenced by one or more other measured attributes and, in consequence, the effect on the criterion of the categorical variables—treatment group assignment in the randomized groups design—is studied with the influence of such other attributes partialed out or controlled. In effect, we use the linear relation between the criterion attribute and the covariate to reduce $\mathbf{e}'\mathbf{e}$ and hence increase our sensitivity to the effects of the treatments.

Error reduction is common to concomitant variable and covariance analyses. If we also seek assurance that criterion scores under the several treatments are not differentially influenced by the covariate, we should choose covariance analysis. The two techniques are illustrated for the Airline Pilots Response Time Study, assuming that response time, in experimental settings, is correlated with tested intelligence: Those who respond more quickly tend to be those who earn higher scores on intelligence measures. There is some variability in tested intelligence among airline pilots, and since our interest in the Response Time Study is in differences among the four stimulus conditions, we want to avoid having stimulus-related differences in response speed affected by intelligence-related differences.

As a methodological control, assume that we obtained an intelligence measure for each of the 48 pilots participating in the study. It is important to note, however, that we shall not use this information in assigning pilots to the four stimulus conditions. We still make random assignments. There are two traditional ways of adding this intelligence measure to the predictor data (design matrix) employed in the analysis. For definiteness we shall assume categorical variable coding for the k-groups design. That is, we start with a design matrix of the form

$$
\left[\begin{array}{cccc}
1 & 0 & 0 & 0 \\
\vdots & \vdots & \vdots & \vdots \\
1 & 0 & 0 & 0 \\
\hdashline
0 & 1 & 0 & 0 \\
\vdots & \vdots & \vdots & \vdots \\
0 & 1 & 0 & 0 \\
\hdashline
0 & 0 & 1 & 0 \\
\vdots & \vdots & \vdots & \vdots \\
0 & 0 & 1 & 0 \\
\hdashline
0 & 0 & 0 & 1 \\
\vdots & \vdots & \vdots & \vdots \\
0 & 0 & 0 & 1 \\
\hdashline
\end{array}\right]
\begin{array}{l}
\\ \\ 12 \\ \\ \\ 24 \\ \\ \\ 36 \\ \\ \\ 48
\end{array}
$$

The first 12 rows correspond to the 12 pilots assigned to the LV condition, the second 12 to those in the RV condition, followed by the LA and finally the RA groups.

Treating intelligence as a *concomitant* variable, we simply add a fifth column to the design matrix containing the test scores. The expectation vector $\mathbf{E(y)} = \mathbf{X\beta}$ has entries of the form

$$
[E(\mathbf{y})]_i = \begin{cases} \beta_1 + \beta_5 X_{i5} & \text{if} \quad \text{the } i\text{th observation is in the first group} \\ \beta_2 + \beta_5 X_{i5} & \text{if} \quad \text{in the second group} \\ \beta_3 + \beta_5 X_{i5} & \text{if} \quad \text{in the third group} \\ \beta_4 + \beta_5 X_{i5} & \text{if} \quad \text{in the fourth group} \end{cases}
$$

The usual constraints on $\beta_1, \ldots, \beta_4$ are then used to compute and evaluate, by the F ratio test, increases in SSE. The only difference between the concomitant variable analysis and the parallel fixed effects analysis of variance is that the parameter β_5 remains a part of each model and is always unconstrained. To the extent that the concomitant variable and the criterion are correlated, the SSE should be smaller in the concomitant variable design.

If we wish to use intelligence test scores as a *covariable*, then we add to the k-column randomized groups design matrix not one column but an additional k columns. If we start, again, with a categorical variable coding of the design these additional columns are simply the k direct products of each of the categorical variables with a vector of scores on the measured covariate. That is, for our four group design, we add columns with entries

$$X_{i5} = \begin{cases} \text{test score for the } i\text{th pilot} & \text{if assigned to the first group} \\ 0 & \text{otherwise} \end{cases}$$

$$X_{i6} = \begin{cases} \text{test score for the } i\text{th pilot} & \text{if in the second group} \\ 0 & \text{otherwise} \end{cases}$$

$$X_{i7} = \begin{cases} \text{test score for the } i\text{th pilot} & \text{if in the third group} \\ 0 & \text{otherwise} \end{cases}$$

$$X_{i8} = \begin{cases} \text{test score for the } i\text{th pilot} & \text{if in the fourth group} \\ 0 & \text{otherwise} \end{cases}$$

In the analysis of covariance for the four-group design, then, the expectation vector $\mathbf{E(y)} = \mathbf{X\beta}$ has elements

$$[E(\mathbf{y})]_i = \begin{cases} \beta_1 + \beta_5 X_{i5} & \text{if } \; y_i \text{ is in the first group} \\ \beta_2 + \beta_6 X_{i6} & \text{if } \; \text{in the second group} \\ \beta_3 + \beta_7 X_{i7} & \text{if } \; \text{in the third group} \\ \beta_4 + \beta_8 X_{i8} & \text{if } \; \text{in the fourth group} \end{cases}$$

The *covariance* approach to controlling, statistically, for the effects of an outside variable on the criterion customarily involves testing one hypothesis in addition to those already discussed. To further ensure that the covariable does not have a differential effect over the several treatment groups, the regression of the criterion on the covariable should be the same within each of the groups. This hypothesis of homogeneity of within groups regression requires that $\boldsymbol{\beta}$ be constrained so that $\beta_5 = \beta_6 = \beta_7 = \beta_8$. A set of constraints that would do this is

$$_7\mathbf{C}' = \begin{bmatrix} 0 & 0 & 0 & 0 & 1 & -1 & 0 & 0 \\ 0 & 0 & 0 & 0 & 0 & 1 & -1 & 0 \\ 0 & 0 & 0 & 0 & 0 & 0 & 1 & -1 \end{bmatrix}$$

where the hypothesis is that $_7\mathbf{C}'\boldsymbol{\beta} = \mathbf{0}$.

As evidence that the random assignment of pilots to the four treatments did not lead to a piling up of the more intelligent volunteers in one of the stimulus presentation conditions, it is common to accompany the analysis of covariance with a fixed effects analysis of variance in which the criterion attribute is the covariate or concomitant variable. One would expect no significant differences on any of the contrasts or constraints employed with the real criterion.

The format of the concomitant or covariable in the design matrix, incidentally, is as just given, whether the randomized groups design template had been based on categorical variables, dummy variables, or effect coding and whether the k groups defined a factorial experiment or not.

Table 9.3 Analysis of Covariance for the Response Time Study

Source of variation	Sum of squares	df	Mean square	F
Covariate: intelligence	331,227.45	1	331,227.45	23.13
Modality	5896.33	1	5896.33	0.42
Direction	74,734.08	1	74,734.08	5.22
Modality × direction	1534.18	1	1534.18	0.11
$\mathbf{e}'\mathbf{e}$	615,798.61	43	14,320.90	

Data from the Response Time Study can now be used to illustrate the analysis of covariance. Comparing the mean reaction time in milliseconds at time 1 for the four treatment groups with intelligence as a covariate results in the summary statistics given in Table 9.3. By comparing the error term in these results with the value in the analysis reported in Table 9.2, which did not include the score on the intelligence test as a covariate, it is clear that taking individual differences in intelligence into account reduced $\mathbf{e}'\mathbf{e}$. None of the substantive conclusions, however, are altered. (The hypothesis of equal slopes for the regression of the dependent variable on the covariate, over the four groups, cannot be rejected, and Table 9.3 reports the subsequent concomitant variable analysis.)

9.3 Multiple Linear Regression Applications

In this section we discuss four applications of multiple linear regression to studies including only measured predictors. These are the interrelated topics of *curve-fitting*, *polynomial regression*, testing for *linearity of a predictor attribute*, and the use of *moderator variables* in linear regression. The four will be illustrated with data from the Preemployment Study.

Three attributes will be used in the illustrations. The criterion is self-reported *job satisfaction* at the end of one year's employment and is obtained from a 200-point scale where 1 denotes "very dissatisfied" and 200 "very satisfied." The two predictors of interest are *years of education* and previous *work experience*. The latter is measured on a 10-point scale with the following definitions of the scale points:

1. no previous full-time work experience;
2. some but no more than 3 months;
3. at least 3 but fewer than 6 months;
4. at least 6 months but less than 1 year;
5. at least 1 but less than 2 years;
6. at least 2 but less than 3 years;
7. at least 3 but less than 5 years;
8. at least 5 but fewer than 7 years;

9. at least 7 but fewer than 10 years;
10. 10 or more years full-time work experience.

Measures are for 150 applicants hired, trained, and appointed as underwriters in a large insurance firm.

9.3.1 Curve Fitting

The work experience scale is not linearly related to months of prior full-time employment, and it should be anticipated that the regression of job satisfaction on work experience also might not be linear. That is, a hypothesis that, on the average, job satisfaction will change as much between those in categories 2 and 3 as between those in categories 8 and 9 would have to be empirically evaluated before its acceptance would be reasonable. In Section 9.3.3 we illustrate a technique for testing for such linearity. Here, however, we concentrate on using linear regression to fit the work experience scale to the criterion observations.

In terms of the linear regression model, curve fitting uses the notion that there exists (at least ideally) a separate, normally distributed population of criterion observations for each of the scale points. In the Preemployment Study, there would be a normal distribution of job satisfaction ratings, for example, for those who had been previously employed for 5–7 years. These theoretical distributions are the ones that would result if we were able to hire, train, and employ all potential applicants with these job experience characteristics.

By assuming that our hired underwriters were randomly and independently chosen from those with the same, scaled, work experience and that each of these normal distributions of job satisfaction ratings would have the same variance, we can compute maximum likelihood estimates of the means of each distribution. We do this using the same technique as for the k-randomized-groups design. That is, by taking the design matrix to be made up of a set of *categorical variables*, coding membership in one of the ten groups corresponding to amounts of work experience, $\boldsymbol{\beta}$ consists, in turn, of the population job satisfaction means, and $\hat{\boldsymbol{\beta}} = (\mathbf{X}'\mathbf{X})^{-1}\mathbf{X}'\mathbf{y}$ estimates those means.

These maximum likelihood estimators, of course, will equal the sample criterion means for the ten groups, paralleling the results in the earlier k-groups analyses. Similarly, the population means could as easily be estimated using a dummy variable, effect coding, or any other *design template* ${}_{\mathrm{T}}\mathbf{X}$ based on the relation among the design template $\boldsymbol{\beta}$ and the *expectation template* ${}_{\mathrm{T}}\mathbf{X}\boldsymbol{\beta} = {}_{\mathrm{T}}\mathbf{E}(\mathbf{y})$. We estimate the expectation template by substituting $\hat{\boldsymbol{\beta}}$ for $\boldsymbol{\beta}$: ${}_{\mathrm{T}}\mathbf{X}\hat{\boldsymbol{\beta}} = {}_{\mathrm{T}}\hat{\mathbf{E}}(\mathbf{y})$.

Carrying through the computations for the work experience scale categories and the job satisfaction criterion, we have, using categorical variable coding or ${}_{\mathrm{T}}\mathbf{X} = \mathbf{I}$, $\mathbf{1}'(\mathbf{X}'\mathbf{X}) = \mathbf{1}'\mathbf{D} = (4, 7, 6, 18, 30, 23, 26, 16, 12, 8)$ and

$$\begin{aligned}\mathbf{D}^{-1}\mathbf{X}'\mathbf{y} &= \hat{\boldsymbol{\beta}} = {}_{\mathrm{T}}\hat{\mathbf{E}}(\mathbf{y}) \\ &= (80.00, 105.71, 116.67, 116.11, 119.00, \\ &\qquad 106.96, 133.85, 133.75, 150.00, 115.00)\end{aligned}$$

The *point estimates* of $[{}_{\mathrm{T}}\mathbf{E}(\mathbf{y})]_i$ can be converted to *interval estimates* of a certain confidence level. These latter clearly depend on the *sampling distribution* of ${}_{\mathrm{T}}\hat{\mathbf{E}}(\mathbf{y})$. In developing this sampling distribution it is usual to adopt another linear regression assumption, that $\mathbf{X}$ is *fixed*. That is, our interval estimates reflect what we would expect to happen if we were to repeat the study, again and again, each time randomly and independently drawing n_1, $n_2, \ldots, n_k$ observations from the first, second, ..., kth category, where these are the numbers used in the sample upon which the estimates were based. The widths of the intervals reflect these subsample sizes. We shall have greater confidence in estimates based on larger samples. Figure 9.1, which plots ${}_{\mathrm{T}}\hat{\mathbf{E}}(\mathbf{y})$ against values of the work experience scale, provides an example of curve fitting. The samples drawn from the different experience categories, given in $\mathbf{1}'\mathbf{D}$, are not all of the same size.

The common population variance for the criterion σ^2 is estimated from the sample data based on $\mathbf{y}'\mathbf{y} - \mathbf{y}'\mathbf{X}\hat{\boldsymbol{\beta}} = \mathbf{e}'\mathbf{e} = 222{,}679.88$ as

$$\hat{\sigma}^2 = \mathbf{e}'\mathbf{e}/(n-m) = \mathbf{e}'\mathbf{e}/(150-10) = 1590.57$$

Based on Eq. (8.4.33), confidence intervals for estimates of the mean job satisfaction are obtained from

$$L_{[\hat{E}(\mathbf{y})]_i} = [\hat{E}(\mathbf{y})]_i \pm [\hat{\sigma}^2(F_{0.95,1,n-m})/n_i]^{1/2} \tag{9.3.1}$$

Figure 9.1 ${}_{\mathrm{T}}\mathbf{E}(\mathbf{y})$ versus work experience.

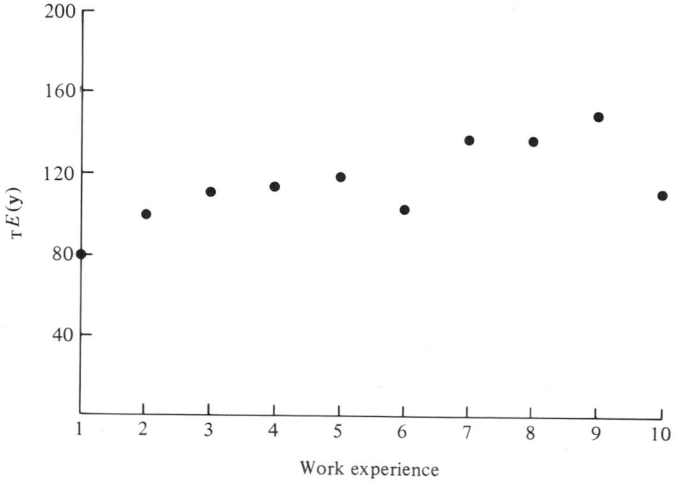

Table 9.4 Curve Fitting to Work Experience Levels

Work experience level	95% confidence interval
1	28.03–131.96
2	78.04–133.39
3	81.13–152.20
4	97.18–134.04
5	103.33–134.67
6	87.47–126.45
7	117.61–150.09
8	111.41–156.09
9	125.32–174.68
10	93.57–136.43

where $\hat{\sigma}^2$ is the unbiased estimator of σ^2, n_i is the number of observations at the ith scale point, and $F_{0.95, 1, n-m}$ is the point below which 95% of the observations making up the $F_{1, n-m}$ distribution are located. The larger n_i, the narrower the confidence interval for $\hat{E}(y_i)$.

For the data of the example, the results are given in Table 9.4.

9.3.2 Polynomial Regression

As an alternative to assuming the regression of job satisfaction on scaled work experience is linear, we substituted for the work experience scale (one measured variable) a set of work experience *categories* (ten categorical variables). In Section 9.3.3 we show how to test the hypothesis of linearity of a predictor attribute by comparing a second set of estimates with this categorical variable solution. Before we do that, however, we present an alternative approach that allows a nonlinear relation between predictor and criterion.

In the Preemployment Study another measured attribute of applicants is the number of years of education. We might expect, in general, self-assessed job satisfaction to increase with the educational level of the employee, although there may be either a diminishing returns effect at higher levels of education or perhaps a downward trend in satisfaction among those substantially "overeducated" for the job they hold. Rather than being linear, the regression of job satisfaction on education might take one of the forms shown in Figure 9.2.

There is no single alternative to the linear regression of a criterion on a measured attribute, $[E(\mathbf{y})]_i = k + az_i$. Rather, the alternatives include all possible nonlinear relations. In some rather limited contexts we may know the alternatives of interest. Competing theories may tell us the regression is linear or logarithmic, linear or exponential, linear or quadratic. More often, however, we cannot specify the alternative to linearity; yet we want to evaluate such an alternative.

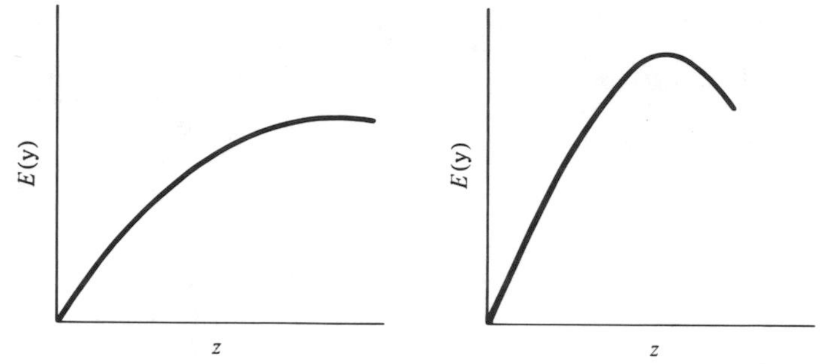

Figure 9.2 Expected job satisfaction versus years of education: two possibilities.

Because a measured attribute may define a very large number of categories, the categorical variable approach to curve fitting in Section 9.3.1 can lead to very large design matrices. Since the variance of the sampling distribution for $\hat{\hat{\sigma}}^2$ increases as $N - m$ decreases, making m—the number of columns in the design matrix—large leads to relatively unstable estimates of σ^2 and hypothesis tests that lack power.

What can be substituted for categorical variables as a general approach to nonlinear attribute relations? Most commonly, regression analysis takes advantage of the fact that almost any continuous functional relation $[E(\mathbf{y})]_i = F(z_i)$ can be reasonably well approximated by a polynomial in z_i:

$$[E(\mathbf{y})]_i = k + az_i + bz_i^2 + cz_i^3 + \cdots + jz_i^j$$

Although very complicated relations may require z_i^j to be a fairly high power of z_i, such complexities are not common with attributes in the behavioral sciences. In general, cubic and quartic equations, those involving z_i^3 or z_i^4, provide a sufficient contrast to a linear regression.

The design matrix for the regression of job satisfaction of insurance underwriters on their years of education might consist of five columns:

- $\mathbf{X}_{.1}$, the unit vector;
- $\mathbf{X}_{.2}$, a vector in which X_{i2} is the years of education of the ith underwriter;
- $\mathbf{X}_{.3}$, the direct product, $\mathbf{X}_{.2} * \mathbf{X}_{.2}$, years squared;
- $\mathbf{X}_{.4}$, the direct product $\mathbf{X}_{.2} * \mathbf{X}_{.2} * \mathbf{X}_{.2}$, years cubed;
- $\mathbf{X}_{.5}$, the direct product $\mathbf{X}_{.2} * \mathbf{X}_{.2} * \mathbf{X}_{.2} * \mathbf{X}_{.2}$, the fourth power of years of education.

If, as in the present case, the attribute with which we start takes only positive values, we can and should improve upon this set of predictors. As defined, $\mathbf{X}_{.2}$, $\mathbf{X}_{.3}$, $\mathbf{X}_{.4}$, and $\mathbf{X}_{.5}$ would be strongly positively intercorrelated. Large z_i give rise to large z_i^2, z_i^3, and z_i^4. If a set of predictor attributes is linked by a number of high correlations, they are said to be *multicollinear*. This multicollinearity can be so pronounced that one or more of the predictors is perfectly reproducible as a linear function of other predictors; the design matrix then has rank less than its column order, and $\mathbf{X}'\mathbf{X}$ lacks a regular inverse. Should this occur, we know we are in trouble.

More insidious are multicollinear design matrices that, while still basic, yield *ill-conditioned* minor product moments. Although $\mathbf{X}'\mathbf{X}$ has an inverse and $\hat{\boldsymbol{\beta}}$ an apparently unique solution, one or more of the $\hat{\beta}_i$ may have an unacceptably large sampling variance. We may be unable to judge, for example, whether $\hat{\beta}_3$ is positive or negative, let alone its magnitude. (Worse, we may forget to inspect the sampling variance and interpret $\hat{\beta}_2$ at face value!)

Highly correlated predictors are to be avoided. One fairly effective (and easily implemented) way of reducing the correlations among predictors in polynomial regression is to begin with a measured attribute in *deviation* form. Odd and even powers of the attribute will then be uncorrelated, although the odd powers will still be correlated among themselves, as will the even powers. If the degree of the polynomial is only three (cubic) or even four (quartic), this reduction in intercorrelation is usually sufficient.

This point can be illustrated by further analysis of data from the Preemployment Study. Assume that we wanted to explain variability in job satisfaction y on the basis of years of education x or some power of years of education, for example, x^2, x^3, or x^4. Based on analyzing raw scores on x and powers of raw scores in this example we have the following values:

$$\mathbf{R}_x = \begin{array}{c} \\ x \\ x^2 \\ x^3 \\ x^4 \end{array}\!\!\begin{array}{c} \begin{array}{cccc} x & x^2 & x^3 & x^4 \end{array} \\ \begin{bmatrix} 1.000 & 0.997 & 0.987 & 0.972 \\ 0.997 & 1.000 & 0.997 & 0.988 \\ 0.987 & 0.997 & 1.000 & 0.997 \\ 0.972 & 0.988 & 0.997 & 1.000 \end{bmatrix} \end{array}, \qquad \hat{\boldsymbol{\beta}} = \begin{bmatrix} -5066.93 \\ 1758.07 \\ -222.02 \\ 12.34 \\ -0.25 \end{bmatrix}$$

In this analysis $R^2_{y.x,\,x^2,\,x^3,\,x^4} = 0.092$, and only years of education x contributed to the prediction beyond a chance level ($R^2_{y.x} = 0.091$).

In comparison, on analyzing $x - \bar{x}$, $(x - \bar{x})^2$, $(x - \bar{x})^3$, and $(x - \bar{x})^4$, we obtain

$$\mathbf{R}_{(x-\bar{x})} = \begin{bmatrix} 1.000 & 0.111 & 0.869 & 0.204 \\ 0.111 & 1.000 & 0.261 & 0.926 \\ 0.869 & 0.261 & 1.000 & 0.422 \\ 0.204 & 0.926 & 0.422 & 1.000 \end{bmatrix}, \qquad \hat{\boldsymbol{\beta}} = \begin{bmatrix} 117.37 \\ 6.34 \\ 3.05 \\ 0.16 \\ -0.25 \end{bmatrix}$$

In this analysis, $R^2_{y.x,\,x^2,\,x^3,\,x^4} = 0.0982$, and years of education x remains the only predictor contributing, beyond the chance level, to the prediction of the criterion. However, in this second analysis we have a much more accurate estimate of the $\boldsymbol{\beta}$ for years of education. The reader may verify this by computing the inverses of the two $\mathbf{R}$ matrices displayed and comparing the magnitudes of the two diagonal elements $(\mathbf{R}^{-1})_{11}$ in each case.

9.3.3 Testing for Linearity

By replacing a measured attribute in the design matrix either with a set of categorical variables or with a set of powers of variables, we can determine the fit of a set of data. A principal reason for doing so is to contrast the fit with that obtained from constraining the maximum likelihood solution so that the regression on the original measured attribute is linear. These comparisons result in F ratios and can be considered as straightforward applications of the constraint matrix approach.

For the categorical variable illustration, involving the work experience scale, we have a full model or unconstrained fit that yields $\mathbf{e}'\mathbf{e} = \mathbf{y}'\mathbf{y} - \mathbf{y}'\mathbf{X}\hat{\boldsymbol{\beta}}$. Recalling that the elements of $\boldsymbol{\beta}$ are the population means for the ten ordered categories, the hypothesis of linearity is a hypothesis of equal spacing to the categories and can be expressed

$$
\begin{aligned}
\beta_3 - \beta_2 &= \beta_2 - \beta_1\\
\beta_4 - \beta_3 &= \beta_3 - \beta_2\\
\beta_5 - \beta_4 &= \beta_4 - \beta_3\\
\beta_6 - \beta_5 &= \beta_5 - \beta_4\\
\beta_7 - \beta_6 &= \beta_6 - \beta_5\\
\beta_8 - \beta_7 &= \beta_7 - \beta_6\\
\beta_9 - \beta_8 &= \beta_8 - \beta_7\\
\beta_{10} - \beta_9 &= \beta_9 - \beta_8
\end{aligned}
$$

The hypothesis of linearity imposes eight constraints, which may be phrased $\mathbf{C}'\boldsymbol{\beta} = \mathbf{0}$, on the ten elements of $\boldsymbol{\beta}$. The rows of $\mathbf{C}'$ are easily developed from the above equalities: $\mathbf{C}_1'$, for example, is (1, −2, 1, 0, 0, 0, 0, 0, 0, 0). The increase in SSE associated with these eight constraints is $\mathbf{f}'\mathbf{f} - \mathbf{e}'\mathbf{e} = 11,$ 741.10. The test of linearity, then, is given by

$$F = \frac{(\mathbf{f}'\mathbf{f} - \mathbf{e}'\mathbf{e})/8}{\mathbf{e}'\mathbf{e}/(N-10)} = \frac{11{,}741.1/8}{222{,}679.88/140} = 0.92$$

and the hypothesis of linearity cannot be rejected.

In the deviation score polynomial example, the full model SSE for the quartic equation is $\mathbf{e}'\mathbf{e} = 228{,}578.80$. The hypothesis of linearity calls for contrasting this with the SSE resulting from an equation in which years of education appears only in the first degree. This hypothesis constrains the full model by requiring that $\beta_3 = \beta_4 = \beta_5 = 0$. The constraint matrix is

$$\mathbf{C}' = \begin{bmatrix} 0 & 0 & 1 & 0 & 0 \\ 0 & 0 & 0 & 1 & 0 \\ 0 & 0 & 0 & 0 & 1 \end{bmatrix}$$

and the increase in SSE associated with these constraints is $\mathbf{h}'\mathbf{h} - \mathbf{e}'\mathbf{e} = 2045.62$. The ratio

$$\frac{(\mathbf{h}'\mathbf{h} - \mathbf{e}'\mathbf{e})/3}{\mathbf{e}'\mathbf{e}/(N-5)} = \frac{2045.62/3}{228{,}578.8/145} = 0.43$$

is an observation from the F variate with 3 and 145 degrees of freedom and, in this instance, we would not reject the hypothesis of linearity.

In evaluating polynomial regression, rather than simply compare the rth-degree polynomial with a linear model it is frequently more informative to conduct several tests, dropping first the highest degree term, then the next highest, and continuing until each polynomial term, including the linear term, has been evaluated. The hierarchical constraints approach provides a convenient way of performing the sequential tests.

To test the quartic component to the regression of job satisfaction on years of work experience, we obtain the increase in SSE associated with the constraint ${}_1\mathbf{C}'\boldsymbol{\beta} = \mathbf{0}$, where ${}_1\mathbf{C}' = (0, 0, 0, 0, 1)$. This increase provides the numerator of the F ratio test:

$$F = \frac{(\mathbf{f}'\mathbf{f} - \mathbf{e}'\mathbf{e})/1}{\mathbf{e}'\mathbf{e}/(N-5)} = \frac{1666.55/1}{228{,}578.8/145} = 1.057$$

To test the cubic component the constraint ${}_2\mathbf{C}'\boldsymbol{\beta} = \mathbf{0}$ is imposed with

$${}_2\mathbf{C}' = \begin{bmatrix} 0 & 0 & 0 & 0 & 1 \\ 0 & 0 & 0 & 1 & 0 \end{bmatrix}$$

The associated increase in SSE is $\mathbf{g}'\mathbf{g} - \mathbf{e}'\mathbf{e} = 1855.5$, but the numerator of the F test, to ensure quasi-independence, is $\mathbf{g}'\mathbf{g} - \mathbf{f}'\mathbf{f} = (\mathbf{g}'\mathbf{g} - \mathbf{e}'\mathbf{e}) - (\mathbf{f}'\mathbf{f} - \mathbf{e}'\mathbf{e}) = 188.95$:

$$\frac{(\mathbf{g}'\mathbf{g} - \mathbf{f}'\mathbf{f})/1}{\mathbf{e}'\mathbf{e}/(N-5)} = \frac{188.95/1}{228{,}578.8/145} = 0.120$$

Similarly, we test the quadratic component for significance by first finding the overall increase in SSE for ${}_3\mathbf{C}'\boldsymbol{\beta} = \mathbf{0}$: $\mathbf{h}'\mathbf{h} - \mathbf{e}'\mathbf{e} = 2045.62$, where

$${}_3\mathbf{C}' = \begin{bmatrix} 0 & 0 & 0 & 0 & 1 \\ 0 & 0 & 0 & 1 & 0 \\ 0 & 0 & 1 & 0 & 0 \end{bmatrix}$$

and then computing the incremental SSE increase by subtraction: $\mathbf{h}'\mathbf{h} - \mathbf{g}'\mathbf{g} = (\mathbf{h}'\mathbf{h} - \mathbf{e}'\mathbf{e}) - (\mathbf{g}'\mathbf{g} - \mathbf{e}'\mathbf{e}) = 190.12$. The F statistic, then, becomes

$$F = \frac{(\mathbf{h}'\mathbf{h} - \mathbf{g}'\mathbf{g})/1}{\mathbf{e}'\mathbf{e}/(N-5)} = \frac{(190.12)/1}{(228{,}578.8)/145} = 0.121$$

Finally, the linear component is evaluated by the ratio

$$F = \frac{(\mathbf{i}'\mathbf{i} - \mathbf{h}'\mathbf{h})/1}{\mathbf{e}'\mathbf{e}/(N-5)} = \frac{(22{,}852.91)/1}{(228{,}578.8)/145} = 14.50$$

where $\mathbf{i}'\mathbf{i} - \mathbf{e}'\mathbf{e} = 24{,}898.53$ is the SSE occasioned by imposing the constraints ${}_4\mathbf{C}'\boldsymbol{\beta} = \mathbf{0}$, and

$${}_4\mathbf{C}' = \begin{bmatrix} 0 & 0 & 0 & 0 & 1 \\ 0 & 0 & 0 & 1 & 0 \\ 0 & 0 & 1 & 0 & 0 \\ 0 & 1 & 0 & 0 & 0 \end{bmatrix}$$

In practice, sequential hypothesis testing would continue only until a significant component is identified. In the final polynomial regression equation all nonsignificant higher-degree terms are eliminated. In our example, the accepted regression equation would be linear since none of the higher-degree terms contributed significantly.

9.3.4 *Moderator Variables*

Consider a linear regression model based on the design matrix with columns $\mathbf{X}_{.1}$, the unit vector; $\mathbf{X}_{.2}$, a vector of work experience scale scores; and $\mathbf{X}_{.3}$, a vector containing years of education entries; so that the regression or expectation equation $\mathbf{E}(\mathbf{y}) = \mathbf{X}\boldsymbol{\beta}$ for the job satisfaction criterion yields, for a typical observation,

$$[E(\mathbf{y})]_i = \beta_1 + \beta_2 X_{i2} + \beta_3 X_{i3}$$

where β_1 is an additive constant and β_2 and β_3 are weights associated with, respectively, the two predictors experience and education.

Note that the weight given the education measure is a constant. Each underwriter's years of education is multiplied by the same quantity, β_3, in arriving at an expected job satisfaction score. It might be argued that the importance of a given amount of education (in determining job satisfaction) ought to vary as a function of the amount of the individual's work experience; the more work experience, perhaps, the less important education is to work satisfaction. The effect of an individual's education on job satisfaction could be *moderated* by the extent of that person's work experience. If we want the weight assigned one attribute to vary as a function of the value of a second attribute, we call the second a *moderator* variable.

The moderating of a weighting function could take any one of several forms. Here we limit our discussion to linear moderating functions. That is, we substitute for the constant β_3 in the regression equation a linear function of the ith underwriter's work experience score:

$$\begin{aligned} [E(\mathbf{y})]_i &= \beta_1 + \beta_2 X_{i2} + (a + bX_{i2}) X_{i3} \\ &= \beta_1 + \beta_2 X_{i2} + aX_{i3} + bX_{i2} X_{i3} \\ &= \beta_1 + \beta_2 X_{i2} + \beta_3 X_{i3} + \beta_4 (X_{i2} X_{i3}) \end{aligned}$$

The moderator variable concept, for measured attributes, is directly analogous, in form and interpretation, to the interaction introduced in the analysis of variance for categorical attributes. As in polynomial regression it is suggested that both the moderating and moderated attributes be transformed to deviation form for the computation of the direct product and for the subsequent regression analysis. This transformation reduces additional collinearities introduced into the design matrix.

Polynomial and product terms can both be accommodated in the same regression model. Also, of course, the direct product may be defined between a categorical and a measured attribute. We saw one such application in the analysis of covariance. In multiple regression studies with, for example, gender as one of the predictor attributes, a possible hypothesis might be that the regression of the criterion on some second predictor is the same for the two genders. An approach to this hypothesis would begin with a full model that includes gender interactions or moderations of the measured predictor attributes.

In the analysis of moderated regression, hierarchical constraints are again useful. Ordinarily the product terms (interactions) are eliminated first. We illustrate with the Preemployment Study data.

The full model has education, experience, and their direct product as predictors and job satisfaction as the criterion and yields a sum of squares of errors (SSE) of $\mathbf{e}'\mathbf{e} = 217{,}707.796$. Increases in SSE for this sequence of hierarchical constraint matrices

$$ {}_1\mathbf{C} = \begin{bmatrix} 0 & 0 & 0 & 1 \end{bmatrix} $$

$$ {}_2\mathbf{C} = \begin{bmatrix} 0 & 0 & 0 & 1 \\ 0 & 1 & 0 & 0 \end{bmatrix} $$

$$ {}_3\mathbf{C} = \begin{bmatrix} 0 & 0 & 0 & 1 \\ 0 & 1 & 0 & 0 \\ 0 & 0 & 1 & 0 \end{bmatrix} $$

are, respectively, $\mathbf{f}'\mathbf{f} - \mathbf{e}'\mathbf{e} = 1{,}994.81$, $\mathbf{g}'\mathbf{g} - \mathbf{e}'\mathbf{e} = 10{,}415.17$, and $\mathbf{h}'\mathbf{h} - \mathbf{e}'\mathbf{e} = 16{,}396.33$.

Three hypotheses may then be tested:

1. no incremental effect owing to an interaction between education and experience,

$$ F = \frac{(\mathbf{f}'\mathbf{f} - \mathbf{e}'\mathbf{e})/1}{\mathbf{e}'\mathbf{e}/(N - 4)} = 1.34 $$

2. no incremental effect for work experience,

$$ F = \frac{(\mathbf{g}'\mathbf{g} - \mathbf{f}'\mathbf{f})/1}{\mathbf{e}'\mathbf{e}/(N - 4)} = 5.57 $$

3. no education effect,

$$F = \frac{(\mathbf{h'h} - \mathbf{g'g})/1}{\mathbf{e'e}/(N - 4)} = 3.96$$

While there is no evidence for an interaction between the two measured attributes, both education and experience (for $\alpha = 0.05$) influence self-reported job satisfaction.

9.4 Summary

Chapter 9 has illustrated several common applications of the GLM with maximum likelihood estimation of parameters. First, we discussed testing hypotheses arising from experimental designs. In such analyses, the treatment information is coded into the predictor matrix using design templates and categorical, dummy variable, effect, or othogonal coding. Subsequent tests of hypotheses about the population treatment means are expressed either as *contrasts* on the expectation template or as *constraints* on the elements of $\boldsymbol{\beta}$.

With these procedures and the theoretical developments of Chapter 8, data from the Pilot Response Time Study were analyzed as a four-group randomized groups design, then as a 2×2 factorial experiment randomized groups design. This factorial experiment was balanced and the effects orthogonal, but the use of contrasts on the expectation template allows a direct extension to the unbalanced design with unequal and nonproportional numbers of participants in each of the factorially designed treatments. Our approach clarifies the problems with coding procedures that force orthogonality on unbalanced designs by explicitly representing the tested hypotheses by appropriate coding schemes. An emphasis on hierarchical contrasts allows the testing of independent effects that correspond typically to comparisons of means of theoretical interest.

The GLM was also used to analyze hypotheses relevant to designs incorporating both categorical treatment attributes and measured covariate attributes. Often the researcher's interest in such designs leads to the analysis of covariance. Additional data on the pilots' tested intelligence in the Response Time Study were analyzed to illustrate the analysis of covariance as an application of the GLM.

Applications of the GLM to studies incorporating only measured predictor attributes were illustrated for hypotheses concerning curve fitting, trend analysis, and interactions between measured predictors. The ability to formulate such hypotheses in terms of constraints on the elements of $\boldsymbol{\beta}$ was again explored in these applications.

Together, Chapters 8 and 9 have presented the theory and application of the general linear regression approach to the analysis of hypotheses about univariate criterion. In Part III we turn to the development of procedures for testing hypotheses involving multiple criteria—the multivariate general linear model (MGLM).

Exercises

1. Given the data in Table 1.2 on vocabulary and quantitative and spatial ability scores and on English, mathematics, and physics grade point averages, find the contribution of the three test scores to the prediction of mathematics grade point average when they are entered hierarchically in the order spatial ability (first), quantitative skills (second), and vocabulary (third). Contrast this with the results for forward selection regression, selecting first the predictor with largest zero order correlation with the criterion, then the one with the largest first order partial, and so on.

2. Given the data in Table 1.3 on the effects of a career education class on females' attitudes toward a career, find the regression equation for predicting level of career aspiration from SES using a single predictor (assuming the values 1, 2, or 3). Then find the regression equation for predicting level of aspiration using SES as three 0–1 attributes. Compare the two results with respect to R^2, $\mathbf{f}'\mathbf{f} - \mathbf{e}'\mathbf{e}$, and the value and significance of the F ratio.

3. Given the data in Table 1.8 on the relation between fertility, education, mobility, and income, calculate R^2, the regression equation, and the overall F ratio for predicting fertility from the other three measures. What are the F ratios for the addition of each predictor for your choice of order? Why did you choose the particular order?

4. A researcher is interested in the effects of three different methods of teaching geometry. She randomly assigns 20 students to each method of teaching. At the end of the period of instruction, she obtains geometry achievement scores. She also measures (prior to instruction) each student's spatial aptitude. The data are given in Table 9.5.

- **a.** What is the common slope for the regression of achievement on spatial ability?
- **b.** What is the value of the F ratio for testing the homogeneity of within group regression of achievement on spatial ability?
- **c.** What proportions, overall, of achievement variance are accounted for by spatial ability and the teaching methods?
- **d.** What is the F ratio for the three methods of instruction as a determiner of achievement (ignoring the spatial ability data)?
- **e.** What is the F ratio for the three methods when spatial ability is used as a covariate?
- **f.** What is the F ratio for the three methods when spatial ability is used as a concomitant variable?
- **g.** What different interpretations do you give to the last three results?

Table 9.5 Spatial Ability and Methods of Teaching Geometry

Method 1		Method 2		Method 3	
Spatial ability	Achievement	Spatial ability	Achievement	Spatial ability	Achievement
12	46	12	52	12	64
12	46	12	50	12	62
11	44	11	50	11	58
11	43	11	48	11	56
11	43	11	48	11	55
10	40	10	46	10	55
10	38	10	46	10	56
10	38	10	45	10	54
10	36	10	44	10	54
9	36	9	44	10	55
9	36	9	44	10	53
9	36	9	46	9	50
9	40	9	46	9	52
8	40	8	44	9	50
8	40	8	44	9	50
7	38	7	43	8	50
7	34	7	42	7	48
6	32	7	40	6	44
5	32	5	40	5	44
4	28	5	40	4	40

5. One theory of the relation between the time required for a psychomotor performance and anxiety about that performance suggests a quadratic relation: mean performance is a quadratic function of anxiety. Test this model (against a linear one) for the data in Table 9.6.

6. Thornton investigated the relation of reported marital happiness to the frequency of sexual intercourse and the frequency of arguments, interviewing 28 married couples. For a period of 35 consecutive days couples recorded the frequency of

Table 9.6 Psychomotor Performance and Anxiety

Performance (sec)	Anxiety test score	Performance (sec)	Anxiety text score
3	35	6	70
4	35	7	65
4	40	7	65
5	45	8	55
5	90	9	55
5	95	10	60
6	85	10	60
6	80		

sexual intercourse and of arguments. They also rated marital happiness on a seven point scale from very unhappy (1) to very happy (7). Some of the results may be summarized as follows:

	Mean	S.D.
Marital happiness	5.32	1.66
Sexual intercourse	13.46	7.32
Arguments	6.15	4.19

The correlation of happiness and intercourse was 0.705, of happiness and arguments −0.740, and of intercourse and arguments −0.448.

a. Determine the linear regression equation for predicting marital happiness from frequency of sexual intercourse and arguments.

b. Does adding the frequency of arguments to the prediction equation add significantly to the prediction of marital happiness over what is obtained using intercourse frequency as a single predictor?

c. Does adding intercourse frequency to argument frequency as the sole predictor of happiness increase significantly the predictability of happiness?

Additional Reading

A wide range of applications of the general linear model is touched on by

Draper, N., and Smith, H. (1981). *Applied Regression Analysis*, 2nd ed. New York: Wiley.

The two references following were written specifically for behavioral scientists and at a fairly basic level of statistical sophistication:

Cohen, J., and Cohen, P. (1975). *Applied Multiple Regression/Correlation Analysis for the Behavioral Sciences*. Hillsdale, NJ: Lawrence Erlbaum Associates.

Pedhazur, E. J. (1982), *Multiple Regression in Behavioral Research*, 2nd ed. New York: Holt, Rinehart, Winston.

Problem 6 refers to

Thornton, B. (1977). *Personality and Social Psychology Bulletin* 3:674–676.

We have not discussed the analysis of data with characteristics that violate the assumptions of the linear regression. The following book is a compendium of procedures for diagnosing and transforming data to meet the assumptions of the model. It also discusses the problem of collinearity of the predictors.

Belsley, D. A., Kuh, E., and Welsch, R. E. (1980). *Regression Diagnostics: Identifying Influential Data and Sources of Collinearity*. New York: Wiley.

PART III
BASIC STRUCTURE AND THE MULTIVARIATE CRITERION

In Part III we extend the general linear model to research in which more than one measured attribute make up the criterion observation. This multivariate general linear model (MGLM), with its structural and sampling assumptions, provides the basis for many classical multivariate analyses. The development concentrates on showing typical multivariate analyses to be natural extensions of more familiar univariate techniques.

To facilitate the extension of the GLM, Chapter 10 provides the reader with an introduction to the multivariate normal distribution. We make use of basic structure concepts to develop multivariate generalizations of sampling distributions that were important to the GLM. In Chapter 11 the MGLM is presented. When introducing multivariate general linear hypotheses and their principal statistical tests we show that a number of separately named multivariate techniques are examples of the MGLM. Chapter 12 provides examples of the application of the MGLM. Hotelling's T^2, the multivariate analysis of variance (and covariance), profile analysis, and multivariate linear regression are featured. We also develop and illustrate the linear discriminant function in a form designed to show its close relation to the MGLM.

vectors =
list of scalars

Chapter 10

Basic Structure and
Multivariate Normal Samples

scalar = element of
a field

Introduction

When we jointly observe *several* measured criterion attributes, the criterion observation for each entity is a vector rather than a scalar. A statistical model that would allow us the same freedom in working with multivariate criteria as the GLM allowed us for a univariate criterion would be quite useful. In fact, the GLM can be extended, as we shall see in the following chapters. As the univariate GLM is based on the assumption that the scalar criterion is sampled from *univariate* normal distributions, its multivariate extension assumes that the criterion vector is drawn from some *multivariate normal population*. In this chapter we present those properties of the multivariate normal distribution that are useful to understanding techniques for the analysis of multivariate measured attributes.

We use basic structure to establish the continuity from univariate through bivariate to multivariate normal distributions. Central to this discussion is the concept of the *generalized variance*, defined as the product of the basic diagonal elements of a variance–covariance matrix. Population or parametric versions of the partial and multiple correlation, introduced in Chapter 7 on a descriptive basis, are also developed. Additionally, we generalize the maximum likelihood estimation of parameters of the univariate normal distribution to the multivariate case and, in Section 10.8, develop the multivariate extension of the family of chi-squared distributions, a result that provides a basis for hypothesis testing in the multivariate general linear model.

10.1 Jointly Distributed Observations

In the development of the general linear model, in Chapter 8, a distinction was drawn between the fixed character of the predictor observations and the sampled character of the criterion. The design matrix, containing the predictor data, was often taken to be set by the researcher in accordance with the hypotheses to be tested or the precision of estimation desired. Observations of the criterion attribute were assumed to have resulted from a random and independent sampling of the populations described by the design matrix, and it was this sampling that permitted hypothesis testing and assessment of the accuracy of estimation of the model or population parameters.

The models in Part III consider observations from more than one sampled criterion attribute. Rather than treat each of the attributes separately—repeating, for example, the linear regression analysis for each of several criteria in a study—we treat the entity's observations on the several attributes as occurring together. In the College Prediction Study, for example, the random selection of a student from among the population of those having completed course work in English, mathematics, and physics carries with it the simultaneous observation of a grade point average in each of these three areas of study. The attributes, occurring together, are said to be *jointly distributed* or, alternatively, to define jointly a *multivariate attribute*.

When we observe a multivariate attribute, we obtain not a scalar but a vector. If we observe that multivariate attribute a second time, a second vector results. In order to extend our ideas of hypothesis testing and parameter estimation to the study of the multivariate attribute, we need to understand and be able to work with the *sampling distribution* of vectors as well as of scalars.

The same concepts are used for multivariate as for univariate distributions. Some vectors are more likely to be observed than others, there is an expected value—or, if you wish, a vector of expected values—for the distribution, and we can describe numerically the variability of observed vectors about this expected value.

10.2 Multivariate Normal Distribution

For many multivariate analyses the set of criterion attributes is assumed to follow a *multivariate normal distribution*. That is, we frequently assume that our observations have been sampled from one or more multivariate normal populations, just as, in regression analyses for univariate criteria, we assumed samples from univariate normal distributions. Important properties of the multivariate normal distribution can be seen to stem directly from the probability density function for that distribution, and we begin by discussing that function.

Consider an observation x_i from a univariate normal attribute. It has associated with it the density function

$$p(x_i) = (1/2\pi\sigma^2)^{1/2}\exp\{-\tfrac{1}{2}([x_i - E(x_i)]/\sigma)^2\} \qquad (10.2.1)$$

where $E(x_i)$ is the expected value (or mean) and σ^2 is the variance of the normal distribution from which x_i was sampled. The function $p(x_i)$ is a maximum when $x_i = E(x_i)$ and decreases symmetrically as x_i becomes smaller or larger than its expected value. The height of the normal curve is given by $p(x_i)$. For convenience we write this function again for an observation drawn from a normally distributed attribute with unit variance, $\sigma^2 = 1.0$,

$$p(Z_{i1}) = (1/2\pi)^{1/2}\exp\{-\tfrac{1}{2}[Z_{i1} - E(Z_1)]^2\} \qquad (10.2.2)$$

We have named the attribute from which the observation was drawn Z_1. This attribute has a mean of $E(Z_1)$ and, as we noted, a variance of 1.0.

Now, let $Z_2, Z_3, \ldots, Z_m$ be additional attributes, each normally distributed and with unit variance. An observation from any one of these m normal attributes has a probability density function of the form

$$p(Z_{ij}) = (1/2\pi)^{1/2}\exp\{-\tfrac{1}{2}[Z_{ij} - E(Z_j)]^2\} \qquad (10.2.3)$$

where $j = 1, 2, \ldots, m$.

Assume we have sampled exactly one observation from each of these m attributes and arranged the result as a vector $\mathbf{Z}'_{i.} = (Z_{i1}, Z_{i2}, \ldots, Z_{im})$. Furthermore, let us assume the m observations were made *independently*; the observation from Z_2, for example, did not influence the outcome of sampling from Z_3.

As in Chapter 8, the assumption of independence can be used to produce a new density function which is a density function for the vector observation (of m independent scalar observations). It will be the product of m densities

$$p(\mathbf{Z}_{i.}) = \prod_{j=1}^{m} p(Z_{ij}) \qquad (10.2.4)$$

each of the form of Eq. (10.2.3):

$$p(\mathbf{Z}_{i.}) = (1/2\pi)^{m/2}\exp\left\{-\tfrac{1}{2}\sum_{j=1}^{m}[Z_{ij} - E(Z_j)]^2\right\} \qquad (10.2.5)$$

or, in matrix notation,

$$p(\mathbf{Z}_{i.}) = (2\pi)^{-m/2}\exp\{-\tfrac{1}{2}[\mathbf{Z}_{i.} - \mathbf{E}(\mathbf{Z}_{i.})]'[\mathbf{Z}_{i.} - \mathbf{E}(\mathbf{Z}_{i.})]\} \qquad (10.2.6)$$

where $\mathbf{E}(\mathbf{Z}_{i.})' = (E(Z_1), E(Z_2), \ldots, E(Z_m))$.

Equation (10.2.6) expresses the likelihood of obtaining a particular m-element vector by independently sampling, once each, from m normal distributions all with unit variance but with different means. Although we shall not prove this result, $p(\mathbf{Z}_{i.})$ is a maximum when $\mathbf{Z}_{i.} = \mathbf{E}(\mathbf{Z}_{i.})$, when each of the m

attributes is represented by its mean value. $\mathbf{Z}_{i.}$ is a vector observation from an attribute with mean vector $\mathbf{E}(\mathbf{Z}_{i.})$ and with $(m \times m)$ variance–covariance matrix **I**. (Diagonal elements of the variance–covariance matrix are all 1; off-diagonal elements, because of independent sampling, are zero.)

Equation (10.2.6) is a special form of the multivariate normal density function. The values it assumes describe the sampling distribution of m-element vectors when each of the m-element vectors results from independently sampling from m univariate normal populations all with common, unit variance. Most attributes whose joint occurrence interests us cannot fit these specifications. Reconsidering the College Prediction Study, grade point averages in English, mathematics, and physics, even if separately normally distributed, are not likely to have the same variance or be distributed independently. We need a more general form of multivariate normal density function.

We let **W** be any square basic matrix of order m. Then the m-element vector of transformed observations

$$\mathbf{Y}_{i.} = \mathbf{W}\mathbf{Z}_{i.} \tag{10.2.7}$$

is a *sample* from a population of vectors with means

$$\mathbf{E}(\mathbf{Y}_{i.}) = \mathbf{W}\mathbf{E}(\mathbf{Z}_{i.}) \tag{10.2.8}$$

and with variance–covariance

$$\mathbf{Var}(\mathbf{Y}_{i.}) = \mathbf{W}\,\mathbf{Var}(\mathbf{Z}_{i.})\,\mathbf{W}' = \mathbf{W}\mathbf{I}\mathbf{W}' = \mathbf{W}\mathbf{W}' \tag{10.2.9}$$

Furthermore, as we now show by deriving $p(\mathbf{Y}_{i.})$, this transformed vector observation is also distributed as a (general) multivariate normal observation.

Since **W** is square and basic, there is an *inverse transformation* to Eq. (10.2.7),

$$\mathbf{W}^{-1}\mathbf{Y}_{i.} = \mathbf{W}^{-1}\mathbf{W}\mathbf{Z}_{i.} = \mathbf{Z}_{i.} \tag{10.2.10}$$

that returns the independent observations of unit variance. It is useful to express **W** and matrices derived from **W** in terms of the basic structure of **W**. If **W** has basic structure

$$\mathbf{W} = \mathbf{P}\boldsymbol{\Delta}\mathbf{Q}' \tag{10.2.11}$$

then $\mathbf{W}'$ and $\mathbf{W}\mathbf{W}'$ have basic structures

$$\mathbf{W}' = \mathbf{Q}\boldsymbol{\Delta}\mathbf{P}' \tag{10.2.12}$$

and

$$\mathbf{W}\mathbf{W}' = \mathbf{Var}(\mathbf{Y}_{i.}) = \mathbf{P}\boldsymbol{\Delta}^2\mathbf{P}' \tag{10.2.13}$$

while $\mathbf{W}^{-1}$ and $[\mathbf{Var}(\mathbf{Y}_{i.})]^{-1}$ can be written

$$\mathbf{W}^{-1} = \mathbf{Q}\boldsymbol{\Delta}^{-1}\mathbf{P}' \tag{10.2.14}$$

and

$$\left[\mathbf{Var}(\mathbf{Y}_{i.})\right]^{-1} = \mathbf{P}\boldsymbol{\Delta}^{-2}\mathbf{P}' \tag{10.2.15}$$

Our goal is to describe the density function for $\mathbf{Y}_{i.}$ in terms of the known density function for $\mathbf{Z}_{i.}$ and the properties of the transformation matrix $\mathbf{W}$. We can use these basic structure representations together with a general theorem in probability on the transformation of continuous variables to do that. The probability theorem requires that we first find the basic structure of the $m \times m$ matrix containing the derivatives of the elements of the original vector observation $\mathbf{Z}_{i.}$ taken with respect to the elements of the transformed vector $\mathbf{Y}_{i.}$. The product of the elements of the basic diagonal of this matrix of derivatives is called the *Jacobian* of the inverse of the transformation from $\mathbf{Z}_{i.}$ to $\mathbf{Y}_{i.}$. Then, the probability density function for the transformed observation is the product of the density function for the original observation—written as the inverse transformation of the new observation—and the Jacobian.

In the application of this general result we can use the rules on matrix differentiation given in Chapter 6 to establish that, based on Eq. (10.2.10), $d\mathbf{Z}_{i.}/d\mathbf{Y}_{i.} = \mathbf{W}^{-1}$. Based on Eq. (10.2.14), the Jacobian of the inverse of the transformation is

$$J = \prod_{j=1}^{m} \Delta_j^{-1}$$

and the density function for the transformed vector can be written

$$p(\mathbf{Y}_{i.}) = p\left[\mathbf{Z}_{i.}(= \mathbf{W}^{-1}\mathbf{Y}_{i.})\right] \prod_{j=1}^{m} \Delta_j^{-1} \tag{10.2.16}$$

Here we have taken into account our earlier basic structure finding that the basic diagonal of W^{-1} consists of a rearrangement of the diagonal elements of $\boldsymbol{\Delta}^{-1}$.

On the right of Eq. (10.2.16) we need to substitute $\mathbf{W}^{-1}\mathbf{Y}_{i.}$ for $\mathbf{Z}_{i.}$ from Eq. (10.2.6). Doing this while noting that $\mathbf{E}[\mathbf{W}^{-1}\mathbf{Y}_{i.}] = \mathbf{W}^{-1}\mathbf{E}(\mathbf{Y}_{i.})$ gives

$$\begin{aligned} p(\mathbf{Z}_{i.}) &= (2\pi)^{-m/2} \exp\{-\tfrac{1}{2}\left[\mathbf{W}^{-1}\mathbf{Y}_{i.} - \mathbf{W}^{-1}\mathbf{E}(\mathbf{Y}_{i.})\right]' \\ &\quad \times \left[\mathbf{W}^{-1}\mathbf{Y}_{i.} - \mathbf{W}^{-1}\mathbf{E}(\mathbf{Y}_{i.})\right]\} \\ &= (2\pi)^{-m/2} \exp\{-\tfrac{1}{2}\left[\mathbf{Y}_{i.} - \mathbf{E}(\mathbf{Y}_{i.})\right]'\mathbf{W}^{-1\prime}\mathbf{W}^{-1}\left[\mathbf{Y}_{i.} - \mathbf{E}(\mathbf{Y}_{i.})\right]\} \end{aligned} \tag{10.2.17}$$

As $\mathbf{W}$ (and $\mathbf{W}'$) are square basic matrices, we can use Eq. (10.2.15) to rewrite this

$$p(\mathbf{Z}_{i.}) = (2\pi)^{-m/2} \exp\{-\tfrac{1}{2}\left[\mathbf{Y}_{i.} - \mathbf{E}(\mathbf{Y}_{i.})\right]'\left[\mathbf{Var}(\mathbf{Y}_{i.})\right]^{-1}\left[\mathbf{Y}_{i.} - \mathbf{E}(\mathbf{Y}_{i.})\right]\} \tag{10.2.18}$$

We substitute this result into Eq. (10.2.16). The probability density for $\mathbf{Y}_{i.}$ is

$$p(\mathbf{Y}_{i.}) = (2\pi)^{-m/2}\left(\prod_{j=1}^{m}\Delta_j^2\right)^{-1/2}$$

$$\times \exp\{-\tfrac{1}{2}[\mathbf{Y}_{i.} - \mathbf{E}(\mathbf{Y}_{i.})]'[\mathbf{Var}(\mathbf{Y}_{i.})]^{-1}[\mathbf{Y}_{i.} - \mathbf{E}(\mathbf{Y}_{i.})]\} \quad (10.2.19)$$

This is the usual form for the density function for a multivariate normal vector observation: $\mathbf{Y}_{i.}$ is a multivariate normal observation with mean (vector) $\mathbf{E}(\mathbf{Y}_{i.})$ and variance–covariance matrix $\mathbf{Var}(\mathbf{Y}_{i.})$. By Eq. (10.2.13) the variance–covariance matrix has basic structure $\mathbf{Var}(\mathbf{Y}_{i.}) = \mathbf{P}\mathbf{\Delta}^2\mathbf{P}'$, so $\prod_{j=1}^{m}\Delta_m^2$ is the product of the basic diagonal elements of $\mathbf{Var}(\mathbf{Y}_{i.})$.

One way of interpreting this result is to say that any m-variate normal observation (an observation from a multivariate normal distribution) can be considered to be the result of applying m linear transformations to a set of m independent (univariate and unit variance) normal observations. This also has a corollary. If $\mathbf{Y}_{i.}$ is a vector observation from a multivariate normal population whose variance–covariance matrix is of full rank and has basic structure $\mathbf{Var}(\mathbf{Y}_{i.}) = \mathbf{P}\mathbf{\Delta}^2\mathbf{P}'$, then the transformed vector observation $\mathbf{Z}_{i.} = \mathbf{\Delta}^{-1}\mathbf{P}'\mathbf{Y}_{i.}$ is a multivariate normal observation with a variance–covariance matrix $\mathbf{Var}(\mathbf{Z}_{i.}) = \mathbf{I}$. That is, $\mathbf{Y}_{i.}$ can be transformed into a set of m independent univariate normal observations, all with unit variance. Later we shall use this ability to decompose the multivariate normal into independent normals.

For now, though, we note that Eq. (10.2.19) enjoys twin roles:

1. a *probability density function*, taking on different values as the value of the observation vector $\mathbf{Y}_{i.}$ changes but the parameters, $\mathbf{E}(\mathbf{Y}_{i.})$ and $\mathbf{Var}(\mathbf{Y}_{i.})$ remain fixed; and
2. a *likelihood function*, taking on different values as the value of the observation vector is fixed and $\mathbf{E}(\mathbf{Y}_{i.})$ and $\mathbf{Var}(\mathbf{Y}_{i.})$ are allowed to vary.

Because of our concern with problems of estimation and hypothesis testing—in the absence of knowledge about population parameters—the likelihood role predominates and usually we write the left-hand side of (10.2.19) as $L(\mathbf{Y}_{i.})$ rather than as $p(\mathbf{Y}_{i.})$.

10.3 Generalized Variance

The likelihood function for the multivariate normal attribute developed in Section 10.2 can be used to show the continuity of the univariate normal distribution (the workhorse of much elementary inferential statistics) with the bivariate normal (the basis of two-attribute normal correlation theory and applications) and with the multivariate normal, source of the bulk of multivariate analysis. We can rewrite the univariate normal likelihood from

(10.2.1) in the form

$$L(\mathbf{y}_i) = (1/\sqrt{2\pi})(1/\sqrt{\sigma^2})\exp\{-\tfrac{1}{2}[\mathbf{y}_i - \mathbf{E}(\mathbf{y}_i)]'(\sigma^2)^{-1}[\mathbf{y}_i - \mathbf{E}(\mathbf{y}_i)]\} \tag{10.3.1}$$

to compare it with the multivariate likelihood [see Eq. (10.2.19)]:

$$L(\mathbf{Y}_{i.}) = (1/\sqrt{2\pi})^m\left(1\Big/\sqrt{\prod_j \Delta_j^2}\right)$$

$$\times \exp\{-\tfrac{1}{2}[\mathbf{Y}_{i.} - \mathbf{E}(\mathbf{Y}_{i.})]'[\mathbf{Var}(\mathbf{Y}_{i.})]^{-1}[\mathbf{Y}_{i.} - \mathbf{E}(\mathbf{Y}_{i.})]\}$$

The *univariate normal likelihood* is just that for the multivariate normal observation with $m = 1$. Parts of that conclusion are clear: $(1/\sqrt{2\pi})^m$ becomes $1/\sqrt{2\pi}$ and, since $\mathrm{Var}(Y_{i.}) = \sigma^2$ for scalar $Y_{i.}$, $[\mathbf{Var}(\mathbf{Y}_{i.})]^{-1}$ in the exponent can be replaced by $(\sigma^2)^{-1}$. What, however, is the univariate equivalent to $\prod_j \Delta_j^2$? Recall that $\boldsymbol{\Delta}^2$ is the basic diagonal of $\mathbf{Var}(\mathbf{Y}_{i.})$ [see Eq. (10.2.15)]: $\mathbf{Var}(\mathbf{Y}_{i.}) = \mathbf{P}\boldsymbol{\Delta}^2\mathbf{P}'$.

The positive-valued scalar σ^2 is a 1×1 square basic matrix, and we saw in Chapter 4 that its basic structure is $\sigma^2 = 1\sigma^2 1$, where the left and right orthonormals are scalar ones and the basic diagonal is the positive scalar itself. Thus, for $\mathbf{Var}(\mathbf{Y}_i) = \sigma^2$, we have $\prod_j \Delta_j^2 = \sigma^2$. The univariate normal distribution likelihood in (10.3.1) is indeed a special case of the multivariate likelihood.

The scalar quantity that is the product of the basic diagonal elements of a variance–covariance matrix is regarded in multivariate analysis as a *generalized variance measure*, a scalar summary of the variances in a set of attributes and their relatedness through their covariances. We have just seen how $\prod_j \Delta_j^2 = \sigma^2$ when $m = 1$. Further, we also deduce that if $\mathbf{Var}(\mathbf{Y}_{i.}) = \mathbf{D}_{\sigma^2}$, a diagonal matrix (if all of the covariances are zero), then $\prod_j \Delta_j^2 = \prod_j \sigma_j^2$, where σ_j^2 is the variance of the jth attribute. The generalized variance measure for a set of independently distributed attributes is the product of their variances. (We have used here the property that the basic diagonal of a diagonal matrix with positive elements is obtained by rearranging the diagonal elements in descending order of magnitude. The product of the diagonal elements is the same for the two matrices.)

If we consider a *bivariate* attribute, consisting of two observations, the effect upon the generalized variance measure of a lack of independence can be understood. For the bivariate criterion attribute the variance–covariance matrix can be written

$$\mathbf{Var}(\mathbf{Y}_{i.}) = \begin{bmatrix} \sigma_1^2 & \sigma_{12} \\ \sigma_{12} & \sigma_2^2 \end{bmatrix}$$

where σ_{12} is the covariance between the two attributes. In Chapter 4 we proved that for a product moment matrix such as $\mathbf{Var}(\mathbf{Y}_{i.})$, the trace of the matrix equals the sum of the basic diagonal elements:

$$\sum_{j=1}^{m} \Delta_j^2 = \mathrm{tr}[\mathbf{Var}(\mathbf{Y}_{i.})]$$

or, in the case of the bivariate covariance matrix,

$$\Delta_1^2 + \Delta_2^2 = \sigma_1^2 + \sigma_2^2 \tag{10.3.2}$$

The inverse of the 2×2 $\mathbf{Var}(\mathbf{Y}_{i.})$ can be written directly:

$$[\mathbf{Var}(\mathbf{Y}_{i.})]^{-1} = \frac{1}{\sigma_1^2\sigma_2^2 - \sigma_{12}^2}\begin{bmatrix} \sigma_2^2 & -\sigma_{12} \\ -\sigma_{12} & \sigma_1^2 \end{bmatrix}$$

Because the basic diagonal of $[\mathbf{Var}(\mathbf{Y}_{i.})]^{-1}$ has as elements the reciprocals of the elements of the basic diagonal of $\mathbf{Var}(\mathbf{Y}_{i.})$, we can write $\mathrm{tr}\{[\mathbf{Var}(\mathbf{Y}_{i.})]^{-1}\} = \Sigma_{j=1}^{m}(1/\Delta_j^2)$ or, for the 2×2 case,

$$1/\Delta_1^2 + 1/\Delta_2^2 = (\sigma_1^2 + \sigma_2^2)/(\sigma_1^2\sigma_2^2 - \sigma_{12}^2) \tag{10.3.3}$$

The left side of this equation can be rewritten $(\Delta_2^2 + \Delta_1^2)/\Delta_1^2\Delta_2^2$, permitting us to replace (10.3.3) with

$$(\Delta_2^2 + \Delta_1^2)/\Delta_1^2\Delta_2^2 = (\sigma_1^2 + \sigma_2^2)/(\sigma_1^2\sigma_2^2 - \sigma_{12}^2) \tag{10.3.4}$$

By Eq. (10.3.2), $\Delta_2^2 + \Delta_1^2$ may be replaced with $\sigma_1^2 + \sigma_2^2$ in Eq. (10.3.4), resulting in

$$\Delta_1^2\Delta_2^2 = \sigma_1^2\sigma_2^2 - \sigma_{12}^2 \tag{10.3.5}$$

Thus, the generalized variance for two attributes is the product of their variances if the two are independently distributed and the product of their variances less the square of their covariance otherwise. This principle generalizes to the multivariate attribute; linear relations among the m attributes reduce the generalized variance. Put another way, the generalized variance measure takes into account the variance shared among the attributes.

10.4 Population Correlation and Bivariate Normal Distributions

In earlier chapters the observed or sample linear or product moment correlation was defined. Computationally, $r_{\mathbf{xy}}$ is the ratio of the observed covariance between $\mathbf{x}$ and $\mathbf{y}$ to the product of the observed standard deviations of the two attributes. In considering certain aspects of multivariate attributes statisticians have found it convenient to define, in parallel, a *population correlation*. We also shall do that here, but in the context of the concept of *statistical expectation*.

The variances and covariance for the bivariate attribute of Section 10.3 can be expressed as the expectations

$$\sigma_j^2 = E\{[y_j - E(y_j)]^2\}, \qquad j = 1,2 \tag{10.4.1}$$

and

$$\sigma_{12} = E\{[y_1 - E(y_1)][y_2 - E(y_2)]\} \tag{10.4.2}$$

while we define the population correlation for the two attributes as the expectation

$$\rho_{12} = E\{[y_1 - E(y_1)]/\sigma_1\}\{[y_2 - E(y_2)]/\sigma_2\}$$

each deviation being divided by the population standard deviation (positive square root of the variance). As these standard deviations are constants in the expectation, the correlation may be expressed as

$$\rho_{12} = (1/\sigma_1\sigma_2)E\{[y_1 - E(y_1)][y_2 - E(y_2)]\}$$

or, from Eq. (10.4.2), as $\rho_{12} = \sigma_{12}/\sigma_1\sigma_2$, which can be rearranged in the form

$$\sigma_{12} = \rho_{12}\sigma_1\sigma_2 \tag{10.4.3}$$

With the population covariance expressed as this triple product, the variance–covariance matrix for a multivariate attribute may be easier to understand. First, we collect scalar relations of the form of Eqs. (10.4.1) and (10.4.2) to write the variance–covariance matrix as the expected value of a product moment matrix:

$$\mathbf{Var}(\mathbf{Y}_{i.}) = \mathbf{E}\{[\mathbf{Y}_{i.} - \mathbf{E}(\mathbf{Y}_{i.})][\mathbf{Y}_{i.} - \mathbf{E}(\mathbf{Y}_{i.})]'\} \tag{10.4.4}$$

That is, this matrix has as its jth diagonal element

$$[\mathrm{Var}(\mathbf{Y}_{i.})]_{jj} = E\{[Y_{ij} - E(Y_{ij})]^2\} = \sigma_j^2$$

and as the element at the intersection of the jth row and kth column

$$[\mathrm{Var}(\mathbf{Y}_{i.})]_{jk} = E\{[Y_{ij} - E(Y_{ij})][Y_{ik} - E(Y_{ik})] = \sigma_{jk} = \rho_{jk}\sigma_j\sigma_k$$

These scalar equations, in turn, suggest that we can write the variance–covariance matrix as the triple product

$$\mathbf{Var}(\mathbf{Y}_{i.}) = \mathbf{D}_\sigma \boldsymbol{\rho} \mathbf{D}_\sigma \tag{10.4.5}$$

where $\mathbf{D}_\sigma$ is a diagonal matrix of population standard deviations and $\boldsymbol{\rho}$ is a symmetric matrix with population correlations in the off-diagonal elements and 1s along the diagonal. For a bivariate observation this gives

$$\mathbf{Var}(\mathbf{Y}_{i.}) = \begin{bmatrix} \sigma_1^2 & \rho_{12}\sigma_1\sigma_2 \\ \rho_{12}\sigma_1\sigma_2 & \sigma_2^2 \end{bmatrix} \tag{10.4.6}$$

while, from Eq. (10.3.5), the bivariate generalized variance is

$$\Delta_1^2\Delta_2^2 = \sigma_1^2\sigma_2^2(1 - \rho_{12}^2) \tag{10.4.7}$$

From Eq. (10.4.5) the inverse of the variance–covariance matrix is $[\mathbf{Var}(\mathbf{Y}_{i.})]^{-1} = \mathbf{D}_\sigma^{-1}\boldsymbol{\rho}^{-1}\mathbf{D}_\sigma^{-1}$, where for the bivariate case

$$\rho^{-1} = \frac{1}{1 - \rho_{12}^2}\begin{bmatrix} 1 & -\rho_{12} \\ -\rho_{12} & 1 \end{bmatrix}$$

Thus, the inverse of the bivariate variance–covariance matrix is

$$[\mathbf{Var}(\mathbf{Y}_{i.})]^{-1} = \frac{1}{1 - \rho_{12}^2}\begin{bmatrix} 1/\sigma_1^2 & -\rho_{12}/\sigma_1\sigma_2 \\ -\rho_{12}/\sigma_1\sigma_2 & 1/\sigma_2^2 \end{bmatrix}$$

Based on this result, the form

$$K = [\mathbf{Y}_{i.} - \mathbf{E}(\mathbf{Y}_{i.})]'[\mathbf{Var}(\mathbf{Y}_{i.})]^{-1}[\mathbf{Y}_{i.} - \mathbf{E}(\mathbf{Y}_{i.})]$$

which appears in the exponent of the multivariate normal likelihood function, Eq. (10.2.19), becomes in the bivariate case

$$\begin{aligned} K = {} & [1/(1 - \rho_{12}^2)][\{[Y_{i1} - E(Y_{i1})]/\sigma_1\}^2 + \{[Y_{i2} - E(Y_{i2})]/\sigma_2\}^2 \\ & -2\rho_{12}\{[Y_{i1} - E(Y_{i1})]/\sigma_1\}\{[Y_{i2} - E(Y_{i2})]/\sigma_2\}] \end{aligned}$$

or

$$K = [1/(1 - \rho_{12}^2)](Z_{i1}^2 + Z_{i2}^2 - 2\rho_{12}Z_{i1}Z_{i2}) \tag{10.4.8}$$

where $Z_{ij} = [Y_{ij} - E(Y_{ij})]/\sigma_j$.

Substituting from Eqs. (10.4.7) and (10.4.8) into the likelihood function for the multivariate normal, we have the bivariate normal likelihood as a special case:

$$\begin{aligned} L(Y_{i1}, Y_{i2}) &= (2\pi)^{-1}[\sigma_1^2\sigma_2^2(1 - \rho_{12}^2)]^{-1/2}\exp(-\tfrac{1}{2}K) \\ &= [2\pi\sigma_1\sigma_2(1 - \rho_{12}^2)^{1/2}]^{-1} \\ &\quad \times \exp\{-\tfrac{1}{2}[1/(1 - \rho_{12}^2)](Z_{i1}^2 + Z_{i2}^2 - 2\rho_{12}Z_{i1}Z_{i2})\} \end{aligned} \tag{10.4.9}$$

10.4.1 Marginal and Conditional Distributions

What does it mean for two attributes to be *jointly distributed* as a *bivariate normal* observation? To answer this, we first review some properties of probability density functions. Figure 10.1 depicts a bivariate normal probability *surface*. The function, $L(Y_{i1}, Y_{i2})$ gives the height of this surface above the (Y_{i1}, Y_{i2}) plane. (Here we treat the function $L(\mathbf{Y}_{i.})$ as a density function. That is, ρ_{12}, σ_1^2, σ_2^2, $E(Y_{i1})$, and $E(Y_{i2})$ are fixed and we are interested in

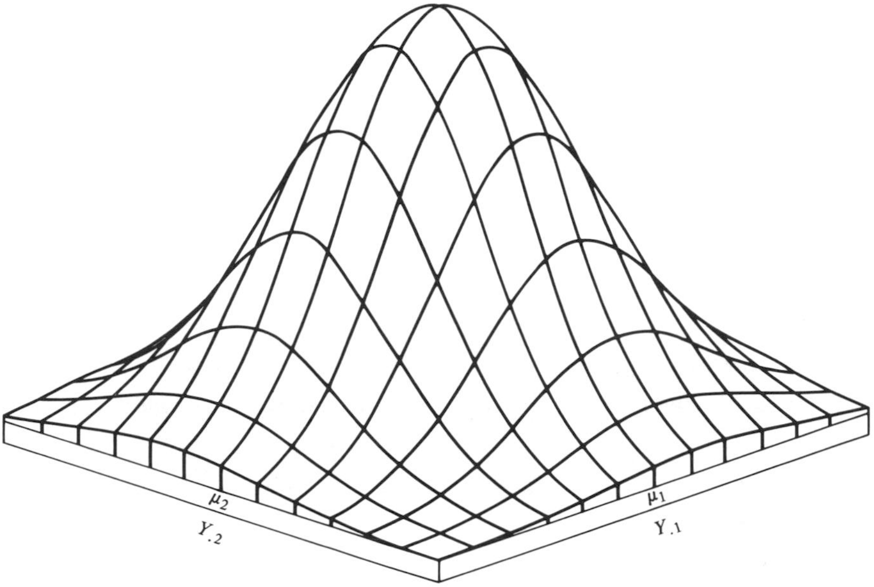

Figure 10.1 A bivariate normal probability surface.

how $L(\mathbf{Y}_{i.})$ changes as $Y_{i.}$ changes.) Some pairs of values, some joint occurrences of $Y_{.1}$ and $Y_{.2}$ are more likely than are others.

The surface above the plane is a three-dimensional analog of the bell-shaped curve we associate with the univariate normal distribution. Two points of comparison are easily seen. First, the density function, whether univariate or bivariate, is always positive; the univariate normal curve and the bivariate normal "tent" is always some distance above the base line or basal plane. Second, just as the normal curve reaches its peak when the univariate observation takes its mean or expected value, so the bivariate normal surface is farthest above the plane when both $Y_{i1} = E(Y_{i1})$ and $Y_{i2} = E(Y_{i2})$.

The numerical values taken on by a univariate density function are such that we interpret the *area* under the function or under a segment of that function as a *probability*. We can illustrate this with the normal probability function Pr illustrated in Figure 10.2. The probability that the normal random variable will take a value between the points a and b is the area enclosed between the normal curve, the base line, and the verticals erected at a and b. Because it is a probability, this area must be a value bounded between 0 and 1. Indeed, since the normal variable can take, in principle, any value along the whole real line between negative infinity and infinity, the total area under the curve is 1.

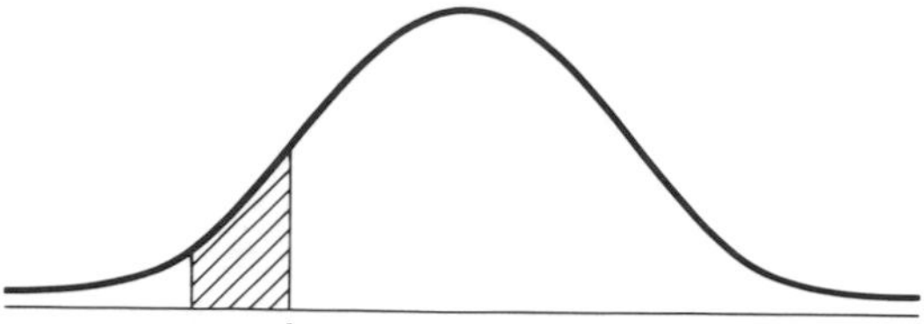

Figure 10.2 Probabilities under the normal curve.

Because the normal density function is a continuous function over the set of real numbers, areas under the curve or probabilities are found by evaluating a definite integral. Three examples from the illustration of Figure 10.2 are

$$\Pr(y_i \leqslant a) = \int_{-\infty}^{a} L(y_i)\, dy$$

$$\Pr(a < y_i \leqslant b) = \int_{a}^{b} L(y_i)\, dy$$

$$\Pr(b < y_i) = \int_{b}^{\infty} L(y_i)\, dy$$

For those not familiar with the integral calculus, these three equations may be read as saying the following:

1. The probability that the variable y will take a value no greater than a is given by the area under the density function to the left of the vertical erected at a.
2. The probability that the variable y will take a value larger than a but no larger than b equals the area under the curve between a and b.
3. The probability that the variable y will take a value exceeding b is equivalent to the area under the density function to the right of b.

Since y_i must be a value in one of the three nonoverlapping intervals, the sums of these three probabilities, the total area under the curve, is 1.

Just as there is one unit of area under the univariate density function, so there is one unit of volume under the bivariate density. Further, this volume also can be divided to describe the probability that the bivariate observation (Y_{i1}, Y_{i2}) will take on certain sets of values. Figure 10.3 illustrates several of these probabilities.

The volume enclosed by the bivariate density "roof" above the rectangle described by lines drawn perpendicular to the points c_1 and c_2 along the $Y_{.1}$ dimension and to the points k_1 and k_2 along the $Y_{.2}$ dimension equals the probability that Y_{i1} will be a value between c_1 and c_2 and that Y_{i2} will simultaneously be a value between k_1 and k_2. Similarly, the volume above

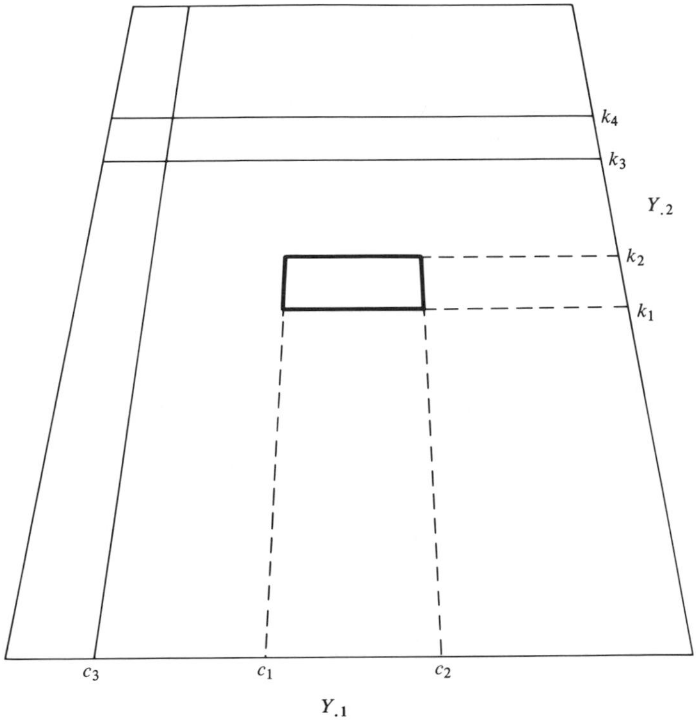

Figure 10.3 The bivariate basal plane.

the region of the basal plane bounded by the lines intersecting the $Y_{.2}$ axis at k_3 and k_4 is the probability that Y_{i2} will be a value between k_3 and k_4 (and that Y_{i1} will be a value between $-\infty$ and ∞). Finally, the volume contained beneath the bivariate density "tent" to the left of the wall erected along the line that intersects the $Y_{.1}$ axis at c_3 tells us the probability that Y_{i1} will assume a value no greater than c_3.

These bivariate probabilities are also formally defined by a set of definite integrals, now double integrals:

$$\Pr(c_1 < Y_{i1} \leqslant c_2, k_1 \leqslant Y_{i2} < k_2) = \int_{k_1}^{k_2}\int_{c_1}^{c_2} L(Y_{i1}, Y_{i2})\, dY_{.1}\, dY_{.2}$$

$$\Pr(k_3 < Y_{i2} \leqslant k_4) = \int_{k_3}^{k_4}\int_{-\infty}^{\infty} L(Y_{i1}, Y_{i2})\, dY_{.1}\, dY_{.2}$$

$$\Pr(Y_{i1} \leqslant c_3) = \int_{-\infty}^{\infty}\int_{-\infty}^{c_3} L(Y_{i1}, Y_{i2})\, dY_{.1}\, dY_{.2}$$

These last two probability statements describe how $Y_{.1}$ and $Y_{.2}$ are each distributed if jointly they are bivariate normal. Statisticians refer to these distributions as marginal distributions, as if in Figure 10.1 or 10.3 we collapsed our three-dimensional representation into a two-dimensional one, allowing the populations to pile up along the $Y_{.1}$ or $Y_{.2}$ dimensions or margins. In fact, we find the marginal densities through calculus, by integrating over the full range of the other variable, the one we are collapsing. The two marginal densities are defined by

$$L(Y_{i1}) = \int_{-\infty}^{\infty} L(Y_{i1}, Y_{i2})\, dY_{.2} \tag{10.4.10}$$

and

$$L(Y_{i2}) = \int_{-\infty}^{\infty} L(Y_{i1}, Y_{i2})\, dY_{.1} \tag{10.4.11}$$

A little algebraic manipulation shows these marginal densities. We first rewrite (10.4.8)

$$\begin{aligned}(1 - \rho_{12}^2)K = {} & [(Y_{i1} - \mu_1)/\sigma_1]^2 + [(Y_{i2} - \mu_2)/\sigma_2]^2 \\ & -2\rho_{12}[(Y_{i1} - \mu_1)/\sigma_1][(Y_{i2} - \mu_2)/\sigma_2]\end{aligned} \tag{10.4.12}$$

where for notational simplicity

$$E(Y_{.1}) = \mu_1 \qquad \text{and} \qquad E(Y_{.2}) = \mu_2$$

The right side of (10.4.12) can be rewritten

$$\begin{aligned}(1 - \rho_{12}^2)K = {} & [(Y_{i2} - \mu_2)/\sigma_2 - \rho_{12}(Y_{i1} - \mu_1)/\sigma_1]^2 \\ & +(1 - \rho_{12}^2)[(Y_{i1} - \mu_1)/\sigma_1]^2\end{aligned} \tag{10.4.13}$$

If we now define

$$M = \mu_2 + \sigma_2\rho_{12}(Y_{i1} - \mu_1)/\sigma_1 \tag{10.4.14}$$

and

$$S = \sigma_2(1 - \rho_{12}^2)^{1/2} \tag{10.4.15}$$

then Eq. (10.4.13) gives rise to

$$K = [(Y_{i2} - M)/S]^2 + [(Y_{i1} - \mu_1)/\sigma_1]^2 \tag{10.4.16}$$

Our interest in getting K in this form is that we know from (10.4.8) and (10.4.9) that the bivariate normal density can be expressed

$$L(Y_{i1}, Y_{i2}) = \left[2\pi\sigma_1\sigma_2(1 - \rho_{12}^2)^{1/2}\right]^{-1} e^{-(1/2)K}$$

and if we substitute the S and K of (10.4.15) and (10.4.16) into this, we can write the bivariate density as

$$\begin{aligned}L(Y_{i1}, Y_{i2}) = {} & \left[(\sigma_1\sqrt{2\pi})^{-1}\exp\{-\tfrac{1}{2}[(Y_{i1} - \mu_1)/\sigma_1]^2\}\right] \\ & \times (S\sqrt{2\pi})^{-1}\exp\{-\tfrac{1}{2}[(Y_{i2} - M)/S]^2\}\end{aligned} \tag{10.4.17}$$

Then the marginal density of Y_{i1}, from (10.4.10) and (10.4.17), is

$$L(Y_{i1}) = \left(\sigma_1\sqrt{2\pi}\right)^{-1}\exp\{-\tfrac{1}{2}[(Y_{i1}-\mu_1)/\sigma_1]^2\}$$
$$\times\int_{-\infty}^{\infty}\left(S\sqrt{2\pi}\right)^{-1}\exp\{-\tfrac{1}{2}[(Y_{i2}-M)/S]^2\}\,dY_{.2} \tag{10.4.18}$$

where we have factored outside the integral the term that does not rely on the variable of integration, $Y_{.2}$.

Equation (10.4.18) calls for integrating over the entire real number line from $-\infty$ to $+\infty$ a function that is in the form of the univariate normal random variable with mean M and variance S^2, e.g., Eq. (10.2.1). We now know that this integral must equal 1, the probability that Y_{i2} takes some value between $-\infty$ and $+\infty$. As a result, we can write the marginal likelihood

$$L(Y_{i1}) = \left(\sigma_1\sqrt{2\pi}\right)^{-1}\exp\{-\tfrac{1}{2}[(Y_{i1}-\mu_1)/\sigma_1]^2\} \tag{10.4.19}$$

The importance of the development of (10.4.12)–(10.4.19) is that it establishes that when (Y_{i1}, Y_{i2}) is a bivariate normal observation the marginal distribution of $Y_{.1}$ is that of a univariate normal. Essentially the same argument could be repeated to show that the marginal distribution of $Y_{.2}$ is also normal:

$$L(Y_{i2}) = \left(\sigma_2\sqrt{2\pi}\right)^{-1}\exp\{-\tfrac{1}{2}[(Y_{i2}-\mu_2)/\sigma_2]^2\} \tag{10.4.20}$$

If two variables are jointly bivariate normal, then *each* variable is normally distributed.

A second consequence of the bivariate normality of (Y_{i1}, Y_{i2}) is the form of the conditional distributions of $Y_{.1}$ and $Y_{.2}$. Returning to the bivariate distribution pictured in Figure 10.1, we might ask not how $Y_{.1}$ is distributed overall (the marginal distribution) but when $Y_{.2}$ is held constant at, say, k_3. That is, what is the distribution of Y_{i1} given that $Y_{i2} = k_3$? This is termed the conditional distribution of $Y_{.1}$—conditional on Y_{i2} assuming a particular value: $L(Y_{i1} | Y_{i2} = k_3)$. $Y_{.2}$ also has a distribution conditional on Y_{i1}: $L(Y_{i2} | Y_{i1} = c_3)$, for example.

The conditional density function is defined in general in terms of the bivariate density and one of the marginal distributions:

$$L(Y_{i1}|Y_{i2}) = L(Y_{i1}, Y_{i2})/L(Y_{i2}), \qquad L(Y_{i2}|Y_{il}) = L(Y_{i1}, Y_{i2})/L(Y_{i1}) \tag{10.4.21}$$

The reader may notice that conditional densities are defined in essentially the same fashion as conditional probabilities: $P(A|B) = P(AB)/P(B)$; in selecting a card at random from a standard bridge deck, the probability that a card will be an ace given that we know it to be a spade $[P(A|B)]$ can be obtained by dividing the probability of the joint occurrence of an ace and a spade $[P(AB) = \frac{1}{52}]$ by the probability of a spade $[P(B) = \frac{13}{52}]$.

If we take the joint distribution as given in (10.4.17) and divide it by the marginal density of Y_{i1} as given in (10.4.19), we see that the conditional density of Y_{i2}, given from (10.4.21), is

$$L(Y_{i2}|Y_{i1}) = \left(\sigma_{2.1}\sqrt{2\pi}\right)^{-1}\exp\left\{-\tfrac{1}{2}\left[(Y_{i2} - \mu_{2.1})/\sigma_{2.1}\right]^2\right\} \quad (10.4.22)$$

where we have substituted new names for the M and S of (10.4.14) and (10.4.15):

$$\mu_{2.1} = \mu_2 + \sigma_2\rho_{12}(Y_{i1} - \mu_1)/\sigma_1 \quad (10.4.23)$$

and

$$\sigma_{2.1} = \sigma_2\left[\left(1 - \rho_{12}^2\right)\right]^{1/2} \quad (10.4.24)$$

From Eq. (10.4.22), $L(Y_{i2}|Y_{i1})$ is also a normal likelihood, here with a mean $\mu_{2.1}$ and a variance $\sigma_{2.1}^2$. These latter constants are referred to as the conditional mean and variance of $Y_{.2}$ (conditional on Y_{i1}). When introducing the linear regression model in Chapter 8 we encountered conditional distributions: $E(y_i)$ was conditional on the vector $\mathbf{X}_{i.}$. Equations (10.4.22)–(10.4.24) indicate that when Y_{i1} and Y_{i2} are jointly bivariate normal, the regression of $Y_{.2}$ on $Y_{.1}$ is linear: $(Y_{i2}|Y_{i1})$ is normally distributed with a mean that depends, linearly, on the value Y_{i1} and with a variance that is constant.

Because the form of the bivariate normal density is symmetric in Y_{i1} and Y_{i2}, we can conclude that the conditional distribution of $Y_{.1}$ (conditional on Y_{i2}) is also normal:

$$L(Y_{i1}|Y_{i2}) = \left(\sigma_{1.2}\sqrt{2\pi}\right)^{-1}\exp\left\{-\tfrac{1}{2}\left[(Y_{i1} - \mu_{1.2})/\sigma_{1.2}\right]^2\right\} \quad (10.4.25)$$

where

$$\mu_{1.2} = \mu_1 + \sigma_1\rho_{12}(Y_{i2} - \mu_2)/\sigma_2 \quad (10.4.26)$$

and

$$\sigma_{1.2} = \sigma_1\left(1 - \rho_{12}^2\right)^{1/2} \quad (10.4.27)$$

Note that the regression of $Y_{.1}$ on $Y_{.2}$ is also linear.

In summary, if two variables are jointly bivariate normal, it is also true that each is normally distributed, each has a conditional distribution (relative to the other) that is normal, and—closely linked with this latter property—the regression of each on the other is linear. While these results all stem from bivariate normality, the opposite is not true: Marginal and conditional normality are not sufficient to ensure bivariate normality.

10.5 Marginal and Conditional Distributions Based on the Multivariate Normal

In Section 10.4 we introduced the notions of marginal and conditional distributions when two attributes are jointly distributed. These ideas generalize to those instances where three or more attributes are considered to be jointly distributed. In this section we present some results, again for the special case where the m attributes are multivariate normal in their joint distribution. The derivations of these results are more involved when we move from the bivariate case, and we shall not offer proofs for all of the generalizations.

10.5.1 Marginal Distributions

Suppose $\mathbf{Y}_{i.}' = (Y_{i1}, Y_{i2}, \ldots, Y_{im})$ is an observation from a multivariate normal distribution with mean vector $\boldsymbol{\mu}' = (\mu_1, \mu_2, \ldots, \mu_m)$ and variance–covariance matrix $\boldsymbol{\Sigma}$, where $\Sigma_{ii} = \sigma_i^2$ and $\Sigma_{jk} = \text{Cov}(\mathbf{Y}_{.j}, \mathbf{Y}_{.k})$.

1. $Y_{i1}, Y_{i2}, \ldots, Y_{im}$ are each an observation from a univariate normal distribution with mean and variance given by the appropriate elements of μ and the diagonals of $\boldsymbol{\Sigma}$.
2. Any subset of k elements from the vector $\mathbf{Y}_{i.}$ $(1 \leqslant k \leqslant m)$ is an observation from a multivariate normal distribution. This k-variate normal has mean vector and variance–covariance matrix, respectively, of the appropriate elements of μ and the appropriate symmetric partition of $\boldsymbol{\Sigma}$.

This second property ensures that every pair of scores from the multivariate normal distribution is a bivariate normal observation. $\mathbf{Y}_{.i}$ has a normal distribution not only overall but at each partitioned level of $\mathbf{Y}_{.i}$.

10.5.2 Conditional Distributions

The notion of a conditional distribution of one univariate attribute (relative to another) extends to *partitions* of the multivariate attribute. For example, we may consider the multivariate observation $\mathbf{Y}_{i.}$ to be partitioned

$$\mathbf{Y}_{i.}' = (\mathbf{Y}_{iA}' \quad \mathbf{Y}_{iB}') \tag{10.5.1}$$

where $\mathbf{Y}_{iA}$ contains k scores and $\mathbf{Y}_{iB}$ the remaining $m - k$. It is convenient to think of the elements of each partition as grouped together, although this is not necessary. The partitions need not include all of the m variates (though frequently they do), but they are nonoverlapping.

A partition of frequent interest is that between k criteria (dependent variables) and $m - k$ predictors. In the College Prediction Study, for example, we are interested in the joint distribution of the three grade point averages (in English, mathematics, and physics) conditional on the values of the scores on the three tests (vocabulary, mathematics achievement, and spatial ability). Similarly, we could be interested in the conditional distribution of spatial ability scores given vocabulary and mathematics achievement test performance.

If $\mathbf{Y}_{i.}'$ is an observation from a multivariate normal attribute, then $(Y_{ij} | Y_{i1}, Y_{i2}, \ldots, Y_{i(j-1)}, Y_{i(j+1)}, \ldots, Y_{im})$ is an observation from a univariate normal distribution and $(Y_{iB} | Y_{iA})$ is an observation from an $(m-k)$-variate normal distribution.

Each of the m dimensions has a linear regression on every subset of the remaining $m-1$ dimensions. Further, each set of k dimensions will have a linear regression on every subset of the remaining $m-k$ dimensions.

Specifying the mean vector and variance–covariance matrix for a set of attributes conditional on a second set of attributes is involved. The results, however, stem directly from the definitions of the multivariate normal density and of the conditional density as the ratio of the joint to the marginal densities.

If we consider the multivariate observation vector partitioned as in (10.5.1), we can write the associated expectation vector and variance–covariance matrix

$$\boldsymbol{\mu}' = (\boldsymbol{\mu}_A' \quad \boldsymbol{\mu}_B') \tag{10.5.2}$$

and

$$\boldsymbol{\Sigma} = \begin{bmatrix} \boldsymbol{\Sigma}_A & \boldsymbol{\Sigma}_{AB} \\ \boldsymbol{\Sigma}_{AB}' & \boldsymbol{\Sigma}_B \end{bmatrix} \tag{10.5.3}$$

To define the conditional distributions, we use two properties of the partitioned $\boldsymbol{\Sigma}$. First we use the rule on the inverse of a partitioned matrix to express

$$\boldsymbol{\Sigma}^{-1} = \begin{bmatrix} \mathbf{F} & \mathbf{G} \\ \mathbf{G}' & \mathbf{M}^{-1} \end{bmatrix} \tag{10.5.4}$$

where

$$\begin{aligned} \mathbf{F} &= \boldsymbol{\Sigma}_A^{-1} + \boldsymbol{\Sigma}_A^{-1}\boldsymbol{\Sigma}_{AB}\mathbf{M}^{-1}\boldsymbol{\Sigma}_{AB}'\boldsymbol{\Sigma}_A^{-1} \\ \mathbf{G} &= -\boldsymbol{\Sigma}_A^{-1}\boldsymbol{\Sigma}_{AB}\mathbf{M}^{-1} \\ \mathbf{M}^{-1} &= \left(\boldsymbol{\Sigma}_B - \boldsymbol{\Sigma}_{AB}'\boldsymbol{\Sigma}_A^{-1}\boldsymbol{\Sigma}_{AB}\right)^{-1} \end{aligned}$$

We next transform the vector $\mathbf{Y}_{i.}'$ into $\mathbf{W}_{i.}' = \mathbf{Y}_{i.}'\mathbf{T}$:

$$\left(\mathbf{Y}_{iA}' \quad \mathbf{Y}_{iB}' - \mathbf{Y}_{iA}'\boldsymbol{\Sigma}_A^{-1}\boldsymbol{\Sigma}_{AB}\right) = (\mathbf{Y}_{iA}' \quad \mathbf{Y}_{iB}')\begin{bmatrix} \mathbf{I} & -\left(\boldsymbol{\Sigma}_A^{-1}\boldsymbol{\Sigma}_{AB}\right) \\ \mathbf{0} & \mathbf{I} \end{bmatrix} \tag{10.5.5}$$

This transformation is chosen such that $\mathbf{W}_{i.}$ has a variance–covariance matrix

$$\mathbf{Var}(\mathbf{W}_{i.}) = \mathbf{T}'\boldsymbol{\Sigma}\mathbf{T}$$

$$= \begin{bmatrix} \boldsymbol{\Sigma}_A & \mathbf{0} \\ \mathbf{0} & \boldsymbol{\Sigma}_B - \boldsymbol{\Sigma}'_{AB}\boldsymbol{\Sigma}_A^{-1}\boldsymbol{\Sigma}_{AB} \end{bmatrix} = \begin{bmatrix} \boldsymbol{\Sigma}_A & \mathbf{0} \\ \mathbf{0} & \mathbf{M} \end{bmatrix} \quad (10.5.6)$$

The transformed observation vector $\mathbf{W}'_{i.}$ is such that the first k elements have zero covariances with the final $m - k$ elements. Since the first k elements have been carried over directly from $\mathbf{Y}$, it is equally clear to say that we have transformed the latter $m - k$ elements so that they are, as a set, orthogonal to the first k.

We now note something about the basic structure of $\mathbf{Var}(\mathbf{W}_{i.})$. Since $\boldsymbol{\Sigma}_A$ and $\mathbf{M}$ are each variance–covariance matrices, they are product moment matrices and have basic structures that can be expressed

$$\boldsymbol{\Sigma}_A = \mathbf{Q}\mathbf{D}^2\mathbf{Q}', \qquad \mathbf{M} = \mathbf{q}\mathbf{d}^2\mathbf{q}' \quad (10.5.7)$$

Because the two sets of attributes are orthogonal one to the other they have nonoverlapping orthonormal bases; as a result, $\mathbf{Var}(\mathbf{W}_{i.})$ can be written

$$\mathbf{Var}(\mathbf{W}_{i.}) = \begin{bmatrix} \mathbf{Q} & \mathbf{0} \\ \mathbf{0} & \mathbf{q} \end{bmatrix} \begin{bmatrix} \mathbf{D}^2 & \mathbf{0} \\ \mathbf{0} & \mathbf{d}^2 \end{bmatrix} \begin{bmatrix} \mathbf{Q}' & \mathbf{0} \\ \mathbf{0} & \mathbf{q}' \end{bmatrix} \quad (10.5.8)$$

The right side is not quite the basic structure of $\mathbf{Var}(\mathbf{W}_{i.})$ since the elements of the diagonal matrix may not be in descending order. The product of the basic diagonals of $\mathbf{Var}(\mathbf{W}_{i.})$, however, is the product of the products of the basic diagonals of $\boldsymbol{\Sigma}_A$ and $\mathbf{M}$. The *generalized variance* of the attribute vector $\mathbf{W}_{i.}$ is equal to the product of the generalized variances of the attribute vector $\mathbf{Y}_{iA}$ and of the attribute vector whose variance–covariance matrix is given by $\mathbf{M}$.

Before further studying this generalized variance, we trace through certain implications of our formal definition of a conditional distribution or density function. As a notational convenience we drop from the observation vector and its partition the matrix subscript notation: $\mathbf{y}' = (\mathbf{y}'_A, \mathbf{y}'_B)$. If we assume $\boldsymbol{\Sigma}$, the $m \times m$ variance–covariance matrix, has basic structure

$$\boldsymbol{\Sigma} = \mathbf{P}\boldsymbol{\Delta}^2\mathbf{P}' \quad (10.5.9)$$

then $\mathbf{y}'$, as a multivariate normal observation has a density function

$$L(\mathbf{y}') = (2\pi)^{-m/2}\left(\prod_{i=1}^{m} \Delta_i\right)^{-1} \exp\{-\tfrac{1}{2}[(\mathbf{y} - \boldsymbol{\mu})'\boldsymbol{\Sigma}^{-1}(\mathbf{y} - \boldsymbol{\mu})]\} \quad (10.5.10)$$

Similarly, if $\mathbf{y}'_A$ must also be a multivariate normal observation, its likelihood function, based on (10.5.2), (10.5.3), and (10.5.7), must be

$$L(\mathbf{y}'_A) = (2\pi)^{-k/2}\left(\prod_{j=1}^{k} D_j\right)^{-1} \exp\{-\tfrac{1}{2}[(\mathbf{y}_A - \boldsymbol{\mu}_A)'\boldsymbol{\Sigma}_A^{-1}(\mathbf{y}_A - \boldsymbol{\mu}_A)]\} \tag{10.5.11}$$

Using these two forms and the conditional relation $L(\mathbf{y}'_B|\mathbf{y}'_A) = L(\mathbf{y}'_A, \mathbf{y}'_B)/L(\mathbf{y}'_A)$, we can write

$$\begin{aligned} L(\mathbf{y}'_B|\mathbf{y}'_A) = {} & (2\pi)^{-(m-k)/2}\left(\prod_{i=1}^{m} \Delta_i \Big/ \prod_{j=1}^{k} D_j\right)^{-1} \\ & \times \exp\{-\tfrac{1}{2}[(\mathbf{y} - \boldsymbol{\mu})'\boldsymbol{\Sigma}^{-1}(\mathbf{y} - \boldsymbol{\mu}) - (\mathbf{y}_A - \boldsymbol{\mu}_A)\boldsymbol{\Sigma}_A^{-1}(\mathbf{y}_A - \boldsymbol{\mu}_A)]\} \end{aligned} \tag{10.5.12}$$

Our goal is to get this last expression into a recognizable form, namely that of another multivariate normal distribution function. To begin, we write, from (10.5.1) and (10.5.4),

$$\begin{aligned} (\mathbf{y} - \boldsymbol{\mu})'\boldsymbol{\Sigma}^{-1}(\mathbf{y} - \boldsymbol{\mu}) = {} & (\mathbf{y}_A - \boldsymbol{\mu}_A)'\mathbf{F}(\mathbf{y}_A - \boldsymbol{\mu}_A) + (\mathbf{y}_A - \boldsymbol{\mu}_A)'\mathbf{G}(\mathbf{y}_B - \boldsymbol{\mu}_B) \\ & + (\mathbf{y}_B - \boldsymbol{\mu}_B)'\mathbf{G}'(\mathbf{y}_A - \boldsymbol{\mu}_A) \\ & + (\mathbf{y}_B - \boldsymbol{\mu}_B)'\mathbf{M}^{-1}(\mathbf{y}_B - \boldsymbol{\mu}_B) \end{aligned} \tag{10.5.13}$$

Based on the definition in (10.5.4) of **F**,

$$\begin{aligned} (\mathbf{y}_A - \boldsymbol{\mu}_A)'\mathbf{F}(\mathbf{y}_A - \boldsymbol{\mu}_A) = {} & (\mathbf{y}_A - \boldsymbol{\mu}_A)\boldsymbol{\Sigma}_A^{-1}(\mathbf{y}_A - \boldsymbol{\mu}_A) \\ & + (\mathbf{y}_A - \boldsymbol{\mu}_A)\boldsymbol{\Sigma}_A^{-1}\boldsymbol{\Sigma}_{AB}\mathbf{M}^{-1}\boldsymbol{\Sigma}'_{AB}\boldsymbol{\Sigma}_A^{-1}(\mathbf{y}_A - \boldsymbol{\mu}_A) \end{aligned}$$

or

$$\begin{aligned} (\mathbf{y}_A - \boldsymbol{\mu}_A)'\mathbf{F}(\mathbf{y}_A - \boldsymbol{\mu}_A) = {} & (\mathbf{y}_A - \boldsymbol{\mu}_A)\boldsymbol{\Sigma}_A^{-1}(\mathbf{y}_A - \boldsymbol{\mu}_A) \\ & + [\boldsymbol{\Sigma}'_{AB}\boldsymbol{\Sigma}_A^{-1}(\mathbf{y}_A - \boldsymbol{\mu}_A)]'\mathbf{M}^{-1}[\boldsymbol{\Sigma}'_{AB}\boldsymbol{\Sigma}_A^{-1}(\mathbf{y}_A - \boldsymbol{\mu}_A)] \end{aligned} \tag{10.5.14}$$

Similarly, given the definition of **G** in (10.5.4),

$$(\mathbf{y}_B - \boldsymbol{\mu}_B)'\mathbf{G}'(\mathbf{y}_A - \boldsymbol{\mu}_A) = -(\mathbf{y}_B - \boldsymbol{\mu}_B)'\mathbf{M}^{-1}[\boldsymbol{\Sigma}'_{AB}\boldsymbol{\Sigma}_A^{-1}(\mathbf{y}_A - \boldsymbol{\mu}_A)]$$

Substituting these two results into (10.5.14) gives

$$\begin{aligned} (\mathbf{y} - \boldsymbol{\mu})'\boldsymbol{\Sigma}^{-1}(\mathbf{y} - \boldsymbol{\mu}) = {} & (\mathbf{y}_A - \boldsymbol{\mu}_A)'\boldsymbol{\Sigma}_A^{-1}(\mathbf{y}_A - \boldsymbol{\mu}_A) \\ & + [(\mathbf{y}_B - \boldsymbol{\mu}_B) - \boldsymbol{\Sigma}'_{AB}\boldsymbol{\Sigma}_A^{-1}(\mathbf{y}_A - \boldsymbol{\mu}_A)]' \\ & \times \mathbf{M}^{-1}[(\mathbf{y}_B - \boldsymbol{\mu}_B) - \boldsymbol{\Sigma}'_{AB}\boldsymbol{\Sigma}_A^{-1}(\mathbf{y}_A - \boldsymbol{\mu}_A)] \end{aligned}$$

or, on rearranging terms in the multipliers of $\mathbf{M}^{-1}$,

$$\begin{aligned}(\mathbf{y}-\boldsymbol{\mu})'\boldsymbol{\Sigma}^{-1}(\mathbf{y}-\boldsymbol{\mu}) = {} & (\mathbf{y}_A-\boldsymbol{\mu}_A)'\boldsymbol{\Sigma}_A^{-1}(\mathbf{y}_A-\boldsymbol{\mu}_A) \\ & + \left[(\mathbf{y}_B-\boldsymbol{\Sigma}'_{AB}\boldsymbol{\Sigma}_A^{-1}\mathbf{y}_A)-(\boldsymbol{\mu}_B-\boldsymbol{\Sigma}'_{AB}\boldsymbol{\Sigma}_A^{-1}\boldsymbol{\mu}_A)\right]' \\ & \times\mathbf{M}^{-1}\left[(\mathbf{y}_B-\boldsymbol{\Sigma}'_{AB}\boldsymbol{\Sigma}_A^{-1}\mathbf{y}_A)-(\boldsymbol{\mu}_B-\boldsymbol{\Sigma}'_{AB}\boldsymbol{\Sigma}_A^{-1}\boldsymbol{\mu}_A)\right]\end{aligned} \tag{10.5.15}$$

Next, we find the exponential term required in (10.5.12) by subtracting from this result $(\mathbf{y}_A-\boldsymbol{\mu}_A)\boldsymbol{\Sigma}_A^{-1}(\mathbf{y}_A-\boldsymbol{\mu}_A)$, so that the exponential becomes

$$\exp\left[-\tfrac{1}{2}(\mathbf{y}_{B.A}-\boldsymbol{\mu}_{B.A})'\mathbf{M}^{-1}(\mathbf{y}_{B.A}-\boldsymbol{\mu}_{B.A})\right] \tag{10.5.16}$$

where the transformed observation vector

$$\mathbf{y}_{B.A} = \mathbf{y}_B - \boldsymbol{\Sigma}'_{AB}\boldsymbol{\Sigma}_A^{-1}\mathbf{y}_A \tag{10.5.17}$$

has as its expectation, given the joint sampling of $\mathbf{y}_A$ and $\mathbf{y}_B$,

$$\mathbf{E}(\mathbf{y}_{B.A}) \equiv \boldsymbol{\mu}_{B.A} = \boldsymbol{\mu}_B - \boldsymbol{\Sigma}'_{AB}\boldsymbol{\Sigma}_A^{-1}\boldsymbol{\mu}_A \tag{10.5.18}$$

You may note that the transformed vector defined by Eq. (10.5.17) is identically the one created in the transformation of (10.5.5): $\mathbf{Y}'_{i.}\mathbf{T} = \mathbf{W}' = (\mathbf{y}'_A, \mathbf{y}'_{B.A})$. The vector $\mathbf{y}_{B.A}$, we saw in (10.5.6), is orthogonal to the vector $\mathbf{y}_A$. From Eq. (10.5.6), we also learned that this transformed vector has variance–covariance matrix

$$\boldsymbol{\Sigma}_{B.A} = \boldsymbol{\Sigma}_B - \boldsymbol{\Sigma}'_{AB}\boldsymbol{\Sigma}_A^{-1}\boldsymbol{\Sigma}_{AB} = \mathbf{M} \tag{10.5.19}$$

We can now rewrite the conditional likelihood function of Eq. (10.5.12) to include the results of (10.5.16)–(10.5.19):

$$\begin{aligned}L(\mathbf{y}_B|\mathbf{y}_A) = {} & (2\pi)^{-(m-k)/2}\left(\prod_{i=1}^{m}\Delta_i\Big/\prod_{j=1}^{k}D_j\right)^{-1} \\ & \times\exp\{-\tfrac{1}{2}(\mathbf{y}_{B.A}-\boldsymbol{\mu}_{B.A})'\boldsymbol{\Sigma}_{B.A}^{-1}(\mathbf{y}_{B.A}-\boldsymbol{\mu}_{B.A})\}\end{aligned}$$

This would be the conditional likelihood function of an $(m-k)$-variate normal observation if we could show that

$$\left(\prod_{i=1}^{m}\Delta_i\Big/\prod_{j=1}^{k}D_j\right)^{-1} = \left(\prod_{p=1}^{m-k}d_p\right)^{-1} \tag{10.5.20}$$

where, as noted in (10.5.7), $\boldsymbol{\Sigma}_{B.A}$ has basic structure $\mathbf{qd}^2\mathbf{q}'$.

We shall not provide a proof of this equivalence but merely offer an analogy based on the bivariate distribution. Let the bivariate observation $\mathbf{y}' = (y_1, y_2)$ have variance–covariance matrix

$$\boldsymbol{\Sigma} = \begin{bmatrix}\Sigma_{11} & \Sigma_{12}\\ \Sigma_{12} & \Sigma_{22}\end{bmatrix}$$

Then, from Eq. (10.3.5), $\boldsymbol{\Sigma}$ provides a generalized variance $\Delta_1^2\Delta_2^2 = \Sigma_{11}\Sigma_{22} - \Sigma_{12}^2$. Now consider a transformation of $\mathbf{y}'$ that yields y_1 and the modified attribute

$$y_{2.1} = y_2 - \Sigma_{12}\Sigma_{11}^{-1}y_1$$

the scalar or univariate form of Eq. (10.5.5), $\mathbf{y}'\mathbf{T} = \mathbf{w}'$.

$$(y_1 \quad y_2)\begin{bmatrix} 1 & -\Sigma_{12}\Sigma_{11}^{-1} \\ 0 & 1 \end{bmatrix} = \begin{bmatrix} y_1 & (y_2 - \Sigma_{12}\Sigma_{11}^{-1}y_1) \end{bmatrix}$$

Based on Eq. (10.5.6) the variance–covariance matrix for $\mathbf{w}$ is

$$\mathbf{Var(w)} = \begin{bmatrix} \Sigma_{11} & 0 \\ 0 & \Sigma_{22} - \Sigma_{11}^{-1}\Sigma_{12}^2 \end{bmatrix}$$

Again, based on Eq. (10.3.5), the vector $\mathbf{w}$ has generalized variance

$$D_1^2 d_1^2 = \Sigma_{11}(\Sigma_{22} - \Sigma_{11}^{-1}\Sigma_{12}^2) = \Sigma_{11}\Sigma_{22} - \Sigma_{12}^2$$

Thus, a transformation of the bivariate $\mathbf{y}$ into two orthogonal attributes has the following property: The generalized variance of the original bivariate attribute is equal to the product of the variances of the two orthogonal attributes. We now assert that this result extends to (1) the multivariate $\mathbf{y}$ and (2) any transformation $\mathbf{y}'\mathbf{T} = \mathbf{w}'$ that transforms the m-variate attribute vector $\mathbf{y}$ into two vector attributes that are, element by element, mutually orthogonal; one a subset of k of the original attributes and the second an $(m-k)$-element vector. (Roughly speaking, this ensures that we neither gain nor lose *generalized variance* when we transform a set of attributes into orthogonal subsets. Later we consider other transformations of an m-variate observation vector with the resulting conservation of variance.)

Applying this principle to the present problem, we know that $\mathbf{Var(y)} = \boldsymbol{\Sigma}$ has generalized variance $\prod_{i=1}^m \Delta_i^2$ and that $\mathbf{Var(w)} = \mathbf{T}'\boldsymbol{\Sigma}\mathbf{T}$ has generalized variance $\prod_{j=1}^k D_j^2 \prod_{p=1}^{m-k} d_p^2$; we can now assert their equivalence,

$$\prod_{i=1}^m \Delta_i^2 = \left[\prod_{j=1}^k D_j^2 \prod_{p=1}^{m-k} d_p^2\right]$$

and, in consequence, establish that

$$\prod_{i=1}^m \Delta_i^2 \Big/ \prod_{j=1}^k D_j^2 = \prod_{p=1}^{m-k} d_p^2 \qquad (10.5.21)$$

This last result, substituted in Eq. (10.5.12), tells us that the multivariate normal observation leads to conditional observations that are also multivariate normal in form:

$$L(\mathbf{y}_B|\mathbf{y}_A) = (2\pi)^{-(m-k)/2}\left(\prod_{p=1}^{m-k} d_p^2\right)^{-1/2}$$
$$\times \exp\{-\tfrac{1}{2}(\mathbf{y}_{B.A} - \boldsymbol{\mu}_{B.A})'\boldsymbol{\Sigma}_{B.A}^{-1}(\mathbf{y}_{B.A} - \boldsymbol{\mu}_{B.A})\} \quad (10.5.22)$$

where $\boldsymbol{\Sigma}_{B.A}$ has $\mathbf{d}^2$ as its basic diagonal.

Although we defined this multivariate conditional on the density or likelihood relation, $L(\mathbf{y}_B|\mathbf{y}_A) = L(\mathbf{y}_A, \mathbf{y}_B)/L(\mathbf{y}_A)$, it has the same substantive interpretation as in the bivariate case: If we hold constant the values of the first k elements of an m-variate normal observation, the remaining $m - k$ elements will also have a multivariate normal distribution. If $\mathbf{y}'$ has expectation vector $\boldsymbol{\mu}' = (\boldsymbol{\mu}'_A, \boldsymbol{\mu}'_B)$ and variance–covariance matrix

$$\boldsymbol{\Sigma} = \begin{bmatrix} \boldsymbol{\Sigma}_A & \boldsymbol{\Sigma}_{AB} \\ \boldsymbol{\Sigma}'_{AB} & \boldsymbol{\Sigma}_B \end{bmatrix}$$

then the conditional attribute $(\mathbf{y}_B|\mathbf{y}_A) = \mathbf{y}_{B.A} = \mathbf{y}_B - \boldsymbol{\Sigma}'_{AB}\boldsymbol{\Sigma}_A^{-1}\mathbf{y}_A$ has expectation vector $\boldsymbol{\mu}_{B.A} = \boldsymbol{\mu}_B - \boldsymbol{\Sigma}'_{AB}\boldsymbol{\Sigma}_A^{-1}\boldsymbol{\mu}_A$ and variance–covariance matrix $\boldsymbol{\Sigma}_{B.A} = \boldsymbol{\Sigma}_B - \boldsymbol{\Sigma}'_{AB}\boldsymbol{\Sigma}_A^{-1}\boldsymbol{\Sigma}_{AB}$. Further, if $\mathbf{y}'$ is multivariate normal, $\mathbf{y}_{B.A}$ is an $(m - k)$-variate normal observation.

10.6 Multivariate Partial and Multiple Correlation

Thc partitioning, developed in Section 10.5, of the multivariate normal observation y' into (y'_A, y'_B) can be used to introduce the notions of population partial and multiple correlations.

10.6.1 Population Partial Correlation Matrix

The vector $\mathbf{y}_{B.A}$ was introduced in Section 10.5 as a transformation of some initial observation vector $\mathbf{y}_B$. In fact, our goal was solely to illuminate the distribution of $\mathbf{y}_B$ when $\mathbf{y}_A$ is fixed; the vector $\mathbf{y}_{B.A}$ allowed us to describe this conditional distribution.

We may, however, treat $\mathbf{y}_{B.A}$ as an actual vector or set of scores based on the observation vector $\mathbf{y}' = (\mathbf{y}'_A, \mathbf{y}'_B)$. Assuming $\boldsymbol{\Sigma}$ is known, the sampled vector $\mathbf{y}'$ could be transformed according to (10.5.17). From this perspective, with $\mathbf{y}_A$ sampled together with $\mathbf{y}_B$ rather than regarded as fixed, $\mathbf{y}_{B.A}$ is a set of observations on $m - k$ *partial variables*. That is, Eq. (10.5.17) partials from each of the last $m - k$ elements of $\mathbf{y}'$ the population effects of the first k attributes included in $\mathbf{y}'$.

The variance–covariance matrix $\boldsymbol{\Sigma}_{B.A}$ is a matrix of population variances and covariances for these $m - k$ partial variables. Alternatively, the diagonal elements of $\boldsymbol{\Sigma}_{B.A}$ are population partial (or residual) variances and the off-diagonals can be termed partial covariances. We shall illustrate these notions for the three-variable case, $k = 1$ attribute assigned to $\mathbf{y}_A$ and $m - k = 2$ to $\mathbf{y}_B$. In this instance, we have

$$\boldsymbol{\Sigma} = \left[\begin{array}{c|cc} \sigma_1^2 & \rho_{12}\sigma_1\sigma_2 & \rho_{13}\sigma_1\sigma_3 \\ \hline \rho_{12}\sigma_1\sigma_2 & \sigma_2^2 & \rho_{23}\sigma_2\sigma_3 \\ \rho_{13}\sigma_1\sigma_3 & \rho_{23}\sigma_2\sigma_3 & \sigma_3^2 \end{array}\right] = \begin{bmatrix} \boldsymbol{\Sigma}_A & \boldsymbol{\Sigma}_{AB} \\ \boldsymbol{\Sigma}'_{AB} & \boldsymbol{\Sigma}_B \end{bmatrix}$$

and

$$\begin{aligned}\boldsymbol{\Sigma}_{B.A} &= \begin{bmatrix} \sigma_2^2 & \rho_{23}\sigma_2\sigma_3 \\ \rho_{23}\sigma_2\sigma_3 & \sigma_3^2 \end{bmatrix} - \frac{1}{\sigma_1^2}\begin{bmatrix} \rho_{12}\sigma_1\sigma_2 \\ \rho_{13}\sigma_1\sigma_3 \end{bmatrix}(\rho_{12}\sigma_1\sigma_2 \quad \rho_{13}\sigma_1\sigma_3) \\ &= \begin{bmatrix} \sigma_2^2(1 - \rho_{12}^2) & (\rho_{23} - \rho_{12}\rho_{13})\sigma_2\sigma_3 \\ (\rho_{23} - \rho_{12}\rho_{13})\sigma_2\sigma_3 & \sigma_3^2(1 - \rho_{13}^2) \end{bmatrix}\end{aligned}$$

We recognize the diagonal elements of this matrix as the population forms of the partial variances (where the effect of attribute 1 has been partialed from attributes 2 and 3). As $\boldsymbol{\Sigma}_{B.A}$ is a variance–covariance matrix, we may form from it an intercorrelation matrix by pre- and postmultiplying by the inverse root diagonal:

$$\boldsymbol{\rho}_{B.A} = \mathbf{D}_{\boldsymbol{\Sigma}_{B.A}}^{-1/2}\boldsymbol{\Sigma}_{B.A}\mathbf{D}_{\boldsymbol{\Sigma}_{B.A}}^{-1/2} \tag{10.6.1}$$

For the three-variable example we have

$$\boldsymbol{\rho}_{B.A} = \begin{bmatrix} 1 & \dfrac{(\rho_{23} - \rho_{12}\rho_{13})}{[(1 - \rho_{12}^2)(1 - \rho_{13}^2)]^{1/2}} \\ \dfrac{(\rho_{23} - \rho_{12}\rho_{13})}{[(1 - \rho_{12}^2)(1 - \rho_{13}^2)]^{1/2}} & 1 \end{bmatrix}$$

The symmetric matrix $\boldsymbol{\rho}_{B.A}$ has as its off-diagonal elements population partial correlations. In this illustration it contains first-order partial correlations (only one attribute partialed from the pair being correlated), but Eq. (10.6.1) generalizes to partitions of $\mathbf{y}$ where $\mathbf{y}_A$ is a vector.

10.6.2 Population Multiple Correlation

If we think of the vector $\mathbf{y}$ as partitioned between $\mathbf{y}_A$ containing observations on p predictor attributes and $\mathbf{y}_B$ containing observations on k criterion attributes, then the *diagonal* elements of $\boldsymbol{\Sigma}_{B.A}$ can be represented

$$\mathbf{D}_{\boldsymbol{\Sigma}_{B.A}} = \begin{bmatrix} \sigma_1^2(1 - \rho_{1.\mathbf{y}_A}^2) & & & 0 \\ & \sigma_2^2(1 - \rho_{2.\mathbf{y}_A}^2) & & \\ 0 & & \ddots & \\ & & & \sigma_k^2(1 - \rho_{k.\mathbf{y}_A}^2) \end{bmatrix} \tag{10.6.2}$$

where $\rho_{i.\mathbf{y}_A}$ symbolizes the population multiple correlation between the ith criterion attribute and the set of p predictor attributes.

If we wish, we can write a diagonal matrix of squared population multiple correlations from Eqs. (10.5.3) and (10.6.2):

$$\mathbf{D}_{\rho^2_{B.A}} = \mathbf{I} - \left(\mathbf{D}_{\boldsymbol{\Sigma}_B}\right)^{-1}\mathbf{D}_{\boldsymbol{\Sigma}_{B.A}} \tag{10.6.3}$$

We reiterate that the model in this section assumes that $\mathbf{y}_A$ and $\mathbf{y}_B$ are simultaneously sampled from an m-variate distribution with mean $\boldsymbol{\mu}' = (\boldsymbol{\mu}'_A, \boldsymbol{\mu}'_B)$ and variance–covariance

$$\boldsymbol{\Sigma} = \begin{bmatrix} \boldsymbol{\Sigma}_A & \boldsymbol{\Sigma}_{AB} \\ \boldsymbol{\Sigma}'_{AB} & \boldsymbol{\Sigma}_B \end{bmatrix}$$

Assuming $\boldsymbol{\Sigma}$ is known, the population partial and multiple correlations are defined, regardless of whether $\mathbf{y}$ is multivariate normal.

10.7 Parameter Estimation for the Multivariate Normal Distribution

In general, of course $\boldsymbol{\mu}$ and $\boldsymbol{\Sigma}$ are not known. They consist of arrays of population parameters, and a principal task of multivariate analysis is to estimate, restrict, and test hypotheses related to these parameters. To begin this task we develop point estimation of the elements in $\boldsymbol{\mu}$ and $\boldsymbol{\Sigma}$ using the maximum likelihood principle introduced in Chapter 8. We assume for the remainder of this chapter that $\mathbf{y} = \mathbf{Y}_{i.}$ is a vector of observations from the m-variate *normal* distribution with parameters $\boldsymbol{\mu}$ and $\boldsymbol{\Sigma}$. That is, $\mathbf{Y}_{i.}$ has a density or, we would now emphasize, *likelihood function* given by Eq. (10.2.19):

$$L(\mathbf{Y}_{i.}) = (2\pi)^{-m/2}\left(\prod_{j=1}^{m}\Delta_j^2\right)^{-1/2} \exp\{-\tfrac{1}{2}(\mathbf{Y}_{i.} - \boldsymbol{\mu})'\boldsymbol{\Sigma}^{-1}(\mathbf{Y}_{i.} - \boldsymbol{\mu})\}$$

where $\boldsymbol{\Delta}^2$ is the basic diagonal of $\boldsymbol{\Sigma}$. The value of this likelihood (for a fixed vector $\mathbf{Y}_{i.}$) depends upon the unknown $\boldsymbol{\mu}$ and $\boldsymbol{\Sigma}$. To estimate these we independently sample n times from this m-variate normal distribution. The result can be represented as the matrix

$$\mathbf{Y} = \begin{bmatrix} \mathbf{Y}'_{1.} \\ \vdots \\ \mathbf{Y}'_{i.} \\ \vdots \\ \mathbf{Y}'_{n.} \end{bmatrix} \tag{10.7.1}$$

in which each row corresponds to a vector observation.

Given our assumption that the rows in **Y** have been independently sampled, we can express the likelihood of the sample (or of the matrix **Y**) as the product of the likelihoods of the n rows: $L(\mathbf{Y}) = \prod_{i=1}^{n} L(\mathbf{Y}_{i.})$ or

$$L(\mathbf{Y}) = (2\pi)^{-nm/2}\left(\prod_{j=1}^{m} \Delta_j^2\right)^{-n/2} \exp\{-\tfrac{1}{2}\operatorname{tr}[(\mathbf{Y} - \mathbf{1}\boldsymbol{\mu}')\boldsymbol{\Sigma}^{-1}(\mathbf{Y} - \mathbf{1}\boldsymbol{\mu}')']\} \tag{10.7.2}$$

As in Chapter 8, we estimate the arrays $\boldsymbol{\mu}$ and $\boldsymbol{\Sigma}$ by finding values that maximize the likelihood of the sample. Again, it is more straightforward to maximize not $L(\mathbf{Y})$ but its logarithm:

$$\log L(\mathbf{Y}) = -\tfrac{1}{2}\left\{nm\log 2\pi + n\sum_{j=1}^{m}\log \Delta_j^2 + \operatorname{tr}\left[(\mathbf{Y} - \mathbf{1}\boldsymbol{\mu}')\boldsymbol{\Sigma}^{-1}(\mathbf{Y} - \mathbf{1}\boldsymbol{\mu}')'\right]\right\} \tag{10.7.3}$$

10.7.1 Maximum Likelihood Estimator of $\boldsymbol{\mu}$

We note that $\boldsymbol{\mu}$ appears in the log sample likelihood expression only in the trace term, which we may rewrite

$$\operatorname{tr}\left[(\mathbf{Y} - \mathbf{1}\boldsymbol{\mu}')\boldsymbol{\Sigma}^{-1}(\mathbf{Y} - \mathbf{1}\boldsymbol{\mu}')'\right] = \operatorname{tr}(\mathbf{Y}\boldsymbol{\Sigma}^{-1}\mathbf{Y}') - 2\operatorname{tr}(\mathbf{Y}\boldsymbol{\Sigma}^{-1}\boldsymbol{\mu}\mathbf{1}') + \operatorname{tr}(\mathbf{1}\boldsymbol{\mu}'\boldsymbol{\Sigma}^{-1}\boldsymbol{\mu}\mathbf{1}')$$

and then isolate the two parts dependent upon $\boldsymbol{\mu}$

$$\phi = n\operatorname{tr}(\boldsymbol{\mu}'\boldsymbol{\Sigma}^{-1}\boldsymbol{\mu}) - 2\operatorname{tr}(\mathbf{1}'\mathbf{Y}\boldsymbol{\Sigma}^{-1}\boldsymbol{\mu}) \tag{10.7.4}$$

In writing down this equation, we made use of the rule $\operatorname{tr}(\mathbf{AB}) = \operatorname{tr}(\mathbf{BA})$ to give terms that are more easily differentiated.

Choosing $\boldsymbol{\mu}$ to minimize ϕ gives the same result as if we set out to maximize $\log L(\mathbf{Y})$. Setting equal to zero the derivative of ϕ with respect to $\boldsymbol{\mu}$ (from the symbolic matrix differentiation rules of Chapter 6), we obtain

$$\partial\phi/\partial\boldsymbol{\mu} = \mathbf{0} = 2n\hat{\boldsymbol{\mu}}'\boldsymbol{\Sigma}^{-1} - 2(\mathbf{1}'\mathbf{Y}\boldsymbol{\Sigma}^{-1})$$

which yields the maximum likelihood estimator

$$\hat{\boldsymbol{\mu}}' = (1/n)\mathbf{1}'\mathbf{Y} \tag{10.7.5}$$

The sample means for each of the m columns of **Y** provide the maximum likelihood estimators for the corresponding population means.

10.7.2 Maximum Likelihood Estimator of $\boldsymbol{\Sigma}$

We must evaluate two derivatives to find the value of $\boldsymbol{\Sigma}$ that maximizes $\log L(\mathbf{Y})$. Because of the form of Eq. (10.7.3), we maximize with respect to

$\boldsymbol{\Sigma}^{-1}$ rather than $\boldsymbol{\Sigma}$. The first derivative is that of the trace term:

$$\partial\{\mathrm{tr}[(\mathbf{Y} - \mathbf{1}\boldsymbol{\mu}')\boldsymbol{\Sigma}^{-1}(\mathbf{Y} - \mathbf{1}\boldsymbol{\mu}')']\}/\partial\boldsymbol{\Sigma}^{-1} = (\mathbf{Y} - \mathbf{1}\boldsymbol{\mu}')'(\mathbf{Y} - \mathbf{1}\boldsymbol{\mu}') \quad (10.7.6)$$

To obtain the derivative of the term in the log likelihood expression based on the log of the *generalized variance*, we first rewrite that term

$$n \sum_{j=1}^{m} \log \Delta_j^2 = -n \sum_{j=1}^{m} \log \Delta_j^{-2}$$

Now, the derivative of the term on the left with respect to $\boldsymbol{\Sigma}^{-1}$ is equivalent to

$$\partial\left(n \sum_{j=1}^{m} \log \Delta_j^2\right)/\partial\boldsymbol{\Sigma}^{-1} = -n \sum_{j=1}^{m} \left[\partial\left(\log \Delta_j^{-2}\right)/\partial\boldsymbol{\Sigma}^{-1}\right] \quad (10.7.7)$$

To continue this development, we note the basic structure $\boldsymbol{\Sigma} = \mathbf{Q}\boldsymbol{\Delta}^2\mathbf{Q}'$ allows us to write $\boldsymbol{\Sigma}^{-1} = \mathbf{Q}\boldsymbol{\Delta}^{-2}\mathbf{Q}'$ or $\boldsymbol{\Delta}^{-2} = \mathbf{Q}'\boldsymbol{\Sigma}^{-1}\mathbf{Q}$ or $\Delta_j^{-2} = \mathbf{Q}'_{.j}\boldsymbol{\Sigma}^{-1}\mathbf{Q}_{.j}$. Given this last form and the rules of differentiation of logs and matrices, we can write

$$\begin{aligned}
\partial\left(\log \Delta_j^{-2}\right)/\partial\boldsymbol{\Sigma}^{-1} &= \partial\left(\log \mathbf{Q}'_{.j}\boldsymbol{\Sigma}^{-1}\mathbf{Q}_{.j}\right)/\partial\boldsymbol{\Sigma}^{-1} \\
&= \left(\mathbf{Q}'_{.j}\boldsymbol{\Sigma}^{-1}\mathbf{Q}_{.j}\right)^{-1}\left[\partial\left(\mathbf{Q}'_{.j}\boldsymbol{\Sigma}^{-1}\mathbf{Q}_{.j}\right)/\partial\boldsymbol{\Sigma}^{-1}\right] \\
&= \Delta_j^2(\mathbf{Q}_{.j}\mathbf{Q}'_{.j}) \\
&= \mathbf{Q}_{.j}\Delta_j^2\mathbf{Q}'_{.j}
\end{aligned}$$

The right side of Eq. (10.7.7) directs us to sum terms of this form for $j = 1, 2, \ldots, m$, which results in

$$\sum_{j=1}^{m} \left[\frac{\partial\left(\log \Delta_j^{-2}\right)}{\partial\boldsymbol{\Sigma}^{-1}}\right] = \mathbf{Q}_{.1}\Delta_1^2\mathbf{Q}'_{.1} + \cdots + \mathbf{Q}_{.m}\Delta_m^2\mathbf{Q}'_{.m} = \mathbf{Q}\boldsymbol{\Delta}^2\mathbf{Q}' = \boldsymbol{\Sigma}$$

and we obtain for the left side of Eq. (10.7.7)

$$\partial\left(n \sum_{j=1}^{m} \log \Delta_j^2\right)\bigg/\partial\boldsymbol{\Sigma}^{-1} = -n\boldsymbol{\Sigma}$$

Putting this result together with that in Eq. (10.7.6), we can write from the log likelihood of Eq. (10.7.3)

$$\partial \log L(\mathbf{Y})/\partial\boldsymbol{\Sigma}^{-1} = -\tfrac{1}{2}[(\mathbf{Y} - \mathbf{1}\boldsymbol{\mu}')'(\mathbf{Y} - \mathbf{1}\boldsymbol{\mu}') - n\boldsymbol{\Sigma}]$$

Setting this equal to zero and substituting for $\boldsymbol{\mu}$ the maximum likelihood estimator, we obtain

$$\hat{\boldsymbol{\Sigma}} = (1/n)(\mathbf{Y} - \mathbf{1}\hat{\boldsymbol{\mu}}')'(\mathbf{Y} - \mathbf{1}\hat{\boldsymbol{\mu}}')$$

or, based on Eq. (10.7.5),

$$\hat{\boldsymbol{\Sigma}} = (1/n)\mathbf{Y}'\left[\mathbf{I} - \mathbf{1}(\mathbf{1}'\mathbf{1})^{-1}\mathbf{1}'\right]\mathbf{Y} \tag{10.7.8}$$

where we recognize $\mathbf{Y}'[\mathbf{I} - \mathbf{1}(\mathbf{1}'\mathbf{1})^{-1}\mathbf{1}']\mathbf{Y}$ as the matrix of sums of squares and cross products of deviations from the sample means.

The maximum likelihood estimator of $\boldsymbol{\Sigma}$ is simply the sample covariance matrix:

$$\hat{\Sigma}_{jj} = \left(\frac{1}{n}\right)\sum_{i=1}^{n}(Y_{ij} - \hat{\mu}_j)^2$$

and

$$\hat{\Sigma}_{jk} = \left(\frac{1}{n}\right)\sum_{i=1}^{n}(Y_{ij} - \hat{\mu}_j)(Y_{ik} - \hat{\mu}_k)$$

where $\hat{\mu}_j$, the maximum likelihood estimator of μ_j, is the average of the values in the jth column of $\mathbf{Y}$.

10.7.3 Distribution of the Maximum Likelihood Estimator for $\boldsymbol{\mu}$

The maximum likelihood estimator (MLE) of $\boldsymbol{\mu}$ is itself a multivariate normal observation. We shall not prove this, but the defining equation, (10.7.5),

$$\hat{\boldsymbol{\mu}} = [(1/n)\mathbf{1}']\mathbf{Y}$$

establishes that each element of $\hat{\boldsymbol{\mu}}$ is a weighted sum (here the mean) of n independent normal observations. As a result, each $\hat{\mu}_j$ must be a normal observation as well.

To state further characteristics of the distribution of $\hat{\boldsymbol{\mu}}$ we need to find an expression for the parameters of the m-variate normal sampling distribution for $\hat{\boldsymbol{\mu}}$. From Eq. (10.7.5), we can write $\mathbf{E}(\hat{\boldsymbol{\mu}}') = (1/n)\mathbf{1}'\mathbf{E}(\mathbf{Y})$ or, because each row of $\mathbf{Y}$ is an observation from the same m-variate distribution,

$$E(\hat{\boldsymbol{\mu}}') = (1/n)\mathbf{1}'\mathbf{1}\mathbf{E}(\mathbf{Y}_{i.}') = (1/n)n\boldsymbol{\mu}' = \boldsymbol{\mu}' \tag{10.7.9}$$

The MLE of $\boldsymbol{\mu}$ is an *unbiased estimator*.

The sampling variance of $\hat{\boldsymbol{\mu}}$ is defined as the $m \times m$ matrix

$$\mathbf{Var}(\boldsymbol{\mu}) = \mathbf{E}\{[\hat{\boldsymbol{\mu}} - \mathbf{E}(\hat{\boldsymbol{\mu}})][\hat{\boldsymbol{\mu}} - \mathbf{E}(\hat{\boldsymbol{\mu}})]'\}$$

Substituting from (10.7.5) and (10.7.9) gives

$$\begin{aligned}
\mathbf{Var}(\hat{\boldsymbol{\mu}}) &= \mathbf{E}\{[(1/n)\mathbf{Y}'\mathbf{1} - \boldsymbol{\mu}][(1/n)\mathbf{Y}'\mathbf{1} - \boldsymbol{\mu}]'\} \\
&= (1/n)^2\mathbf{E}[(\mathbf{Y}'\mathbf{1} - \boldsymbol{\mu}\mathbf{1}'\mathbf{1})(\mathbf{Y}'\mathbf{1} - \boldsymbol{\mu}\mathbf{1}'\mathbf{1})'] \\
&= (1/n)^2\mathbf{E}[(\mathbf{Y}' - \boldsymbol{\mu}\mathbf{1}')\mathbf{1}\mathbf{1}'(\mathbf{Y} - \mathbf{1}\boldsymbol{\mu}')]
\end{aligned}$$

where the expectation is now of a major product moment of a vector of sums. It will be helpful to examine a typical element of this matrix:

$$[\text{Var}(\hat{\boldsymbol{\mu}})]_{jk} = \left(\frac{1}{n}\right)^2 E\left[\sum_{i=1}^{n}(Y_{ij} - \mu_{.j})\sum_{i=1}^{n}(Y_{ik} - \mu_{.k})\right]$$

On the right, the expectation is of a sum of n^2 terms. Of these, $n^2 - n$ are of the form $(Y_{pj} - \mu_{.j})(Y_{qk} - \mu_{.k})$, involving observations from different rows of $\mathbf{Y}$. Because the rows are independent samples, each of these terms has zero expectation. The remaining n terms may be summed:

$$[\text{Var}(\hat{\boldsymbol{\mu}})]_{jk} = \left(\frac{1}{n}\right)^2 E\left[\sum_{i=1}^{n}(Y_{ij} - \mu_{.j})(Y_{ik} - \mu_{.k})\right]$$

From this result we can construct the matrix representation

$$\mathbf{Var}(\hat{\boldsymbol{\mu}}) = (1/n)^2\mathbf{E}[(\mathbf{Y} - \mathbf{1}\boldsymbol{\mu}')'(\mathbf{Y} - \mathbf{1}\boldsymbol{\mu}')]$$

where the expectation is of a matrix of sums of squares and cross products of deviations from the population means; that is, the expectation is $n\boldsymbol{\Sigma}$, or

$$\mathbf{Var}(\hat{\boldsymbol{\mu}}) = (1/n)^2 n\boldsymbol{\Sigma} = (1/n)\boldsymbol{\Sigma} \tag{10.7.10}$$

In summary, the MLE $\hat{\boldsymbol{\mu}}$ has an m-variate normal sampling distribution with mean $\mathbf{E}(\hat{\boldsymbol{\mu}}) = \boldsymbol{\mu}$ and variance $\mathbf{Var}(\hat{\boldsymbol{\mu}}) = (1/n)\boldsymbol{\Sigma}$. $\hat{\boldsymbol{\mu}}$ is an unbiased and consistent estimator of $\boldsymbol{\mu}$. Consistent estimators have sampling variances that decrease as sample sizes increase.

10.7.4 Expected Value of $\hat{\boldsymbol{\Sigma}}$

In the next two sections we discuss properties of the sampling distribution of $\hat{\boldsymbol{\Sigma}}$. The MLE for $\boldsymbol{\Sigma}$ is defined as

$$\begin{aligned}\hat{\boldsymbol{\Sigma}} &= (1/n)(\mathbf{Y} - \mathbf{1}\hat{\boldsymbol{\mu}}')'(\mathbf{Y} - \mathbf{1}\hat{\boldsymbol{\mu}}') \\ &= (1/n)[(\mathbf{Y} - \mathbf{1}\boldsymbol{\mu}') + (\mathbf{1}\boldsymbol{\mu}' - \mathbf{1}\hat{\boldsymbol{\mu}}')]'[(\mathbf{Y} - \mathbf{1}\boldsymbol{\mu}') + (\mathbf{1}\boldsymbol{\mu}' - \mathbf{1}\hat{\boldsymbol{\mu}}')]\end{aligned}$$

which yields, upon multiplying through,

$$\begin{aligned}\hat{\boldsymbol{\Sigma}} = (1/n)[&(\mathbf{Y} - \mathbf{1}\boldsymbol{\mu}')'(\mathbf{Y} - \mathbf{1}\boldsymbol{\mu}') + (\mathbf{Y} - \mathbf{1}\boldsymbol{\mu}')'\mathbf{1}(\boldsymbol{\mu}' - \hat{\boldsymbol{\mu}}') \\ &+ (\boldsymbol{\mu} - \hat{\boldsymbol{\mu}})\mathbf{1}'(\mathbf{Y} - \mathbf{1}\boldsymbol{\mu}') + (\boldsymbol{\mu} - \hat{\boldsymbol{\mu}})\mathbf{1}'\mathbf{1}(\boldsymbol{\mu} - \hat{\boldsymbol{\mu}})']\end{aligned}$$

Noting, however, that

$$\mathbf{1}'(\mathbf{Y} - \mathbf{1}\boldsymbol{\mu}') = n\hat{\boldsymbol{\mu}}' - n\boldsymbol{\mu}' = n(\hat{\boldsymbol{\mu}} - \boldsymbol{\mu})' = -n(\boldsymbol{\mu} - \hat{\boldsymbol{\mu}})'$$

and substituting the result into the preceding expression, we obtain

$$\hat{\boldsymbol{\Sigma}} = (1/n)(\mathbf{Y} - \mathbf{1}\boldsymbol{\mu}')'(\mathbf{Y} - \mathbf{1}\boldsymbol{\mu}') - (\boldsymbol{\mu} - \hat{\boldsymbol{\mu}})(\boldsymbol{\mu} - \hat{\boldsymbol{\mu}})'$$

Taking expectations, we can use Section 10.7.3 to obtain

$$\mathbf{E}(\hat{\boldsymbol{\Sigma}}) = \boldsymbol{\Sigma} - (1/n)\boldsymbol{\Sigma} = [(n-1)/n]\boldsymbol{\Sigma} \tag{10.7.11}$$

The MLE of $\boldsymbol{\Sigma}$ is not unbiased.

10.8 A Multivariate Generalization of Chi-Squared

The maximum likelihood estimator of $\mathbf{\Sigma}$, based on a sample of n independent observations from a multivariate normal distribution, is an $m \times m$ matrix $\hat{\mathbf{\Sigma}}$. If a second sample of n observations were drawn, a second estimator $\hat{\mathbf{\Sigma}}$ would be available. Similarly, additional samples would yield additional matrix estimators. The sampling distribution of $\hat{\mathbf{\Sigma}}$ is a *distribution of matrices*. The reader may sense the difficulty of describing a distribution of matrices and deciding how a particular matrix fits into that distribution. To avoid these difficulties, statisticians have found it useful to develop both theory and techniques for working not with distributions of sample variance–covariance matrices themselves, but with the *generalized variance* measures associated with such matrices.

The sample estimator $\hat{\mathbf{\Sigma}}$, of course, has a basic structure just as certainly as does $\mathbf{\Sigma}$ itself. The product of the basic diagonals of $\hat{\mathbf{\Sigma}}$ is a *generalized sample variance* measure. Indeed, each $\hat{\mathbf{\Sigma}}$ in the sampling distribution has a basic structure and a generalized sample variance measure. The importance of this is that the generalized variance is a scalar, with a scalar rather than matrix sampling distribution.

The distribution of generalized variances for sample $\hat{\mathbf{\Sigma}}$ matrices, those estimating a particular $\mathbf{\Sigma}$, is representative of a larger class of distributions, all of products of basic diagonal elements, heavily relied upon for inference in multivariate statistics. This class of distributions has much in common with the chi-squared distributions, based on quadratic forms as discussed in Chapter 8. Before describing these distributions of basic diagonal products, we review a part of the discussion of χ^2.

If $\mathbf{y}$ is a vector of observations and $\mathbf{A}$ is an $n \times n$ symmetric matrix, then the scalar $\mathbf{y}'\mathbf{A}\mathbf{y}$ is a quadratic form in the observations and $\mathbf{A}$ is the matrix of the form. Further, if the vector $\mathbf{y}$ consists of normally distributed observations with expectation vector $\mathbf{E}(\mathbf{y})$ and variance–covariance matrix $\mathbf{Var}(\mathbf{y})$, then the quadratic form

$$w = [\mathbf{y} - \mathbf{E}(\mathbf{y})]'\mathbf{A}[\mathbf{y} - \mathbf{E}(\mathbf{y})] \tag{10.8.1}$$

is distributed as (central) chi-squared with $r(\mathbf{A})$ degrees of freedom provided that $\mathbf{A}\,\mathbf{Var}(\mathbf{y})\,\mathbf{A} = \mathbf{A}$ and that $r(\mathbf{A})$ is the rank of $\mathbf{A}$. (We note that when the n normally distributed observations making up $\mathbf{y}$ are sampled independently from populations with a common variance σ^2, then $\mathbf{Var}(\mathbf{y}) = \sigma^2\mathbf{I}$ and $\mathbf{A}\,\mathbf{Var}(\mathbf{y})\,\mathbf{A} = \mathbf{A}$ is satisfied for $\mathbf{A} = (\sigma^2)^{-1}\boldsymbol{\alpha}$, where $\boldsymbol{\alpha}$ is any symmetric $n \times n$ idempotent matrix. In this case $r(\mathbf{A})$ is also the rank of $\boldsymbol{\alpha}$.)

The scalar w has a distribution because each vector $\mathbf{y}$, each sample of n normally distributed observations, can be converted to a scalar by Eq. (10.8.1). We saw in Chapter 8 that, given these assumptions, w is a chi-squared observation since it is a sum of squares of $r(\mathbf{A})$ independent, normally distributed observations, each with unit variance and zero mean.

Two quadratic forms

$$w = [\mathbf{y} - \mathbf{E}(\mathbf{y})]'\mathbf{A}[\mathbf{y} - \mathbf{E}(\mathbf{y})], \qquad u = [\mathbf{y} - \mathbf{E}(\mathbf{y})]'\mathbf{B}[\mathbf{y} - \mathbf{E}(\mathbf{y})]$$

are *independently distributed* as chi-squared observations (their sampling covariance is zero) if each meets the above assumptions and if, additionally, $\mathbf{A}\,\mathbf{Var}(\mathbf{y})\,\mathbf{B} = \mathbf{0}$. The ratio of two independently distributed chi-squared observations, each a quadratic form in the same set of normally distributed observations, is heavily used in the general linear model to test linear regression and analysis of variance hypotheses.

10.8.1 Wishart Matrices

Now we develop similar statistical techniques where $\mathbf{Y}$ is an $n \times m$ matrix rather than an n-element vector. If $\mathbf{E}(\mathbf{Y})$ is the expectation matrix for $\mathbf{Y}$ (so that $[\mathbf{E}(\mathbf{Y})]_{i.} = \mathbf{E}(\mathbf{Y}_{i.})$), and if $\mathbf{A}$ is an $n \times n$ symmetric matrix with rank $r(\mathbf{A})$, we can write, in parallel with Eq. (10.8.1),

$$\mathbf{W} = [\mathbf{Y} - \mathbf{E}(\mathbf{Y})]'\mathbf{A}[\mathbf{Y} - \mathbf{E}(\mathbf{Y})] \tag{10.8.2}$$

$\mathbf{W}$ is a symmetric matrix. The diagonal elements of $\mathbf{W}$ are quadratic forms in the respective (centered) columns of $\mathbf{Y}$,

$$W_{jj} = [\mathbf{Y}_{.j} - \mathbf{E}(\mathbf{Y}_{.j})]'\mathbf{A}[\mathbf{Y}_{.j} - \mathbf{E}(\mathbf{Y}_{.j})]$$

while the off-diagonals are bilinear forms,

$$W_{jk} = [\mathbf{Y}_{.j} - \mathbf{E}(\mathbf{Y}_{.j})]'\mathbf{A}[\mathbf{Y}_{.k} - \mathbf{E}(\mathbf{Y}_{.k})]$$

involving two (centered) columns of $\mathbf{Y}$.

We now assume that the rows of $\mathbf{Y}$ are independent vector observations and that each row is a multivariate normal observation, drawn from a distribution with variance–covariance matrix $\mathbf{Var}(\mathbf{Y}_{i.}) = \mathbf{\Sigma}$. (Each multivariate observation has a common variance–covariance matrix, although the expectation vector for a row may vary from row to row.) Finally, if the matrix $\mathbf{A}$ is idempotent, then the matrix $\mathbf{W}$ is called a *Wishart matrix*. Wishart matrices provide a multivariate analog to quadratic forms and their distribution (actually, the distribution of functions of their basic diagonals) plays the same role as does the distribution of chi-squared observations when $\mathbf{Y}_{i.}$ is univariate normal. With $\mathbf{Y}$ written in deviation- or column-centered form, the resulting $\mathbf{W}$ is a *central Wishart matrix*. Not centering $\mathbf{Y}$ yields a noncentral Wishart matrix, just as not centering $\mathbf{y}$ in Eq. (10.8.1) produces a noncentral chi-squared observation. In this text, we limit our discussion to central chi-squared and Wishart distributions.

The Wishart matrix in Eq. (10.8.2) is an observation from a particular Wishart distribution, and we indicate this by tagging $\mathbf{W}$ with three parameters:

$$[\mathbf{Y} - \mathbf{E}(\mathbf{Y})]'\mathbf{A}[\mathbf{Y} - \mathbf{E}(\mathbf{Y})] = \mathbf{W} \sim \mathbf{W}_m[r(\mathbf{A}), \mathbf{\Sigma}] \tag{10.8.3}$$

The matrix on the left is a Wishart matrix of order m, the rank of the symmetric idempotent matrix of the form is $r(\mathbf{A})$, and each row of $\mathbf{Y}$ is an observation from an m-variate normal distribution with variance–covariance matrix $\mathbf{\Sigma}$.

Similar to what is true for chi-squared observations, two Wishart matrices $\mathbf{W} = [\mathbf{Y} - \mathbf{E}(\mathbf{Y})]'\mathbf{A}[\mathbf{Y} - \mathbf{E}(\mathbf{Y})]$ and $\mathbf{U} = [\mathbf{Y} - \mathbf{E}(\mathbf{Y})]'\mathbf{B}[\mathbf{Y} - \mathbf{E}(\mathbf{Y})]$ have independent sampling distributions if $\mathbf{AB} = \mathbf{0}$.

We shall illustrate these points for the two symmetric matrices

$$\mathbf{G} = \left[\mathbf{Y} - \hat{\mathbf{E}}(\mathbf{Y})\right]'\left[\mathbf{Y} - \hat{\mathbf{E}}(\mathbf{Y})\right] \tag{10.8.4}$$

and

$$\mathbf{H} = \left[\hat{\mathbf{E}}(\mathbf{Y}) - \mathbf{E}(\mathbf{Y})\right]'\left[\hat{\mathbf{E}}(\mathbf{Y}) - \mathbf{E}(\mathbf{Y})\right] \tag{10.8.5}$$

where

$$\mathbf{E}(\mathbf{Y}) = \mathbf{1E}(\mathbf{Y}_{i.})' = \mathbf{1u}' \tag{10.8.6}$$

and

$$\hat{\mathbf{E}}(\mathbf{Y}) = \mathbf{1}\hat{\boldsymbol{\mu}}' = \mathbf{1}(\mathbf{1}'\mathbf{1})^{-1}\mathbf{1}'\mathbf{Y} \tag{10.8.7}$$

This is simply a rephrasing of the concepts developed in Section 10.7. Each of the $\mathbf{1}'\mathbf{1} = n$ independent rows of $\mathbf{Y}$ is assumed to be drawn from a distribution with mean $\boldsymbol{\mu}'$ and variance $\mathbf{\Sigma}$. The maximum likelihood estimator of $\boldsymbol{\mu}'$ is $(1/n)\mathbf{1}'\mathbf{Y}$. As a result, the matrix $\mathbf{G}$ is proportional to the maximum likelihood estimator of $\mathbf{\Sigma}$:

$$\mathbf{G} = (\mathbf{Y} - \mathbf{1}\hat{\boldsymbol{\mu}}')'(\mathbf{Y} - \mathbf{1}\hat{\boldsymbol{\mu}}) = n\hat{\mathbf{\Sigma}} \tag{10.8.8}$$

Similarly, the matrix $\mathbf{H}$ can be written

$$\begin{aligned}\mathbf{H} &= (\mathbf{1}\hat{\boldsymbol{\mu}}' - \mathbf{1}\boldsymbol{\mu}')'(\mathbf{1}\hat{\boldsymbol{\mu}}' - \mathbf{1}\boldsymbol{\mu}') = (\hat{\boldsymbol{\mu}}' - \boldsymbol{\mu}')'\mathbf{1}'\mathbf{1}(\hat{\boldsymbol{\mu}}' - \boldsymbol{\mu}') \\ &= n(\hat{\boldsymbol{\mu}} - \boldsymbol{\mu})(\hat{\boldsymbol{\mu}} - \boldsymbol{\mu})' \end{aligned} \tag{10.8.9}$$

Diagonal elements of $\mathbf{H}$ are squared discrepancies between elements of $\boldsymbol{\mu}$ and their respective estimators, each scaled by n. Each off-diagonal entry in $\mathbf{H}$ is a product of the estimation discrepancies for two elements of $\boldsymbol{\mu}$, again scaled by n.

To show that $\mathbf{G}$ and $\mathbf{H}$ are independently distributed Wishart matrices, we first write $\mathbf{G}$, substituting from (10.8.7), in the form

$$\begin{aligned}\mathbf{G} &= \left[\mathbf{Y} - \mathbf{1}(\mathbf{1}'\mathbf{1})^{-1}\mathbf{1}'\mathbf{Y}\right]'\left[\mathbf{Y} - \mathbf{1}(\mathbf{1}'\mathbf{1})^{-1}\mathbf{1}'\mathbf{Y}\right] \\ &= \mathbf{Y}'\left[\mathbf{I} - \mathbf{1}(\mathbf{1}'\mathbf{1})^{-1}\mathbf{1}'\right]\left[\mathbf{I} - \mathbf{1}(\mathbf{1}'\mathbf{1})^{-1}\mathbf{1}'\right]\mathbf{Y} \\ &= \mathbf{Y}'\left[\mathbf{I} - \mathbf{1}(\mathbf{1}'\mathbf{1})^{-1}\mathbf{1}'\right]\mathbf{Y}\end{aligned} \tag{10.8.10}$$

and recognize the familiar idempotent form. Further, compare this result with

$$\mathbf{J} = [\mathbf{Y} - \mathbf{E}(\mathbf{Y})]'\left[\mathbf{I} - \mathbf{1}(\mathbf{1}'\mathbf{1})^{-1}\mathbf{1}'\right][\mathbf{Y} - \mathbf{E}(\mathbf{Y})]$$

which, on expansion, yields

$$\mathbf{J} = \mathbf{G} + \mathbf{Y}'\left[\mathbf{I} - \mathbf{1}(\mathbf{1}'\mathbf{1})^{-1}\mathbf{1}'\right]\mathbf{E}(\mathbf{Y}) + \mathbf{E}(\mathbf{Y})'\left[\mathbf{I} - \mathbf{1}(\mathbf{1}'\mathbf{1})^{-1}\mathbf{1}'\right]\mathbf{Y}$$
$$+\mathbf{E}(\mathbf{Y})'\left[\mathbf{I} - \mathbf{1}(\mathbf{1}'\mathbf{1})^{-1}\mathbf{1}'\right]\mathbf{E}(\mathbf{Y})$$

By substituting [from Eq. (10.8.6)] $\mathbf{1}\boldsymbol{\mu}'$ for $\mathbf{E}(\mathbf{Y})$, the second, third, and fourth terms are each null. Hence, (10.8.10) may be written

$$\mathbf{G} = [\mathbf{Y} - \mathbf{E}(\mathbf{Y})]'\left[\mathbf{I} - \mathbf{1}(\mathbf{1}'\mathbf{1})^{-1}\mathbf{1}'\right][\mathbf{Y} - \mathbf{E}(\mathbf{Y})] \qquad (10.8.11)$$

The matrix $\mathbf{G}$ is in the form of a Wishart matrix. The matrix $\mathbf{A} = \mathbf{I} - \mathbf{1}(\mathbf{1}'\mathbf{1})^{-1}\mathbf{1}'$ we know to be idempotent. We also know $\mathbf{A}$ has rank $n - 1$. These results are sufficient to establish that the $\mathbf{G}$ of (10.7.8) is Wishart distributed:

$$\mathbf{G} = (\mathbf{Y} - \mathbf{1}\hat{\boldsymbol{\mu}}')'(\mathbf{Y} - \mathbf{1}\hat{\boldsymbol{\mu}}) \sim W_m(n - 1, \boldsymbol{\Sigma}) \qquad (10.8.12)$$

If we substitute from Eq. (10.8.7) for $\hat{\mathbf{E}}(\mathbf{Y})$, the matrix $\mathbf{H}$ can be rewritten

$$\mathbf{H} = \left[\mathbf{1}(\mathbf{1}'\mathbf{1})^{-1}\mathbf{1}'\mathbf{Y} - \mathbf{E}(\mathbf{Y})\right]'\left[\mathbf{1}(\mathbf{1}'\mathbf{1})^{-1}\mathbf{1}'\mathbf{Y} - \mathbf{E}(\mathbf{Y})\right] \qquad (10.8.13)$$

This may be expressed in recognizable Wishart form by first noticing that from Eq. (10.8.6) we can write $\mathbf{1}'\mathbf{E}(\mathbf{Y}) = \mathbf{1}'\mathbf{1}\boldsymbol{\mu}'$ or

$$\mathbf{1}(\mathbf{1}'\mathbf{1})^{-1}\mathbf{1}'\mathbf{E}(\mathbf{Y}) = \mathbf{1}\boldsymbol{\mu}'$$

and, from (10.8.6),

$$\mathbf{1}(\mathbf{1}'\mathbf{1})^{-1}\mathbf{1}'\mathbf{E}(\mathbf{Y}) = \mathbf{E}(\mathbf{Y})$$

If we substitute the term on the left for $\mathbf{E}(\mathbf{Y})$ in (10.8.13), $\mathbf{H}$ becomes

$$\mathbf{H} = [\mathbf{Y} - \mathbf{E}(\mathbf{Y})]'\left[\mathbf{1}(\mathbf{1}'\mathbf{1})^{-1}\mathbf{1}'\right]\left[\mathbf{1}(\mathbf{1}'\mathbf{1})^{-1}\mathbf{1}'\right][\mathbf{Y} - \mathbf{E}(\mathbf{Y})]$$
$$= [\mathbf{Y} - \mathbf{E}(\mathbf{Y})]'\left[\mathbf{1}(\mathbf{1}'\mathbf{1})^{-1}\mathbf{1}'\right][\mathbf{Y} - \mathbf{E}(\mathbf{Y})] \qquad (10.8.14)$$

The form of $\mathbf{H}$ is now that of a Wishart matrix. The matrix $\mathbf{B} = [\mathbf{1}(\mathbf{1}'\mathbf{1})^{-1}\mathbf{1}']$ is idempotent and a major product of vectors, so $r(\mathbf{B}) = 1$. As a result, $\mathbf{H}$ in Eq. (10.8.9) is also a Wishart observation:

$$\mathbf{H} = n(\hat{\boldsymbol{\mu}} - \boldsymbol{\mu})(\hat{\boldsymbol{\mu}} - \boldsymbol{\mu})' \sim W_m(1, \boldsymbol{\Sigma}) \qquad (10.8.15)$$

Finally, the two Wishart matrices are independently distributed:

$$\mathbf{AB} = \left[\mathbf{I} - \mathbf{1}(\mathbf{1}'\mathbf{1})^{-1}\mathbf{1}'\right]\left[\mathbf{1}(\mathbf{1}'\mathbf{1})^{-1}\mathbf{1}'\right] = \mathbf{0}$$

Given the sampling model of Section 10.7, the maximum likelihood estimator of $\boldsymbol{\Sigma}$ is independent of (the major product moment of) the discrepancy between $\boldsymbol{\mu}$ and its maximum likelihood estimator: Scaling each by n yields independently distributed Wishart matrices. The reader may sense that this result is not unlike the finding in Section 8.4 that $\mathbf{e}'\mathbf{e}/\sigma^2$ and $(1/\sigma^2)[\hat{\mathbf{E}}(\mathbf{y}) - \mathbf{E}(\mathbf{y})]'[\hat{\mathbf{E}}(\mathbf{y}) - \mathbf{E}(\mathbf{y})]$ are independently distributed chi-squared observations.

Though they are specialized forms, Wishart matrices are nonetheless still two-dimensional arrays of numbers. It is not the distribution of these matrices themselves but the distribution of one or more scalar functions of these matrices that is useful in multivariate analysis.

10.8.2 Distribution of Products of Basic Diagonals

To facilitate the development of the distribution of scalars based on Wishart matrices, assume $\mathbf{Y}_{i.}$ is a vector observation from the m-variate normal with expectation $\boldsymbol{\mu}$ and variance $\boldsymbol{\Sigma} = \mathbf{I}$. The matrix $\mathbf{Y}$ consists of n such independent observations. The basic structure of the minor product moment of $\mathbf{Y} - \mathbf{1}\boldsymbol{\mu}'$ can be symbolized $(\mathbf{Y} - \mathbf{1}\boldsymbol{\mu}')'(\mathbf{Y} - \mathbf{1}\boldsymbol{\mu}') = \mathbf{Q}\Delta^2\mathbf{Q}'$. Our first goal is to describe the distribution of $\prod_{j=1}^{2}\Delta_j^2$ over repeated samples of n vector observations.

We begin by transforming the sample $n \times m$ matrix $\mathbf{Y}$ into a second matrix $\mathbf{Z}$ of the same order according to the following rules:

$$\mathbf{Z}_{.1} = \mathbf{Y}_{.1}$$

$$\mathbf{Z}_{.2} = (\mathbf{Y}_{.2}|\mathbf{Y}_{.1}) = (\mathbf{Y}_{.2}|\mathbf{Z}_{.1})$$

$$\mathbf{Z}_{.3} = (\mathbf{Y}_{.3}|\mathbf{Y}_{.1}, \mathbf{Y}_{.2}) = (\mathbf{Y}_{.3}|\mathbf{Z}_{.1}, \mathbf{Z}_{.2})$$

and, in general,

$$\mathbf{Z}_{.k} = (\mathbf{Y}_{.k}|\mathbf{Y}_{.1}, \ldots, \mathbf{Y}_{.(k-1)}) = (\mathbf{Y}_{.k}|\mathbf{Z}_{.1}, \ldots, \mathbf{Z}_{.(k-1)}) \quad (10.8.16)$$

Each successive column of $\mathbf{Z}$ is *conditioned* on all preceding columns.

We shall make use of the Chapter 7 discussion of partial variables to describe these conditional attributes. If we use the notation ${}_{(k-1)}\mathbf{Y}$ for the matrix consisting of the first $k - 1$ columns of $\mathbf{Y}$, we can write $\mathbf{Z}_{.k}$ as

$$\mathbf{Z}_{.k} = \left\{\mathbf{I} - {}_{(k-1)}\mathbf{Y}\left[{}_{(k-1)}\mathbf{Y}'{}_{(k-1)}\mathbf{Y}\right]^{-1}{}_{(k-1)}\mathbf{Y}'\right\}\mathbf{Y}_{.k} \quad (10.8.17)$$

This equation has the same general form as $\mathbf{y.X} = [\mathbf{I} - \mathbf{X}(\mathbf{X}'\mathbf{X})^{-1}\mathbf{X}']\mathbf{y}$, which gives the residual part of $\mathbf{y}$, the result of partialing the columns of $\mathbf{X}$ from $\mathbf{y}$. In the same way we have partialed from $\mathbf{Y}_{.k}$ the effects of $\mathbf{Y}_{.1}$ through $\mathbf{Y}_{.(k-1)}$ in producing $\mathbf{Z}_{.k}$. The vector $\mathbf{y.X}$ is orthogonal to the columns of $\mathbf{X}$ and $\mathbf{Z}_{.k}$ is orthogonal to $\mathbf{Z}_{.j}$ for $j = 1, 2, \ldots, k - 1$.

The transformed $\mathbf{Z}_{.k}$ may also be expressed $\mathbf{Z}_{.k} = \mathbf{Y}\mathbf{B}_{.k}$, where $\mathbf{B}_{.k}$ is the vector

$$\mathbf{B}_{.k} = \left[\begin{array}{c} -\left[{}_{(k-1)}\mathbf{Y}'{}_{(k-1)}\mathbf{Y}\right]^{-1}{}_{(k-1)}\mathbf{Y}'\mathbf{Y}_{.k} \\ \hdashline 1 \\ \hdashline 0 \\ \hdashline \end{array}\right] \begin{array}{c} k-1 \\ k \\ m \end{array}$$

This permits us to write the collection of transformations as $\mathbf{Z} = \mathbf{YB}$ or

$$\mathbf{Z} - \mathbf{E}(\mathbf{Z}) = [\mathbf{Y} - \mathbf{E}(\mathbf{Y})]\mathbf{B} \tag{10.8.18}$$

We next examine the minor product moment of this matrix. Because each successive column of $\mathbf{Z} - \mathbf{E}(\mathbf{Z})$ is conditioned on all earlier columns, the matrix is orthogonal by column and its minor product moment is diagonal: $\mathbf{D} = [\mathbf{Z} - \mathbf{E}(\mathbf{Z})]'[\mathbf{Z} - \mathbf{E}(\mathbf{Z})]$. If we assume $\mathbf{Y}$ to be of full rank, the diagonal elements of $\mathbf{D}$ are all positive. In fact, $\mathbf{D}$ is a permutation of the basic diagonal of $[\mathbf{Z} - \mathbf{E}(\mathbf{Z})]'[\mathbf{Z} - \mathbf{E}(\mathbf{Z})]$ and $\prod_{j=1}^{m} D_j$ is the product of these basic diagonal elements.

For $m = 2$ we were able to show in Section 10.5 that the product of the basic diagonals of $[\mathbf{Y} - \mathbf{E}(\mathbf{Y})]'[\mathbf{Y} - \mathbf{E}(\mathbf{Y})] = \mathbf{Q}\boldsymbol{\Delta}^2\mathbf{Q}'$ is identical to the product of the basic diagonals of $[\mathbf{Z} - \mathbf{E}(\mathbf{Z})]'[\mathbf{Z} - \mathbf{E}(\mathbf{Z})] = \mathbf{qDq}'$. This result generalizes to $m > 2$, and we can write

$$\prod_{j=1}^{m} \Delta_j^2 = \prod_{j=1}^{m} D_j = \prod_{j=1}^{m} [\mathbf{Z} - \mathbf{E}(\mathbf{Z})]'_{.j}[\mathbf{Z} - \mathbf{E}(\mathbf{Z})]_{.j} \tag{10.8.19}$$

We may now deduce some properties of the terms on the right. Each column of $\mathbf{Y} - \mathbf{E}(\mathbf{Y})$ consists of n independent observations from a normal distribution with mean 0 and variance 1. We use this to show that each of the m terms on the right of Eq. (10.8.19) is a chi-squared observation. First, as

$$[\mathbf{Z} - \mathbf{E}(\mathbf{Z})]_{.1} = [\mathbf{Y} - \mathbf{E}(\mathbf{Y})]_{.1}$$

$D_1 = [\mathbf{Z} - \mathbf{E}(\mathbf{Z})]'_{.1}[\mathbf{Z} - \mathbf{E}(\mathbf{Z})]_{.1}$ is a sum of squares of n independent normally distributed observations each with mean 0 and variance 1. Consequently, D_1 is an observation from the χ^2 distribution with n degrees of freedom.

Each successive D_j, based on Eq. (10.8.17), can be expressed

$$D_j = [\mathbf{Y} - \mathbf{E}(\mathbf{Y})]'_{.j}\left\{\mathbf{I} - {}_{(j-1)}\mathbf{Y}\left[{}_{(j-1)}\mathbf{Y}'{}_{(j-1)}\mathbf{Y}\right]^{-1}{}_{(j-1)}\mathbf{Y}'\right\}[\mathbf{Y} - \mathbf{E}(\mathbf{Y})]_{.j}$$

That is, D_j is a *quadratic transformation* of a vector, taking the form $D_j = [\mathbf{y} - \mathbf{E}(\mathbf{y})]'\mathbf{A}[\mathbf{y} - \mathbf{E}(\mathbf{y})]$. The vector $\mathbf{y} = \mathbf{Y}_{.j}$ consists of n independent and normally distributed observations. Each observation is assumed to be from a population with variance of 1 although the populations sampled may have different means. Thus, associated with the vector $\mathbf{y} = \mathbf{Y}_{.j}$ is an expectation vector $\mathbf{E}(\mathbf{y}) = \mathbf{E}(\mathbf{Y}_{.j})$ and a variance–covariance matrix $\mathbf{Var}(\mathbf{y}) = \mathbf{Var}(\mathbf{Y}_{.j}) = \mathbf{I}$. Also, the matrix of the quadratic form $\mathbf{A} = [\mathbf{I} - \mathbf{X}(\mathbf{X}'\mathbf{X})^{-1}\mathbf{X}']$ is idempotent. We have $\mathbf{A}\,\mathbf{Var}(\mathbf{y})\,\mathbf{A} = \mathbf{AIA} = \mathbf{A}$ and the conditions of the theorem of Section 8.3 are satisfied: D_j is an observation from the chi-squared distribution with $r(\mathbf{A})$ degrees of freedom. The idempotent

$$\mathbf{A} = \left\{\mathbf{I} - {}_{(j-1)}\mathbf{Y}\left[{}_{(j-1)}\mathbf{Y}'{}_{(j-1)}\mathbf{Y}\right]^{-1}{}_{(j-1)}\mathbf{Y}\right\}$$

has a rank $n - (j - 1)$. Each D_j contributing to the product in Eq. (10.8.19) is a chi-squared observation with $n - (j - 1)$ degrees of freedom, $j = 1, 2, \ldots, m$.

Further, as the columns of $\mathbf{Z} - \mathbf{E(Z)}$ are mutually orthogonal, these m chi-squared observations are independently distributed. We conclude that the right side of Eq. (10.8.19) is a product of m independently distributed chi-squared observations.

So far we are able to say that

$$\prod_{j=1}^{m} \Delta_j^2 \sim \prod_{j=1}^{m} \chi^2_{n-(j-1)\,\mathrm{df}} \tag{10.8.20}$$

where Δ^2 is the basic diagonal of a sample $[\mathbf{Y} - \mathbf{E(Y)}]'[\mathbf{Y} - \mathbf{E(Y)}]$ when $\mathbf{Var}(\mathbf{Y}_{i.}) = \mathbf{\Sigma}$ is taken to be $\mathbf{I}$. We have found a way to relate a particular Wishart matrix to a scalar with a specifiable sampling distribution. Based on a sequence of sample $\mathbf{Y}$ matrices, the sampling distribution of the product of the basic diagonal elements of the Wishart matrix $\mathbf{W} = [\mathbf{Y} - \mathbf{E(Y)}]'\mathbf{I}[\mathbf{Y} - \mathbf{E(Y)}]$ is given by the right-hand side of Eq. (10.8.20).

In applied multivariate analysis we need to extend our approach to the distribution of basic diagonal products to the general Wishart matrix:

$$\mathbf{W} = [\mathbf{Y} - \mathbf{E(Y)}]'\mathbf{A}[\mathbf{Y} - \mathbf{E(Y)}] \sim \mathbf{W}_m[r(\mathbf{A}), \mathbf{\Sigma}]$$

$\mathbf{Var}(\mathbf{Y}_{i.}) = \mathbf{\Sigma}$, and $\mathbf{AA} = \mathbf{A}$. How can we generalize our result?

First, if we express the basic structure of the arbitrary $\mathbf{\Sigma}$ as $\mathbf{\Sigma} = \mathbf{qd}^2\mathbf{q}'$, then the product

$$(\mathbf{d}^{-1}\mathbf{q}')\mathbf{W}(\mathbf{qd}^{-1}) = (\mathbf{d}^{-1}\mathbf{q}')[\mathbf{Y} - \mathbf{E(Y)}]'\mathbf{A}[\mathbf{Y} - \mathbf{E(Y)}](\mathbf{qd}^{-1})$$

is also a Wishart matrix:

$$(\mathbf{d}^{-1}\mathbf{q}')\mathbf{W}(\mathbf{qd}^{-1}) = [\mathbf{Z} - \mathbf{E(Z)}]'\mathbf{A}[\mathbf{Z} - \mathbf{E(Z)}] \tag{10.8.21}$$

in the multivariate normal sample

$$\mathbf{Z} - \mathbf{E(Z)} = [\mathbf{Y} - \mathbf{E(Y)}](\mathbf{qd}^{-1}) \tag{10.8.23}$$

where the transformation from $\mathbf{Y}$ to $\mathbf{Z}$ ensures that $\mathbf{Var}(\mathbf{Z}_{i.}) = \mathbf{I}$. That is, in the population our transformation would have been a spherizing one.

Thus, for $\mathbf{A} = \mathbf{I}$ at least, we know that the product of the basic diagonals of $(\mathbf{d}^{-1}\mathbf{q}')\mathbf{W}(\mathbf{qd}^{-1})$ is distributed as the product of m independent chi-squared observations. We can broaden this result by recalling from Chapter 4 that the general symmetric idempotent matrix has a basic structure of the form $\mathbf{A} = \mathbf{PIP}' = \mathbf{PP}'$ where, of course, $\mathbf{P}'\mathbf{P} = \mathbf{I}$. Substituting this basic structure for $\mathbf{A}$ in Eq. (10.8.21) gives

$$(\mathbf{d}^{-1}\mathbf{q}')\mathbf{W}(\mathbf{qd}^{-1}) = [\mathbf{Z} - \mathbf{E(Z)}]'\mathbf{PP}'[\mathbf{Z} - \mathbf{E(Z)}]$$

The matrix $\mathbf{P}$ is $n \times r(\mathbf{A})$, so the product $\mathbf{P}'[\mathbf{Z} - \mathbf{E(Z)}]$ is $r(\mathbf{A}) \times m$.

What is the nature of $\mathbf{P}'[\mathbf{Z} - \mathbf{E(Z)}]$? First, we rename it

$$\mathbf{X} = \mathbf{P}'[\mathbf{Z} - \mathbf{E(Z)}] \tag{10.8.24}$$

The matrix $\mathbf{X}$ represents transformations of $\mathbf{Z} - \mathbf{E}(\mathbf{Z})$, but these are transformations of the rows rather than the columns. Each row of $\mathbf{X}$ is a weighed sum of n independent multivariate normal observations each with mean $\mathbf{0}'$ and variance $\mathbf{I}$. This is sufficient to ensure that each of the $r(\mathbf{A})$ rows of $\mathbf{X}$ is itself a multivariate normal observation.

What means, variances, and covariances can be assigned the multivariate rows of $\mathbf{X}$? Because the rows being transformed are independent and expressed as deviations from the population mean, the expected values of the transformed rows must also be the expectations of deviation vectors or null vectors. Equivalently the rows of $\mathbf{X}$ are in population deviation form, and we may write the transformation of Eq. (10.8.24)

$$\mathbf{X} - \mathbf{E}(\mathbf{X}) = \mathbf{P}'[\mathbf{Z} - \mathbf{E}(\mathbf{Z})] \tag{10.8.25}$$

The variance–covariance structure of $\mathbf{X} - \mathbf{E}(\mathbf{X})$ is seen more clearly if we rewrite this matrix and $\mathbf{Z} - \mathbf{E}(\mathbf{Z})$ as vectors. In particular, express $\mathbf{Z} - \mathbf{E}(\mathbf{Z})$ as an $n \times m$ vector $\mathbf{v_Z}$ in which the first m elements, say, are the first row of $\mathbf{Z} - \mathbf{E}(\mathbf{Z})$ arranged in column form:

$$\mathbf{v_Z} = \begin{bmatrix} [\mathbf{Z} - \mathbf{E}(\mathbf{Z})]_{1.} \\ \hdashline [\mathbf{Z} - \mathbf{E}(\mathbf{Z})]_{2.} \\ \hdashline \vdots \\ [\mathbf{Z} - \mathbf{E}(\mathbf{Z})]_{n.} \\ \hdashline \end{bmatrix} \begin{matrix} m \\ 2m \\ \\ nm \end{matrix}$$

In the same fashion, $\mathbf{X} - \mathbf{E}(\mathbf{X})$ can be put in the vector form

$$\mathbf{v_X} = \begin{bmatrix} [\mathbf{X} - \mathbf{E}(\mathbf{X})]_{1.} \\ \hdashline [\mathbf{X} - \mathbf{E}(\mathbf{X})]_{2.} \\ \hdashline \vdots \\ [\mathbf{X} - \mathbf{E}(\mathbf{X})]_{r(\mathbf{A}).} \\ \hdashline \end{bmatrix} \begin{matrix} m \\ 2m \\ \\ r(A)m \end{matrix}$$

Based on these representations, we can express the variance of $\mathbf{v_Z}$ as an $nm \times nm$ matrix, $\mathbf{Var}(\mathbf{v_Z})$. In this matrix, the covariance between scalar observations drawn from different rows of $\mathbf{Z} - \mathbf{E}(\mathbf{Z})$ must be zero, as the rows were selected independently. As a result $\mathbf{Var}(\mathbf{v_Z})$ has the form

$$\mathbf{Var}(\mathbf{v_Z}) = \begin{bmatrix} \mathbf{Var}(\mathbf{Z}_{1.}) & & & 0 \\ & \mathbf{Var}(\mathbf{Z}_{2.}) & & \\ & & \ddots & \\ 0 & & & \mathbf{Var}(\mathbf{Z}_{n.}) \end{bmatrix}$$

where each of the elements is an $m \times m$ matrix. However, we know $\mathbf{Z} - \mathbf{E}(\mathbf{Z})$ is defined so that $\mathbf{Var}(Z_{i.}) = \mathbf{I}$, and as a result $\mathbf{Var}(\mathbf{v_Z}) = \mathbf{I}$ as well.

The variance of $\mathbf{v_X}$, of interest to us, is given by the product

$$\mathbf{Var}(\mathbf{v_X}) = \mathbf{G'Var}(\mathbf{v_Z})\mathbf{G} = \mathbf{G'G} \tag{10.8.26}$$

where $\mathbf{G'}$ is the matrix needed for the transformation

$$\mathbf{v_X} = \mathbf{G'v_Z}$$

This is the same transformation as given by Eq. (10.8.25) so $\mathbf{G'}$ must be some form of $\mathbf{P'}$. Dimensional considerations alone tell us that $\mathbf{G'}$ must have $r(\mathbf{A})$ rows and nm columns. The typical row of $\mathbf{G'}$ consists of the n elements of a particular column of $\mathbf{P}$ and $(m - 1)n$ zeros. As examples, $\mathbf{G'_{.1}}$, $\mathbf{G'_{.2}}$ and $\mathbf{G'_{.4}}$ are as follows, illustrating for $m = 3$, $n = 4$:

$$(P_{11}, 0, 0, P_{21}, 0, 0, P_{31}, 0, 0, P_{41}, 0, 0)$$

$$(0, P_{11}, 0, 0, P_{21}, 0, 0, P_{31}, 0, 0, P_{41}, 0)$$

$$(P_{12}, 0, 0, P_{22}, 0, 0, P_{32}, 0, 0, P_{42}, 0, 0)$$

respectively.

Now $\mathbf{P'P} = \mathbf{I}$, so the computation of $\mathbf{Var}(\mathbf{v_X}) = \mathbf{G'G}$ from Eq. (10.8.26) yields

$$\mathbf{Var}(\mathbf{v_X}) = \begin{bmatrix} \mathbf{I} & & & & \\ & \mathbf{I} & & 0 & \\ & & \mathbf{I} & & \\ & 0 & & \ddots & \\ & & & & \mathbf{I} \end{bmatrix}$$

where each of the identity matrices is $m \times m$ and can be interpreted as the variance of a row of $\mathbf{X} - \mathbf{E(X)}$; that is,

$$\mathbf{Var}(\mathbf{X}_{i.}) = \mathbf{I} \tag{10.8.27}$$

Each off-diagonal matrix element in $\mathbf{Var}(\mathbf{v_X})$ is also $m \times m$ and is a covariance matrix between rows of $\mathbf{X} - \mathbf{E(X)}$.

The importance of this result is that, based on (10.8.23) and (10.8.25), we may now write

$$[\mathbf{X} - \mathbf{E(X)}]'[\mathbf{X} - \mathbf{E(X)}] = (\mathbf{d}^{-1}\mathbf{q'})\mathbf{W}(\mathbf{q}\mathbf{d}^{-1}) \tag{10.8.28}$$

where

1. $\mathbf{X}_{i.}$ has been shown to have variance, $\mathbf{Var}(\mathbf{X}_{i.}) = \mathbf{I}$;
2. the rows of $\mathbf{X} - \mathbf{E(X)}$ are independent, and
3. $\mathbf{X} - \mathbf{E(X)}$ consists of $r(\mathbf{A})$ rows.

All the conditions have now been met for the product of the basic diagonals of $(\mathbf{d}^{-1}\mathbf{q'})\mathbf{W}(\mathbf{q}\mathbf{d}^{-1})$ to be distributed as the product of independent chi-squared observations. If the basic diagonal elements of

$(\mathbf{d}^{-1}\mathbf{q}')\mathbf{W}(\mathbf{q}\mathbf{d}^{-1})$ are $g_j, j = 1, 2, \ldots, m$, then

$$\prod_{j=1}^{m} g_j \sim \prod_{j=1}^{m} \chi^2_{r(\mathbf{A}) - (j-1)\,\mathrm{df}}$$

provided that $r(\mathbf{A})$ is not smaller than m.

In summary, if $\mathbf{W}$ is a Wishart form in the m-variate normal $\mathbf{Y}$,

$$\mathbf{W} = [\mathbf{Y} - \mathbf{E}(\mathbf{Y})]'\mathbf{A}[\mathbf{Y} - \mathbf{E}(\mathbf{Y})] \sim \mathbf{W}_m[r(\mathbf{A}), \Sigma]$$

if the rank $r(\mathbf{A})$ of the symmetric idempotent matrix $\mathbf{A}$ is not less than m, and if the basic structure of $\mathbf{Var}(\mathbf{Y}_{i.}) = \boldsymbol{\Sigma} = \mathbf{q}\mathbf{d}^2\mathbf{q}'$, then the product of the basic diagonals of $[(\mathbf{d}^{-1}\mathbf{q}')\mathbf{W}(\mathbf{q}\mathbf{d}^{-1})]$ is distributed as the product of m independent chi-squared observations:

$$\prod_{j=1}^{m} \chi^2_{r(\mathbf{A}) - (j-1)\,\mathrm{df}}$$

We now have a principle that gives information about the sampling distribution of the product of the basic diagonals of any Wishart form. It can be more conveniently stated, however, if we apply two properties of the product of basic diagonals. For this purpose let $\mathbf{A}$ and $\mathbf{B}$ be square basic matrices of the same order with basic diagonals ${}_{\mathbf{A}}\mathbf{D}$ and ${}_{\mathbf{B}}\mathbf{D}$. The first principle, already familiar, is that the product of the basic diagonals of $\mathbf{A}^{-1}$ is the reciprocal of the product of the basic diagonals of $\mathbf{A}$:

$$\prod_{j=1}^{m} {}_{(\mathbf{A}^{-1})}D_j = 1 \Big/ \prod_{j=1}^{m} {}_{\mathbf{A}}D_j$$

The second principle states that the matrix $\mathbf{C} = \mathbf{AB}$ has basic diagonal ${}_{\mathbf{C}}\mathbf{D}$ related to those of $\mathbf{A}$ and $\mathbf{B}$ by

$$\prod_{j=1}^{m} {}_{\mathbf{C}}D_j = \left(\prod_{j=1}^{m} {}_{\mathbf{A}}D_j\right)\left(\prod_{j=1}^{m} {}_{\mathbf{B}}D_j\right)$$

By this second principle, the product of the basic diagonals of $(\mathbf{d}^{-1}\mathbf{q}')\mathbf{W}(\mathbf{q}\mathbf{d}^{-1})$ can be expressed as the product of three basic diagonal products:

$$\left(\prod_{j=1}^{m} {}_{(\mathbf{d}^{-1}\mathbf{q}')}D_j\right)\left(\prod_{j=1}^{m} {}_{\mathbf{W}}D_j\right)\left(\prod_{j=1}^{m} {}_{(\mathbf{q}\mathbf{d}^{-1})}D_j\right)$$

The first and third matrices, however, are inverses of the matrices $\mathbf{qd}$ and $\mathbf{dq}'$, each of the latter with basic diagonal $\mathbf{d}$. Applying the first principle, then, the product of the basic diagonals of $(\mathbf{d}^{-1}\mathbf{q}')\mathbf{W}(\mathbf{q}\mathbf{d}^{-1})$ can be written

$$\left(1 \Big/ \prod_{j=1}^{m} d_j\right) \prod_{j=1}^{m} {}_{\mathbf{W}}D_j \left(1 \Big/ \prod_{j=1}^{m} d_j\right) = \left(1 \Big/ \prod_{j=1}^{m} d_j^2\right) \prod_{j=1}^{m} {}_{\mathbf{W}}D_j$$

The distribution of basic diagonals of Wishart matrices may now be restated. Let **W** be a Wishart form in the m-variate normal **Y**:

$$\mathbf{W} = [\mathbf{Y} - \mathbf{E}(\mathbf{Y})]'\mathbf{A}[\mathbf{Y} - \mathbf{E}(\mathbf{Y})] \sim \mathbf{W}_m[r(\mathbf{A}), \boldsymbol{\Sigma}]$$

If the rank $r(\mathbf{A})$ of the symmetric idempotent **A** is not less than **m** and if the basic diagonal of $\boldsymbol{\Sigma}$ is $\mathbf{d}^2$, then the product of the basic diagonal of **W** is distributed as

$$\prod_{j=1}^{m} {}_{\mathbf{W}}D_j \sim \prod_{j=1}^{m} d_j^2 \prod_{j=1}^{m} \chi^2_{r(\mathbf{A})-(j-1)\,\mathrm{df}}$$

where the m chi-squared observations are mutually independent.

This principle can be illustrated for the sample variance–covariance matrix maximum likelihood estimator $\hat{\boldsymbol{\Sigma}}$. Beginning with Eq. (10.8.12), which established that

$$n\hat{\boldsymbol{\Sigma}} = (\mathbf{Y} - \mathbf{1}\hat{\boldsymbol{\mu}}')'(\mathbf{Y} - \mathbf{1}\hat{\boldsymbol{\mu}}') \sim \mathbf{W}_m[(n-1), \boldsymbol{\Sigma}]$$

we are able to state that the product of the basic diagonals of $n\hat{\boldsymbol{\Sigma}}$ is distributed as

$$(1/n)^m \prod_{j=1}^{m} d_j^2 \prod_{j=1}^{m} \chi^2_{n-j\,\mathrm{df}}$$

Because each of the basic diagonals of $n\hat{\boldsymbol{\Sigma}}$ is exactly n times the magnitude of the corresponding basic diagonal of $\hat{\boldsymbol{\Sigma}}$, it is also true that the product of the basic diagonals of $\hat{\boldsymbol{\Sigma}}$ is distributed as

$$(1/n)^m \prod_{j=1}^{m} d_j^2 \prod_{j=1}^{m} \chi^2_{n-j\,\mathrm{df}}$$

As the distribution of the product of basic diagonals for the general Wishart matrix involves the basic diagonal $\mathbf{d}^2$ of the usually unknown $\boldsymbol{\Sigma}$, it is clear we must either have a hypothesis about $\boldsymbol{\Sigma}$ sufficient to establish at least the product of its basic diagonals or else we must find a way of canceling d_j^2 from the expression. The situation is comparable, indeed, to the result in Chapter 8 that each of the relevant chi-squared terms in the GLM was a ratio involving the unknown σ^2. There we took advantage of the theorem that the ratio of two independently distributed chi-squared observations, a ratio now cleared of the unknown σ^2, had a well-known sampling distribution. To look ahead, we shall find in the next chapter a comparable solution for evaluating basic diagonal products.

10.9 Summary

We began Chapter 10 by emphasizing the interpretation of the data as jointly distributed observations on a multivariate attribute. The multivariate attribute was assumed to define a multivariate normal distribution. Basic to exploring such multivariate attributes is an understanding of this distribution. We generated the expression for the multivariate normal likelihood by assuming we have independent observations on m attributes. Univariate and bivariate normal likelihood distributions were then shown to be special cases of the multivariate normal likelihood.

Discussion of the bivariate normal likelihood led to the definition of population zero-order partial and multiple correlations. Within the context of bivariate normal random variables, marginal and conditional distributions were derived and illustrated. We showed that if attributes are multivariate normal, then the marginal distribution of each attribute is univariate normal and that the distribution of an attribute conditional on any partition of the multivariate attribute is conditionally normal. We also defined the generalized variance as the product of the basic diagonal elements of the variance–covariance matrix of the set of attributes.

Further discussion of the sampling distribution of the maximum likelihood estimators of $\boldsymbol{\mu}$ and $\boldsymbol{\Sigma}$ indicated that to avoid working with the sampling distribution of $\hat{\boldsymbol{\Sigma}}$, which is a distribution of matrices, statisticians rely on generalized variance measures to judge the degree of fit of hypothesized and constrained variance–covariance matrices. The discussion of sampling distributions in the case of univariate criteria was extended to show that the MLEs of $n\hat{\boldsymbol{\Sigma}}$ and $n(\hat{\boldsymbol{\mu}} - \boldsymbol{\mu})'(\hat{\boldsymbol{\mu}} - \boldsymbol{\mu})$ are independently distributed Wishart matrices. Further, we showed that the product of the basic diagonals of $\hat{\boldsymbol{\Sigma}}$ is distributed as the product of m independent chi-squared observations. These results prepare us to consider hypothesis testing in the multivariate general linear model.

Exercises

1. Let

$$\mathbf{S} = \begin{bmatrix} S_{11} & S_{12} \\ S_{21} & S_{22} \end{bmatrix}$$

be a basic variance–covariance matrix. Find a nonsingular matrix **G** such that $\mathbf{G}'\mathbf{S}\mathbf{G} = \mathbf{I}$. Show that any number of matrices can be formed to satisfy this condition. [*Hint:* Consider the basic structure of **S**.]

2. Generalize the results of exercise 1 to a $k \times k$ basic variance–covariance matrix.

3. Let

$$\mathbf{S} = \begin{bmatrix} 25 & 10 \\ 10 & 16 \end{bmatrix}$$

be a variance–covariance matrix. Find the generalized variance of **S**. What is the product of the basic diagonals of $\mathbf{S}^{-1}$?

4. Let

$$\mathbf{C} = \begin{bmatrix} 25 & 10 & 5 \\ 10 & 20 & 0 \\ 5 & 0 & 15 \end{bmatrix}$$

be a variance–covariance matrix among the three variates **x**, **y**, and **z** (in that order).

a. Construct the partial correlation matrix for **y**.**x** and **z**.**x**.

b. Find the variance of the partial variable **x**.**yz**.

Additional Reading

The following two references discuss sampling from the multivariate normal at a more detailed and somewhat more technical level:

Hogg, R. V., and Craig, A. T. (1970). *Introduction to Mathematical Statistics*, 3rd ed. New York: Macmillan.

Srivastava, M. S., and Khatri, C. G. (1979). *An Introduction to Multivariate Statistics*. New York: North-Holland.

Chapter 11

The Multivariate General Linear Model: Some Principles

Introduction

In this chapter we extend systematically the general linear model to the multivariate criterion. We focus, as we did in Chapter 8, on the use of sets of linear constraints to phrase hypotheses, previewing the use of the multivariate general linear hypothesis in profile analysis, the multivariate analysis of variance, and multivariate linear regression.

Hypothesis testing in the multivariate general linear model can take several forms, and in the latter part of this chapter we trace the extension of the univariate approach in two different fashions, giving rise to the two most frequently encountered multivariate test statistics—one based on a product of basic diagonal elements, the other on the largest element in a basic diagonal.

11.1 Sampling from Several Multivariate Normal Populations

In Section 10.7 we considered the problem of estimating, based on a maximum likelihood approach, the mean vector $\boldsymbol{\mu}$ and variance–covariance matrix $\boldsymbol{\Sigma}$ of a multivariate normal distribution. We assumed n independent samplings, each an m-element vector, and found the maximum likelihood estimators to be, from Eqs. (10.7.5) and (10.7.8),

$$\hat{\boldsymbol{\mu}}' = (1/n)\mathbf{1}'\mathbf{Y}, \qquad \hat{\boldsymbol{\Sigma}} = (1/n)(\mathbf{Y} - \mathbf{1}\hat{\boldsymbol{\mu}}')'(\mathbf{Y} - \mathbf{1}\hat{\boldsymbol{\mu}}')$$

where $\mathbf{Y}$ is the $n \times m$ matrix of sample observations.

In this section we extend this estimation problem. Quite often in multivariate analysis we need to decide whether subsets of multivariate observations are drawn from different populations. For example, in the Airline Pilot Response Time Study a vector of response times (two average times based on an initial and later block of trials) is associated with each participant. The vectors observed under the four different conditions, visual or aural stimulation to the left or right stimulus fields, are bivariate observations that might be samples from *different* populations—different in form, in expectation, or in variability.

In the Preemployment Study we have a vector of job performance scores for each employed applicant. Are different vectors produced between those with at least some college education and those with less education? Does the number of years of prior employment affect all or only some of the job performance attributes?

We develop in this chapter a statistical and design framework through which many multivariate research questions may be answered. In this *multivariate general linear model* (MGLM), an extension of the general linear model of Chapter 8, the *criterion* or dependent variable observations are multivariate, organized into an $n \times m$ matrix rather than an n-element vector. As preparation for the extension, we review the general linear model as developed for a univariate criterion attribute.

11.1.1 Review of the Univariate General Linear Model

The GLM assumes n criterion observations, sampled independently from normally distributed populations with a common variance σ^2, but with potentially different means. These population means, in turn, are assumed to be linearly based in the fixed values of a set of *predictor* attributes. If $\mathbf{E}(\mathbf{y})$ is an n-element vector containing the means of the populations from which the corresponding elements of $\mathbf{y}$ were sampled, this linear relation is expressed by Eq. (8.1.1), $\mathbf{E}(\mathbf{y}) = \mathbf{X}\boldsymbol{\beta}$, where $\mathbf{X}$ is an $n \times p$ matrix of fixed predictor or design scores and $\boldsymbol{\beta}$ is a p-element parameter vector.

The mean of the population from which the ith element of $\mathbf{y}$ was selected is thus given by $E(y_i) = \mathbf{X}'_{i.}\boldsymbol{\beta}$ and is a linear function of the p predictor scores making up the ith row of $\mathbf{X}$. The matrix $\mathbf{X}$ is called the *design* matrix, as it represents the researcher's (or nature's) design of the study, the choice of values of the predictor attributes to be studied. $\boldsymbol{\beta}$ consists of weights to be given the p predictors in reproducing the criterion means.

Predictor attributes may be *categorical*, as in the Response Time Study, *measured*, as in the College Prediction Study, or a mixture of the two types, as in our covariance or concomitant variable examples. Predictor values are regarded as known, however, and $\mathbf{y}$ values sampled in the course of the study. The general linear model analysis estimates the unknown vector $\boldsymbol{\beta}$ and σ^2 and tests hypotheses about the values, usually, of one or more elements of $\boldsymbol{\beta}$. These hypotheses most commonly are based on hypotheses about the means of populations from which the $\mathbf{y}$ observations were drawn.

The maximum likelihood estimator of $\boldsymbol{\beta}$ was found to be [see Eq. (8.2.5)]

$$\hat{\boldsymbol{\beta}} = (\mathbf{X}'\mathbf{X})^{-1}\mathbf{X}'\mathbf{y}$$

for the *full rank* or basic design matrix, while the common population variance for the sampled normal distributions was estimated, not unbiasedly, by Eq. (8.2.10):

$$\hat{\sigma}^2 = (1/n)\mathbf{e}'\mathbf{e} = (1/n)(\mathbf{y} - \mathbf{X}\hat{\boldsymbol{\beta}})'(\mathbf{y} - \mathbf{X}\hat{\boldsymbol{\beta}})$$

Hypotheses about the elements of $\boldsymbol{\beta}$ were tested by first phrasing these as *constraints* on $\boldsymbol{\beta}$ of the form given in Eq. (8.4.18):

$$\mathbf{C}'\boldsymbol{\beta} = \boldsymbol{\Gamma}$$

where $\mathbf{C}$ and $\boldsymbol{\Gamma}$ are $p \times k$ and $k \times 1$ arrays, respectively, chosen so as to impose $k \leqslant p$ constraints on the p elements of $\boldsymbol{\beta}$. $\boldsymbol{\beta}$ can then be reestimated subject to these k constraints by Eq. (8.4.20),

$$_{\mathrm{c}}\hat{\boldsymbol{\beta}} = \hat{\boldsymbol{\beta}} - \left\{(\mathbf{X}'\mathbf{X})^{-1}\mathbf{C}\left[\mathbf{C}'(\mathbf{X}'\mathbf{X})^{-1}\mathbf{C}\right]^{-1}\mathbf{C}'\right\}(\mathbf{C}'\hat{\boldsymbol{\beta}} - \boldsymbol{\Gamma})$$

and the two *sums of squares of errors*,

$$\mathbf{e}'\mathbf{e} = (\mathbf{y} - \mathbf{X}\hat{\boldsymbol{\beta}})'(\mathbf{y} - \mathbf{X}\hat{\boldsymbol{\beta}})$$

for the full model, and

$$\mathbf{f}'\mathbf{f} = (\mathbf{y} - \mathbf{X}\,_{\mathrm{c}}\hat{\boldsymbol{\beta}})'(\mathbf{y} - \mathbf{X}\,_{\mathrm{c}}\hat{\boldsymbol{\beta}})$$

for the constrained or *restricted* model, then compared.

Since $\mathbf{e}'\mathbf{e}/\sigma^2$ and $(\mathbf{f}'\mathbf{f} - \mathbf{e}'\mathbf{e})/\sigma^2$ are, under the hypothesis $\mathbf{C}'\boldsymbol{\beta} = \boldsymbol{\Gamma}$, distributed independently of one another as chi-squared observations with $n - p$ and k degrees of freedom, respectively, the ratio

$$F = \frac{(\mathbf{f}'\mathbf{f} - \mathbf{e}'\mathbf{e})/k}{\mathbf{e}'\mathbf{e}/(n - p)} \tag{11.1.1}$$

is an observation from the F variable with k and $n - p$ degrees of freedom and provides a likelihood ratio based test of that hypothesis.

11.1.2 Framework of the Multivariate General Linear Model

To generalize to the multivariate case, we assume $\mathbf{Y}$ to be an $n \times m$ matrix. Each row of $\mathbf{Y}$, each $\mathbf{Y}_{i.}$, is a multivariate normal observation sampled independently of other rows of $\mathbf{Y}$ from a distribution with variance–covariance matrix $\boldsymbol{\Sigma}$ and expectation vector $\mathbf{E}(\mathbf{Y}_{i.}') = \mathbf{X}_{i.}'\boldsymbol{\beta}$, where $\mathbf{X}_{i.}'$ is a p-element vector of predictor scores and $\boldsymbol{\beta}$ is a $p \times m$ matrix. That is, $\mathbf{Y}$ consists of n independent m-variate normal observations with a common variance–covariance structure.

$$\mathbf{Var}(\mathbf{Y}_{i.}) = \boldsymbol{\Sigma} \tag{11.1.2}$$

but potentially different expectations, linearly based on the values of the associated p predictor attributes:

$$\mathbf{E}(\mathbf{Y}_{i.}') = \mathbf{X}_{i.}'\boldsymbol{\beta} \tag{11.1.3}$$

Putting the expectation structure into matrix form,

$$\mathbf{E}(\mathbf{Y}) = \mathbf{X}\boldsymbol{\beta} \tag{11.1.4}$$

we note not only the similarities of the univariate and multivariate general linear models but two distinctions: $\boldsymbol{\Sigma}$, the common variance from row to row of $\mathbf{Y}$, is now a symmetric $m \times m$ matrix rather than a scalar, and $\boldsymbol{\beta}$, the set of weights linking predictor values to criterion expectations, is a $p \times m$ matrix rather than a p-element vector.

The matrix $\mathbf{X}$ may be, as before, a set of fixed or design quantities. It specifies the multivariate populations to be sampled. The matrix $\mathbf{Y}$ contains the sampled multivariate criterion observations. The matrices $\boldsymbol{\Sigma}$ and $\boldsymbol{\beta}$ consist of population parameters, usually unknown and requiring estimation.

11.1.3 Maximum Likelihood Estimation of $\boldsymbol{\beta}$

The likelihood of any row of $\mathbf{Y}$ can be written, from Eq. (10.2.19),

$$L(\mathbf{Y}_{i.}) = (2\pi)^{-m/2}\left[\prod_{j=1}^{m} \Delta_j^2\right]^{-1/2} \exp\left[-\tfrac{1}{2}(\mathbf{Y}_{i.}' - \mathbf{X}_{i.}'\boldsymbol{\beta})\boldsymbol{\Sigma}^{-1}(\mathbf{Y}_{i.}' - \mathbf{X}_{i.}'\boldsymbol{\beta})'\right] \tag{11.1.5}$$

where the variance covariance matrix $\boldsymbol{\Sigma}$ has basic structure

$$\boldsymbol{\Sigma} = \mathbf{Q}\boldsymbol{\Delta}^2\mathbf{Q}' \tag{11.1.6}$$

The likelihood of the sample of n independent observations is the product of n terms of this form:

$$L(\mathbf{Y}) = (2\pi)^{-nm/2}\left(\prod_{j=1}^{m} \Delta_j^2\right)^{-n/2} \exp\left\{-\tfrac{1}{2}\operatorname{tr}\left[(\mathbf{Y} - \mathbf{X}\boldsymbol{\beta})\boldsymbol{\Sigma}^{-1}(\mathbf{Y} - \mathbf{X}\boldsymbol{\beta})'\right]\right\} \tag{11.1.7}$$

Rather than maximize this likelihood directly, it again will be easier to maximize the logarithm:

$$\log[L(\mathbf{Y})] = \left(-\frac{nm}{2}\right)\log(2\pi) - \left(\frac{n}{2}\right)\sum_{j=1}^{m}\log\left(\Delta_j^2\right) - \tfrac{1}{2}\operatorname{tr}\left[(\mathbf{Y} - \mathbf{X}\boldsymbol{\beta})\boldsymbol{\Sigma}^{-1}(\mathbf{Y} - \mathbf{X}\boldsymbol{\beta})'\right] \tag{11.1.8}$$

As the regression matrix appears only in the final term, optimizing with respect to $\boldsymbol{\beta}$ requires only that we minimize

$$\phi = \operatorname{tr}(\mathbf{X}\boldsymbol{\beta}\boldsymbol{\Sigma}^{-1}\boldsymbol{\beta}'\mathbf{X}') - 2\operatorname{tr}(\mathbf{Y}\boldsymbol{\Sigma}^{-1}\boldsymbol{\beta}'\mathbf{X}') \tag{11.1.9}$$

By the rules presented in Chapter 6, the derivative of ϕ with respect to $\boldsymbol{\beta}$ is found by differentiating in turn each of the two matrix traces on the right.

This yields

$$\partial\phi/\partial\boldsymbol{\beta} = 2\mathbf{X}'\mathbf{X}\boldsymbol{\beta}\boldsymbol{\Sigma}^{-1} - 2\mathbf{X}'\mathbf{Y}\boldsymbol{\Sigma}^{-1}$$

Setting this derivative equal to the null matrix and pre- and postmultiplying both sides of the resultant equation by $(\mathbf{X}'\mathbf{X})^{-1}$ and $\boldsymbol{\Sigma}$, respectively, gives the solution:

$$\hat{\boldsymbol{\beta}} = (\mathbf{X}'\mathbf{X})^{-1}\mathbf{X}'\mathbf{Y} \tag{11.1.10}$$

In this we assume $\mathbf{X}$ is a vertical basic matrix. The maximum likelihood estimator of the matrix $\boldsymbol{\beta}$ is algebraically similar to that for the vector $\boldsymbol{\beta}$ in the single criterion GLM.

11.1.4 Maximum Likelihood Estimation of $\boldsymbol{\Sigma}$

We use the results of Section 10.7.2 to estimate $\boldsymbol{\Sigma}$ from Eq. (11.1.8). First, parallel to Eq. (10.7.6), the derivative of $\mathrm{tr}[(\mathbf{Y} - \mathbf{X}\boldsymbol{\beta})\boldsymbol{\Sigma}^{-1}(\mathbf{Y} - \mathbf{X}\boldsymbol{\beta})']$ with respect to $\boldsymbol{\Sigma}^{-1}$ is $(\mathbf{Y} - \mathbf{X}\boldsymbol{\beta})'(\mathbf{Y} - \mathbf{X}\boldsymbol{\beta})$. Then, again following Section 10.7.2, the derivative of $[-n\Sigma_{j=1}^{m} \log \Delta_j^2)]$ with respect to $\boldsymbol{\Sigma}^{-1}$ is found to be $n\boldsymbol{\Sigma}$. Combining these two results gives

$$\partial\{\log[L(\mathbf{Y})]\}/\partial(\boldsymbol{\Sigma}^{-1}) = n\boldsymbol{\Sigma} - (\mathbf{Y} - \mathbf{X}\boldsymbol{\beta})'(\mathbf{Y} - \mathbf{X}\boldsymbol{\beta})$$

Setting this equal to the null matrix, substituting $\hat{\boldsymbol{\beta}}$ for $\boldsymbol{\beta}$, and multiplying through by $1/n$ gives the estimator

$$\hat{\boldsymbol{\Sigma}} = (1/n)(\mathbf{Y} - \mathbf{X}\hat{\boldsymbol{\beta}})'(\mathbf{Y} - \mathbf{X}\hat{\boldsymbol{\beta}}) \tag{11.1.11}$$

The result is analogous again to the univariate solution.

11.1.5 Some Sampling Distribution Results for the Estimators of $\boldsymbol{\beta}$ *and* $\boldsymbol{\Sigma}$

Each row of $\hat{\boldsymbol{\beta}}$,

$$\hat{\boldsymbol{\beta}}_{i.}' = \left[(\mathbf{X}'\mathbf{X})^{-1}\mathbf{X}'\right]_{i.}'\mathbf{Y} \tag{11.1.12}$$

is a multivariate normal observation, as it is a weighted sum of the independent multivariate normal observations that are the rows of $\mathbf{Y}$. Further, as in the univariate case, $\hat{\boldsymbol{\beta}}$ is an unbiased estimator of $\boldsymbol{\beta}$:

$$\mathbf{E}(\hat{\boldsymbol{\beta}}) = (\mathbf{X}'\mathbf{X})^{-1}\mathbf{X}'\mathbf{E}(\mathbf{Y}) = (\mathbf{X}'\mathbf{X})^{-1}\mathbf{X}'\mathbf{X}\boldsymbol{\beta} = \boldsymbol{\beta} \tag{11.1.13}$$

The variance–covariance matrix for the sample $\hat{\boldsymbol{\beta}}_{i.}$ can be deduced using the technique we developed in Chapter 10 for finding the variance of

Wishart-centered attributes. Writing $\mathbf{Y}$ as a single long nm-element vector,

$$\mathbf{v_Y} = \begin{bmatrix} \mathbf{Y}_{1.} \\ \mathbf{Y}_{2.} \\ \vdots \\ \mathbf{Y}_{n.} \end{bmatrix}$$

we obtain a variance–covariance matrix,

$$\mathbf{Var}(\mathbf{v_Y}) = \begin{bmatrix} \boldsymbol{\Sigma} & & 0 \\ & \boldsymbol{\Sigma} & \\ 0 & & \ddots \\ & & & \boldsymbol{\Sigma} \end{bmatrix}$$

where each element is an $m \times m$ matrix. Using (11.1.12) we let $\mathbf{G}_I$ and $\mathbf{G}_J$ be the transformation matrices necessary to yield $\hat{\boldsymbol{\beta}}'_{i.} = \mathbf{G}'_I \mathbf{v_Y}$ and $\hat{\boldsymbol{\beta}}'_{j.} = \mathbf{G}'_J \mathbf{v_Y}$. This tells us that the variance–covariance for the sample $\hat{\boldsymbol{\beta}}_{i.}$ can be expressed

$$\mathbf{Var}(\hat{\boldsymbol{\beta}}_{i.}) = \mathbf{G}'_I \mathbf{Var}(\mathbf{v_Y}) \mathbf{G}_I$$

and the covariance between $\hat{\boldsymbol{\beta}}_{i.}$ and $\hat{\boldsymbol{\beta}}_{j.}$ as

$$\mathbf{Cov}(\hat{\boldsymbol{\beta}}_{i.}, \hat{\boldsymbol{\beta}}_{j.}) = \mathbf{G}'_I \mathbf{Var}(\mathbf{v_Y}) \mathbf{G}_J$$

Writing, from Eq. (11.1.12) $(\mathbf{X'X})^{-1}\mathbf{X'}$ as $\mathbf{g'}$, the appropriate choice of $\mathbf{G}_I$ can be shown to be $\mathbf{G}'_I = (g_{1i}\mathbf{I}, g_{2i}\mathbf{I}, \ldots, g_{ni}\mathbf{I})$, where the identity matrices are each $m \times m$. Thus, the sampling variances and covariances for rows of $\boldsymbol{\beta}$ become

$$\mathbf{Var}(\hat{\boldsymbol{\beta}}_{i.}) = \mathbf{g}'_{.i}\mathbf{g}_{.i}\boldsymbol{\Sigma}, \qquad \mathbf{Cov}(\hat{\boldsymbol{\beta}}_{i.}, \hat{\boldsymbol{\beta}}_{j.}) = \mathbf{g}'_{.i}\mathbf{g}_{.j}\boldsymbol{\Sigma}$$

However, $\mathbf{g}'_{.i}\mathbf{g}_{.j} = (\mathbf{g'g})_{ij} = [(\mathbf{X'X})^{-1}\mathbf{X'X}(\mathbf{X'X})^{-1}]_{ij} = [(\mathbf{X'X})^{-1}]_{ij}$. As a result,

$$\mathbf{Var}(\hat{\boldsymbol{\beta}}_{i.}) = \left[(\mathbf{X'X})^{-1}\right]_{ii}\boldsymbol{\Sigma} \tag{11.1.14}$$

and

$$\mathbf{Cov}(\hat{\boldsymbol{\beta}}_{i.}, \hat{\boldsymbol{\beta}}_{j.}) = \left[(\mathbf{X'X})^{-1}\right]_{ij}\boldsymbol{\Sigma} \tag{11.1.15}$$

The expressions on the right of (11.1.14) and (11.1.15), although $m \times m$ matrices, are extensions of the univariate model finding, Eq. (8.3.3), that $\text{Var}(\hat{\beta}_i) = \sigma^2[(\mathbf{X'X})^{-1}]_{ii}$ and $\text{Cov}(\hat{\beta}_i, \hat{\beta}_j) = \sigma^2[(\mathbf{X'X})^{-1}]_{ij}$.

The maximum likelihood estimator of $\boldsymbol{\Sigma}$ is not unbiased. However, as in the univariate case, it is possible to show that

$$\hat{\boldsymbol{\Sigma}} = [1/(n-p)](\mathbf{Y} - \mathbf{X}\hat{\boldsymbol{\beta}})'(\mathbf{Y} - \mathbf{X}\hat{\boldsymbol{\beta}})$$

is unbiased.

11.1.6 Wishart Distributions in the MGLM

For the univariate GLM we found the two components,

$$\frac{1}{\sigma^2}(\mathbf{y} - \mathbf{X}\boldsymbol{\beta})'(\mathbf{y} - \mathbf{X}\boldsymbol{\beta}) = \frac{1}{\sigma^2}(\mathbf{y} - \mathbf{X}\hat{\boldsymbol{\beta}})'(\mathbf{y} - \mathbf{X}\hat{\boldsymbol{\beta}}) + \frac{1}{\sigma^2}(\mathbf{X}\hat{\boldsymbol{\beta}} - \mathbf{X}\boldsymbol{\beta})'(\mathbf{X}\hat{\boldsymbol{\beta}} - \mathbf{X}\boldsymbol{\beta})$$

to be independently distributed as chi-squared observations:

$$\begin{aligned}(1/\sigma^2)(\mathbf{y} - \mathbf{X}\hat{\boldsymbol{\beta}})'(\mathbf{y} - \mathbf{X}\hat{\boldsymbol{\beta}}) &= (\mathbf{y} - \mathbf{X}\boldsymbol{\beta})'\left\{(1/\sigma^2)\left[\mathbf{I} - \mathbf{X}(\mathbf{X}'\mathbf{X})^{-1}\mathbf{X}'\right]\right\}(\mathbf{y} - \mathbf{X}\boldsymbol{\beta}) \\ &\sim \chi^2_{n-p} \text{ df}\end{aligned}$$

and

$$\begin{aligned}(1/\sigma^2)(\mathbf{X}\hat{\boldsymbol{\beta}} - \mathbf{X}\boldsymbol{\beta})'(\mathbf{X}\hat{\boldsymbol{\beta}} - \mathbf{X}\boldsymbol{\beta}) &= (\mathbf{y} - \mathbf{X}\boldsymbol{\beta})'\left\{(1/\sigma^2)\left[\mathbf{X}(\mathbf{X}'\mathbf{X})^{-1}\mathbf{X}'\right]\right\}(\mathbf{y} - \mathbf{X}\boldsymbol{\beta}) \\ &\sim \chi^2_{p} \text{ df}\end{aligned}$$

Will similar terms, now matrices, have independent Wishart distributions, the multivariate analog to the chi-squared distribution, in the MGLM? First, we define

$$\mathbf{E}'\mathbf{E} = (\mathbf{Y} - \mathbf{X}\hat{\boldsymbol{\beta}})'(\mathbf{Y} - \mathbf{X}\hat{\boldsymbol{\beta}}) \tag{11.1.16}$$

or, substituting for $\hat{\boldsymbol{\beta}}$,

$$\mathbf{E}'\mathbf{E} = \mathbf{Y}'\left[\mathbf{I} - \mathbf{X}(\mathbf{X}'\mathbf{X})^{-1}\mathbf{X}'\right]\mathbf{Y} \tag{11.1.17}$$

Because $\mathbf{X}$ is orthogonal to the idempotent matrix on the right, we can write, as in the univariate case,

$$\mathbf{E}'\mathbf{E} = (\mathbf{Y} - \mathbf{X}\boldsymbol{\beta})'\left[\mathbf{I} - \mathbf{X}(\mathbf{X}'\mathbf{X})^{-1}\mathbf{X}'\right](\mathbf{Y} - \mathbf{X}\boldsymbol{\beta}) \tag{11.1.18}$$

which establishes that $\mathbf{E}'\mathbf{E}$ is a Wishart matrix:

$$\mathbf{E}'\mathbf{E} \sim \mathbf{W}_m[(n-p), \boldsymbol{\Sigma}] \tag{11.1.19}$$

Similarly, the algebraic argument of Chapter 8 allows us to write

$$\mathbf{X}\hat{\boldsymbol{\beta}} - \mathbf{X}\boldsymbol{\beta} = \mathbf{X}(\mathbf{X}'\mathbf{X})^{-1}\mathbf{X}'\mathbf{Y} - \mathbf{X}(\mathbf{X}'\mathbf{X})^{-1}\mathbf{X}'\mathbf{X}\boldsymbol{\beta} = \left[\mathbf{X}(\mathbf{X}'\mathbf{X})^{-1}\mathbf{X}'\right](\mathbf{Y} - \mathbf{X}\boldsymbol{\beta}) \tag{11.1.20}$$

and, from this,

$$(\mathbf{X}\hat{\boldsymbol{\beta}} - \mathbf{X}\boldsymbol{\beta})'(\mathbf{X}\hat{\boldsymbol{\beta}} - \mathbf{X}\boldsymbol{\beta}) = (\mathbf{Y} - \mathbf{X}\boldsymbol{\beta})'\left[\mathbf{X}(\mathbf{X}'\mathbf{X})^{-1}\mathbf{X}'\right](\mathbf{Y} - \mathbf{X}\boldsymbol{\beta}) \tag{11.1.21}$$

The right side is in Wishart form, so

$$(\mathbf{X}\hat{\boldsymbol{\beta}} - \mathbf{X}\boldsymbol{\beta})'(\mathbf{X}\hat{\boldsymbol{\beta}} - \mathbf{X}\boldsymbol{\beta}) \sim \mathbf{W}_m(p, \boldsymbol{\Sigma}) \tag{11.1.22}$$

Since the two idempotent matrices $\mathbf{I} - \mathbf{X}(\mathbf{X}'\mathbf{X})^{-1}\mathbf{X}'$ and $\mathbf{X}(\mathbf{X}'\mathbf{X})^{-1}\mathbf{X}'$ are mutually orthogonal, the two Wishart matrices, $\mathbf{E}'\mathbf{E}$ and $(\mathbf{X}\hat{\boldsymbol{\beta}} - \mathbf{X}\boldsymbol{\beta})'(\mathbf{X}\hat{\boldsymbol{\beta}} - \mathbf{X}\boldsymbol{\beta})$, are independent.

An additional note on the distribution of the product of basic diagonals will aid in establishing further the correspondence of $\mathbf{E}'\mathbf{E}$ to the univariate $\mathbf{e}'\mathbf{e}$. If $n - p \geqslant m$, then, taking the basic structure of $\boldsymbol{\Sigma}$ as $\mathbf{Q}\boldsymbol{\Delta}^2\mathbf{Q}'$, the results of Section 10.8 can be applied to show that the product of the basic diagonals of

$$\mathbf{Q}_{\mathbf{E}} = (\boldsymbol{\Delta}^{-1}\mathbf{Q}')\mathbf{E}'\mathbf{E}(\mathbf{Q}\boldsymbol{\Delta}^{-1}) \tag{11.1.23}$$

is distributed as the product of m independent chi-squared observations:

$$\prod_{j=1}^{m} \chi^2_{(n-p)-j+1 \text{ df}}$$

or, equivalently,

$$\prod_{(n-p)-(m-1)}^{n-p} \chi^2_{j \text{ df}} \tag{11.1.24}$$

When $m = 1$, $\mathbf{E}'\mathbf{E} = \mathbf{e}'\mathbf{e}$, a scalar; $\Sigma = \sigma^2 = 1\sigma^2 1$, also a scalar; and the Wishart form of Eq. (11.1.23) becomes

$$Q_{\mathbf{e}} = (1/\sigma)\mathbf{E}'\mathbf{E}(1/\sigma) = \mathbf{e}'\mathbf{e}/\sigma^2$$

a nonnegative scalar with basic diagonal product $\mathbf{e}'\mathbf{e}/\sigma^2$. The sampling distribution for this term, according to (11.1.24), is

$$\mathbf{e}'\mathbf{e}/\sigma^2 \sim \chi^2_{n-p \text{ df}}$$

the univariate result of Chapter 8.

11.2 Effects of Linear Constraints

Now that we can estimate parameters of the MGLM, we can develop techniques for testing hypotheses about relations among those parameters. In parallel with our development of the univariate GLM we again focus on the effects of linear constraints on the $\boldsymbol{\beta}$ matrix or, equivalently, on contrasts among the population means (now vectors). Such constraints cover a wide range of multivariate questions. We preview some of these before developing the hypothesis testing procedure.

11.2.1 Two-Group T^2

We begin with the multivariate equivalent of the hypothesis of equal means for two populations. In the Women's Career Study, criterion data were collected for two groups (ignoring for the present other refinements of the design): those who had participated in a career development class and those who had not. For each woman three attributes were measured: level of career *aspiration*, *certainty* of career choice, and the *centrality* of work in the woman's future.

The design template for the study could be of the form

$$_{\mathrm{T}}\mathbf{X} = \begin{bmatrix} 1 & 0 \\ 0 & 1 \end{bmatrix}$$

such that a row of the resulting $n \times 2$ design matrix is

(1, 0) if the corresponding woman was a class participant

and

(0, 1) if the woman was not a class participant

The (sampled) criterion matrix is $n \times 3$ with $\mathbf{Y}_{.1}$, $\mathbf{Y}_{.2}$, and $\mathbf{Y}_{.3}$ carrying career aspiration, certainty, and centrality scores, respectively.

The $\boldsymbol{\beta}$ matrix must have $p = 2$ rows and $m = 3$ columns, and we extend the relation of the design template to the expectation template for the multivariate case:

$$\begin{bmatrix} 1 & 0 \\ 0 & 1 \end{bmatrix} \begin{bmatrix} \beta_{11} & \beta_{12} & \beta_{13} \\ \beta_{21} & \beta_{22} & \beta_{23} \end{bmatrix} = \begin{bmatrix} E(Y_{.1}|\mathrm{P}) & E(Y_{.2}|\mathrm{P}) & E(Y_{.3}|\mathrm{P}) \\ E(Y_{.1}|\mathrm{NP}) & E(Y_{.2}|\mathrm{NP}) & E(Y_{.3}|\mathrm{NP}) \end{bmatrix}$$

With our choice of design template, $\boldsymbol{\beta}$ is identical to the expectation template:

$$\boldsymbol{\beta} = {}_{\mathrm{T}}\mathbf{E}(\mathbf{Y}) = \begin{bmatrix} \mathbf{E}(\mathbf{Y}_{i.}'|\mathrm{P}) \\ \mathbf{E}(\mathbf{Y}_{i.}'|\mathrm{NP}) \end{bmatrix}$$

(Here P and NP code participation or nonparticipation in the career class.)

A common hypothesis for this design, termed the two-group T^2 design, is that the two expectation vectors are identical:

$$\mathbf{E}(\mathbf{Y}_{i.}|\mathrm{P}) = \mathbf{E}(\mathbf{Y}_{i.}|\mathrm{NP})$$

This translates, again given our choice of design template, into the linear constraint,

$$\mathbf{C}'\boldsymbol{\beta} = (1 \quad -1) \begin{bmatrix} \mathbf{E}(\mathbf{Y}_{i.}'|\mathrm{P}) \\ \mathbf{E}(\mathbf{Y}_{i.}'|\mathrm{NP}) \end{bmatrix} = (0 \quad 0 \quad 0) = \boldsymbol{\Gamma}$$

Note that the hypothesis is that the population means are identical for each of the three criterion attributes.

11.2.2 Randomized Groups Multivariate Analysis of Variance

The design just discussed can be generalized to randomized groups designs with more than two populations sampled. Pursuing further the Women's Career Study recall that six groups of women contributed criterion data.

The populations from which they were sampled were defined by crossing the two levels of participation with three levels of socioeconomic status (high, middle, and low.) For an effect coding of this factorial design, one version of the 6 × 6 design template is

$$
{}_T\mathbf{X} = \begin{bmatrix} 1 & 1 & 0 & 1 & 1 & 0 \\ 1 & 1 & 0 & -1 & -1 & 0 \\ 1 & 0 & 1 & 1 & 0 & 1 \\ 1 & 0 & 1 & -1 & 0 & -1 \\ 1 & -1 & -1 & 1 & -1 & -1 \\ 1 & -1 & -1 & -1 & 1 & 1 \end{bmatrix}
$$

where the second and third columns code SES levels, the fourth codes class participation, and the fifth and sixth are the two interactions, the direct products of columns 2 and 3 with column 4.

However, based on the relation ${}_T\mathbf{E}(\mathbf{Y}) = {}_T\mathbf{X}\boldsymbol{\beta}$ and the choice of design template, the hypothesis of additivity or of no interaction between the effects of participation and SES on the multivariate criterion requires that the fifth and sixth rows of $\boldsymbol{\beta}$ each be null vectors. This hypothesis imposes two constraints on the rows of $\boldsymbol{\beta}$. These constraints can be expressed

$$
{}_1\mathbf{C}'\boldsymbol{\beta} = \begin{bmatrix} 0 & 0 & 0 & 0 & 1 & 0 \\ 0 & 0 & 0 & 0 & 0 & 1 \end{bmatrix} \boldsymbol{\beta} = \begin{bmatrix} 0 & 0 & 0 \\ 0 & 0 & 0 \end{bmatrix} = \boldsymbol{\Gamma}
$$

The hypotheses of no participation effect and of no SES effect are similarly accommodated by the pair of linear constraints ${}_2\mathbf{C}'\boldsymbol{\beta} = \mathbf{0}$ and ${}_3\mathbf{C}'\boldsymbol{\beta} = \mathbf{0}$, where

$$
{}_2\mathbf{C}' = (0, 0, 1, 0, 0), \qquad {}_3\mathbf{C}' = \begin{bmatrix} 0 & 1 & 0 & 0 & 0 & 0 \\ 0 & 0 & 1 & 0 & 0 & 0 \end{bmatrix}
$$

We return to the *multivariate analysis of variance* later in this chapter. For now, we note that the constraints developed here are identical to those developed for the univariate linear model in Section 9.1. In both cases constraints are imposed as linear functions of the rows of $\boldsymbol{\beta}$. The sole difference thus far is that in the multivariate model the rows of $\boldsymbol{\beta}$ are vectors rather than scalars.

11.2.3 Multivariate Linear Regression

In the same vein, the multiple linear regression techniques of Chapter 9 suggest analogous constraints when the criterion is a set of multivariate normal observations. This is true both for standard regression designs (where the usual hypotheses or constraints are those associated with dropping one or more of the potential predictors) and for such special applications as were illustrated in Section 9.3. We preview multivariate linear regression with a *moderator variable* application.

In the Preemployment Study, the three outcome of employment measures—time to complete *training*, supervisor's *performance* rating, and self-reported job *satisfaction*—together constitute a trivariate observation on each of the insurance underwriters. Two preemployment predictors of these outcomes are extent of previous *work experience* (scaled as noted in Section 9.3) and *gender* of applicant. A design matrix, permitting us to ask whether gender *moderates* the regression of the (multivariate) employment outcome on amount of previous work, is given by the $(n \times 4)$ $\mathbf{X}$ in which

$$X_{i1} = \begin{cases} 0 & \text{if } i\text{th applicant is male} \\ 1 & \text{if } i\text{th applicant is female} \end{cases}$$

$$X_{i2} = \begin{cases} 0 & \text{if } i\text{th applicant is female} \\ 1 & \text{if } i\text{th applicant is male} \end{cases}$$

$$X_{i3} = \begin{cases} 0 & \text{if } i\text{th applicant is male} \\ \text{work experience score} & \text{if } i\text{th applicant is female} \end{cases}$$

$$X_{i4} = \begin{cases} 0 & \text{if } i\text{th applicant is female} \\ \text{work experience score} & \text{if } i\text{th applicant is male} \end{cases}$$

The ith applicant is that underwriter whose criterion vector makes up the ith row of the 150×3 matrix $\mathbf{Y}$.

A 4×3 $\boldsymbol{\beta}$ matrix is necessary to satisfy the equation $\mathbf{X}\boldsymbol{\beta} = \mathbf{E}(\mathbf{Y})$. The hypothesis that the slope of the regression of any one of the three measures on work experience is the same for male as for female underwriters is equivalent to requiring that the third and fourth rows of $\boldsymbol{\beta}$ be identical, giving the constraint

$${}_1\mathbf{C}'\boldsymbol{\beta} = (0 \quad 0 \quad 1 \quad -1)\begin{bmatrix} \boldsymbol{\beta}'_{1.} \\ \boldsymbol{\beta}'_{2.} \\ \boldsymbol{\beta}'_{3.} \\ \boldsymbol{\beta}'_{4.} \end{bmatrix} = (0 \quad 0 \quad 0) = \boldsymbol{\Gamma}$$

Further, the hypothesis of no difference between male and female underwriters in the intercept of the regression of any one of the three criteria translates into the constraint ${}_2\mathbf{C}'\boldsymbol{\beta} = 0$ where ${}_2\mathbf{C}' = (1, -1, 0, 0)$. For the multivariate criterion, the hypothesis of an identical slope or intercept is a hypothesis of an identical slope or intercept for each of the m outcome attributes.

11.2.4 Known Population Mean Vectors

A multivariate analog to the one sample design is one in which it is hypothesized that the sample of criterion observations, each a multivariate normal vector, was obtained from a population with a known expectation vector. For the College Prediction Study we may define a **Y** matrix whose ith row contains, in order, the scores of the ith student on the three tests of vocabulary, mathematics achievement, and spatial ability. A relevant question might be whether the sample can be regarded as having been selected from the multivariate population with expectation $\mathbf{E}(\mathbf{Y}_{i.}') = (50, 50, 50)$.

We may take the design matrix here to be simply the unit vector, so that $\boldsymbol{\beta}$ is the row expectation vector $\boldsymbol{\beta} = \mathbf{E}(\mathbf{Y}_{i.}')$, and the preceding hypothesis can be expressed as a constraint on the rows of $\boldsymbol{\beta}$ of the form

$$1\boldsymbol{\beta} = (50, 50, 50)$$
$$C'\boldsymbol{\beta} = \boldsymbol{\Gamma}$$

where C' is simply the scalar 1.

As in the univariate GLM, weighted sums of the rows of $\boldsymbol{\beta}$ may be hypothesized to yield other than the null vector.

11.2.5 Constraining β *by Column*

In these examples we have traced parallels between the univariate GLM and the MGLM. The rows of $\boldsymbol{\beta}$ may be constrained whether those rows are scalars or vectors. However, $\boldsymbol{\beta}$ is a matrix in the multivariate case, so constraints, again linear, may be applied to the columns as well. What meaning would we attach to such constraints?

We approach an answer by considering an example. In the Response Time Study, two average response times were recorded for each pilot, one based on an initial and a second based on a later block of trials. The criterion is bivariate. The design was a randomized groups one with four groups of pilots, each group assigned a different stimulus condition. The multivariate linear expectation equation $\mathbf{E}(\mathbf{Y}) = \mathbf{X}\boldsymbol{\beta}$ involves matrices of the following orders:

$$(\mathbf{X}_{.1} \quad \mathbf{X}_{.2} \quad \mathbf{X}_{.3} \quad \mathbf{X}_{.4})\begin{bmatrix} \beta_{11} & \beta_{12} \\ \beta_{21} & \beta_{22} \\ \beta_{31} & \beta_{32} \\ \beta_{41} & \beta_{42} \end{bmatrix} = [\mathbf{E}(\mathbf{Y}_{.1}) \quad \mathbf{E}(\mathbf{Y}_{.2})]$$

Both **X** and **E(Y)** have n rows, one for each participating pilot.

Consider the hypothesis that the average response times on the first block of trials and the second block of trials should be the same. Symbolically, $\mathbf{E}(\mathbf{Y}_{.1}) = \mathbf{E}(\mathbf{Y}_{.2})$. We have, then, a hypothesis about the columns of **E(Y)**. Heretofore our hypotheses have been about the rows of **E(Y)**. Note that, based on the linear expectation equation, each column of **E(Y)** is the result of postmultiplying the design matrix by the corresponding column of $\boldsymbol{\beta}$. The columns of **E(Y)** are produced, in this sense, by the columns of $\boldsymbol{\beta}$. As a result, we should be able to translate hypotheses about the columns of **E(Y)** into restrictions or constraints on the columns of $\boldsymbol{\beta}$. In the present example the hypothesis requires that $\mathbf{E}(\mathbf{Y}_{.1}) = \mathbf{E}(\mathbf{Y}_{.2})$ or that $\mathbf{X}\boldsymbol{\beta}_{.1} = \mathbf{X}\boldsymbol{\beta}_{.2}$. This will be true, certainly, if $\boldsymbol{\beta}_{.1} = \boldsymbol{\beta}_{.2}$. This last represents a linear constraint $\boldsymbol{\beta}\mathbf{A} = \boldsymbol{\Gamma} = \mathbf{0}$ on the columns of $\boldsymbol{\beta}$:

$$(\boldsymbol{\beta}_{.1} \quad \boldsymbol{\beta}_{.2})\begin{bmatrix} 1 \\ -1 \end{bmatrix} = \mathbf{0}$$

Constraints on the columns of $\boldsymbol{\beta}$ frequently can be interpreted as changes in the criterion. In this instance the effect is to substitute as the criterion a univariate vector of difference scores, $\mathbf{Y}_{.1} - \mathbf{Y}_{.2}$, and to hypothesize that this new attribute has a mean of zero (perhaps over the four populations from which the randomized groups were drawn).

The constraint $\boldsymbol{\beta}\mathbf{A} = \boldsymbol{\Gamma}$ or the contrast ${}_{\mathrm{T}}\mathbf{E}(\mathbf{Y}) = \mathbf{A} = \boldsymbol{\Gamma}$ is incomplete. As we shall see when we develop tests for multivariate general linear model hypotheses, we need to develop and include $\mathbf{C}$ as well, yielding multivariate constraints of the form

$$\mathbf{C}'\boldsymbol{\beta}\mathbf{A} = \boldsymbol{\Gamma} \tag{11.2.1}$$

The necessity for specifying $\mathbf{C}$ may be motivated in the present example by asking, "For whom do we hypothesize that response times will be the same on the two blocks of trials?"

To respond to that question with a constraint on $\boldsymbol{\beta}$ requires, of course, that we know the design matrix associated with $\boldsymbol{\beta}$. If we assume *categorical variable* coding, the expectation equation ${}_{\mathrm{T}}\mathbf{X}\boldsymbol{\beta} = {}_{\mathrm{T}}\mathbf{E}(\mathbf{Y})$ corresponds to

$$\begin{bmatrix} 1 & 0 & 0 & 0 \\ 0 & 1 & 0 & 0 \\ 0 & 0 & 1 & 0 \\ 0 & 0 & 0 & 1 \end{bmatrix}\begin{bmatrix} \beta_{11} & \beta_{12} \\ \beta_{21} & \beta_{22} \\ \beta_{31} & \beta_{32} \\ \beta_{41} & \beta_{42} \end{bmatrix} = \begin{bmatrix} E(Y_{.1}|\mathrm{LV}) & E(Y_{.2}|\mathrm{LV}) \\ E(Y_{.1}|\mathrm{RV}) & E(Y_{.2}|\mathrm{RV}) \\ E(Y_{.1}|\mathrm{LA}) & E(Y_{.2}|\mathrm{LA}) \\ E(Y_{.1}|\mathrm{RA}) & E(Y_{.2}|\mathrm{RA}) \end{bmatrix}$$

If the hypothesis of equal response times is intended only in those instances where stimulation is to the left visual field, then we could begin with the *contrast*

$${}_{1}\mathbf{H}'\,{}_{\mathrm{T}}\mathbf{E}(\mathbf{Y})\mathbf{A} = (1 \quad 0 \quad 0 \quad 0)_{\mathrm{T}}\mathbf{E}(\mathbf{Y})\begin{bmatrix} 1 \\ -1 \end{bmatrix} = 0 = {}_{1}\Gamma$$

and from this develop a *constraint* ${}_{1}\mathbf{C}'\boldsymbol{\beta}\mathbf{A} = {}_{1}\Gamma$, based on the contrast–constraint relationship developed in Chapter 8, $\mathbf{C}' = \mathbf{H}'\,{}_{\mathrm{T}}\mathbf{X}$. (Given our present choice of design matrix, ${}_{\mathrm{T}}\mathbf{X} = \mathbf{I}$ and $\mathbf{C}' = \mathbf{H}'$.)

Alternatively, our hypothesis might have been that response times for the two blocks are the same in each of the four populations from which we sampled:

$${}_{2}\mathbf{H}'\,{}_{\mathrm{T}}\mathbf{E}(\mathbf{Y})\mathbf{A} = \begin{bmatrix} 1 & 0 & 0 & 0 \\ 0 & 1 & 0 & 0 \\ 0 & 0 & 1 & 0 \\ 0 & 0 & 0 & 1 \end{bmatrix}{}_{\mathrm{T}}\mathbf{E}(\mathbf{Y})\begin{bmatrix} 1 \\ -1 \end{bmatrix} = \begin{bmatrix} 0 \\ 0 \\ 0 \\ 0 \end{bmatrix} = {}_{2}\boldsymbol{\Gamma}$$

or, a weaker version of this, that the difference in response times between the two blocks of trials, *averaged* over the four populations, is zero:

$$ {}_3\mathbf{H}'\,{}_T\mathbf{E}(\mathbf{Y})\mathbf{A} = (1 \quad 1 \quad 1 \quad 1)\,{}_T\mathbf{E}(\mathbf{Y})\begin{bmatrix} 1 \\ -1 \end{bmatrix} = 0 = {}_3\Gamma $$

We use the Response Time Study to see how to express, as multivariate contrasts, some other potential hypotheses. First, consider the hypothesis that the *change* in average response time from first block to later block is the same for each of the four groups. (We do not require this time the change to be zero.) As a start, note that the product

$$ {}_T\mathbf{E}(\mathbf{Y})\mathbf{A} = {}_T\mathbf{E}(\mathbf{Y})\begin{bmatrix} 1 \\ -1 \end{bmatrix} = \begin{bmatrix} \mathbf{E}[(\mathbf{Y}_{.1} - \mathbf{Y}_{.2})|\mathrm{LV}] \\ \mathbf{E}[(\mathbf{Y}_{.1} - \mathbf{Y}_{.2})|\mathrm{RV}] \\ \mathbf{E}[(\mathbf{Y}_{.1} - \mathbf{Y}_{.2})|\mathrm{LA}] \\ \mathbf{E}[(\mathbf{Y}_{.1} - \mathbf{Y}_{.2})|\mathrm{RA}] \end{bmatrix} $$

provides a vector of expected changes in response times for the four treatment groups. One way to assert that the four changes are the same is to assert that the difference between any pair of changes is always zero. This we know we can represent by premultiplying ${}_T\mathbf{E}(\mathbf{Y})\mathbf{A}$ by

$$ {}_4\mathbf{H}' = \begin{bmatrix} 1 & -1 & 0 & 0 \\ 1 & 0 & -1 & 0 \\ 1 & 0 & 0 & -1 \end{bmatrix} $$

for example, and then setting the result equal to the null vector:

$$ {}_4\mathbf{H}'\,{}_T\mathbf{E}(\mathbf{Y})\mathbf{A} = \begin{bmatrix} 0 \\ 0 \\ 0 \end{bmatrix} $$

Alternatively, the hypothesis might be modified slightly to require change of the same magnitude for the two visual stimulation conditions (LV and RV) but change of only half that magnitude for each of the two aural conditions (RA and LA). For this situation the matrix ${}_4\mathbf{H}'$ could be replaced with

$$ {}_5\mathbf{H}' = \begin{bmatrix} 1 & -1 & 0 & 0 \\ 1 & 0 & -2 & 0 \\ 1 & 0 & 0 & -2 \end{bmatrix} $$

in the constraining equation.

Finally, the contrasts need not equate to zero or the null matrix. A hypothesis that response times in the visually stimulated groups should decrease by 50 msec from the first to later trials, for instance, gives rise to the contrasts

$$ {}_6\mathbf{H}'\,{}_T\mathbf{E}(\mathbf{Y})\mathbf{A} = \begin{bmatrix} 1 & 0 & 0 & 0 \\ 0 & 1 & 0 & 0 \end{bmatrix}\boldsymbol{\beta}\begin{bmatrix} 1 \\ -1 \end{bmatrix} = \begin{bmatrix} 50 \\ 50 \end{bmatrix} = {}_6\Gamma $$

Given categorical variable coding of the four groups, **H** of course becomes **C** and ${}_{\mathrm{T}}\mathbf{E}(\mathbf{Y})$ becomes $\boldsymbol{\beta}$. With other choices of ${}_{\mathrm{T}}\mathbf{X}$ the two will be given by $\mathbf{C}' = \mathbf{H}'\,{}_{\mathrm{T}}\mathbf{X}$ and $\boldsymbol{\beta} = ({}_{\mathrm{T}}\mathbf{X})^{-1}\,{}_{\mathrm{T}}\mathbf{E}(\mathbf{Y})$.

We have illustrated the most general form of linear constraint on the $\boldsymbol{\beta}$ matrix, and hence on the expectations of the criterion attribute. It takes the form of (11.2.1), $\mathbf{C}'\boldsymbol{\beta}\mathbf{A} = \boldsymbol{\Gamma}$. Some examples of how to phrase these constraints have been given. More examples appear in this and the following chapter. It is not enough, of course, simply to phrase the constraints. We must have some way of testing such hypotheses based on a sample of criterion observations.

11.3 Hypothesis Testing in the Multivariate General Linear Model

We follow as closely as possible the hypothesis testing paradigm established for the univariate GLM. Already [Eq. (11.1.16)] we have shown how to obtain $\mathbf{E}'\mathbf{E} = (\mathbf{Y} - \mathbf{X}\hat{\boldsymbol{\beta}})'(\mathbf{Y} - \mathbf{X}\hat{\boldsymbol{\beta}})$, the multivariate analog to $\mathbf{e}'\mathbf{e}$, the *unconstrained* or full model error sum of squares in the univariate model. Further [see Eq. (11.1.19)], we have shown $\mathbf{E}'\mathbf{E}$ to have a Wishart distribution,

$$\mathbf{E}'\mathbf{E} \sim \mathbf{W}_m[(n-p), \boldsymbol{\Sigma}]$$

when each of the n rows of $\mathbf{Y}$ is an independent m-variate normal observation with $\mathbf{Var}(\mathbf{Y}_{i.}) = \boldsymbol{\Sigma}$ and $\mathbf{E}(\mathbf{Y}_{i.}') = \mathbf{X}_{i.}'\boldsymbol{\beta}$. The design matrix $\mathbf{X}$ is assumed to be basic; $r(\mathbf{X}) = p$, its column order.

A *constrained* model sum of squares (and sum of cross products) of errors can now be defined, again in parallel with the univariate development:

$$\mathbf{F}'\mathbf{F} = (\mathbf{Y} - \mathbf{X}\,{}_{c}\hat{\boldsymbol{\beta}})'(\mathbf{Y} - \mathbf{X}\,{}_{c}\hat{\boldsymbol{\beta}}) \tag{11.3.1}$$

where ${}_{c}\hat{\boldsymbol{\beta}}$ is the maximum likelihood estimator of $\boldsymbol{\beta}$ subject to the constraints $\mathbf{C}'\boldsymbol{\beta}\mathbf{A} = \boldsymbol{\Gamma}$ [Eq. (11.2.1)]. We need now to see how we can

1. find ${}_{c}\hat{\boldsymbol{\beta}}$,
2. compute $\mathbf{F}'\mathbf{F}$, and
3. determine the sampling distribution of $\mathbf{F}'\mathbf{F}$.

11.3.1 Maximum Likelihood Estimation for $\mathbf{C}'\boldsymbol{\beta}\mathbf{A} = \boldsymbol{\Gamma}$

To estimate $\boldsymbol{\beta}$ under the constraints we begin with the log likelihood expression for the sample, as given in (11.1.8), and build in the side conditions using Lagrangian multipliers. This gives a new expression to be maximized:

$$\phi = -n \sum_{j=1}^{m} \log(\Delta_j^2) - \mathrm{tr}[(\mathbf{Y} - \mathbf{X}\boldsymbol{\beta})\boldsymbol{\Sigma}^{-1}(\mathbf{Y} - \mathbf{X}\boldsymbol{\beta})'] - \mathrm{tr}[\boldsymbol{\lambda}'(\mathbf{C}'\boldsymbol{\beta}\mathbf{A} - \boldsymbol{\Gamma})] \tag{11.3.2}$$

We take $\mathbf{C}$ to be $p \times s \leqslant p$ and $\mathbf{A}$ to be $m \times t \leqslant m$ and both to be basic. $\mathbf{C}'\boldsymbol{\beta}\mathbf{A}$ and $\boldsymbol{\Gamma}$ are, then $s \times t$ matrices. The $p \times m$ matrix $\boldsymbol{\beta}$ is subject to $s \times t$ constraints. [The form in which the side condition is added ensures that differentiation with respect to any element λ_{ij} of $\boldsymbol{\lambda}$ gives a result of the form $\mathbf{C}'_{.i}\boldsymbol{\beta}\mathbf{A}_{.j} - \Gamma_{ij}$ so that setting the result equal to zero will yield $\mathbf{C}_{.i}\hat{\boldsymbol{\beta}}\mathbf{A}_{.j} = \Gamma_{ij}$. That is, upon optimizing ϕ we shall have each of the $s \times t$ elements of $\mathbf{C}'\,{}_{c}\boldsymbol{\beta}\mathbf{A}$ equal to the corresponding element of $\boldsymbol{\Gamma}$, as (11.2.1) requires.]

The derivative of ϕ with respect to $\boldsymbol{\lambda}$, based on the differentiation rules for traces presented in Chapter 6, is $\partial\phi/\partial\boldsymbol{\lambda} = \mathbf{C}'\boldsymbol{\beta}\mathbf{A} - \boldsymbol{\Gamma}$. Setting this equal to the null matrix, we retrieve the side conditions

$$\mathbf{C}'\,{}_{c}\hat{\boldsymbol{\beta}}\mathbf{A} = \boldsymbol{\Gamma} \tag{11.3.3}$$

The side condition terms of (11.3.2) do not involve $\boldsymbol{\Sigma}$, so the constrained maximum likelihood estimator of that matrix can be written in the same form as the unconstrained estimator:

$${}_{c}\hat{\boldsymbol{\Sigma}} = \left(\mathbf{Y} - \mathbf{X}\,{}_{c}\hat{\boldsymbol{\beta}}\right)'\left(\mathbf{Y} - \mathbf{X}\,{}_{c}\hat{\boldsymbol{\beta}}\right) \tag{11.3.4}$$

The derivative of ϕ with respect to $\boldsymbol{\beta}$ can be written, drawing on the unconstrained differentiation demonstrated in Section 11.1,

$$\partial\phi/\partial\boldsymbol{\beta} = (\mathbf{X}'\mathbf{Y})\boldsymbol{\Sigma}^{-1} - (\mathbf{X}'\mathbf{X})\boldsymbol{\beta}\boldsymbol{\Sigma}^{-1} - \tfrac{1}{2}\mathbf{C}\boldsymbol{\lambda}\mathbf{A}' \tag{11.3.5}$$

Setting this derivative to nullity and rearranging terms gives

$$(\mathbf{X}'\mathbf{X})\,{}_{c}\hat{\boldsymbol{\beta}}\,{}_{c}\hat{\boldsymbol{\Sigma}}^{-1} = (\mathbf{X}'\mathbf{Y})\,{}_{c}\hat{\boldsymbol{\Sigma}}^{-1} - \mathbf{C}\tfrac{1}{2}\boldsymbol{\lambda}\mathbf{A}'$$

Premultiplying by $(\mathbf{X}'\mathbf{X})^{-1}$ and postmultiplying by ${}_{c}\hat{\boldsymbol{\Sigma}}$ isolates the $\boldsymbol{\beta}$ estimator:

$${}_{c}\hat{\boldsymbol{\beta}} = (\mathbf{X}'\mathbf{X})^{-1}\mathbf{X}'\mathbf{Y} - (\mathbf{X}'\mathbf{X})^{-1}\mathbf{C}\tfrac{1}{2}\boldsymbol{\lambda}\mathbf{A}'\,{}_{c}\hat{\boldsymbol{\Sigma}} \tag{11.3.6}$$

We can begin to eliminate the unknown $\boldsymbol{\lambda}$ by premultiplying by $\mathbf{C}'$ and postmultiplying by $\mathbf{A}$:

$$\mathbf{C}'\,{}_{c}\hat{\boldsymbol{\beta}}\mathbf{A} = \mathbf{C}'\hat{\boldsymbol{\beta}}\mathbf{A} - \left[\mathbf{C}'(\mathbf{X}'\mathbf{X})^{-1}\mathbf{C}\right]\tfrac{1}{2}\boldsymbol{\lambda}\left(\mathbf{A}'\,{}_{c}\hat{\boldsymbol{\Sigma}}\mathbf{A}\right)$$

the left side of which is $\boldsymbol{\Gamma}$. Next we premultiply by $[\mathbf{C}'(\mathbf{X}'\mathbf{X})^{-1}\mathbf{C}]^{-1}$, postmultiply by $(\mathbf{A}'\,{}_{c}\hat{\boldsymbol{\Sigma}}\mathbf{A})^{-1}$, and isolate the term involving $\boldsymbol{\lambda}$:

$$\tfrac{1}{2}\boldsymbol{\lambda} = \left[\mathbf{C}'(\mathbf{X}'\mathbf{X})^{-1}\mathbf{C}\right]^{-1}\left(\mathbf{C}'\hat{\boldsymbol{\beta}}\mathbf{A} - \boldsymbol{\Gamma}\right)\left(\mathbf{A}'\,{}_{c}\hat{\boldsymbol{\Sigma}}\mathbf{A}\right)^{-1}$$

Substituting this result for $\tfrac{1}{2}\boldsymbol{\lambda}$ in (11.3.6) gives

$${}_{c}\hat{\boldsymbol{\beta}} = \hat{\boldsymbol{\beta}} - (\mathbf{X}'\mathbf{X})^{-1}\mathbf{C}\left[\mathbf{C}'(\mathbf{X}'\mathbf{X})^{-1}\mathbf{C}\right]^{-1}\left(\mathbf{C}'\hat{\boldsymbol{\beta}}\mathbf{A} - \boldsymbol{\Gamma}\right)\left(\mathbf{A}'\,{}_{c}\hat{\boldsymbol{\Sigma}}\mathbf{A}\right)^{-1}\mathbf{A}'\,{}_{c}\hat{\boldsymbol{\Sigma}} \tag{11.3.7}$$

For $\mathbf{A} = \mathbf{I}$ this result is identical in form to the constrained estimator of the univariate $\boldsymbol{\beta}$, found in Eq. (8.4.20):

$${}_{c}\hat{\boldsymbol{\beta}} = \hat{\boldsymbol{\beta}} - (\mathbf{X}'\mathbf{X})^{-1}\mathbf{C}\left[\mathbf{C}'(\mathbf{X}'\mathbf{X})^{-1}\mathbf{C}\right]^{-1}\left(\mathbf{C}'\hat{\boldsymbol{\beta}} - \boldsymbol{\Gamma}\right)$$

For $\mathbf{A} \neq \mathbf{I}$, however, (11.3.7), presenting ${}_c\hat{\boldsymbol{\beta}}$ in terms of ${}_c\hat{\boldsymbol{\Sigma}}$ is of little help: As (11.3.4) reminds us ${}_c\hat{\boldsymbol{\Sigma}}$ is a function of ${}_c\hat{\boldsymbol{\beta}}$. How we can escape the circularity?

In Section 11.2 we noted that constraints on the columns of $\boldsymbol{\beta}$ could be thought of as changing the criterion vector. We explore this further. Our MGLM is marked by the expectation equation (11.1.2), $\mathbf{E}(\mathbf{Y}_{i.}') = \mathbf{X}_{i.}'\boldsymbol{\beta}$. When we constrain $\boldsymbol{\beta}$ by columns, $\boldsymbol{\beta}\mathbf{A} = \boldsymbol{\varepsilon}$, Eq. (11.1.2) becomes $\mathbf{X}_{i.}'(\boldsymbol{\beta}\mathbf{A}) = \mathbf{E}(\mathbf{Y}_{i.}')\mathbf{A}$ $\mathbf{X}_{i.}'\boldsymbol{\varepsilon} = \mathbf{E}(\mathbf{Z}_{i.}')$, and we are now able to act as if we were interested not in the m-variate normal attribute, $\mathbf{Y}_{i.} \sim \mathbf{N}_m(\mathbf{X}_{i.}'\boldsymbol{\beta}, \boldsymbol{\Sigma})$, but in the t-variate normal attribute, $\mathbf{Z}_{i.} \sim \mathbf{N}_t(\mathbf{X}_{i.}'\boldsymbol{\varepsilon}, \boldsymbol{\Phi})$, where $\boldsymbol{\varepsilon} = \boldsymbol{\beta}\mathbf{A}$ and $\boldsymbol{\Phi} = \mathbf{A}'\boldsymbol{\Sigma}\mathbf{A}$ reflect the relation between the two—each $\mathbf{Z}$ attribute is a linear transformation of $\mathbf{Y}$:

$$\mathbf{Z}_{i.}' = \mathbf{Y}_{i.}'\mathbf{A}$$

or, in matrix form, $\mathbf{Z} = \mathbf{YA}$. $\mathbf{Z}_{i.}$ is normally distributed, since it is a linear transformation of the multivariate normal $\mathbf{Y}_{i.}$.

We can adapt the results of this chapter to the multivariate $\mathbf{Z}_{i.}$. We can write directly the unconstrained maximum likelihood estimators of $\boldsymbol{\varepsilon}$ and $\boldsymbol{\Phi}$:

$$\hat{\boldsymbol{\varepsilon}} = (\mathbf{X}'\mathbf{X})^{-1}\mathbf{X}'\mathbf{Z} = (\mathbf{X}'\mathbf{X})^{-1}\mathbf{X}'\mathbf{YA} = \hat{\boldsymbol{\beta}}\mathbf{A} \qquad (11.3.8)$$

and

$$\begin{aligned}\hat{\boldsymbol{\Phi}} &= (1/n)(\mathbf{Z} - \mathbf{X}\hat{\boldsymbol{\varepsilon}})'(\mathbf{Z} - \mathbf{X}\hat{\boldsymbol{\varepsilon}}) = (1/n)(\mathbf{YA} - \mathbf{X}\hat{\boldsymbol{\beta}}\mathbf{A})'(\mathbf{Y} - \mathbf{X}\hat{\boldsymbol{\beta}}\mathbf{A}) \\ &= \mathbf{A}'(1/n)(\mathbf{Y} - \mathbf{X}\hat{\boldsymbol{\beta}})'(\mathbf{Y} - \mathbf{X}\hat{\boldsymbol{\beta}})\mathbf{A} = \mathbf{A}'\hat{\boldsymbol{\Sigma}}\mathbf{A} = (1/n)\mathbf{A}'\mathbf{E}'\mathbf{E}\mathbf{A}\end{aligned} \qquad (11.3.9)$$

Furthermore, if we impose the restriction $\mathbf{C}'\boldsymbol{\varepsilon} = \boldsymbol{\Gamma}$, we can write the constrained MLE of $\boldsymbol{\varepsilon}$ directly from (8.4.20):

$$\begin{aligned}{}_c\hat{\boldsymbol{\varepsilon}} &= \hat{\boldsymbol{\varepsilon}} - (\mathbf{X}'\mathbf{X})^{-1}\mathbf{C}\left[\mathbf{C}'(\mathbf{X}'\mathbf{X})^{-1}\mathbf{C}\right]^{-1}(\mathbf{C}'\boldsymbol{\varepsilon} - \boldsymbol{\Gamma}) \\ &= \hat{\boldsymbol{\beta}}\mathbf{A} - (\mathbf{X}'\mathbf{X})^{-1}\mathbf{C}\left[\mathbf{C}'(\mathbf{X}'\mathbf{X})^{-1}\mathbf{C}\right]^{-1}(\mathbf{C}'\hat{\boldsymbol{\beta}}\mathbf{A} - \boldsymbol{\Gamma})\end{aligned} \qquad (11.3.10)$$

Looking back, we can see that the right side is the equivalent of having postmultiplied the right side of (11.3.7) by $\mathbf{A}$. Thus we can write ${}_c\hat{\boldsymbol{\varepsilon}} = {}_c\hat{\boldsymbol{\beta}}\mathbf{A}$.

Finally, the constrained MLE of the covariance matrix is

$$\begin{aligned}{}_c\hat{\boldsymbol{\Phi}} &= (1/n)(\mathbf{Z} - \mathbf{X}\,{}_c\hat{\boldsymbol{\varepsilon}})'(\mathbf{Z} - \mathbf{X}\,{}_c\hat{\boldsymbol{\varepsilon}}) = \mathbf{A}'(1/n)(\mathbf{Y} - \mathbf{X}\,{}_c\hat{\boldsymbol{\beta}})'(\mathbf{Y} - \mathbf{X}\,{}_c\hat{\boldsymbol{\beta}})\mathbf{A} \\ &= \mathbf{A}'\,{}_c\hat{\boldsymbol{\Sigma}}\mathbf{A} = (1/n)\mathbf{A}'\mathbf{F}'\mathbf{F}\mathbf{A}\end{aligned} \qquad (11.3.11)$$

This, again, is a straightforward extension of the univariate GLM result. We have taken $\mathbf{F}'\mathbf{F}$ to be

$$\mathbf{F}'\mathbf{F} = (\mathbf{Y} - \mathbf{X}\,{}_c\hat{\boldsymbol{\beta}})'(\mathbf{Y} - \mathbf{X}\,{}_c\hat{\boldsymbol{\beta}}) \qquad (11.3.12)$$

the *constrained* sum of squares (and cross products) of errors.

11.3.2 Sums of Squares of Errors for the Effective Criterion

For the MGLM the relevant matrices of sums of squares of errors, both constrained and unconstrained, are for the *effective* multivariate criterion attribute, the t-variate $\mathbf{Y}_{i.}'\mathbf{A}$. The unconstrained error matrix, by (11.3.9), is

$$\mathbf{A'E'EA} = (\mathbf{YA} - \mathbf{X}\hat{\boldsymbol{\beta}}\mathbf{A})'(\mathbf{Y} - \mathbf{X}\hat{\boldsymbol{\beta}}\mathbf{A})$$

which, because it is of the form $\mathbf{A'E'EA} = [\mathbf{YA} - \hat{\mathbf{E}}(\mathbf{YA})]'[\mathbf{YA} - \hat{\mathbf{E}}(\mathbf{YA})]$, can be shown, by the argument of Section 11.1.6, to have a Wishart form [see Eq. (11.1.18)]:

$$\mathbf{A'E'EA} = [\mathbf{YA} - \mathbf{E}(\mathbf{YA})]'\left[\mathbf{I} - \mathbf{X}(\mathbf{X'X})^{-1}\mathbf{X'}\right][\mathbf{YA} - \mathbf{E}(\mathbf{YA})]$$

and

$$\mathbf{A'E'EA} \sim \mathbf{W}_t[(n-p), \mathbf{A'\Sigma A}] \qquad (11.3.13)$$

A common computational form for $\mathbf{A'E'EA}$ is

$$\mathbf{A'E'EA} = \mathbf{A'}\left[\mathbf{Y'Y} - \mathbf{Y'X}(\mathbf{X'X})^{-1}\mathbf{X'Y}\right]\mathbf{A} \qquad (11.3.14)$$

While it is not unusual to obtain $\mathbf{A'E'EA}$ from $\mathbf{E'E}$—indeed, different hypotheses may imply different $\mathbf{A}$ matrices—the sum of squares matrix $\mathbf{F'F}$ and the ${}_c\hat{\boldsymbol{\beta}}$ from which it is derived are seldom directly obtained. Rather, we are most likely to obtain ${}_c\hat{\boldsymbol{\beta}}\mathbf{A}$ first from (11.3.10) as

$${}_c\hat{\boldsymbol{\beta}}\mathbf{A} = \hat{\boldsymbol{\beta}}\mathbf{A} - (\mathbf{X'X})^{-1}\mathbf{C}\left[\mathbf{C'}(\mathbf{X'X})^{-1}\mathbf{C}\right]^{-1}(\mathbf{C'}\hat{\boldsymbol{\beta}}\mathbf{A} - \boldsymbol{\Gamma}) \qquad (11.3.15)$$

and base our constrained sum of squares for the effective criterion on this. We can write $\mathbf{FA} = \mathbf{YA} - \mathbf{X}\,{}_c\hat{\boldsymbol{\beta}}\mathbf{A}$, substituting from the right side of (11.3.15) and, on taking the minor product moment of $\mathbf{FA}$, show that

$$\mathbf{A'F'FA} = \mathbf{A'E'EA} + (\mathbf{C'}\hat{\boldsymbol{\beta}}\mathbf{A} - \boldsymbol{\Gamma})'\left[\mathbf{C'}(\mathbf{X'X})^{-1}\mathbf{C}\right]^{-1}(\mathbf{C'}\hat{\boldsymbol{\beta}}\mathbf{A} - \boldsymbol{\Gamma}) \qquad (11.3.16)$$

This, too, is a common computational form. Note the right side; it suggests, reasonably enough, that $\mathbf{A'F'FA}$ becomes "larger" than $\mathbf{A'E'EA}$ as $\mathbf{C'}\hat{\boldsymbol{\beta}}\mathbf{A} - \boldsymbol{\Gamma}$ departs from the null matrix.

Equation (11.3.16) permits us to write the multivariate equivalent of the *hypothesis sum of squares*, the increase in the sum of squares of errors occasioned by imposing the constraints $\mathbf{C'}\boldsymbol{\beta}\mathbf{A} = \boldsymbol{\Gamma}$:

$$\mathbf{A'F'FA} - \mathbf{A'E'EA} = (\mathbf{C'}\hat{\boldsymbol{\beta}}\mathbf{A} - \boldsymbol{\Gamma})'\left[\mathbf{C'}(\mathbf{X'X})^{-1}\mathbf{C}\right]^{-1}(\mathbf{C'}\hat{\boldsymbol{\beta}}\mathbf{A} - \boldsymbol{\Gamma}) \qquad (11.3.17)$$

11.3.3 Sampling Distributions of Multivariate Error Sums of Squares

We have established that $\mathbf{A'E'EA}$ has a Wishart distribution just as in the univariate GLM $\mathbf{e'e}/\sigma^2$ has a chi-squared distribution. We can establish a similar link between the distributions of $\mathbf{A'F'FA} - \mathbf{A'E'EA}$ and $(\mathbf{f'f} -$

$\mathbf{e'e})/\sigma^2$: The chi-squared form of the latter also generalizes to a Wishart form.

The algebraic argument of Chapter 8 that established that $(\mathbf{f'f} - \mathbf{e'e})/\sigma^2$ is distributed as chi-squared can be repeated almost exactly. Beginning with the outer term in Eq. (11.3.17), we rewrite it

$$\begin{aligned}\mathbf{C'\hat{\beta}A} - \mathbf{\Gamma} &= \mathbf{C'\hat{\beta}A} - \mathbf{C'\beta A} = \mathbf{C'(X'X)^{-1}X'YA} - \mathbf{C'\beta A}\\ &= \mathbf{C'(X'X)^{-1}X'YA} - \mathbf{C'(X'X)^{-1}(X'X)\beta A}\\ &= \mathbf{C'(X'X)^{-1}X'(YA - X\beta A)} = \mathbf{C'(X'X)^{-1}X'[YA - E(YA)]}\end{aligned}$$

Substituting this last form into (11.3.17) gives

$$\mathbf{A'F'FA} - \mathbf{A'E'EA} = [\mathbf{YA} - \mathbf{E(YA)}]'\mathbf{M}[\mathbf{YA} - \mathbf{E(YA)}] \qquad (11.3.18)$$

where

$$\mathbf{M} = \mathbf{X(X'X)^{-1}C}\left[\mathbf{C'(X'X)^{-1}C}\right]^{-1}\mathbf{C'(X'X)^{-1}X'} \qquad (11.3.19)$$

is an idempotent matrix.

Since the rank of $\mathbf{M}$ is equal to the rank of $\mathbf{C}$, or s (the number of restrictions imposed on the p rows of $\boldsymbol{\beta}$), we have established that

$$\mathbf{A'F'FA} - \mathbf{A'E'EA} \sim \mathbf{W}_t[s, \mathbf{A'\Sigma A}] \qquad (11.3.20)$$

By paralleling the univariate results further, the matrix $\mathbf{M}$ of (11.3.19) may be shown to be orthogonal to $\mathbf{I} - \mathbf{X(X'X)^{-1}X'}$ the idempotent matrix of the Wishart form of $\mathbf{A'E'EA}$ as given in (11.1.18). As a result, $\mathbf{A'E'EA}$ and $\mathbf{A'F'FA} - \mathbf{A'E'EA}$ have independent Wishart distributions.

In summary, if each of n m-element vectors $\mathbf{Y}_{i.}$ is an independent multivariate normal observation with $\mathbf{E(Y}_{i.}'\mathbf{)} = \mathbf{X}_{i.}'\boldsymbol{\beta}$ and $\mathbf{Var(Y}_{i.}\mathbf{)} = \boldsymbol{\Sigma}$, and if $\mathbf{X}$ is an $n \times p$ matrix of fixed constants with rank equal to its column order, then under the system of linear constraints on the $p \times m$ parameter matrix $\boldsymbol{\beta}$ given by $\mathbf{C'\beta A} = \mathbf{\Gamma}$—where $\mathbf{C}_{p\times s\leqslant p}$, $\mathbf{A}_{m\times t\leqslant m}$, and $\mathbf{\Gamma}_{s\times t}$ are *hypothesis* matrices of fixed constants with $r(\mathbf{C}) = s$ and $r(\mathbf{A}) = t$—the following hold:

1. $\mathbf{A'E'EA} = \mathbf{A'(Y - X\hat{\beta})'(Y - X\hat{\beta})A}$ has the Wishart distribution
$$\mathbf{W}_t[(n-p), \mathbf{A'\Sigma A}]$$
where $\hat{\boldsymbol{\beta}} = \mathbf{(X'X)^{-1}X'Y}$ is the unconstrained MLE of $\boldsymbol{\beta}$.
2. $\mathbf{A'F'FA} - \mathbf{A'E'EA} = (\mathbf{C'\hat{\beta}A} - \mathbf{\Gamma})'[\mathbf{C'(X'X)^{-1}C}]^{-1}(\mathbf{C'\hat{\beta}A} - \mathbf{\Gamma})$ has Wishart distribution
$$\mathbf{W}_t(s, \mathbf{A'\Sigma A})$$
where $\mathbf{A'F'FA} = \mathbf{A'(Y - X\,{}_c\hat{\beta})'(Y - X\,{}_c\hat{\beta})A}$ and ${}_c\hat{\boldsymbol{\beta}}\mathbf{A}$ is the MLE of $\boldsymbol{\beta}\mathbf{A}$ constrained by $\mathbf{C'\beta A} = \mathbf{\Gamma}$.
3. The two Wishart matrices are distributed independently.

11.3.4 Likelihood Ratio Basis for Hypothesis Testing

In Eq. (11.1.7) we developed an expression for the likelihood of a sample of n independent observations under the MGLM:

$$L(\mathbf{Y}) = (2\pi)^{-(nm/2)}\left(\prod_{j=1}^{m}\Delta_j^2\right)^{-(n/2)} \exp\{-\tfrac{1}{2}\operatorname{tr}[(\mathbf{Y}-\mathbf{X}\boldsymbol{\beta})\boldsymbol{\Sigma}^{-1}(\mathbf{Y}-\mathbf{X}\boldsymbol{\beta}')]\}$$

The matrices $\boldsymbol{\beta}$ and $\boldsymbol{\Sigma}$, together with the basic diagonal of the latter, $\boldsymbol{\Delta}^2$, are unknown parameters; hence, the likelihood of the sample is not known. We have estimated $\boldsymbol{\beta}$ and $\boldsymbol{\Sigma}$, however, specifically to maximize the likelihood of the sample. If we substitute the MLE for $\boldsymbol{\beta}$ and $\boldsymbol{\Sigma}$ in (11.1.7), we have an unconstrained likelihood for the sample:

$$L(\mathbf{Y}|\hat{\boldsymbol{\beta}}, \hat{\boldsymbol{\Sigma}}) = (2\pi)^{-(nm/2)}\left(\prod_{j=1}^{m}D_j^2\right)^{-(n/2)} \times \exp\{-\tfrac{1}{2}\operatorname{tr}[(\mathbf{Y}-\mathbf{X}\hat{\boldsymbol{\beta}})\hat{\boldsymbol{\Sigma}}^{-1}(\mathbf{Y}-\mathbf{X}\hat{\boldsymbol{\beta}})']\} \quad (11.3.21)$$

where $\hat{\boldsymbol{\Sigma}}$ has basic structure $\mathbf{qD}^2\mathbf{q}'$.

If we write $\hat{\boldsymbol{\Sigma}}^{-1}$ as $\mathbf{qD}^{-2}\mathbf{q}'$, or $(\mathbf{qD}^{-1})(\mathbf{D}^{-1}\mathbf{q}')$, and recall that the trace of the minor and major product moments of a matrix are the same, we can simplify the exponential term in this equation:

$$\begin{aligned}
&-\tfrac{1}{2}\operatorname{tr}[(\mathbf{Y}-\mathbf{X}\hat{\boldsymbol{\beta}})(\mathbf{qD}^{-1})(\mathbf{D}^{-1}\mathbf{q}')(\mathbf{Y}-\mathbf{X}\hat{\boldsymbol{\beta}})'] \\
&\quad = -\tfrac{1}{2}\operatorname{tr}[\mathbf{D}^{-1}\mathbf{q}'(\mathbf{Y}-\mathbf{X}\hat{\boldsymbol{\beta}})'(\mathbf{Y}-\mathbf{X}\hat{\boldsymbol{\beta}})\mathbf{qD}^{-1}] \\
&\quad = -\tfrac{1}{2}\operatorname{tr}[\mathbf{D}^{-1}\mathbf{q}'(n\hat{\boldsymbol{\Sigma}})\mathbf{qD}^{-1}] \\
&\quad = (-n/2)\operatorname{tr}[\mathbf{D}^{-1}\mathbf{q}'(\mathbf{qD}^2\mathbf{q}')\mathbf{qD}^{-1}] \\
&\quad = (-n/2)\operatorname{tr}(\mathbf{I}) = -nm/2
\end{aligned} \quad (11.3.22)$$

As a result, the likelihood of the sample based upon unconstrained estimates is

$$L(\mathbf{Y}|\hat{\boldsymbol{\beta}}, \hat{\boldsymbol{\Sigma}}) = (2\pi e)^{-nm/2}\left(\prod_{j=1}^{m}D_j^2\right)^{-n/2} \quad (11.3.23)$$

Remember that $\boldsymbol{\beta}$ and $\boldsymbol{\Sigma}$ were estimated so as to maximize this quantity.

Although (11.3.23) was developed for the observation matrix $\mathbf{Y}$, it holds equally well for the observation matrix $\mathbf{YA}$. We shift to that interpretation now. All we need keep in mind is that $\mathbf{D}^2$ is now the basic diagonal of the "effective" criterion variance–covariance matrix, $\mathbf{A}'\hat{\boldsymbol{\Sigma}}\mathbf{A}$.

We also obtained MLEs of $\boldsymbol{\beta}\mathbf{A}$ and $\mathbf{A}'\boldsymbol{\Sigma}\mathbf{A}$ constrained by the requirement that $\mathbf{C}'\boldsymbol{\beta}\mathbf{A} = \boldsymbol{\Gamma}$. Substituting these into (11.1.17) gives a constrained estimated likelihood for the sample. If $\mathbf{A}'{}_c\hat{\boldsymbol{\Sigma}}\mathbf{A} = (1/n)\mathbf{A}'\mathbf{F}'\mathbf{F}\mathbf{A}$ has basic structure $\mathbf{pd}^2\mathbf{p}'$, then the trace argument may be repeated to show that this constrained estimated likelihood can be expressed

$$L(\mathbf{Y}|\mathbf{C}'\boldsymbol{\beta}\mathbf{A} = \boldsymbol{\Gamma}) = (2\pi e)^{-nt/2}\left(\prod_{j=1}^{t} d_j^2\right)^{-n/2} \tag{11.3.24}$$

The likelihood has been maximized in (11.3.24) over all values of $\boldsymbol{\beta}$ that satisfy $\mathbf{C}'\boldsymbol{\beta}\mathbf{A} = \boldsymbol{\Gamma}$, whereas in Eq. (11.2.23) the likelihood has been maximized over all values of $\boldsymbol{\beta}$ (or $\boldsymbol{\beta}\mathbf{A}$), whether $\mathbf{C}'\boldsymbol{\beta}\mathbf{A} = \boldsymbol{\Gamma}$ is satisfied or not. As a result, the ratio

$$\text{LR} = L(\mathbf{Y}|\mathbf{C}'\hat{\boldsymbol{\beta}}\mathbf{A} = \boldsymbol{\Gamma})/L(\mathbf{Y}|\hat{\boldsymbol{\beta}}, \hat{\boldsymbol{\Sigma}}) \leqslant 1.0 \tag{11.3.25}$$

LR can never exceed 1.0. It is another *likelihood ratio*, the ratio of the likelihood of the sample, given that a particular hypothesis is true, to the likelihood of the sample, given all possible hypotheses. It can serve as the basis for testing statistical hypotheses.

To use LR, however, requires that the ratio have some known sampling distribution, at least under the hypothesis. We need to know how large LR is likely to be. For the MGLM, substituting from (11.3.23) and (11.3.24) provides

$$\begin{aligned}\text{LR} &= \left(\prod_{j=1}^{t} d_j^2\right)^{-n/2} \Big/ \left(\prod_{j=1}^{t} D_j^2\right)^{-n/2} \\ &= \left(\prod_{j=1}^{t} D_j^2 \Big/ \prod_{j=1}^{t} d_j^2\right)^{n/2}\end{aligned}$$

As in Chapter 8, the likelihood ratio may be replaced with a simpler form, eliminating the positive exponent:

$$\lambda = \prod_{j=1}^{t} D_j^2 \Big/ \prod_{j=1}^{t} d_j^2 \tag{11.3.26}$$

since there is a monotonic relation between the two ratios.

To this point we have taken $\mathbf{D}^2$ and $\mathbf{d}^2$ to be the basic diagonals of $(1/n)\mathbf{A}'\mathbf{E}'\mathbf{E}\mathbf{A}$ and $(1/n)\mathbf{A}'\mathbf{F}'\mathbf{F}\mathbf{A}$. Because the multiplication by $1/n$ does not alter the ratio of the diagonal products, we take the ratio to be that of the products of the basic diagonals of $\mathbf{A}'\mathbf{E}'\mathbf{E}\mathbf{A}$ and of $\mathbf{A}'\mathbf{F}'\mathbf{F}\mathbf{A} = (\mathbf{A}'\mathbf{F}'\mathbf{F}\mathbf{A} - \mathbf{A}'\mathbf{E}'\mathbf{E}\mathbf{A}) + \mathbf{A}'\mathbf{E}'\mathbf{E}\mathbf{A}$:

$$\lambda = \frac{\prod_{j=1}^{t} {}_{(\mathbf{A}'\mathbf{E}'\mathbf{E}\mathbf{A})}D_j}{\prod_{j=1}^{t} {}_{[(\mathbf{A}'\mathbf{F}'\mathbf{F}\mathbf{A} - \mathbf{A}'\mathbf{E}'\mathbf{E}\mathbf{A}) + \mathbf{A}'\mathbf{E}'\mathbf{E}\mathbf{A}]}D_j} \tag{11.3.27}$$

Writing $\mathbf{A}'\mathbf{F}'\mathbf{F}\mathbf{A}$ in this complicated fashion serves a purpose, as we shall see in the following section.

11.3.5 Wilks's Lambda and the U Distribution

The ratio given in (11.3.27) is one of the most frequently used test statistics in multivariate analysis. Known as Wilks's lambda, it is based on the following distributional properties of Wishart matrices: If $\mathbf{A} \sim \mathbf{W}_m(a, \sigma)$ and $\mathbf{B} \sim \mathbf{W}_m(b, \sigma)$ are independently distributed, and $b \geqslant m$, then the ratio

$$\prod_{j=1}^{m} {}_{\mathbf{B}}D_j \Big/ \prod_{j=1}^{m} {}_{\mathbf{A}+\mathbf{B}}D_j \sim U(m, a, b)$$

follows the distribution of the U statistic, depending upon the three parameters. *Smaller* values of U make up a *critical region* for the MGLM likelihood ratio test. They appear in tabular form as an appendix.

In Section 11.3.3 it was established that $\mathbf{A'E'EA}$ and $\mathbf{A'F'FA} - \mathbf{A'E'EA}$ are independently distributed Wishart matrices:

$$\mathbf{A'E'EA} \sim \mathbf{W}_t[(n-p), \mathbf{A'\Sigma A}], \qquad \mathbf{A'F'FA} - \mathbf{A'E'EA} \sim \mathbf{W}_t[s, \mathbf{A'\Sigma A}]$$

As long as $n - p \geqslant t$, the ratio in (11.3.27) is distributed under the hypothesis $\mathbf{C'\beta A} = \mathbf{\Gamma}$ as $U[t, s, (n-p)]$. In developing this test statistic we have assumed $\mathbf{X}_{n \times p}$, $\mathbf{C}_{p \times s}$, and $\mathbf{A}_{m \times t}$ to be basic and each row of $\mathbf{Y}_{n \times m}$ to be a multivariate normal observation.

When $t = 1$, the criterion is univariate, $\mathbf{A'E'EA} = \mathbf{e'e}$, $\mathbf{A'F'FA} - \mathbf{A'E'EA} = \mathbf{f'f} - \mathbf{e'e}$ in GLM notation, and the λ of Eq. (11.3.27) is related to the univariate F ratio test:

$$\lambda = \frac{\mathbf{e'e}}{(\mathbf{f'f} - \mathbf{e'e}) + \mathbf{e'e}} = \frac{1}{1 + (\mathbf{f'f} - \mathbf{e'e})/\mathbf{e'e}}$$

whereas the F ratio test of the general linear hypothesis is of the form

$$F_{s,(n-p)} = \frac{(\mathbf{f'f} - \mathbf{e'e})/s}{\mathbf{e'e}/(n-p)}$$

Thus we can write

$$\lambda = U[1, s, (n-p)] = \{1 + [s/(n-p)]F_{s,(n-p)}\}^{-1}$$

Large values of F are equivalent to small values of U. In Chapter 8 we could have developed $U[1, s, (n-p)]$ as the likelihood ratio test statistic for the univariate GLM hypothesis as readily, although $F_{s,(n-p)}$ is better known.

In obtaining a value for the likelihood ratio statistic, many computational algorithms use a product of basic diagonals for a single matrix rather than a ratio of two such products. The basis for this is an application of the two rules given in Section 10.8 relating to the products of basic diagonals of square basic matrices:

1. For $\mathbf{C} = \mathbf{AB}$, $\prod_{j=1}^{m} {}_{\mathbf{C}}D_j = (\prod_{j=1}^{m} {}_{\mathbf{A}}D_j)(\prod_{j=1}^{m} {}_{\mathbf{B}}D_j)$ and
2. $\prod_{j=1}^{m} {}_{\mathbf{A}^{-1}}D_j = 1/\prod_{j=1}^{m} {}_{\mathbf{A}}D_j$.

If we write the basic structure $\mathbf{A'E'EA} = \mathbf{QD^2Q'} = (\mathbf{QDQ'})(\mathbf{QDQ'})$, then

$$\lambda = \left(\prod_{j=1}^{t} {}_{(\mathbf{QDQ'})}D_j\right)\left(\prod_{j=1}^{t} {}_{(\mathbf{QDQ'})}D_j\right) \Big/ \prod_{j=1}^{t} {}_{(\mathbf{A'F'FA})}D_j$$

or

$$1/\lambda = \left(\prod_{j=1}^{t} {}_{(\mathbf{QD^{-1}Q'})}D_j\right)\left(\prod_{j=1}^{t} {}_{(\mathbf{A'F'FA})}D_j\right)\left(\prod_{j=1}^{t} {}_{(\mathbf{QD^{-1}Q'})}D_j\right)$$

In this form, lambda is identical to the reciprocal of the product of the basic diagonals of the symmetric product moment matrix,

$$\mathbf{L} = (\mathbf{QD^{-1}Q'})(\mathbf{A'F'FA})(\mathbf{QD^{-1}Q'}) \tag{11.3.28}$$

That is,

$$\left(\prod_{j=1}^{t} {}_{\mathbf{L}}D_j\right)^{-1} \sim U[t, s, (n-p)] \tag{11.3.29}$$

An approximation to the U statistic based on the F distribution is widely reported by computer programs for the multivariate analysis of variance or other multivariate techniques. The quantity

$$[(1 - U^{1/g})/(U^{1/g})](h/st) \sim F_{ts,h}$$

approximately, where

$$g = \begin{cases} (s^2t^2 - 4)/(s^2 + t^2 - 5) & \text{if } s^2 + t^2 \neq 5 \\ 1 & \text{if } s^2 + t^2 = 5 \end{cases}$$

and

$$h = [(n-p) - \tfrac{1}{2}(t - s + 1)]g - \tfrac{1}{2}(ts - 2)$$

The computation of h is likely to produce a noninteger degree of freedom and call for interpolation in standard tables of F.

This F test is exact when t (the number of effective criteria) or s (the number of constraints on the rows of $\boldsymbol{\beta}$) is 1. Thus, for the 1 df hypothesis ($\mathbf{C'}$ a row vector, $s = 1$) we have $g = 1$, $h = (n-p) - (t-1)$, and

$$[(1 - U)/U]\{[(n-p) - (t-1)]/t\} \sim F_{t,[(n-p)-(t-1)]}$$

For $t = 1$ the approximate F statistic reduces, as we noted earlier, to the conventional univariate F ratio.

11.3.6 Union–Intersection Basis for Hypothesis Testing

The likelihood ratio test developed in the last two sections, though widely used in multivariate analysis, is not the only approach to testing hypotheses in the MGLM. The principal alternative is based on the *union–intersection principle*. We outline this principle and develop from it a test of the MGLM hypothesis $\mathbf{C}'\boldsymbol{\beta}\mathbf{A} = \boldsymbol{\Gamma}$.

The union–intersection principle applied to the multivariate hypothesis requires

1. that the hypothesis be regarded as tenable (not to be rejected) if the *intersection* of all composite univariate hypotheses is tenable, and
2. that the hypothesis be rejected if the *union* of all composite hypotheses is not tenable.

That is, the multivariate hypothesis is not rejected only if every composite univariate hypothesis is not rejected, and the multivariate hypothesis is rejected if at least one of the composite univariate hypotheses is rejected. What are these composite univariate hypotheses?

The MGLM hypothesis $\mathbf{C}'\boldsymbol{\beta}\mathbf{A} = \mathbf{0}$ giving rise to an $s \times t$ null matrix has as its composite univariate hypotheses the set of equations

$$\mathbf{C}'\boldsymbol{\beta}\mathbf{A}\mathbf{w} = \mathbf{0} \tag{11.3.30}$$

where $\mathbf{w}$ is *any* nonnull t-element vector. Each such univariate hypothesis requires evaluating an s-element null vector for the s constraints imposed by $\mathbf{C}$. The union–intersection principle requires that we reject $\mathbf{C}'\boldsymbol{\beta}\mathbf{A} = \mathbf{0}$ if we can find any nonnull vector $\mathbf{w}$ such that we would have to reject the univariate hypothesis $\mathbf{C}'(\boldsymbol{\beta}\mathbf{A}\mathbf{w}) = \mathbf{0}$.

Each row of $\mathbf{YA}$, under MGLM assumptions, is an observation from a t-variate normal distribution with $\mathbf{E}[(\mathbf{YA})'_{i.}] = \mathbf{X}'_{i.}\boldsymbol{\beta}\mathbf{A}$ and $\mathbf{Var}[(\mathbf{YA})_{i.}] = \mathbf{A}'\boldsymbol{\Sigma}\mathbf{A}$. Each row of $\mathbf{YA}$, further, is sampled independently. Under these conditions, for any nonnull t-element vector $\mathbf{w}$, the n-element vector $\mathbf{YAw}$ consists of n independent observations from *univariate* normal distributions such that $\mathbf{E}[(\mathbf{YAw})_{i}] = \mathbf{X}'_{i.}\boldsymbol{\beta}\mathbf{A}\mathbf{w}$ and $\mathbf{Var}[(\mathbf{YAw})_{i}] = \mathbf{w}'\mathbf{A}'\boldsymbol{\Sigma}\mathbf{A}\mathbf{w}$.

The test of a composite univariate hypothesis, as a result, should be the test developed for the univariate GLM. That is, by the union–intersection principle we are led to reject the multivariate hypothesis $\mathbf{C}'\boldsymbol{\beta}\mathbf{A} = \mathbf{0}$ if there is any univariate normal attribute $\mathbf{YAw}$ with expectation $\mathbf{E}(\mathbf{YAw}) = \mathbf{X}\boldsymbol{\beta}\mathbf{A}\mathbf{w}$ for which we would reject (at the agreed upon significance level) the hypothesis $\mathbf{C}'\boldsymbol{\beta}\mathbf{A}\mathbf{w} = \mathbf{0}$. From Chapter 8 we know an appropriate form of the univariate test to be

$$F = \frac{(\mathbf{f}'\mathbf{f} - \mathbf{e}'\mathbf{e})/(n-p)}{\mathbf{e}'\mathbf{e}/s}$$

Based on the results of Section 11.3.5 this test statistic may be stated in MGLM terms:

$$[s/(n-p)]F = \frac{\mathbf{w}'(\mathbf{A}'\mathbf{F}'\mathbf{F}\mathbf{A} - \mathbf{A}'\mathbf{E}'\mathbf{E}\mathbf{A})\mathbf{w}}{\mathbf{w}'(\mathbf{A}'\mathbf{E}'\mathbf{E}\mathbf{A})\mathbf{w}} \tag{11.3.31}$$

where, as earlier, $\mathbf{A}'\mathbf{E}'\mathbf{E}\mathbf{A}$ and $\mathbf{A}'\mathbf{F}'\mathbf{F}\mathbf{A}$ are the multivariate sums of squares of errors matrices.

We wish to reject the multivariate hypothesis if we can find a vector **w** that will cause the right side of (11.3.31) to exceed

$$\tau_\alpha = [s/(n-p)]F_{\alpha,s,(n-p)}$$

where the proportion of values in the distribution of $F_{s,(n-p)}$ that are greater than $F_{\alpha,s,(n-p)}$ is α, the level of significance of the test.

Rather than search by trial and error for a vector that, substituted in (11.3.31), might produce a value greater than τ_α, we can use matrix differentiation techniques to find **w** to *maximize* the function. If this maximum value exceeds the critical value, we know there is at least one vector for which the univariate hypothesis of (11.3.27) would be rejected. Under these circumstances, we would reject the multivariate hypothesis. On the other hand, if the maximum value of (11.3.31) is not greater than τ_α, we are assured that there is no vector **w** for which the univariate hypothesis would be rejected; hence we would not, under the union–intersection principle, be led to reject the multivariate hypothesis.

11.3.7 Roy's Greatest Basic Diagonal Test

A ratio of matrix forms, as in (11.3.31) is difficult to optimize, so we manipulate it a bit. As in presenting an alternative form for the *U* statistic, we want to "reduce" the ratio. First, we write the basic structure of the denominator sums of squares.

$$\mathbf{A'E'EA} = \mathbf{QD^2Q'} = (\mathbf{QDQ'})(\mathbf{QDQ'})$$

Then we write $\mathbf{w} = (\mathbf{QD^{-1}Q'})\mathbf{v}$. This expression does not restrict the number of **w** vectors considered, since every nonnull **w** has a corresponding **v**: $(\mathbf{QDQ'})\mathbf{w} = \mathbf{v}$. By considering all possible vectors **v**, we shall be considering all possible nonnull vectors.

Substituting for **w** in (11.3.31) gives

$$[s,(n-p)]F = \frac{\mathbf{v'(QD^{-1}Q')(A'F'FA - A'E'EA)(QD^{-1}Q')v}}{\mathbf{v'(QD^{-1}Q')(QD^2Q')(QD^{-1}Q')v}}$$
$$= \frac{\mathbf{v'(QD^{-1}Q')(A'F'FA - A'E'EA)(QD^{-1}Q')v}}{\mathbf{v'v}} \tag{11.3.32}$$

This equation does not have a unique maximum. If two vectors are proportional, they give the same value to (11.3.32) when substituted for **v**.

A useful way to overcome this lack of uniqueness is to restrict the search for optimizing vectors **v** to those for which $\mathbf{v'v} = 1$. This, again, is not limiting, as any nonnull vector is proportional to another with sum of squares of 1. The vector that maximizes (11.3.32) is among those whose sum of squares is 1. Using a Lagrangian multiplier to build in this side condition, we write from (11.3.32) a new function to be maximized:

$$\phi = \mathbf{v}'(\mathbf{QD}^{-1}\mathbf{Q}')(\mathbf{A}'\mathbf{F}'\mathbf{FA} - \mathbf{A}'\mathbf{E}'\mathbf{EA})(\mathbf{QD}^{-1}\mathbf{Q}')\mathbf{v} - \lambda(\mathbf{v}'\mathbf{v} - 1) \tag{11.3.33}$$

Differentiating gives

$$\partial\phi/\partial\mathbf{v} = 2(\mathbf{QD}^{-1}\mathbf{Q}')(\mathbf{A}'\mathbf{F}'\mathbf{FA} - \mathbf{A}'\mathbf{E}'\mathbf{EA})(\mathbf{QD}^{-1}\mathbf{Q}')\mathbf{v} - 2\lambda\mathbf{v}$$

and

$$\partial\phi/\partial\lambda = -(\mathbf{v}'\mathbf{v} - 1)$$

Setting the former equal to zero, we get

$$(\mathbf{QD}^{-1}\mathbf{Q}')(\mathbf{A}'\mathbf{F}'\mathbf{FA} - \mathbf{A}'\mathbf{E}'\mathbf{EA})(\mathbf{QD}^{-1}\mathbf{Q}')\mathbf{v} = \lambda\mathbf{v} \tag{11.3.34}$$

a result of the form $\mathbf{Mv} = s\mathbf{v}$, where $\mathbf{M}$ is a product moment matrix, $\mathbf{v}$ a nonnull vector, and s a scalar. Sometimes known as the characteristic equation (of the matrix $\mathbf{M}$), we encountered a form of it in Section 6.5 when we previewed principal components analysis. The solution vector $\mathbf{v}$ can be stated in terms of the basic structure of the matrix $\mathbf{M}$. If $\mathbf{M}$ has basic structure $\mathbf{PdP}'$, then

$$\mathbf{MP}_{.i} = \mathbf{PdP}'\mathbf{P}_{.i} = \mathbf{Pde}_i = \mathbf{P}_{.i}d_i = d_i\mathbf{P}_{.i}$$

and we see that any column of $\mathbf{P}$ is a solution vector.

Equation (11.3.34) is satisfied for any column $\mathbf{v}$ of the left orthonormal to the matrix

$$\mathbf{M} = (\mathbf{QD}^{-1}\mathbf{Q}')(\mathbf{A}'\mathbf{F}'\mathbf{FA} - \mathbf{A}'\mathbf{E}'\mathbf{EA})(\mathbf{QD}^{-1}\mathbf{Q}') = \mathbf{PdP}' \tag{11.3.35}$$

Further, λ, the scalar constant on the right of (11.3.34), takes the value of the basic diagonal element associated with that orthonormal column.

For our MGLM test statistic example, $r(\mathbf{M}) = t$ and there are t solutions to (11.3.34). We still seek the one that maximizes (11.3.32), which we can rewrite $[s(n-p)]F = \mathbf{P}'_{.i}(\mathbf{PdP}')\mathbf{P}_{.i} = d_i$. Because $\mathbf{d}$ is a basic diagonal, it is now clear that this is a maximum when $\mathbf{v} = \mathbf{P}_{.1}$, the first column of the left orthonormal of $\mathbf{M}$. In this case, we have $[s,(n-p)]F = d_1$.

What we have found, then, is that the multivariate hypothesis will be rejected, according to the union–intersection principle, if $d_1 > \tau_\alpha$, where

$$\tau_\alpha = [s,(n-p)]F_{\alpha,s,(n-p)}$$

and d_1 is the first (largest) basic diagonal of

$$\mathbf{M} = (\mathbf{QD}^{-1}\mathbf{Q}')(\mathbf{A}'\mathbf{F}'\mathbf{FA} - \mathbf{A}'\mathbf{E}'\mathbf{EA})(\mathbf{QD}^{-1}\mathbf{Q}')$$

$\mathbf{QD}^2\mathbf{Q}'$ being the basic structure of $\mathbf{A}'\mathbf{E}'\mathbf{EA}$.

The largest basic diagonal is an optimized property of each sample matrix, and the sampling distribution of d_1 under the hypothesis is quite complex. Nevertheless, critical values are available in tabular form. The Appendix gives values not of d_1 itself but of

$$\theta = d_1/(1 + d_1) \tag{11.3.36}$$

The distribution of θ is dependent on three parameters:

$$d_1/(1 + d_1) \sim \theta_{p_1, p_2, p_3} \tag{11.3.37}$$

where $p_1 = \min(s, t)$, $p_2 = \frac{1}{2}(|s - t| - 1)$, and $p_3 = \frac{1}{2}[(n - p) - (t + 1)]$.

For $p_1 = 1$ the largest diagonal d_1 itself has a sampling distribution closely related to that of F:

$$d_1[(p_3 + 1)/(p_2 + 1)] \sim F_{2(p_2+1), 2(p_3+1)} \tag{11.3.38}$$

This special case includes not only the univariate effective criterion ($t = 1$) but also the single constraint on the rows of $\boldsymbol{\beta}$, $s = 1$. This is not uncommon in multivariate analysis.

We have developed two hypothesis testing techniques for the MGLM. Both the likelihood ratio based test (Wilks's lambda) and the union intersection based test (Roy's largest diagonal) converge to the usual F test when $t = 1$. In the multivariate case their resemblances and differences may be highlighted by reexpressing Wilks's lambda. We have seen in (11.3.29) and (11.3.28) that

$$\lambda = \left(\prod_{j=1}^{t} {}_{\mathrm{L}}D_j\right)^{-1} \qquad \text{for} \quad \mathbf{L} = \left[(\mathbf{QD}^{-1}\mathbf{Q}')(\mathbf{A}'\mathbf{F}'\mathbf{FA})(\mathbf{QD}^{-1}\mathbf{Q}')\right]$$

where $\mathbf{A}'\mathbf{E}'\mathbf{EA}$ has basic structure $\mathbf{QD}^2\mathbf{Q}'$. The matrix $\mathbf{L}$ may be rewritten successively

$$\begin{aligned}
\mathbf{L} &= (\mathbf{QD}^{-1}\mathbf{Q}')[(\mathbf{A}'\mathbf{F}'\mathbf{FA} - \mathbf{A}'\mathbf{E}'\mathbf{EA}) + \mathbf{A}'\mathbf{E}'\mathbf{EA}](\mathbf{QD}^{-1}\mathbf{Q}') \\
&= (\mathbf{QD}^{-1}\mathbf{Q}')(\mathbf{A}'\mathbf{F}'\mathbf{FA} - \mathbf{A}'\mathbf{E}'\mathbf{EA})(\mathbf{QD}^{-1}\mathbf{Q}') \\
&\quad + (\mathbf{QD}^{-1}\mathbf{Q}')(\mathbf{A}'\mathbf{E}'\mathbf{EA})(\mathbf{QD}^{-1}\mathbf{Q}') \\
&= (\mathbf{QD}^{-1}\mathbf{Q}')(\mathbf{A}'\mathbf{F}'\mathbf{FA} - \mathbf{A}'\mathbf{E}'\mathbf{EA})(\mathbf{QD}^{-1}\mathbf{Q}') + \mathbf{I}
\end{aligned}$$

or as $\mathbf{L} = \mathbf{M} + \mathbf{I}$ where, from Eq. (11.3.35),

$$\mathbf{M} = (\mathbf{QD}^{-1}\mathbf{Q}')(\mathbf{A}'\mathbf{F}'\mathbf{FA} - \mathbf{A}'\mathbf{E}'\mathbf{EA})(\mathbf{QD}^{-1}\mathbf{Q}')$$

has basic structure $\mathbf{PdP}'$.

Though we shall not prove this, the basic diagonal of $\mathbf{L} = \mathbf{M} + \mathbf{I}$ can be expressed

$${}_{\mathrm{L}}\mathbf{D} = {}_{\mathrm{M}}\mathbf{D} + \mathbf{I}$$

and, as a consequence, the lambda statistic can be shown to be

$$\lambda = \left[\prod_{j=1}^{t} (d_j + 1)\right]^{-1} \tag{11.3.39}$$

Thus both hypothesis testing strategies depend on the magnitude of the basic diagonal of $\mathbf{M} = \mathbf{PdP}'$. The largest of these diagonals, d_1, is evaluated for significance in Roy's test and the product of all t of them gives us the substance of Wilks's test. In both instances, a large basic diagonal leads to rejection of the multivariate general linear hypothesis.

The matrix **M** may be given an interpretation that suggests its importance to multivariate hypothesis testing. At the core of **M** is the matrix $\mathbf{A'F'FA} - \mathbf{A'E'EA}$, which contains the increase in the sum of squares and sum of products of errors occasioned by imposing the constraints of the multivariate general linear hypothesis. However, the magnitudes of the elements of this matrix and how they are arranged reflect not only the effects of the hypothesis but also the sizes of within population variances and covariances of $\mathbf{YA} - \mathbf{E(YA)}$.

The matrix $\mathbf{QD}^{-1}\mathbf{Q'}$, in turn, can be thought of as the sample (or estimated) spherizing transformation matrix for the multivariate observations making up $\mathbf{YA} - \mathbf{E(YA)}$. That is, the result of this transformation

$$[\mathbf{YA} - \hat{\mathbf{E}}(\mathbf{YA})](\mathbf{QD}^{-1}\mathbf{Q'}) = \mathbf{Z} - \hat{\mathbf{E}}(\mathbf{Z})$$

has minor product moment

$$\begin{aligned}[\mathbf{Z} - \hat{\mathbf{E}}(\mathbf{Z})]'[\mathbf{Z} - \hat{\mathbf{E}}(\mathbf{Z})] &= (\mathbf{QD}^{-1}\mathbf{Q'})[\mathbf{YA} - \hat{\mathbf{E}}(\mathbf{YA})]' \\ &\quad \times [\mathbf{YA} - \hat{\mathbf{E}}(\mathbf{YA})](\mathbf{QD}^{-1}\mathbf{Q'}) \\ &= (\mathbf{QD}^{-1}\mathbf{Q'})(\mathbf{A'E'EA})(\mathbf{QD}^{-1}\mathbf{Q'}) = \mathbf{I}\end{aligned}$$

and the transformed sample variates have common variance and zero covariance.

In effect, then, the matrix **M** is a standardized version of the increase in the sums of squares and cross products of errors resulting from the hypothesis. The spherizing transformation counters the influence of disparate variances and nonzero covariances among the criteria.

11.4 Summary

In this chapter we extended to the multivariate criterion the general linear model developed in Chapter 8. We considered independent samples from multivariate normal populations with a common variance–covariance matrix $\boldsymbol{\Sigma}$ but potentially different expectation vectors. By taking the expectation vector to be linearly dependent on a vector of predictor attribute scores $\mathbf{E}(\mathbf{Y}_{i.})' = \mathbf{X}'_{i.}\boldsymbol{\beta}$, we found the maximum likelihood estimates of $\boldsymbol{\Sigma}$ and $\boldsymbol{\beta}$ to have the same form as for the univariate case.

We then developed the multivariate general linear hypothesis, $\mathbf{C}'\boldsymbol{\beta}\mathbf{A} = \boldsymbol{\Gamma}$, permitting us to constrain $\boldsymbol{\beta}$ [or $_{\mathrm{T}}\mathbf{E}(\mathbf{Y})$] not only by row but by column. Columnar constraints yield the concept of an effective number of criteria, which may be fewer than the observed criterion attributes. Two of the most frequently used tests of the multivariate general linear hypothesis were deduced. On the likelihood ratio principle, Wilks's lambda, a generalized variance, was obtained and related to the U statistic. An alternative test criterion, Roy's greatest basic diagonal, resulted from application of the union–intersection principle. This result, in turn, was related to the tabulated θ statistic.

These findings are applied in the next chapter to a series of hypotheses based on the data sets of this text.

Exercises

1. Show that the expression in (11.3.19) is idempotent.

2. *Profile analysis* is the name given to one kind of analysis of a set of multivariate or vector observations. There are typically three hypotheses addressed in profile analysis. Here we assume that three samples of students—high school seniors, college freshmen, and college seniors—have been administered a set of four ability measures.

Hypothesis 1: The three population mean vectors are mutually parallel.

Hypothesis 2: The three population mean vectors (profiles) are at the same level. That is, if we average across the four ability measures for each group, we should get about the same value for each group.

Hypothesis 3: The combined profile is flat. If we average over the groups, the resulting profile is flat.

a. Construct an appropriate design template and $\boldsymbol{\beta}$ matrix.
b. Construct the constraint matrices appropriate to testing hypothesis 1, and describe the degrees of freedom associated with the test.
c. Construct the constraint matrices appropriate to testing hypothesis 2, and describe the degrees of freedom associated with the test.
d. Construct the constraint matrices appropriate to hypothesis 3.
e. In which order would you test these three hypotheses? How would the selection of later hypotheses depend upon the outcome of earlier tests?

Additional Reading

More statistical or mathematical detail on the MGLM is provided in the following two sources:

Morrison, D. F. (1976). *Multivariate Statistical Methods* (2nd ed.). New York: McGraw-Hill.

Rao, C. R. (1973). *Linear Statistical Inference and Its Applications* (2nd ed.). New York: Wiley.

An alternative development of the multivariate normal model is provided by

Timm, N. H. (1975). *Multivariate Analysis with Applications in Education and Psychology*. Belmont, CA: Wadsworth.

An interesting recent reference on the multivariate analysis of variance is

Stevens, J. P. (1980). Power of the multivariate analysis of variance tests, *Psychological Bulletin* 88:728–737.

Chapter 12

The Multivariate General Linear Model: Some Applications

Introduction

In the preceding two chapters we developed the rationale for classical multivariate analysis, testing hypotheses about the mean vectors of several sampled multivariate normal populations. Here we illustrate some of the more common applications of this multivariate general linear model.

In Section 12.1 we draw on the data sets presented earlier in the book to provide examples of Hotelling's T^2 (contrasting the mean vectors for two populations), the multivariate analysis of variance (MANOVA) for randomized groups, and multivariate linear regression. In Section 12.2 we develop a technique of multivariate analysis closely associated with the MGLM, but one that has been elaborated to answer somewhat different questions—the linear discriminant function. We illustrate the discriminant function and discuss its several applications. Finally, the MGLM presumes that the several normally distributed populations contributing criterion observations share a common variance covariance matrix. In Section 12.3 we present a likelihood ratio approach to tests of that assumption.

12.1 Examples of MGLM Problems

In this section we provide computational examples of a number of applications of the MGLM approach. The solutions reflect a series of problems raised in Chapter 11. The examples use the data matrices reported or described in Chapter 1.

12.1.1 Two-Group T^2 Example

The first example is based on the Women's Career Study data. Two populations of women are sampled, those who have participated in a career development course and those who have not, to yield 30 multivariate observations from each. The particular observation, in turn, consists of three values: level of occupational aspiration (values of 1–10 assigned to occupations aspired to, with higher values assigned to occupations requiring more training, having difficult entry criteria, or commanding higher status), career centrality (1 or 0, depending on whether the woman reports her career is expected to be the center of her adult life), and certainty of career choice (a self-rating of 1–7 of the certainty of the woman's occupational choice, higher values indicating greater certainty.)

The means and standard deviations of the multivariate observations are given, together with the pooled within groups deviation sum of squares matrix, in Table 12.1. Women who participated in the career development classes, as sampled, had higher scores on the average on each of the three measures.

Testing the hypothesis that the mean vector is the same in the two populations sampled yielded a Wilks's lambda statistic of $U = 0.36280$, significant beyond the 0.01 level. For this example, the U statistic obtained is from a distribution with parameters $t = 3$, $s = 1$, and $n - p = 58$, where n is the total number of observations, p the number of populations sampled, t the effective number of criteria, and s the number of constraints on the rows of the $p \times t$ $\boldsymbol{\beta}$ matrix. For $s = 1$ (i.e., $\boldsymbol{\beta}_{1.} - \boldsymbol{\beta}_{2.} = \mathbf{0}$ is the sole constraint) the F approximation to the multivariate test statistic presented in Section 11.3 is an exact test of the hypothesis. We calculate

$$\begin{aligned} F &= [(1 - U)/U][(n - p) - (t - 1)]/t \\ &= (0.6372/0.3628)(56/3) = 32.78 \end{aligned}$$

an observation, under the hypothesis, from the F distribution with $t = 3$

Table 12.1 Average Criterion Scores for Women Participants and Nonparticipants in High School Career Development Classes[a]

	Level of aspiration	Career centrality	Choice certainty
Participants	5.83 (1.78)	0.63 (0.49)	5.07 (1.51)
Nonparticipants	3.77 (2.37)	0.43 (0.50)	1.70 (1.09)
Within groups deviation scores sum of squares matrix $\mathbf{E}'\mathbf{E}$			
Level of aspiration	255.5333	34.2000	45.2333
Career centrality	34.2000	14.3333	8.6333
Choice certainty	45.2333	8.6333	100.1667
Hypothesis sum of squares matrix $\mathbf{F}'\mathbf{F} - \mathbf{E}'\mathbf{E}$			
Level of aspiration	64.0667	6.2000	104.3667
Career centrality	6.2000	0.6000	10.1000
Choice certainty	104.3667	10.1000	170.0167

[a]Standard deviations in parentheses.

and $(n - p) - (t - 1) = 56$ degrees of freedom (df). The hypothesis would be rejected, of course, at the same level using this F test as using the U statistic. Table 12.1 also reports the hypothesis sum of squares and cross products matrix $\mathbf{F'F} - \mathbf{E'E}$.

12.1.2 Randomized Groups Multivariate Analysis of Variance

We expand the analysis of these data by treating the 60 observations as samples not from two populations (participants and nonparticipants) but from six (participation crossed with three levels of socioeconomic status, SES). Ten observations have been drawn independently from each of the six populations. The hypotheses of interest are, first, no interaction between SES and class participation in determining level of criterion scores and, second, no impact of SES or (as in the previous analysis) class participation, taken separately, on the three dependent variables.

Sample means for the six groups are presented in Table 12.2 (together with the pooled within groups deviation scores sum of squares and cross products matrix) and plotted in Figure 12.1. The sample data are contrived with population means chosen, as Figure 12.1 suggests, to reflect

1. an interaction between SES and class participation in the determination of level of aspiration,
2. the additive contribution of both SES and participation to the determination of career centrality, and
3. the relevance of class participation but not of SES to ratings of certainty of career choice.

Table 12.2 Average Criterion Scores (SES × Participation), Women's Career Study, Six Groups

	Level of aspiration	Career centrality	Choice certainty
Participants			
High SES	6.00	0.90	5.20
Medium SES	5.60	0.70	5.20
Low SES	6.00	0.30	4.80
Nonparticipants			
High SES	6.40	0.80	1.70
Medium SES	3.10	0.30	1.60
Low SES	1.80	0.20	1.80
Within group deviation scores sum of squares matrix **E′E**			
Level of aspiration	142.2000	19.6000	47.2000
Career centrality	19.6000	10.4000	7.4000
Choice certainty	47.2000	7.4000	98.9000

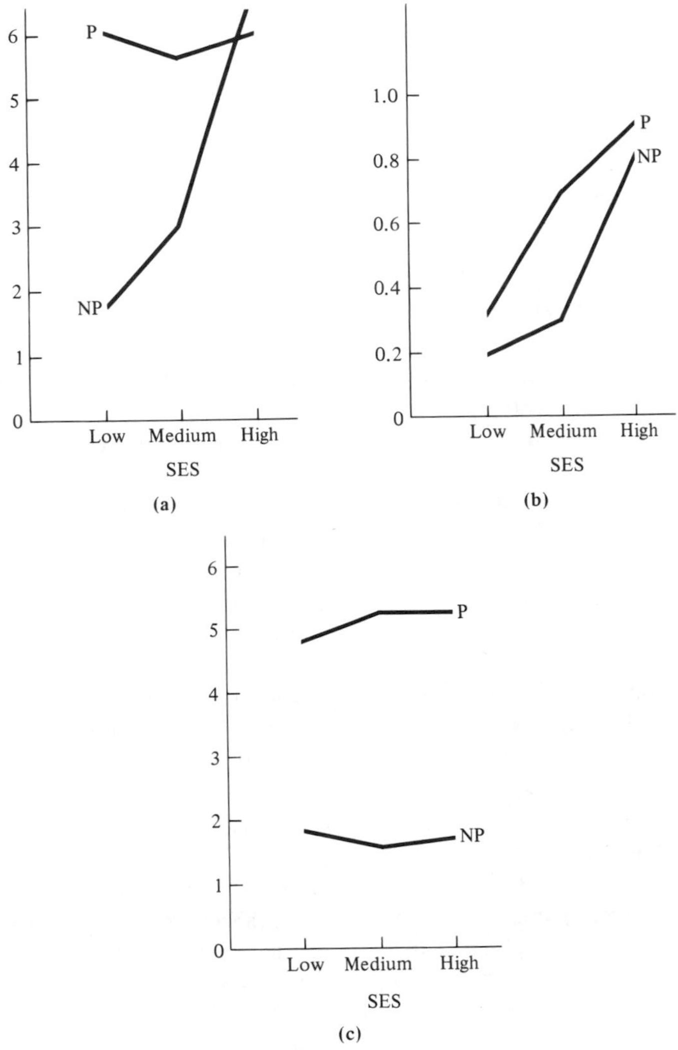

Figure 12.1 Mean scores for SES and P/NP (participants/nonparticipants) groups, Women's Career Study: **(a)** level of aspiration, **(b)** career centrality, and **(c)** choice certainty.

(Note that $\mathbf{E}'\mathbf{E}$ is smaller here than for the T^2 example, reflecting the mean differences between SES levels.)

How closely is the underlying structure revealed by the sample MANOVA results? The multivariate no interaction hypothesis yielded a Wilks's lambda or likelihood ratio statistic of $U = 0.59113$. With parameters $t = 3$, $s = 2$, and $n - p = 54$, this value is within the lower 1% of the distribution of this statistic. The multivariate hypothesis of no interaction must be rejected.

Table 12.3 Women's Career Study: Hypothesis of No Interaction

	Hypothesis sum of squares $\mathbf{F'F} - \mathbf{E'E}$			Discriminant function $\mathbf{P}_{.1}$
Level of aspiration	56.6333	0.6500	−5.1167	0.89
Career centrality	0.6500	0.3000	0.3500	−0.35
Choice certainty	−5.1167	0.3500	1.0333	−0.29

Similarly, the largest diagonal test of Roy yields a value of the statistic $\theta = 0.39141$. With parameters $p_1 = \min(s, t) = 2$, $p_2 = \frac{1}{2}(|s - t| - 1) = 0$, and $p_3 = \frac{1}{2}[(n - p) - (t + 1)] = 25$, this value is in the upper 1% of the statistic's sampling distribution.

Table 12.3 gives the hypothesis sums of squares matrix, $\mathbf{F'F} - \mathbf{E'E}$, for the no interaction hypothesis and the *discriminant function*, the first column of the left orthonormal to the standardized hypothesis sum of squares matrix, $\mathbf{M}$ of (11.3.35). Both suggest the relative importance of level of aspiration to the multivariate tests: Consider first the diagonal elements of $\mathbf{F'F} - \mathbf{E'E}$ as increases relative to the corresponding elements of $\mathbf{E'E}$. The increase for level of aspiration is much the largest of the three. Second, consider the discriminant function as a set of weights to be applied to the three (standardized) criteria to detect optimally an interaction. Much the greatest weight attaches to level of aspiration. (Follow-up univariate tests of the hypothesis of no interaction would lead to rejection only for the dependent variable level of aspiration.)

The hypothesis of no SES influence on the multivariate criterion yielded a likelihood ratio statistic of $U = 0.63249$ and a largest diagonal statistic of $\theta = 0.35291$. For $t = 3$, $s = 2$, and $n - p = 54$, both values are significant at the 0.01 level. Table 12.4 gives the hypothesis sums of squares matrix $\mathbf{G'G} - \mathbf{E'E}$ and discriminant function vector for this multivariate hypothesis. The hypothesis of no SES effect does not increase the sum of squares of errors for certainty of career choice, but it does so for the other two criteria. Follow-up tests of univariate hypotheses would find significant SES effects for level of aspiration and career centrality.

The multivariate hypothesis of no class participation effect on the multivariate dependent variable yielded $U = 0.36091$ and $\theta = 0.63909$ for the

Table 12.4 Women's Career Study: Hypothesis of No SES Effect

	Hypothesis sum of squares $\mathbf{G'G} - \mathbf{E'E}$			Discriminant function $\mathbf{P}_{.1}$
Level of aspiration	56.7000	13.9500	3.1500	−0.74
Career centrality	13.9500	3.6333	0.8833	−0.49
Choice certainty	3.1500	0.8833	0.2333	0.35

two test statistics. Here $t = 3$, $s = 1$, and $n - p = 54$, and the U statistic is significant. With $s = 1$ the F approximation to U is exact: $F = 30.69$, degrees of freedom 3 and 52, and highly significant.

The largest diagonal test when $s = 1$ is of a special form, given by (11.3.38). An F statistic is defined by

$$d_1[(p_3 + 1)/(p_2 + 1)] \sim F_{[2(p_2+1),2(p_3+1)]\text{ df}}$$

where $d_1 = \theta/(1 - \theta)$ is the largest diagonal and p_2 and p_3 are defined as they were for the distribution of θ itself. In the present example, $d_1 = 0.63909/0.36091 = 1.7708$, $p_2 = \frac{1}{2}$, $p_3 = 25$, and $F = 1.7708(26/1.5) = 30.69$, which (with 3 and 52 degrees of freedom) is in the upper 1% of the distribution. (Note that the two hypothesis tests yield the same statistic in this instance.) Follow-up univariate analyses of variance would show that class participation had an effect on two of the three criteria, level of aspiration, and certainty of career choice. The hypothesis sum of squares matrix $\mathbf{H'H - E'E}$ and discriminant function are reported in Table 12.5, and the relative magnitudes of the diagonals of $\mathbf{H'H - E'E}$ confirm our test statistic findings. Owing to the correlations among the dependent variables, the discriminant function, as the weighting vector in multiple regression, can be somewhat less informative about which criterion attributes are strongly affected by a hypothesis.

A word of warning to the reader. The Women's Career Study data provided a handy example for a MANOVA illustration. The results, even for contrived data, need to be treated with caution. Given the definition of the criterion measures, particularly that of career centrality, the effect of violations of the assumption of multivariate normality needs to be investigated further.

12.1.3 Multivariate Linear Regression Example

We turn to the Preemployment Study for an example of a test for a multivariate *moderator variable*. The multivariate dependent variable for a hypothetical sample of 150 beginning insurance underwriters consists of three criterion measures: training time, job productivity, and supervisor's proficiency rating. We are interested in two predictors, amount of prior

Table 12.5 Women's Career Study: Hypothesis of No Participation Effect

	Hypothesis sum of squares $\mathbf{H'H - E'E}$			Discriminant function $\mathbf{P}_{.1}$
Level of aspiration	60.0667	6.2000	104.3667	0.20
Career centrality	6.2000	0.6000	10.1000	−0.14
Choice certainty	104.3667	10.1000	170.0167	0.94

work experience and years of education. The question is whether their effects on the dependent measures are additive or interactive. Is predictability of the multivariate criterion increased when the two experience measures are complemented by a third predictor, the (deviation) product of the two?

Table 12.6 gives the means and standard deviations for the sample of 150 agents. Work experience is reported in years of earlier full-time work, training time in months required to obtain job skills, job productivity as percentage of sales quota attained, and supervisor's rating on a seven-point scale.

In Table 12.7 we have the full model sum of squares matrix $\mathbf{E'E}$ and the associated matrix $\hat{\boldsymbol{\beta}}$. The product term appears to carry little weight. The hypothesis that this vector of weights is the null vector yields the hypothesis sum of squares matrix $\mathbf{F'F} - \mathbf{E'E}$ and the constrained vector ${}_{c}\hat{\boldsymbol{\beta}}$, also reported in Table 12.7. The reader may note that the diagonals of $\mathbf{F'F'} - \mathbf{E'E}$ are relatively small and that the nonzero elements of ${}_{c}\hat{\boldsymbol{\beta}}$ are substantially identical to the corresponding elements of $\hat{\boldsymbol{\beta}}$. While these results are consistent with a failure to reject the hypothesis of zero weights for the predictor interaction or product term, they do not provide a formal test of that hypothesis.

Representing the basic structure of the full model sum of squares as $\mathbf{E'E} = \mathbf{QD}^2\mathbf{Q'}$, the standardized form of the hypothesis matrix given by (11.3.35),

$$\mathbf{M} = (\mathbf{QD}^{-1}\mathbf{Q'})(\mathbf{F'F} - \mathbf{E'E})(\mathbf{QD}^{-1}\mathbf{Q'}) = \mathbf{PdP'}$$

yields for these data a basic diagonal with elements $d_1 = 0.01776991$, $d_2 = 0.0000092$, and $d_3 = 0.0000009$. The greatest root or diagonal test is typically based on the statistic $\theta = d_1/(1 + d_1) = 0.017459$, with parameters $p_1 = \min(s = 1,\ t = 3) = 1$, $p_2 = \frac{1}{2}(|s - t| - 1) = 1$, and $p_3 = \frac{1}{2}[\{(n - p) = 147\} - \{(t + 1) = 4\}] = 71.5$. For $p_1 = 1$ we can use the sampling distribution of d_1 directly [see Eq. (11.3.38)], as then

$$d_1[(p_3 + 1)/(p_2 + 1)] \sim F_{2(p_2+1),\,2(p_3+1)\ \mathrm{df}}$$

Table 12.6 Preemployment Study: Means and Standard Deviations, Predictors, and Criteria

	Mean	SD
Predictors		
Prior work experience	5.96	2.16
Years of education	12.01	1.79
Product of deviations	0.41	3.80
Criteria		
Training time	8.23	1.83
Productivity	99.85	8.45
Supervisor's rating	6.15	1.85

Table 12.7 Preemployment Study

Full model sum of squares matrix **E′E**

Training time	419.12	−154.11	66.69
Productivity	−154.11	9612.11	555.33
Supervisor's rating	66.69	555.33	446.99

Full model predictor weight matrix $\hat{\boldsymbol{\beta}}$

	Training time	Productivity	Supervisor's ratings
Prior work	0.14	0.30	0.28
Education	0.32	1.38	−0.00
Product (interaction)	0.05	−0.01	0.04

Hypothesis sum of squares matrix **F′F** − **E′E**

Training time	4.56	−1.04	4.31
Productivity	−1.04	0.23	−0.98
Supervisor's rating	4.31	−0.98	4.08

Constrained weight matrix ${}_c\hat{\boldsymbol{\beta}}$

	Training time	Productivity	Supervisor's ratings
Prior work	0.14	0.30	0.29
Education	0.34	1.38	0.02
Product (interaction)	0.00	0.00	0.00

Our test statistic is $0.0177691(72.5/2) = 0.6$, and, as this value is less than 1, it clearly is in the lower half of the referenced F distribution, not in the upper tail. The hypothesis of no product or no interaction effect cannot be rejected.

Similarly, applying the likelihood ratio test with only one constraint, for $s = 1$, provides an exact F ratio test:

$$[(1 - U)/U]\{[(n - p) - (t - 1)]/t\} \sim F_{t, \{(n-p)-(t-1)\}\ \mathrm{df}}$$

In this instance the U statistic is the value [see Eq. (11.3.39)]

$$U = \left[\prod_{i=1}^{t} (d_i + 1)\right]^{-1} = [(1.0178)(1.0)(1.0)]^{-1} = 0.9825$$

yielding the F distributed quantity, $[(1 - 0.9825)/0.9825][(147 - 2)/3] = 0.8609$, which is again less than 1, and the interaction of the two predictors is found to be nonsignificant.

12.1.4 Combining Criterion Variables

We use some results from the simulated Airline Pilots Reaction Time Study to illustrate the constraint by columns of $\boldsymbol{\beta}$, or, equivalently, the formation of an effective criterion as a linear combination of the measured criteria.

Table 12.8 Airline Pilot Reaction Time Study

	Mean reaction time by condition and block[a]	
	Block 1	Block 2
Auditory, right	764.500 (110.151)	666.333 (82.958)
Auditory, left	827.000 (151.916)	739.083 (89.549)
Visual, right	725.917 (141.303)	624.583 (87.650)
Visual, left	821.250 (175.387)	710.583 (105.897)

Full model sum of squares matrix **E′E**

Block 1	945,326.167	519,516.333
Block 2	519,516.333	371,794.750

Full model sum of squares for time reduction

$$\mathbf{e}'\mathbf{e} = \mathbf{A}'\mathbf{E}'\mathbf{E}\mathbf{A} = 278{,}068.256$$

[a]Standard deviations reported in parentheses.

Table 12.8 reports the means (and standard deviations) of the pilots' average reaction times in milliseconds. It is clear that reaction time decreases overall from block 1 to block 2. The question of interest is whether this improvement in performance is differential for modality, direction, or their interaction. To test these hypotheses a new criterion or dependent variable is defined, (block 1) − (block 2) = time reduction. In the notation of Chapter 11, $\mathbf{a}' = (1, -1)$ is a transformation vector that creates this new variable $\mathbf{z} = \mathbf{Y}\mathbf{a}$. Table 12.8 also reports the full model sum of squares for the original bivariate criterion $\mathbf{E}'\mathbf{E}$ and the corresponding value for the newly defined or effective criterion $\mathbf{e}'\mathbf{e} = \mathbf{a}'\mathbf{E}'\mathbf{E}\mathbf{a}$.

In Table 12.9 the sums of squares for the three hypotheses are reported. As the criterion is now univariate, the hypothesis testing techniques of

Table 12.9 Airline Pilot Reaction Time Study: Time Reduction—Hypothesis Sums of Squares

Interaction

$$\mathbf{a}'\mathbf{F}'\mathbf{F}\mathbf{a} - \mathbf{a}'\mathbf{E}'\mathbf{E}\mathbf{a} = \mathbf{f}'\mathbf{f} - \mathbf{e}'\mathbf{e} = 1131.0208$$

Modality

$$\mathbf{a}'\mathbf{G}'\mathbf{G}\mathbf{a} - \mathbf{a}'\mathbf{E}'\mathbf{E}\mathbf{a} = \mathbf{g}'\mathbf{g} - \mathbf{e}'\mathbf{e} = 1989.1867$$

Direction

$$\mathbf{a}'\mathbf{H}'\mathbf{H}\mathbf{a} - \mathbf{a}'\mathbf{E}'\mathbf{E}\mathbf{a} = \mathbf{h}'\mathbf{h} - \mathbf{e}'\mathbf{e} = 3.5200$$

Denominator of F test statistic

$$\mathbf{e}'\mathbf{e}/(n - p) = 278{,}068.256/(48 - 4) = 6319.73$$

Chapter 8 are appropriate:

$$\frac{(\mathbf{f}'\mathbf{f} - \mathbf{e}'\mathbf{e})/s}{\mathbf{e}'\mathbf{e}/(n-p)} \sim F_{s,(n-p)\,\mathrm{df}}$$

For each hypothesis $s = 1$ and the denominator of the test statistic is $\mathbf{e}'\mathbf{e}/(n-p) = 6319.73$. From this it is clear that any one of the computed test statistics is <1.0 and the gain in speed of responding was not a function of sensory modality, direction of stimulation, or their interaction.

12.2 Linear Discriminant Function

In this section we consider a common problem in multivariate analysis, that of differentiating samples or groups on the basis of sets of multivariate observations contributed by the samples. The approach we illustrate here permits us to subsume the discrimination problem under MGLM, although the motivation for discrimination is somewhat different than for the MGLM techniques developed earlier.

Suppose, for example, that an investigator has Wechsler Adult Intelligence Scale (WAIS) data for samples from three adult populations. Ten WAIS subtest scores are available for a group of persons 62–68 years old who have sustained no brain damage (group 1) as well as for groups of similarly aged persons who are known to have suffered injury to the left (group 2) or the right cerebral hemisphere (group 3). The general question of interest is whether the WAIS results are different for the three groups. In *linear discriminant function* (LDF) analysis this question is posed more specifically, usually in one of the following senses:

1. "How should the WAIS subtest scores be combined to give the best separation of the three groups?"
2. "Can the three populations of WAIS profiles be reliably discriminated?"
3. "Given a profile of WAIS scores for a person not in the original sample, what brain damage (right, left, or none) should be predicted?"

12.2.1 Linear Discriminant Problem

We structure the problem in terms drawn from the MGLM, beginning with an $n \times m$ matrix $\mathbf{Y}$:

$$\mathbf{Y} = \begin{bmatrix} --- \\ --- \\ \vdots \\ --- \end{bmatrix} \begin{matrix} n_1 \\ n_1 + n_2 \\ \vdots \\ n = n_1 + n_2 + \cdots + n_k \end{matrix}$$

The matrix $\mathbf{Y}$ has been constructed, for ease of illustration, such that the first n_1 rows contain a set of independent m-variate observations from population 1, the next following n_2 rows are independent m-variate observations from population 2, and successive rows are similarly defined, the final n_k rows consisting of observations from the kth population sampled. We have samples from k multivariate populations.

Next we assume that each of these multivariate populations is distributed normally and that all k share a common variance–covariance structure. Each row of $\mathbf{Y}$ is a sample from a multivariate normal population with common variance matrix but potentially unique expectation vector:

$$\mathbf{Y}_{i.} \sim \mathbf{N}_m\left[\boldsymbol{\mu}_j = \mathbf{E}(\mathbf{Y}_{i.}|i\text{th observation from } j\text{th population}), \boldsymbol{\Sigma}\right]$$

The m-element mean vector $\boldsymbol{\mu}_j$ may differ from population to population, and it will be convenient to array these expectations for the k populations in an expectation template:

$$_{\mathrm{T}}\mathbf{E}(\mathbf{Y})' = (\boldsymbol{\mu}_1, \boldsymbol{\mu}_2, \ldots, \boldsymbol{\mu}_k) \tag{12.2.1}$$

To complete the representation of the data in MGLM terms, we assume a design template $_{\mathrm{T}}\mathbf{X}$ that codes the population source of each observation in terms of a set of k mutually exclusive categorical variables. That is, we take $_{\mathrm{T}}\mathbf{X}$ to be the kth-order identity matrix.

In the LDF we seek to combine linearly the m elements of the multivariate observation, producing a set of n univariate scores that, in some sense, best differentiate among or discriminate between the k contributing populations. Designate the vector $\mathbf{a}$ as a set of k weights; then

$$z_i = \mathbf{Y}_{i.}'\mathbf{a} \tag{12.2.2}$$

is an univariate score corresponding to the ith multivariate observation. Given our assumptions about the source of the vector $\mathbf{Y}_i$, each z_i must be an independent *univariate normal* observation,

$$z_i \sim N_1(\mathbf{a}'\boldsymbol{\mu}_j, \mathbf{a}'\boldsymbol{\Sigma}\mathbf{a}),$$

where $\mathbf{Y}_{i.}$ was sampled from population j. Each z_i is an observation from one of k normal distributions, all with the same variance $\mathbf{a}'\boldsymbol{\Sigma}\mathbf{a}$, but with potentially different means $\mathbf{a}'\boldsymbol{\mu}_j$.

To discriminate best among the contributing populations, the LDF seeks to identify that weighting vector $\mathbf{a}$ for which the within group variability, $\mathbf{a}'\boldsymbol{\Sigma}\mathbf{a}$, is as small as possible while the between groups variability—that among the elements of the vector $_{\mathrm{T}}\mathbf{E}(\mathbf{z})' = \mathbf{a}'[_{\mathrm{T}}\mathbf{E}(\mathbf{Y})'] = (\mathbf{a}'\boldsymbol{\mu}_1, \mathbf{a}'\boldsymbol{\mu}_2, \ldots, \mathbf{a}'\boldsymbol{\mu}_k)$ —is as large as possible. The LDF seeks a set of scores that are very close in value within any one of the k samples and very different in value between any two samples.

12.2.2 *Sample Solution to the Linear Discriminant Function*

The parameter matrices $\mathbf{\Sigma}$ and ${}_{\mathrm{T}}\mathbf{E}(\mathbf{Y})$ on which the above definitions depend are unknown in most cases, and we must use their estimators in searching for the desired weighting vector.

Given our choice of the identity design template, we have in MGLM terms $\boldsymbol{\beta} = {}_{\mathrm{T}}\mathbf{E}(\mathbf{Y})$, where ${}_{\mathrm{T}}\mathbf{E}(\mathbf{Y}) = {}_{\mathrm{T}}\mathbf{X}\boldsymbol{\beta}$, or, by extension, $\mathbf{E}(\mathbf{Y}) = \mathbf{X}\boldsymbol{\beta}$. The full design matrix consists of the appropriate number of replications n_i of each row of ${}_{\mathrm{T}}\mathbf{X}$.

From our earlier development of the MGLM, we can write the unconstrained maximum likelihood estimator of $\boldsymbol{\beta}$, or ${}_{\mathrm{T}}\mathbf{E}(\mathbf{Y})$,

$$ {}_{\mathrm{T}}\hat{\mathbf{E}}(\mathbf{Y}) = \hat{\boldsymbol{\beta}} = (\mathbf{X}'\mathbf{X})^{-1}\mathbf{X}'\mathbf{Y} $$

Given the nature of $\mathbf{X}$, this matrix must have as its elements the within group sample means, $\hat{\beta}_{jp}$ is the average score on the pth variate for the sample drawn from the jth population, and $\hat{\boldsymbol{\beta}}_{j.} = \overline{\mathbf{Y}}_{j}$. The latter is a sample estimator of the population expectation vector $\boldsymbol{\mu}_{j}$.

We usually express the unconstrained maximum likelihood estimator of $\mathbf{\Sigma}$

$$ \hat{\mathbf{\Sigma}} = (1/n)\mathbf{E}'\mathbf{E} $$

where $\mathbf{E} = \mathbf{Y} - \mathbf{X}\hat{\boldsymbol{\beta}}$. The matrix $\mathbf{E}$ consists in deviations of the $\mathbf{Y}$ observations about the sample means computed separately for each of the k groups:

$$ \mathbf{E}_{i.} = \mathbf{Y}_{i.} - \overline{\mathbf{Y}}_{j} \tag{12.2.3} $$

assuming $\mathbf{Y}_{i.}$ was sampled from the jth population.

For any choice of weighting vector $\mathbf{a}$, the maximum likelihood estimator of $\mathbf{a}'\mathbf{\Sigma}\mathbf{a}$ is

$$ \mathbf{a}'\hat{\mathbf{\Sigma}}\mathbf{a} = (1/n)\mathbf{a}'\mathbf{E}'\mathbf{E}\mathbf{a} \tag{12.2.4} $$

The goal of LDF is to determine $\mathbf{a}$ so as to minimize this estimated within group variability.

To assess the between groups variability, we begin with a maximum likelihood estimator of $\boldsymbol{\beta}$ constrained to have k identical rows. This would be our estimator if we were to hypothesize that all k populations had the same mean vector. That is, it is $\boldsymbol{\beta}$ estimated subject to a set of $k-1$ linear constraints $\mathbf{C}'\boldsymbol{\beta} = \mathbf{0}$, where $\mathbf{C}'$ could be the $(k-1) \times k$ matrix

$$ \mathbf{C}' = \begin{bmatrix} 1 & -1 & 0 & \cdots & 0 \\ 1 & 0 & -1 & \cdots & 0 \\ \vdots & \vdots & \vdots & & \vdots \\ 1 & 0 & 0 & \cdots & -1 \end{bmatrix} $$

Each row of the constrained estimator ${}_{\mathrm{c}}\hat{\boldsymbol{\beta}}$ is the total sample mean vector ${}_{\mathrm{c}}\hat{\boldsymbol{\beta}}_{j.} = \overline{\mathbf{Y}}$ for all $j = 1, 2, \ldots, k$. The corresponding matrix of deviations $\mathbf{F} = \mathbf{Y} - \mathbf{X}\,{}_{\mathrm{c}}\hat{\boldsymbol{\beta}} = \mathbf{Y} - \mathbf{1}\overline{\mathbf{Y}}'$ gives deviations of the $\mathbf{Y}_{i.}$ observations about this total sample mean.

For any choice of weighting vector **a**, the vector

$$\mathbf{Fa} = (\mathbf{Y} - \mathbf{1}\bar{\mathbf{Y}}')\mathbf{a} = \mathbf{Ya} - \mathbf{1}\bar{\mathbf{Y}}'\mathbf{a} = \mathbf{z} - \mathbf{1}\bar{\mathbf{z}} \tag{12.2.5}$$

is a vector of discrepancies of the linear composite scores about their overall mean. The minor product moment of this vector, $\mathbf{a'F'Fa}$, is a sum of squares of these discrepancies about an overall mean. We shall compare these interpretations with those of $\mathbf{Ea}$ and $\mathbf{a'E'Ea}$. From Eq. (12.2.3),

$$\mathbf{E}'_{i.}\mathbf{a} = \left(\mathbf{Y}_{i.} - \bar{\mathbf{Y}}_j\right)'\mathbf{a} = \mathbf{Y}'_{i.}\mathbf{a} - \bar{\mathbf{Y}}'_j\mathbf{a} = z_i - \bar{z}_j \tag{12.2.6}$$

where $\bar{z}_j$ is the composite score mean for the jth group. The vector $\mathbf{Ea}$, then, consists of discrepancies between composite scores and their appropriate group means. The minor product $\mathbf{a'E'Ea}$ is a sum of squares of these discrepancies.

The difference between these two sums of squares,

$$\mathbf{a'F'Fa} - \mathbf{a'E'Ea} = \mathbf{a'(F'F - E'E)a}$$

is a function of the variability of the group sample means for the linear composite. If $\bar{z}_1, \bar{z}_2, \ldots, \bar{z}_k$ are all nearly the same in value, $\mathbf{a'(F'F - E'E)a}$ is small. If these k sample means are highly variable, the hypothesis sum of squares is large.

We are ready now to define the optimization problem for the LDF. A set of weights is sought that minimizes $\mathbf{a'E'Ea}$ (or $\mathbf{a}'\hat{\boldsymbol{\Sigma}}\mathbf{a} = (1/n)\mathbf{a'E'Ea}$) and maximizes $\mathbf{a'(F'F - E'E)a}$. The two criteria are neatly combined in one function:

$$\phi = \mathbf{a'(F'F - E'E)a/a'E'Ea} \tag{12.2.7}$$

What choice of **a** results in ϕ assuming the largest possible value?

With (12.2.7), the optimization task for the LDF is identical to that already faced in Sections 11.3.6 and 11.3.7 in the development of the union–intersection test of the MGLH, $\mathbf{C}'\boldsymbol{\beta}\mathbf{A} = \boldsymbol{\Gamma}$. There we established that ϕ is a maximum for $\mathbf{a} = (\mathbf{QD}^{-1}\mathbf{Q}')\mathbf{P}_{.1}$, where $\mathbf{P}_{.1}$ is the first column of the left orthonormal of $\mathbf{M} = (\mathbf{QD}^{-1}\mathbf{Q}')(\mathbf{F'F} - \mathbf{E'E})(\mathbf{QD}^{-1}\mathbf{Q}')$ and where $\mathbf{E'E}$ itself has basic structure $\mathbf{QD}^2\mathbf{Q}'$.

12.2.3 Significance of the Linear Discriminant Function

While **a** provides the best set of weights for defining a linear composite, the amount of group separation or discrimination provided by the resulting LDF scores may be quite small. The significance of the LDF can be assessed, as the derivation hints, with the greatest diagonal test provided by the union–intersection approach to the MGLH. If d_1 is the largest basic diagonal to $\mathbf{M}$, then by Eq. (11.3.37) the statistic

$$d_1/(1 + d_1) \sim \theta_{p_1, p_2, p_3}$$

where $p_1 = \min(s, t)$, $p_2 = \frac{1}{2}(|s - t| - 1)$, $p_3 = \frac{1}{2}[(n - p) - (t - 1)]$. In turn, s is the number of constraints imposed on the rows of $\boldsymbol{\beta}$, t is the number of effective criteria, p is the number of columns to the (basic) design matrix, and n is the (total) sample size.

If we adapt this general test to the LDF situation, we have $\theta = d_1/(1 + d_1)$ distributed with parameters $p_1 = \min[(k - 1), m]$, $p_2 = \frac{1}{2}(|k - m| - 2)$, and $p_3 = \frac{1}{2}[n - (k + m + 1)]$, where k groups contributing a total of n observations are to be discriminated by a linear function of m measures.

12.2.4 Additional Linear Discriminant Functions Beyond the First

The LDF spreads the k groups out along a single dimension defined by the LDF scores. The results for three hypothetical groups are presented in the top half of Figure 12.2. There observations sampled from population 1 yielded low LDF scores, whereas observations from populations 2 and 3 yielded high LDF scores. The reader will remember that the LDF scores are the result of using **a** as the vector of weights for the m elements of the typical multivariate observation. The average of the LDF scores for each group is termed the *centroid* for that group. These are represented in Figure 12.2 as $\mathbf{C}_1$, $\mathbf{C}_2$, and $\mathbf{C}_3$.

Where more than two groups are to be discriminated, however, the best separation may be in a higher-dimensional space. The LDF approach generalizes to successive linear discriminant functions, defined much as was the first LDF.

We begin by considering the addition of a second LDF. What we seek is a second set of weights, forming a vector **b**, that makes

$$\phi_2 = \mathbf{b}'(\mathbf{F}'\mathbf{F} - \mathbf{E}'\mathbf{E})\mathbf{b}/\mathbf{b}'\mathbf{E}'\mathbf{E}\mathbf{b} \tag{12.2.8}$$

Figure 12.2 Distribution of members of three groups on one and two discriminant functions.

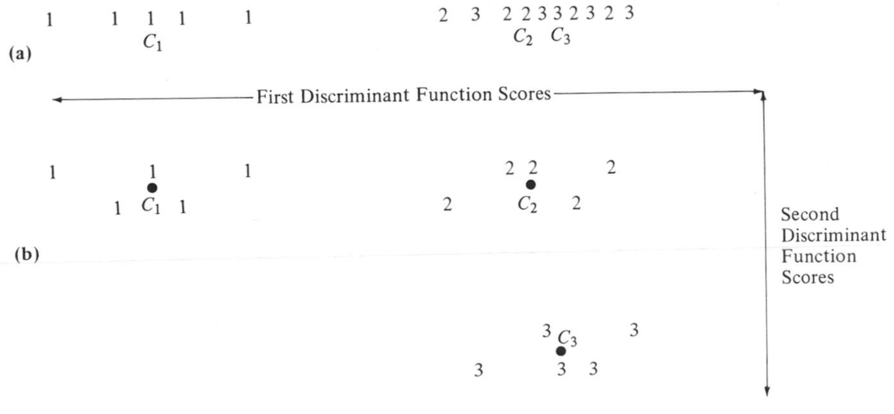

as large as possible subject to the constraint that this second LDF is to be "orthogonal" to the first. The quotes signal that the independence required is not what might first come to mind. It is required that the two LDFs be orthogonal not in the sample space but in a standardized between groups space.

Orthogonality in the sample space would imply that the two sets of LDF scores $\mathbf{Ya} = \mathbf{z}_1$ and $\mathbf{Yb} = \mathbf{z}_2$ are orthogonal; that is, $\mathbf{z}_1'\mathbf{z}_2 = 0$. This would be a natural orthogonality, perhaps, but it relates to the interobservation variability in $\mathbf{Y}$ rather than to the intergroup or intersample variability. How should we look at intergroup variability?

In Section 11.3.7 we mentioned briefly an interpretation of the matrix of Eq. (11.3.35),

$$\mathbf{M} = (\mathbf{E}'\mathbf{E})^{-1/2}(\mathbf{F}'\mathbf{F} - \mathbf{E}'\mathbf{E})(\mathbf{E}'\mathbf{E})^{-1/2}$$

where the inverse square root matrix $(\mathbf{E}'\mathbf{E})^{-1/2}$ is $\mathbf{QD}^{-1}\mathbf{Q}'$ based on the basic structure of $\mathbf{E}'\mathbf{E}$, $\mathbf{QD}^2\mathbf{Q}'$ [$= (\mathbf{QDQ}')(\mathbf{QDQ}') = (\mathbf{E}'\mathbf{E})^{1/2}(\mathbf{E}'\mathbf{E})^{1/2}$]. This interpretation will be of further interest here.

The matrices $\mathbf{E}'\mathbf{E}$, $\mathbf{F}'\mathbf{F}$, and $\mathbf{F}'\mathbf{F} - \mathbf{E}'\mathbf{E}$ are sums of squares and cross products matrices. Because in the LDF formulation the particular MGLH that yields $\mathbf{F}'\mathbf{F}$ is that the k populations sampled have a common multivariate mean, it is possible to show that $\mathbf{F}'\mathbf{F} - \mathbf{E}'\mathbf{E}$ can be expressed as the minor product moment of an $n \times m$ matrix $\mathbf{H} = \mathbf{X}(\hat{\boldsymbol{\beta}} - {}_c\hat{\boldsymbol{\beta}})$ with $H_{ip} = [\bar{Y}_{j(p)} - \bar{Y}_{(p)}]$ as its typical element. Here $\bar{Y}_{(p)}$ is the *total* sample mean for the pth variable and $\bar{Y}_{j(p)}$ is the sample mean in the jth *group* for that same variable, provided that the ith observation was sampled from the jth population. Although $\mathbf{H}$ contains a row for each observation, any row is a set of differences between means.

The hypothesis sum of squares matrix $\mathbf{F}'\mathbf{F} - \mathbf{E}'\mathbf{E} = \mathbf{H}'\mathbf{H}$ consists of sums of squares and cross products of discrepancies between group and total means. Further, the columns of $\mathbf{H}$ are in deviation form ($\mathbf{1}'\mathbf{H} = \mathbf{0}$), with the result that $\mathbf{F}'\mathbf{F} - \mathbf{E}'\mathbf{E}$ is proportional to a variance–covariance matrix.

As we noted in Section 11.3.7, the magnitudes of the entries of $\mathbf{F}'\mathbf{F} - \mathbf{E}'\mathbf{E}$ reflect (differences in) the metrics of the m variables and their covariation as well as, in this instance, the sample to sample variability of their means. Also, we know that the matrix $\mathbf{E}'\mathbf{E}$ (scaled by a reciprocal of sample size) is an estimator of the common within population variance–covariance structure for the m variates. In consequence of this, we may induce a transformation on $\mathbf{H}$,

$$\mathbf{G} = \mathbf{H}(\mathbf{E}'\mathbf{E})^{-1/2} = \mathbf{H}(\mathbf{QD}^{-1}\mathbf{Q}')$$

which is the multivariate analog to standardizing a vector: If $\mathbf{h}$ is vector $\mathbf{H}_{.1}$, then the transformed vector $\mathbf{g} = \mathbf{h}[(\mathbf{E}'\mathbf{E})_{11}]^{-1/2}$ consists of standardized mean differences on the first variable, these mean differences each expressed as fractions of the estimated within groups standard deviation for that variable (again except for a sample size constant).

The matrix $\mathbf{G}$ remains in deviation form, and its minor product moment is a variance–covariance matrix:

$$\begin{aligned}\mathbf{G'G} &= (\mathbf{E'E})^{-1/2}\mathbf{H'H}(\mathbf{E'E})^{-1/2} \\ &= (\mathbf{QD}^{-1}\mathbf{Q'})(\mathbf{F'F} - \mathbf{E'E})(\mathbf{QD}^{-1}\mathbf{Q'}) = \mathbf{M}\end{aligned}$$

(The sample size constants for $\mathbf{E'E}$ and $\mathbf{H'H}$ cancel in the computation of $\mathbf{G'G} = \mathbf{M}$.)

The matrix $\mathbf{M}$ of Eq. (11.3.35) for the LDF problem can be interpreted as a variance–covariance matrix over a collection of n observations for m variables each of which assesses a standardized or corrected mean deviation for each observation. The mean deviation for an observation is the discrepancy between a total mean for some variable and the mean for that variable in the group of which that observation is a member. The standardization of these mean deviations serves to correct for the within group variances and covariances of these same variables.

The orthogonality required of a pair of LDFs is in the space of these corrected mean deviation vectors: in the space of the columns of $\mathbf{G} = \mathbf{H}(\mathbf{QD}^{-1}\mathbf{Q'})$, where $\mathbf{H} = \mathbf{X}(\hat{\boldsymbol{\beta}} - {}_c\hat{\boldsymbol{\beta}})$ and $\mathbf{E'E} = \mathbf{QD}^2\mathbf{Q'}$. If $\mathbf{Gv} = \mathbf{Ha} = \mathbf{s}$ is the vector transformation of the corrected mean deviations induced by the first set of LDF weights, then we want the second set of LDF weights, the vector $\mathbf{b}$, to be defined so that for the product $\mathbf{Gw} = \mathbf{Hb} = \mathbf{t}$ we find that $\mathbf{t's} = 0$. That is, $\mathbf{b}$ is to be chosen so that

$$\mathbf{w'}(\mathbf{G'G})\mathbf{v} = \mathbf{w'Mv} = 0 \tag{12.2.9}$$

In summary, the first LDF, $\mathbf{z}_1 = \mathbf{Ya}$, requires that we find $\mathbf{a}$ both to minimize the estimated within group variability of $\mathbf{z}_1$, $\mathbf{Var}(\mathbf{z}_1) = \mathbf{a'}\hat{\boldsymbol{\Sigma}}\mathbf{a}$ (or, equivalently, to minimize $\mathbf{a'E'Ea}$), and to maximize the (weighted) among groups variability of $\mathbf{z}_1$, $\Sigma_{j=1}^{k}[n_j(\bar{z}_{j(1)} - \bar{z}_{(1)})^2] = \mathbf{a'}(\mathbf{F'F} - \mathbf{E'E})\mathbf{a}$. The second LDF, $\mathbf{z}_2 = \mathbf{Yb}$, requires a weight vector $\mathbf{b}$ that must minimize $\mathbf{b'E'Eb}$ and maximize $\mathbf{b'}(\mathbf{F'F} - \mathbf{E'E})\mathbf{b}$. Further, $\mathbf{b}$ must be chosen so that the corrected among groups variability accounted for by the second transformation does not overlap that accounted for by the first transformation. That is, we want $\mathbf{v'Mw} = \mathbf{a'}(\mathbf{F'F} - \mathbf{E'E})\mathbf{b} = 0$. In more intuitive terms, we want the second LDF to account for sources of among groups differences that were not tapped by the first LDF.

We now discuss the solution satisfying these requirements. The triple product $\mathbf{v'Mw}$ may be rewritten $\mathbf{v'}(\mathbf{PdP'})\mathbf{w}$— if we replace $\mathbf{M}$ with its basic structure—or as $\mathbf{P'}_{.1}(\mathbf{PdP'})\mathbf{w}$—if we replace $\mathbf{v}$ with its solution, the first column to the orthonormal $\mathbf{P}$. With what choice of $\mathbf{w}$ is $\mathbf{P'}_{.1}(\mathbf{PdP'})\mathbf{w} = 0$? A little algebra,

$$\mathbf{P}'_{.1}(\mathbf{PdP}')\mathbf{w} = \mathbf{e}'_1\mathbf{dP}'\mathbf{w} = d_1\mathbf{e}'_1\mathbf{P}'\mathbf{w} = d_1\mathbf{P}'_{.1}\mathbf{w}$$

establishes that $\mathbf{w}$ must be orthogonal to $\mathbf{P}_{.1}$. Thus any vector proportional to one of the latter columns of $\mathbf{P}$ (any column except $\mathbf{P}_{.1}$) satisfies the condition $\mathbf{v}'\mathbf{M}\mathbf{w} = 0$.

Among the vectors orthogonal to $\mathbf{P}_{.1}$ we seek that one that gives the largest ratio

$$\phi_2 = \mathbf{b}'(\mathbf{F}'\mathbf{F} - \mathbf{E}'\mathbf{E})\mathbf{b}/\mathbf{b}'\mathbf{E}'\mathbf{E}\mathbf{b}$$

The same argument used to establish that $\mathbf{a} = (\mathbf{E}'\mathbf{E})^{-1/2}\mathbf{P}_{.1}$ provides the unconstrained maximization of $[\mathbf{a}'(\mathbf{F}'\mathbf{F} - \mathbf{E}'\mathbf{E})\mathbf{a}]/(\mathbf{a}'\mathbf{E}'\mathbf{E}\mathbf{a})$ suggests that, among those vectors satisfying $\mathbf{w}'\mathbf{P}_{.1} = 0$, the one that maximizes ϕ_2 is

$$\mathbf{w} = \mathbf{P}_{.2} \quad \text{or} \quad \mathbf{b} = (\mathbf{E}'\mathbf{E})^{1/2}\mathbf{P}_{.2} \tag{12.2.10}$$

That is, the weights for the second LDF are based on the second column to the left orthonormal of the $\mathbf{M}$ of Eq. (11.3.35).

The lower half of Figure 12.2 illustrates the addition of a second set of LDF scores. The effect in this illustration is to separate group 3 from groups 1 and 2. The centroid of each group is that point with coordinates equaling the mean first and second LDF scores for members of that group.

Often, more than two LDFs may be found. Reiterating our argument and requiring, for example, that the third LDF be orthogonal to each of the first two in the corrected among groups space leads to identifying $(\mathbf{E}'\mathbf{E})^{1/2}\mathbf{P}_{.3}$ as the appropriate third set of weights. Generalizing from this leads to the conclusion that the number of LDFs that can be constructed equals the rank of the matrix $\mathbf{M}$. In turn, the rank of $\mathbf{M}$ can be shown to be the minimum of the number of variables (the rank of $\mathbf{E}'\mathbf{E}$) and the number of constraints imposed on the rows of $\boldsymbol{\beta}$ (the rank of $\mathbf{F}'\mathbf{F} - \mathbf{E}'\mathbf{E}$), For m-variate observations distributed over k groups, then, the number of LDFs is $r = \min[m, (k - 1)]$.

The number of discriminant functions may be large, but are they all useful? To answer this we examine the transformations induced on the corrected among groups variables, the columns of $\mathbf{G}$. First, we find the sample variance–covariance matrix for the full set of these transformations:

$$\mathrm{Var}(\mathbf{GP}) = \mathbf{P}'\mathbf{G}'\mathbf{G}\mathbf{P} = \mathbf{P}'\mathbf{M}\mathbf{P} = \mathbf{P}'\mathbf{PdP}'\mathbf{P} = \mathbf{d}$$

The covariances are all zero, as they should be. Each successive LDF does account for additional among groups variability. It is apparent, however, that later LDFs may account for relatively little of this variability. Because $\mathbf{d}$ is a basic diagonal, $d_1 \geqslant d_2 \geqslant \cdots \geqslant d_r$.

We have seen how to test whether d_1 is large enough—that is, whether the first LDF differentiates significantly among the k groups. We can evaluate the contribution of the subsequent, likely smaller, LDFs in very much the same fashion. The θ test statistic for the largest diagonal generalizes to the smaller elements of d. The ratio

$$\theta_i = d_i/(1 + d_i)$$

has the same distributional form for $i > 1$ as for $i = 1$. We can use the θ tables for this statistic, after adjusting p_1, the first of the three distributional parameters. When we were interested in testing the largest diagonal, for $i = 1$, we defined this parameter as $p_1 = \min(s, t)$. Now we can define it more generally. To test the ith basic diagonal with the statistic θ_i we compute that first parameter as $p_1 = \min[(s + 1 - i), (t + 1 - i)]$. The other two parameters are unchanged.

In the LDF application of the MGLM with m variables and k groups, we have $s = (k - 1)$, $t = m$, $p = k$, and the significance of the ith LDF is tested by the statistic θ_i distributed with parameters

$$\begin{aligned} p_1 &= \min[(k - i), (m + 1 - i)] \\ p_2 &= \tfrac{1}{2}(|k - m| - 2) \\ p_3 &= \tfrac{1}{2}[n - (k + m + 1)] \end{aligned}$$

12.2.5 Classifying New Observations

Assume that for some sample of observations from k multivariate populations, we have determined a set of q LDFs. How can we use these q functions to classify future observations?

To answer this question, recall that the multivariate observations are assumed to be normally distributed:

$$\mathbf{Y}_{i.} \sim \mathbf{N}_m(\boldsymbol{\mu}_j, \boldsymbol{\Sigma})$$

where the ith observation was taken from the jth population. The parameters $\boldsymbol{\mu}_j$ and $\boldsymbol{\Sigma}$ are unknown but were estimated from the original or calibration sample: $\hat{\boldsymbol{\mu}}_j = \overline{\mathbf{Y}}_j$, the sample mean for the jth group, and $\hat{\boldsymbol{\Sigma}} = (1/n)\mathbf{E}'\mathbf{E}$, where $\mathbf{E}'\mathbf{E}$ is the sum of squares and cross products of the original observations taken about their group means. This sum of squares matrix has basic structure $\mathbf{Q}\mathbf{D}^2\mathbf{Q}'$.

A sample sum of squares and cross products among groups was defined as $\mathbf{F}'\mathbf{F} - \mathbf{E}'\mathbf{E}$, where $\mathbf{F}'\mathbf{F}$ is itself the sum of squares and cross products of the calibration observations about the total sample mean. From the basic structure of the corrected among groups matrix,

$$\mathbf{M} = (\mathbf{E}'\mathbf{E})^{-1/2}(\mathbf{F}'\mathbf{F} - \mathbf{E}'\mathbf{E})(\mathbf{E}'\mathbf{E})^{-1/2} = \mathbf{P}\mathbf{d}\mathbf{P}'$$

we obtained weights for the q LDFs, $\mathbf{Z} = \mathbf{Y}\mathbf{A}$, where $\mathbf{A}$ consists of the first q columns of $(\mathbf{E}'\mathbf{E})^{-1/2}\mathbf{P}$.

Each row of $\mathbf{Z}$, whether computed for a calibration sample observation or a new observation, is a q-variate normal observation:

$$\mathbf{Z}_{i.} \sim \mathbf{N}_q(\mathbf{A}'\boldsymbol{\mu}_j, \mathbf{A}'\boldsymbol{\Sigma}\mathbf{A}) \qquad (12.2.11)$$

We can express the basic structure of the variance–covariance matrix for $\mathbf{Z}_{i.}$ as $\mathbf{A}'\boldsymbol{\Sigma}\mathbf{A} = \mathbf{RC}^2\mathbf{R}'$ and use these basic structure terms to define a second transformation, $\mathbf{W}_{i.}' = (\mathbf{Z}_{i.} - \mathbf{A}'\boldsymbol{\mu}_j)'(\mathbf{RC}^{-1}\mathbf{R}')$. This $\mathbf{W}_{i.}$ (which we need not actually compute) is an observation from the q-variate normal with a null mean vector and variance–covariance matrix equal to the identity matrix:

$$\mathbf{W}_{i.} \sim \mathbf{N}_q(\mathbf{0}, \mathbf{I})$$

inasmuch as

$$\mathbf{E}(\mathbf{W}_{i.}) = \mathbf{E}(\mathbf{Z}_{i.} - \mathbf{A}'\boldsymbol{\mu}_j) = \mathbf{A}'\boldsymbol{\mu}_j - \mathbf{A}'\boldsymbol{\mu}_j = \mathbf{0}$$

and

$$\begin{aligned}\mathbf{Var}(\mathbf{W}_{i.}) &= (\mathbf{RC}^{-1}\mathbf{R}')(\mathbf{A}'\boldsymbol{\Sigma}\mathbf{A})(\mathbf{RC}^{-1}\mathbf{R}') \\ &= (\mathbf{RC}^{-1}\mathbf{R}')(\mathbf{RC}^2\mathbf{R}')(\mathbf{RC}^{-1}\mathbf{R}') = \mathbf{I}\end{aligned}$$

Further, as $\mathbf{W}_{i.}$ consists of q independently distributed normal observations each with mean 0 and variance 1, the minor product moment of that vector is a chi-squared observation: $\mathbf{W}_{i.}'\mathbf{W}_{i.} \sim \chi^2_{q\ \mathrm{df}}$

If we did not know to which of k populations a vector observation $\mathbf{Y}_{i.}$ belonged, we could compute this sum of squares for each possibility:

$$\begin{aligned}D^2_{i(j)} &= {}_j\mathbf{W}_{i.j}'\mathbf{W}_{i.} = (\mathbf{Z}_{i.} - \mathbf{A}'\boldsymbol{\mu}_j)'(\mathbf{RC}^{-1}\mathbf{R}')(\mathbf{Z}_{i.} - \mathbf{A}'\boldsymbol{\mu}_j) \\ &= (\mathbf{Y}_{i.} - \boldsymbol{\mu}_j)'\mathbf{A}(\mathbf{A}'\boldsymbol{\Sigma}\mathbf{A})^{-1}\mathbf{A}'(\mathbf{Y}_{i.} - \boldsymbol{\mu}_j) \qquad (12.2.12)\end{aligned}$$

We could then assign the $\mathbf{Y}_{i.}$ observation to that population that yielded the smallest $D^2_{i(j)}$. The assignment would be based on determining for the new observation the closest group centroid in the space of the $\mathbf{W}$ variables.

A problem arises in the computation of $D^2_{i(j)}$. The matrix $\boldsymbol{\Sigma}$ and vectors $\boldsymbol{\mu}_j$ are unknown. In practice, the maximum likelihood estimators of these can be substituted:

$$n\hat{D}^2_{i(j)} = \left[(\mathbf{Y}_{i.} - \hat{\boldsymbol{\mu}}_j)'\mathbf{A}\right]\left[\mathbf{A}'(\mathbf{Y}_{i.} - \hat{\boldsymbol{\mu}}_j)\right] \qquad (12.2.13)$$

and the same search for a minimum conducted. [The computations implied by Eq. (12.2.13) are not as forbidding as might be thought. The number of LDFs that are either significant in the calibration sample or retained for use in classifying new observations is normally quite small. Very infrequently are more than three LDFs reported in the literature.]

12.2.6 LDF Example: Vocational Interest and College Major

Among those taking their bachelor's degrees at the University of Washington in June 1977 were 1405 who had, as high school juniors, completed the

Vocational Interest Inventory (VII). Table 12.10 reports the average score profiles for these students, categorized by their majors at the time of university graduation. Scales on the VII each have a mean of 50 and a standard deviation of 10 for all high school juniors.

The within major estimated variance–covariance matrix, $\hat{\mathbf{\Sigma}} = (1/n)\mathbf{E}'\mathbf{E}$, is given in Table 12.11, as are the eight elements of the basic diagonal of **M**. Because of the large sample size, the first five of the LDFs are significant. Not all are equally important, however; taking the sum of the eight d_i as the total amount of corrected among groups variability that can be accounted for (this is the trace of **M**), we find the first LDF, the first two LDFs, and the first three LDFs cumulatively account for, in turn, 42%, 73%, and 84% of the maximum achievable.

The first few LDFs, then, are of the greatest interest. In Table 12.12 we present the weights for the first three LDFs, the first three columns to the left orthonormal of **M**. The centroids of the 15 groups, based on the first two of these LDFs, are displayed in Figure 12.3. The coordinates of these points are the results of applying the two sets of weights, $\mathbf{A}_{.1}$ and $\mathbf{A}_{.2}$ to the 15 VII mean profiles listed in Table 12.10. In this example the first LDF arranges engineering graduates at one end and history and foreign language and literature majors at the other, while the second LDF ranges the graduates from nursing and the biological sciences at one end and political

Table 12.10 High School Vocational Interests of University Graduates: Mean Interest Scale Profiles by Major

		Interest scale[a]							
Major	n	SER	BUS	ORG	TEC	OUT	SCI	CUL	ART
Communications	102	47.8	52.9	50.6	46.3	45.7	47.0	54.8	51.1
History	61	49.5	49.5	48.1	46.2	49.7	46.9	59.8	51.1
Accounting	122	45.5	51.5	56.6	50.4	45.4	50.9	52.7	46.1
Art	55	46.8	47.2	44.6	49.4	50.6	49.7	48.9	62.2
Economics	85	46.0	51.9	53.0	50.3	47.0	50.2	54.7	45.8
Home economics	51	52.1	46.5	50.6	46.4	49.6	51.0	50.5	53.6
Political science	60	47.2	55.0	53.4	47.5	46.4	47.1	56.8	47.0
Nursing	105	53.5	42.6	48.9	45.9	51.3	58.6	49.9	48.6
Biological science	213	47.1	43.8	46.7	50.2	50.8	61.7	49.8	48.1
Urban design	50	44.9	49.3	51.8	51.1	46.7	51.1	50.4	53.9
Social services	68	54.8	49.0	49.8	46.0	48.0	50.6	54.9	47.2
Health professions	79	53.7	47.1	48.2	46.6	48.7	55.1	50.3	50.0
Fisheries/forestry	68	45.1	45.6	45.6	50.4	59.1	58.2	46.4	46.9
Engineering	239	49.9	45.2	49.2	45.2	51.4	49.4	54.5	54.6

[a] Key: SER, service; BUS, business contact; ORG, organization; TEC, technical; OUT, outdoors; SCI, science; CUL, general cultural; and ART, arts/entertainment.

Table 12.11 High School Vocational Interests of University Graduates[a]

	Within majors variance–covariance matrix							
Service	78.3	−0.8	−16.5	−15.3	−2.9	−6.9	4.5	−7.6
Business contact	−0.8	102.3	18.1	−4.8	−25.3	−32.3	−9.6	−12.5
Organization	−16.5	18.1	98.1	−1.5	−39.7	−14.6	10.9	−16.7
Technical	−15.3	−4.8	−1.5	64.4	−4.6	6.5	−23.6	0.6
Outdoors	−2.9	−25.3	−39.7	−4.6	103.1	7.0	−18.1	−6.2
Science	−6.9	−32.3	−14.6	6.5	7.0	94.4	−14.9	−13.9
General cultural	4.5	−9.6	10.9	−23.6	−18.1	−14.9	102.3	−8.1
Arts/entertainment	−7.6	−12.5	−16.7	0.6	−6.2	−13.9	−8.1	94.0

	Basic diagonals of $\mathbf{M} = (\mathbf{E}'\mathbf{E})^{-1/2}(\mathbf{F}'\mathbf{F} - \mathbf{E}'\mathbf{E})(\mathbf{E}'\mathbf{E})^{-1/2}$		
i	d_i	θ_i	p_1
1	0.41745	0.2945	8
2	0.30519	0.2339	7
3	0.11974	0.1069	6
4	0.06821	0.0639	5
5	0.03767	0.0363	4
6	0.02417	0.0236	3
7	0.01495	0.0147	2
8	0.00426	0.0042	1

[a] $k = 15$, $m = 8$, $n = 1405$. $p_2 = \frac{1}{2}[(15 - 8) - 2] = 3.5$, $p_3 = \frac{1}{2}(1405 - 24) = 690.5$.

science and economics majors at the other. In Table 12.12 the first LDF assigned largest weights to technical and science interests (negative weights) and art and service interests (positive), whereas the second LDF weighted science and service negatively and technical and cultural interests positively. Because of the interrelatedness of the eight scales to the VII, weights such as these should be treated with some caution.

Table 12.12 High School Vocational Interests of University Graduates: LDF Weights, First Three Functions

	$\mathbf{P}_{.1}$	$\mathbf{P}_{.2}$	$\mathbf{P}_{.3}$
Service	0.393	−0.432	0.517
Business contact	0.187	0.277	−0.012
Organization	0.006	0.078	0.310
Technical	−0.652	0.385	−0.223
Outdoors	0.158	−0.242	−0.102
Science	−0.376	−0.590	0.137
General cultural	0.177	0.279	0.106
Arts/entertainment	0.427	−0.215	−0.701

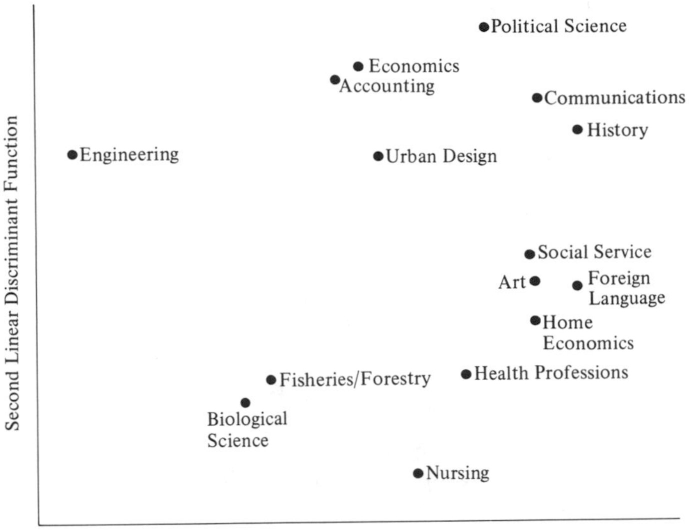

Figure 12.3 Group centroids on the first two discriminant functions.

12.3 Homogeneity of Variance–Covariance Matrices

One of the assumptions of the MGLM is that each of the multivariate normal populations sampled has a common variance–covariance matrix $\mathbf{\Sigma}$. In this section we develop a test of that assumption, one closely based on the likelihood ratio principle.

We work in the context of the k-groups design. That is, we assume $n_1, n_2, \ldots, n_k$ independent observations from k m-variate normal distributions. Not only may these k populations have different expectation vectors, but now we want to consider the possibility that they have different variance–covariance matrices. If we designate the k variance–covariance matrices ${}_1\mathbf{\Sigma}, {}_2\mathbf{\Sigma}, \ldots, {}_k\mathbf{\Sigma}$, a hypothesis of interest is that

$${}_1\mathbf{\Sigma} = {}_2\mathbf{\Sigma} = \cdots = {}_k\mathbf{\Sigma}$$

The alternative to this hypothesis is that at least one of the matrices differs from the others.

In Section 11.3.4 we developed a likelihood ratio test of the MGLM hypothesis $\mathbf{C}'\boldsymbol{\beta}\mathbf{A} = \mathbf{\Gamma}$. We exploit part of that development here. For the unconstrained maximum likelihood estimators of $\boldsymbol{\beta}$ and $\mathbf{\Sigma}$ (the latter common across the k populations sampled), we obtained in Eq. (11.3.23) an expression for the estimated likelihood of the sample of $n = n_1 + n_2 + \cdots + n_k$ independent observations that we can rewrite

$$L(\mathbf{Y}|\hat{\mathbf{\Sigma}}) = (2\pi e)^{-nm/2}\left(\prod_{j=1}^{m} D_j^2\right)^{-n/2} \tag{12.3.1}$$

where the maximum likelihood estimator of the common $\boldsymbol{\Sigma}$ has basic structure

$$\hat{\boldsymbol{\Sigma}} = \mathbf{Q}\mathbf{D}^2\mathbf{Q}'$$

Consider now the likelihood of the same sample of observations given that we must estimate the variance–covariance matrix separately for each of the k populations. Each of the estimators, however, is still maximum likelihood. To see the result we partition the observation matrix $\mathbf{Y}$ by rows,

$$\mathbf{Y} = \begin{bmatrix} {}_1\mathbf{Y} \\ {}_2\mathbf{Y} \\ \vdots \\ {}_k\mathbf{Y} \end{bmatrix}$$

where ${}_i\mathbf{Y}$ contains n_i independent observations from the ith m-variate normal population. The likelihood of the sample from the ith population can be written, from (11.1.7),

$$L({}_i\mathbf{Y}) = (2\pi)^{-mn_i/2}\left(\prod_{j=1}^{m} {}_i\Delta_j^2\right)^{-n_i/2} \exp\{-\tfrac{1}{2}\operatorname{tr}[({}_i\mathbf{Y} - {}_i\mathbf{X}\boldsymbol{\beta})_i\boldsymbol{\Sigma}^{-1}({}_i\mathbf{Y} - {}_i\mathbf{X}\boldsymbol{\beta})']\} \tag{12.3.2}$$

${}_i\mathbf{X}$ is the horizontal partitioning of the design matrix corresponding to the sample of observations ${}_i\mathbf{Y}$; ${}_i\boldsymbol{\Sigma}$, the variance–covariance matrix for the ith population, has basic structure

$${}_i\boldsymbol{\Sigma} = {}_i\mathbf{Q}\,{}_i\boldsymbol{\Delta}^2\,{}_i\mathbf{Q}' \tag{12.3.3}$$

The maximum likelihood estimator of ${}_i\boldsymbol{\Sigma}$ is

$${}_i\hat{\boldsymbol{\Sigma}} = (1/n_i)({}_i\mathbf{Y} - {}_i\mathbf{X}\hat{\boldsymbol{\beta}})'({}_i\mathbf{Y} - {}_i\mathbf{X}\hat{\boldsymbol{\beta}}) \tag{12.3.4}$$

where $\hat{\boldsymbol{\beta}} = (\mathbf{X}'\mathbf{X})^{-1}\mathbf{X}'\mathbf{Y}$. Following the same basic structure argument as in Section 11.3.4, we replace $\boldsymbol{\beta}$ with $\hat{\boldsymbol{\beta}}$, ${}_i\boldsymbol{\Sigma}$ with ${}_i\hat{\boldsymbol{\Sigma}}$, and ${}_i\boldsymbol{\Delta}^2$ with the basic diagonal ${}_i\mathbf{D}^2$ of ${}_i\hat{\boldsymbol{\Sigma}}$, and then write from (12.3.2) another estimate of the likelihood of the sample in the same form as (11.3.23):

$$L({}_i\mathbf{Y}|\hat{\boldsymbol{\beta}}, {}_i\hat{\boldsymbol{\Sigma}}) = (2\pi e)^{-mn_i/2}\left(\prod_{j=1}^{m} {}_iD_j^2\right)^{-n_i/2} \tag{12.3.5}$$

Assuming the k samples are independent, we can write the likelihood of the total sample, given potentially different variance–covariance matrices, as the product of k terms of the form given in Eq. (12.3.5):

$$L(\mathbf{Y}|\hat{\boldsymbol{\beta}}, {}_i\hat{\boldsymbol{\Sigma}}, i = 1, 2, \ldots, k) = (2\pi e)^{-nm/2}\prod_{i=1}^{k}\left(\prod_{j=1}^{m} {}_iD_j^2\right)^{-n_i/2} \tag{12.3.6}$$

We now have two estimated likelihoods for the sample, one based on an assumption of a common variance–covariance structure across the populations and one based on an assumption of possibly different variance–covariance matrices. Since the first is subject to restrictions or constraints beyond those imposed on the second, we can construct from them a likelihood ratio, a ratio of the more restricted to the less restricted

$$\Lambda = L(\mathbf{Y}|\hat{\boldsymbol{\beta}}, \hat{\boldsymbol{\Sigma}})/L(\mathbf{Y}|\hat{\boldsymbol{\beta}}, {}_i\boldsymbol{\Sigma}, i = 1, 2, \ldots, k) \leqslant 1 \qquad (12.3.7)$$

(The reader may note that what was the unconstrained sample likelihood when we developed a test of the mean structure of our k populations is here the constrained likelihood. It is more constraining to require that all k populations have a common variance–convariance matrix than to permit them different variance–covariance structures.) On substituting from (12.3.1) and (12.3.6), the likelihood ratio becomes

$$\Lambda = \left(\prod_{j=1}^{m} D_j^2\right)^{-(n/2)} \bigg/ \prod_{i=1}^{k}\left(\prod_{j=1}^{m} {}_iD_j^2\right)^{-(n_i/2)}$$

$$\lambda = \prod_{i=1}^{k}\left(\prod_{j=1}^{m} {}_iD_j^2\right)^{n_i} \bigg/ \left(\prod_{j=1}^{m} D_j^2\right)^{-n} \qquad (12.3.8)$$

Small values of λ are associated with instances in which the k variance–covariance matrices are not all identical. By the same logic, we reject the hypothesis of a common variance–covariance structure if the *reciprocal* of λ is large:

$$1/\lambda = \left(\prod_{j=1}^{m} D_j^2\right)^{n} \bigg/ \prod_{i=1}^{k}\left(\prod_{j=1}^{m} {}_iD_j^2\right)^{n_i}$$

As λ and its reciprocal are always positive, the logarithm of $1/\lambda$ can also serve as an indicator:

$$\log(1/\lambda) = n\left(\log \prod_{j=1}^{m} D_j^2\right) - \sum_{i=1}^{k}\left[n_i\left(\log \prod_{j=1}^{m} {}_iD_j^2\right)\right] \qquad (12.3.9)$$

Again, large values would be inconsistent with a common variance–covariance matrix.

In fact, a statistic very closely related to $\log(1/\lambda)$—and, hence, to the likelihood ratio—is used in hypothesis testing. (The trick, again, is to find some function of the likelihood ratio with a known sampling distribution.) First, n and n_i are replaced in Eq. (12.3.9) by their corresponding *degrees of freedom*:

$$L = (n - k)\left(\log \prod_{j=1}^{m} D_j^2\right) - \sum_{i=1}^{k}\left[(n_i - 1)\left(\log \prod_{j=1}^{m} {}_iD_j^2\right)\right] \qquad (12.3.10)$$

Two test statistics are based on L. For $m < 6$, $k < 6$, and each $n_i > 20$, the product $LG \sim \chi^2_{g\text{ df}}$, where

$$G = 1 - \frac{2m^2 + 3p - 1}{6(m+1)(k-1)}\left[\sum_{i=1}^{k}\left(\frac{1}{n_i - 1}\right) - \frac{1}{n-k}\right] \quad (12.3.11)$$

and

$$g = \tfrac{1}{2}[m(m+1)(k-1)] \quad (12.3.12)$$

For smaller subsample sizes (down to about 10) or larger values of m or k, an F distribution approximation is thought to provide a better test of the hypothesis of homogeneity of variance–covariance matrices: $L[(G/g) - h] \sim F_{g,\,h\text{ df}}$, where L, G, and g are as described and

$$h = \frac{1}{k+2}\left\{\frac{(m-1)(m+2)}{6(k-1)}\left[\sum_{i=1}^{k}(n_i - 1)^{-2} - (n-k)^{-2}\right] - (1-G)^2\right\} \quad (12.3.13)$$

For either test, large values of the statistic lead to a rejection of the hypothesis of identical variance–covariance matrices.

12.4 Summary

Examples of several applications of the multivariate general linear hypothesis were presented. These included a two-group T^2 problem extended to illustrate the multivariate analysis of variance, the study of a moderator variable in multivariate linear regression, and an exercise in combining criterion attributes to produce a difference score.

The philosophy of the linear discriminant function was developed in the context of the MGLM and illustrated with data for tested interest patterns among college majors. The discussion touched on the use of more than one discriminant function.

Finally, the likelihood ratio technique was employed to develop a test for the multivariate general linear model assumption of common variance–covariance matrices for each of the sampled multivariate normal populations. Chapter 12 completed the discussion of the MGLM.

Exercises

1. Describe the design matrix and beta matrix for the Airline Pilot Study if three covariate measures had been collected on each pilot.

2. Describe the constraint matrices you would use to evaluate whether any one of the covariates contributed to the multivariate dependent variable.

3. How would you go about evaluating whether there is homogeneity of regression for the covariates in each of the experimental groups?

4. Outline the tests you would make of the experimental effects and their interaction. Develop the constraint matrices assuming that all three covariate measures had been retained.

Additional Reading

Applications of the MGLM are featured in the following two references:

Harris, R. J. (1975). *A Primer of Multivariate Statistics*. New York: Academic.

Tatsuoka, M. M. (1971). *Multivariate Analysis: Techniques for Education and Psychology*. New York: Wiley.

A computation related treatment of some common multivariate designs is given by

Cooley, W. W., and Lohnes, P. R. (1971). *Multivariate Data Analysis*. New York: Wiley.

An informative review of the kinds of questions addressed by the linear discriminant function is provided by

Huberty, C. J. (1975). Discriminant analysis, *Review of Educational Research* 45:543–598.

PART IV
MULTIVARIATE APPROACHES AND OTHER APPLICATIONS

Parts II and III have focused primarily on answering questions about the mean structures of a family of populations or measured observations, either univariate or multivariate. In Part IV we address some other issues in the analysis of multivariate data.

In Chapter 13 we direct attention to observations that are categorical rather than measured. A form of log linear analysis is presented that is very similar to the GLM in the way in which hypotheses about frequencies (rather than means) may be phrased and tested. Applications of the approach are included.

The GLM and its multivariate extensions assume that some attributes are predictors and others criteria; we then analyze the dependence of the criteria on the predictors. Some multivariate problems, however, do not presume a partitioning of the attributes. Rather, we want to understand the interdependence of the set of attributes as a whole. To develop such an approach we return to basic structure. In Chapter 14 we establish the equivalence of the basic structure representation of a data matrix to its principal components, discuss some applications of principal components, and compare principal components or basic structure to the closely related techniques of factor analysis.

The factors of factor analysis, unlike principal components, are not observable. They are *latent* attributes. They help to determine but are not identical to any observed attributes or any function of these observed (or, manifest) attributes. In Chapter 15 we discuss how such latent variables might be useful in dependence as well as in interdependence (factor analysis) studies. Joreskog's technique for the analysis of covariance structures, as realized in the computer program LISREL (linear structural relations), provides a way of testing, against an observed covariance matrix, models that specify patterns of covariation among such latent attributes.

Chapter 13

An Analysis Model for Categorical Data

Introduction

The general linear model, extended in the last two chapters to those cases in which the criterion attribute is a vector of observations, is the workhorse of multivariate inferential analysis. Yet, despite its breadth of applications, the GLM has some inherent limitations. It requires that the criterion, univariate or multivariate, have a normal sampling distribution. Here we look at a model of similar generality but suited to a different kind of observation. In particular, we are concerned with observations that are not in the form of measurements at all and can only be assigned to one or another of several nonoverlapping categories.

An approach now widely used for the analysis of such categorized, cross classified, or contingency table frequencies is the *log linear model.* We develop a form of this model and illustrate some of its uses. To make the development of the log linear approach as broadly comparable to the GLM as possible, we pay particular attention to the information analytic orientation provided by D. V. Gokhale and Solomon Kullback. Although their presentation of the log linear model is less well known among behavioral scientists than the work of Stephen Fienberg, for example, it is more in the tradition of the GLM. Computational programs that exploit the information approach, however, are not yet widely available, and our major examples will be consistent with the more restricted log linear approaches that can be pursued with such generally distributed programs as the BMDP routine P3F, Goodman's ECTA (Everyman's Contingency Table Analyzer), and the British package GLIM.

We think it important that the reader understand that, just as the analysis of the randomized *k*-groups design is but one application of the GLM, so too the analysis of *hierarchical cross classification models* through the iterative proportional fitting of selected marginal frequencies is but one application of the log linear approach. It is, at the time of our writing, the problem most frequently approached by log linear analyses, but as behavioral scientists gain more experience, this will change. We hope that the presentation of a quite general log linear model at this time—in advance of any widespread realization of its range of applications—will help the reader prepare for such later developments.

13.1 Three Sampling Models for Obtaining Frequencies

In this section we consider three models that explain how a set of frequencies has come to be observed. These three models describe the types of population samplings that give rise to the usual observed category counts.

To provide specificity, we assume that we are sampling within the context of our Preemployment Study. In particular, we have observed for each sampled underwriter three characteristics: *education* (categorized as some education beyond high school or no education beyond high school), *work experience* (two or more years previous full-time employment or a shorter period of experience), and *training time* (training required a minimum of 12 weeks or training completed in a shorter period). In this instance each attribute could have been measured. For present purposes, however, we wish to treat the observations simply as categorical. Taking the three attributes together, there are eight possible categories in which our underwriters may fall. We represent the number of observations in a particular category by f_{ijk}, where the indices i, j, and k correspond to the levels of the three attributes. As a bit of further notation we indicate the number of levels of the three attributes by I, J, and K. In general, for categories defined by such cross classification of observations, the total number of categories is given by the product $W = IJK$.

Note that this example yields a set of eight categories, cross classified by three attributes or dimensions. We should keep in mind that the sampling schemes to be described will apply as well where the W categories are defined in some fashion other than the full crossing of the levels of two or more categorized attributes. In Section 13.2 we consider the more general case.

13.1.1 Sampling with No Marginal Restrictions

Assume our sampling experiment for the Preemployment Study involved collecting data for every trainee underwriter hired during the first three months of 1980. No restrictions are imposed by this scheme on any of the frequencies f_{ijk} (or any combination of them). The sampling model most frequently used to describe frequencies obtained under these circumstances is that of *independent Poisson distributions*. The Poisson distribution gives for the probability of a certain frequency in cell *ijk*

$$\Pr(f_{ijk}) = \left(m_{ijk}^{f_{ijk}}\right)\left(e^{-m_{ijk}}\right)/f_{ijk}! \tag{13.1.1}$$

(The notation $f_{ijk}!$ is read "f_{ijk} factorial" and indicates the product of the positive integers from 1 through f_{ijk}. By convention, $0! = 1$.) The Poisson distribution has a single parameter, m_{ijk}, which is also the mean of the sampling distribution of f_{ijk}:

$$E(f_{ijk}) = m_{ijk} \tag{13.1.2}$$

The n independent observations making up the sample of underwriters' performances have a *sample probability* equal to the product of the (independent) cell probabilities over the *IJK* cells in the three-way table:

$$\Pr(f|\text{Poisson}) = \prod_{k=1}^{K}\prod_{j=1}^{J}\prod_{i=1}^{I} m_{ijk}^{f_{ijk}} e^{-m_{ijk}}/f_{ijk}! \tag{13.1.3}$$

Except when all cells have the same expected value and all values of m_{ijk} are identical, this expression does not simplify much.

13.1.2 Sampling to Obtain a Fixed Total Number of Observations

A second sampling model would be appropriate to the Preemployment Study example if we had chosen at random n of the underwriters hired in calendar year 1980. Although we do not fix the value of any f_{ijk} in this way, we do fix the sum of these over the *IJK* cells at n. Here a useful model is that of sampling from a *multinomial population*. The total number of observations n is fixed in multinomial sampling, but variable in independent Poisson sampling. If the probability that any one observation will fall in cell *ijk* is symbolized P_{ijk}, the multinomial model requires

$$\sum_{k=1}^{K}\sum_{j=1}^{J}\sum_{i=1}^{I} P_{ijk} = 1$$

In independently sampling n underwriters, then, the probability of obtaining a particular sample of f_{ijk} entries (summing to n) is given by

$$\Pr(f|\text{multinomial}) = n! \prod_{i,j,k}\left(P_{ijk}^{f_{ijk}}\right) \Big/ \prod_{i,j,k} f_{ijk}! \tag{13.1.4}$$

Again, unless all P_{ijk} values are the same (the probability that a randomly chosen underwriter will fall into any one of the eight cells is the same for all eight cells), this equation does not simplify. With n independent samples from the *IJK* multinomial distribution (on each sampling one and only one of the *IJK* possible outcomes is observed), the *expected* total number of outcomes in cell *ijk* is

$$E(f_{ijk}) = nP_{ijk} \tag{13.1.5}$$

13.1.3 Sampling to Obtain Fixed Marginal Totals

We might have sampled to ensure not only a particular total number of observations but particular *marginal* totals in our frequency table as well. In the underwriter training example, we could have decided in advance to sample $n_1 = f_{1..}$ from among those hired with no postsecondary education and $n_2 = f_{2..}$ from among those hired with at least some college training. In this instance not only is the total number of observations fixed at $n = n_1 + n_2$, but so are the total numbers in each of the two education levels of our three-way table.

One sampling model for the situation just described would be to assume the n_1 high-school-educated underwriters are independently chosen from one JK multinomial population and the n_2 college-educated underwriters from a second JK multinomial. Assuming the high-school- and college-educated underwriters are chosen independently, the probability of the overall sample of f_{ijk} observations is a *product of multinomial* probabilities:

$$\Pr(f|\text{product multinomial}) = \prod_i \left[f_{i..}! \left(\prod_{j,k} {}_i p_{jk}^{f_{ijk}} \Big/ \prod_{j,k} f_{ijk}! \right) \right] \tag{13.1.6}$$

where ${}_i p_{jk}$ is the probability of obtaining an occupant of cell jk when sampling from the ith multinomial distribution ($\Sigma_{j,k}\, {}_i p_{jk} = 1$, for $i = 1, 2, \ldots, I$). The expected number of observations for cell ijk under product multinomial sampling is

$$E(f_{ijk}) = f_{i..}({}_i p_{jk}) \tag{13.1.7}$$

Marginal totals may be fixed for other levels of the multiway table. In our Preemployment Study illustration, for example, $f_{11.}$, $f_{12.}$, $f_{21.}$, and $f_{22.}$ might all have been fixed in advance, corresponding to sampling known numbers (perhaps the same numbers) of low education–low experience, low education–high experience, high education–low experience, and high education–high experience underwriter trainees for study. Our sampling model could still be represented as involving independent samples from each of several multinomials, there now being a multinomial distribution with K outcomes for each of the IJ fixed size samples. (This design has some similarities to factorial design ANOVA, in which we might require the same number of observations in each cell; again, we would be interested in the sampling of criterion observations from each of IJ distributions.)

Of course, we cannot fix *all* of the frequencies in a multiway table, for then we have nothing left to vary so as to test hypotheses. Also, depending on the hypotheses to be tested, some marginal frequencies are more reasonably fixed than others. If training time were the criterion in the present instance, it would be awkward to consider a design in which we fixed the number of slow and fast trainees. The reasons for these distinctions will be clearer after the log linear *structural model* has been introduced.

13.2 General Log Linear Model for Cell Frequencies

In each of the three sampling models the frequency of occurrence, that is, the number of observations in a particular cell, is dependent on some *cell parameter*—a cell probability in the multinomial and product multinomial instances and an expected cell frequency for the Poisson process—as well as on the experimenter manipulated sample sizes or, for the Poisson, sampling interval. The three are referred to as sampling models because, for fixed or frozen values of cell parameters and sampling factors, they explain the trial-to-trial variability in the observed frequencies that would occur if we were to repeat our experiment over and over. In this role they correspond to the normal sampling distribution postulated for criterion observations in the GLM.

In the GLM we have not only a sampling model but also a *structural* model,

$$\mathbf{E}(\mathbf{Y}) = \mathbf{X}\boldsymbol{\beta}$$

That is, the *expected values* of the criterion observations are presumed to have a structure, each is a linear function of a set of predictor values. It is the structural part of the GLM that allows us to phrase hypotheses or contrasts, $\mathbf{C}'\boldsymbol{\beta} = \boldsymbol{\Gamma}$, whereas the sampling model part allows us to test those hypotheses and to put confidence intervals about the contrasts.

If we are conveniently to phrase hypotheses about what we expect as cell frequencies, it will be of assistance to have a structural model of about the same generality as we have for the GLM.

13.2.1 Some Likelihood Ratios

We begin our development of the general log linear model by constructing likelihood ratios similar to those we used to establish hypothesis-testing techniques for the GLM. Both to simplify the notation and to make the construction more general, we assume W categories overall or, in the product multinomial case, ${}_iW$ categories for the ith multinomial. For now we shall not be concerned with how the W categories relate to any cross classifying dimensions. The estimated likelihood of a sample set of frequencies f can be rewritten from Eqs. (13.1.3), (13.1.4), and (13.1.6) as

$$L(f|\text{Poisson}) = \prod_w \frac{\tilde{m}_w^{f_w} e^{-\tilde{m}_w}}{f_w!} \tag{13.2.1}$$

$$L(f|\text{multinomial}) = n!\left(\prod_w \tilde{p}_w^{f_w} \Big/ \prod_w f_w!\right) \tag{13.2.2}$$

$$L(f|\text{product multinomial}) = \prod_i \left[f_i!\left(\prod_{{}_iw} {}_i\tilde{p}_w^{f_w} \Big/ \prod_{{}_iw} f_w!\right)\right] \tag{13.2.3}$$

where $\tilde{m}_w$, $\tilde{p}_w$, and ${}_i\tilde{p}_w$ are estimates of the corresponding parameters, and the index ${}_iw$ indicates that the product includes only terms for the ith multinomial.

As in the GLM, we are interested in finding both unconstrained and constrained estimators. First, we define the estimators $\tilde{m}_w$, $\tilde{p}_w$, and ${}_i\tilde{p}_w$ as the unconstrained maximum likelihood estimators. Without formally deriving these, each has an intuitively appealing form:

$$\tilde{m}_w = f_w \tag{13.2.4}$$

$$\tilde{p}_w = f_w/n \tag{13.2.5}$$

$${}_i\tilde{p}_w = f_w/{}_in \tag{13.2.6}$$

where $n = \Sigma_w f_w$, the total sample size; ${}_in = \Sigma_{{}_iw} f_w$, the size of the sample drawn from the ith multinomial; and m_w is the expected frequency of the wth category when that refers to sampling from the wth Poisson process.

Our second set of estimators, ${}_A\tilde{m}_w$, ${}_A\tilde{p}_w$, and ${}_{Ai}\tilde{p}_w$ consists of the maximum likelihood estimators *constrained by some model*, an as yet unspecified model A. The ratio of the constrained to unconstrained likelihoods, each obtained by substituting the appropriate set of estimators into (13.2.1), (13.2.2), or (13.2.3), is a *likelihood ratio*. As we know, the constrained likelihood cannot exceed the unconstrained likelihood, and the ratio cannot exceed 1.

From (13.2.1)–(13.2.3), the three likelihood ratios may be written

$$\text{LR(Poisson)} = \exp\left(\sum_w \tilde{m}_w - \sum_w {}_A\tilde{m}_w\right)\prod_w \left(\frac{{}_A\tilde{m}_w}{\tilde{m}_w}\right)^{f_w} \tag{13.2.7}$$

$$\text{LR(multinomial)} = \prod_w \left(\frac{{}_A\tilde{p}_w}{\tilde{p}_w}\right)^{f_w} \tag{13.2.8}$$

$$\text{LR(product multinomial)} = \prod_i \prod_{{}_iw} \left(\frac{{}_{Ai}\tilde{p}_w}{{}_i\tilde{p}_w}\right)^{f_w} \tag{13.2.9}$$

For Poisson samples, the constrained models we shall be interested in are such that

$$\sum_w {}_A\tilde{m}_w = \sum_w \tilde{m}_w = n$$

and Eq. (13.2.7) simplifies. The *logarithms* of the three likelihood ratios are

$$\log \text{LR(Poisson)} = \sum_w \left[f_w \log\left(\frac{{}_A\tilde{m}_w}{\tilde{m}_w}\right)\right] \tag{13.2.10}$$

$$\log \text{LR(multinomial)} = \sum_w \left[f_w \log\left(\frac{{}_A\tilde{p}_w}{\tilde{p}_w}\right)\right] \tag{13.2.11}$$

$$\log \text{LR(product multinomial)} = \sum_w \left[f_w \log\left(\frac{{}_{Ai}\tilde{p}_w}{{}_i\tilde{p}_w}\right)\right] \tag{13.2.12}$$

Substituting the unconstrained estimators from (13.2.4)–(13.2.6) into these results gives

$$\log \mathrm{LR}(\text{Poisson}) = \sum_w \left[f_w \log\left(\frac{{}_A\tilde{m}_w}{f_w} \right) \right] \tag{13.2.13}$$

$$\log \mathrm{LR}(\text{multinomial}) = \sum_w \left[f_w \log\left(\frac{n_A \tilde{p}_w}{f_w} \right) \right] \tag{13.2.14}$$

$$\log \mathrm{LR}(\text{product multinomial}) = \sum_w \left[f_w \log\left(\frac{{}_i n_{Ai} \tilde{p}_w}{f_w} \right) \right] \tag{13.2.15}$$

Any one of the three log likelihood ratios now is of the form

$$\log \mathrm{LR} = \sum_w \left\{ f_w \log\left[\frac{E(f_w|A)}{f_w} \right] \right\} \tag{13.2.16}$$

where $E(f_w|A)$ is the *expected frequency* in the wth cell given the constraints of model A.

13.2.2 Kullback's Discrimination Information Statistic

These likelihood ratios assume values in the interval $0 < \mathrm{LR} \leqslant 1$, and consequently the $\log \mathrm{LR} \leqslant 0$. In developing their version of the log linear model, Gokhale and Kullback (see the suggested reading at the end of this chapter) use a simple transformation of log LR. The quantity

$$\mathrm{DI}[f: E(f|A)] = \sum_w \left\{ f_w \log\left[\frac{f_w}{E(f_w|A)} \right] \right\} \tag{13.2.17}$$

is the negative of log LR. This nonnegative quantity is one of Kullback's *discrimination information* (DI) statistics.

Roughly speaking, the DI provides an index of the degree of discrimination provided by two descriptions of a probability or frequency distribution. Typically, the two descriptions are such that one is simpler than the other and the question is how much less information about the distribution is captured by the simpler description. In Eq. (13.2.17) the comparison is between a full or unconstrained description of the distribution—we know the value of each f_w—and one which permits us only to estimate f_w subject to the constraints of model A. The DI is zero only if the two descriptions, the two sets of estimates, are identical. Because the estimates under model A usually diverge from the full description—they provide less information about the distribution—the DI takes on larger, positive values.

The discrepancy between two probability or frequency distributions can be indexed by a variety of statistics other than the DI. We use the DI because under an important class of constraining models or hypotheses and for sufficiently large samples $2\mathrm{DI}[f: E(f|A)] \sim \chi^2_{r\,df}$. That is, twice the DI is distributed, asymptotically and under the constraining model, as a central chi-squared observation with degrees of freedom that depend on the sampling design and the constraining model or hypothesis. "Sufficiently large" may be interpreted here as meaning that n, the number of observations, should be at least ten times W, the number of categories.

The DI of Eq. (13.2.17) provides a measure of *goodness of fit*. If the values of $E(f_w|A)$ are close to the observed values of f_w, the resulting χ^2 will be small, reflecting that model A provides a good fit or description of the observations.

13.2.3 A Class of Models

What can model A look like? How might estimates of the cell frequencies be constrained? We follow the approach of Gokhale and Kullback, which provides a way of defining constraints that is easily understood and can be applied extensively. Before introducing the class of constraints we can work with, we simplify our notation slightly. Let ${}_A f_w$ stand for $E(f_w|A)$, the expected frequency in cell w under model A. Further, we write $\mathbf{f}$ and ${}_A\mathbf{f}$, taking them to be W-element vectors of observed and expected frequencies.

The models that restrict, constrain, or limit our expectations of $\mathbf{f}$ and that are consistent with the DI statistic are those that can be expressed in the form

$$ {}_A\mathbf{T}'\,{}_A\mathbf{f} = {}_A\boldsymbol{\phi} \qquad (13.2.18) $$

where ${}_A\mathbf{T}$ and ${}_A\boldsymbol{\phi}$ are $W \times r$ and $r \times 1$ arrays specified by the investigator. What (13.2.18) tells us is that under model A our estimates of the W cell frequencies must satisfy the r linear relations ${}_A\mathbf{T}'_{.i}\,{}_A\mathbf{f} = {}_A\phi_i$. Each row of ${}_A\mathbf{T}'$, together with the corresponding element of the vector ${}_A\boldsymbol{\phi}$, describes a restriction or constraint on the elements of ${}_A\mathbf{f}$.

In later examples, we consider a number of linear constraints on ${}_A\mathbf{f}$. For now we mention a commonly employed and easily interpreted one. The constraint $\mathbf{1}'\,{}_A\mathbf{f} = n$ ensures that the sum of the expected frequencies under model A equals the total observed sample size n.

13.2.4 Internal and External Constraints

Two classes of estimation problems are encountered when finding expectations for the cell frequencies, subject to a set of linear constraints. These constraints can be wholly *internal*, or they may include some that are *external*. "Internal" and "external" refer to the vector of observed frequencies $\mathbf{f}$. If all of the constraints are chosen to ensure that ${}_A\mathbf{f}$ possesses some of the characteristics of $\mathbf{f}$, then the constraints are internal. For internal constraints (13.2.18) may be written in the specialized form

$$ {}_A\mathbf{T}'\,{}_A\mathbf{f} = {}_A\mathbf{T}'\mathbf{f} \qquad (13.2.19) $$

or, for our simple example, $\mathbf{1}'\,{}_A\mathbf{f} = \mathbf{1}'\mathbf{f}$. The latter says that the vector of estimates ${}_A\mathbf{f}$ is to have the same sum as the observed vector $\mathbf{f}$.

By way of contrast, we may impose constraints that do not depend upon the observed frequencies. For example, two estimated frequencies can be required to be the same,

$$ {}_A\mathbf{T}'\,{}_A\mathbf{f} = (0,\ldots,0,1,\ldots,0,-1,0,\ldots,0)\,{}_A\mathbf{f} = {}_Af_i - {}_Af_j = 0 = {}_A\phi $$

whether or not f_i and f_j, the corresponding observed frequencies, are the same. This constraint is *external* to the data. It does not require that the vector ${}_A\mathbf{f}$ look like $\mathbf{f}$. Rather, it requires that ${}_A\mathbf{f}$ look like an external (or ideal) distribution in some particular regard.

In either case we are interested in assessing the goodness of fit. For the internal constraints problem (ICP), we assess how well the data at hand, $\mathbf{f}$, are *fit by* a set of expectations constrained only partially to resemble $\mathbf{f}$. For the external constraints problem (ECP), we ask how well the data at hand *fit to* a set of expectations constrained to take into account restrictions outside that data.

The differences between the two classes of problem are sufficient that they merit separate attention. We take up first the ICP, which, as indicated in the introduction, has been more widely treated by the log linear approach. Following that we discuss the ECP in a way that highlights the unity of the two.

13.2.5 Internal Constraints Problem and Minimum Discrimination Information

At first glance, Eq. (13.2.18), ${}_A\mathbf{T}'\,{}_A\mathbf{f} = {}_A\boldsymbol{\phi}$, appears similar to the familiar GLM constraint $\mathbf{C}'\boldsymbol{\beta} = \boldsymbol{\Gamma}$; we should point out some differences. The reader will recall that as we increased the number of constraints on $\boldsymbol{\beta}$ in the GLM, as we added rows to $\mathbf{C}'$, we could expect progressively poorer fits of the model to the data. That is, $\mathbf{e}'\mathbf{e} \leqslant \mathbf{f}'\mathbf{f} \leqslant \mathbf{g}'\mathbf{g}$, where $\mathbf{e}'\mathbf{e}$, $\mathbf{f}'\mathbf{f}$, and $\mathbf{g}'\mathbf{g}$ are, respectively, the unconstrained sum of squares of errors and the sum of squares of errors for $\boldsymbol{\beta}$ constrained in turn by ${}_1\mathbf{C}'\boldsymbol{\beta} = {}_1\boldsymbol{\Gamma}$ and ${}_2\mathbf{C}'\boldsymbol{\beta} = {}_2\boldsymbol{\Gamma}$ with ${}_2\mathbf{C} = ({}_1\mathbf{C}, {}_2\mathbf{K})$ and ${}_2\boldsymbol{\Gamma}' = ({}_1\boldsymbol{\Gamma}', {}_2\mathbf{G}')$.

For the ICP, where the constraints are designed to make ${}_A\mathbf{f}$ look like $\mathbf{f}$, the reverse will be true: The more ways in which we require that ${}_A\mathbf{f}$ be like $\mathbf{f}$—the more internal constraints we impose on ${}_A\mathbf{f}$—the better the fit will be of ${}_A\mathbf{f}$ to $\mathbf{f}$. In fact, it is helpful to think of the observed frequency distribution $\mathbf{f}$ as a *fully internally constrained* set of expectations. We can express this in terms of Eq. (13.2.19) as the result of imposing W constraints on ${}_A\mathbf{f}$: ${}_A\mathbf{f} = \mathbf{I}\,{}_A\mathbf{f} = \mathbf{I}\mathbf{f} = \mathbf{f}$. This full set of internal constraints requires that each ${}_A\mathbf{f}_w = \mathbf{f}_w$. The fit is perfect, and uninteresting. In most ICP analyses we are interested in how few constraints we can impose on ${}_A\mathbf{f}$ and still have a close fit to $\mathbf{f}$.

To help develop ICP analysis, consider the following table, containing the (hypothetical) results of stopping 50 consecutive passersby and inquiring as to their gender and marital status:

	Female	Male
Married	4	16
Single	16	14

Next, write these results as a vector $\mathbf{f}' = (4, 16, 16, 14)$. The first two elements are for the married samples, the second two for the single respondents. These observed frequencies, we just noted, are also a set of *fully constrained* estimators. In what follows we use the notation ${}_{\mathrm{F}}\mathbf{f}$ to indicate the vector of fully constrained estimators.

Additionally, the sampling plan provides a set of *least constrained* estimators. In the example, assume we questioned 50 consecutive passersby. The total number of observations is fixed by that decision. We want *any* set of estimators also to have that property. Thus the *least constrained* estimator, written ${}_{\mathrm{L}}\mathbf{f}$, must satisfy the condition $\mathbf{1}'\,{}_{\mathrm{L}}\mathbf{f} = 50$. For *multinomial* sampling, illustrated by the preceding example, the *least constrained* estimator must satisfy the condition

$$\mathbf{1}'\,{}_{\mathrm{L}}\mathbf{f} = n\,(= \mathbf{1}'\mathbf{f}) \tag{13.2.20}$$

The expression in parentheses reminds us that this constraint is internal.

If the table of observed frequencies had been the result of *product multinomial* sampling—if, for example, we had questioned not 50 consecutive passersby but 20 consecutive female passersby and 30 consecutive male passersby—the least constrained estimator would need to satisfy slightly different restrictions. In this case, any set of estimators must reflect the decision to query 20 women and 30 men. The least constrained estimator satisfies

$$\begin{bmatrix} 1 & 0 & 1 & 0 \\ 0 & 1 & 0 & 1 \end{bmatrix} {}_{\mathrm{L}}\mathbf{f} = \begin{bmatrix} 20 \\ 30 \end{bmatrix}$$

Written more generally, the least constrained set of estimators under product multinomial sampling must satisfy

$$ {}_{I}\mathbf{U}'\,{}_{\mathrm{L}}\mathbf{f} = {}_{I}\mathbf{n}\,(= {}_{I}\mathbf{U}'\mathbf{f}) \tag{13.2.21}$$

where the ith row of ${}_{I}\mathbf{U}'$ serves to sum over those categories associated with the ith multinomial distribution. This sum is constrained to be equal to ${}_{I}\mathbf{n}_i$, the total number of observations sampled from the ith multinomial.

Finally, if an obtained frequency vector is assumed to be the result of sampling from *W independent Poisson* processes—as would have been reasonable if our data above had resulted from surveying every passerby over a half-hour interval—a set of least constrained estimators of cell frequencies would need to satisfy the restriction

$$\mathbf{1}'\,{}_{\mathrm{L}}\mathbf{f} = \mathbf{1}'\mathbf{f} \tag{13.2.22}$$

(The best guess of the number of passersby in any half-hour period is the number we observed.)

Equations (13.2.20)–(13.2.22) are not sufficient to tell us exactly the least constrained estimator ${}_{\mathrm{L}}\mathbf{f}$. Many vectors satisfy each of these equations. For 50 observations sampled from a four category multinomial, the vectors (5, 20, 15, 10), (10, 15, 10, 15), (0, 25, 0, 25), and (12.5, 12.5, 12.5, 12.5) all satisfy the least constrained estimator condition $\mathbf{1}'\,{}_{\mathrm{L}}\mathbf{f} = 50$. Is there one among them, or some other, that is *the* least constrained estimator? What we ask of ${}_{\mathrm{L}}\mathbf{f}$ is that it resemble $\mathbf{f}$ only through the relation $\mathbf{1}'\,{}_{\mathrm{L}}\mathbf{f} = n$. We do not want ${}_{\mathrm{L}}\mathbf{f}$ inadvertently to carry more information about $\mathbf{f}$ than is implied by this least constraint. How can we ensure this?

Again taking our lead from Gokhale and Kullback, we assume that ${}_{\mathrm{L}}\mathbf{f}$ is composed in ignorance. Except for conforming to one of Eqs. (13.2.20)–(13.2.22), the elements of ${}_{\mathrm{L}}\mathbf{f}$ are equal, or *uniform*. In this fashion they will be least informative about $\mathbf{f}$. Specifically, the least constrained estimator for independent Poisson sampling is

$${}_{\mathrm{L}}\mathbf{f} = \left[\left(\frac{1}{W}\right)\sum_{w} f_w\right]\mathbf{1} \tag{13.2.23}$$

while for multinomial sampling it is

$${}_{\mathrm{L}}\mathbf{f} = (n/W)\mathbf{1} \tag{13.2.24}$$

and for product multinomial

$${}_{\mathrm{L}}f_{(i)} = {}_{I}n_i/{}_{i}W \tag{13.2.25}$$

where $f_{(i)}$ is any one of the categories containing observations from the ith multinomial.

Thus, the least and fully constrained estimators for our 2×2 table, assuming multinomial (or independent Poisson) sampling, are

$${}_{\mathrm{L}}\mathbf{f} = (12.5 \quad 12.5 \quad 12.5 \quad 12.5), \qquad {}_{\mathrm{F}}\mathbf{f}' = (4 \quad 16 \quad 16 \quad 14)$$

[Had we assumed separate sampling of our male and female respondents, the least constrained estimator would have been ${}_{\mathrm{L}}\mathbf{f}' = (10, 15, 10, 15)$, as given by Eq. (13.2.25).]

We now know how to write the least and fully (internally) constrained estimators. Most hypotheses of interest, however, imply estimators somewhere between the two. That is, we often want to test the fit of models that assume neither that all of the cell frequencies should be identical nor that all should be independently determined. For our gender $\times$ marital status example, we might be interested in the fit of a model that takes explicitly into account the numbers of males and females questioned but nothing about their marital status. That is, we find an estimator satisfying

$$\begin{bmatrix} 1 & 0 & 1 & 0 \\ 0 & 1 & 0 & 1 \end{bmatrix} {}_A\mathbf{f} = \begin{bmatrix} 20 \\ 30 \end{bmatrix}$$

Many four-element vectors can satisfy this pair of constraints. Choosing one to compare with the observed frequencies emphasizes the role of least and fully constrained estimators.

Equation (13.2.17) defined a discrimination information statistic,

$$\mathrm{DI}(\mathbf{f} : {}_A\mathbf{f}) = \sum_w \left[f_w \log\left(\frac{f_w}{{}_Af_w} \right) \right]$$

which was described as a measure of the extent to which the model constrained estimator ${}_A\mathbf{f}$ carries less information about the frequency distribution than does $\mathbf{f}$ itself. Recalling that $\mathbf{f}$ is also the fully constrained estimator, we can rewrite Eq. (13.2.17)

$$\mathrm{DI}({}_\mathrm{F}\mathbf{f} : {}_A\mathbf{f}) = \sum_w \left[{}_\mathrm{F}f_w \log\left(\frac{{}_\mathrm{F}f_w}{{}_Af_w} \right) \right]$$

Written this way, the DI measures how well the fully constrained estimators ${}_\mathrm{F}\mathbf{f}$ are fit by the partially constrained ones ${}_A\mathbf{f}$.

The importance of reexpressing the DI in this form is to point up a broader range of application. In comparing two estimates of a set of frequencies, we need not require that one of these be the set of observed frequencies. Assume that we have three constrained sets of estimators, ${}_\mathrm{L}\mathbf{f}$, ${}_A\mathbf{f}$, and ${}_\mathrm{F}\mathbf{f}$, where ${}_\mathrm{L}\mathbf{f}$ and ${}_\mathrm{F}\mathbf{f}$ are the previously defined least and fully constrained estimators and ${}_A\mathbf{f}$ is between the two in the sense that it is subject to the same constraints as ${}_\mathrm{L}\mathbf{f}$ plus one or more additional constraints, but it is not subject to all of the constraints imposed on ${}_\mathrm{F}\mathbf{f}$. (The pair of constraints that would ensure that the estimated numbers of male and female respondents correspond to the observed numbers is of this intermediate type. It would ensure that the total of the estimated frequencies is 50, thus incorporating the least constraint, but it would not require that all cells be estimated by their observed frequencies, as would the fully constrained set of estimators.) For these three estimated frequency distributions we can compute three DI statistics. Two of them,

$$\mathrm{DI}({}_\mathrm{F}\mathbf{f} : {}_A\mathbf{f}) = \sum_w \left[{}_\mathrm{F}f_w \log\left(\frac{{}_\mathrm{F}f_w}{{}_Af_w} \right) \right], \qquad \mathrm{DI}({}_\mathrm{F}\mathbf{f} : {}_\mathrm{L}\mathbf{f}) = \sum_w \left[{}_\mathrm{F}f_w \log\left(\frac{{}_\mathrm{F}f_w}{{}_\mathrm{L}f_w} \right) \right]$$

are measures of how well an observed frequency distribution (or a fully constrained set of estimators of those frequencies) is fit by a partly constrained set of estimators. The third DI,

$$\mathrm{DI}({}_A\mathbf{f} : {}_\mathrm{L}\mathbf{f}) = \sum_w \left[{}_Af_w \log\left(\frac{{}_Af_w}{{}_\mathrm{L}f_w} \right) \right] \tag{13.2.26}$$

provides a measure of how well a set of (partially) constrained estimators ${}_A\mathbf{f}$ is fit by the least constrained estimator ${}_L\mathbf{f}$.

In fact, Eqs. (13.2.17) and (13.2.26) are special forms of a more general DI statistic. Let ${}_A\mathbf{f}$ and ${}_B\mathbf{f}$ be two estimators of a W-element vector $\mathbf{f}$ of frequencies constrained, respectively, by ${}_A\mathbf{T}'\,{}_A\mathbf{f} = {}_A\mathbf{T}'\mathbf{f}$ and ${}_B\mathbf{T}'\,{}_B\mathbf{f} = {}_B\mathbf{T}'\mathbf{f}$. Then,

$$\mathrm{DI}({}_B\mathbf{f} : {}_A\mathbf{f}) = \sum_W \left[{}_Bf_w \log\left(\frac{{}_Bf_w}{{}_Af_w}\right)\right] \tag{13.2.27}$$

is a nonnegative discrimination information statistic if ${}_A\mathbf{T}$ and ${}_B\mathbf{T}$ are such that any nonnull vector $\mathbf{v}$ that satisfies ${}_B\mathbf{T}'\mathbf{v} = {}_B\mathbf{T}'\mathbf{f}$ also satisfies ${}_A\mathbf{T}'\mathbf{v} = {}_A\mathbf{T}'\mathbf{f}$. That is, (13.2.27) describes a DI statistic if ${}_B\mathbf{f}$ has to satisfy all the constraints met by ${}_A\mathbf{f}$ plus one or more additional constraints.

Furthermore, twice the DI of (13.2.27) has, at least asymptotically, a chi-squared distribution with $q - p$ degrees of freedom, where ${}_A\mathbf{T}$ and ${}_B\mathbf{T}$ have ranks p and q satisfying $p < q \leqslant W$:

$$2\mathrm{DI}({}_B\mathbf{f} : {}_A\mathbf{f}) \sim \chi^2_{q-p\ \mathrm{df}} \tag{13.2.28}$$

whenever the sampling distribution of $\mathbf{f}$ is such that ${}_A\mathbf{T}'\mathbf{f}$ is fixed but ${}_B\mathbf{T}'\mathbf{f}$ is not. Broadly, this means that 2DI has a central chi-squared distribution when the hypothesis implied by ${}_A\mathbf{T}'\,{}_A\mathbf{f} = {}_A\mathbf{T}'\mathbf{f}$ is *true* and the hypothesis implied by ${}_B\mathbf{T}'\,{}_B\mathbf{f} = {}_B\mathbf{T}'\mathbf{f}$ is *not true*. The interpretation of this will be clearer after we have considered further the problem of estimation.

We return now to the task of finding the appropriate ${}_A\mathbf{f}$ to satisfy ${}_A\mathbf{T}'\,{}_A\mathbf{f} = {}_A\mathbf{T}'\mathbf{f}$ given by

$$\begin{bmatrix} 1 & 0 & 1 & 0 \\ 0 & 1 & 0 & 1 \end{bmatrix} {}_A\mathbf{f} = \begin{bmatrix} 20 \\ 30 \end{bmatrix}$$

Just as we found that vector that satisfies

$${}_L\mathbf{T}'\,{}_L\mathbf{f} = (1 \quad 1 \quad 1 \quad 1)\,{}_L\mathbf{f} = 50 = {}_L\mathbf{T}'\mathbf{f}$$

without capturing any additional characteristics of $\mathbf{f}$, so too the selected ${}_A\mathbf{f}$ is to be free of any unintended resemblance to $\mathbf{f}$. The choice is the ${}_A\mathbf{f}$—among all those satisfying the constraints ${}_A\mathbf{T}'\,{}_A\mathbf{f} = {}_A\mathbf{T}'\mathbf{f}$—that increases by the smallest amount possible the information about $\mathbf{f}$ carried by the least constrained estimator ${}_L\mathbf{f}$. The mechanism for conducting the search involves the DI of Eq. (13.2.26),

$$\mathrm{DI}({}_A\mathbf{f} : {}_L\mathbf{f}) = \sum_w \left[{}_Af_w \log\left(\frac{{}_Af_w}{{}_Lf_w}\right)\right]$$

This DI provides a measure of the information increase (discrimination) between ${}_L\mathbf{f}$ and ${}_A\mathbf{f}$. We seek the ${}_A\mathbf{f}$ satisfying ${}_A\mathbf{T}'\,{}_A\mathbf{f} = {}_A\mathbf{T}'\mathbf{f}$ that minimizes Eq. (13.2.26).

The side conditions or constraints that must be satisfied can be built into the minimization by writing a composite function,

$$\phi = \sum_w {}_Af_w \log({}_Af_w/{}_Lf_w) - \sum_{i=1}^{p} \lambda_i \left[\sum_w {}_AT_{wi}\,{}_Af_w - \sum_w {}_AT_{wi}\, f_w \right] \tag{13.2.29}$$

where the λ_i are the unknown Lagrangian multipliers associated with the p scalar side conditions or linear constraints on ${}_A\mathbf{f}$. The vector ${}_A\mathbf{f}$ that minimizes Eq. (13.2.29) is described by Gokhale and Kullback as a *minimum discrimination information* (MDI) estimator of $\mathbf{f}$.

To find the MDI, we differentiate this composite function with respect to ${}_A\mathbf{f}$ and set the result equal to zero. This yields

$$\log\left(\frac{{}_Af_w}{{}_Lf_w}\right) = \sum_{i=1}^{p} \lambda_i\,{}_AT_{wi} \tag{13.2.30}$$

or

$$\log({}_Af_w) = \log({}_Lf_w) + \sum_{i=1}^{p} \lambda_i\,{}_AT_{wi} \tag{13.2.31}$$

or

$${}_Af_w = \exp\left(\log {}_Lf_w + \sum_{i=1}^{p} \lambda_i\,{}_AT_{wi} \right) \tag{13.2.32}$$

It is Eq. (13.2.31) that gives rise to the name *log linear* model: The natural logarithm of each of the expected cell frequencies $\log({}_Af_w)$ is linear in a set of parameters, the λ_i. For our 2×2 example,

$${}_L\mathbf{f}' = \left(\tfrac{50}{4}, \tfrac{50}{4}, \tfrac{50}{4}, \tfrac{50}{4}\right), \qquad {}_A\mathbf{T}' = \begin{bmatrix} 1 & 0 & 1 & 0 \\ 0 & 1 & 0 & 1 \end{bmatrix}$$

and (13.2.31) provides

$$\log({}_A\mathbf{f}) = \begin{bmatrix} \log \frac{50}{4} + \lambda_1 \\ \log \frac{50}{4} + \lambda_2 \\ \log \frac{50}{4} + \lambda_1 \\ \log \frac{50}{4} + \lambda_2 \end{bmatrix}$$

13.2.6 Iterative Proportional Fitting of Marginals

Although the MDI estimator of (13.2.32) has the desired minimization property and is properly constrained, it is dependent on the as yet unknown λ_i. While there is no direct, general solution for ${}_A\mathbf{f}$, an iterative computational procedure—called *iterative proportional fitting* (IPF)—converges to the required solution when the constraints on ${}_A\mathbf{f}$ are internal. We offer no proof of this convergence or of the finding that, under the sampling and constraint conditions we have been discussing, the IPF estimators are the appropriately constrained maximum likelihood estimators of the cell frequency expectations. The interested reader may learn more of the estimation theory from Bishop et al. or Andersen (see the list of additional reading at the end of this chapter).

In effect, IPF requires that we fit each of the p constraints one at a time, beginning with ${}_{\mathrm{L}}\mathbf{f}$, adjusting elements of the current estimator of ${}_{A}\mathbf{f}$ to fit that constraint and going on to the next. Depending on the model, IPF may require us to cycle through all p constraints several times before an ${}_{A}\mathbf{f}$ is found that satisfies all constraints. Some sets of constraints, as the following simple example illustrates, can be satisfied by a single cycle of adjustments. We begin with ${}_{\mathrm{L}}\mathbf{f}$ and move away from it as little as possible to satisfy the constraints.

We illustrate IPF with our 2×2 example. For notational convenience we write ${}_{0}\mathbf{f} = {}_{\mathrm{L}}\mathbf{f}$ and $\mathbf{T}_j = {}_{A}\mathbf{T}_{.j}$, which gives at the outset

$$\mathbf{f}' = (4, 16, 16, 14), \qquad {}_{0}\mathbf{f}' = \left(\tfrac{50}{4}, \tfrac{50}{4}, \tfrac{50}{4}, \tfrac{50}{4}\right)$$
$$\mathbf{T}_1' = (1, 0, 1, 0), \qquad \mathbf{T}_2' = (0, 1, 0, 1)$$

First cycle. Take ${}_{0}\mathbf{f}$ as the initial estimator of ${}_{A}\mathbf{f}$.

1. Compute $\mathbf{T}_1' {}_{0}\mathbf{f} = \phi_1$ and $\mathbf{T}_1'\mathbf{f} = P_1$. If $\phi_1 = P_1$, set ${}_{1}\mathbf{f} = {}_{0}\mathbf{f}$ and go to step 2. If $\phi_1 \neq P_1$, write ${}_{1}\mathbf{f}$ as follows:

$${}_{1}f_w = \begin{cases} {}_{0}f_w & \text{if } T_{w1} = 0 \\ {}_{0}f_w(P_1/\phi_1) & \text{if } T_{w1} \neq 0 \end{cases}$$

   For the example, $\phi_1 = 25$, $P_1 = 20$ and we compute

$${}_{1}\mathbf{f}' = \left(\left(\tfrac{50}{4}\right)\left(\tfrac{20}{25}\right), \tfrac{50}{4}, \left(\tfrac{50}{4}\right)\left(\tfrac{20}{25}\right), \tfrac{50}{4}\right) = \left(10, \tfrac{50}{4}, 10, \tfrac{50}{4}\right)$$

   and we go to step 2.

2. Compute $\mathbf{T}_2' {}_{1}\mathbf{f} = \phi_2$ and $\mathbf{T}_2'\mathbf{f} = P_2$. If $\phi_2 = P_2$, set ${}_{2}\mathbf{f} = {}_{1}\mathbf{f}$ and go to second cycle. If they disagree, write ${}_{2}\mathbf{f}$ as follows:

$${}_{2}f_w = \begin{cases} {}_{1}f_w & \text{if } T_{w2} = 0 \\ {}_{1}f_w(P_2/\phi_2) & \text{if } T_{w2} \neq 0 \end{cases}$$

   Here, we have $\phi_2 = 25$ and $P_2 = 30$, so we compute ${}_{2}\mathbf{f}$ as

$${}_{2}\mathbf{f}' = \left(10, \left(\tfrac{50}{4}\right)\left(\tfrac{30}{25}\right), 10, \left(\tfrac{50}{4}\right)\left(\tfrac{30}{25}\right)\right) = (10, 15, 10, 15)$$

   and we have finished the first cycle.

Second cycle. Does ${}_{p}\mathbf{f} = {}_{0}\mathbf{f}$? That is, was our estimate of ${}_{A}\mathbf{f}$ modified by the cycle of p steps? If no change has occurred, we are finished. If the two vectors differ, we start a second cycle with ${}_{p}\mathbf{f}$. (For our example $p = 2$, the number of constraints.)

1. Compute $\mathbf{T}_1' {}_{p}\mathbf{f} = \phi_1$ and $\mathbf{T}_1'\mathbf{f} = P_1$. If the two agree, set ${}_{p+1}\mathbf{f} = {}_{p}\mathbf{f}$ and proceed to step 2. If the two are not equal, compute ${}_{p+1}\mathbf{f}$ as in step 1

above. For our example $P_1 = \phi_1 = 20$, and the estimator is not changed:

$$_{p+1}\mathbf{f} = {}_{p}\mathbf{f}$$

2. Compute $\mathbf{T}_2'\,{}_{p+1}\mathbf{f} = \phi_2$ and $\mathbf{T}_2'\mathbf{f} = P_2$ and compare. If they agree, set $_{p+2}\mathbf{f} = {}_{p+1}\mathbf{f}$; otherwise compute $_{p+2}\mathbf{f}$ as in step 2, then go to the third cycle. For our data both products are 30 and $_4\mathbf{f} = {}_3\mathbf{f}$.

Third cycle. Does $_{2p}\mathbf{f} = {}_{p}\mathbf{f}$? If the second cycle did not modify our vector of estimates, we have completed the estimation process. For our 2×2 example, this is true and $_A\mathbf{f}' = (10,\ 15,\ 10,\ 15)$. Had the vector been modified, we would cycle through the p steps once again.

As noted, for the gender $\times$ marital status data, convergence was achieved by the end of the first cycle. The second cycle served simply to verify that all constraints were satisfied by the vector produced in the first cycle.

13.2.7 Log Linear Computations

We have obtained an estimator that, among the class of those satisfying $_A\mathbf{T}'\,{}_A\mathbf{f} = {}_A\mathbf{T}'\mathbf{f}$, is closest to the least constrained estimator. Normally, our interest in this estimator is twofold. First, and perhaps foremost, we want to know whether or how well a model including only these constraints fits an observed set of frequencies.

The DI statistic could be computed from (13.2.28):

$$\begin{aligned} \mathrm{DI}(\mathbf{f} : {}_A\mathbf{f}) &= \sum_w f_w \log(f_w / {}_A f_w) \\ &= 4 \log \tfrac{4}{10} + 16 \log \tfrac{16}{15} + 16 \log \tfrac{16}{10} + 14 \log \tfrac{14}{15} \\ &= 3.9216113 \end{aligned}$$

Under the null hypothesis that fixing $_A\mathbf{T}'\mathbf{f}$ is sufficient to describe the populations from which we sampled, $2\mathrm{DI}(\mathbf{f} : {}_A\mathbf{f})$ may be taken as an observation from the chi-squared distribution with $W - p = 4 - 2 = 2$ degrees of freedom. Because $2\mathrm{DI} = 7.84$ and the value of chi-squared with 2 df significant at the 0.05 level is only 5.99, we conclude that a model that constrains only the numbers of male and female respondents is not sufficient to account for the observed gender $\times$ marital status frequencies.

A second computation gives us the log linear parameters from (13.2.31),

$$\log({}_A f_w) = \log({}_L f_w) + \sum_{i=1}^{p} \lambda_i T_{wi}$$

For our 2×2 example,

$$\begin{bmatrix} \log 10 \\ \log 15 \\ \log 10 \\ \log 15 \end{bmatrix} = \begin{bmatrix} \log \frac{50}{4} + \lambda_1 \\ \log \frac{50}{4} + \lambda_2 \\ \log \frac{50}{4} + \lambda_1 \\ \log \frac{50}{4} + \lambda_2 \end{bmatrix}$$

or $\lambda_1 = \log \frac{40}{50} = -0.22314$ and $\lambda_2 = \log \frac{60}{50} = 0.18232$. If we redefine $\log({}_{\text{L}}f_w) = \log \frac{50}{4} = 2.52573$ as λ_0, we can write

$$\log({}_A\mathbf{f}) = \begin{bmatrix} \lambda_0 + \lambda_1 \\ \lambda_0 + \lambda_2 \\ \lambda_0 + \lambda_1 \\ \lambda_0 + \lambda_2 \end{bmatrix} = \begin{bmatrix} 2.30259 \\ 2.70805 \\ 2.30259 \\ 2.70805 \end{bmatrix}$$

A glance at Eq. (13.2.32) tells us that the log linear model provides not only for representing the log of an estimated frequency as the weighted sum of parameters, but also for representing the estimated frequency itself as a product of parametric terms. In the present notation,

$${}_A\mathbf{f} = \begin{bmatrix} e^{\lambda_0}e^{\lambda_1} \\ e^{\lambda_0}e^{\lambda_2} \\ e^{\lambda_0}e^{\lambda_1} \\ e^{\lambda_0}e^{\lambda_2} \end{bmatrix} = \begin{bmatrix} (\frac{50}{4})(\frac{40}{50}) \\ (\frac{50}{4})(\frac{60}{50}) \\ (\frac{50}{4})(\frac{40}{50}) \\ (\frac{50}{4})(\frac{60}{50}) \end{bmatrix} = \begin{bmatrix} 10 \\ 15 \\ 10 \\ 15 \end{bmatrix}$$

We call the λ_i the additive or linear parameters and the e^{λ_i} the multiplicative parameters of the log linear model. The linear parameters model the logarithms of expected cell frequencies, and the multiplicative parameters relate directly to these expectations. This distinction suggests that the e^{λ_i} might be the more attractive. However, because of the way in which constraints on expected frequencies are phrased in the log linear model, the hypotheses tested by this approach are, effectively, hypotheses about the λ_i.

13.2.8 Partitioning the Discrimination Information Statistic

Before summarizing the internal constraints version of the log linear model and illustrating its classical use with cross classified frequencies, we need to say a bit more about the generalized discrimination information statistic. Consider anew a W-element sample vector of observed frequencies $\mathbf{f}$ and four internally constrained estimators of the expectations of those frequencies, ${}_{\text{L}}\mathbf{f}$, ${}_A\mathbf{f}$, ${}_B\mathbf{f}$, and ${}_{\text{F}}\mathbf{f}$ $(= \mathbf{f})$, where ${}_{\text{L}}\mathbf{f}$ and ${}_{\text{F}}\mathbf{f}$ are the least and fully constrained estimators and ${}_A\mathbf{f}$ and ${}_B\mathbf{f}$ are two partially constrained estimators, which satisfy

1. ${}_B\mathbf{T}'\,{}_B\mathbf{f} = {}_B\mathbf{T}'\mathbf{f}$ (implies ${}_A\mathbf{T}'\,{}_B\mathbf{f} = {}_A\mathbf{T}'\mathbf{f}$) and
2. ${}_A\mathbf{T}'\,{}_A\mathbf{f} = {}_A\mathbf{T}'\mathbf{f}$ (implies $\mathbf{U}'\,{}_A\mathbf{f} = \mathbf{U}'\mathbf{f}$),

where ${}_B\mathbf{T}$, ${}_A\mathbf{T}$, and $\mathbf{U} = {}_{\text{L}}\mathbf{T}$ are the defining constraints for model B, model A, and the least constrained model, respectively. In other words, the four estimators are ordered, left to right, from least to most constrained, with each successive estimator satisfying all of the lower constraints plus some new ones. In particular, model B requires that the constraints of model A be satisfied plus one or more in addition.

If we specify further that the ranks of $\mathbf{U}$, ${}_A\mathbf{T}$, and ${}_B\mathbf{T}$ are u, p, and q, respectively, satisfying the inequalities $u < p < q < W$, then we know that we can form a number of DI statistics, each of which has asymptotically a central chi-squared distribution:

a. $2\mathrm{DI}({}_{\mathrm{F}}\mathbf{f} : {}_B\mathbf{f}) \sim \chi^2_{(W-q)\text{ df}}$ when ${}_B\mathbf{T}'\mathbf{f}$ is fixed;
b. $2\mathrm{DI}({}_{\mathrm{F}}\mathbf{f} : {}_A\mathbf{f}) \sim \chi^2_{(W-p)\text{ df}}$ and $2\mathrm{DI}({}_B\mathbf{f} : {}_A\mathbf{f}) \sim \chi^2_{(q-p)\text{ df}}$ when ${}_A\mathbf{T}'\mathbf{f}$ is fixed;
c. $2\mathrm{DI}({}_{\mathrm{F}}\mathbf{f} : {}_{\mathrm{L}}\mathbf{f}) \sim \chi^2_{(W-u)\text{ df}}$, $2\mathrm{DI}({}_B\mathbf{f} : {}_{\mathrm{L}}\mathbf{f}) \sim \chi^2_{(q-u)\text{ df}}$, and $2\mathrm{DI}({}_A\mathbf{f} : {}_{\mathrm{L}}\mathbf{f}) \sim \chi^2_{(p-u)\text{ df}}$ when $\mathbf{U}'\mathbf{f}$ is fixed.

We now develop the relations among these statistics through a numerical example. For the gender × marital status example, ${}_{\mathrm{L}}\mathbf{f}' = (12.5, 12.5, 12.5, 12.5)$, ${}_A\mathbf{f}' = (10, 15, 10, 15)$, and ${}_{\mathrm{F}}\mathbf{f}' = \mathbf{f}' = (4, 16, 16, 14)$. To complement these, we add a fourth estimator based on a model B that imposes more constraints than did model A. Model A required that the estimators provide the same totals by gender of respondent as did the observed frequencies. A more constrained model B would be one that requires that the estimators replicate the observed totals not only by gender but by marital status as well. That is, we look for a set of estimators that have these fixed marginal sums:

	Married	Single	
Female	${}_Bf_1$	${}_Bf_2$	20
Male	${}_Bf_3$	${}_Bf_4$	30
	20	30	50

We can represent this by the set of constraints ${}_B\mathbf{T}'\,{}_B\mathbf{f} = {}_B\mathbf{T}'\mathbf{f}$ given by

$$\begin{bmatrix} 1 & 0 & 1 & 0 \\ 0 & 1 & 0 & 1 \\ 1 & 1 & 0 & 0 \\ 0 & 0 & 1 & 1 \end{bmatrix} {}_B\mathbf{f} = \begin{bmatrix} 1 & 0 & 1 & 0 \\ 0 & 1 & 0 & 1 \\ 1 & 1 & 0 & 0 \\ 0 & 0 & 1 & 1 \end{bmatrix} \mathbf{f} = \begin{bmatrix} 20 \\ 30 \\ 20 \\ 30 \end{bmatrix}$$

The reader should apply the iterative proportional fitting algorithm to this set of constraints, beginning with ${}_{\mathrm{L}}\mathbf{f}$, to obtain ${}_B\mathbf{f}$. The resulting vector will be ${}_B\mathbf{f}' = (18, 12, 12, 18)$. From these results we compute the DI statistic for our data

$$\begin{aligned} \mathrm{DI}(\mathbf{f} : {}_B\mathbf{f}) &= \sum_w \left[f_w \log\left(\frac{f_w}{{}_Bf_w} \right) \right] \\ &= 4\log\tfrac{4}{8} + 16\log\tfrac{16}{12} + 16\log\tfrac{16}{12} + 14\log\tfrac{14}{18} \\ &= 2.9148357 \end{aligned}$$

which provides a measure of the fit of our model B to the observed frequencies.

It is important to compare this fit with that of model A, which constrained only the gender of respondent totals. Because model B contains more of the information in $\mathbf{f}$, it provides a closer fit—but how much closer?

We answer this question by first finding the difference between the two DI statistics. Algebraically, we form

$$\begin{aligned}
\mathrm{DI}(f : {}_A\mathbf{f}) - \mathrm{DI}(\mathbf{f} : {}_B\mathbf{f}) &= \sum_w f_w \log\left(\frac{f_w}{{}_Af_w}\right) - \sum_w f_w \log\left(\frac{f_w}{{}_Bf_w}\right) \\
&= \sum_w f_w \left[\log\left(\frac{f_w}{{}_Af_w}\right) - \log\left(\frac{f_w}{{}_Bf_w}\right)\right] \\
&= \sum_w f_w \log\left[\left(\frac{f_w}{{}_Af_w}\right)\left(\frac{{}_Bf_w}{f_w}\right)\right] \\
&= \sum_w \left[f_w \log\left(\frac{{}_Bf_w}{{}_Af_w}\right)\right] \qquad (13.2.33)
\end{aligned}$$

For the 2×2 example,

$$\mathrm{DI}(\mathbf{f} : {}_A\mathbf{f}) - \mathrm{DI}(\mathbf{f} : {}_B\mathbf{f}) = 4 \log \tfrac{8}{10} + 16 \log \tfrac{12}{15} + 16 \log \tfrac{12}{10} + 14 \log \tfrac{18}{15}$$

which can be written

$$\begin{aligned}
\mathrm{DI}(\mathbf{f} : {}_A\mathbf{f}) - \mathrm{DI}(\mathbf{f} : {}_B\mathbf{f}) &= 8 \log \tfrac{8}{10} + 12 \log \tfrac{12}{15} + 12 \log \tfrac{12}{10} + 18 \log \tfrac{18}{15} \\
&\quad - 4 \log \tfrac{8}{10} + 4 \log \tfrac{12}{15} + 4 \log \tfrac{12}{10} - 4 \log \tfrac{18}{15}
\end{aligned}$$

The purpose of expressing it in this form is that the first four terms on the right are now

$$\mathrm{DI}({}_B\mathbf{f} : {}_A\mathbf{f}) = \sum_w \left[{}_Bf_w \log\left(\frac{{}_Bf_w}{{}_Af_w}\right)\right]$$

which allows us to write

$$\begin{aligned}
\mathrm{DI}(\mathbf{f} : {}_A\mathbf{f}) - \mathrm{DI}(\mathbf{f} : {}_B\mathbf{f}) &= \mathrm{DI}({}_B\mathbf{f} : {}_A\mathbf{f}) + 4 \log\left[\left(\tfrac{10}{8}\right)\left(\tfrac{12}{15}\right)\left(\tfrac{12}{10}\right)\left(\tfrac{15}{18}\right)\right] \\
&= \mathrm{DI}({}_B\mathbf{f} : {}_A\mathbf{f}) + 4 \log 1 \\
&= \mathrm{DI}({}_B\mathbf{f} : {}_A\mathbf{f}) \qquad (13.2.34)
\end{aligned}$$

We have illustrated Eq. (13.2.34) for our numerical example, but it is a quite general result. The constraints imposed by model B are sufficient to ensure that

$$\sum_w \left[f_w \log\left(\frac{{}_Bf_w}{{}_Af_w}\right)\right] = \sum_w \left[{}_Bf_w \log\left(\frac{{}_Bf_w}{{}_Af_w}\right)\right]$$

whenever the model A constraints are contained in the model B constraints in the sense described earlier in this section.

The terms of (13.2.34) may be rearranged to give an important log linear analysis relation:

$$\mathrm{DI}(\mathbf{f} : {}_A\mathbf{f}) = \mathrm{DI}(\mathbf{f} : {}_B\mathbf{f}) + \mathrm{DI}({}_B\mathbf{f} : {}_A\mathbf{f}) \qquad (13.2.35)$$

Equation (13.2.35) indicates that the fit of model A to the data may be partitioned into two parts: the fit of (the more constrained) model B to the data and the fit of model A to model B. We need to keep in mind that model B is closer to the data than is model A. The following sketch illustrates the decomposition:

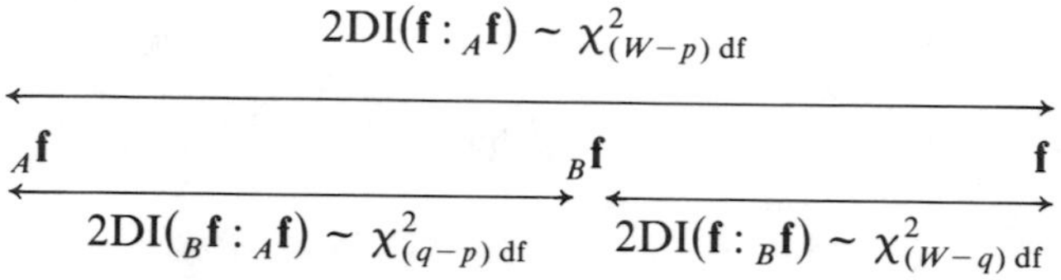

Since each of the 2DI statistics is an observation from a chi-squared distribution, we have here a partitioning of tests of significance of the model fit as well. Equation (13.2.35) generalizes to deeper hierarchies. For the present example we could write

$$\mathrm{DI}(\mathbf{f} : {}_L\mathbf{f}) = \mathrm{DI}(\mathbf{f} : {}_B\mathbf{f}) + \mathrm{DI}({}_B\mathbf{f} : {}_A\mathbf{f}) + \mathrm{DI}({}_A\mathbf{f} : {}_L\mathbf{f})$$

partitioning the (lack of) fit of the least constrained estimators to the observed frequencies into three nonoverlapping parts. Each of these three parts might be the subject of a hypothesis test.

So far we have discussed the internal constraints model in quite general terms. Ways in which alternative models can be developed to provide useful tests of hypotheses are discussed in Section 13.3.

13.3 Hierarchical Internal Constraints Problem

If $\mathbf{f}$ is a W-element vector of observed frequencies generated by one of the three sampling models described earlier, the log linear internal contraints analysis allows us to investigate how well these data are fit by an estimation vector $\tilde{\mathbf{f}} = {}_A\mathbf{f}$ whose natural logarithm, another W-element vector, is structured as

$$\log(\tilde{\mathbf{f}}) = \mathbf{T}\boldsymbol{\lambda} \qquad (13.3.1)$$

where $\boldsymbol{\lambda}$ is a p-element vector of (usually) unknown parameters and $\mathbf{T}$ is a $W \times p$ *design matrix* selected by the researcher: The estimation vector $\tilde{\mathbf{f}}$ must satisfy the internal constraints

$$\mathbf{T}'\tilde{\mathbf{f}} = \mathbf{T}'\mathbf{f} \qquad (13.3.2)$$

Equation (13.3.1) is the scalar equation (13.2.31) written in matrix form, and (13.3.2) is simply a restatement of the internal constraints condition of, for example, (13.2.19). Written in this compact form, the p constraints on $\tilde{\mathbf{f}}$ imposed by (13.3.2) include both *sample constraints* (those necessary for the least constrained estimator) and *hypothesis constraints* (the additional constraints necessary for model A).

The design matrix $\mathbf{T}$ is so named because it describes the number and kind of internal constraints to be imposed on the vector of cell frequency estimates; equivalently, it describes the number of $\boldsymbol{\lambda}$ parameters and the linear weightings of these necessary to determine the logarithm of the vector of frequency estimates. There is a certain parallel with the design matrix of the general linear model, the matrix $\mathbf{X}$ of $\mathbf{E}(\mathbf{Y}) = \mathbf{X}\boldsymbol{\beta}$.

When we explored the GLM in Chapter 8, we considered several choices of the design matrix for categorical predictors, where, for example, the independent variables assigned entities to treatments. Depending on the choice of coding, we saw that the elements of the vector $\boldsymbol{\beta}$ had different interpretations. Here too the matrix $\mathbf{T}$ and the vector $\boldsymbol{\lambda}$ are interdependent. Rather than explore different choices for $\mathbf{T}$, however, we focus on the particular class of design matrices and their associated $\boldsymbol{\lambda}$ that are appropriate to fully crossed or cross classified observations. This tabular arrangement of frequencies is studied more often than other designs by log linear analysis.

13.3.1 Factorial Design

Log linear analysis of cross classified frequencies is better understood, perhaps, by comparison to the ANOVA approach to the factorial experimental design.

For discussion purposes we assume a three-factor design with the factors A, B, and C fully crossed with I, J, and K levels, respectively. An independent sample is drawn from each of the IJK populations, and we wish to test hypotheses about or estimate contrasts among the means of these populations. A typical population mean is symbolized $E(\mathbf{Y}|A_i, B_j, C_k)$. The full ANOVA linear model assumes this to mean to be a linear function of a set of underlying parameters:

$$E(\mathbf{Y}|A_i, B_j, C_k) = \mu_{...} + \mu_{i..} + \mu_{.j.} + \mu_{..k} + \mu_{ij.} + \mu_{i.k} + \mu_{.jk} + \mu_{ijk} \tag{13.3.3}$$

where

$$\mu_{...} = \left(\frac{1}{IJK}\right) \sum_i \sum_j \sum_k E(\mathbf{Y}|A_i, B_j, C_k) \tag{13.3.4}$$

$$\mu_{i..} = \left(\frac{1}{JK}\right) \sum_j \sum_k E(\mathbf{Y}|A_i, B_j, C_k) - \mu_{...} \tag{13.3.5}$$

$$\mu_{ij.} = \left(\frac{1}{K}\right)\sum_k E(\mathbf{Y}|A_i, B_j, C_k) - (\mu_{...} + \mu_{i..} + \mu_{.j.}) \tag{13.3.6}$$

$$\mu_{ijk} = E(\mathbf{Y}|A_i, B_j, C_k) - (\mu_{...} + \mu_{i..} + \mu_{.j.} + \mu_{..k} + \mu_{ij.} + \mu_{i.k} + \mu_{.jk}) \tag{13.3.7}$$

with corresponding definitions for the other μ terms.

These defined *effects* provide sufficient restrictions on the μ,

$$\sum_i \mu_{i..} = \sum_j \mu_{.j.} = \sum_k \mu_{..k} = \sum_i \mu_{ij.} = \sum_j \mu_{ij.} = \sum_i \mu_{i.k}$$
$$= \sum_k \mu_{i.k} = \sum_j \mu_{.jk} = \sum_k \mu_{.jk} = \sum_i \mu_{ijk} = \sum_j \mu_{ijk} = \sum_k \mu_{ijk} = 0 \tag{13.3.8}$$

that there remain, in the full model, exactly *IJK* indeterminate μ elements to be estimated, based on the criterion data for the *IJK* cells of the factorial design. In this and subsequent sections we discuss *explicitly* the *IJK* factorial design or the *IJK* frequency table. Unless we indicate otherwise, these discussions extend to lower- and higher-order designs with appropriate changes in notation.

Restricted models for the factorial design are models of the form of (13.3.3) but with one or more of the μ terms dropped (in GLM terms, set equal to zero). A large number of models are possible. Here, to facilitate comparison with the log linear analysis, we consider only restricted models that are *hierarchical* in the sense that Fienberg uses the term: A hierarchical model is one that includes all effects that are "nested within" each of its highest-order interactions. A model that includes $\mu_{ij.}$, for example, must also include $\mu_{i..}$, $\mu_{.j.}$, and $\mu_{...}$. (The definition of $\mu_{ij.}$ suggests its dependence on these lower-order μ terms.)

An additional advantage of considering only hierarchical models is that such restricted models are relatively easy to denote. It is common to specify only the highest-order interactions to be included, all nested terms being inferred. For our *IJK* factorial design, these are the possible hierarchical models:

1. [*ABC*], the full model;
2. [*AB*, *AC*, *BC*], which omits μ_{ijk};
3. [*AB*, *AC*], which omits μ_{ijk} and $\mu_{.jk}$;
4. [*AB*, *BC*], which omits μ_{ijk} and $\mu_{i.j}$;
5. [*AC*, *BC*], which omits μ_{ijk} and $\mu_{.jk}$;
6. [*AB*, *C*], which includes only one interaction term, $\mu_{ij.}$;
7. [*AC*, *B*], a second one-interaction model;
8. [*BC*, *A*], the third one-interaction model;
9. [*A*, *B*, *C*], with no interactions but all three factors;

10. $[AB]$, with no C factor but an AB interaction postulated;
11. $[AC]$, with no B factor but an AC interaction;
12. $[BC]$, the third two-factor with interaction model;
13. $[A, B]$, with A and B in an additive model;
14. $[A, C]$, with A and C additive;
15. $[B, C]$, with B and C additive;
16. $[A]$, $\mu_{...} + \mu_{i..}$;
17. $[B]$, $\mu_{...} + \mu_{.j.}$;
18. $[C]$, $\mu_{...} + \mu_{..k}$; and
19. [.], $\mu_{...}$ alone in the model.

These 19 models cannot be ordered, of course, in a single hierarchical order, either in the GLM sense of a higher-order model that includes all of the effects in lower-order models or in the closely related sense, given in Section 13.2, of nested constraints. Such hierarchies of hierarchical models can be constructed, however. One such hierarchy for the three-factor design is $[ABC]$, $[AB, AC, BC]$, $[AB, AC]$, $[AB, C]$, $[A, B, C]$, $[A, B]$, $[A]$, and [.]. As noted in Chapters 8 and 9, the choice of a particular hierarchy should reflect the sequence of hypotheses that is most justified theoretically.

A sequence of F ratio tests associated with this particular hierarchy of restricted models has a common error term or denominator $\mathbf{e}'\mathbf{e}/(n - IJK)$, where $\mathbf{e}'\mathbf{e}$ is the sum of squares of errors for the full model, and numerator terms as shown in Table 13.1. This sequence of hypothesis tests is discontinued, of course, whenever the proposal of a more restricted model leads to a significant increase in the sum of squares of errors. This would indicate a worsening fit of the data by the model.

One GLM representation of this design is provided by *effect coding*. Assume factor A at three levels ($I = 3$) and factors B and C each at two levels ($J = K = 2$); then an appropriate design template ${}_{\mathrm{T}}\mathbf{X}$ would be

$$
\begin{array}{l}
A_1B_1C_1\\ A_1B_1C_2\\ A_1B_2C_1\\ A_1B_2C_2\\ A_2B_1C_1\\ A_2B_1C_2\\ A_2B_2C_1\\ A_2B_2C_2\\ A_3B_1C_1\\ A_3B_1C_2\\ A_3B_2C_1\\ A_3B_2C_2
\end{array}
\begin{bmatrix}
1 & 1 & 0 & 1 & 1 & 1 & 0 & 1 & 0 & 1 & 1 & 0\\
1 & 1 & 0 & 1 & -1 & 1 & 0 & -1 & 0 & -1 & -1 & 0\\
1 & 1 & 0 & -1 & 1 & -1 & 0 & 1 & 0 & -1 & -1 & 0\\
1 & 1 & 0 & -1 & -1 & -1 & 0 & -1 & 0 & 1 & 1 & 0\\
1 & 0 & 1 & 1 & 1 & 0 & 1 & 0 & 1 & 1 & 0 & 1\\
1 & 0 & 1 & 1 & -1 & 0 & 1 & 0 & -1 & -1 & 0 & -1\\
1 & 0 & 1 & -1 & 1 & 0 & -1 & 0 & 1 & -1 & 0 & -1\\
1 & 0 & 1 & -1 & -1 & 0 & -1 & 0 & -1 & 1 & 0 & 1\\
1 & -1 & -1 & 1 & 1 & -1 & -1 & -1 & -1 & 1 & -1 & -1\\
1 & -1 & -1 & 1 & -1 & -1 & -1 & 1 & 1 & -1 & 1 & 1\\
1 & -1 & -1 & -1 & 1 & 1 & 1 & -1 & -1 & -1 & 1 & 1\\
1 & -1 & -1 & -1 & -1 & 1 & 1 & 1 & 1 & 1 & -1 & -1
\end{bmatrix}
$$

Table 13.1

Model	Error SS	Numerator of F ratio	Tests
$[AB, AC, BC]$	$\mathbf{f'f}$	$(\mathbf{f'f} - \mathbf{e'e})/(I-1)(J-1)(K-1)$	μ_{ijk}
$[AB, AC]$	$\mathbf{g'g}$	$(\mathbf{g'g} - \mathbf{f'f})/(J-1)(K-1)$	$\mu_{.jk}$
$[AB, C]$	$\mathbf{h'h}$	$(\mathbf{h'h} - \mathbf{g'g})/(I-1)(K-1)$	$\mu_{i.k}$
$[A, B, C]$	$\mathbf{j'j}$	$(\mathbf{j'j} - \mathbf{h'h})/(I-1)(J-1)$	$\mu_{ij.}$
$[A, B]$	$\mathbf{k'k}$	$(\mathbf{k'k} - \mathbf{j'j})/(K-1)$	$\mu_{..k}$
$[A]$	$\mathbf{m'm}$	$(\mathbf{m'm} - \mathbf{k'k})/(J-1)$	$\mu_{.j.}$
$[.]$	$\mathbf{n'n}$	$(\mathbf{n'n} - \mathbf{m'm})/(I-1)$	$\mu_{i..}$

where, for each row, the population sampled is identified to the left. Based on the relation ${}_T\mathbf{X}\boldsymbol{\beta} = {}_T\mathbf{E}(\mathbf{Y})$, elements of the vector are

$$\beta_1 = \mu_{...}$$

$$\beta_2 = \mu_{1..}$$

$$\beta_3 = \mu_{2..}, \qquad \mu_{3..} = -(\mu_{1..} + \mu_{2..}) = -(\beta_2 + \beta_3)$$

$$\beta_4 = \mu_{.1.}, \qquad \mu_{.2.} = -\mu_{.1.} = -\beta_4$$

$$\beta_5 = \mu_{..1}, \qquad \mu_{..2} = -\mu_{..1} = -\beta_5$$

$$\beta_6 = \mu_{11.}, \qquad \mu_{12.} = -\beta_6$$

$$\beta_7 = \mu_{21.}, \qquad \mu_{22.} = -\beta_7, \quad \mu_{31.} = -(\mu_{11.} + \mu_{21.}) = -(\beta_6 + \beta_7)$$

$$\mu_{32.} = \beta_6 + \beta_7$$

$$\beta_8 = \mu_{1.1}, \qquad \mu_{1.2} = -\beta_8$$

$$\beta_9 = \mu_{2.1}, \qquad \mu_{2.2} = -\beta_9, \quad \mu_{3.1} = -(\mu_{1.1} + \mu_{2.1}) = -(\beta_8 + \beta_9)$$

$$\mu_{3.2} = \beta_8 + \beta_9$$

$$\beta_{10} = \mu_{.11}, \qquad \mu_{.21} = \mu_{.12} = -\beta_{10}, \quad \mu_{.22} = \beta_{10}$$

$$\beta_{11} = \mu_{111}, \qquad \mu_{121} = \mu_{112} = -\beta_{11}, \quad \mu_{122} = \beta_{11}$$

$$\beta_{12} = \mu_{211}, \qquad \mu_{221} = \mu_{212} = -\beta_{12}, \quad \mu_{222} = \beta_{12}$$

$$\mu_{311} = -(\beta_{11} + \beta_{12})$$

$$\mu_{321} = \mu_{312} = \beta_{11} + \beta_{12}, \quad \mu_{322} = -(\beta_{11} + \beta_{12})$$

The series of sums of squares of errors given for these models involves the constraints on $\boldsymbol{\beta}$ given in Table 13.2. The $\mathbf{C}$ matrices required for each of the sets of constraints $\mathbf{C}'\boldsymbol{\beta} = \boldsymbol{\Gamma}$ are easily specified.

Table 13.2

Sum of squares of errors	Constraint on $\boldsymbol{\beta}$
$\mathbf{f}'\mathbf{f}$	$\beta_{11} = \beta_{12} = 0$
$\mathbf{g}'\mathbf{g}$	β_{10}–β_{12}, all 0
$\mathbf{h}'\mathbf{h}$	β_8–β_{12}, all 0
$\mathbf{j}'\mathbf{j}$	β_6–β_{12}, all 0
$\mathbf{k}'\mathbf{k}$	β_5–β_{12}, all 0
$\mathbf{m}'\mathbf{m}$	β_4–β_{12}, all 0
$\mathbf{n}'\mathbf{n}$	β_2–β_{12}, all 0

13.3.2 Log Linear Model for Cross Classified Frequencies

The *full log linear model* for observations cross classified into a set of $I \times J \times K$ frequencies can be written

$$\log E(f_{ijk}) = \lambda_{...} + \lambda_{i..} + \lambda_{.j.} + \lambda_{..k} + \lambda_{ij.} + \lambda_{i.k} + \lambda_{.jk} + \lambda_{ijk} \tag{13.3.9}$$

where

$$\lambda_{...} = \left(\frac{1}{IJK}\right) \sum_{i,j,k} \log E(f_{ijk}) \tag{13.3.10}$$

$$\lambda_{i..} = \left(\frac{1}{JK}\right) \sum_{j,k} \log E(f_{ijk}) - \lambda_{...} \tag{13.3.11}$$

$$\lambda_{ij.} = \left(\frac{1}{K}\right) \sum_{k} \log E(f_{ijk}) - (\lambda_{...} + \lambda_{i..} + \lambda_{.j.}) \tag{13.3.12}$$

$$\lambda_{ijk} = \log E(f_{ijk}) - (\lambda_{...} + \lambda_{i..} + \lambda_{.j.} + \lambda_{..k} + \lambda_{ij.} + \lambda_{i.k} + \lambda_{.jk}) \tag{13.3.13}$$

with corresponding definitions for the other λ parameters.

This full log linear model is nearly identical to the full model for the factorial design as given in Eqs. (13.3.3)–(13.3.7). The expectations (of a family of normal distributions) have been replaced by the logarithms of expectations (of frequencies generated by one of the three processes described earlier in the chapter) and the μ parameters replaced by λ. The coding of effects is essentially the same, λ being constrained in the full model by the equivalent of the restrictions given for the μ by Eq. (13.3.8).

The full model log linear representation of cross classified frequencies, Eq. (13.3.9), indicates that the *logarithm* of the expected frequency for any cell is *linear* in the effects of the cross classifying dimensions and their interactions. Appropriate restricted models are those in which some of the dimensions and their interactions are dropped from this log expectation equation, maintaining a *hierarchy* of effects. All hierarchical restricted

Table 13.3

Model	λ terms
$[ABC]$	the full model
$[AB, AC, BC]$	no λ_{ijk} term
$[AB, AC]$	neither λ_{ijk} nor $\lambda_{.jk}$
$[AB, C]$	$\lambda_{ij.}$ the only interaction term
$[A, B, C]$	no interaction terms
$[A, B]$	$\lambda_{...} + \lambda_{i..} + \lambda_{.j.}$
$[A]$	$\lambda_{...} + \lambda_{i..}$
[.]	$\lambda_{...}$ alone in the model

models are consistent with the Poisson and multinomial sampling models as the presence of $\lambda_{...}$ in each model is sufficient to ensure that the least constrained model is included in each restricted model. If the observed frequencies are the result of product multinomial sampling, certain marginal frequencies are fixed and the corresponding effects must be retained in each restricted model. For example, if $f_{i..} = \Sigma_j \Sigma_k f_{ijk}$ is fixed for $i = 1, 2, \ldots, I$ by the choice of sampling design, then the $\lambda_{i..}$ terms must be present in each restricted model. This ensures the least constrained model will be included in each restricted model.

Assuming multinomial sampling, the 19 models listed in Section 13.3.1 for the IJK factorial design are also the hierarchical restricted models for the IJK cross classification of frequencies. We need only replace references to the μ with references to λ. In particular, we could study the same hierarchical arrangement of these (hierarchical) restricted models. (See Table 13.3.) A sequence of tests of goodness of fit to the observations of these restricted models is given by the series of DI statistics in Table 13.4.

If, to further the comparison with the factorial design, we assume that factors A, B, and C are represented again at 3, 2, and 2 levels, the general

Table 13.4

Model	Goodness of fit	Sequential χ^2 test
$[ABC]$	perfect	$\mathbf{f} = {}_{\mathrm{F}}\mathbf{f}$
$[AB, AC, BC]$	$\mathrm{DI}(\mathbf{f} : {}_A\mathbf{f})$	$2\mathrm{DI}(\mathbf{f} : {}_A\mathbf{f}) \sim \chi^2_{(I-1)(J-1)(K-1)\ \mathrm{df}}$
$[AB, AC]$	$\mathrm{DI}(\mathbf{f} : {}_B\mathbf{f})$	$2\mathrm{DI}(\mathbf{f} : {}_B\mathbf{f}) - 2\mathrm{DI}(\mathbf{f} : {}_A\mathbf{f}) = 2\mathrm{DI}({}_A\mathbf{f} : {}_B\mathbf{f}) \sim \chi^2_{(J-1)(K-1)\ \mathrm{df}}$
$[AB, C]$	$\mathrm{DI}(\mathbf{f} : {}_C\mathbf{f})$	$2\mathrm{DI}({}_B\mathbf{f} : {}_C\mathbf{f}) \sim \chi^2_{(I-1)(K-1)\ \mathrm{df}}$
$[A, B, C]$	$\mathrm{DI}(\mathbf{f} : {}_D\mathbf{f})$	$2\mathrm{DI}({}_C\mathbf{f} : {}_D\mathbf{f}) \sim \chi^2_{(I-1)(J-1)\ \mathrm{df}}$
$[A, B]$	$\mathrm{DI}(\mathbf{f} : {}_E\mathbf{f})$	$2\mathrm{DI}({}_D\mathbf{f} : {}_E\mathbf{f}) \sim \chi^2_{(K-1)\ \mathrm{df}}$
$[A]$	$\mathrm{DI}(\mathbf{f} : {}_G\mathbf{f})$	$2\mathrm{DI}({}_E\mathbf{f} : {}_G\mathbf{f}) \sim \chi^2_{(J-1)\ \mathrm{df}}$
[.]	$\mathrm{DI}(\mathbf{f} : {}_{\mathrm{L}}\mathbf{f})$	$2\mathrm{DI}({}_G\mathbf{f} : {}_{\mathrm{L}}\mathbf{f}) \sim \chi^2_{(I-1)\ \mathrm{df}}$

log linear relation Eq. (13.3.1), $\log(\tilde{\mathbf{f}}) = \mathbf{T}\boldsymbol{\lambda}$, can be satisfied by the following choices for $\mathbf{T}$ and $\boldsymbol{\lambda}$. First, we take $\mathbf{T}$, the full model design matrix, to be identical to ${}_T\mathbf{X}$, the design template we used for effect coding the factorial design of Section 13.3.1:

$$
\begin{array}{l}
A_1B_1C_1\\ A_1B_1C_2\\ A_1B_2C_1\\ A_1B_2C_2\\ A_2B_1C_1\\ A_2B_1C_2\\ A_2B_2C_1\\ A_2B_2C_2\\ A_3B_1C_1\\ A_3B_1C_2\\ A_3B_2C_1\\ A_3B_2C_2\\ \\
\end{array}
\begin{array}{c}
\left[\begin{array}{rrrrrrrrrrrr}
1 & 1 & 0 & 1 & 1 & 1 & 0 & 1 & 0 & 1 & 1 & 0\\
1 & 1 & 0 & 1 & -1 & 1 & 0 & -1 & 0 & -1 & -1 & 0\\
1 & 1 & 0 & -1 & 1 & -1 & 0 & 1 & 0 & -1 & -1 & 0\\
1 & 1 & 0 & -1 & -1 & -1 & 0 & -1 & 0 & 1 & 1 & 0\\
1 & 0 & 1 & 1 & 1 & 0 & 1 & 0 & 1 & 1 & 0 & 1\\
1 & 0 & 1 & 1 & -1 & 0 & 1 & 0 & -1 & -1 & 0 & -1\\
1 & 0 & 1 & -1 & 1 & 0 & -1 & 0 & 1 & -1 & 0 & -1\\
1 & 0 & 1 & -1 & -1 & 0 & -1 & 0 & -1 & 1 & 0 & 1\\
1 & -1 & -1 & 1 & 1 & -1 & -1 & -1 & -1 & 1 & -1 & -1\\
1 & -1 & -1 & 1 & -1 & -1 & -1 & 1 & 1 & -1 & 1 & 1\\
1 & -1 & -1 & -1 & 1 & 1 & 1 & -1 & -1 & -1 & 1 & 1\\
1 & -1 & -1 & -1 & -1 & 1 & 1 & 1 & 1 & 1 & -1 & -1
\end{array}\right]\\
\begin{array}{cccccccccccc}
T_{.1} & T_{.2} & T_{.3} & T_{.4} & T_{.5} & T_{.6} & T_{.7} & T_{.8} & T_{.9} & T_{.10} & T_{.11} & T_{.12}
\end{array}
\end{array}
$$

where the cell corresponding to each row is identified to the left. The vector $\boldsymbol{\lambda}$ corresponding to this full model design has elements

$$
\begin{array}{lll}
\lambda_1 = \lambda_{...} & & \\
\lambda_2 = \lambda_{1..} & & \\
\lambda_3 = \lambda_{2..}, & \lambda_{3..} = -(\lambda_2 + \lambda_3) & \\
\lambda_4 = \lambda_{.1.}, & \lambda_{.2.} = -\lambda_4 & \\
\lambda_5 = \lambda_{..1}, & \lambda_{..2} = -\lambda_5 & \\
\lambda_6 = \lambda_{11.}, & \lambda_{12.} = -\lambda_6 & \\
\lambda_7 = \lambda_{21.}, & \lambda_{22.} = -\lambda_7, \quad \lambda_{31.} = -(\lambda_6 + \lambda_7), \quad \lambda_{32.} = \lambda_6 + \lambda_7 & \\
\lambda_8 = \lambda_{1.1}, & \lambda_{1.2} = -\lambda_8 & \\
\lambda_9 = \lambda_{2.1}, & \lambda_{2.2} = -\lambda_9, \quad \lambda_{3.1} = -(\lambda_8 + \lambda_9), \quad \lambda_{3.2} = \lambda_8 + \lambda_9 & \\
\lambda_{10} = \lambda_{.11}, & \lambda_{.12} = \lambda_{.21} = -\lambda_{10}, \quad \lambda_{.22} = \lambda_{10} & \\
\lambda_{11} = \lambda_{111}, & \lambda_{121} = \lambda_{112} = -\lambda_{11}, \quad \lambda_{122} = \lambda_{11} & \\
\lambda_{12} = \lambda_{211}, & \lambda_{221} = \lambda_{212} = -\lambda_{12}, \quad \lambda_{222} = \lambda_{12} & \\
 & \lambda_{311} = -(\lambda_{11} + \lambda_{12}) & \\
 & \lambda_{312} = \lambda_{321} = \lambda_{11} + \lambda_{12}, \quad \lambda_{322} = -(\lambda_{11} + \lambda_{12}) &
\end{array}
$$

Table 13.5 Estimators and Design Matrices

Estimator ${}_M\mathbf{f}$	Design matrix ${}_M\mathbf{T}$
${}_F\mathbf{f}\ (= \mathbf{f})$	full matrix **T**
${}_A\mathbf{f}$	columns 1–10 of **T**
${}_B\mathbf{f}$	columns 1–9 of **T**
${}_C\mathbf{f}$	columns 1–7 of **T**
${}_D\mathbf{f}$	columns 1–5 of **T**
${}_E\mathbf{f}$	columns 1–4 of **T**
${}_G\mathbf{f}$	columns 1–3 of **T**
${}_L\mathbf{f}$	column 1 of **T**

The *internal constraints* to be imposed on successive estimators ${}_M\mathbf{f}$ for model M must satisfy Eq. (13.2.19),

$$ {}_M\mathbf{T}'\,{}_M\mathbf{f} = {}_M\mathbf{T}'\mathbf{f} $$

where ${}_M\mathbf{T}$ describes a sequence of design matrices, all based on **T** (see Table 13.5).

The reader may note that column 1 of **T** imposes, through Eq. (13.2.19), the condition that the sum of the estimated cell frequencies must equal the sum of the observed cell frequencies, the *sample* or *least constraint* for multinomial sampling. Columns 1–3, similarly, ensure that the estimated frequencies at the three levels of A match the observed frequencies at those three levels. Reducing the number of columns to ${}_M\mathbf{T}$ is equivalent to reducing the number of λ parameters in the restricted log linear models. Corresponding to the hierarchy of models, interaction terms are dropped prior to major effect terms and in the prescribed sequence.

There is considerable similarity, then, of the GLM approach to factorial designs and the hierarchical internal constraints approach to the log linear analysis of cross classified frequencies.

13.3.3 Job Satisfaction Example

V. Wilson (M.S. project, University of Washington, 1980) provides an instance of the use of the log linear approach. Data are for 595 graduates of a baccalaureate program in psychology, surveyed one year following graduation. For the analyses of interest, the respondents are cross classified by *gender* ($S = 1, 2$ for male, female), rated *satisfaction with degree* studies ($D = 1, 2, 3$ for no, some, and great satisfaction), and rated *satisfaction with job* presently held ($J = 1, 2, 3$ for no, some, and great satisfaction.) The cross classified frequencies are given in Table 13.6.

The full log linear model is of the form

$$ \log({}_F f_{sdj}) = \lambda_{...} + \lambda_{s..} + \lambda_{.d.} + \lambda_{..j} + \lambda_{sd.} + \lambda_{s.j} + \lambda_{.dj} + \lambda_{sdj} $$

Table 13.6

	Gender (S)					
	males			females		
Degree satisfaction (D)	no	some	great	no	some	great
Job satisfaction (J)						
no	3	14	1	9	11	4
some	4	57	17	5	84	26
great	4	75	44	3	98	58

briefly represented $[SDJ]$. For our purposes we take job satisfaction as a criterion measure. The questions of interest are the extent to which the gender of respondent and respondent's rated satisfaction with degree influence, separately or in interaction, the criterion. Since only the interaction terms involving J reflect these influences, there is little to be gained from restricting the log linear model beyond

$$[SD, J]: \qquad \log({}_{L}f_{sdj}) = \lambda_{...} + \lambda_{s..} + \lambda_{.d.} + \lambda_{..j} + \lambda_{sd.}$$

Between these fully and minimally constrained models lie three others that are of interest:

$$[SD, SJ, DJ]: \quad \log({}_{A}f_{sdj}) = \lambda_{...} + \lambda_{s..} + \lambda_{.d.} + \lambda_{..j} + \lambda_{sd.} + \lambda_{s.j} + \lambda_{.dj}$$

$$[SD, DJ]: \quad \log({}_{B}f_{sdj}) = \lambda_{...} + \lambda_{s..} + \lambda_{.d.} + \lambda_{..j} + \lambda_{sd.} + \lambda_{.dj}$$

$$[SD, SJ]: \quad \log({}_{C}f_{sdj}) = \lambda_{...} + \lambda_{s..} + \lambda_{.d.} + \lambda_{..j} + \lambda_{sd.} + \lambda_{s.j}$$

The first of these restricted models drops the second-order interaction term λ_{sdj}, and each of the latter two eliminates as well an interaction with job satisfaction, either $\lambda_{.dj}$ or $\lambda_{s.j}$.

The null hypothesis associated with the first restricted model can be phrased in either of these ways:

1. The influence of degree satisfaction on job satisfaction is the same for men as it is for women.
2. Any differences between the genders on job satisfaction are the same at all levels of degree satisfaction.

The alternative, in either event, is that gender and degree satisfaction interact in influencing job satisfaction. For the model $[SD, SJ, DJ]$, Wilson found, as noted in Table 13.7, $2\mathrm{DI}(\mathbf{f} : {}_{A}\mathbf{f}) = 1.59$, where ${}_{A}\mathbf{f}$ is the estimator of $\mathbf{f}$ for this model. Under the null hypothesis, this can be considered an observation from the central chi-squared distribution with $(S - 1)(D - 1)(J - 1) = (2 - 1)(3 - 1)(3 - 1) = 4$ degrees of freedom. The small value of the DI statistic obtained does not permit rejection of the null hypothesis,

Table 13.7 Job Satisfaction DI Statistics

Model: ${}_M\mathbf{f}$	$2\text{DI}(\mathbf{f} : {}_M\mathbf{f})$	df
$[SDJ]$: ${}_F\mathbf{f}$	$2\text{DI}(\mathbf{f} : {}_F\mathbf{f}) = 0$	
$[SD, SJ, DJ]$: ${}_A\mathbf{f}$	$2\text{DI}(\mathbf{f} : {}_A\mathbf{f}) = 1.59$	4
$[SD, DJ]$: ${}_B\mathbf{f}$	$2\text{DI}(\mathbf{f} : {}_B\mathbf{f}) = 32.46$	6
$[SD, SJ]$: ${}_C\mathbf{f}$	$2\text{DI}(\mathbf{f} : {}_C\mathbf{f}) = 27.20$	8

and $[SD, SJ, DJ]$ provides an acceptable fit; the data (Table 13.6) offer no evidence that an interaction between gender and degree satisfaction influences rated job satisfaction.

Next, we evaluate the result, in effect, of dropping gender as a predictor of job satisfaction. The further restricted model $[SD, DJ]$ provides an estimator ${}_B\mathbf{f}$ and a discrimination information statistic (relative to the full model). The difference $2\text{DI}(\mathbf{f} : {}_B\mathbf{f}) - 2\text{DI}(\mathbf{f} : {}_A\mathbf{f}) = 2\text{DI}({}_A\mathbf{f} : {}_B\mathbf{f})$ has a chi-squared distribution under the more restricted model with $[(S-1)\times(D-1)(J-1) + (S-1)(J-1)] - (S-1)(D-1)(J-1) = (S-1)(J-1) = (2-1)(3-1) = 2$ degrees of freedom. The value obtained, $32.46 - 1.59 = 30.87$, is sufficiently large that the null hypothesis is rejected and we return to the model $[SD, SJ, DJ]$.

Gender, then, is a significant determiner of job satisfaction for this sample. What of the contribution of rated extent of degree satisfaction? The model $[SD, SJ]$ yields an estimator ${}_C\mathbf{f}$ such that any observed degree satisfaction–job satisfaction relation is nullified because the relevant marginal totals have been pulled back to the uniform distribution of the least constrained model. The increase in the DI statistic from $\text{DI}(\mathbf{f} : {}_A\mathbf{f})$ to $\text{DI}(\mathbf{f} : {}_C\mathbf{f})$ yields a chi-squared distribution observation: $2\text{DI}({}_A\mathbf{f} : {}_C\mathbf{f}) = 2\text{DI}(\mathbf{f} : {}_C\mathbf{f}) - 2\text{DI}(\mathbf{f} : {}_A\mathbf{f}) = 27.20 - 1.59 = 25.61$, which, with $8 - 4 = 4$ degrees of freedom, is sufficiently large that we reject as well the null hypothesis of an absence of relation between ratings of degree and job satisfaction.

Had we not first rejected the hypothesis of no gender differences in job satisfaction, the more appropriate test of the hypothesis of no degree satisfaction effect would have been to contrast the models $[SD, DJ]$ and $[SD, J]$. That is, given that $[SD, DJ]$ provides an acceptable fit to the data, we would want to know whether $[SD, J]$ is also close. For this example, however, we can drop neither SJ nor DJ without affecting the fit adversely.

13.4 Log Linear External Constraints Problem

In Section 13.2.4 we noted that Gokhale and Kullback distinguished two classes of log linear problem: internal and external constraints problems (ICP and ECP). For the ICP we have a vector $\mathbf{f}$ of W observed frequencies, and we seek a set of estimates of the expectancies of these frequencies, $\tilde{\mathbf{f}}$, that satisfy the constraints of Eq. (13.3.2), $\mathbf{T}'\tilde{\mathbf{f}} = \mathbf{T}'\mathbf{f}$, while departing as little as possible from the least constrained estimator (${}_{\mathrm{L}}\mathbf{f}$, the *uniform distribution*, or something close to it depending on the particular sampling strategy employed in data collection). Having found $\tilde{\mathbf{f}}$, we then evaluate the goodness of fit of these estimates to the data obtained as measured by Eq. (13.2.17),

$$2\mathrm{DI}(\mathbf{f} : \tilde{\mathbf{f}}) = 2\sum_{w}\left[f_w \log\left(\frac{f_w}{\tilde{f}_w}\right)\right]$$

a statistic that, under the hypothesis that $\mathbf{T}'\tilde{\mathbf{f}} = \mathbf{T}'\mathbf{f}$, is distributed as chi-squared with $W - r(\mathbf{T})$ degrees of freedom, $r(\mathbf{T})$ being the rank of the design matrix $\mathbf{T}$.

The constraints of (13.3.2) are internal to the data at hand: As the right side suggests, we constrain our estimates to be like the observed frequencies in certain ways. In short, in the ICP analysis we first change the estimates as little as possible, and in a carefully described manner, from a flat distribution towards the observed distribution, and then assess whether we have come close enough to the observed distribution. In developing this argument, it was convenient to think of the observed frequencies as a set of fully internally constrained estimators ${}_{\mathrm{F}}\mathbf{f}$ and to check the closeness to ${}_{\mathrm{F}}\mathbf{f}$ of *partially* constrained estimator ${}_{A}\mathbf{f} = \tilde{\mathbf{f}}$,

$${}_{\mathrm{L}}\mathbf{f} \leftrightarrow {}_{A}\mathbf{f} \leftrightarrow {}_{\mathrm{F}}\mathbf{f}$$

Where two (or more) sets of internal constraints could be nested, one in the other, we saw that the goodness of fit statistics, all distributed as chi-squared variates, could be partitioned:

$$\overleftrightarrow{{}_{\mathrm{L}}\mathbf{f} \longleftrightarrow {}_{A}\mathbf{f} \longleftrightarrow {}_{B}\mathbf{f} \longleftrightarrow {}_{\mathrm{F}}\mathbf{f}}$$

$$2\mathrm{DI}({}_{A}\mathbf{f} : {}_{\mathrm{L}}\mathbf{f}) + 2\mathrm{DI}({}_{B}\mathbf{f} : {}_{A}\mathbf{f}) + 2\mathrm{DI}({}_{\mathrm{F}}\mathbf{f} : {}_{B}\mathbf{f}) = 2\mathrm{DI}({}_{\mathrm{F}}\mathbf{f} : {}_{\mathrm{L}}\mathbf{f})$$

The *external constraints problem* also begins with a vector $\mathbf{f}$ of W observed frequencies and with an interest in obtaining an expected value for that vector $\tilde{\mathbf{f}}$ that is linearly constrained, now by Eq. (13.2.18):

$$\mathbf{T}'\tilde{\mathbf{f}} = \boldsymbol{\phi}$$

The important difference is that here the term on the right need not depend upon the observed frequencies. Indeed, the hallmark of the ECP is that at least one of the constraints imposed by Eq. (13.2.18) does not refer to the observed frequencies.

As an example, consider the 2×2 illustration used in Section 13.2.5:

	Female	Male
Married	4	16
Single	16	14

the two-way table resulting from asking 50 consecutive passersby their gender and marital status. One external constraint we might entertain here is based on the hypothesis that the population from which we sampled consists of equal numbers of males and females. That is, if the observed frequencies are ordered in the vector $\mathbf{f}' = (4, 16, 16, 14)$, the first two entries being the number of female and male married respondents, then the equal-numbers hypothesis could be written as the set of constraints

$$\mathbf{T}'\tilde{\mathbf{f}} = \begin{bmatrix} 1 & 0 & 1 & 0 \\ 0 & 1 & 0 & 1 \end{bmatrix}\tilde{\mathbf{f}} = \begin{bmatrix} 25 \\ 25 \end{bmatrix} = \boldsymbol{\phi}$$

where $\boldsymbol{\phi}$ reflects the hypothesis that we expect 25 women and 25 men in our sample of 50 respondents.

Imposing these external constraints allows us to ask whether the obtained frequencies, 20 women and 30 men, are consistent with sampling from a population in which males and females are equally represented.

13.4.1 MDI Solution for External Constraints

In solving the ICP we extracted certain features from the observed distribution and added these to a uniform or least constrained estimator; we employed only some of the information in the observed frequency distribution to develop the constrained estimator. For the ECP the goal is a constrained estimator in which we have added to the observations some information external to those observations.

Gokhale and Kullback approach the ECP question, then, by taking the observed frequencies to constitute a least constrained estimator and assessing the information discriminating those frequencies and the elements of the externally constrained estimator ${}_a\mathbf{f}$, using the DI statistic

$$\mathrm{DI}\left({}_a\mathbf{f} : \mathbf{f}\right) = \sum_w \left[{}_af_w \log\left(\frac{{}_af_w}{f_w}\right)\right] \tag{13.4.1}$$

The reader will note the difference between this statistic and the similar statistic (13.2.17) used for the ICP,

$$\mathrm{DI}(\mathbf{f} : {}_A\mathbf{f}) = \sum_w \left[f_w \log\left(\frac{f_w}{{}_Af_w}\right)\right]$$

In effect, the internally constrained ${}_A\mathbf{f}$ satisfying

$${}_A\mathbf{T}'\,{}_A\mathbf{f} = {}_A\mathbf{T}'\mathbf{f}$$

is *less* constrained than $\mathbf{f}$, whereas the externally constrained ${}_a\mathbf{f}$ satisfying

$${}_a\mathbf{T}'\,{}_a\mathbf{f} = \boldsymbol{\phi} \tag{13.4.2}$$

is constrained *beyond* $\mathbf{f}$.

In going beyond $\mathbf{f}$, however, we need to ensure that we move no further than is necessary. We do not want to get away from the observations any more than is required to have the estimator satisfy the external constraints. To this end, Gokhale and Kullback again use the principle of minimum discrimination information estimation. It is required that, among the vectors satisfying the constraints of Eq. (13.4.2), ${}_a\mathbf{f}$ *minimize* (13.4.1). Using Lagrangian multipliers, these goals can be expressed by the composite function,

$$\sum_w \left[{}_af_w \log\left(\frac{{}_af_w}{f_w}\right)\right] - \sum_i \left\{\lambda_i \left[\sum_w ({}_aT_{wi\,a}f_w) - \phi_i\right]\right\} \qquad (13.4.3)$$

Differentiating Eq. (13.4.3) with respect to ${}_af_w$ and setting the result equal to zero yields

$$\log\left(\frac{{}_af_w}{f_w}\right) = \sum_i \lambda_i T_{wi} \qquad (13.4.4)$$

$$\log {}_af_w = \log f_w + \sum_i \lambda_i T_{wi} \qquad (13.4.5)$$

$${}_af_w = \exp(\log f_w) \exp\left(\sum_i \lambda_i T_{wi}\right) \qquad (13.4.6)$$

13.4.2 Iterative Solution to the External Constraints Problem

As in the ICP, minimizing the discrimination information statistic subject to the linear constraints does not lead to a direct solution either for ${}_a\mathbf{f}$ or for $\boldsymbol{\lambda}$, the vector of log linear parameters. Both are present in Eqs. (13.4.4)–(13.4.6). An iterative solution must be sought. The iterative proportional fitting of marginals, useful for ICP, will not work here, as not all of the constraints are phrased in terms of observed marginal frequencies.

An alternative approach, suggested by Gokhale and Kullback, is a Newton–Raphson iteration procedure. Briefly, each cycle of the estimation begins with an approximation to $\boldsymbol{\lambda}$, which is used to estimate ${}_a\mathbf{f}$ by (13.4.6). The error in the approximation is then determined by seeing how close ${}_a\mathbf{T}'\,{}_a\mathbf{f}$ is to $\boldsymbol{\phi}$, the set of external constraints of (13.4.2). Finally, the approximation to $\boldsymbol{\lambda}$ is modified by a term proportional to the error. Cycles are repeated until the error is of negligible magnitude.

In this section we outline the flow of this iterative approach and in Section 13.4.3 apply it to the 2×2 example. The following sequence is appropriate to multinomial or product multinomial sampling. It is also useful for independent Poisson sampling, so long as the external constraints are derived from statements about the relative (rather than absolute) expected frequencies in different categories. In Section 13.4.4 we show the tests of goodness of fit for ECP problems.

The vector $\mathbf{f}$ of W observed frequencies is estimated by an ${}_a\mathbf{f}$ satisfying Eq. (13.4.2), where ${}_a\mathbf{T}$ and $\boldsymbol{\phi}$ are specified in advance. Here we further assume that ${}_a\mathbf{T}$ and $\boldsymbol{\phi}$ are partitioned between a set of k *sampling* constraints and a set of r *hypothesis* constraints. The former are internal constraints necessary to ensure that the estimator ${}_a\mathbf{f}$ is consistent with our sampling scheme. In particular, the $W \times (k + r)$ ${}_a\mathbf{T}$ and the $(k + r)$-element $\boldsymbol{\phi}$ are to be partitioned as

a. ${}_a\mathbf{T} = ({}_k\mathbf{T}, {}_a\mathbf{t})$,
b. $\boldsymbol{\phi}' = ({}_I\mathbf{n}', {}_2\boldsymbol{\phi}')$.

The form of the sampling constraints ensures, for example, that the estimated frequencies in the W_i categories arising from the ith multinomial sum to ${}_I n_i$, the number of observations sampled from that multinomial:

$$ {}_k\mathbf{T}'\,{}_a\mathbf{f} = {}_I\mathbf{n} \tag{13.4.7}$$

The number of multinomial distributions contributing to the W frequencies, then, is k. For Poisson or multinomial samples ${}_k\mathbf{T} = \mathbf{1}$, the unit vector, and ${}_I n = n$, a scalar.

Corresponding to this partitioning of the constraints, the log linear relation of Eq. (13.4.5) can be written

$$\log {}_a\mathbf{f} = \log \mathbf{f} + {}_k\mathbf{T}\mathbf{L} + {}_a\mathbf{t}\boldsymbol{\lambda} \tag{13.4.8}$$

where $\mathbf{L}$ provides a set of standardizing constants associated with the sampling from the k populations, while $\boldsymbol{\lambda}$ remains a vector of log linear parameters associated with the r external hypothesis constraints.

Initial Approximations to $\boldsymbol{\lambda}$, $\mathbf{L}$, *and* ${}_a\mathbf{f}$. To obtain ${}_a\mathbf{f}$ satisfying (13.4.2) and closest to the observed $\mathbf{f}$, begin with a version of Eq. (13.4.8) in which

c. ${}_1\mathbf{L} = \mathbf{0}$,
d. ${}_1\boldsymbol{\lambda} = \mathbf{0}$.

As a result, the initial approximation to ${}_a\mathbf{f}$ is the vector of observed frequencies,

e. ${}_1\mathbf{f} = \mathbf{f}$.

Closeness of Fit. The closeness of fit of ${}_a\mathbf{f}$ to ${}_j\mathbf{f}$ is checked using the criterion of (13.4.2), by computing

f. ${}_j\mathbf{F} = {}_a\mathbf{t}'\,{}_j\mathbf{f}$

and comparing this with the target $\boldsymbol{\phi}$:

g. ${}_j\mathbf{d} = {}_2\boldsymbol{\phi} - {}_j\mathbf{F}$.

If ${}_j\mathbf{d}$ is close enough to the null vector, the iterative part is finished and the solution is completed by going to step t; otherwise, $\boldsymbol{\lambda}$ is modified.

Modification of $\boldsymbol{\lambda}$. To modify $\boldsymbol{\lambda}$ we define a diagonal matrix form of the ${}_j\mathbf{f}$ vector,

h. ${}_j\mathbf{D} = \mathbf{D}_{{}_j\mathbf{f}}$,

and use this to obtain the symmetric matrix

i. ${}_j\mathbf{S} = {}_a\mathbf{T}' \, {}_j\mathbf{D} \, {}_a\mathbf{T}$.

As a result of the partitioning of ${}_a\mathbf{T}$, ${}_j\mathbf{S}$ is partitioned symmetrically:

$$\begin{bmatrix} {}_j\mathbf{S}_{11} & {}_j\mathbf{S}_{12} \\ {}_j\mathbf{S}_{21} & {}_j\mathbf{S}_{22} \end{bmatrix} = {}_j\mathbf{S}$$

where ${}_j\mathbf{S}_{11}$ is $(k \times k)$ and ${}_j\mathbf{S}_{22}$ is $(r \times r)$. From ${}_j\mathbf{S}$ compute

j. ${}_j\mathbf{S}_{2.1} = {}_j\mathbf{S}_{22} - {}_j\mathbf{S}_{21} \, {}_j\mathbf{S}_{11}^{-1} \, {}_j\mathbf{S}_{12}$

and its inverse,

k. $({}_j\mathbf{S}_{2.1})^{-1}$.

This $r \times r$ matrix standardizes the error of the jth approximation,

l. ${}_j\boldsymbol{\Delta} = ({}_j\mathbf{S}_{2.1})^{-1} \, {}_j\mathbf{d}$,

which is then used in the adjustment

m. ${}_{j+1}\boldsymbol{\lambda} = {}_j\boldsymbol{\lambda} + {}_j\boldsymbol{\Delta}$.

Modification of **L** *and* ${}_a\mathbf{f}$. A temporary version of the log linear relation (13.4.8) is next computed:

n. $\log({}_{j+1}\mathbf{y}) = \log \mathbf{f} + {}_a\mathbf{t} \, {}_{j+1}\boldsymbol{\lambda}$.

The vector $\mathbf{y}$ on the left may differ from an acceptable approximation to ${}_a\mathbf{f}$ by not giving the appropriate marginal sums. Steps o–r scale $\mathbf{y}$ appropriately. First, find $\mathbf{y}$,

o. ${}_{j+1}\mathbf{y} = \text{antilog}(\log \mathbf{f} + {}_a\mathbf{t} \, {}_{j+1}\boldsymbol{\lambda})$,

and then compute k sums, each over the appropriate W_i elements of $\mathbf{y}$:

p. ${}_{j+1}\boldsymbol{\sigma} = {}_k\mathbf{T}' \, {}_{j+1}\mathbf{y}$.

These k sums contribute to a k-element vector

q. ${}_{j+1}\mathbf{L}' = [\log({}_I n_1/\sigma_1), \ldots, \log({}_I n_k/\sigma_k)]$

used to scale $\log \mathbf{y}$:

r. $\log({}_{j+1}\mathbf{f}) = \log({}_{j+1}\mathbf{y}) + {}_k\mathbf{T} \, {}_{j+1}\mathbf{L}$.

The next approximation to ${}_a\mathbf{f}$ is

s. ${}_{j+1}\mathbf{f} = \text{antilog}[\log({}_{j+1}\mathbf{f})]$;

return to step f to check its adequacy.

Exiting from the Iterative Solution. When a difference sufficiently close to the r-element null vector is obtained at step g, the iterations are finished. The last approximations are taken as the convergent solutions satisfying Eq. (13.4.8):

t. ${}_a\mathbf{f} = {}_j\mathbf{f}$,
u. $\boldsymbol{\lambda} = {}_j\boldsymbol{\lambda}$,
v. $\mathbf{L} = {}_{j+1}\mathbf{L}$.

13.4.3 *Application of the Newton–Raphson Iteration*

For our example we have ${}_kT = 1$, ${}_In = 50$,

$$ {}_a\mathbf{t}' = (1, 0, 1, 0), \qquad {}_2\phi = 25, \qquad {}_1\mathbf{f} = (4, 16, 16, 14)$$

Cycle 1

f. ${}_1F = {}_a\mathbf{t}'\,{}_1\mathbf{f} = 20.$
g. ${}_1d = {}_2\phi - {}_1F = 5$, which is not null.
i.

$$ {}_1\mathbf{S} = \begin{bmatrix} 1 & 1 & 1 & 1 \\ 1 & 0 & 1 & 0 \end{bmatrix} \begin{bmatrix} 4 & & & 0 \\ & 16 & & \\ & & 16 & \\ 0 & & & 14 \end{bmatrix} \begin{bmatrix} 1 & 1 \\ 1 & 0 \\ 1 & 1 \\ 1 & 0 \end{bmatrix} = \begin{bmatrix} 50 & 20 \\ 20 & 20 \end{bmatrix}$$

j. ${}_1S_{2.1} = 20 - (20)(\frac{1}{50})(20) = 12.$
k. $({}_1S_{2.1})^{-1} = \frac{1}{12}.$
l. ${}_1\Delta = \frac{1}{12}(5) = \frac{5}{12}.$
m. ${}_2\lambda = {}_1\lambda + {}_1\Delta = 0 + (\frac{5}{12}) = \frac{5}{12}.$
n. $\log \mathbf{f} + {}_a\mathbf{t}\lambda = \log \mathbf{y}$:

$$\begin{bmatrix} 1.3862944 \\ 2.7725887 \\ 2.7725887 \\ 2.6390573 \end{bmatrix} + \begin{bmatrix} \frac{5}{12} \\ 0 \\ \frac{5}{12} \\ 0 \end{bmatrix} = \begin{bmatrix} 1.802961 \\ 2.775887 \\ 3.1892554 \\ 2.6390573 \end{bmatrix}$$

o.

$$\mathbf{y} = \begin{bmatrix} 6.067587 \\ 16 \\ 24.270349 \\ 14 \end{bmatrix}$$

p. $\mathbf{1}'\mathbf{y} = 60.337936.$
q. $\log(50/60.337936) = -0.187938.$

r. $\log {}_2\mathbf{f} = \log \mathbf{y} + (-0.187938)\mathbf{1}$:

$$\log {}_2\mathbf{f} = \begin{bmatrix} 1.6150230 \\ 2.5846507 \\ 3.0013174 \\ 2.4511193 \end{bmatrix}$$

s.

$${}_2\mathbf{f} = \begin{bmatrix} 5.0280036 \\ 13.258657 \\ 20.112015 \\ 11.601325 \end{bmatrix}$$

Cycle 2

f. ${}_2F = {}_a\mathbf{t}'\,{}_2\mathbf{f} = 5.0280036 + 20.112015 = 25.140019.$

g. ${}_2d = 25 - 25.140019 = -0.140019.$

i.

$${}_2\mathbf{S} = \begin{bmatrix} 50 & 25.140019 \\ 25.140019 & 25.140019 \end{bmatrix}$$

j. ${}_2S_{2.1} = 25.140019(1 - 25.140019/50) = 13.499608.$

l. ${}_2\Delta = -0.140019/13.499608 = -0.0112019.$

m. ${}_3\lambda = {}_2\lambda + {}_2\Delta = \frac{5}{12} - 0.0112019 = 0.4054648.$

n.

$$\log \mathbf{y} = \begin{bmatrix} 1.3862944 + 0.4054648 \\ 2.7725887 \\ 2.7725887 + 0.4054648 \\ 2.6390573 \end{bmatrix} = \begin{bmatrix} 1.7917592 \\ 2.7725887 \\ 3.1780535 \\ 2.6390573 \end{bmatrix}$$

o.

$$\mathbf{y} = \begin{bmatrix} 5.9999984 \\ 16 \\ 23.9999920 \\ 14 \end{bmatrix}$$

p. $\mathbf{1}'\mathbf{y} = 59.99999.$

q. $\log(50/59.99999) = -0.1823214.$

r.

$$\log {}_3\mathbf{f} = \begin{bmatrix} 1.6094378 \\ 2.5902673 \\ 2.9457321 \\ 2.4567359 \end{bmatrix}$$

s.

$${}_3\mathbf{f} = \begin{bmatrix} 4.999999 \\ 13.333335 \\ 19.999997 \\ 11.666668 \end{bmatrix}$$

Cycle 3

f. ${}_3F = {}_a\mathbf{t}'\,{}_3\mathbf{f} = 24.999996.$

g. ${}_3d = 25 - {}_3F = 0.000004$, declared to be null and the solution is completed: ${}_a\mathbf{f}' = (5, \frac{40}{3}, 20, \frac{35}{3})$, and $\lambda = 0.4054648$.

13.4.4 Goodness of Fit for the External Constraint Problem

Under the ECP null hypothesis, when the *expected* frequencies are constrained by ${}_a\mathbf{T}'\mathbf{E}(\mathbf{f}) = \boldsymbol{\phi}$, then twice the DI statistic,

$$2\mathrm{DI}({}_a\mathbf{f} : \mathbf{f}) = 2\sum_w \left[{}_af_w \log\left(\frac{{}_af_w}{f_w} \right) \right]$$

has as its sampling distribution, again for large n, that of a chi-squared variate with k degrees of freedom, where k is the rank of ${}_a\mathbf{T}$. This quantity is used to test whether a set of observed frequencies fits an external set of constraints or model.

For the 2 × 2 example we obtained $\mathbf{f}' = (4, 16, 16, 14)$ and ${}_a\mathbf{f}' = (5, \frac{40}{3}, 20, \frac{35}{3})$, from which we compute the DI statistic, 1.0205499. This is multiplyed by 2, producing 2.04, which may be regarded as an observation from a chi-squared distribution with 2 degrees of freedom. In this instance we do not reject the hypothesis of sampling from a population containing equal numbers of males and females.

Just as the statistic for lack of fit for the ICP can be partitioned when the constraints ${}_B\mathbf{T}'\tilde{\mathbf{f}} = {}_B\mathbf{T}'\mathbf{f}$ imply a smaller set of constraints ${}_A\mathbf{T}'\tilde{\mathbf{f}} = {}_A\mathbf{T}'\mathbf{f}$, the ECP goodness of fit statistic $\mathrm{DI}({}_b\mathbf{f} : \mathbf{f})$ also can be partitioned. The circumstances are much the same. Consider the following diagram:

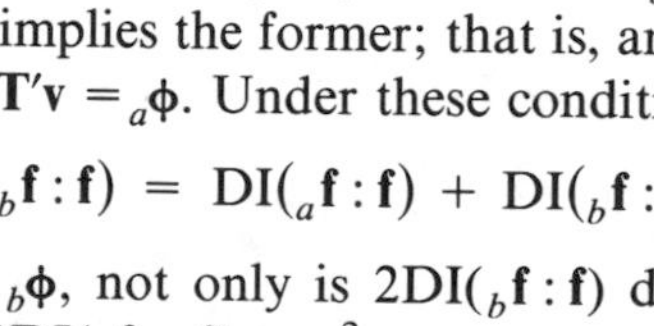

Here we assume that the estimator ${}_a\mathbf{f}$ is subject to p constraints ${}_a\mathbf{T}'\,{}_a\mathbf{f} = {}_a\boldsymbol{\phi}$ while the estimator ${}_b\mathbf{f}$ is subject to $q > p$ constraints ${}_b\mathbf{T}'\,{}_b\mathbf{f} = {}_b\boldsymbol{\phi}$, and that the latter set of constraints implies the former; that is, any vector $\mathbf{v}$ that satisfies ${}_b\mathbf{T}'\mathbf{v} = {}_b\boldsymbol{\phi}$ also satisfies ${}_a\mathbf{T}'\mathbf{v} = {}_a\boldsymbol{\phi}$. Under these conditions,

$$\mathrm{DI}({}_b\mathbf{f} : \mathbf{f}) = \mathrm{DI}({}_a\mathbf{f} : \mathbf{f}) + \mathrm{DI}({}_b\mathbf{f} : {}_a\mathbf{f})$$

When in fact ${}_b\mathbf{T}'\mathbf{E}(\mathbf{f}) = {}_b\boldsymbol{\phi}$, not only is $2\mathrm{DI}({}_b\mathbf{f} : \mathbf{f})$ distributed as $\chi^2_{q\,\mathrm{df}}$, but $2\mathrm{DI}({}_a\mathbf{f} : \mathbf{f}) \sim \chi^2_{p\,\mathrm{df}}$ and $2\mathrm{DI}({}_b\mathbf{f} : {}_a\mathbf{f}) \sim \chi^2_{(q-p)\,\mathrm{df}}$.

13.5 Summary

In this chapter we introduced an approach to the analysis of frequency data. The approach treats observations that are counts in each of a number of categories. The categories may be cross classified by two or more substantively interpretable dimensions.

The techniques of analysis were shown to be appropriate for three different sampling models: independent Poisson processes (a time-limited set of counts, for instance), multinomial categories (fixed total number of observations), and two or more independent multinomial samples (where row, column, or other marginal totals are fixed by the sampling design).

We presented one particular version of log linear analysis, the information theory approach of Gokhale and Kullback, to emphasize the continuity with the general linear model for the measured criterion attribute. Exploiting this continuity, we illustrated the development of hypotheses using constraints and a design matrix.

Problems involving both internal and external constraints were developed, and iterative techniques for arriving at numerical solutions were provided.

Exercises

Ries and Smith obtained the data given in Table 13.8 from a survey of consumer preferences for brands of laundry soap.

1. Assuming that preference is a criterion variable, describe these data in terms of their cross classifications.

2. Are previous use and water softness additive or interactive in their effects on brand preference?

3. How does the choice of laundering temperature reflect water softness?

4. Given the possible additive models to determine brand preference, which seems best?

Table 13.8 Cross Classification of 1008 Consumers[a]

		Previous M user		Previous non-M user	
Water softness	Brand preference	high temperature	low temperature	high temperature	low temperature
soft	X	19	57	29	63
	M	29	49	27	53
medium	X	23	47	33	66
	M	47	55	23	50
hard	X	24	37	42	68
	M	43	52	30	42

Source: From Ries and Smith (1963).

[a]According to (1) the softness of the laundry water used, (2) the previous use of detergent brand M, (3) the temperature of the laundry water used, and (4) the preference for detergent brand X or brand M in a blind test.

Additional Reading

The main reference is

Gokhale, D. V., and Kullback, S. (1978). *The Information in Contingency Tables*. New York: Dekker.

The maximum likelihood approach to log linear models is well explained by

Fienberg, S. E. (1980). *The Analysis of Cross-Classified Categorical Data*, 2nd ed. Cambridge, MA: MIT Press.

Application of yet a third approach to contingency table analysis, weighted least squares, is featured in

Forthofer, R. N., and Lehnen, R. G. (1981). *Public Program Analysis: A New Categorical Data Approach*. Belmont, CA: Wadsworth.

Two rigorous presentations of the statistical basis of estimation for categorical designs are

Andersen, E. B. (1980). *Discrete Statistical Models with Social Science Applications*. Amsterdam: North-Holland.

Bishop, Y. M. M., Fienberg, S. E., and Holland, P. W. (1975). *Discrete Multivariate Analysis: Theory and Practice*. Cambridge, MA: MIT Press.

When there are but two categories to the criterion, *probit analysis* can be an effective technique:

Amemiya, T. (1975). Qualitative response models, *Annals of Economic and Social Measurement* 4:363–372.

The data in Table 13.8, used in the exercises, are from

Ries, P. N., and Smith, H. (1963). The use of chi square for preference testing in multidimensional problems, *Chemical Engineering Progress* 59:39–43.

Chapter 14

Basic Structure and the Analysis of Interdependence

Introduction

In previous chapters we stressed the value of the basic structure representation of matrices as an intermediate step in the statistical analysis of predictor–criterion relations. This is appropriate when the researcher has identified some attributes as predictors and others as criteria. Many research problems, however, make no distinction between predictor and criterion attributes and are concerned instead with analyzing the *interdependence* of the attributes. The analysis of interdependence is directed toward such goals as

1. reducing the dimensionality of a set of multivariate data,
2. interpreting the resultant dimensions in terms of a theory,
3. developing taxonomies of attributes or entities, and
4. scaling attributes and entities.

Historically, researchers have either analyzed the interdependence of the measured attributes or assumed a set of unmeasured *latent attributes* in order to account for the covariance or correlations of the measured attributes. The former approach is exemplified by principal components analysis, the latter by factor analysis.

14.1 Basic Structure and Principal Components Analysis

In 1901 Karl Pearson (see the list of additional reading) was among the first to suggest an analytical solution to the analysis of the interdependence of multivariate data. His discussion employed a hypothetical scatter diagram, similar to Figure 14.1, of attributes correlated in two dimensions. In Part II our concern was with finding either the linear transformation of $\mathbf{x}$ that fits $\mathbf{y}$

and minimizes $\mathbf{e}'\mathbf{e}$ (the least squares approach) or the linear transformation that maximizes the likelihood of the sample of normally distributed $\mathbf{y}$ observations given fixed values of $\mathbf{x}$ (the linear regression approach). In either case, these solutions are useful and appropriate when the researcher conceptualizes the problem in terms of the dependence of the criteria on the predictors and accepts the assumptions of the approach. Pearson, however focusing on the interdependence of $\mathbf{x}$ and $\mathbf{y}$, suggested as a solution the *line of best fit* (actually a plane in three or more dimensions) indicated in Figure 14.1. Using theorems of geometry, Pearson proved that the line of best fit minimizes the squares of the perpendicular projections of the points in the scatter diagram. In terms of the residuals for point (x_i, y_i) shown in Figure 14.1, Pearson's line of best fit minimizes $\mathbf{c}'\mathbf{c}$, whereas the regression line of $\mathbf{y}$ on $\mathbf{x}$ minimizes $\mathbf{a}'\mathbf{a}$ and the regression line of $\mathbf{x}$ on $\mathbf{y}$ minimizes $\mathbf{b}'\mathbf{b}$.

Maximizing these projections is equivalent to requiring the variance accounted for (by the line of best fit) in $\mathbf{x}$ and $\mathbf{y}$ to be a maximum or the residual variance of $\mathbf{x}$ and $\mathbf{y}$ to be a minimum. Pearson showed that the best-fitting line in this sense passes through the centroid of the system of points, is the principal axis of the correlation ellipse, and is perpendicular to the greatest axis of the ellipsoid of residuals.

Figure 14.1 Scatter diagram of correlated variables in two dimensions.

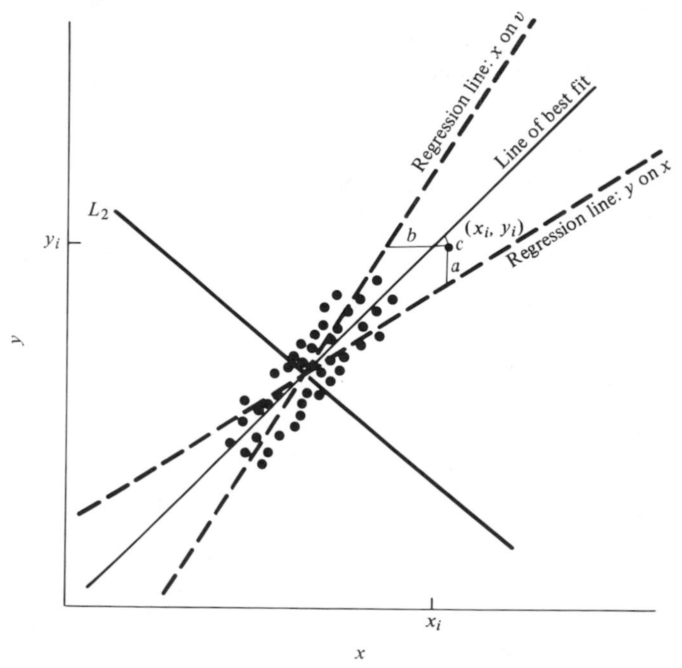

As shown in Section 6.5, $\mathbf{Q}_{.1}$ and Δ_1^2 in the basic structure of $\mathbf{X}'\mathbf{X} = \mathbf{C}$, maximize the variance accounted for by a linear combination of the measured attributes and minimize the remaining variance by providing the best rank 1 approximation to $\mathbf{X}$ and $\mathbf{C}$. Thus the solution proposed by Pearson is parallel to finding $\mathbf{Q}_{.1}$ in the basic structure of $\mathbf{C}$. We also showed in Section 6.5 that the square of the basic diagonal is the sum of squares accounted for by $\mathbf{Q}_{.1}$. The value Δ_1^2 is also called the first *eigenvalue* or first *latent root* of $\mathbf{C}$.

Thus in the analysis of interdependence the basic structure solution provides the orthogonal linear combination of $\mathbf{x}$ that best accounts for the variance in $\mathbf{C}$. Historically, the use of basic structure to this end has been termed *principal components analysis*.

In principal components analysis the basic structure $\mathbf{X} = \mathbf{P}\mathbf{\Delta}\mathbf{Q}'$ is interpreted in terms of $\mathbf{P}$ and $\mathbf{Q}\mathbf{\Delta}$, where $\mathbf{P}$ is taken to be the matrix of scores for the n entities on the m principal components. The matrix $\mathbf{\Delta}\mathbf{Q}'$ is, then, a transformation or weighting matrix, and when $\mathbf{X}$ is scaled so that $\mathbf{X}'\mathbf{X} = \mathbf{R}$ (an intercorrelation matrix), $\mathbf{Q}\mathbf{\Delta}$ is the matrix of correlations between scores on the components and scores on the attributes. Much research has been based on an analysis of the correlation matrix $\mathbf{R}$. Recall that $\mathbf{R}$ may be expressed

$$\mathbf{R} = \mathbf{Q}\mathbf{\Delta}^2\mathbf{Q}' = \mathbf{Q}\mathbf{\Delta}(\mathbf{\Delta}\mathbf{Q}') = \mathbf{Q}\mathbf{\Delta}(\mathbf{Q}\mathbf{\Delta})'$$

If we let $\mathbf{F} = \mathbf{Q}\mathbf{\Delta}$, the intercorrelation matrix $\mathbf{R}$ may then be represented as the major product moment

$$\mathbf{R} = \mathbf{F}\mathbf{F}' \tag{14.1.1}$$

The matrix $\mathbf{F}$ is referred to as the *loading matrix*.

As shown in Section 6.5, $\mathbf{Q}_{.1}\Delta_1^2\mathbf{Q}'_{.1}$ provides the best rank 1 approximation to the variance–covariance matrix $\mathbf{C}$, in that the residual has smallest variance. Similarly, the first q components of $\mathbf{C}$ provide the best rank q approximation to $\mathbf{C}$. Based on this property, researchers often interpret only the first q components as of theoretical value.

Other interpretations of the properties of the principal components of an intercorrelation matrix arise from the property that $\mathbf{Q}\mathbf{\Delta}$ is the matrix of correlations between scores on $\mathbf{X}$ and scores on $\mathbf{P}$. If $(\mathbf{Q}\mathbf{\Delta})_{ij}$ is the correlation of X_i with P_j, then $(\mathbf{Q}\mathbf{\Delta})_{ij}^2$ is the proportion of variance that the jth component accounts for in X_i. The sum

$$\sum_{j=1}^{q} (\mathbf{Q}\mathbf{\Delta})_{ij}^2 = \mathbf{Q}'_{i.}\mathbf{\Delta}^2\mathbf{Q}_{i.}$$

represents the variance accounted for in X_i by the first q components. This sum has been referred to as the communality of the attribute based on the first q components. Similarly, $\Sigma_{i=1}^{m}(\mathbf{Q\Delta})_{ij}^2 = \Delta_j^2$ represents the total variance that the jth component, P_j accounts for in the set of m attributes.

Recall, too, that the columns of component loadings are orthogonal, $\mathbf{F'F} = \Delta^2$, as are the columns of $\mathbf{P}$, since, from Chapter 4, $\mathbf{P'P} = \mathbf{I}$.

14.2 Solving for the Principal Components

The solution for the principal components of $\mathbf{R}$ is the same as the solution for the basic structure of $\mathbf{R}$. Since $\mathbf{R} = \mathbf{Q\Delta}^2\mathbf{Q}'$, one could solve for $\mathbf{Q}$ and $\mathbf{\Delta}$ by one of the methods discussed in Section 4.7. The matrix of scores of the n entities on the components is then obtained as shown in Section 4.3, based on

$$\mathbf{P} = \mathbf{XQ\Delta}^{-1} \tag{14.2.1}$$

14.3 Application of Principal Components

Much of principal components analysis arose as a technique for assessing dimensions of individual differences in cognitive skills. One early example of such an application is a 1939 study by Holzinger and Swineford (listed at the end of this chapter). They studied individual differences in the performance of 145 seventh and eighth graders on 24 psychological tests. A central issue in their research was whether a single dimension, interpreted as general intelligence, could explain the variance in the tests. If one dimension does account for all the variance in the tests, then we should find that the intercorrelation matrix has rank 1 and that the best fitting linear combination of the test scores with weights given by $\mathbf{Q}_{.1}$ accounts for all the variance in the test scores. Table 14.1 shows the means and standard deviations of

Table 14.1 Means and Standard Deviations of 145 Children on 12 Psychological Tests[a]

Test	Mean	S.D.
Visual perception	29.60	6.90
Cubes	24.84	4.50
Paper form board	15.65	3.07
Flags	36.31	8.38
General information	44.92	11.75
Paragraph comprehension	9.95	3.36
Sentence completion	18.79	4.63
Word meaning	17.24	7.89
Addition	90.16	23.60
Code	68.41	16.84
Counting dots	109.83	21.04
Straight–curved capitals	191.81	37.03

[a]From Holzinger and Swineford (1939). Copyright 1939 by University of Chicago. Reprinted by permission of the publisher.

the scores on 12 of the 24 psychological tests used by Holzinger and Swineford. Table 14.2 shows the intercorrelations of the 12 tests.

In most studies the researcher decides that the means and standard deviations are arbitrary and applies principal components analysis to the matrix of correlations based on the standardized attributes. In some applications, however, the researcher considers it inappropriate to correct for differences in standard deviations among the measured variables; principal components analysis of the variance–covariance matrix is then appropriate. We present the solutions for these two alternatives for the data from the 12 psychological tests.

Table 14.3 shows the correlation of each of the 12 principal components with each attribute, the amount (Δ_i^2) of the total variance in the 12 tests accounted for by each component, and the cumulative percentage of the total variance accounted for by successive components.

The first component is correlated with each test, but it accounts only for 38% of the variance in the 12 tests. Whether this is sufficient to account for the important variance depends on one's definition of "important." Many definitions have been offered, based on maximum-likelihood estimation, Monte Carlo simulation evidence, graphical procedures, and measurement theory. Probably the most widely used rule assumes that a principal component to be important should account for at least as much variance as a single attribute. Consequently, by this rule, only those principal components of a correlation matrix with $\Delta_i^2 > 1$ are deemed important. Based on this rule, Table 14.3 indicates that one component is not sufficient. Three components would be judged to be of importance.

Table 14.4 shows, in the second and third columns, the results of the principal components analysis of the variance–covariance matrix of the 12 psychological tests. Since this matrix is no longer based on standardized measures, only the relative magnitude of the components is interpreted. The results clearly show that the addition and straight–curved capitals tests, which had the largest standard deviations, greatly influenced the positions

Table 14.2 Intercorrelations of 12 Psychological Tests for 145 Children

Visual perception	1.000											
Cubes	0.318	1.000										
Paper form board	0.403	0.317	1.000									
Flags	0.468	0.230	0.305	1.000								
General information	0.321	0.285	0.247	0.227	1.000							
Paragraph comprehension	0.335	0.234	0.268	0.327	0.622	1.000						
Sentence completion	0.304	0.157	0.223	0.335	0.656	0.722	1.000					
Word meaning	0.326	0.195	0.184	0.325	0.723	0.714	0.685	1.000				
Addition	0.116	0.057	−0.075	0.099	0.311	0.203	0.346	0.170	1.000			
Code	0.308	0.150	0.091	0.110	0.344	0.353	0.232	0.280	0.484	1.000		
Counting dots	0.314	0.145	0.140	0.160	0.215	0.095	0.181	0.113	0.585	0.428	1.000	
Straight–curved capitals	0.489	0.239	0.321	0.327	0.344	0.309	0.345	0.280	0.408	0.535	0.512	

Table 14.3 Principal Component Loadings of 12 Psychological Tests

Test	Components											
	1	2	3	4	5	6	7	8	9	10	11	12
Visual perception	0.63	0.04	0.49	−0.18	−0.04	−0.24	−0.45	0.15	0.11	−0.20	0.00	0.02
Cubes	0.41	−0.08	0.44	0.66	0.41	−0.08	0.04	−0.08	0.06	0.01	0.00	0.04
Paper form board	0.43	−0.18	0.60	0.11	−0.43	0.37	0.21	0.16	−0.03	−0.03	0.02	0.06
Flags	0.52	−0.15	0.40	−0.52	0.41	−0.06	0.27	0.06	−0.12	0.09	−0.06	−0.04
General information	0.77	−0.24	−0.29	0.20	−0.01	0.11	−0.15	0.08	−0.35	−0.06	−0.16	−0.20
Paragraph comprehension	0.76	−0.38	−0.24	0.01	−0.07	−0.08	0.15	0.04	0.32	−0.00	0.17	−0.24
Sentence completion	0.75	−0.34	−0.29	−0.11	0.01	0.18	−0.01	−0.21	0.22	−0.06	−0.28	0.16
Word meaning	0.74	−0.41	−0.30	−0.02	0.02	−0.04	−0.12	0.05	−0.15	0.19	0.26	0.22
Addition	0.47	0.62	−0.36	0.00	0.23	0.20	0.14	0.15	−0.01	−0.33	0.11	0.08
Code	0.58	0.48	−0.16	0.13	−0.25	−0.46	0.21	0.17	−0.01	0.13	−0.15	0.07
Counting dots	0.49	0.68	0.13	−0.03	0.08	0.33	−0.20	0.04	0.13	0.34	−0.01	−0.07
Straight–curved capitals	0.68	0.41	0.20	−0.08	−0.19	−0.08	0.04	−0.48	−0.14	−0.06	0.11	−0.04
Δ_i^2	4.55	1.79	1.46	0.83	0.69	0.62	0.49	0.41	0.37	0.33	0.25	0.20
Percentage of total Δ_i^2	38.0	14.9	12.2	6.9	5.8	5.2	4.1	3.4	3.0	2.8	2.1	1.6
Cumulative percentage	38.0	52.9	65.1	72.0	77.8	83.0	87.1	90.5	93.5	96.3	98.4	100

Table 14.4 Unrotated and Rotated Principal Component Loading Matrices for the Variance–Covariance Matrix of 12 Psychological Tests

	Unrotated components		Rotated components	
Test	1	2	1	2
Visual perception	3.34	−1.47	3.45	1.19
Cubes	1.08	−0.48	1.12	0.38
Paper form board	8.14	−0.90	1.21	−0.10
Flags	2.66	−1.40	2.91	0.78
General information	4.97	0.67	3.19	3.87
Paragraph comprehension	1.17	−1.70	0.98	0.67
Sentence completion	1.78	−0.12	1.39	1.13
Word meaning	2.55	−0.40	2.14	1.44
Addition	15.10	16.40	−0.06	22.30
Code	11.20	2.86	6.24	9.68
Counting dots	14.70	9.01	4.63	16.60
Straight–curved capitals	35.00	−11.50	33.50	15.30
Δ_i^2	1851.61	498.28	1226.43	1123.46
Percentage of total variance	61.01	16.42	40.41	37.02

of the principal components. It is also clear that the results of the principal components analysis are dependent on the scale of measurement and the resultant type of cross products matrix that is analyzed.

14.4 Further Interpretation of Principal Components

The principal components shown in Tables 14.3 and 14.4 have elegant mathematical properties, but they are difficult to interpret psychologically. When every attribute is correlated with the first component, theoretical interpretation of the principal component may be unclear.

The educational psychologist L. L. Thurstone made one of the first suggestions for improving the interpretability of analyses of interdependence. He argued that the matrix of correlations of the attributes and substantively interpretable components should have a *simple structure* characterized as follows:

1. Each component should be highly correlated with only a few attributes and relatively uncorrelated with the other attributes.
2. Each attribute should be correlated strongly with only one component and uncorrelated with the other components.
3. Pairs of components should exhibit different patterns of correlations with the attributes.

By these criteria, the principal components shown in Table 14.3 are not characterized by a simple structure. In the earliest attempts to obtain simple structure, the loadings for a pair of components were plotted against one another and the component axes rotated. Later work attempted to define a mathematical function or analytic criterion that could be minimized or maximized and satisfy the criteria for a simple structure.

One of the most widely used analytic criteria is the VARIMAX scheme. Thurstone's suggestions imply that we want each loading to be either large or small. The VARIMAX approach therefore maximizes the sum of squares of the loadings for each component by multiplying $\mathbf{Q\Delta}$ by an appropriately chosen square orthonormal rotation matrix $\mathbf{H}$. That is, VARIMAX finds the matrix $\mathbf{H}$ from among those satisfying $\mathbf{H'H} = \mathbf{I}$ that maximizes $\mathrm{tr}(\mathbf{B'B})$, where $\mathbf{B} = \mathbf{Q\Delta H}$. The scores of the entities on these rotated components are then found by $\mathbf{M} = \mathbf{PH'} = \mathbf{XQ\Delta}^{-1}\mathbf{H'}$. The scores on these rotated components remain orthogonal, $\mathbf{M'M} = \mathbf{P'P} = \mathbf{I}$, and VARIMAX is an orthogonal rotation.

For the first three components in Table 14.3, the rows of the loading matrix $\mathbf{F}$ have different sums of squares or communalities, and so the attributes with the largest communalities would contribute most to the VARIMAX solution. To alleviate this, the loading matrix is first normalized by rows; each row of $\mathbf{F}$ is divided by the square root of that row's communality. After finding the $\mathbf{H}$ that maximizes the VARIMAX criterion for this normalized loading matrix, the resulting matrix $\mathbf{B}$ is denormalized to restore the original communality. This is the normalized VARIMAX solution. Table 14.5 shows the normalized VARIMAX solution for the first three components in Table 14.3. We can see that the rotated components account for the same total amount of variance as the unrotated components, but this variance is split up differently among the components. We no longer have a basic structure or principal components solution, but the pattern of high and low loadings in Table 14.5 is more easily interpretable, which was the goal of the rotation.

Most applied work has used orthogonal rotations of principal component loadings. However, some researchers have argued that nonorthogonal or oblique transformations are more likely to produce simple structure. In oblique transformations we find a square basic transformation matrix $\mathbf{J}$ of the $m \times q$ loadings matrix $\mathbf{F}$. Unlike $\mathbf{H}$ in the VARIMAX solution, the oblique transformation matrix is not orthonormal. Because the obliquely rotated components are no longer uncorrelated, we also need to take into account the matrix of correlations among pairs of transformed components. As a result obliquely rotated components may be examined from either of two perspectives: First, a *structure* or loading matrix giving the perpendicular projections of the attributes on the oblique axes and interpretable as the set of correlations between attributes and oblique dimensions when $\mathbf{F}$ is a correlation matrix; second, a *pattern* matrix containing the regression weights used to fit the attributes with the obliquely rotated components.

Table 14.5 Normalized VARIMAX Loadings of First Three Principal Components from Table 14.3[a]

	VARIMAX rotated components			
	1	2	3	h^2
Visual perception	0.18	0.27	0.73	0.63
Cubes	0.11	0.06	0.59	0.37
Paper form board	0.10	−0.04	0.76	0.58
Flags	0.24	0.06	0.63	0.46
General information	0.80	0.23	0.17	0.73
Paragraph comprehension	0.84	0.10	0.23	0.77
Sentence completion	0.84	0.14	0.17	0.76
Word meaning	0.88	0.08	0.17	0.81
Addition	0.18	0.83	−0.15	0.74
Code	0.24	0.72	0.09	0.59
Counting dots	−0.03	0.82	0.19	0.70
Straight–curved capitals	0.17	0.66	0.46	0.68
$\sum_{i=1}^{12} (Q_j \Delta_{ij})^2$	3.070	2.478	2.262	
Percentage of total variance	25.6	20.7	18.9	
Cumulative percentage	25.6	46.3	65.2	

[a]The transformation matrix is

$$\mathbf{H} = \begin{bmatrix} 0.70 & 0.50 & 0.51 \\ -0.52 & 0.85 & -0.12 \\ -0.49 & -0.17 & 0.85 \end{bmatrix}$$

Table 14.6 Oblimin Obliquely Rotated Principal Components for Table 14.3[a]

	1		2		3	
	pattern	structure	pattern	structure	pattern	structure
Visual perception	−0.31	0.56	0.05	0.55	0.99	0.78
Cubes	−0.22	0.36	−0.17	0.31	0.88	0.57
Paper form board	−0.31	0.38	−0.39	0.28	1.21	0.67
Flags	−0.02	0.48	−0.26	0.37	0.86	0.65
General information	1.17	0.83	−0.12	0.58	−0.31	0.56
Paragraph comprehension	1.29	0.83	−0.36	0.51	−0.20	0.58
Sentence completion	1.30	0.82	−0.29	0.52	−0.30	0.55
Word meaning	1.40	0.83	−0.40	0.48	−0.31	0.54
Addition	−0.05	0.40	1.27	0.72	−0.69	0.23
Code	−0.04	0.51	0.99	0.74	−0.28	0.43
Counting dots	−0.63	0.36	1.24	0.75	0.01	0.42
Straight–curved capitals	−0.38	0.57	0.78	0.79	0.40	0.68

[a]Based upon a direct oblimin solution (delta of 0.50). The matrix of factor pattern correlations is

$$\begin{bmatrix} 1.00 & 0.79 & 0.82 \\ 0.79 & 1.00 & 0.75 \\ 0.82 & 0.75 & 1.00 \end{bmatrix}$$

Although many analytical oblique transformations have been suggested, the most popular include a class of rotations termed *oblimin*. In general, these procedures attempt to minimize a function of the sum of cross products of squared loadings for pairs of attributes. Table 14.6 shows the obliquely transformed loadings of the first three components in Table 14.3. The similarities between the results of the oblique rotation in Table 14.6 and the VARIMAX solution in Table 14.5 are apparent. While this is not always the case, comparisons of various rotations on the same set of data suggest that oblique and orthogonal transformations often lead to similar theoretical interpretations.

14.5 Principal Components as a Method of Orthogonalization

We noted that finding the basic structure of correlation matrices has historically been termed principal components analysis. In principal components analysis we find an orthogonal basis for the interdependences among the measured attributes. We have discussed other orthogonalizations, such as the triangular factoring procedures. These could be, and historically have been, used to identify the orthogonal basis for a set of measured attributes.

However, principal components analysis, like other basic structure solutions, has the advantage of providing, for a given number of components, the best possible approximation to the correlation matrix $\mathbf{R}$ or the data matrix $\mathbf{X}$. The goal of much research is to reduce the number of attributes with least loss of information. Principal components analysis, as a rank-reducing variance-maximizing basic structure solution, does this best. In contrast, the usefulness of triangular factoring procedures is limited by an arbitrariness of the order in which linear combinations of attributes are extracted.

14.6 Computer Solutions and Principal Components Analysis

Two widely used statistical packages, SPSS and BMDP, have principal components analysis programs. BMDP allows the analysis of minor-product moment and variance–covariance matrices as well as correlation matrices. SPSS provides analyses only of correlation matrices. Both SPSS and BMDP provide orthogonal and oblique rotations.

14.7 Principal Components and Factor Analysis

Thus far we have discussed the basic structure application known as principal components and its role in the interpretation of interdependence. Other techniques, known as *common factor analysis*, have also been used for this purpose. Procedures of common factor analysis differ from principal component approaches (and among themselves) in the scaling of the attributes and the loss function minimized in the solution.

Differences in the choice of scaling of the attributes are based in part on philosophical differences. Principal components analysis seeks to account for the correlation matrix (less commonly the variance–covariance matrix) for a set of measured or observed attributes. A basic equation for the principal components of a correlation matrix

$$\mathbf{R} = \mathbf{F}\mathbf{F}' \tag{14.7.1}$$

implies that the observed attributes are standardized or scaled to have the same variance. Equivalently, Eq. (14.7.1) tells us that it is the observed correlation matrix that is to be reproduced from the components.

Common factor analysis, by contrast, seeks to explain not the observed correlation matrix, but only that part of it that reflects what is *common* to the attributes or is due to latent attributes. Parallel to Eq. (14.7.1), the fundamental factor analysis equation is

$$\mathbf{R} = \mathbf{F}_c\mathbf{F}_c' + \mathbf{U}$$

or

$$\mathbf{R} - \mathbf{U} = \mathbf{F}_c\mathbf{F}_c' \tag{14.7.2}$$

where $\mathbf{U}$ is a diagonal matrix of uniquenesses for the observed attributes.

Viewed from a scaling perspective, Eq. (14.7.2) tells us that each attribute should be scaled to reflect the amount of common variance or communality of that attribute. [In Eqs. (14.7.1) and (14.7.2), $\mathbf{F}$ is the component loading matrix for a set of orthogonal components and $\mathbf{F}_c$, similarly, is a factor loading matrix for a set of orthogonal factors.] Equation (14.7.2) implies that the uniquenesses (or communalities) must be known before the factors can be sought.

Most computer programs have two options for determining the communalities in factor analysis. The researcher, acting on prior information, may determine the values of the communalities or a computer algorithm may estimate the communalities.

Having elected a common factor approach and having determined (or estimated) the communalities, the researcher must next decide what loss function is to be minimized in finding the factors. The researcher may accept the same loss function as in the principal components solution, so that the variance accounted for in $\mathbf{R} - \mathbf{U}$ by the first and each succeeding factor is maximized. Principal axis factor analysis uses this loss function. Adoption of the loss function minimizing the off-diagonals in $\mathbf{R} - \mathbf{U}$ at each step leads to *minimum residual factor analysis*. Still other loss functions lead to *maximum likelihood factor analysis* and *alpha factor analysis*. These procedures all share the problems of estimating the communality of each variable, deciding on the number of underlying factors, selecting a method of transformation to a simple structure, and finally estimating the scores of the entities on the factors. This last issue is important, because whereas scores on components may be calculated exactly, scores on factors can only be estimated imprecisely.

14.8 Principal Components and Factor Analysis: A Comparison

Table 14.7 shows the unrotated and VARIMAX rotated principal axis and maximum likelihood factors for the intercorrelation matrix in Table 14.2. Three factors were assumed and, when appropriate, communalities were estimated using the iterative solution discussed in Section 14.7. The psychological interpretations of all solutions are comparable. It has been suggested that this correspondence holds for correlation matrices with widely differing properties. This correspondence in the interpretation of the results of principal components and factor analysis, the parsimonious interpretation of the unrotated or initial solution, and the exactness with which component scores can be calculated all support the continued use of basic structure or principal components analysis for the analysis of interdependence.

14.9 Regression on Principal Components

In Chapter 8 we showed that the sampling variability of the maximum likelihood estimator of β_i is given by

$$\operatorname{Var}(\hat{\beta}_i) = \sigma^2\left[(\mathbf{X}'\mathbf{X})^{-1}\right]_{ii} \tag{14.9.1}$$

so that the sum of these sampling variances over the p elements of $\boldsymbol{\beta}$ is

$$\sum_{i=1}^{p} \operatorname{Var}(\hat{\beta}_i) = \sigma^2 \operatorname{tr}(\mathbf{X}'\mathbf{X})^{-1} \tag{14.9.2}$$

Table 14.7 Unrotated and VARIMAX Rotated Principal Axis and Maximum Likelihood Factors[a]

	Principal axis loadings							Maximum likelihood loadings						
	unrotated			VARIMAX				unrotated			VARIMAX			
	1	2	3	1	2	3	h^2	1	2	3	1	2	3	h^2
Visual perception	59	08	46	17	23	70	57	52	15	52	17	21	70	57
Cubes	35	−02	27	14	08	41	20	32	03	29	14	08	40	19
Paper form board	39	−08	47	12	01	60	38	34	−03	50	12	−01	59	37
Flags	46	−07	31	23	08	50	31	43	00	36	23	09	50	32
General information	75	−22	−20	74	24	21	64	78	−15	−13	75	24	20	65
Paragraph comprehension	75	−36	−14	79	11	26	71	79	−28	−06	79	10	26	70
Sentence completion	73	−31	−19	78	15	21	67	78	−23	−11	78	16	21	67
Word meaning	74	−40	−21	84	08	20	75	79	−33	−11	83	08	21	75
Addition	46	56	−40	18	80	−12	68	43	60	−37	18	80	−12	69
Code	53	37	−13	23	60	16	43	51	41	−06	23	59	18	43
Counting dots	46	60	−02	00	72	21	56	40	65	00	00	74	21	59
Straight–curved capitals	65	39	18	16	59	49	60	58	45	25	16	57	49	60

[a]Decimals have been omitted, and h^2 is the estimated communality of the variable $\mathbf{R} - \mathbf{U}$.

The trace of a symmetric matrix is the sum of its basic diagonal elements, and Eq. (14.9.2) can be rewritten

$$\sum_{i=1}^{p} \mathrm{Var}(\hat{\beta}_i) = \sigma^2 \sum_{i=1}^{p} \frac{1}{\Delta_i^2} \tag{14.9.3}$$

where the design matrix and its minor product moment have basic structures $\mathbf{X} = \mathbf{P\Delta Q'}$ and $\mathbf{X'X} = \mathbf{Q\Delta^2 Q'}$, respectively.

The importance of (14.9.3) is that it suggests that the sampling variance of the regression parameters, summed across those parameters, can be quite large if $\mathbf{X}$ or $\mathbf{X'X}$ has very small basic diagonal elements. One attempt to reduce the sampling variability or uncertainty of regression parameters could be to replace $\hat{\boldsymbol{\beta}}$ with an estimator based on a subset of the principal components of the design matrix. Since the full set of orthogonal or principal components provides a basis for the set of predictor attributes, we can regard the components themselves as predictor attributes.

Consider the transformed (principal component) attributes $\mathbf{Z} = \mathbf{XQ} = \mathbf{P\Delta Q'Q} = \mathbf{P\Delta}$. If the dependent attribute $\mathbf{y}$ regresses linearly on $\mathbf{X}$, then it also regresses linearly on $\mathbf{Z}$. That is,

$$\mathrm{E}(\mathbf{y}) = \mathbf{X}\boldsymbol{\beta} = \mathbf{P\Delta Q'}\boldsymbol{\beta} = \mathbf{Z}(\mathbf{Q'}\boldsymbol{\beta}) = \mathbf{Z}\boldsymbol{\gamma}$$

Thus the $\mathbf{y}$ population means are linear functions of $\mathbf{Z}$ as much as of $\mathbf{X}$. To estimate $\boldsymbol{\gamma}$ in a maximum likelihood fashion leads to

$$\hat{\boldsymbol{\gamma}} = (\mathbf{Z'Z})^{-1}\mathbf{Z'y} = (\mathbf{\Delta P'P\Delta})^{-1}\mathbf{\Delta P'y} = (\mathbf{\Delta})^{-1}\mathbf{P'y} \tag{14.9.4}$$

Each of these estimators has sampling variability, again generalizing from Chapter 8, of

$$\mathrm{Var}(\hat{\gamma}_i) = \sigma^2\left[(\mathbf{Z'Z})^{-1}\right]_{ii} = \sigma^2/\Delta_i^2 \tag{14.9.5}$$

From Eq. (14.9.5) it is clear that it is the later principal components that have regression parameters with large sampling variabilities. This is true particularly when the columns of $\mathbf{X}$ are strongly intercorrelated. The sums of the sampling variabilities are the same for $\hat{\boldsymbol{\gamma}}$ as for $\hat{\boldsymbol{\beta}}$:

$$\sum_{i=1}^{p} \mathrm{Var}(\hat{\gamma}_i) = \sigma^2 \sum_{i=1}^{p} \frac{1}{\Delta_i^2} \tag{14.9.6}$$

Equation (14.9.6) suggests that we might reduce the overall uncertainty in the estimates of $\boldsymbol{\gamma}$ by, in effect, not weighting or eliminating the smaller principal components.

Consider the result of regressing on only the first k (among p) principal components. Partitioning the basic structure of $\mathbf{X}$ as

$$\mathbf{X} = (\mathbf{P}_k \quad \mathbf{p})\begin{bmatrix} \mathbf{\Delta}_k & \mathbf{0} \\ \mathbf{0} & \mathbf{d} \end{bmatrix}\begin{bmatrix} \mathbf{Q}'_k \\ \mathbf{q}' \end{bmatrix}$$

we can write the vector of k maximum likelihood estimators, from (14.9.4),

$$\hat{\boldsymbol{\gamma}}_k = (\boldsymbol{\Delta}_k)^{-1}\mathbf{P}'_k\mathbf{y} \tag{14.9.7}$$

This vector has sampling variance–covariance matrix

$$\mathbf{Var}(\hat{\boldsymbol{\gamma}}_k) = \sigma^2(\boldsymbol{\Delta}_k)^{-2} \tag{14.9.8}$$

The estimate of the expectation vector, based on Eq. (14.9.7), is

$$\hat{\mathbf{E}}(\mathbf{y}) = \mathbf{P}_k\boldsymbol{\Delta}_k(\boldsymbol{\Delta}_k)^{-2}\boldsymbol{\Delta}_k\mathbf{P}'_k\mathbf{y} = \mathbf{P}_k\mathbf{P}'_k\mathbf{y}$$

The first k principal components can be obtained from $\mathbf{X}$ by $\mathbf{X}\mathbf{Q}_k(\boldsymbol{\Delta}_k)^{-1} = \mathbf{P}_k$, and substituting this into the expectation equation gives $\hat{\mathbf{E}}(\mathbf{y}) = \mathbf{X}\mathbf{Q}_k(\boldsymbol{\Delta}_k)^{-2}\mathbf{Q}'_k\mathbf{X}'\mathbf{y} = \mathbf{X}\boldsymbol{\beta}^*$ where

$$\boldsymbol{\beta}^* = \mathbf{Q}_k(\boldsymbol{\Delta}_k)^{-2}\mathbf{Q}'_k\mathbf{X}'\mathbf{y} = \left[\mathbf{Q}_k(\boldsymbol{\Delta}_k)^2\mathbf{Q}'_k\right]^{-1}\mathbf{X}'\mathbf{y} \tag{14.9.9}$$

is a set of weights for the $\mathbf{X}$ attributes.

These alternative weights can be shown to have, overall, less sampling variability than the maximum likelihood estimators. First, the relation between $\boldsymbol{\beta}^*$ and $\hat{\boldsymbol{\gamma}}$, obtained from Eqs. (14.9.7) and (14.9.9), is $\boldsymbol{\beta}^* = \mathbf{Q}_k\hat{\boldsymbol{\gamma}}_k$, and we can write from Eq. (14.9.8)

$$\mathbf{Var}(\boldsymbol{\beta}^*) = \mathbf{Q}_k\,\mathbf{Var}(\hat{\boldsymbol{\gamma}})\,\mathbf{Q}'_k = \sigma^2\mathbf{Q}_k(\boldsymbol{\Delta}_k)^{-2}\mathbf{Q}'_k$$

The sum of the sampling variabilities for the k weights is

$$\sum_{i=1}^{k}\mathrm{Var}(\beta_i^*) = \sigma^2\,\mathrm{tr}\left[\mathbf{Q}_k(\boldsymbol{\Delta}_k)^{-2}\mathbf{Q}'_k\right]$$

or, since the traces of a major and minor product moment are equal,

$$\sum_{i=1}^{k}\mathrm{Var}(\beta_i^*) = \sigma^2\,\mathrm{tr}\left[(\boldsymbol{\Delta}_k)^{-1}\mathbf{Q}'_k\mathbf{Q}_k(\boldsymbol{\Delta}_k)^{-1}\right]$$

$$= \sigma^2\,\mathrm{tr}\left[(\boldsymbol{\Delta}_k)^{-2}\right] = \sigma^2\sum_{i=1}^{k}\left(\frac{1}{\Delta_i^2}\right)$$

This is less than the sum of the sampling variabilities of the γ_i given in (14.9.3). Although the weights based on the first few principal components may be more stable than the usual maximum likelihood estimators, care must be exercised. The latter components may have important influences on the criterion that are masked by the uncertainty of their regression weights.

14.10 Summary

We introduced in this chapter the two major approaches to the analysis of interdependence among a set of attributes. Principal components analysis is equivalent to finding the basic structure of a correlation or variance–covariance matrix. It is appropriate when the pattern observed in such a matrix is to be accounted for parsimoniously.

Common factor analysis emphasizes constructing latent variables or attributes. By rotation of an initial solution, a simple structure can be obtained. Rotated and unrotated components and factors for an example involving a set of cognitive psychological measures were developed and compared.

Finally, regression on principal components was outlined as a technique for controlling the large sampling variability of regression estimates commonly observed when intercorrelated predictors are employed.

Exercises

1. Given the data shown in Table 1.2 on ability test scores and grade point averages for 50 university students, compute a VARIMAX rotated principal components analysis and a VARIMAX rotated principal axis factor analysis. Rotate on two factors and compare the solutions.

2. Briefly discuss the theoretical difference between the goals of a principal components analysis and a factor analysis in terms of the assumptions that are made about the scaling of the data.

3. Consider the following correlation matrix of eight physical variables for 305 girls, taken from a study by Mullen in 1939:

	1	2	3	3	5	6	7	8
1	—							
2	0.846	—						
3	0.805	0.881	—					
4	0.859	0.826	0.801	—				
5	0.473	0.376	0.380	0.436	—			
6	0.398	0.326	0.319	0.329	0.762	—		
7	0.301	0.277	0.237	0.327	0.730	0.583	—	
8	0.382	0.415	0.345	0.365	0.629	0.577	0.539	—

where variable 1 is height, 2 arm span, 3 forearm length, 4 length of lower leg, 5 weight, 6 size of the bitrochanteric chamber, 7 chest girth, and 8 chest width.

a. Find the component loadings of the intercorrelation matrix for the first two components.

b. Rotate the component loadings by VARIMAX and an oblique solution of your choice. Compare the interpretations.

c. Is it possible to find the component scores in this problem? Why or why not?

d. Do a maximum likelihood factor analysis of the intercorrelation matrix. What are the unrotated factor loadings? What are the values of the iterated communalities? Rotate the solution using VARIMAX. How does the rotated factor loading matrix compare with the rotated component loading matrix of (b).

4. Suppose you are interested in the determinants of mathematical ability. You have 50 measures, many of which can be assumed to influence mathematical ability. What are the disadvantages of doing a multiple regression analysis with 50 predictors? How can component analysis be used to help you in this problem?

Additional Reading

The following three references provide good coverage of factor analysis. Harman is a classic compendium, Horst is close in development to the present work, and Mulaik is somewhat more recent.

Harman, H. H. (1967). *Modern Factor Analysis*, 2nd ed. Chicago: University of Chicago Press.

Horst, P. (1965). *Factor Analysis of Data Matrices*. New York: Holt, Rinehart, Winston.

Mulaik, S. A. (1972). *The Foundations of Factor Analysis*. New York: McGraw-Hill.

The Holzinger and Swineford data are from the following reference:

Holzinger, K. J., and Swineford, F. (1939). A study in factor analysis: The stability of a bi-factor solution, in *Supplementary Educational Monographs*, No. 48. Chicago: Department of Education, University of Chicago.

Karl Pearson's original development of principal components analysis was contained in the following article:

Pearson, K. (1901). On lines and planes of closest fit to a system of points in space, *Philosophical Magazine* (*Ser.* 6) 2:557–572.

Chapter 15
Latent Variable Structural Equation Models

Introduction

In analyses of dependence, such as the GLM of Parts II and III, the attributes are partitioned into predictors and criteria, and it is assumed that the observed values of the predictors are without (measurement) error. In much research, however, this is a questionable assumption. It is partly as a result of this concern that *latent variable structural equation models* of dependence have been developed.

The linear models presented in this chapter assume that the observed attributes are fallible indicators of underlying unobservable latent factors. Relations among these latent variables then form the basis for estimating and testing hypotheses concerning dependence. Thus the model discussed in this chapter combines elements of analyses of dependence and factor analysis.

Whereas most of the hypotheses tested in Parts II and III were about means, latent variable structural equation models have been used primarily to test hypotheses about the structure of covariances of attributes. For this reason, some authors call the approach the *analysis of covariance structures*.

Several statisticians have proposed latent variable structural equation models. We have identified a number of sources in the selected references at the end of this chapter. To illustrate the approach, we discuss a model that has been implemented by Karl Jöreskog, a Swedish statistician, in the LISREL V computer program. Following the suggestion of Peter Bentler, we shall speak of the JKW model (after Jöreskog, Ward Keesling, and David Wiley).

Table 15.1 summarizes the steps taken by the researcher in applying latent variable structural models to data. The chapter is organized according to this list of topics. We first define a general latent structural equation

Table 15.1 Latent Variable Structural Equation Modeling

1. Define the general substantive model.
2. Construct a path diagram of relations among the observed and latent variables.
3. Develop a set of hierarchical, theoretically based models. (These will be compared in subsequent steps.)
4. Translate into matrix equations the hypothesized relations among variables.
5. Test the system of equations in a model for identifiability.[a]
6. Constrain or fix parameters to achieve a testable model.
7. Estimate the goodness of fit of the hypothesized models to the covariance matrix of observed variates.
8. Based on the hierarchical set of theoretical models, compare the goodness of fit index for the models, taking into account the statistical assumptions, sample size, and theory.
9. If model fitting is exploratory in purpose, use theory and diagnostic information about the model fit to suggest further model changes. Then return to steps 2–8 and cross-validate the model using additional data.

[a] For example, if an inadequate number of indicators are measured for a particular latent construct, many parameters involving that construct may fail to be identified and many interesting models may not be testable. Clearly this issue should be considered prior to data collection.

model using the notation of the LISREL V implementation. We then illustrate this approach with two applications, which are discussed more fully in the LISREL V manual. In the final section we indicate how the techniques of previous chapters can be discussed as specific applications of the model described in Section 15.1.

Our focus in this chapter is on the ideas and equations of latent variable structural equation models. Analyzing data using these approaches requires access to an appropriate computer program and a thorough understanding of that particular program. Nevertheless, we hope this chapter provides a useful introduction to these techniques.

15.1 JKW Latent Variable Structural Equation Model

We develop the JKW model by

1. specifying the measurement models defining the latent factors;
2. identifying the structural equation *models* that specify the dependence relations among the latent factors;
3. developing the covariance matrices of the latent variables; and
4. expressing the covariance of the observed attributes in terms of the latent factors. We depend heavily on the notation adopted by Jöreskog for LISREL V. Because the matrix symbols may be confusing, we present a summary in Table 15.2.

Table 15.2 Matrices Used in the LISREL V Model

	Dimensions	Mean	Covariance	Interpretation
$\mathbf{y}$	$p \times 1$	$\boldsymbol{\mu}_y$	(15.1.6)	observed criterion variates
$\mathbf{x}$	$q \times 1$	$\boldsymbol{\mu}_x$	(15.1.6)	observed predictor variates
$\boldsymbol{\eta}$	$m \times 1$	$\mathbf{0}^a$	(15.1.5)	latent criterion factors
$\boldsymbol{\xi}$	$n \times 1$	$\mathbf{0}^a$	$\boldsymbol{\Phi}$	latent predictor factors
$\boldsymbol{\Lambda}_y$	$p \times m$	—	—	factor matrix of $\mathbf{y}$ on $\boldsymbol{\eta}$
$\boldsymbol{\Lambda}_x$	$q \times n$	—	—	factor matrix of $\mathbf{x}$ on $\boldsymbol{\xi}$
$\boldsymbol{\varepsilon}$	$p \times 1$	$\mathbf{0}$	$\boldsymbol{\Theta}_\varepsilon$	errors in measurement for criterion variates
$\boldsymbol{\delta}$	$q \times 1$	$\mathbf{0}$	$\boldsymbol{\Theta}_\delta$	errors in measurement for predictor variates
$\boldsymbol{\beta}$	$m \times m$	—	—	coefficient matrix for latent criterion variables
$\boldsymbol{\Gamma}$	$m \times n$	—	—	coefficient matrix for latent predictor variables
$\boldsymbol{\zeta}$	$m \times 1$	$\mathbf{0}$	$\boldsymbol{\Psi}$	errors in structural equations

Basic equations in the

measurement models	structural equation model
$\mathbf{y} = \boldsymbol{\mu}_y + \boldsymbol{\Lambda}_y\boldsymbol{\eta} + \boldsymbol{\varepsilon}$	$\boldsymbol{\eta} + \boldsymbol{\beta}\boldsymbol{\eta} + \boldsymbol{\Gamma}\boldsymbol{\xi} + \boldsymbol{\zeta}$
$\mathbf{x} = \boldsymbol{\mu}_x + \boldsymbol{\Lambda}_x\boldsymbol{\xi} + \boldsymbol{\delta}$	

[a]Assuming that the observed data are expressed as deviation scores.

15.1.1 Measurement Models

As suggested in the introduction to this chapter, there are two aspects to the use of a linear structural equation model in the analysis of dependence. First, the observed criterion attributes Y and predictor attributes X are interpreted by *measurement models* as indicators of latent or common factors:

$$\mathbf{y} = \boldsymbol{\Lambda}_y\boldsymbol{\eta} + \boldsymbol{\varepsilon} \tag{15.1.1}$$

$$\mathbf{x} = \boldsymbol{\Lambda}_x\boldsymbol{\xi} + \boldsymbol{\delta} \tag{15.1.2}$$

where $\mathbf{y}$ is a $p \times 1$ vector of observed criterion scores, $\boldsymbol{\Lambda}_y$ is a $p \times m$ matrix of factor loadings, $\boldsymbol{\eta}$ is an $m \times 1$ vector of common factor scores, $\boldsymbol{\varepsilon}$ is a $p \times 1$ vector of residuals (where $p > m$), $\mathbf{x}$ is a $q \times 1$ vector of observed predictor scores, $\boldsymbol{\Lambda}_x$ is a $q \times n$ matrix of factor loadings, $\boldsymbol{\xi}$ is an $n \times 1$ vector of common factor scores, and $\boldsymbol{\delta}$ is a $q \times 1$ vector of residuals (where $q > n$). The elements of $\boldsymbol{\varepsilon}$ and $\boldsymbol{\delta}$ are referred to as the *errors of measurement*. It is assumed, first, that these errors of measurement are uncorrelated with each other and with the latent factors and, second, that in (15.1.1) and (15.1.2) scores on the observed attributes and latent factors are expressed as deviations from population means. In analyses based on single samples, this last assumption is not restrictive; however, multiple samples require defining the measurement models in terms of raw scores, as noted in Table 15.1. $\mathbf{\Phi}$, $\mathbf{\Theta}_\varepsilon$, and $\mathbf{\Theta}_\delta$ are the variance–covariance matrices of $\boldsymbol{\xi}$, $\boldsymbol{\varepsilon}$, and $\boldsymbol{\delta}$, respectively.

These measurement models are just the factor analytic models that we briefly considered in Chapter 14. Here our emphasis will be on their confirmatory rather than exploratory use. We can see that these measurement models are factor models by considering the variance–covariance of $\mathbf{x}$ obtained from Eq. (15.1.2):

$$\begin{aligned}\mathbf{Var(x)} &= \mathbf{E(xx')} = E[(\mathbf{\Lambda_x}\boldsymbol{\xi} + \boldsymbol{\delta})(\mathbf{\Lambda_x}\boldsymbol{\xi} + \boldsymbol{\delta})'] \\ &= \mathbf{E}(\mathbf{\Lambda_x}\boldsymbol{\xi}\boldsymbol{\xi}'\mathbf{\Lambda_x'} + \mathbf{\Lambda_x}\boldsymbol{\xi}\boldsymbol{\delta}' + \boldsymbol{\delta}\boldsymbol{\xi}'\mathbf{\Lambda_x'} + \boldsymbol{\delta}\boldsymbol{\delta}') \\ &= \mathbf{\Lambda_x}\mathbf{E}(\boldsymbol{\xi}\boldsymbol{\xi}')\mathbf{\Lambda_x'} + \mathbf{\Lambda_x}\mathbf{E}(\boldsymbol{\xi}\boldsymbol{\delta}') + \mathbf{E}(\boldsymbol{\delta}\boldsymbol{\xi}')\mathbf{\Lambda_x'} + \mathbf{E}(\boldsymbol{\delta}\boldsymbol{\delta}') \\ &= \mathbf{\Lambda_x}\mathbf{\Phi}\mathbf{\Lambda_x'} + \mathbf{\Theta}_\delta \end{aligned} \tag{15.1.3}$$

Equation (15.1.3) corresponds to (14.7.2) under the assumption that $\mathbf{\Phi}$ is an identity matrix. Equation (15.1.3) expresses the $\frac{1}{2}[q(q + 1)]$ elements of the symmetric population variance–covariance matrix as functions of the $qn + \frac{1}{2}[n(n + 1)] + q$ unknown independent parameters in the matrices $\mathbf{\Lambda_x}$, $\mathbf{\Phi}$, and $\mathbf{\Theta}_\delta$. These three unknown matrices are estimated using equations analogous to (15.1.3) but with $\mathbf{Var(x)}$ replaced by the sample variance–covariance matrix for the observed predictor attributes.

In Chapter 14 we discussed exploratory factor analytic solutions to an equation like Eq. (15.1.3). There, $\mathbf{\Phi}$ was typically constrained to be $\mathbf{I}$—at least initially—and $\mathbf{\Theta}_\delta$ was obtained by unconstrained communality estimation. The proeedures of confirmatory factor analysis, which are useful here, constrain the parameters in $\mathbf{\Lambda_x}$, $\mathbf{\Phi}$, and $\mathbf{\Theta}_\delta$ in ways that allow the researcher to incorporate prior theoretical and empirical findings into a statistical model and to subject these findings to confirmation. Usually this takes the form of constraints specifying that certain elements, or a certain number of elements, of $\mathbf{\Lambda_x}$ and $\mathbf{\Phi}$ are zero. In the confirmatory factor model, sufficient constraints must be imposed that the parameters are identified uniquely.

15.1.2 Structural Equation Models

Models for the analysis of dependence specify how certain criteria are functions of certain predictors. The analysis of covariance structure models based on latent variables requires that the latent criterion factors be expressed as a function of the latent predictor factors. Further, these latent criterion variables may directly influence one another. Equation (15.1.4) expresses these relations in terms of the matrices identified in Table 15.2:

$$\boldsymbol{\eta} = \boldsymbol{\beta}\boldsymbol{\eta} + \mathbf{\Gamma}\boldsymbol{\xi} + \boldsymbol{\zeta} \tag{15.1.4}$$

where $\boldsymbol{\eta}$ is an $m \times 1$ vector of latent criterion factors, $\boldsymbol{\xi}$ is an $n \times 1$ vector of latent predictor factors, $\mathbf{\Gamma}$ is an $m \times n$ dependence coefficient matrix, and $\boldsymbol{\beta}$

is an $m \times m$ coefficient matrix (with zeros along the diagonal) expressing the direct effects of each η_i on other η_i. (Note that this $\boldsymbol{\beta}$ is related to the $\boldsymbol{\beta}_{IV}$ in LISREL IV by $\boldsymbol{\beta}_{IV} = \mathbf{I} - \boldsymbol{\beta}$.) $\boldsymbol{\zeta}$ is an $m \times 1$ vector of residuals, as η_i may not be completely determined by $\boldsymbol{\xi}$ and the other components of $\boldsymbol{\eta}$. Equation (15.1.4) is called a *structural equation model*, since each choice of zero and nonzero elements expresses a hypothesis about the structure linking predictors and criteria.

It is assumed that $\mathbf{E}(\boldsymbol{\eta}) = \mathbf{E}(\boldsymbol{\zeta}) = \mathbf{E}(\boldsymbol{\xi}) = \mathbf{0}$ and that $\boldsymbol{\beta}$ is chosen such that $(\mathbf{I} - \boldsymbol{\beta})^{-1}$ exists. It is also assumed that the predictor latent factors are uncorrelated with the errors in the structural equations: $\mathbf{E}(\boldsymbol{\xi}\boldsymbol{\zeta}') = \mathbf{0}$. $\boldsymbol{\Psi}$ is the covariance matrix of $\boldsymbol{\zeta}$; that is, $\mathbf{E}(\boldsymbol{\zeta}\boldsymbol{\zeta}') = \boldsymbol{\Psi}$. We have now defined the matrices in the measurement and structural equation models, their expected values, and (with the exception of $\boldsymbol{\eta}$) their variance–covariance matrices. With these it is possible to derive the variance–covariance matrix of $\boldsymbol{\eta}$.

15.1.3 Variance–Covariance Matrix of the Observed Attributes

Since $\mathbf{I} - \boldsymbol{\beta}$ has an inverse, Eq. (15.1.4) can be rewritten $\boldsymbol{\eta} = (\mathbf{I} - \boldsymbol{\beta})^{-1}\boldsymbol{\Gamma}\boldsymbol{\xi} + (\mathbf{I} - \boldsymbol{\beta})^{-1}\boldsymbol{\zeta}$. As a result,

$$\begin{aligned}\mathbf{Var}(\boldsymbol{\eta}) &= \mathbf{E}(\boldsymbol{\eta}\boldsymbol{\eta}') \\ &= \mathbf{E}\left\{\left[(\mathbf{I} - \boldsymbol{\beta})^{-1}\boldsymbol{\Gamma}\boldsymbol{\xi} + (\mathbf{I} - \boldsymbol{\beta})\boldsymbol{\zeta}\right]\left[(\mathbf{I} - \boldsymbol{\beta})^{-1}\boldsymbol{\Gamma}\boldsymbol{\xi} + (\mathbf{I} - \boldsymbol{\beta})\boldsymbol{\zeta}\right]'\right\}\end{aligned}$$

Multiplying, distributing the expectation operator, and recalling that $\mathbf{E}(\boldsymbol{\xi}\boldsymbol{\xi}') = \boldsymbol{\Phi}$ and $\mathbf{E}(\boldsymbol{\xi}\boldsymbol{\zeta}') = \mathbf{0}$, we obtain

$$\begin{aligned}\mathbf{Var}(\boldsymbol{\eta}) &= (\mathbf{I} - \boldsymbol{\beta})^{-1}\boldsymbol{\Gamma}\boldsymbol{\Phi}\boldsymbol{\Gamma}'(\mathbf{I} - \boldsymbol{\beta}')^{-1} + (\mathbf{I} - \boldsymbol{\beta})^{-1}\boldsymbol{\Psi}(\mathbf{I} - \boldsymbol{\beta}')^{-1} \\ &= (\mathbf{I} - \boldsymbol{\beta})^{-1}(\boldsymbol{\Gamma}\boldsymbol{\Phi}\boldsymbol{\Gamma}' + \boldsymbol{\Psi})(\mathbf{I} - \boldsymbol{\beta}')^{-1} \qquad (15.1.5)\end{aligned}$$

From Eqs. (15.1.1) and (15.1.5), the variance–covariance matrix for the observed criterion attributes can be expressed

$$\mathbf{Var}(\mathbf{y}) = \boldsymbol{\Lambda}_y(\mathbf{I} - \boldsymbol{\beta})^{-1}(\boldsymbol{\Gamma}\boldsymbol{\Phi}\boldsymbol{\Gamma}' + \boldsymbol{\Psi})(\mathbf{I} - \boldsymbol{\beta}')^{-1}\boldsymbol{\Lambda}_y' + \boldsymbol{\theta}_\varepsilon \qquad (15.1.6)$$

since $\mathbf{E}(\boldsymbol{\eta}\boldsymbol{\varepsilon}') = \mathbf{0}$ and $\mathbf{E}(\boldsymbol{\varepsilon}\boldsymbol{\varepsilon}') = \boldsymbol{\theta}_\varepsilon$.

For $\boldsymbol{\eta} = (\mathbf{I} - \boldsymbol{\beta})^{-1}\boldsymbol{\Gamma}\boldsymbol{\xi} + (\mathbf{I} - \boldsymbol{\beta})^{-1}\boldsymbol{\zeta}$ we can obtain the covariance between the $\mathbf{x}$ and $\mathbf{y}$ of Eqs. (15.1.2) and (15.1.1):

$$\begin{aligned}\mathbf{Cov}(\mathbf{x}, \mathbf{y}) &= \mathbf{E}(\mathbf{x}\mathbf{y}') \\ &= \mathbf{E}\left\{(\boldsymbol{\Lambda}_x\boldsymbol{\xi} + \boldsymbol{\delta})\left[\boldsymbol{\Lambda}_y(\mathbf{I} - \boldsymbol{\beta})^{-1}\boldsymbol{\Gamma}\boldsymbol{\xi} + \boldsymbol{\Lambda}_y(\mathbf{I} - \boldsymbol{\beta})^{-1}\boldsymbol{\zeta} + \boldsymbol{\varepsilon}\right]'\right\} \\ &= \boldsymbol{\Lambda}_x\mathbf{E}(\boldsymbol{\xi}\boldsymbol{\xi}')\boldsymbol{\Gamma}'(\mathbf{I} - \boldsymbol{\beta}')^{-1}\boldsymbol{\Lambda}_y' \\ &= \boldsymbol{\Lambda}_x\boldsymbol{\Phi}\boldsymbol{\Gamma}'(\mathbf{I} - \boldsymbol{\beta}')^{-1}\boldsymbol{\Lambda}_y' \qquad (15.1.7)\end{aligned}$$

where we have used the property that the error terms are everywhere orthogonal. Equations (15.1.3), (15.1.6), and (15.1.7) contribute to the partitioned variance–covariance matrix for the full set of observed attributes:

$$\boldsymbol{\Sigma} = \mathbf{Var}[(\mathbf{y}, \mathbf{x})] = \begin{bmatrix} \mathbf{Var}(\mathbf{y}) & \mathbf{Cov}(\mathbf{y}, \mathbf{x}) \\ \mathbf{Cov}(\mathbf{x}, \mathbf{y}) & \mathbf{Var}(\mathbf{x}) \end{bmatrix}$$

$$= \left[\begin{array}{c|c} \boldsymbol{\Lambda}_y(\mathbf{I} - \boldsymbol{\beta})^{-1}(\boldsymbol{\Gamma\Phi\Gamma}' + \boldsymbol{\Psi})(\mathbf{I} - \boldsymbol{\beta}')^{-1}\boldsymbol{\Lambda}_y' + \boldsymbol{\Theta}_\varepsilon & \boldsymbol{\Lambda}_y(\mathbf{I} - \boldsymbol{\beta})^{-1}\boldsymbol{\Gamma\Phi\Lambda}_x' \\ \hline \boldsymbol{\Lambda}_x\boldsymbol{\Phi\Gamma}'(\mathbf{I} - \boldsymbol{\beta}')^{-1}\boldsymbol{\Lambda}_y' & \boldsymbol{\Lambda}_x\boldsymbol{\Phi\Lambda}_x' + \boldsymbol{\Theta}_\delta \end{array}\right] \tag{15.1.8}$$

The variance–covariance matrix in Eq. (15.1.8) contains four submatrices. The diagonal submatrices reflect the relations *within* each of the observed attribute sets expressed in terms of the latent predictor factors. The off-diagonal submatrices reflect the relations *between* the two sets of observed attributes, again expressed in terms of the latent predictor variables. The elements of the variance–covariance matrix $\boldsymbol{\Sigma}$ are then functions of the elements of the matrices $\boldsymbol{\Lambda}_y$, $\boldsymbol{\Lambda}_x$, $\boldsymbol{\beta}$, $\boldsymbol{\Gamma}$, $\boldsymbol{\Phi}$, $\boldsymbol{\Psi}$, $\boldsymbol{\Theta}_\varepsilon$, and $\boldsymbol{\Theta}_\delta$. Model testing and estimation is based on Eq. (15.1.8).

In applying this model to data, elements of these matrices, and hence of $\boldsymbol{\Sigma}$ are of three types:

1. fixed parameters that have been assigned certain values by theory;
2. constrained parameters that are unknown but are constrained in some way; and
3. free parameters that are unknown and unconstrained.

The constraints placed on the second type of parameter depend on the model being estimated.

15.1.4 Identifiability in Latent Variable Models

In order to obtain unique estimates of the free parameters in the variance–covariance matrix of Eq. (15.1.8), the parameters must be identified. Establishing whether a parameter is identified is not an easy task.

In practice, identification depends on the structure: the specific model under consideration and the specification of fixed, constrained, and free parameters. Although a particular structure leads to a unique covariance matrix, several different structures may generate the same variance–covariance matrix; if so, we say the structures are equivalent. Only if a parameter has the same value in all equivalent structures is that parameter identified.

Several necessary conditions for a model to be identified have been suggested, but counterexamples have subsequently been constructed for each. Because a parameter is identifiable if it can be expressed in terms of the covariances (or correlations) of the observed variables, a sufficient condition for identifiability is that the equations can be specified and manipulated algebraically so that each free parameter can be obtained in terms of the observed variances and covariances. For complex structural models, this is difficult—if not impossible.

The LISREL V program does not claim to solve this problem; indeed, it assumes that the model being estimated has been identified. Researchers overlooking this have assumed the LISREL solution for their parameters to be unique, only later to discover that some parameters in the model were not identified.

In complex structural models it is possible for only some of the parameters to be identified.

In general, useful models must be overidentified. Loosely speaking, there must be fewer free parameters than observed variances and covariances, so that the degree of fit can be compared to other models. Although no general conditions for identifiability have been discovered, the researcher typically relies on heuristic aids. We shall return to this issue after discussing estimation.

15.1.5 Estimation in Latent Variable Models

The covariance structure of (15.1.8) specifies that the population variances and covariances of the observed attributes are certain functions of parameter matrices. Let $\boldsymbol{\theta}$ be defined as the $r \times 1$ vector of unknown parameters, the free or constrained elements of $\boldsymbol{\Lambda}_x$, $\boldsymbol{\Lambda}_y$, $\boldsymbol{\beta}$, $\boldsymbol{\Gamma}$, $\boldsymbol{\Phi}$, $\boldsymbol{\Psi}$, $\boldsymbol{\Theta}_\varepsilon$, and $\boldsymbol{\Theta}_\delta$. Although models contain fixed parameters, by convention these are not included in $\boldsymbol{\theta}$. Also, if two or more parameters are constrained to be equal, $\boldsymbol{\theta}$ contains only one of the set. If we let $\hat{\boldsymbol{\Sigma}}(\boldsymbol{\theta})$ represent the estimated covariance matrix obtained from (15.1.8) using a particular $\boldsymbol{\theta}$, then the estimation problem can be stated as follows: We want to find that $\boldsymbol{\theta}$ that takes $\hat{\boldsymbol{\Sigma}}(\boldsymbol{\theta})$ closest to $\mathbf{S}$, the observed or sample variance–covariance matrix.

Three methods of fitting $\hat{\boldsymbol{\Sigma}}(\boldsymbol{\theta})$ to $\mathbf{S}$ have been widely considered. Unweighted least squares minimizes the function $U(\boldsymbol{\theta}) = \frac{1}{2}\operatorname{tr}[\mathbf{S} - \hat{\boldsymbol{\Sigma}}(\boldsymbol{\theta})]^2$. Generalized least squares minimizes the function

$$G(\boldsymbol{\theta}) = \tfrac{1}{2}\operatorname{tr}\left[\mathbf{I} - \mathbf{S}^{-1}\hat{\boldsymbol{\Sigma}}(\boldsymbol{\theta})\right]^2$$

Finally, the maximum likelihood method minimizes

$$F(\boldsymbol{\theta}) = \operatorname{tr}\left[\hat{\boldsymbol{\Sigma}}(\boldsymbol{\theta})^{-1}\mathbf{S}\right] - \prod_{j=1}^{p+q} D_j^2 \qquad (15.1.9)$$

where $\mathbf{D}^2$ is the basic diagonal of $\hat{\boldsymbol{\Sigma}}(\boldsymbol{\theta})$; the criterion function $F(\boldsymbol{\theta})$ is obtained by assuming $\mathbf{x}$ and $\mathbf{y}$ to be jointly multivariate normal and then writing the likelihood of N independent $(\mathbf{x}', \mathbf{y}')$ vector observations. [This should be compared with the results in Chapter 10, notably Eq. (10.7.3).] Minimizing $F(\boldsymbol{\theta})$ is equivalent to maximizing the likelihood of $\mathbf{S}$, the observed variance–covariance matrix.

Equation (15.1.9) defies minimization by analytic techniques, so LISREL V minimizes $F(\boldsymbol{\theta})$ numerically, by Fisher's scoring method and the method of Fletcher and Powell. The procedure uses the first derivatives $\partial F(\boldsymbol{\theta})/\partial\boldsymbol{\theta}$ and approximations to the second derivatives $\partial^2 F(\boldsymbol{\theta})/\partial\boldsymbol{\theta}^2$. Convergence to a local minimum of $F(\boldsymbol{\theta})$ from most starting points is rapid, but there is no guarantee that the method converges to an absolute minimum if there are several local minima.

Of the several properties of maximum likelihood estimators discussed in Chapters 10 and 11, the most important for our purposes is that relating $F(\hat{\boldsymbol{\theta}})$ to the chi-squared distribution. Consider

$$\kappa = n\left[F(\hat{\boldsymbol{\theta}}) - \prod_{j=1}^{p+q} d_j^2 - (p+q)\right] \tag{15.1.10}$$

which is based on a likelihood ratio. It has an asymptotic chi-squared distribution with degrees of freedom equal to the number of sample variances and covariances minus the number of parameters estimated. This is typically

$$\mathrm{df} = \tfrac{1}{2}(p+q)(p+q+1) - r \tag{15.1.11}$$

where p is the number of observed dependent attributes, q is the number of observed independent attributes, r is the total number of independent parameters estimated under the model, and $\mathbf{d}^2$ is the basic diagonal of the observed variance–covariance matrix $\mathbf{S}$.

As a result of the minimization, a matrix of the expected values of the second partial derivatives of $F(\boldsymbol{\theta})$ with respect to $\boldsymbol{\theta}$ is formed. This *information matrix* is helpful in assessing the identifiability of parameters in the model. In most cases, if this matrix has an inverse, the parameters in the model are identified. If not, the information matrix may indicate which parameters are not identified.

15.1.6 Interpreting the Results of Estimation

The results of estimation can be viewed from several perspectives. The researcher interested in the absolute fit of $\hat{\boldsymbol{\Sigma}}(\boldsymbol{\theta})$ to $\mathbf{S}$ can use the results of the likelihood ratio test in Eq. (15.1.10). A large chi-squared value is an indication that more information can be extracted from the data. A chi-squared value that is small or approximately equal to the degrees of freedom may be an indication that the model fits "too well," and such a model is likely to be unstable in future samples. However, the chi-squared value is very dependent on the sample size. With a large sample size, almost any model will be rejected; with a small sample size, the test has little power for most alternatives and the hypothesized model is unlikely to be rejected.

In an exploratory study several sources of information are available to direct model revision. For example, the standard errors of the estimated elements of $\boldsymbol{\theta}$ may help screen out poorly identified parameters. Similarly, the agreement between the observed attributes and their estimates provided by either the measurement model or a combination of the measurement and structural models (usually assessed by a correlation or squared correlation) can point to deficiencies in aspects of these models. All these and other guides are available in LISREL V. We strongly urge that model changes have external theoretical or empirical support and that any final model be cross-validated with new data.

Confirmatory research using latent variable structural equation models assumes that a series of theoretically meaningful models are to be compared. The comparisons are most informative if the models are hierarchically arranged. A nested or hierarchical model, say M_2, can be obtained from M_1 by constraining one or more parameters that are free in M_1, either assigning values to free parameters in M_1 or by equating previously free parameters. In such cases M_2 is a subset of the more general model M_1. This process of defining hierarchical models is conceptually and statistically similar to the process of defining constraints in Parts II and III, as well as in Chapter 13 for log linear models.

The fit of two models within the hierarchy is assessed by the difference in their κ statistics. This difference is distributed as chi-squared with number of degrees of freedom equal to the difference in the number of parameters constrained in the two models. Several other incremental fit indices have been proposed, but none as yet is in general use by researchers.

It should also be noted that the maximum likelihood estimation procedure and the interpretation of standard errors of the parameter estimates must take into account the assumptions of the procedure. In particular, estimates of standard errors may be inaccurate for analyses based on correlation matrices or data that violate the assumption of multivariate normality. In the examples in the next section, our analysis of correlation matrices is consistent with the LISREL V manual.

15.2 Some Applications of the JKW Model

The specification of latent variable models is generally facilitated by constructing path diagrams of the relations expressed in the measurement and structural equation models. Figure 15.1 shows a path diagram for one possible model for the attributes observed in a study by Calsyn and Kenny.

Calsyn and Kenny (see the list of additional reading) were interested in the relations between a student's ability and educational aspirations. They collected ratings for 556 white eighth-grade students on

x_1, an eight-item-scale measure of the student's evaluation of the student's own academic ability;

x_2, a five-item-scale measure of the student's perception of the parent's evaluation of the student's present school ability;

x_3, a five-item-scale measure of the student's perception of the teacher's evaluation of the student's present ability;

x_4, a five-item-scale measure of the student's perception of a friend's evaluation of the student's present school ability;

x_5, a single-item measure of how far the student desires to go in school; and

x_6, a single-item measure of how far the student expects to go in school.

The measurement models hypothesize that x_1, x_2, x_3, and x_4 are indicators of a latent factor called *ability* and x_5 and x_6 are indicators of a latent factor called *aspiration*. This is represented in the path diagram of Figure 15.1 by the lines from the latent factors (represented by circles) to the indicators or observed attributes (represented by boxes). The errors of measurement in the observed attributes are represented by the lines from the appropriate δ_i. The elements of the coefficient matrix Λ are represented by the λ_{ij}. The double-headed arrow between the two latent factors represents the assumed recursive (mutual causation) correlation between the latent factors.

Figure 15.1 Path diagram for ability and acquisition.

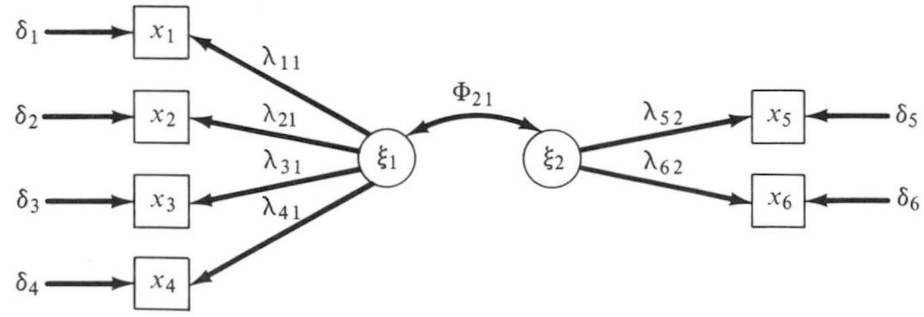

$$\mathbf{x}' = \Lambda_x \xi + \delta : \quad \begin{pmatrix} x_1 \\ x_2 \\ x_3 \\ x_4 \\ x_5 \\ x_6 \end{pmatrix} = \begin{pmatrix} \lambda_{11} & 0 \\ \lambda_{21} & 0 \\ \lambda_{31} & 0 \\ \lambda_{41} & 0 \\ 0 & \lambda_{52} \\ 0 & \lambda_{62} \end{pmatrix} \begin{pmatrix} \xi_1 \\ \xi_2 \end{pmatrix} + \begin{pmatrix} \delta_1 \\ \delta_2 \\ \delta_3 \\ \delta_4 \\ \delta_5 \\ \delta_6 \end{pmatrix}$$

Table 15.3 Correlations Among Ability and Aspiration Measures

x_1	1.00					
x_2	0.73	1.00				
x_3	0.70	0.68	1.00			
x_4	0.58	0.61	0.57	1.00		
x_5	0.46	0.43	0.40	0.37	1.00	
x_6	0.56	0.52	0.48	0.41	0.72	1.00

From Calsyn and Kenny (1977). Copyright 1977 by the American Psychological Association. Adapted by permission of the authors.

To investigate the correlation between the latent variables aspiration and ability, we need only three matrices of parameters in the JKW model: $\mathbf{\Lambda}_\mathbf{x}$ (a 6×2 matrix of factor loadings), $\mathbf{\Phi}$ (a 2×2 correlation matrix of factors), and $\mathbf{\Theta}_\delta$ (a 6×6 diagonal matrix of error variances). Table 15.3 gives the intercorrelations of the six ratings, and Table 15.4 the LISREL V maximum likelihood estimates of the factor loadings $\mathbf{\Lambda}_\mathbf{x}$, the correlation $\mathbf{\Phi}$ between the factors, the error variances $\mathbf{\Theta}_\delta$ in the measurement model, and other information useful in judging the degree of fit of the model. These statistics are defined in the LISREL V manual. LISREL also provides a printed table of standard errors of the parameter estimates, residual variances and covariances $\mathbf{S} - \hat{\mathbf{\Sigma}}$, normalized residuals, and several modification indices that point to possible parameters that might be defined as free parameters in a subsequent model. Inspection of Table 15.4 indicates that the diagonals of $\mathbf{\Phi}$ were fixed at unity; that certain elements of Λ, such as λ_{51}, λ_{61}, λ_{12}, λ_{22}, λ_{32}, and λ_{42}, were constrained to be 0; and that the measurement errors δ_i were assumed to be uncorrelated. The goodness of fit indices indicate that

Table 15.4 Model Estimates for Ability and Aspiration[a]

	Λ_1	Λ_2	$\mathbf{\Theta}_\delta$
	$\mathbf{\Lambda}_\mathbf{x}$		
x_1	0.863	0.000	0.255
x_2	0.849	0.000	0.279
x_3	0.805	0.000	0.352
x_4	0.695	0.000	0.517
x_5	0.000	0.775	0.399
x_6	0.000	0.929	0.137
	$\mathbf{\Phi}$		
Λ_1	1.000	0.666	
Λ_2	0.666	1.000	

[a]LISREL estimates maximum likelihood. The measures of goodness of fit for the model as a whole are $\chi^2 = 9.26$ with 8 df (probability level of 0.321), goodness of fit index 0.992, and root mean square residual 0.012. Adapted from Calsyn and Kenny (1977).

further constraints are unnecessary. Models with fewer constraints could be compared in fit with that of Figure 15.1, but the theory should be considered in formulating other models.

The example from Calsyn and Kenny was selected because it is simple enough to provide an introduction to operationalizing distinctions between observed attributes and latent variables and relating path diagrams to LISREL matrices. As a second example, we consider a study by Wheaton and co-workers, published in 1977, that was concerned with the stability over time of certain attitudes, including alienation, and the relation of such attitudes to background variables, including occupational status and education. The path diagram of the initial theoretical model is shown as model A in Figure 15.2. Model A assumes that the ratings of anomie and powerlessness in 1967 are indicators of the latent variable *alienation 1967*, that education and a socioeconomic index (SEI) are indicators of the latent variable *socioeconomic status* (SES), and that the ratings of anomie and powerlessness in 1971 are indicators of the latent variable *alienation 1971*. These are the three measurement models. In order to achieve identifiability of the parameters in Λ_x and Λ_y, the factor loading of anomie in 1967 and 1971 and the factor loading of SEI were fixed at 1. This is a common heuristic rule for achieving identifiability in parameters of the measurement model.

In fitting this model to the variance–covariance matrix of observed variables, $\chi^2 = 71.54$ with 6 df, an unacceptable fit. As in many studies where the measurements are repeated on entities over time, there is a tendency for measurement errors (ε_1 and ε_3) to be correlated. This possibility is supported by the LISREL V modification indices for model A in Figure 15.2, suggesting that allowing ε_1 and ε_3 to be correlated would lead to an improved fit for the overall model. Model B in Figure 15.2 allows these two measurement errors to be correlated. For the overall model, $\chi^2 = 6.33$ with 5 df. The drop in χ^2 is therefore $71.54 - 6.33 = 65.21$ with $6 - 5 = 1$ degree of freedom.

A more complete analysis of these models is reported in the LISREL V manual, where many other examples are also analyzed. One example examines the effects of assigning discrete values to data generated from a multivariate normal population.

The two examples in this section are single-sample analyses in which interest is in the covariances and correlations of the latent variables. By analyzing the product moment matrix of raw scores rather than deviations from the mean, latent variable structural equation models allow the researcher to test hypotheses about the mean structure of the data in the different groups. The researcher can also test the equality of factor patterns or error variances and of factor covariance matrices in various groups. It therefore provides an extension of the multivariate general linear model discussed in Part III, which assumed equality of covariance matrices, since

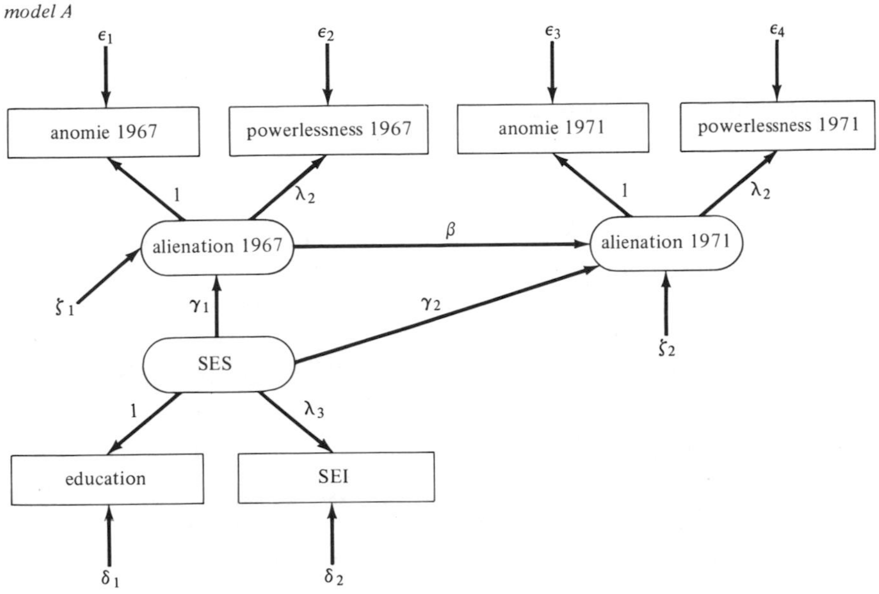

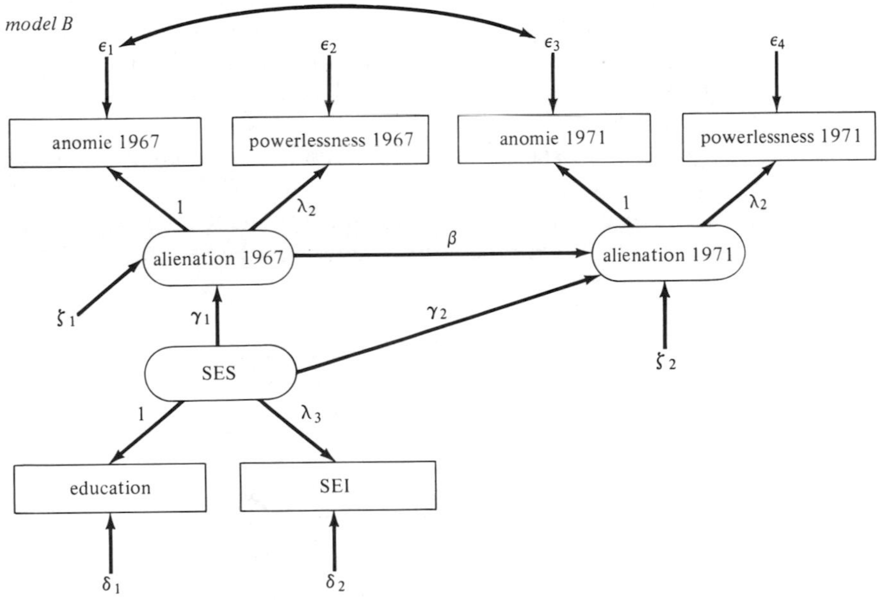

Figure 15.2 Models for the study of stability of alienation.

in the LISREL V implementation hypotheses about mean structures can be tested in the presence of heterogeneous covariance matrices. In the LISREL V manual Jöreskog and Sorbom provide a discussion, with an example, of testing models with structured means.

15.3 Special Cases of the JKW Model

Equations (15.1.1) and (15.1.2) define the measurement models, Eq. (15.1.4) the structural equation model, and Eq. (15.1.8) the covariance structure of the observed attributes **x** and **y**. With proper assumptions about the latent factors and measured attributes, the models of Parts II and III as well as Chapter 14 can be expressed as special cases of the JKW model.

15.3.1 Factor Analysis

We characterized factor analysis as the analysis of interdependence. Thus, to express the factor analysis model as a special case of the JKW model, we do not need to assume a structural equation model such as (15.1.4). Additionally, we have already shown that the exploratory factor analysis models of Chapter 14 are a special case of the general JKW model for which only measurement equation (15.1.3) needs to be specified. We also suggested that most initial exploratory solutions implicitly assume that the correlations among factors are zero and $\mathbf{\Phi}$ is an identity matrix. Figure 15.3 shows the path diagram for a factor analysis model with five measured variables and two latent variables.

Model B in Figure 15.3 shows a competing model for which five measured variables are indicators of a single factor model. The JKW model allows us to compare statistically the fit of the two measurement models. The constraint tested in model B fixes the correlation of the two latent variables at 1.00. This defines a hierarchy of models.

Figure 15.3 Path diagrams of two latent variable factor analysis models.

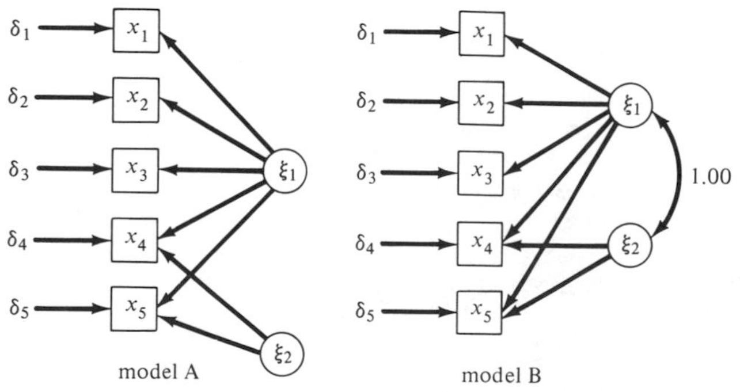

15.3.2 Linear Regression on Observed Attributes

In the analysis of dependence using a linear regression model, discussed in Part II, we identify some attribute as a criterion and others as predictors. Such a model can be treated as a special case of the JKW model by assuming $\boldsymbol{\beta} = \mathbf{0}$ in the structural equation model and expressing $\mathbf{y}$ as a function of the observed values of $\mathbf{x}$:

$$\mathbf{y} = \boldsymbol{\Gamma}\mathbf{x} + \boldsymbol{\zeta}$$

In this case $\boldsymbol{\xi}$ and $\mathbf{x}$ are equated: there are no latent variables.

15.3.3 Population Correlation Models

If we define a totally recursive system in which each observed attribute is related to every other observed attribute, we have a system that allows us to calculate the multiple R^2 for each $\mathbf{y}$ and partition it into direct relations between attributes and indirect relations through an association with other attributes. Such a model can be generated as a special case of the JKW model by assuming a measurement model in which each attribute is a perfect indicator ($r = 1.00$) of $\boldsymbol{\eta}$ and $\boldsymbol{\xi}$, and expressing the structural model $\mathbf{y} = \boldsymbol{\beta}\mathbf{y} + \boldsymbol{\Gamma}\mathbf{x} + \boldsymbol{\zeta}$.

15.3.4 Finding Principal Components with LISREL

By working only with the measurement model for $\mathbf{x}$, fixing $\boldsymbol{\Phi} = \mathbf{I}$ and $\boldsymbol{\Theta}_\delta = \mathbf{0}$, and freely estimating all the elements in just the first column of $\boldsymbol{\Lambda}_x$, one can find the first principal component of the correlation matrix. Subsequent principal components can be found in several ways; for example, we could constrain the elements of the first column of $\boldsymbol{\Lambda}_x$ to be the values in the first principal component and freely estimate the elements in the second column.

Other factoring procedures, such as the triangular procedure, can also be represented in terms of LISREL. Triangular factors can be derived by setting the pivotal element of $\boldsymbol{\Lambda}_x$ equal to 1.00 and freely estimating the other elements in that column of $\boldsymbol{\Lambda}_x$. Subsequent triangular factors can be obtained by inputting the residual matrix and fixing the pivotal element at 1.00.

15.4 Summary

With the multivariate general linear model developed in Chapters 11 and 12, the dependence of a set of criterion attributes on a set of predictor attributes can be evaluated. In Chapter 14 we presented models of interdependence by which interrelations among a set of observed attributes or their dependence on a common set of unobserved or latent factors could be studied. The present chapter integrated the two approaches. The analysis of covariance structures approach developed by Jöreskog and others provided the vehicle: the presence of both measurement (or factor analysis) models and structural equation (or dependence) models in the same analysis permitted the integration.

Following development of the general model, problems in obtaining a satisfactory solution were discussed. Several applications were presented, and, in conclusion, the techniques discussed earlier in this book were reinterpreted as special instances of the analysis of covariance structures approach.

Exercises

To analyze complex structural models it is necessary to use a computer program such as LISREL V. Working out the exercises in the appropriate manual will clarify many of the issues in latent variable structural modeling of covariance and mean structures.

1. Construct the path diagram for predicting a single criterion attribute from three predictor attributes. Label the paths using the appropriate LISREL V notation. Construct an alternative model with paths for only two of the predictor attributes to the single criterion attribute. Write the structural equations for these two models. What is the difference in their degrees of freedom? Solve for the paths in terms of the variances and covariances of the observed attributes.

2. Assume that two of the predictors in Exercise 1 are indicators of a single latent factor and the third predictor is an indicator of another latent factor. Draw the path diagram for this model. Are all the parameters identified? What constraints might be reasonable?

3. Construct a path diagram for the zero-order correlations among x_1, x_2, and x_3.

4. Draw a path diagram for the following partial correlations for three attributes: $r_{x_1 x_2 \cdot x_3}$ and $r_{x_1 x_3 \cdot x_2}$

5. Construct a path diagram for the semipartial correlations $r_{x_1(x_2 \cdot x_3)}$ and $r_{x_1(x_3 \cdot x_2)}$. How do these diagrams differ from the diagram in Exercise 4? If you interpreted these correlations within the context of a structural model, what is the difference in structural interpretation of the partial and semipartial correlations in Exercises 4 and 5?

6. For the data in the College Prediction Study, consider the three grade point averages as indicators of a latent factor and the three precollege test scores as indicators of a second latent factor. Define the measurement models in terms of

the LISREL V notation. Specify the structural equation to test the hypothesis that the two latent variables are uncorrelated. How does this differ theoretically from doing a multivariate regression as discussed in Chapter 12 with the three grade point averages as criteria attributes and the three precollege test scores as predictor attributes? If you have access to LISREL, test the model of no correlation between the two latent factors.

7. This chapter has presented the latent variable model following Jöreskog. Express the JKW model as an extension of the representation presented in Chapter 12.

8. Some authors have argued that the distinction between measurement and structural models is artificial. Construct a measurement model that will encompass both the LISREL V measurement and structural models.

Additional Reading

The following two papers present excellent reviews of the theory and application of latent variable models:

Bentler, P. M. (1980). Multivariate analysis with latent variables: Causal modelling, *Annual Review of Psychology* 31:419–456.

Bielby, W. T., and Hauser, R. M. (1977). Structural equation models, *Annual Review of Sociology* 3:137–161.

The following papers present theoretically oriented discussions of covariance structure models:

Bentler, P. M., and Weeks, D. G. (1980). Linear structural equations with latent variables, *Psychometrika* 45:289–308.

McDonald, R. P. (1978). A simple comprehensive model for the analysis of covariance structures, *British Journal of Mathematical and Statistical Psychology* 31:59–72.

The following books present elementary introductions to causal modeling with latent variables:

Heise, D. R. (1975). *Causal Analysis*. New York: Wiley.

Kenny, D. A. (1979). *Correlation and Causality*. New York: Wiley.

The following books present collections of papers focusing on more advanced aspects of latent variable structural equation models:

Goldberger, A. S., and Duncan, O. D., eds. (1973). *Structural Equation Models in the Social Sciences*. New York: Seminar Press.

Jöreskog, K. G., and Sorbom, D. (1979). *Advances in Factor Analysis and Structural Equation Models*. Cambridge, MA: Abt Books.

Jöreskog, K. G., and Wold, H., eds. (1983). *Systems Under Indirect Observation: Causality, Structure, Prediction*. Amsterdam: North-Holland.

The following works are referred to in the text:

Calsyn, R. J., and Kenny, D. A. (1977). Self-concept of ability and perceived evaluation of others: Cause or effect of academic achievement?, *Journal of Educational Psychology* 69:136–145.

Jöreskog, K. G., and Sorbom, D. (1981). *LISREL V Manual*. Chicago: National Educational Resources.

Wheaton, B., Muthen, B., Alwin, D., and Summers, G. (1977). In *Sociological Methodology* (D. R. Heise, ed.), San Francisco: Josey Bass.

Appendix

Table 1 Upper Percentage Points of the Chi-Squared Distributions

$[\chi^2_{\mathrm{df},\,\alpha}]$

	α						
df	0.250	0.100	0.050	0.025	0.010	0.005	0.001
1	1.32330	2.70554	3.84146	5.02389	6.63490	7.87944	10.828
2	2.77259	4.60517	5.99146	7.37776	9.21034	10.5966	13.816
3	4.10834	6.21539	7.81473	9.34840	11.3449	12.8382	16.266
4	5.38527	7.77944	9.48773	11.1433	13.2767	14.8603	18.467
5	6.62568	9.23636	11.0705	12.8325	15.0863	16.7496	20.515
6	7.84080	10.6446	12.5916	14.4494	16.8119	18.5476	22.458
7	9.03715	12.0170	14.0671	16.0128	18.4753	20.2777	24.322
8	10.2189	13.3616	15.5073	17.5345	20.0902	21.9550	26.125
9	11.3888	14.6837	16.9190	19.0228	21.6660	23.5894	27.8777
10	12.5489	15.9872	18.3070	20.4832	23.2093	25.1882	29.588
11	13.7007	17.2750	19.6751	21.9200	24.7250	26.7568	31.264
12	14.8454	18.5493	21.0261	23.3367	26.2170	28.2995	32.909
13	15.9839	19.8119	22.3620	24.7356	27.6882	29.8195	34.528
14	17.1169	21.0641	23.6848	26.1189	29.1412	31.3194	36.123
15	18.2451	22.3071	24.9958	27.4884	30.5779	32.8013	37.697
16	19.3689	23.5418	26.2962	28.8454	31.9999	34.2672	39.252
17	20.4887	24.7690	27.5871	30.1910	33.4087	35.7185	40.790
18	21.6049	25.9894	28.8693	31.5264	34.8053	37.1565	42.312
19	22.7178	27.2036	30.1435	32.8523	36.1909	38.5823	43.820
20	23.8277	28.4120	31.4104	34.1696	37.5662	39.9968	45.315
21	24.9348	29.6151	32.6706	35.4789	38.9322	41.4011	46.797
22	26.0393	30.8133	33.9244	36.7807	40.2894	42.7957	48.268
23	27.1413	32.0069	35.1725	38.0756	41.6384	44.1813	49.728
24	28.2412	33.1962	36.4150	39.3641	42.9798	45.5585	51.179
25	29.3389	34.3816	37.6525	40.6465	44.3141	46.9279	52.618
26	30.4346	35.5632	38.8851	41.9232	45.6417	48.2899	54.052
27	31.5284	36.7412	40.1133	43.1945	46.9629	49.6449	55.476
28	32.6205	37.9159	41.3371	44.4608	48.2782	50.9934	56.892
29	33.7109	39.0875	42.5570	45.7223	49.5879	52.3356	58.301
30	34.7997	40.2560	43.7730	46.9792	50.8922	53.6720	59.703
40	45.6160	51.8051	55.7585	59.3417	63.6907	66.7660	73.402
50	56.3336	63.1671	67.5048	71.4202	76.1539	79.4900	86.661
60	66.9815	74.3970	79.0819	83.2977	88.3794	91.9517	99.607
70	77.5767	85.5270	90.5312	95.0232	100.425	104.215	112.317
80	88.1303	96.5782	101.879	106.629	112.329	116.321	124.839
90	98.6499	107.565	113.145	118.136	124.116	128.299	137.208
100	109.141	118.498	124.342	129.561	135.807	140.169	149.449

From Table 8, E. S. Pearson and H. O. Hartley (eds.), *Biometrika Tables for Statisticians*, Vol. 1 (3rd ed.). New York: Cambridge, 1966. Reproduced by permission of the editors and trustees of *Biometrika*.

Table 2 Upper Percentage Points of the F Distributions
$[F_{\text{df numerator, df denominator}, \alpha}]$

df for denominator	α	df for numerator 1	2	3	4	5	6	7	8	9	10	11
1	0.25	5.83	7.50	8.20	8.58	8.82	8.98	9.10	9.19	9.26	9.32	9.36
	0.10	39.9	49.5	53.6	55.8	57.2	58.2	58.9	59.4	59.9	60.2	60.5
	0.05	161	200	216	225	230	234	237	239	241	242	243
2	0.25	2.57	3.00	3.15	3.23	3.28	3.31	3.34	3.35	3.37	3.38	3.39
	0.10	8.53	9.00	9.16	9.24	9.29	9.33	9.35	9.37	9.38	9.39	9.40
	0.05	18.5	19.0	19.2	19.2	19.3	19.3	19.4	19.4	19.4	19.4	19.4
	0.01	98.5	99.0	99.2	99.2	99.3	99.3	99.4	99.4	99.4	99.4	99.4
3	0.25	2.02	2.28	2.36	2.39	2.41	2.42	2.43	2.44	2.44	2.44	2.45
	0.10	5.54	5.46	5.39	5.34	5.31	5.28	5.27	5.25	5.24	5.23	5.22
	0.05	10.1	9.55	9.28	9.12	9.01	8.94	8.89	8.85	8.81	8.79	8.76
	0.01	34.1	30.8	29.5	28.7	28.2	27.9	27.7	27.5	27.3	27.2	27.1
4	0.25	1.81	2.00	2.05	2.06	2.07	2.08	2.08	2.08	2.08	2.08	2.08
	0.10	4.54	4.32	4.19	4.11	4.05	4.01	3.98	3.95	3.94	3.92	3.91
	0.05	7.71	6.94	6.59	6.39	6.26	6.16	6.09	6.04	6.00	5.96	5.94
	0.01	21.2	18.0	16.7	16.0	15.5	15.2	15.0	14.8	14.7	14.5	14.4
5	0.25	1.69	1.85	1.88	1.89	1.89	1.89	1.89	1.89	1.89	1.89	1.89
	0.10	4.06	3.78	3.62	3.52	3.45	3.40	3.37	3.34	3.32	3.30	3.28
	0.05	6.61	5.79	5.41	5.19	5.05	4.95	4.88	4.82	4.77	4.74	4.71
	0.01	16.3	13.3	12.1	11.4	11.0	10.7	10.5	10.3	10.2	10.1	9.96
6	0.25	1.62	1.76	1.78	1.79	1.79	1.78	1.78	1.78	1.77	1.77	1.77
	0.10	3.78	3.46	3.29	3.18	3.11	3.05	3.01	2.98	2.96	2.94	2.92
	0.05	5.99	5.14	4.76	4.53	4.39	4.28	4.21	4.15	4.10	4.06	4.03
	0.01	13.7	10.9	9.78	9.15	8.75	8.47	8.26	8.10	7.98	7.87	7.79
7	0.25	1.57	1.70	1.72	1.72	1.71	1.71	1.70	1.70	1.69	1.69	1.69
	0.10	3.59	3.26	3.07	2.96	2.88	2.83	2.78	2.75	2.72	2.70	2.68
	0.05	5.59	4.74	4.35	4.12	3.97	3.87	3.79	3.73	3.68	3.64	3.60
	0.01	12.2	9.55	8.45	7.85	7.46	7.19	6.99	6.84	6.72	6.62	6.54
8	0.25	1.54	1.66	1.67	1.66	1.66	1.65	1.64	1.64	1.63	1.63	1.63
	0.10	3.46	3.11	2.92	2.81	2.73	2.67	2.62	2.59	2.56	2.54	2.52
	0.05	5.32	4.46	4.07	3.84	3.69	3.58	3.50	3.44	3.39	3.35	3.31
	0.01	11.3	8.65	7.59	7.01	6.63	6.37	6.18	6.03	5.91	5.81	5.73
9	0.25	1.51	1.62	1.63	1.63	1.62	1.61	1.60	1.60	1.59	1.59	1.58
	0.10	3.36	3.01	2.81	2.69	2.61	2.55	2.51	2.47	2.44	2.42	2.40
	0.05	5.12	4.26	3.86	3.63	3.48	3.37	3.29	3.23	3.18	3.14	3.10
	0.01	10.6	8.02	6.99	6.42	6.06	5.80	5.61	5.47	5.35	5.26	5.18
10	0.25	1.49	1.60	1.60	1.59	1.59	1.58	1.57	1.56	1.56	1.55	1.55
	0.10	3.29	2.92	2.73	2.61	2.52	2.46	2.41	2.38	2.35	2.32	2.30
	0.05	4.96	4.10	3.71	3.48	3.33	3.22	3.14	3.07	3.02	2.98	2.94
	0.01	10.0	7.56	6.55	5.99	5.64	5.39	5.20	5.06	4.94	4.85	4.77

From Table 18, E. S. Pearson and H. O. Hartley (eds.), *Biometrika Tables for Statisticians*, Vol. 1 (3rd ed.). New York: Cambridge, 1966. Reproduced by permission of the editors and trustees of *Biometrika*.

Table 2 (*continued*)

df for numerator												
12	15	20	24	30	40	50	60	100	120	200	500	∞
9.41	9.49	9.58	9.63	9.67	9.71	9.74	9.76	9.78	9.80	9.82	9.84	9.85
60.7	61.2	61.7	62.0	62.3	62.5	62.7	62.8	63.0	63.1	63.2	63.3	63.3
244	246	248	249	250	251	252	252	253	253	254	254	254
3.39	3.41	3.43	3.43	3.44	3.45	3.45	3.46	3.47	3.47	3.48	3.48	3.48
9.41	9.42	9.44	9.45	9.46	9.47	9.47	9.47	9.48	9.48	9.49	9.49	9.49
19.4	19.4	19.4	19.5	19.5	19.5	19.5	19.5	19.5	19.5	19.5	19.5	19.5
99.4	99.4	99.4	99.5	99.5	99.5	99.5	99.5	99.5	99.5	99.5	99.5	99.5
2.45	2.46	2.46	2.46	2.47	2.47	2.47	2.47	2.47	2.47	2.47	2.47	2.47
5.22	5.20	5.18	5.18	5.17	5.16	5.15	5.15	5.14	5.14	5.14	5.14	5.13
8.74	8.70	8.66	8.64	8.62	8.59	8.58	8.57	8.55	8.55	8.54	8.53	8.53
27.1	26.9	26.7	26.6	26.5	26.4	26.4	26.3	26.2	26.2	26.2	26.1	26.1
2.08	2.08	2.08	2.08	2.08	2.08	2.08	2.08	2.08	2.08	2.08	2.08	2.08
3.90	3.87	3.84	3.83	3.82	3.80	3.80	3.79	3.78	3.78	3.77	3.76	3.76
5.91	5.86	5.80	5.77	5.75	5.72	5.70	5.69	5.66	5.66	5.65	5.64	5.63
14.4	14.2	14.0	13.9	13.8	13.7	13.7	13.7	13.6	13.6	13.5	13.5	13.5
1.89	1.89	1.88	1.88	1.88	1.88	1.88	1.87	1.87	1.87	1.87	1.87	1.87
3.27	3.24	3.21	3.19	3.17	3.16	3.15	3.14	3.13	3.12	3.12	3.11	3.10
4.68	4.62	4.56	4.53	4.50	4.46	4.44	4.43	4.41	4.40	4.39	4.37	4.36
9.89	9.72	9.55	9.47	9.38	9.29	9.24	9.20	9.13	9.11	9.08	9.04	9.02
1.77	1.76	1.76	1.75	1.75	1.75	1.75	1.74	1.74	1.74	1.74	1.74	1.74
2.90	2.87	2.84	2.82	2.80	2.78	2.77	2.76	2.75	2.74	2.73	2.73	2.72
4.00	3.94	3.87	3.84	3.81	3.77	3.75	3.74	3.71	3.70	3.69	3.68	3.67
7.72	7.56	7.40	7.31	7.23	7.14	7.09	7.06	6.99	6.97	6.93	6.90	6.88
1.68	1.68	1.67	1.67	1.66	1.66	1.66	1.65	1.65	1.65	1.65	1.65	1.65
2.67	2.63	2.59	2.58	2.56	2.54	2.52	2.51	2.50	2.49	2.48	2.48	2.47
3.57	3.51	3.44	3.41	3.38	3.34	3.32	3.30	3.27	3.27	3.25	3.24	3.23
6.47	6.31	6.16	6.07	5.99	5.91	5.86	5.82	5.75	5.74	5.70	5.67	5.65
1.62	1.62	1.61	1.60	1.60	1.59	1.59	1.59	1.58	1.58	1.58	1.58	1.58
2.50	2.46	2.42	2.40	2.38	2.36	2.35	2.34	2.32	2.32	2.31	2.30	2.29
3.28	3.22	3.15	3.12	3.08	3.04	3.02	3.01	2.97	2.97	2.95	2.94	2.93
5.67	5.52	5.36	5.28	5.20	5.12	5.07	5.03	4.96	4.95	4.91	4.88	4.86
1.58	1.57	1.56	1.56	1.55	1.55	1.54	1.54	1.53	1.53	1.53	1.53	1.53
2.38	2.34	2.30	2.28	2.25	2.23	2.22	2.21	2.19	2.18	2.17	2.17	2.16
3.07	3.01	2.94	2.90	2.86	2.83	2.80	2.79	2.76	2.75	2.73	2.72	2.71
5.11	4.96	4.81	4.73	4.65	4.57	4.52	4.48	4.42	4.40	4.36	4.33	4.31
1.54	1.53	1.52	1.52	1.51	1.51	1.50	1.50	1.49	1.49	1.49	1.48	1.48
2.28	2.24	2.20	2.18	2.16	2.13	2.12	2.11	2.09	2.08	2.07	2.06	2.06
2.91	2.85	2.77	2.74	2.70	2.66	2.64	2.62	2.59	2.58	2.56	2.55	2.54
4.71	4.56	4.41	4.33	4.25	4.17	4.12	4.08	4.01	4.00	3.96	3.93	3.91

(*continued*)

Table 2 Upper Percentage Points of the *F* Distributions (*continued*)
$[F_{\text{df numerator, df denominator}, \alpha}]$

df for denominator	α	1	2	3	4	5	6	7	8	9	10	11
11	0.25	1.47	1.58	1.58	1.57	1.56	1.55	1.54	1.53	1.53	1.52	1.52
	0.10	3.23	2.86	2.66	2.54	2.45	2.39	2.34	2.30	2.27	2.25	2.23
	0.05	4.84	3.98	3.59	3.36	3.20	3.09	3.01	2.95	2.90	2.85	2.82
	0.01	9.65	7.21	6.22	5.67	5.32	5.07	4.89	4.74	4.63	4.54	4.46
12	0.25	1.46	1.56	1.56	1.55	1.54	1.53	1.52	1.51	1.51	1.50	1.50
	0.10	3.18	2.81	2.61	2.48	2.39	2.33	2.28	2.24	2.21	2.19	2.17
	0.05	4.75	3.89	3.49	3.26	3.11	3.00	2.91	2.85	2.80	2.75	2.72
	0.01	9.33	6.93	5.95	5.41	5.06	4.82	4.64	4.50	4.39	4.30	4.22
13	0.25	1.45	1.55	1.55	1.53	1.52	1.51	1.50	1.49	1.49	1.48	1.47
	0.10	3.14	2.76	2.56	2.43	2.35	2.28	2.23	2.20	2.16	2.14	2.12
	0.05	4.67	3.81	3.41	3.18	3.03	2.92	2.83	2.77	2.71	2.67	2.63
	0.01	9.07	6.70	5.74	5.21	4.86	4.62	4.44	4.30	4.19	4.10	4.02
14	0.25	1.44	1.53	1.53	1.52	1.51	1.50	1.49	1.48	1.47	1.46	1.46
	0.10	3.10	2.73	2.52	2.39	2.31	2.24	2.19	2.15	2.12	2.10	2.08
	0.05	4.60	3.74	3.34	3.11	2.96	2.85	2.76	2.70	2.65	2.60	2.57
	0.01	8.86	6.51	5.56	5.04	4.69	4.46	4.28	4.14	4.03	3.94	3.86
15	0.25	1.43	1.52	1.52	1.51	1.49	1.48	1.47	1.46	1.46	1.45	1.44
	0.10	3.07	2.70	2.49	2.36	2.27	2.21	2.16	2.12	2.09	2.06	2.04
	0.05	4.54	3.68	3.29	3.06	2.90	2.79	2.71	2.64	2.59	2.54	2.51
	0.01	8.68	6.36	5.42	4.89	4.56	4.32	4.14	4.00	3.89	3.80	3.73
16	0.25	1.42	1.51	1.51	1.50	1.48	1.47	1.46	1.45	1.44	1.44	1.44
	0.10	3.05	2.67	2.46	2.33	2.24	2.18	2.13	2.09	2.06	2.03	2.01
	0.05	4.49	3.63	3.24	3.01	2.85	2.74	2.66	2.59	2.54	2.49	2.46
	0.01	8.53	6.23	5.29	4.77	4.44	4.20	4.03	3.89	3.78	3.69	3.62
17	0.25	1.42	1.51	1.50	1.49	1.47	1.46	1.45	1.44	1.43	1.43	1.42
	0.10	3.03	2.64	2.44	2.31	2.22	2.15	2.10	2.06	2.03	2.00	1.98
	0.05	4.45	3.59	3.20	2.96	2.81	2.70	2.61	2.55	2.49	2.45	2.41
	0.01	8.40	6.11	5.18	4.67	4.34	4.10	3.93	3.79	3.68	3.59	3.52
18	0.25	1.41	1.50	1.49	1.48	1.46	1.45	1.44	1.43	1.42	1.42	1.41
	0.10	3.01	2.62	2.42	2.29	2.20	2.13	2.08	2.04	2.00	1.98	1.96
	0.05	4.41	3.55	3.16	2.93	2.77	2.66	2.58	2.51	2.46	2.41	2.37
	0.01	8.29	6.01	5.09	4.58	4.25	4.01	3.84	3.71	3.60	3.51	3.43
19	0.25	1.41	1.49	1.49	1.47	1.46	1.44	1.43	1.42	1.41	1.41	1.40
	0.10	2.99	2.61	2.40	2.27	2.18	2.11	2.06	2.02	1.98	1.96	1.94
	0.05	4.38	3.52	3.13	2.90	2.74	2.63	2.54	2.48	2.42	2.38	2.34
	0.01	8.18	5.93	5.01	4.50	4.17	3.94	3.77	3.63	3.52	3.43	3.36
20	0.25	1.40	1.49	1.48	1.46	1.45	1.44	1.43	1.42	1.41	1.40	1.39
	0.10	2.97	2.59	2.38	2.25	2.16	2.09	2.04	2.00	1.96	1.94	1.92
	0.05	4.35	3.49	3.10	2.87	2.71	2.60	2.51	2.45	2.39	2.35	2.31
	0.01	8.10	5.85	4.94	4.43	4.10	3.87	3.70	3.56	3.46	3.37	3.29

Table 2 (*continued*)

df for numerator												
12	15	20	24	30	40	50	60	100	120	200	500	∞
1.51	1.50	1.49	1.49	1.48	1.47	1.47	1.47	1.46	1.46	1.46	1.45	1.45
2.21	2.17	2.12	2.10	2.08	2.05	2.04	2.03	2.00	2.00	1.99	1.98	1.97
2.79	2.72	2.65	2.61	2.57	2.53	2.51	2.49	2.46	2.45	2.43	2.42	2.40
4.40	4.25	4.10	4.02	3.94	3.86	3.81	3.78	3.71	3.69	3.66	3.62	3.60
1.49	1.48	1.47	1.46	1.45	1.45	1.44	1.44	1.43	1.43	1.43	1.42	1.42
2.15	2.10	2.06	2.04	2.01	1.99	1.97	1.96	1.94	1.93	1.92	1.91	1.90
2.69	2.62	2.54	2.51	2.47	2.43	2.40	2.38	2.35	2.34	2.32	2.31	2.30
4.16	4.01	3.86	3.78	3.70	3.62	3.57	3.54	3.47	3.45	3.41	3.38	3.36
1.47	1.46	1.45	1.44	1.43	1.42	1.42	1.42	1.41	1.41	1.40	1.40	1.40
2.10	2.05	2.01	1.98	1.96	1.93	1.92	1.90	1.88	1.88	1.86	1.85	1.85
2.60	2.53	2.46	2.42	2.38	2.34	2.31	2.30	2.26	2.25	2.23	2.22	2.21
3.96	3.82	3.66	3.59	3.51	3.43	3.38	3.34	3.27	3.25	3.22	3.19	3.17
1.45	1.44	1.43	1.42	1.41	1.41	1.40	1.40	1.39	1.39	1.39	1.38	1.38
2.05	2.01	1.96	1.94	1.91	1.89	1.87	1.86	1.83	1.83	1.82	1.80	1.80
2.53	2.46	2.39	2.35	2.31	2.27	2.24	2.22	2.19	2.18	2.16	2.14	2.13
3.80	3.66	3.51	3.43	3.35	3.27	3.22	3.18	3.11	3.09	3.06	3.03	3.00
1.44	1.43	1.41	1.41	1.40	1.39	1.39	1.38	1.38	1.37	1.37	1.36	1.36
2.02	1.97	1.92	1.90	1.87	1.85	1.83	1.82	1.79	1.79	1.77	1.76	1.76
2.48	2.40	2.33	2.29	2.25	2.20	2.18	2.16	2.12	2.11	2.10	2.08	2.07
3.67	3.52	3.37	3.29	3.21	3.13	3.08	3.05	2.98	2.96	2.92	2.89	2.87
1.43	1.41	1.40	1.39	1.38	1.37	1.37	1.36	1.36	1.35	1.35	1.34	1.34
1.99	1.94	1.89	1.87	1.84	1.81	1.79	1.78	1.76	1.75	1.74	1.73	1.72
2.42	2.35	2.28	2.24	2.19	2.15	2.12	2.11	2.07	2.06	2.04	2.02	2.01
3.55	3.41	3.26	3.18	3.10	3.02	2.97	2.93	2.86	2.84	2.81	2.78	2.75
1.41	1.40	1.39	1.38	1.37	1.36	1.35	1.35	1.34	1.34	1.34	1.33	1.33
1.96	1.91	1.86	1.84	1.81	1.78	1.76	1.75	1.73	1.72	1.71	1.69	1.69
2.38	2.31	2.23	2.19	2.15	2.10	2.08	2.06	2.02	2.01	1.99	1.97	1.96
3.46	3.31	3.16	3.08	3.00	2.92	2.87	2.83	2.76	2.75	2.71	2.68	2.65
1.40	1.39	1.38	1.37	1.36	1.35	1.34	1.34	1.33	1.33	1.32	1.32	1.32
1.93	1.89	1.84	1.81	1.78	1.75	1.74	1.72	1.70	1.69	1.68	1.67	1.66
2.34	2.27	2.19	2.15	2.11	2.06	2.04	2.02	1.98	1.97	1.95	1.93	1.92
3.37	3.23	3.08	3.00	2.92	2.84	2.78	2.75	2.68	2.66	2.62	2.59	2.57
1.40	1.38	1.37	1.36	1.35	1.34	1.33	1.33	1.32	1.32	1.31	1.31	1.30
1.91	1.86	1.81	1.79	1.76	1.73	1.71	1.70	1.67	1.67	1.65	1.64	1.63
2.31	2.23	2.16	2.11	2.07	2.03	2.00	1.98	1.94	1.93	1.91	1.89	1.88
3.30	3.15	3.00	2.92	2.84	2.76	2.71	2.67	2.60	2.58	2.55	2.51	2.49
1.39	1.37	1.36	1.35	1.34	1.33	1.33	1.32	1.31	1.31	1.30	1.30	1.29
1.89	1.84	1.79	1.77	1.74	1.71	1.69	1.68	1.65	1.64	1.63	1.62	1.61
2.28	2.20	2.12	2.08	2.04	1.99	1.97	1.95	1.91	1.90	1.88	1.86	1.84
3.23	3.09	2.94	2.86	2.78	2.69	2.64	2.61	2.54	2.52	2.48	2.44	2.42

(*continued*)

Table 2 Upper Percentage Points of the *F* Distributions (*continued*)
$[F_{\text{df numerator, df denominator}, \alpha}]$

df for denominator	α	df for numerator 1	2	3	4	5	6	7	8	9	10	11
22	0.25	1.40	1.48	1.47	1.45	1.44	1.42	1.41	1.40	1.39	1.39	1.38
	0.10	2.95	2.56	2.35	2.22	2.13	2.06	2.01	1.97	1.93	1.90	1.88
	0.05	4.30	3.44	3.05	2.82	2.66	2.55	2.46	2.40	2.34	2.30	2.26
	0.01	7.95	5.72	4.82	4.31	3.99	3.76	3.59	3.45	3.35	3.26	3.18
24	0.25	1.39	1.47	1.46	1.44	1.43	1.41	1.40	1.39	1.38	1.38	1.37
	0.10	2.93	2.54	2.33	2.19	2.10	2.04	1.98	1.94	1.91	1.88	1.85
	0.05	4.26	3.40	3.01	2.78	2.62	2.51	2.42	2.36	2.30	2.25	2.21
	0.01	7.82	5.61	4.72	4.22	3.90	3.67	3.50	3.36	3.26	3.17	3.09
26	0.25	1.38	1.46	1.45	1.44	1.42	1.41	1.39	1.38	1.37	1.37	1.36
	0.10	2.91	2.52	2.31	2.17	2.08	2.01	1.96	1.92	1.88	1.86	1.84
	0.05	4.23	3.37	2.98	2.74	2.59	2.47	2.39	2.32	2.27	2.22	2.18
	0.01	7.72	5.53	4.64	4.14	3.82	3.59	3.42	3.29	3.18	3.09	3.02
28	0.25	1.38	1.46	1.45	1.43	1.41	1.40	1.39	1.38	1.37	1.36	1.35
	0.10	2.89	2.50	2.29	2.16	2.06	2.00	1.94	1.90	1.87	1.84	1.81
	0.05	4.20	3.34	2.95	2.71	2.56	2.45	2.36	2.29	2.24	2.19	2.15
	0.01	7.64	5.45	4.57	4.07	3.75	3.53	3.36	3.23	3.12	3.03	2.96
30	0.25	1.38	1.45	1.44	1.42	1.41	1.39	1.38	1.37	1.36	1.35	1.35
	0.10	2.88	2.49	2.28	2.14	2.05	1.98	1.93	1.88	1.85	1.82	1.79
	0.05	4.17	3.32	2.92	2.69	2.53	2.42	2.33	2.27	2.21	2.16	2.13
	0.01	7.56	5.39	4.51	4.02	3.70	3.47	3.30	3.17	3.07	2.98	2.91
40	0.25	1.36	1.44	1.42	1.40	1.39	1.37	1.36	1.35	1.34	1.33	1.32
	0.10	2.84	2.44	2.23	2.09	2.00	1.93	1.87	1.83	1.79	1.76	1.73
	0.05	4.08	3.23	2.84	2.61	2.45	2.34	2.25	2.18	2.12	2.08	2.04
	0.01	7.31	5.18	4.31	3.83	3.51	3.29	3.12	2.99	2.89	2.80	2.73
60	0.25	1.35	1.42	1.41	1.38	1.37	1.35	1.33	1.32	1.31	1.30	1.29
	0.10	2.79	2.39	2.18	2.04	1.95	1.87	1.82	1.77	1.74	1.71	1.68
	0.05	4.00	3.15	2.76	2.53	2.37	2.25	2.17	2.10	2.04	1.99	1.95
	0.01	7.08	4.98	4.13	3.65	3.34	3.12	2.95	2.82	2.72	2.63	2.56
120	0.25	1.34	1.40	1.39	1.37	1.35	1.33	1.31	1.30	1.29	1.28	1.27
	0.10	2.75	2.35	2.13	1.99	1.90	1.82	1.77	1.72	1.68	1.65	1.62
	0.05	3.92	3.07	2.68	2.45	2.29	2.17	2.09	2.02	1.96	1.91	1.87
	0.01	6.85	4.79	3.95	3.48	3.17	2.96	2.79	2.66	2.56	2.47	2.40
200	0.25	1.33	1.39	1.38	1.36	1.34	1.32	1.31	1.29	1.28	1.27	1.26
	0.10	2.73	2.33	2.11	1.97	1.88	1.80	1.75	1.70	1.66	1.63	1.60
	0.05	3.89	3.04	2.65	2.42	2.26	2.14	2.06	1.98	1.93	1.88	1.84
	0.01	6.76	4.71	3.88	3.41	3.11	2.89	2.73	2.60	2.50	2.41	2.34
∞	0.25	1.32	1.39	1.37	1.35	1.33	1.31	1.29	1.28	1.27	1.25	1.24
	0.10	2.71	2.30	2.08	1.94	1.85	1.77	1.72	1.67	1.63	1.60	1.57
	0.05	3.84	3.00	2.60	2.37	2.21	2.10	2.01	1.94	1.88	1.83	1.79
	0.01	6.63	4.61	3.78	3.32	3.02	2.80	2.64	2.51	2.41	2.32	2.25

Table 2 (*continued*)

df for numerator												
12	15	20	24	30	40	50	60	100	120	200	500	∞
1.37	1.36	1.34	1.33	1.32	1.31	1.31	1.30	1.30	1.30	1.29	1.29	1.28
1.86	1.81	1.76	1.73	1.70	1.67	1.65	1.64	1.61	1.60	1.59	1.58	1.57
2.23	2.15	2.07	2.03	1.98	1.94	1.91	1.89	1.85	1.84	1.82	1.80	1.78
3.12	2.98	2.83	2.75	2.67	2.58	2.53	2.50	2.42	2.40	2.36	2.33	2.31
1.36	1.35	1.33	1.32	1.31	1.30	1.29	1.29	1.28	1.28	1.27	1.27	1.26
1.83	1.78	1.73	1.70	1.67	1.64	1.62	1.61	1.58	1.57	1.56	1.54	1.53
2.18	2.11	2.03	1.98	1.94	1.89	1.86	1.84	1.80	1.79	1.77	1.75	1.73
3.03	2.89	2.74	2.66	2.58	2.49	2.44	2.40	2.33	2.31	2.27	2.24	2.21
1.35	1.34	1.32	1.31	1.30	1.29	1.28	1.28	1.26	1.26	1.26	1.25	1.25
1.81	1.76	1.71	1.68	1.65	1.61	1.59	1.58	1.55	1.54	1.53	1.51	1.50
2.15	2.07	1.99	1.95	1.90	1.85	1.82	1.80	1.76	1.75	1.73	1.71	1.69
2.96	2.81	2.66	2.58	2.50	2.42	2.36	2.33	2.25	2.23	2.19	2.16	2.13
1.34	1.33	1.31	1.30	1.29	1.28	1.27	1.27	1.26	1.25	1.25	1.24	1.24
1.79	1.74	1.69	1.66	1.63	1.59	1.57	1.56	1.53	1.52	1.50	1.49	1.48
2.12	2.04	1.96	1.91	1.87	1.82	1.79	1.77	1.73	1.71	1.69	1.67	1.65
2.90	2.75	2.60	2.52	2.44	2.35	2.30	2.26	2.19	2.17	2.13	2.09	2.06
1.34	1.32	1.30	1.29	1.28	1.27	1.26	1.26	1.25	1.24	1.24	1.23	1.23
1.77	1.72	1.67	1.64	1.61	1.57	1.55	1.54	1.51	1.50	1.48	1.47	1.56
2.09	2.01	1.93	1.89	1.84	1.79	1.76	1.74	1.70	1.68	1.66	1.64	1.62
2.84	2.70	2.55	2.47	2.39	2.30	2.25	2.21	2.13	2.11	2.07	2.03	2.01
1.31	1.30	1.28	1.26	1.25	1.24	1.23	1.22	1.21	1.21	1.20	1.19	1.19
1.71	1.66	1.61	1.57	1.54	1.51	1.48	1.47	1.43	1.42	1.41	1.39	1.38
2.00	1.92	1.84	1.79	1.74	1.69	1.66	1.64	1.59	1.58	1.55	1.53	1.51
2.66	2.52	2.37	2.29	2.20	2.11	2.06	2.02	1.94	1.92	1.87	1.83	1.80
1.29	1.27	1.25	1.24	1.22	1.21	1.20	1.19	1.17	1.17	1.16	1.15	1.15
1.66	1.60	1.54	1.51	1.48	1.44	1.41	1.40	1.36	1.35	1.33	1.31	1.29
1.92	1.84	1.75	1.70	1.65	1.59	1.56	1.53	1.48	1.47	1.44	1.41	1.39
2.50	2.35	2.20	2.12	2.03	1.94	1.88	1.84	1.75	1.73	1.68	1.63	1.60
1.26	1.24	1.22	1.21	1.19	1.18	1.17	1.16	1.14	1.13	1.12	1.11	1.10
1.60	1.55	1.48	1.45	1.41	1.37	1.34	1.32	1.27	1.26	1.24	1.21	1.19
1.83	1.75	1.66	1.61	1.55	1.50	1.46	1.43	1.37	1.35	1.32	1.28	1.25
2.34	2.19	2.03	1.95	1.86	1.76	1.70	1.66	1.56	1.53	1.48	1.42	1.38
1.25	1.23	1.21	1.20	1.18	1.16	1.14	1.12	1.11	1.10	1.09	1.08	1.06
1.57	1.52	1.46	1.42	1.38	1.34	1.31	1.28	1.24	1.22	1.20	1.17	1.14
1.80	1.72	1.62	1.57	1.52	1.46	1.41	1.39	1.32	1.29	1.26	1.22	1.19
2.27	2.13	1.97	1.89	1.79	1.69	1.63	1.58	1.48	1.44	1.39	1.33	1.28
1.24	1.22	1.19	1.18	1.16	1.14	1.13	1.12	1.09	1.08	1.07	1.04	1.00
1.55	1.49	1.42	1.38	1.34	1.30	1.26	1.24	1.18	1.17	1.13	1.08	1.00
1.75	1.67	1.57	1.52	1.46	1.39	1.35	1.32	1.24	1.22	1.17	1.11	1.00
2.18	2.04	1.88	1.79	1.70	1.59	1.52	1.47	1.36	1.32	1.25	1.15	1.00

Table 3 Lower Percentage Points of Wilks's Lambda

$[U_{m,a,b,\alpha}]$

$m = 1, \quad \alpha = 0.01$

	a					
b	1	2	3	4	5	6
1	0.006157	0.002501	0.001543	0.001112	0.000868	0.000712
2	0.097504	0.050003	0.033615	0.025322	0.020309	0.016953
3	0.228516	0.135712	0.097321	0.076019	0.062408	0.052963
4	0.341614	0.223602	0.168243	0.135345	0.113373	0.097610
5	0.430725	0.301697	0.235535	0.194031	0.165283	0.144073
6	0.500549	0.368408	0.295990	0.248596	0.214783	0.189255
7	0.555908	0.424896	0.349304	0.298096	0.260620	0.231811
8	0.600708	0.472870	0.396057	0.342590	0.302612	0.271332
9	0.637512	0.513916	0.437164	0.382446	0.340790	0.307770
10	0.668243	0.549286	0.473389	0.418213	0.375519	0.341248
11	0.694275	0.580017	0.505463	0.450317	0.407104	0.372040
12	0.716553	0.606964	0.534027	0.479309	0.435913	0.400299
13	0.735840	0.630737	0.559570	0.505524	0.462189	0.426361
14	0.752686	0.651825	0.582581	0.529327	0.486267	0.450348
15	0.767548	0.670715	0.603333	0.551025	0.508362	0.472534
16	0.780701	0.687653	0.622162	0.570862	0.528717	0.493103
17	0.792480	0.702972	0.639343	0.589081	0.547516	0.512176
18	0.803070	0.716858	0.655029	0.605835	0.564911	0.529907
19	0.812622	0.729553	0.669434	0.621307	0.581024	0.546448
20	0.821320	0.741135	0.682709	0.635651	0.596039	0.561890
21	0.829224	0.751770	0.694977	0.648941	0.610046	0.576355
22	0.836472	0.761597	0.706329	0.661316	0.623108	0.589905
23	0.843140	0.770660	0.716858	0.672867	0.635361	0.602631
24	0.849274	0.779083	0.726685	0.683655	0.646851	0.614609
25	0.854950	0.786896	0.735870	0.693771	0.657639	0.625900
26	0.860199	0.794189	0.744446	0.703278	0.667786	0.636566
27	0.865112	0.800995	0.752487	0.712189	0.677383	0.646637
28	0.869675	0.807373	0.760040	0.720612	0.686432	0.656174
29	0.873947	0.813339	0.767151	0.728546	0.694992	0.665222
30	0.877945	0.818970	0.773865	0.736053	0.703110	0.673798
40	0.907349	0.860886	0.824463	0.793274	0.765594	0.740539
60	0.937485	0.904968	0.878807	0.855911	0.835175	0.816055
80	0.952827	0.927841	0.907471	0.889450	0.872940	0.857590
100	0.962128	0.941845	0.925179	0.910324	0.896637	0.883835
120	0.968363	0.951297	0.937199	0.924578	0.912894	0.901916
140	0.972836	0.958107	0.945890	0.934921	0.924731	0.915131
170	0.977588	0.965370	0.955195	0.946025	0.937478	0.929401
200	0.980926	0.970487	0.961767	0.953893	0.946532	0.939564
240	0.984085	0.975345	0.968024	0.961396	0.955187	0.949296
320	0.988046	0.981451	0.975907	0.970876	0.966145	0.961649
440	0.991295	0.986475	0.982411	0.978715	0.975232	0.971914
600	0.993610	0.990064	0.987067	0.984337	0.981759	0.979301
800	0.995204	0.992539	0.990282	0.988225	0.986279	0.984422
1000	0.996161	0.994026	0.992216	0.990566	0.989003	0.987512
INF	1.000000	1.000000	1.000000	1.000000	1.000000	1.000000

From F. J. Wall, *The Generalized Variance Ratio or the U Statistic*, 1967. Reproduced with permission of the author.

Table 3 (*continued*)

a						
7	8	9	10	11	12	b
0.000603	0.000523	0.000462	0.000413	0.000374	0.000341	1
0.014549	0.012741	0.011333	0.010208	0.009281	0.008512	2
0.046005	0.040672	0.036446	0.033020	0.030182	0.027794	3
0.085724	0.076446	0.068985	0.062851	0.057724	0.053375	4
0.127777	0.114822	0.104279	0.095505	0.088120	0.081787	5
0.169266	0.153168	0.139893	0.128754	0.119278	0.111114	6
0.208893	0.190186	0.174606	0.161423	0.150116	0.140289	7
0.246124	0.225311	0.207825	0.192902	0.180008	0.168747	8
0.280823	0.258362	0.239288	0.222931	0.208679	0.196182	9
0.313019	0.289246	0.268936	0.251373	0.235992	0.222443	10
0.342834	0.318054	0.296768	0.278229	0.261932	0.247467	11
0.370453	0.344940	0.322876	0.303528	0.286469	0.271240	12
0.396057	0.369995	0.347321	0.327362	0.309662	0.293823	13
0.419800	0.393372	0.370239	0.349823	0.331589	0.315247	14
0.441864	0.415222	0.391754	0.370941	0.352325	0.335541	15
0.462433	0.435638	0.411957	0.390869	0.371918	0.354797	16
0.481598	0.454742	0.430939	0.409637	0.390472	0.373077	17
0.499481	0.472687	0.448807	0.427368	0.408020	0.390411	18
0.516235	0.489502	0.465637	0.444138	0.424652	0.406891	19
0.531952	0.505341	0.481506	0.459991	0.440430	0.422546	20
0.546692	0.520264	0.496521	0.475006	0.455414	0.437469	21
0.560562	0.534332	0.510712	0.489258	0.469635	0.451660	22
0.573639	0.547638	0.524139	0.502762	0.483185	0.465179	23
0.585968	0.560211	0.536896	0.515594	0.496078	0.478088	24
0.597626	0.572128	0.548981	0.527817	0.508362	0.490402	25
0.608643	0.583435	0.560486	0.539459	0.520081	0.502167	26
0.619080	0.594147	0.571411	0.550537	0.531281	0.513428	27
0.628998	0.604370	0.581833	0.561127	0.541962	0.524200	28
0.638428	0.614075	0.591766	0.571228	0.552200	0.534515	29
0.647385	0.623321	0.601242	0.580872	0.561996	0.544418	30
0.717575	0.696365	0.676636	0.658188	0.640884	0.624603	40
0.798233	0.781494	0.765686	0.750702	0.736420	0.722809	60
0.843124	0.829437	0.816391	0.803925	0.791962	0.780464	80
0.871696	0.860153	0.849083	0.838455	0.828201	0.818314	100
0.891475	0.881501	0.871901	0.862660	0.853706	0.845045	120
0.905971	0.897200	0.888734	0.880563	0.872625	0.864929	140
0.921669	0.914245	0.907057	0.900101	0.893324	0.886738	170
0.932877	0.926443	0.920200	0.914149	0.908239	0.902486	200
0.943631	0.938171	0.932861	0.927705	0.922660	0.917740	240
0.957311	0.953121	0.949035	0.945058	0.941155	0.937344	320
0.968704	0.965599	0.962561	0.959604	0.956692	0.953846	440
0.976917	0.974611	0.972349	0.970144	0.967969	0.965842	600
0.982619	0.980873	0.979158	0.977487	0.975834	0.974218	800
0.986062	0.984658	0.983276	0.981931	0.980598	0.979296	1000
1.000000	1.000000	1.000000	1.000000	1.000000	1.000000	INF

(*continued*)

Table 3 Lower Percentage Points of Wilks's Lambda (*continued*)
$[U_{m,a,b,\alpha}]$
$m = 1, \quad \alpha = 0.05$

b	a = 1	2	3	4	5	6
1	0.000247	0.000100	0.000062	0.000044	0.000035	0.000028
2	0.019900	0.010000	0.006678	0.005013	0.004012	0.003344
3	0.080827	0.046416	0.032834	0.025458	0.020806	0.017599
4	0.158742	0.100000	0.073958	0.058903	0.049014	0.041999
5	0.235203	0.158489	0.121418	0.098877	0.083563	0.072430
6	0.303867	0.215443	0.169784	0.140867	0.120651	0.105640
7	0.363705	0.268270	0.216358	0.182355	0.158006	0.139585
8	0.415397	0.316227	0.259967	0.222073	0.194363	0.173070
9	0.460089	0.359381	0.300242	0.259453	0.229097	0.205430
10	0.498896	0.398108	0.337189	0.294313	0.261901	0.236323
11	0.532793	0.432877	0.370993	0.326670	0.292708	0.265602
12	0.562582	0.464159	0.401904	0.356635	0.321526	0.293230
13	0.588936	0.492388	0.430204	0.384373	0.348450	0.319237
14	0.612381	0.517948	0.456147	0.410058	0.373579	0.343685
15	0.633365	0.541170	0.479986	0.433866	0.397056	0.366662
16	0.652233	0.562342	0.501931	0.455967	0.418982	0.388257
17	0.669300	0.581709	0.522195	0.476513	0.439506	0.408565
18	0.684789	0.599484	0.540936	0.495647	0.458725	0.427677
19	0.698917	0.615848	0.558319	0.513499	0.476742	0.445681
20	0.711843	0.630958	0.574470	0.530184	0.493561	0.462657
21	0.723730	0.644947	0.589523	0.545805	0.509577	0.478683
22	0.734669	0.657933	0.603568	0.560456	0.524563	0.493830
23	0.744795	0.670019	0.616713	0.574221	0.538693	0.508161
24	0.754176	0.681292	0.629026	0.587173	0.552034	0.521736
25	0.762902	0.691831	0.640594	0.599381	0.564657	0.534611
26	0.771028	0.701704	0.651468	0.610905	0.576603	0.546834
27	0.778625	0.710971	0.661723	0.621798	0.587931	0.558452
28	0.785730	0.719685	0.671391	0.632109	0.598682	0.569507
29	0.792406	0.727896	0.680539	0.641884	0.608900	0.580037
30	0.798670	0.735642	0.689191	0.651161	0.618619	0.590076
40	0.845412	0.794328	0.755603	0.723155	0.694813	0.669500
60	0.894480	0.857696	0.828970	0.804330	0.782305	0.762272
80	0.919918	0.891251	0.868522	0.848784	0.830928	0.814526
100	0.935478	0.912011	0.893219	0.876803	0.861820	0.847989
120	0.945976	0.926119	0.910119	0.896070	0.883183	0.871238
140	0.953532	0.936329	0.922402	0.910129	0.898827	0.888325
170	0.961595	0.947263	0.935601	0.925292	0.915755	0.906869
200	0.967270	0.954993	0.944964	0.936079	0.927834	0.920134
240	0.972661	0.962351	0.953904	0.946399	0.939414	0.932883
320	0.979433	0.971628	0.965202	0.959483	0.954136	0.949127
440	0.985001	0.979285	0.974556	0.970342	0.966383	0.962678
600	0.988980	0.984767	0.981267	0.978151	0.975211	0.972459
800	0.991723	0.988553	0.985913	0.983561	0.981336	0.979256
1000	0.993372	0.990832	0.988711	0.986824	0.985033	0.983362
INF	1.000000	1.000000	1.000000	1.000000	1.000000	1.000000

Table 3 (*continued*)

a						
7	8	9	10	11	12	*b*
0.000024	0.000021	0.000018	0.000017	0.000015	0.000014	1
0.002867	0.002509	0.002231	0.002008	0.001826	0.001674	2
0.015251	0.013458	0.012043	0.010898	0.009951	0.009157	3
0.036755	0.032682	0.029427	0.026763	0.024544	0.022665	4
0.063947	0.057265	0.051854	0.047390	0.043634	0.040434	5
0.094010	0.084728	0.077134	0.070801	0.065439	0.060839	6
0.125112	0.113416	0.103749	0.095628	0.088696	0.082716	7
0.156116	0.142270	0.130724	0.120944	0.112552	0.105261	8
0.186374	0.170658	0.157452	0.146189	0.136452	0.127957	9
0.215512	0.198202	0.183548	0.170965	0.160030	0.150442	10
0.243349	0.224692	0.208793	0.195061	0.183068	0.172501	11
0.269804	0.250027	0.233063	0.218338	0.205413	0.193976	12
0.294872	0.274166	0.256310	0.240729	0.226996	0.214791	13
0.318575	0.297116	0.278506	0.262202	0.247764	0.234893	14
0.340981	0.318908	0.299682	0.282756	0.267719	0.254259	15
0.362133	0.339583	0.319844	0.302404	0.286850	0.272887	16
0.382133	0.359198	0.339049	0.321175	0.305186	0.290785	17
0.401023	0.377807	0.357322	0.339101	0.322737	0.307970	18
0.418900	0.395470	0.374735	0.356217	0.339555	0.324463	19
0.435811	0.412241	0.391308	0.372565	0.355644	0.340290	20
0.451836	0.428178	0.407108	0.388183	0.371060	0.355477	21
0.467022	0.443332	0.422166	0.403108	0.385819	0.370054	22
0.481441	0.457752	0.436534	0.417377	0.399965	0.384048	23
0.495132	0.471485	0.450247	0.431029	0.413515	0.397487	24
0.508160	0.484576	0.463349	0.444097	0.426523	0.410397	25
0.520546	0.497064	0.475866	0.456613	0.438989	0.422804	26
0.532362	0.508986	0.487854	0.468607	0.450973	0.434734	27
0.543615	0.520379	0.499314	0.480110	0.462471	0.446211	28
0.554370	0.531274	0.510313	0.491149	0.473532	0.457257	29
0.564636	0.541702	0.520842	0.501748	0.484160	0.467894	30
0.646550	0.625549	0.606163	0.588188	0.571417	0.555726	40
0.743738	0.726513	0.710318	0.695108	0.680672	0.667012	60
0.799185	0.784809	0.771162	0.758235	0.745861	0.734069	80
0.834952	0.822679	0.810943	0.799788	0.789035	0.778749	100
0.859925	0.849237	0.838971	0.829183	0.819700	0.810616	120
0.878338	0.868886	0.859771	0.851065	0.842600	0.834476	140
0.898385	0.890335	0.882539	0.875082	0.867790	0.860800	170
0.912760	0.905757	0.898951	0.892434	0.886056	0.879907	200
0.926606	0.920640	0.914819	0.909247	0.903766	0.898492	240
0.944294	0.939692	0.935191	0.930865	0.926597	0.922489	320
0.959082	0.955661	0.952291	0.949067	0.945865	0.942783	440
0.969781	0.967231	0.964708	0.962302	0.959899	0.957590	600
0.977220	0.975291	0.973377	0.971545	0.969713	0.967956	800
0.981722	0.980169	0.978621	0.977148	0.975664	0.974250	1000
1.000000	1.000000	1.000000	1.000000	1.000000	1.000000	INF

(*continued*)

Table 3 Lower Percentage Points of Wilks's Lambda (*continued*)
$[U_{m,a,b,\alpha}]$

$m = 2, \quad \alpha = 0.01$

	a					
b	1	2	3	4	5	6
1	0.000000	0.000000	0.000000	0.000000	0.000000	0.000000
2	0.000100	0.000025	0.000011	0.000006	0.000004	0.000003
3	0.010000	0.003470	0.001764	0.001068	0.000716	0.000514
4	0.046416	0.019844	0.011160	0.007179	0.005013	0.003701
5	0.099999	0.049316	0.029953	0.020241	0.014627	0.011080
6	0.158490	0.086620	0.055849	0.039284	0.029229	0.022633
7	0.215443	0.127189	0.085984	0.062513	0.047671	0.037627
8	0.268270	0.168148	0.118119	0.088278	0.068750	0.055175
9	0.316228	0.207906	0.150743	0.115317	0.091448	0.074467
10	0.359382	0.245666	0.182908	0.142738	0.114989	0.094845
11	0.398107	0.281095	0.214051	0.169943	0.138805	0.115797
12	0.432876	0.314111	0.243868	0.196543	0.162496	0.136940
13	0.464159	0.344773	0.272209	0.222298	0.185786	0.157996
14	0.492388	0.373205	0.299027	0.247072	0.208495	0.178764
15	0.517947	0.399561	0.324338	0.270794	0.230506	0.199104
16	0.541170	0.424011	0.348190	0.293441	0.251751	0.218924
17	0.562341	0.446714	0.370654	0.315019	0.272197	0.238163
18	0.581709	0.467823	0.391807	0.335555	0.291830	0.256785
19	0.599484	0.487482	0.411733	0.355085	0.310657	0.274771
20	0.615848	0.505819	0.430515	0.373654	0.328695	0.292119
21	0.630957	0.522953	0.448231	0.391312	0.345965	0.308831
22	0.644947	0.538990	0.464956	0.408104	0.362497	0.324920
23	0.657933	0.554026	0.480762	0.424082	0.378320	0.340402
24	0.670019	0.568146	0.495715	0.439293	0.393467	0.355296
25	0.681292	0.581428	0.509875	0.453782	0.407970	0.369622
26	0.691831	0.593939	0.523299	0.467592	0.421860	0.383403
27	0.701704	0.605746	0.536039	0.480765	0.435169	0.396661
28	0.710971	0.616902	0.548144	0.493339	0.447927	0.409418
29	0.719686	0.627457	0.559655	0.505352	0.460162	0.421696
30	0.727896	0.637459	0.570615	0.516835	0.471903	0.433519
40	0.789652	0.714476	0.656673	0.608581	0.567185	0.530850
60	0.855467	0.799984	0.755573	0.717315	0.683328	0.652617
80	0.889953	0.846188	0.810436	0.779081	0.750765	0.724783
100	0.911163	0.875081	0.845239	0.818780	0.794644	0.772286
120	0.925522	0.894844	0.869263	0.846415	0.825431	0.805865
140	0.935886	0.909213	0.886840	0.866750	0.848208	0.830840
170	0.946959	0.924659	0.905836	0.888839	0.873070	0.858224
200	0.954772	0.935614	0.919375	0.904652	0.890941	0.877988
240	0.962196	0.946071	0.932346	0.919856	0.908184	0.897119
320	0.971540	0.959297	0.948819	0.939239	0.930247	0.921688
440	0.979238	0.970243	0.962512	0.955416	0.948732	0.942346
600	0.984741	0.978097	0.972369	0.967097	0.962118	0.957350
800	0.988539	0.983531	0.979204	0.975215	0.971440	0.967819
1000	0.990823	0.986804	0.983329	0.980120	0.977081	0.974162
INF	1.000000	1.000000	1.000000	1.000000	1.000000	1.000000

Table 3 (*continued*)

a						
7	8	9	10	11	12	b
0.000000	0.000000	0.000000	0.000000	0.000000	0.000000	1
0.000002	0.000002	0.000001	0.000001	0.000000	0.000000	2
0.000386	0.000301	0.000241	0.000198	0.000165	0.000140	3
0.002846	0.002257	0.001834	0.001520	0.001280	0.001093	4
0.008688	0.006999	0.005760	0.004824	0.004099	0.003527	5
0.018059	0.014752	0.012283	0.010389	0.008903	0.007715	6
0.030485	0.025222	0.021222	0.018110	0.015639	0.013643	7
0.045317	0.037913	0.032206	0.027708	0.024097	0.021153	8
0.061901	0.052316	0.044821	0.038849	0.034006	0.030024	9
0.079687	0.067961	0.058684	0.051208	0.045092	0.040019	10
0.098225	0.084458	0.073448	0.064492	0.057100	0.050924	11
0.117163	0.101490	0.088832	0.078445	0.069807	0.062537	12
0.136231	0.118805	0.104603	0.092856	0.083018	0.074689	13
0.155227	0.136206	0.120576	0.107553	0.096574	0.087224	14
0.174003	0.153543	0.136603	0.122393	0.110341	0.100021	15
0.192450	0.170701	0.152570	0.137265	0.124211	0.112976	16
0.210492	0.187598	0.168387	0.152079	0.138096	0.126003	17
0.228078	0.204169	0.183988	0.166764	0.151924	0.139032	18
0.245174	0.220372	0.199322	0.181266	0.165639	0.152007	19
0.261761	0.236177	0.214352	0.195544	0.179196	0.164879	20
0.277829	0.251565	0.229052	0.209566	0.192561	0.177614	21
0.293378	0.266524	0.243403	0.223308	0.205707	0.190182	22
0.308412	0.281051	0.257394	0.236756	0.218614	0.202560	23
0.322939	0.295145	0.271020	0.249897	0.231267	0.214731	24
0.336971	0.308812	0.284279	0.262726	0.243657	0.226681	25
0.350522	0.322057	0.297172	0.275239	0.255776	0.238402	26
0.363607	0.334890	0.309703	0.287436	0.267621	0.249886	27
0.376242	0.347322	0.321877	0.299318	0.279190	0.261129	28
0.388443	0.359363	0.333703	0.310890	0.290483	0.272130	29
0.400227	0.371026	0.345187	0.322156	0.301504	0.282889	30
0.498541	0.469542	0.443323	0.419481	0.397694	0.377702	40
0.624558	0.598723	0.574795	0.552534	0.531746	0.512271	60
0.700697	0.678213	0.657114	0.637235	0.618444	0.600635	80
0.751373	0.731681	0.713048	0.695352	0.678495	0.662399	100
0.787452	0.770011	0.753414	0.737564	0.722385	0.707815	120
0.814421	0.798803	0.783877	0.769567	0.755808	0.742551	140
0.844122	0.830645	0.817710	0.805252	0.793223	0.781586	170
0.865642	0.853805	0.842407	0.831395	0.820731	0.810382	200
0.886539	0.876363	0.866534	0.857011	0.847760	0.838758	240
0.913469	0.905533	0.897838	0.890354	0.883057	0.875930	320
0.936194	0.930234	0.924437	0.918780	0.913248	0.907828	440
0.952745	0.948273	0.943913	0.939649	0.935470	0.931366	600
0.964316	0.960909	0.957581	0.954322	0.951123	0.947977	800
0.971336	0.968584	0.965894	0.963257	0.960665	0.958115	1000
1.000000	1.000000	1.000000	1.000000	1.000000	1.000000	INF

(*continued*)

Table 3 Lower Percentage Points of Wilks's Lambda (*continued*)

$[U_{m,a,b,\alpha}]$

$m = 2, \quad \alpha = 0.05$

	a					
b	1	2	3	4	5	6
1	0.000000	0.000000	0.000000	0.000000	0.000000	0.000000
2	0.002500	0.000641	0.000287	0.000162	0.000104	0.000072
3	0.049998	0.018318	0.009528	0.005844	0.003950	0.002849
4	0.135725	0.061800	0.035817	0.023460	0.016578	0.012346
5	0.223605	0.117368	0.073621	0.050765	0.037211	0.028476
6	0.301715	0.174902	0.116450	0.083663	0.063188	0.049481
7	0.368405	0.229737	0.160239	0.118984	0.092129	0.073571
8	0.424876	0.280167	0.202813	0.154741	0.122376	0.099380
9	0.472866	0.325883	0.243151	0.189781	0.152779	0.125881
10	0.513885	0.367036	0.280602	0.223433	0.182643	0.152421
11	0.549280	0.404052	0.315720	0.255369	0.211592	0.178545
12	0.580029	0.437339	0.347988	0.285511	0.239373	0.203997
13	0.606971	0.467384	0.377744	0.313836	0.265838	0.228568
14	0.630737	0.494599	0.405216	0.340396	0.291016	0.252171
15	0.651851	0.519281	0.430564	0.365263	0.314363	0.274786
16	0.670711	0.541775	0.454003	0.388530	0.337412	0.296391
17	0.687662	0.562317	0.475724	0.410322	0.358763	0.316990
18	0.702981	0.581146	0.495888	0.430784	0.378964	0.336632
19	0.716866	0.598489	0.514629	0.449961	0.398041	0.355335
20	0.729531	0.614483	0.532092	0.467968	0.416109	0.373163
21	0.741124	0.629283	0.548399	0.484925	0.433211	0.390129
22	0.751776	0.643011	0.563622	0.500886	0.449429	0.406286
23	0.761598	0.655775	0.577893	0.515922	0.464800	0.421699
24	0.770680	0.667666	0.591286	0.530135	0.479373	0.436391
25	0.779088	0.678783	0.603884	0.543551	0.493227	0.450412
26	0.786893	0.689182	0.615752	0.556269	0.506409	0.463802
27	0.794192	0.698945	0.626937	0.568306	0.518951	0.476588
28	0.800992	0.708108	0.637517	0.579727	0.530891	0.488822
29	0.807354	0.716737	0.647497	0.590582	0.542291	0.500519
30	0.813343	0.724899	0.656961	0.600899	0.553155	0.511722
40	0.857594	0.786432	0.729818	0.681627	0.639419	0.601870
60	0.903437	0.852599	0.810662	0.773804	0.740586	0.710190
80	0.926967	0.887496	0.854347	0.824736	0.797636	0.772490
100	0.941272	0.909051	0.881684	0.856993	0.834186	0.812834
120	0.950898	0.923673	0.900382	0.879233	0.859569	0.841056
140	0.957812	0.934247	0.913983	0.895493	0.878224	0.861896
170	0.965169	0.945562	0.928606	0.913057	0.898465	0.884603
200	0.970341	0.953554	0.938982	0.925569	0.912940	0.900904
240	0.975243	0.961158	0.948887	0.937554	0.926848	0.916613
320	0.981393	0.970741	0.961415	0.952766	0.944563	0.936691
440	0.986445	0.978644	0.971788	0.965408	0.959337	0.953491
600	0.990047	0.984298	0.979233	0.974507	0.969998	0.965648
800	0.992529	0.988203	0.984384	0.980814	0.977404	0.974108
1000	0.994021	0.990552	0.987487	0.984620	0.981877	0.979224
INF	1.000000	1.000000	1.000000	1.000000	1.000000	1.000000

Table 3 (*continued*)

a						
7	8	9	10	11	12	b
0.000000	0.000000	0.000000	0.000000	0.000000	0.000000	1
0.000053	0.000041	0.000032	0.000026	0.000022	0.000018	2
0.002152	0.001683	0.001352	0.001110	0.000928	0.000787	3
0.009555	0.007615	0.006212	0.005165	0.004362	0.003734	4
0.022507	0.018244	0.015092	0.012695	0.010826	0.009343	5
0.039834	0.032772	0.027440	0.023320	0.020068	0.017453	6
0.060172	0.050155	0.042465	0.036426	0.031600	0.027678	7
0.082397	0.069475	0.059404	0.051386	0.044908	0.039579	8
0.105643	0.089993	0.077615	0.067661	0.059515	0.052772	9
0.129282	0.111138	0.096610	0.084797	0.075044	0.066901	10
0.152898	0.132506	0.116013	0.102453	0.091177	0.081680	11
0.176155	0.153782	0.135511	0.120356	0.107656	0.096885	12
0.198874	0.174774	0.154909	0.138311	0.124284	0.112321	13
0.220930	0.195325	0.174061	0.156149	0.140923	0.127849	14
0.242249	0.215357	0.192837	0.173755	0.157442	0.143350	15
0.262763	0.234782	0.211185	0.191059	0.173755	0.158740	16
0.282502	0.253583	0.229036	0.208000	0.189807	0.173946	17
0.301430	0.271723	0.246366	0.224530	0.205530	0.188918	18
0.319573	0.289225	0.263169	0.240614	0.220915	0.203611	19
0.336951	0.306072	0.279429	0.256249	0.235936	0.218013	20
0.353609	0.322287	0.295146	0.271437	0.250565	0.232083	21
0.369555	0.337873	0.310325	0.286147	0.264800	0.245821	22
0.384810	0.352883	0.324978	0.300409	0.278639	0.259223	23
0.399429	0.367295	0.339116	0.314213	0.292087	0.272280	24
0.413436	0.381165	0.352775	0.327593	0.305127	0.285006	25
0.426867	0.394506	0.365946	0.340539	0.317798	0.297372	26
0.439744	0.407337	0.378645	0.353047	0.330095	0.309407	27
0.452093	0.419700	0.390911	0.365171	0.342019	0.321110	28
0.463948	0.431586	0.402753	0.376900	0.353591	0.332484	29
0.475325	0.443028	0.414182	0.388244	0.364802	0.343537	30
0.568076	0.537426	0.509476	0.483873	0.460296	0.438550	40
0.682157	0.656096	0.631804	0.609029	0.587643	0.567501	60
0.748974	0.726849	0.705927	0.686107	0.667279	0.649328	80
0.792697	0.773596	0.755405	0.738034	0.721395	0.705440	100
0.823491	0.806739	0.790700	0.775302	0.760485	0.746201	120
0.846339	0.831442	0.817125	0.803326	0.789999	0.777105	140
0.871338	0.858581	0.846267	0.834352	0.822797	0.811574	170
0.889349	0.878202	0.867412	0.856939	0.846755	0.836834	200
0.906758	0.897224	0.887968	0.878959	0.870174	0.861593	240
0.929082	0.921692	0.914493	0.907461	0.900579	0.893835	¯20
0.947824	0.942303	0.936908	0.931623	0.926435	0.921337	440
0.961420	0.957293	0.953251	0.949283	0.945380	0.941537	600
0.970900	0.967763	0.964686	0.961662	0.958683	0.955744	800
0.976640	0.974110	0.971627	0.969184	0.966775	0.964397	1000
1.000000	1.000000	1.000000	1.000000	1.000000	1.000000	INF

(*continued*)

Table 3 Lower Percentage Points of Wilks's Lambda (*continued*)

$[U_{m,a,b,\alpha}]$

$m = 3, \quad \alpha = 0.01$

	a					
b	1	2	3	4	5	6
1	0.000000	0.000000	0.000000	0.000000	0.000000	0.000000
2	0.000000	0.000000	0.000000	0.000000	0.000000	0.000000
3	0.000080	0.000021	0.000016	0.000015	0.000015	0.000017
4	0.006763	0.001829	0.000824	0.000484	0.000335	0.000258
5	0.032882	0.011211	0.005326	0.003037	0.001959	0.001383
6	0.073980	0.029981	0.015536	0.009229	0.006018	0.004211
7	0.121426	0.055863	0.031196	0.019423	0.013027	0.009244
8	0.169788	0.085991	0.051041	0.033146	0.022897	0.016575
9	0.216359	0.118124	0.073700	0.049627	0.035223	0.026018
10	0.259966	0.150745	0.098030	0.068087	0.049501	0.037260
11	0.300240	0.182909	0.123161	0.087852	0.065237	0.049951
12	0.337186	0.214052	0.148469	0.108380	0.081998	0.063758
13	0.370989	0.243868	0.173524	0.129256	0.099424	0.078384
14	0.401904	0.272209	0.198043	0.150167	0.117224	0.093576
15	0.430202	0.299027	0.221841	0.170888	0.135171	0.109125
16	0.456147	0.324338	0.244809	0.191257	0.153091	0.124859
17	0.479984	0.348191	0.266888	0.211160	0.170848	0.140644
18	0.501932	0.370654	0.288051	0.230524	0.188345	0.156370
19	0.522191	0.391807	0.308300	0.249300	0.205509	0.171955
20	0.540934	0.411734	0.327644	0.267462	0.222288	0.187333
21	0.558316	0.430515	0.346122	0.284999	0.238647	0.202457
22	0.574470	0.448231	0.363762	0.301910	0.254564	0.217287
23	0.589519	0.464956	0.380593	0.318203	0.270024	0.231801
24	0.603567	0.480761	0.396664	0.333888	0.285024	0.245977
25	0.616709	0.495715	0.412006	0.348987	0.299564	0.259807
26	0.629025	0.509875	0.426661	0.363513	0.313646	0.273282
27	0.640592	0.523299	0.440664	0.377492	0.327281	0.286401
28	0.651469	0.536040	0.454050	0.390942	0.340476	0.299164
29	0.661719	0.548144	0.466858	0.403887	0.353244	0.311575
30	0.671391	0.559656	0.479116	0.416348	0.365597	0.323637
40	0.744674	0.649620	0.577483	0.518712	0.469272	0.426891
60	0.823683	0.751990	0.694679	0.645816	0.602970	0.564801
80	0.865422	0.808282	0.761397	0.720482	0.683828	0.650513
100	0.891201	0.843804	0.804298	0.769332	0.737595	0.708389
120	0.908698	0.868241	0.834163	0.803715	0.775833	0.749959
140	0.921350	0.886074	0.856136	0.829206	0.804388	0.781216
170	0.934886	0.905306	0.880001	0.857075	0.835802	0.815812
200	0.944448	0.918986	0.897083	0.877137	0.858541	0.840986
240	0.953545	0.932073	0.913505	0.896514	0.880601	0.865514
320	0.965006	0.948662	0.934434	0.921336	0.909000	0.897239
440	0.974459	0.962428	0.951897	0.942156	0.932937	0.924109
600	0.981223	0.972324	0.964504	0.957246	0.950353	0.943732
800	0.985892	0.979179	0.973264	0.967760	0.962522	0.957478
1000	0.988702	0.983312	0.978556	0.974124	0.969900	0.965827
INF	1.000000	1.000000	1.000000	1.000000	1.000000	1.000000

Table 3 (*continued*)

a						
7	8	9	10	11	12	*b*
0.000000	0.000000	0.000000	0.000000	0.000000	0.000000	1
0.000000	0.000000	0.000001	0.000002	0.000003	0.000004	2
0.000019	0.000021	0.000023	0.000026	0.000029	0.000031	3
0.000215	0.000188	0.000172	0.000161	0.000154	0.000149	4
0.001047	0.000837	0.000698	0.000602	0.000533	0.000483	5
0.003116	0.002414	0.001943	0.001614	0.001376	0.001200	6
0.006864	0.005293	0.004214	0.003450	0.002892	0.002474	7
0.012459	0.009664	0.007703	0.006286	0.005237	0.004444	8
0.019845	0.015549	0.012468	0.010203	0.008501	0.007198	9
0.028840	0.022851	0.018472	0.015200	0.012705	0.010772	10
0.039203	0.031408	0.025614	0.021217	0.017821	0.015158	11
0.050682	0.041037	0.033759	0.028161	0.023785	0.020317	12
0.063039	0.051550	0.042762	0.035922	0.030516	0.026188	13
0.076062	0.062771	0.052482	0.044385	0.037922	0.032700	14
0.089567	0.074542	0.062783	0.053438	0.045911	0.039779	15
0.103395	0.086724	0.073546	0.062977	0.054394	0.047349	16
0.117419	0.099197	0.084662	0.072908	0.063289	0.055338	17
0.131529	0.111859	0.096037	0.083144	0.072519	0.063680	18
0.145640	0.124624	0.107589	0.093611	0.082016	0.072312	19
0.159680	0.137421	0.119251	0.104243	0.091719	0.081178	20
0.173594	0.150193	0.130962	0.114983	0.101574	0.090228	21
0.187337	0.162890	0.142676	0.125784	0.111534	0.099418	22
0.200875	0.175473	0.154348	0.136602	0.121558	0.108708	23
0.214182	0.187912	0.165948	0.147403	0.131611	0.118064	24
0.227238	0.200180	0.177443	0.158158	0.141663	0.127455	25
0.240029	0.212260	0.188816	0.168840	0.151686	0.136855	26
0.252545	0.224135	0.200042	0.179430	0.161661	0.146240	27
0.264779	0.235795	0.211110	0.189910	0.171566	0.155593	28
0.276730	0.247231	0.222009	0.200265	0.181387	0.164895	29
0.288394	0.258438	0.232727	0.210485	0.191110	0.174132	30
0.390088	0.357822	0.329317	0.303979	0.281338	0.261014	40
0.530443	0.499282	0.470857	0.444810	0.420849	0.398737	60
0.619945	0.591715	0.565509	0.541095	0.518272	0.496881	80
0.681275	0.655947	0.632179	0.609801	0.588667	0.568664	100
0.725743	0.702949	0.681398	0.660958	0.641519	0.622994	120
0.759404	0.738756	0.719127	0.700413	0.682523	0.665388	140
0.796876	0.778843	0.761600	0.745064	0.729169	0.713861	170
0.824284	0.808309	0.792971	0.778202	0.763939	0.750169	200
0.851099	0.837256	0.823911	0.811012	0.798488	0.786374	240
0.885943	0.875039	0.864473	0.854210	0.844184	0.834449	320
0.915592	0.907335	0.899302	0.891467	0.883774	0.876280	440
0.937324	0.931093	0.925012	0.919063	0.913203	0.907480	600
0.952586	0.947819	0.943157	0.938588	0.934077	0.929662	800
0.961872	0.958013	0.954234	0.950525	0.946859	0.943268	1000
1.000000	1.000000	1.000000	1.000000	1.000000	1.000000	INF

(*continued*)

Table 3 Lower Percentage Points of Wilks's Lambda (*continued*)
$[U_{m,a,b,\alpha}]$

$m = 3, \quad \alpha = 0.05$

	a					
b	1	2	3	4	5	6
1	0.000000	0.000000	0.000000	0.000000	0.000000	0.000000
2	0.000000	0.000000	0.000000	0.000000	0.000000	0.000001
3	0.001698	0.000354	0.000179	0.000127	0.000105	0.000095
4	0.033740	0.009612	0.004205	0.002314	0.001479	0.001052
5	0.097355	0.035855	0.017521	0.010010	0.006357	0.004369
6	0.168271	0.073634	0.039672	0.024047	0.015792	0.011018
7	0.235525	0.116476	0.067711	0.043226	0.029433	0.021043
8	0.295976	0.160244	0.098932	0.065947	0.046378	0.033966
9	0.349276	0.202814	0.131378	0.090794	0.065660	0.049161
10	0.396084	0.243139	0.163846	0.116701	0.086448	0.066012
11	0.437147	0.280808	0.195556	0.142927	0.108110	0.083979
12	0.473377	0.315719	0.226090	0.168939	0.130131	0.102644
13	0.505452	0.347981	0.255220	0.194413	0.152159	0.121656
14	0.534017	0.377735	0.282849	0.219113	0.173959	0.140775
15	0.559570	0.405221	0.308951	0.242944	0.195322	0.159796
16	0.582577	0.430566	0.333588	0.265812	0.216138	0.178574
17	0.603338	0.454006	0.356777	0.287689	0.236338	0.197017
18	0.622168	0.475728	0.378631	0.308599	0.255858	0.215044
19	0.639337	0.495908	0.399223	0.328552	0.274710	0.232604
20	0.655028	0.514622	0.418629	0.347546	0.292843	0.249666
21	0.669437	0.532101	0.436898	0.365676	0.310304	0.266216
22	0.682712	0.548393	0.454182	0.382934	0.327083	0.282253
23	0.694960	0.563637	0.470473	0.399402	0.343191	0.297740
24	0.706310	0.577895	0.485889	0.415077	0.358665	0.312738
25	0.716875	0.591311	0.500491	0.430041	0.373523	0.327221
26	0.726681	0.603899	0.514336	0.444332	0.387790	0.341199
27	0.735837	0.615757	0.527453	0.457946	0.401488	0.354711
28	0.744404	0.626944	0.539914	0.470981	0.414658	0.367742
29	0.752437	0.637514	0.551741	0.483430	0.427307	0.380334
30	0.759984	0.647501	0.563023	0.495347	0.439474	0.392490
40	0.816139	0.723938	0.651355	0.590773	0.538846	0.493686
60	0.874843	0.807777	0.752424	0.704238	0.661334	0.622640
80	0.905160	0.852653	0.808266	0.768805	0.732964	0.700026
100	0.923660	0.880557	0.843610	0.810333	0.779746	0.751296
120	0.936178	0.899588	0.867973	0.839253	0.812632	0.787686
140	0.945137	0.913391	0.885776	0.860534	0.836998	0.814819
170	0.954680	0.928199	0.904999	0.883652	0.863624	0.844636
200	0.961395	0.938685	0.918687	0.900202	0.882782	0.866197
240	0.967765	0.948678	0.931793	0.916116	0.901281	0.887100
320	0.975762	0.961296	0.948422	0.936405	0.924971	0.913987
440	0.982336	0.971725	0.962235	0.953337	0.944835	0.936632
600	0.987028	0.979198	0.972173	0.965563	0.959229	0.953099
800	0.990261	0.984364	0.979060	0.974060	0.969257	0.964600
1000	0.992204	0.987475	0.983215	0.979193	0.975326	0.971571
INF	1.000000	1.000000	1.000000	1.000000	1.000000	1.000000

Table 3 (*continued*)

a						
7	8	9	10	11	12	*b*
0.000000	0.000000	0.000000	0.000000	0.000000	0.000000	1
0.000002	0.000004	0.000005	0.000008	0.000010	0.000013	2
0.000091	0.000090	0.000091	0.000092	0.000095	0.000098	3
0.000809	0.000659	0.000562	0.000496	0.000449	0.000416	4
0.003195	0.002458	0.001971	0.001636	0.001397	0.001222	5
0.008067	0.006148	0.004849	0.003939	0.003281	0.002793	6
0.015642	0.012012	0.009485	0.007674	0.006345	0.005347	7
0.025706	0.019990	0.015911	0.012927	0.010697	0.008997	8
0.037855	0.029838	0.023995	0.019637	0.016323	0.013763	9
0.051643	0.041238	0.033514	0.027654	0.023135	0.019593	10
0.066659	0.053876	0.044225	0.036801	0.030993	0.026391	11
0.082534	0.067443	0.055894	0.046882	0.039757	0.034049	12
0.098973	0.081704	0.068298	0.057724	0.049278	0.042437	13
0.115736	0.096413	0.081246	0.069166	0.059407	0.051442	14
0.132619	0.111416	0.094593	0.081052	0.070028	0.060954	15
0.149493	0.126564	0.108178	0.093264	0.081026	0.070875	16
0.166236	0.141728	0.121917	0.105704	0.092299	0.081109	17
0.182762	0.156827	0.135693	0.118273	0.103768	0.091588	18
0.199009	0.171789	0.149445	0.130904	0.115361	0.102241	19
0.214918	0.186544	0.163097	0.143521	0.127018	0.113012	20
0.230467	0.201077	0.176620	0.156088	0.138689	0.123835	21
0.245626	0.215325	0.189969	0.168561	0.150321	0.134680	22
0.260397	0.229291	0.203123	0.180907	0.161896	0.145521	23
0.274743	0.242939	0.216044	0.193091	0.173370	0.156313	24
0.288709	0.256276	0.228718	0.205103	0.184720	0.167023	25
0.302238	0.269280	0.241137	0.216929	0.195944	0.177651	26
0.315386	0.281968	0.253300	0.228535	0.206998	0.188160	27
0.328131	0.294313	0.265188	0.239935	0.217899	0.198546	28
0.340477	0.306326	0.276805	0.251110	0.228615	0.208808	29
0.352461	0.318033	0.288158	0.262062	0.239155	0.218912	30
0.453976	0.418785	0.387401	0.359270	0.333940	0.311045	40
0.587440	0.555224	0.525598	0.498272	0.472957	0.449477	60
0.669520	0.641124	0.614572	0.589678	0.566281	0.544236	80
0.724666	0.699598	0.675935	0.653520	0.632235	0.611999	100
0.764150	0.741841	0.720623	0.700389	0.681054	0.662546	120
0.793780	0.773732	0.754565	0.736197	0.718557	0.701592	140
0.826518	0.809156	0.792465	0.776383	0.760857	0.745847	170
0.850307	0.835018	0.820262	0.805990	0.792160	0.778739	200
0.873459	0.860284	0.847521	0.835131	0.823081	0.811346	240
0.903369	0.893064	0.883033	0.873250	0.863692	0.854341	320
0.928671	0.920913	0.913332	0.905910	0.898630	0.891482	440
0.947133	0.941302	0.935589	0.929978	0.924461	0.919029	600
0.960057	0.955610	0.951243	0.946947	0.942713	0.938538	800
0.967905	0.964310	0.960776	0.957296	0.953863	0.950473	1000
1.000000	1.000000	1.000000	1.000000	1.000000	1.000000	INF

(*continued*)

Table 3 Lower Percentage Points of Wilks's Lambda (*continued*)
$[U_{m,a,b,\alpha}]$

$m = 4, \quad \alpha = 0.01$

b	a: 1	2	3	4	5	6
1	0.000000	0.000000	0.000000	0.000000	0.000000	0.000000
2	0.000000	0.000000	0.000000	0.000000	0.000000	0.000000
3	0.000000	0.000000	0.000000	0.000000	0.000000	0.000000
4	0.000090	0.000026	0.000015	0.000011	0.000009	0.000008
5	0.005218	0.001224	0.000484	0.000250	0.000153	0.000106
6	0.025586	0.007345	0.003037	0.001538	0.000893	0.000574
7	0.058962	0.020352	0.009229	0.004891	0.002885	0.001846
8	0.098904	0.039349	0.019423	0.010860	0.006623	0.004315
9	0.140881	0.062551	0.033146	0.019474	0.012300	0.008211
10	0.182361	0.088300	0.049627	0.030445	0.019865	0.013591
11	0.222076	0.115330	0.068087	0.043357	0.029128	0.020392
12	0.259456	0.142746	0.087852	0.057777	0.039832	0.028478
13	0.294315	0.169948	0.108380	0.073308	0.051709	0.037679
14	0.326670	0.196546	0.129256	0.089607	0.064505	0.047814
15	0.356636	0.222301	0.150167	0.106392	0.077992	0.058712
16	0.384374	0.247074	0.170888	0.123435	0.091973	0.070212
17	0.410058	0.270796	0.191257	0.140556	0.106281	0.082172
18	0.433867	0.293442	0.211160	0.157615	0.120777	0.094469
19	0.455967	0.315021	0.230524	0.174505	0.135348	0.106995
20	0.476513	0.335555	0.249300	0.191144	0.149903	0.119660
21	0.495648	0.355086	0.267462	0.207474	0.164368	0.132389
22	0.513499	0.373655	0.284999	0.223450	0.178686	0.145118
23	0.530184	0.391312	0.301910	0.239044	0.192810	0.157796
24	0.545805	0.408104	0.318203	0.254237	0.206707	0.170381
25	0.560457	0.424083	0.333888	0.269015	0.220349	0.182837
26	0.574221	0.439293	0.348987	0.283374	0.233718	0.195137
27	0.587173	0.453781	0.363513	0.297314	0.246799	0.207259
28	0.599381	0.467592	0.377492	0.310838	0.259584	0.219187
29	0.610904	0.480765	0.390942	0.323953	0.272068	0.230906
30	0.621798	0.493340	0.403887	0.336665	0.284247	0.242408
40	0.704846	0.593044	0.510028	0.444079	0.390022	0.344862
60	0.795314	0.709205	0.641042	0.583746	0.534292	0.490946
80	0.843446	0.774138	0.717496	0.668503	0.625079	0.586056
100	0.873280	0.815461	0.767296	0.724909	0.686729	0.651890
120	0.893573	0.844036	0.802240	0.765030	0.731147	0.699908
140	0.908268	0.864962	0.828089	0.794989	0.764613	0.736396
170	0.924010	0.887597	0.856293	0.827943	0.801710	0.777147
200	0.935142	0.903739	0.876559	0.851792	0.828739	0.807033
240	0.945741	0.919212	0.896104	0.874924	0.855100	0.836335
320	0.959108	0.938870	0.921100	0.904693	0.889230	0.874494
440	0.970142	0.955217	0.942028	0.929777	0.918164	0.907036
600	0.978043	0.966989	0.957176	0.948021	0.939309	0.930927
800	0.983500	0.975154	0.967720	0.960765	0.954127	0.947724
1000	0.986785	0.980081	0.974098	0.968491	0.963130	0.957951
INF	1.000000	1.000000	1.000000	1.000000	1.000000	1.000000

Table 3 (*continued*)

a						
7	8	9	10	11	12	*b*
0.000000	0.000000	0.000000	0.000000	0.000000	0.000000	1
0.000000	0.000000	0.000000	0.000000	0.000000	0.000000	2
0.000000	0.000000	0.000000	0.000000	0.000000	0.000000	3
0.000007	0.000007	0.000007	0.000007	0.000007	0.000007	4
0.000079	0.000063	0.000052	0.000045	0.000040	0.000037	5
0.000398	0.000293	0.000227	0.000183	0.000152	0.000130	6
0.001259	0.000906	0.000681	0.000531	0.000427	0.000353	7
0.002966	0.002131	0.001590	0.001225	0.000971	0.000789	8
0.005732	0.004155	0.003111	0.002396	0.001891	0.001527	9
0.009662	0.007095	0.005357	0.004145	0.003278	0.002644	10
0.014763	0.010993	0.008388	0.006539	0.005197	0.004202	11
0.020973	0.015836	0.012218	0.009607	0.007685	0.006242	12
0.028192	0.021570	0.016825	0.013349	0.010755	0.008785	13
0.036299	0.028118	0.022164	0.017742	0.014400	0.011834	14
0.045168	0.035392	0.028175	0.022747	0.018597	0.015378	15
0.054675	0.043297	0.034790	0.028315	0.023313	0.019396	16
0.064703	0.051742	0.041937	0.034394	0.028510	0.023860	17
0.075148	0.060641	0.049546	0.040928	0.034143	0.028739	18
0.085915	0.069912	0.057551	0.047861	0.040170	0.033996	19
0.096921	0.079482	0.065888	0.055142	0.046546	0.039596	20
0.108094	0.089287	0.074500	0.062720	0.053228	0.045503	21
0.119371	0.099266	0.083333	0.070548	0.060177	0.051683	22
0.130700	0.109371	0.092342	0.078585	0.067354	0.058103	23
0.142036	0.119555	0.101484	0.086790	0.074724	0.064730	24
0.153340	0.129781	0.110720	0.095129	0.082255	0.071537	25
0.164581	0.140015	0.120018	0.103570	0.089917	0.078495	26
0.175732	0.150228	0.129349	0.112085	0.097684	0.085579	27
0.186772	0.160396	0.138688	0.120648	0.105530	0.092768	28
0.197681	0.170499	0.148013	0.129238	0.113435	0.100038	29
0.208447	0.180518	0.157304	0.137834	0.121378	0.107372	30
0.306628	0.273929	0.245739	0.221270	0.199908	0.181164	40
0.452558	0.418305	0.387562	0.359837	0.334734	0.311927	60
0.550669	0.518371	0.488748	0.461472	0.436276	0.412939	80
0.619831	0.590158	0.562571	0.536836	0.512760	0.490184	100
0.670876	0.643745	0.618288	0.594325	0.571711	0.550325	120
0.709985	0.685132	0.661654	0.639410	0.618283	0.598178	140
0.753978	0.731988	0.711113	0.691172	0.672100	0.653829	170
0.786448	0.766791	0.748045	0.730070	0.712779	0.696132	200
0.818447	0.801269	0.784813	0.768957	0.753648	0.738836	240
0.860356	0.846684	0.833511	0.820740	0.808335	0.796267	320
0.896302	0.885863	0.875756	0.865908	0.856294	0.846895	440
0.922811	0.914887	0.907187	0.899657	0.892279	0.885041	600
0.941507	0.935422	0.929493	0.923681	0.917972	0.912357	800
0.952914	0.947976	0.943158	0.938426	0.933773	0.929189	1000
1.000000	1.000000	1.000000	1.000000	1.000000	1.000000	INF

(*continued*)

Table 3 Lower Percentage Points of Wilks's Lambda (*continued*)
$[U_{m,a,b,\alpha}]$

$m = 4, \quad \alpha = 0.05$

	a					
b	1	2	3	4	5	6
1	0.000000	0.000000	0.000000	0.000000	0.000000	0.000000
2	0.000000	0.000000	0.000000	0.000000	0.000000	0.000000
3	0.000000	0.000000	0.000000	0.000000	0.000000	0.000001
4	0.001378	0.000292	0.000127	0.000075	0.000052	0.000040
5	0.025529	0.006091	0.002314	0.001128	0.000647	0.000416
6	0.076071	0.023604	0.010010	0.005073	0.002903	0.001818
7	0.135374	0.050839	0.024047	0.013014	0.007737	0.004938
8	0.194043	0.083695	0.043226	0.024857	0.015415	0.010129
9	0.248619	0.118995	0.065947	0.039919	0.025729	0.017408
10	0.298130	0.154758	0.090794	0.057378	0.038260	0.026586
11	0.342593	0.189778	0.116701	0.076502	0.052524	0.037385
12	0.382448	0.223411	0.142927	0.096664	0.068077	0.049495
13	0.418181	0.255376	0.168939	0.117377	0.084546	0.062632
14	0.450335	0.285511	0.194413	0.138286	0.101586	0.076537
15	0.479286	0.313829	0.219113	0.159131	0.118954	0.090983
16	0.505511	0.340400	0.242944	0.179688	0.136434	0.105779
17	0.529312	0.365253	0.265812	0.199832	0.153891	0.120780
18	0.551035	0.388530	0.287689	0.219490	0.171171	0.135856
19	0.570858	0.410325	0.308599	0.238570	0.188209	0.150905
20	0.589077	0.430766	0.328552	0.257052	0.204926	0.165853
21	0.605832	0.449947	0.347546	0.274909	0.221288	0.180626
22	0.621318	0.467988	0.365676	0.292142	0.237242	0.195197
23	0.635634	0.484922	0.382934	0.308765	0.252783	0.209511
24	0.648934	0.500883	0.399402	0.324767	0.267896	0.223535
25	0.661320	0.515918	0.415077	0.340175	0.282568	0.237277
26	0.672864	0.530124	0.430041	0.355004	0.296810	0.250710
27	0.683663	0.543561	0.444332	0.369254	0.310608	0.263809
28	0.693769	0.556262	0.457946	0.382979	0.323980	0.276602
29	0.703259	0.568303	0.470981	0.396197	0.336947	0.289051
30	0.712188	0.579734	0.483430	0.408914	0.349488	0.301188
40	0.778877	0.668158	0.582817	0.513297	0.455181	0.405867
60	0.849044	0.767047	0.700065	0.642556	0.592126	0.547349
80	0.885442	0.820705	0.766251	0.718260	0.675124	0.635912
100	0.907714	0.854312	0.808614	0.767700	0.730354	0.695928
120	0.922736	0.877325	0.838018	0.802443	0.769650	0.739118
140	0.933554	0.894066	0.859605	0.828175	0.798994	0.771635
170	0.945088	0.912072	0.883006	0.856283	0.831278	0.807661
200	0.953211	0.924848	0.899727	0.876499	0.854647	0.833900
240	0.960919	0.937047	0.915781	0.896012	0.877319	0.859481
320	0.970605	0.952477	0.936212	0.920990	0.906503	0.892593
440	0.978571	0.965253	0.953233	0.941922	0.931100	0.920655
600	0.984259	0.974422	0.965507	0.957084	0.948995	0.941160
800	0.988181	0.980767	0.974028	0.967644	0.961498	0.955529
1000	0.990538	0.984589	0.979173	0.974034	0.969078	0.964257
INF	1.000000	1.000000	1.000000	1.000000	1.000000	1.000000

Table 3 (*continued*)

a						
7	8	9	10	11	12	*b*
0.000000	0.000000	0.000000	0.000000	0.000000	0.000000	1
0.000000	0.000000	0.000000	0.000000	0.000000	0.000000	2
0.000001	0.000001	0.000002	0.000002	0.000002	0.000003	3
0.000033	0.000029	0.000026	0.000025	0.000023	0.000022	4
0.000292	0.000218	0.000172	0.000141	0.000120	0.000105	5
0.001223	0.000872	0.000652	0.000508	0.000409	0.000338	6
0.003338	0.002365	0.001745	0.001333	0.001050	0.000848	7
0.006975	0.004994	0.003698	0.002819	0.002206	0.001766	8
0.012249	0.008907	0.006664	0.005112	0.004009	0.003208	9
0.019107	0.014130	0.010706	0.008288	0.006542	0.005254	10
0.027402	0.020589	0.015806	0.012365	0.009839	0.007948	11
0.036933	0.028170	0.021899	0.017314	0.013895	0.011302	12
0.047493	0.036731	0.028895	0.023075	0.018675	0.015303	13
0.058886	0.046115	0.036676	0.029572	0.024133	0.019917	14
0.070925	0.056188	0.045140	0.036722	0.030208	0.025101	15
0.083443	0.066806	0.054181	0.044440	0.036830	0.030804	16
0.096316	0.077856	0.063688	0.052645	0.043936	0.036980	17
0.109411	0.089236	0.073577	0.061263	0.051456	0.043568	18
0.122643	0.100843	0.083764	0.070213	0.059338	0.050514	19
0.135926	0.112607	0.094180	0.079441	0.067512	0.057782	20
0.149180	0.124462	0.104757	0.088877	0.075938	0.065315	21
0.162364	0.136342	0.115440	0.098474	0.084565	0.073068	22
0.175434	0.148204	0.126185	0.108191	0.093352	0.081008	23
0.188341	0.160009	0.136950	0.117977	0.102254	0.089100	24
0.201067	0.171726	0.147695	0.127818	0.111240	0.097305	25
0.213597	0.183333	0.158399	0.137656	0.120274	0.105608	26
0.225900	0.194794	0.169017	0.147483	0.129346	0.113968	27
0.237971	0.206105	0.179569	0.157274	0.138418	0.122368	28
0.249798	0.217241	0.189991	0.167006	0.147478	0.130784	29
0.261373	0.228198	0.200311	0.176673	0.156516	0.139205	30
0.363565	0.326959	0.295085	0.267163	0.242600	0.220888	40
0.507256	0.471148	0.438462	0.408771	0.381699	0.356960	60
0.600023	0.566986	0.536460	0.508176	0.481887	0.457414	80
0.663968	0.634166	0.606280	0.580111	0.555487	0.532298	100
0.710513	0.683595	0.658183	0.634132	0.611324	0.589657	120
0.745829	0.721386	0.698162	0.676045	0.654943	0.634778	140
0.785224	0.763821	0.743347	0.723717	0.704865	0.686733	170
0.814087	0.795095	0.776838	0.759251	0.742281	0.725885	200
0.842366	0.825881	0.809960	0.794554	0.779622	0.765130	240
0.879164	0.866153	0.853513	0.841211	0.829220	0.817517	320
0.910522	0.900654	0.891022	0.881602	0.872376	0.863330	440
0.933530	0.926075	0.918772	0.911606	0.904563	0.897634	600
0.949702	0.943994	0.938390	0.932877	0.927446	0.922092	800
0.959545	0.954922	0.950376	0.945898	0.941481	0.937120	1000
1.000000	1.000000	1.000000	1.000000	1.000000	1.000000	INF

(*continued*)

Table 3 Lower Percentage Points of Wilks's Lambda (*continued*)
$[U_{m,\,a,\,b,\,\alpha}]$

$m = 5, \quad \alpha = 0.01$

	a					
b	1	2	3	4	5	6
1	0.000000	0.000000	0.000000	0.000000	0.000000	0.000000
2	0.000000	0.000000	0.000000	0.000000	0.000000	0.000000
3	0.000000	0.000000	0.000000	0.000000	0.000000	0.000000
4	0.000000	0.000000	0.000000	0.000000	0.000000	0.000000
5	0.000164	0.000036	0.000015	0.000009	0.000006	0.000004
6	0.004668	0.000962	0.000335	0.000153	0.000084	0.000052
7	0.021333	0.005332	0.001959	0.000893	0.000472	0.000277
8	0.049302	0.014879	0.006018	0.002885	0.001557	0.000918
9	0.083710	0.029395	0.013027	0.006623	0.003709	0.002237
10	0.120729	0.047777	0.022897	0.012300	0.007165	0.004443
11	0.158044	0.068815	0.035223	0.019865	0.012007	0.007658
12	0.194389	0.091490	0.049501	0.029128	0.018203	0.011922
13	0.229107	0.115016	0.065237	0.039832	0.025645	0.017209
14	0.261911	0.138822	0.081993	0.051709	0.034186	0.023452
15	0.292711	0.162507	0.099424	0.064505	0.043665	0.030560
16	0.321529	0.185794	0.117224	0.077992	0.053925	0.038428
17	0.348449	0.208500	0.135171	0.091973	0.064813	0.046952
18	0.373583	0.230510	0.153091	0.106281	0.076195	0.056028
19	0.397053	0.251754	0.170849	0.120777	0.087948	0.065559
20	0.418988	0.272198	0.188345	0.135348	0.099970	0.075458
21	0.439504	0.291832	0.205509	0.149903	0.112170	0.085645
22	0.458721	0.310659	0.222288	0.164368	0.124471	0.096050
23	0.476739	0.328696	0.238647	0.178686	0.136809	0.106612
24	0.493662	0.345966	0.254564	0.192810	0.149131	0.117275
25	0.509575	0.362498	0.270024	0.206707	0.161392	0.127995
26	0.524560	0.378321	0.285024	0.220349	0.173556	0.138732
27	0.538691	0.393468	0.299564	0.233718	0.185593	0.149451
28	0.552034	0.407970	0.313646	0.246799	0.197480	0.160123
29	0.564655	0.421860	0.327281	0.259584	0.209196	0.170724
30	0.576601	0.435170	0.340476	0.272068	0.220728	0.181235
40	0.668249	0.542257	0.451107	0.380670	0.324502	0.278831
60	0.769057	0.669979	0.592611	0.528633	0.474345	0.427572
80	0.823038	0.742525	0.677232	0.621368	0.572440	0.529007
100	0.856606	0.789069	0.733048	0.684125	0.640446	0.600959
120	0.879484	0.821412	0.772515	0.729225	0.690073	0.654241
140	0.896071	0.845177	0.801862	0.763137	0.727789	0.695150
170	0.913861	0.870955	0.834024	0.800661	0.769907	0.741242
200	0.926453	0.889383	0.857222	0.827957	0.800792	0.775304
240	0.938451	0.907082	0.879663	0.854540	0.831068	0.808906
320	0.953594	0.929613	0.908455	0.888903	0.870485	0.852956
440	0.966104	0.948389	0.932642	0.917985	0.904086	0.890771
600	0.975067	0.961932	0.950192	0.939210	0.928746	0.918676
800	0.981261	0.971335	0.962430	0.954071	0.946080	0.938365
1000	0.984990	0.977013	0.969840	0.963094	0.956632	0.950381
INF	1.000000	1.000000	1.000000	1.000000	1.000000	1.000000

Table 3 (*continued*)

a						
7	8	9	10	11	12	*b*
0.000000	0.000000	0.000000	0.000000	0.000000	0.000000	1
0.000000	0.000000	0.000000	0.000000	0.000000	0.000000	2
0.000000	0.000000	0.000000	0.000000	0.000000	0.000000	3
0.000000	0.000000	0.000000	0.000000	0.000000	0.000000	4
0.000003	0.000003	0.000003	0.000002	0.000002	0.000002	5
0.000035	0.000025	0.000020	0.000016	0.000013	0.000011	6
0.000177	0.000121	0.000088	0.000066	0.000052	0.000042	7
0.000582	0.000390	0.000275	0.000202	0.000154	0.000121	8
0.001432	0.000964	0.000677	0.000493	0.000371	0.000287	9
0.002899	0.001974	0.001394	0.001017	0.000763	0.000587	10
0.005103	0.003528	0.002519	0.001850	0.001392	0.001072	11
0.008113	0.005702	0.004122	0.003055	0.002315	0.001790	12
0.011946	0.008535	0.006252	0.004681	0.003576	0.002781	13
0.016584	0.012034	0.008929	0.006758	0.005207	0.004077	14
0.021981	0.016183	0.012158	0.009299	0.007228	0.005702	15
0.028076	0.020951	0.015926	0.012305	0.009648	0.007667	16
0.034797	0.026294	0.020207	0.015764	0.012465	0.009978	17
0.042071	0.032161	0.024971	0.019658	0.015670	0.012633	18
0.049825	0.038499	0.030179	0.023963	0.019247	0.015623	19
0.057989	0.045254	0.035793	0.028649	0.023178	0.018937	20
0.066497	0.052374	0.041771	0.033687	0.027441	0.022558	21
0.075288	0.059809	0.048073	0.039045	0.032012	0.026471	22
0.084307	0.067511	0.054660	0.044692	0.036865	0.030655	23
0.093504	0.075438	0.061497	0.050597	0.041977	0.035091	24
0.102837	0.083549	0.068546	0.056730	0.047322	0.039758	25
0.112264	0.091808	0.075777	0.063064	0.052876	0.044637	26
0.121752	0.100183	0.083160	0.069572	0.058617	0.049707	27
0.131271	0.108643	0.090666	0.076229	0.064523	0.054951	28
0.140795	0.117163	0.098272	0.083013	0.070573	0.060349	29
0.150300	0.125719	0.105955	0.089902	0.076748	0.065885	30
0.241173	0.209788	0.183401	0.161054	0.142005	0.125677	40
0.386853	0.351131	0.319602	0.291636	0.266722	0.244448	60
0.490107	0.455043	0.423278	0.394390	0.368029	0.343906	80
0.564970	0.531981	0.501610	0.473552	0.447557	0.423417	100
0.621196	0.590558	0.562038	0.535406	0.510473	0.487083	120
0.664792	0.636413	0.609782	0.584719	0.561076	0.538731	140
0.714313	0.688960	0.664941	0.642146	0.620465	0.599807	170
0.751191	0.728357	0.706614	0.685839	0.665962	0.646914	200
0.787800	0.767701	0.748450	0.729967	0.712175	0.695028	240
0.836125	0.819982	0.804406	0.789342	0.774746	0.760584	320
0.877902	0.865484	0.853429	0.841699	0.830267	0.819109	440
0.908897	0.899419	0.890178	0.881148	0.872309	0.863646	600
0.930849	0.923543	0.916398	0.909395	0.902519	0.895761	800
0.944279	0.938337	0.932515	0.926799	0.921177	0.915641	1000
1.000000	1.000000	1.000000	1.000000	1.000000	1.000000	INF

(*continued*)

Table 3 Lower Percentage Points of Wilks's Lambda (*continued*)
$[U_{m,a,b,\alpha}]$

$m = 5, \quad \alpha = 0.05$

b	a = 1	2	3	4	5	6
1	0.000000	0.000000	0.000000	0.000000	0.000000	0.000000
2	0.000000	0.000000	0.000000	0.000000	0.000000	0.000000
3	0.000000	0.000000	0.000000	0.000000	0.000000	0.000000
4	0.000000	0.000000	0.000000	0.000000	0.000001	0.000001
5	0.001598	0.000291	0.000105	0.000052	0.000031	0.000021
6	0.021145	0.004391	0.001479	0.000647	0.000335	0.000197
7	0.062770	0.016898	0.006357	0.002903	0.001514	0.000872
8	0.113526	0.037390	0.015792	0.007737	0.004208	0.002479
9	0.165351	0.063279	0.029433	0.015415	0.008787	0.005348
10	0.214794	0.092191	0.046378	0.025729	0.015321	0.009639
11	0.260635	0.122403	0.065660	0.038260	0.023674	0.015360
12	0.302608	0.152793	0.086448	0.052524	0.033618	0.022418
13	0.340813	0.182662	0.108110	0.068077	0.044878	0.030680
14	0.375528	0.211602	0.130131	0.084546	0.057198	0.039965
15	0.407128	0.239373	0.152159	0.101586	0.070324	0.050117
16	0.435899	0.265851	0.173959	0.118954	0.084048	0.060965
17	0.462173	0.291015	0.195322	0.136434	0.098187	0.072367
18	0.486266	0.314859	0.216138	0.153891	0.112582	0.084178
19	0.508362	0.337418	0.236338	0.171171	0.127108	0.096308
20	0.528714	0.358776	0.255858	0.188209	0.141662	0.108634
21	0.547516	0.378956	0.274710	0.204926	0.156176	0.121083
22	0.564905	0.398037	0.292843	0.221288	0.170563	0.133590
23	0.581036	0.416105	0.310304	0.237242	0.184782	0.146095
24	0.596032	0.433216	0.327083	0.252783	0.198795	0.158544
25	0.610030	0.449429	0.343191	0.267896	0.212568	0.170898
26	0.623126	0.464800	0.358665	0.282568	0.226071	0.183129
27	0.635368	0.479382	0.373523	0.296810	0.239294	0.195207
28	0.646832	0.493247	0.387790	0.310608	0.252224	0.207116
29	0.657645	0.506421	0.401488	0.323980	0.264872	0.218828
30	0.667803	0.518945	0.414658	0.336947	0.277200	0.230347
40	0.744009	0.617178	0.521747	0.446045	0.384424	0.333492
60	0.824764	0.729155	0.652037	0.586878	0.530670	0.481578
80	0.866847	0.790730	0.727186	0.671775	0.622536	0.578316
100	0.892643	0.829563	0.775817	0.728040	0.684827	0.645343
120	0.910071	0.856267	0.809790	0.767957	0.729656	0.694256
140	0.922634	0.875748	0.834850	0.797705	0.763400	0.731431
170	0.936039	0.896748	0.862122	0.830370	0.800777	0.772953
200	0.945486	0.911680	0.881674	0.853973	0.827989	0.803406
240	0.954455	0.925960	0.900496	0.876838	0.854512	0.833264
320	0.965732	0.944055	0.924519	0.906224	0.888827	0.872146
440	0.975013	0.959064	0.944590	0.930949	0.917894	0.905302
600	0.981642	0.969850	0.959096	0.948912	0.939124	0.929642
800	0.986214	0.977320	0.969181	0.961450	0.953996	0.946753
1000	0.988963	0.981823	0.975277	0.969047	0.963029	0.957171
INF	1.000000	1.000000	1.000000	1.000000	1.000000	1.000000

Table 3 (*continued*)

a						
7	8	9	10	11	12	*b*
0.000000	0.000000	0.000000	0.000000	0.000000	0.000000	1
0.000000	0.000000	0.000000	0.000000	0.000000	0.000000	2
0.000000	0.000000	0.000000	0.000000	0.000000	0.000000	3
0.000001	0.000001	0.000001	0.000001	0.000001	0.000001	4
0.000015	0.000012	0.000010	0.000008	0.000007	0.000007	5
0.000126	0.000087	0.000064	0.000049	0.000039	0.000032	6
0.000544	0.000361	0.000253	0.000185	0.000141	0.000110	7
0.001557	0.001032	0.000716	0.000516	0.000385	0.000296	8
0.003433	0.002304	0.001607	0.001159	0.000861	0.000657	9
0.006343	0.004335	0.003062	0.002225	0.001660	0.001267	10
0.010358	0.007216	0.005173	0.003802	0.002858	0.002192	11
0.015467	0.010980	0.007991	0.005946	0.004512	0.003486	12
0.021607	0.015611	0.011530	0.008685	0.006659	0.005187	13
0.028683	0.021061	0.015774	0.012024	0.009313	0.007317	14
0.036584	0.027266	0.020687	0.015949	0.012475	0.009884	15
0.045199	0.034145	0.026219	0.020428	0.016129	0.012885	16
0.054409	0.041618	0.032312	0.025427	0.020252	0.016307	17
0.064111	0.049602	0.038909	0.030904	0.024819	0.020133	18
0.074209	0.058024	0.045951	0.036810	0.029790	0.024339	19
0.084619	0.066805	0.053373	0.043100	0.035137	0.028896	20
0.095254	0.075885	0.061122	0.049724	0.040817	0.033782	21
0.106063	0.085203	0.069149	0.056652	0.046803	0.038962	22
0.116974	0.094699	0.077408	0.063832	0.053052	0.044411	23
0.127948	0.104337	0.085849	0.071231	0.059537	0.050103	24
0.138945	0.114058	0.094444	0.078809	0.066222	0.056005	25
0.149909	0.123843	0.103144	0.086536	0.073084	0.062103	26
0.160826	0.133657	0.111931	0.094385	0.080092	0.068358	27
0.171667	0.143454	0.120766	0.102328	0.087220	0.074761	28
0.182408	0.153240	0.129630	0.110336	0.094455	0.081283	29
0.193043	0.162971	0.138499	0.118393	0.101767	0.087901	30
0.290896	0.254963	0.224433	0.198322	0.175874	0.156480	40
0.438367	0.400085	0.365997	0.335520	0.308193	0.283593	60
0.538319	0.501966	0.468774	0.438392	0.410497	0.384827	80
0.609037	0.575508	0.544420	0.515540	0.488629	0.463515	100
0.661341	0.630608	0.601822	0.574793	0.549362	0.525395	120
0.701466	0.673268	0.646653	0.621477	0.597615	0.574968	140
0.746648	0.721687	0.697934	0.675284	0.653648	0.632953	170
0.780024	0.757705	0.736343	0.715856	0.696177	0.677251	200
0.812938	0.793426	0.774647	0.756540	0.739054	0.722148	240
0.856074	0.840535	0.825476	0.810855	0.796641	0.782805	320
0.893096	0.881226	0.869655	0.858357	0.847311	0.836500	440
0.920410	0.911396	0.902572	0.893921	0.885429	0.877084	600
0.939682	0.932756	0.925957	0.919273	0.912693	0.906209	800
0.951441	0.945820	0.940292	0.934848	0.929480	0.924182	1000
1.000000	1.000000	1.000000	1.000000	1.000000	1.000000	INF

(*continued*)

Table 3 Lower Percentage Points of Wilks's Lambda (*continued*)

$[U_{m,a,b,\alpha}]$

$m = 6, \quad \alpha = 0.01$

	a					
b	1	2	3	4	5	6
1	0.000000	0.000000	0.000000	0.000000	0.000000	0.000000
2	0.000000	0.000000	0.000000	0.000000	0.000000	0.000000
3	0.000000	0.000000	0.000000	0.000000	0.000000	0.000000
4	0.000000	0.000000	0.000000	0.000000	0.000000	0.000000
5	0.000000	0.000000	0.000000	0.000000	0.000000	0.000000
6	0.000295	0.000050	0.000017	0.000008	0.000004	0.000003
7	0.004608	0.000839	0.000258	0.000106	0.000052	0.000029
8	0.018808	0.004182	0.001383	0.000574	0.000277	0.000150
9	0.042762	0.011508	0.004211	0.001846	0.000918	0.000503
10	0.072861	0.022948	0.009244	0.004315	0.002237	0.001257
11	0.105882	0.037842	0.016575	0.008211	0.004443	0.002575
12	0.139723	0.055318	0.026013	0.013591	0.007658	0.004578
13	0.173151	0.074563	0.037260	0.020392	0.011922	0.007339
14	0.205478	0.094910	0.049951	0.028478	0.017209	0.010886
15	0.236354	0.115841	0.063758	0.037679	0.023452	0.015207
16	0.265622	0.136970	0.078384	0.047814	0.030560	0.020265
17	0.293243	0.158016	0.093576	0.058712	0.038428	0.026009
18	0.319245	0.178778	0.109125	0.070212	0.046952	0.032373
19	0.343692	0.199115	0.124859	0.082172	0.056028	0.039291
20	0.366666	0.218933	0.140644	0.094469	0.065559	0.046693
21	0.388259	0.238169	0.156370	0.106995	0.075458	0.054514
22	0.408567	0.256789	0.171955	0.119660	0.085645	0.062689
23	0.427679	0.274775	0.187333	0.132389	0.096050	0.071160
24	0.445681	0.292121	0.202457	0.145118	0.106612	0.079873
25	0.462657	0.308833	0.217287	0.157796	0.117275	0.088781
26	0.478684	0.324922	0.231801	0.170381	0.127995	0.097838
27	0.493829	0.340404	0.245977	0.182837	0.138732	0.107006
28	0.508160	0.355297	0.259807	0.195137	0.149451	0.116251
29	0.521737	0.369623	0.273282	0.207259	0.160123	0.125542
30	0.534611	0.383404	0.286401	0.219187	0.170724	0.134852
40	0.633971	0.495984	0.398981	0.326182	0.269778	0.225181
60	0.744292	0.633481	0.548313	0.479114	0.421438	0.372626
80	0.803733	0.712825	0.639827	0.578119	0.524751	0.477984
100	0.840806	0.764135	0.700945	0.646245	0.597871	0.554570
120	0.866117	0.799962	0.744489	0.695702	0.651910	0.612148
140	0.884492	0.826371	0.777033	0.733148	0.693333	0.656808
170	0.904218	0.855095	0.812851	0.774824	0.739929	0.707546
200	0.918192	0.875677	0.838782	0.805291	0.774313	0.745329
240	0.931516	0.895480	0.863940	0.835082	0.808188	0.782831
320	0.948345	0.920741	0.896322	0.873758	0.852531	0.832324
440	0.962259	0.941834	0.923609	0.906632	0.890537	0.875096
600	0.972232	0.957070	0.943458	0.930704	0.918546	0.906818
800	0.979127	0.967660	0.957321	0.947597	0.938291	0.929281
1000	0.983279	0.974060	0.965726	0.957869	0.950332	0.943020
INF	1.000000	1.000000	1.000000	1.000000	1.000000	1.000000

Table 3 (*continued*)

a = 7	8	9	10	11	12	*b*
0.000000	0.000000	0.000000	0.000000	0.000000	0.000000	1
0.000000	0.000000	0.000000	0.000000	0.000000	0.000000	2
0.000000	0.000000	0.000000	0.000000	0.000000	0.000000	3
0.000000	0.000000	0.000000	0.000000	0.000000	0.000000	4
0.000000	0.000000	0.000000	0.000000	0.000000	0.000000	5
0.000002	0.000001	0.000001	0.000000	0.000000	0.000000	6
0.000018	0.000012	0.000008	0.000006	0.000005	0.000004	7
0.000089	0.000057	0.000038	0.000027	0.000020	0.000015	8
0.000297	0.000187	0.000124	0.000086	0.000062	0.000046	9
0.000754	0.000477	0.000317	0.000219	0.000156	0.000115	10
0.001578	0.001014	0.000678	0.000470	0.000336	0.000247	11
0.002873	0.001878	0.001272	0.000889	0.000639	0.000471	12
0.004715	0.003140	0.002158	0.001525	0.001105	0.000818	13
0.007151	0.004851	0.003384	0.002420	0.001769	0.001320	14
0.010199	0.007041	0.004984	0.003608	0.002664	0.002004	15
0.013856	0.009723	0.006980	0.005114	0.003815	0.002894	16
0.018099	0.012897	0.009383	0.006954	0.005240	0.004009	17
0.022896	0.016549	0.012192	0.009135	0.006950	0.005362	18
0.028206	0.020658	0.015398	0.011658	0.008952	0.006963	19
0.033983	0.025196	0.018987	0.014517	0.011245	0.008816	20
0.040181	0.030131	0.022939	0.017701	0.013827	0.010921	21
0.046751	0.035431	0.027233	0.021197	0.016688	0.013275	22
0.053649	0.041060	0.031843	0.024988	0.019819	0.015873	23
0.060830	0.046985	0.036744	0.029056	0.023208	0.018708	24
0.068253	0.053173	0.041911	0.033382	0.026841	0.021769	25
0.075880	0.059592	0.047319	0.037946	0.030703	0.025046	26
0.083676	0.066212	0.052941	0.042728	0.034778	0.028527	27
0.091609	0.073004	0.058756	0.047709	0.039052	0.032201	28
0.099650	0.079944	0.064740	0.052871	0.043508	0.036054	29
0.107773	0.087007	0.070872	0.058193	0.048132	0.040075	30
0.189401	0.160362	0.136572	0.116924	0.100582	0.086906	40
0.330877	0.294885	0.263658	0.236424	0.212564	0.191578	60
0.436634	0.399836	0.366925	0.337368	0.310733	0.286658	80
0.515501	0.480046	0.447731	0.418175	0.391064	0.366135	100
0.575775	0.542326	0.511446	0.482846	0.456292	0.431584	120
0.623066	0.591738	0.562544	0.535260	0.509701	0.485712	140
0.677354	0.649014	0.622338	0.597164	0.573355	0.550798	170
0.718117	0.692405	0.668017	0.644841	0.622774	0.601727	200
0.758861	0.736059	0.714299	0.693473	0.673511	0.654352	240
0.813059	0.794571	0.776777	0.759613	0.743028	0.726975	320
0.860269	0.845937	0.832046	0.818554	0.805427	0.792640	440
0.895499	0.884501	0.873788	0.863330	0.853106	0.843097	600
0.920553	0.912044	0.903725	0.895577	0.887583	0.879732	800
0.935921	0.928984	0.922190	0.915520	0.908964	0.902511	1000
1.000000	1.000000	1.000000	1.000000	1.000000	1.000000	INF

(*continued*)

Table 3 Lower Percentage Points of Wilks's Lambda (*continued*)

$[U_{m,a,b,\alpha}]$

$m = 6, \quad \alpha = 0.05$

	a					
b	1	2	3	4	5	6
1	0.000000	0.000000	0.000000	0.000000	0.000000	0.000000
2	0.000000	0.000000	0.000000	0.000000	0.000000	0.000000
3	0.000000	0.000000	0.000000	0.000000	0.000000	0.000000
4	0.000000	0.000000	0.000000	0.000000	0.000000	0.000000
5	0.000007	0.000002	0.000001	0.000001	0.000001	0.000000
6	0.002045	0.000315	0.000095	0.000040	0.000021	0.000012
7	0.018804	0.003479	0.001052	0.000416	0.000197	0.000106
8	0.053911	0.012883	0.004369	0.001818	0.000872	0.000465
9	0.098038	0.028824	0.011018	0.004938	0.002479	0.001358
10	0.144274	0.049685	0.021043	0.010129	0.005348	0.003035
11	0.189355	0.073697	0.033966	0.017408	0.009639	0.005672
12	0.231866	0.099450	0.049161	0.026586	0.015360	0.009348
13	0.271356	0.125933	0.066012	0.037385	0.022418	0.014071
14	0.307797	0.152453	0.083979	0.049495	0.030680	0.019795
15	0.341285	0.178581	0.102644	0.062632	0.039965	0.026433
16	0.372033	0.204010	0.121656	0.076537	0.050117	0.033893
17	0.400304	0.228568	0.140775	0.090983	0.060965	0.042061
18	0.426364	0.252176	0.159796	0.105779	0.072367	0.050834
19	0.450349	0.274785	0.178574	0.120780	0.084178	0.060119
20	0.472562	0.296393	0.197017	0.135856	0.096308	0.069818
21	0.493091	0.316990	0.215044	0.150905	0.108634	0.079840
22	0.512181	0.336628	0.232604	0.165853	0.121083	0.090122
23	0.529913	0.355328	0.249666	0.180626	0.133590	0.100596
24	0.546452	0.373143	0.266216	0.195197	0.146095	0.111189
25	0.561889	0.390109	0.282253	0.209511	0.158544	0.121873
26	0.576348	0.406285	0.297740	0.223535	0.170898	0.132587
27	0.589899	0.421688	0.312738	0.237277	0.183129	0.143309
28	0.602633	0.436379	0.327221	0.250710	0.195207	0.153998
29	0.614602	0.450416	0.341199	0.263809	0.207116	0.164629
30	0.625896	0.463794	0.354711	0.276602	0.218828	0.175171
40	0.710937	0.569976	0.466792	0.387183	0.324162	0.273470
60	0.801604	0.693451	0.607528	0.536153	0.475641	0.423707
80	0.849063	0.762264	0.690479	0.628610	0.574313	0.526153
100	0.878218	0.805945	0.744748	0.690824	0.642495	0.598763
120	0.897944	0.836112	0.782919	0.735354	0.692128	0.652489
140	0.912172	0.858176	0.811198	0.768751	0.729786	0.693709
170	0.927365	0.882016	0.842092	0.805615	0.771775	0.740119
200	0.938078	0.899001	0.864313	0.832375	0.802523	0.774395
240	0.948255	0.915270	0.885761	0.858391	0.832628	0.808187
320	0.961056	0.935919	0.913212	0.891956	0.871772	0.852459
440	0.971597	0.953076	0.936212	0.920308	0.905097	0.890438
600	0.979129	0.965422	0.952870	0.940969	0.929528	0.918448
800	0.984325	0.973979	0.964469	0.955420	0.946688	0.938203
1000	0.987450	0.979142	0.971487	0.964187	0.957129	0.950256
INF	1.000000	1.000000	1.000000	1.000000	1.000000	1.000000

Table 3 (*continued*)

a						
7	8	9	10	11	12	b
0.000000	0.000000	0.000000	0.000000	0.000000	0.000000	1
0.000000	0.000000	0.000000	0.000000	0.000000	0.000000	2
0.000000	0.000000	0.000000	0.000000	0.000000	0.000000	3
0.000000	0.000000	0.000000	0.000000	0.000000	0.000000	4
0.000000	0.000000	0.000000	0.000000	0.000000	0.000000	5
0.000008	0.000006	0.000004	0.000003	0.000003	0.000002	6
0.000063	0.000040	0.000027	0.000020	0.000015	0.000011	7
0.000270	0.000168	0.000111	0.000076	0.000055	0.000041	8
0.000798	0.000497	0.000325	0.000222	0.000157	0.000115	9
0.001826	0.001155	0.000762	0.000521	0.000369	0.000269	10
0.003507	0.002263	0.001514	0.001046	0.000744	0.000543	11
0.005940	0.003915	0.002664	0.001865	0.001338	0.000983	12
0.009172	0.006173	0.004273	0.003033	0.002200	0.001630	13
0.013205	0.009066	0.006381	0.004592	0.003370	0.002520	14
0.018012	0.012593	0.009005	0.006568	0.004877	0.003682	15
0.023544	0.016741	0.012147	0.008974	0.006740	0.005137	16
0.029737	0.021472	0.015794	0.011811	0.008966	0.006898	17
0.036522	0.026746	0.019924	0.015070	0.011554	0.008971	18
0.043825	0.032520	0.024510	0.018734	0.014503	0.011356	19
0.051576	0.038739	0.029518	0.022785	0.017796	0.014049	20
0.059715	0.045349	0.034906	0.027193	0.021418	0.017040	21
0.068178	0.052311	0.040646	0.031936	0.025354	0.020317	22
0.076899	0.059574	0.046695	0.036988	0.029581	0.023864	23
0.085836	0.067090	0.053016	0.042316	0.034078	0.027670	24
0.094944	0.074824	0.059586	0.047895	0.038825	0.031716	25
0.104168	0.082735	0.066362	0.053696	0.043795	0.035986	26
0.113485	0.090793	0.073318	0.059697	0.048977	0.040460	27
0.122849	0.098970	0.080420	0.065867	0.054339	0.045123	28
0.132250	0.107224	0.087654	0.072196	0.059866	0.049957	29
0.141648	0.115539	0.094994	0.078649	0.065542	0.054951	30
0.232192	0.198251	0.170132	0.146678	0.126985	0.110367	40
0.378774	0.339636	0.305361	0.275238	0.248638	0.225098	60
0.483144	0.444543	0.409736	0.378269	0.349725	0.323787	80
0.558956	0.522538	0.489125	0.458377	0.430004	0.403784	100
0.615927	0.582063	0.550602	0.521300	0.493955	0.468392	120
0.660119	0.628724	0.599296	0.571649	0.545628	0.521100	140
0.710350	0.682254	0.655667	0.630455	0.606507	0.583730	170
0.747758	0.722444	0.698328	0.675308	0.653300	0.632233	200
0.784886	0.762599	0.741229	0.720701	0.700953	0.681935	240
0.833892	0.815985	0.798676	0.781916	0.765666	0.749894	320
0.876249	0.862471	0.849063	0.835995	0.823242	0.810784	440
0.907669	0.897152	0.886868	0.876798	0.866923	0.857233	600
0.929921	0.921812	0.913858	0.906042	0.898354	0.890785	800
0.943532	0.936937	0.930455	0.924073	0.917783	0.911578	1000
1.000000	1.000000	1.000000	1.000000	1.000000	1.000000	INF

(*continued*)

Table 3 Lower Percentage Points of Wilks's Lambda (*continued*)
$[U_{m,a,b,\alpha}]$

$m = 7, \quad \alpha = 0.01$

	a					
b	1	2	3	4	5	6
1	0.000000	0.000000	0.000000	0.000000	0.000000	0.000000
2	0.000000	0.000000	0.000000	0.000000	0.000000	0.000000
3	0.000000	0.000000	0.000000	0.000000	0.000000	0.000000
4	0.000000	0.000000	0.000000	0.000000	0.000000	0.000000
5	0.000000	0.000000	0.000000	0.000000	0.000000	0.000000
6	0.000005	0.000001	0.000000	0.000000	0.000000	0.000000
7	0.000486	0.000068	0.000019	0.000007	0.000003	0.000002
8	0.004798	0.000782	0.000215	0.000079	0.000035	0.000018
9	0.017314	0.003481	0.001047	0.000398	0.000177	0.000089
10	0.038208	0.009312	0.003116	0.001259	0.000582	0.000297
11	0.064845	0.018560	0.006864	0.002966	0.001432	0.000754
12	0.094551	0.030855	0.012459	0.005732	0.002899	0.001578
13	0.125434	0.045575	0.019845	0.009662	0.005103	0.002873
14	0.156314	0.062081	0.028840	0.014763	0.008113	0.004715
15	0.186495	0.079813	0.039203	0.020973	0.011946	0.007151
16	0.215591	0.098313	0.050682	0.028192	0.016584	0.010199
17	0.243398	0.117225	0.063039	0.036299	0.021981	0.013856
18	0.269838	0.136276	0.076063	0.045168	0.028076	0.018099
19	0.294893	0.155259	0.089567	0.054675	0.034797	0.022896
20	0.318593	0.174026	0.103395	0.064703	0.042071	0.028206
21	0.340992	0.192467	0.117419	0.075148	0.049825	0.033983
22	0.362149	0.210506	0.131529	0.085915	0.057989	0.040181
23	0.382138	0.228088	0.145640	0.096921	0.066497	0.046751
24	0.401034	0.245182	0.159680	0.108094	0.075288	0.053649
25	0.418901	0.261767	0.173594	0.119371	0.084307	0.060830
26	0.435815	0.277834	0.187337	0.130700	0.093504	0.068253
27	0.451836	0.293382	0.200875	0.142036	0.102837	0.075880
28	0.467026	0.308414	0.214182	0.153341	0.112264	0.083676
29	0.481442	0.322941	0.227238	0.164581	0.121752	0.091609
30	0.495137	0.336973	0.240029	0.175732	0.131271	0.099650
40	0.601481	0.453452	0.352532	0.279077	0.223835	0.181405
60	0.720662	0.599212	0.507477	0.434307	0.374443	0.324703
80	0.785257	0.684670	0.604805	0.538151	0.481262	0.432066
100	0.825659	0.740370	0.670619	0.610809	0.558448	0.512056
120	0.853291	0.779445	0.717860	0.664095	0.616219	0.573107
140	0.873373	0.808339	0.753345	0.704715	0.660882	0.620947
170	0.894951	0.839849	0.792566	0.750184	0.711466	0.675794
200	0.910250	0.862477	0.821063	0.783589	0.749029	0.716923
240	0.924846	0.884289	0.848791	0.816382	0.786225	0.757986
320	0.943294	0.912166	0.884592	0.859136	0.835186	0.812531
440	0.958556	0.935488	0.914854	0.895630	0.877381	0.859974
600	0.969501	0.952359	0.936918	0.922439	0.908607	0.895334
800	0.977071	0.964096	0.952354	0.941294	0.930682	0.920457
1000	0.981630	0.971194	0.961722	0.952776	0.944171	0.935859
INF	1.000000	1.000000	1.000000	1.000000	1.000000	1.000000

Table 3 (*continued*)

a = 7	8	9	10	11	12	*b*
0.000000	0.000000	0.000000	0.000000	0.000000	0.000000	1
0.000000	0.000000	0.000000	0.000000	0.000000	0.000000	2
0.000000	0.000000	0.000000	0.000000	0.000000	0.000000	3
0.000000	0.000000	0.000000	0.000000	0.000000	0.000000	4
0.000000	0.000000	0.000000	0.000000	0.000000	0.000000	5
0.000000	0.000000	0.000000	0.000000	0.000000	0.000000	6
0.000001	0.000001	0.000001	0.000000	0.000000	0.000000	7
0.000010	0.000006	0.000004	0.000003	0.000002	0.000002	8
0.000049	0.000029	0.000019	0.000012	0.000009	0.000006	9
0.000165	0.000098	0.000061	0.000040	0.000028	0.000020	10
0.000426	0.000255	0.000160	0.000105	0.000072	0.000051	11
0.000912	0.000555	0.000353	0.000233	0.000159	0.000112	12
0.001704	0.001057	0.000682	0.000455	0.000313	0.000221	13
0.002870	0.001817	0.001191	0.000804	0.000558	0.000397	14
0.004460	0.002881	0.001919	0.001314	0.000922	0.000661	15
0.006508	0.004286	0.002902	0.002013	0.001428	0.001034	16
0.009030	0.006056	0.004164	0.002928	0.002101	0.001535	17
0.012027	0.008203	0.005725	0.004077	0.002958	0.002182	18
0.015490	0.010733	0.007595	0.005476	0.004016	0.002991	19
0.019400	0.013641	0.009779	0.007133	0.005285	0.003972	20
0.023733	0.016917	0.012276	0.009053	0.006773	0.005135	21
0.028460	0.020544	0.015079	0.011235	0.008484	0.006487	22
0.033549	0.024505	0.018179	0.013677	0.010420	0.008031	23
0.038968	0.028778	0.021564	0.016372	0.012577	0.009768	24
0.044686	0.033340	0.025219	0.019311	0.014953	0.011698	25
0.050669	0.038170	0.029127	0.022486	0.017542	0.013817	26
0.056888	0.043242	0.033272	0.025883	0.020335	0.016123	27
0.063314	0.048535	0.037637	0.029490	0.023325	0.018609	28
0.069918	0.054026	0.042205	0.033296	0.026502	0.021269	29
0.076676	0.059694	0.046958	0.037285	0.029857	0.024096	30
0.148306	0.122167	0.101312	0.084528	0.070913	0.059797	40
0.282924	0.247545	0.217383	0.191527	0.169255	0.149991	60
0.389140	0.351441	0.318157	0.288649	0.262392	0.238958	80
0.470627	0.433418	0.399849	0.369460	0.341864	0.316744	100
0.533999	0.498338	0.465691	0.435710	0.408105	0.382634	120
0.584311	0.550538	0.519289	0.490293	0.463324	0.438191	140
0.642653	0.611744	0.582816	0.555676	0.530155	0.506117	170
0.686859	0.658572	0.631886	0.606648	0.582733	0.560038	200
0.731332	0.706070	0.682043	0.659149	0.637294	0.616401	240
0.790932	0.770257	0.750414	0.731325	0.712926	0.695173	320
0.843239	0.827091	0.811467	0.796319	0.781611	0.767310	440
0.882498	0.870040	0.857917	0.846097	0.834556	0.823272	600
0.910529	0.900854	0.891402	0.882150	0.873081	0.864181	800
0.927767	0.919864	0.912124	0.904530	0.897070	0.889731	1000
1.000000	1.000000	1.000000	1.000000	1.000000	1.000000	INF

(*continued*)

Table 3 Lower Percentage Points of Wilks's Lambda (*continued*)
$[U_{m,a,b,\alpha}]$

$m = 7, \quad \alpha = 0.05$

	a					
b	1	2	3	4	5	6
1	0.000000	0.000000	0.000000	0.000000	0.000000	0.000000
2	0.000000	0.000000	0.000000	0.000000	0.000000	0.000000
3	0.000000	0.000000	0.000000	0.000000	0.000000	0.000000
4	0.000000	0.000000	0.000000	0.000000	0.000000	0.000000
5	0.000000	0.000000	0.000000	0.000000	0.000000	0.000000
6	0.000043	0.000006	0.000002	0.000001	0.000001	0.000000
7	0.002625	0.000350	0.000091	0.000033	0.000015	0.000008
8	0.017612	0.002953	0.000809	0.000292	0.000126	0.000063
9	0.047835	0.010329	0.003195	0.001223	0.000543	0.000270
10	0.086645	0.023060	0.008067	0.003338	0.001558	0.000798
11	0.128234	0.040186	0.015642	0.006974	0.003433	0.001826
12	0.169506	0.060396	0.025707	0.012249	0.006343	0.003508
13	0.209026	0.082538	0.037857	0.019109	0.010357	0.005940
14	0.246203	0.105734	0.051646	0.027402	0.015466	0.009172
15	0.280861	0.129346	0.066659	0.036933	0.021607	0.013206
16	0.313032	0.152929	0.082533	0.047494	0.028684	0.018013
17	0.342842	0.176179	0.098971	0.058884	0.036586	0.023544
18	0.370455	0.198894	0.115731	0.070921	0.045199	0.029736
19	0.396050	0.220944	0.132623	0.083445	0.054409	0.036520
20	0.419802	0.242252	0.149498	0.096315	0.064111	0.043824
21	0.441876	0.262777	0.166240	0.109415	0.074209	0.051579
22	0.462425	0.282503	0.182765	0.122645	0.084616	0.059717
23	0.481587	0.301432	0.199007	0.135923	0.095257	0.068177
24	0.499486	0.319577	0.214919	0.149181	0.106063	0.076901
25	0.516238	0.336959	0.230467	0.162364	0.116978	0.085838
26	0.531942	0.353606	0.245630	0.175429	0.127951	0.094941
27	0.546689	0.369546	0.260395	0.188339	0.138940	0.104168
28	0.560560	0.384810	0.274752	0.201068	0.149908	0.113482
29	0.573629	0.399430	0.288701	0.213591	0.160826	0.122851
30	0.585961	0.413438	0.302243	0.225894	0.171667	0.132247
40	0.679227	0.525996	0.417050	0.335433	0.272668	0.223571
60	0.779306	0.659576	0.566032	0.489695	0.426135	0.372561
80	0.831906	0.735024	0.655779	0.588321	0.529875	0.478709
100	0.864288	0.783251	0.715144	0.655689	0.602930	0.555673
120	0.886219	0.816680	0.757179	0.704361	0.656738	0.613420
140	0.902052	0.841199	0.798462	0.741086	0.697881	0.658148
170	0.918970	0.867751	0.822764	0.781839	0.744063	0.708913
200	0.930905	0.886705	0.847518	0.811553	0.778074	0.746666
240	0.942249	0.904887	0.871470	0.840546	0.811527	0.784091
320	0.956525	0.928004	0.902213	0.878097	0.855239	0.833417
440	0.968286	0.947243	0.928043	0.909937	0.892635	0.875985
600	0.976693	0.961103	0.946788	0.933208	0.920155	0.907522
800	0.982494	0.970720	0.959861	0.949517	0.939535	0.929836
1000	0.985983	0.976524	0.967778	0.959426	0.951346	0.943478
INF	1.000000	1.000000	1.000000	1.000000	1.000000	1.000000

Table 3 (*continued*)

a: 7	8	9	10	11	12	b
0.000000	0.000000	0.000000	0.000000	0.000000	0.000000	1
0.000000	0.000000	0.000000	0.000000	0.000000	0.000000	2
0.000000	0.000000	0.000000	0.000000	0.000000	0.000000	3
0.000000	0.000000	0.000000	0.000000	0.000000	0.000000	4
0.000000	0.000000	0.000000	0.000000	0.000000	0.000000	5
0.000000	0.000000	0.000000	0.000000	0.000000	0.000000	6
0.000005	0.000003	0.000002	0.000002	0.000001	0.000001	7
0.000034	0.000020	0.000013	0.000009	0.000006	0.000005	8
0.000147	0.000086	0.000053	0.000035	0.000024	0.000017	9
0.000440	0.000259	0.000160	0.000104	0.000070	0.000049	10
0.001035	0.000619	0.000387	0.000252	0.000170	0.000119	11
0.002048	0.001252	0.000796	0.000525	0.000357	0.000249	12
0.003571	0.002234	0.001448	0.000967	0.000665	0.000468	13
0.005668	0.003628	0.002395	0.001625	0.001131	0.000804	14
0.008371	0.005476	0.003682	0.002537	0.001787	0.001285	15
0.011688	0.007801	0.005337	0.003733	0.002664	0.001936	16
0.015606	0.010611	0.007379	0.005235	0.003782	0.002778	17
0.020096	0.013900	0.009814	0.007057	0.005159	0.003829	18
0.025122	0.017653	0.012640	0.009204	0.006805	0.005102	19
0.030640	0.021845	0.015847	0.011676	0.008725	0.006605	20
0.036603	0.026450	0.019422	0.014469	0.010921	0.008342	21
0.042965	0.031435	0.023345	0.017571	0.013387	0.010314	22
0.049678	0.036769	0.027595	0.020971	0.016120	0.012521	23
0.056697	0.042416	0.032148	0.024653	0.019108	0.014956	24
0.063980	0.048346	0.036980	0.028599	0.022341	0.017614	25
0.071488	0.054525	0.042067	0.032794	0.025807	0.020487	26
0.079183	0.060924	0.047385	0.037217	0.029493	0.023565	27
0.087032	0.067514	0.052911	0.041851	0.033384	0.026838	28
0.095005	0.074268	0.058622	0.046678	0.037467	0.030296	29
0.103073	0.081161	0.064496	0.051680	0.041727	0.033928	30
0.184671	0.153533	0.128393	0.107940	0.091192	0.077392	40
0.327012	0.288026	0.254476	0.225471	0.200293	0.178361	60
0.433602	0.393626	0.358051	0.326284	0.297833	0.272287	80
0.513081	0.474521	0.439488	0.407570	0.378421	0.351744	100
0.573796	0.537400	0.503866	0.472893	0.444226	0.417647	120
0.621410	0.587314	0.555578	0.525974	0.498306	0.472408	140
0.676042	0.645194	0.616167	0.588800	0.562955	0.538514	170
0.717058	0.689053	0.662499	0.637274	0.613274	0.590412	200
0.758031	0.733197	0.709478	0.686784	0.665038	0.644178	240
0.812491	0.792362	0.772959	0.754224	0.736112	0.718583	320
0.859892	0.844294	0.829142	0.814403	0.800046	0.786051	440
0.895244	0.883276	0.871588	0.860157	0.848954	0.837994	600
0.920373	0.911114	0.902038	0.893128	0.884371	0.875758	800
0.935782	0.928236	0.920822	0.913527	0.906342	0.899259	1000
1.000000	1.000000	1.000000	1.000000	1.000000	1.000000	INF

(*continued*)

Table 3 Lower Percentage Points of Wilks's Lambda (*continued*)
$[U_{m,a,b,\alpha}]$

$m = 8, \quad \alpha = 0.01$

b	a: 1	2	3	4	5	6
1	0.000000	0.000000	0.000000	0.000000	0.000000	0.000000
2	0.000000	0.000000	0.000000	0.000000	0.000000	0.000000
3	0.000000	0.000000	0.000000	0.000000	0.000000	0.000000
4	0.000000	0.000000	0.000000	0.000000	0.000000	0.000000
5	0.000000	0.000000	0.000000	0.000000	0.000000	0.000000
6	0.000000	0.000000	0.000000	0.000000	0.000000	0.000000
7	0.000021	0.000002	0.000001	0.000000	0.000000	0.000000
8	0.000738	0.000088	0.000021	0.000007	0.000003	0.000001
9	0.005130	0.000759	0.000188	0.000063	0.000025	0.000012
10	0.016457	0.003031	0.000837	0.000293	0.000121	0.000057
11	0.034984	0.007819	0.002414	0.000906	0.000390	0.000187
12	0.058795	0.015460	0.005293	0.002131	0.000964	0.000477
13	0.085701	0.025773	0.009664	0.004155	0.001974	0.001014
14	0.114026	0.038323	0.015549	0.007095	0.003528	0.001878
15	0.142659	0.052612	0.022851	0.010993	0.005702	0.003140
16	0.170909	0.068175	0.031408	0.015836	0.008535	0.004851
17	0.198367	0.084613	0.041037	0.021570	0.012034	0.007041
18	0.224804	0.101603	0.051550	0.028118	0.016183	0.009723
19	0.250102	0.118888	0.062771	0.035392	0.020951	0.012897
20	0.274219	0.136267	0.074542	0.043297	0.026294	0.016549
21	0.297154	0.153588	0.086724	0.051742	0.032161	0.020658
22	0.318934	0.170736	0.099197	0.060641	0.038499	0.025196
23	0.339603	0.187624	0.111859	0.069912	0.045254	0.030131
24	0.359212	0.204189	0.124624	0.079482	0.052374	0.035431
25	0.377817	0.220388	0.137421	0.089287	0.059809	0.041060
26	0.395477	0.236190	0.150193	0.099266	0.067511	0.046985
27	0.412247	0.251575	0.162890	0.109371	0.075438	0.053173
28	0.428183	0.266532	0.175473	0.119555	0.083549	0.059592
29	0.443335	0.281057	0.187912	0.129781	0.091808	0.066212
30	0.457755	0.295150	0.200180	0.140015	0.100183	0.073004
40	0.570452	0.414141	0.310972	0.238227	0.185181	0.145635
60	0.697943	0.566845	0.469642	0.393584	0.332536	0.282759
80	0.767444	0.657821	0.571844	0.501042	0.441431	0.390577
100	0.811035	0.717584	0.641821	0.577506	0.521783	0.472932
120	0.840895	0.759705	0.692428	0.634154	0.582694	0.536747
140	0.862619	0.790948	0.730631	0.677631	0.630186	0.587267
170	0.885982	0.825107	0.773033	0.726554	0.684388	0.645690
200	0.902559	0.849693	0.803953	0.762678	0.724868	0.689844
240	0.918384	0.873433	0.834123	0.798284	0.765147	0.734170
320	0.938398	0.903831	0.873198	0.844911	0.818448	0.793424
440	0.954965	0.929310	0.906326	0.884882	0.864628	0.845292
600	0.966853	0.947766	0.930536	0.914342	0.898941	0.884139
800	0.975077	0.960620	0.947501	0.935108	0.923267	0.911833
1000	0.980031	0.968398	0.957807	0.947773	0.938158	0.928847
INF	1.000000	1.000000	1.000000	1.000000	1.000000	1.000000

Table 3 (*continued*)

a						
7	8	9	10	11	12	*b*
0.000000	0.000000	0.000000	0.000000	0.000000	0.000000	1
0.000000	0.000000	0.000000	0.000000	0.000000	0.000000	2
0.000000	0.000000	0.000000	0.000000	0.000000	0.000000	3
0.000000	0.000000	0.000000	0.000000	0.000000	0.000000	4
0.000000	0.000000	0.000000	0.000000	0.000000	0.000000	5
0.000000	0.000000	0.000000	0.000000	0.000000	0.000000	6
0.000000	0.000000	0.000000	0.000000	0.000000	0.000000	7
0.000001	0.000000	0.000000	0.000000	0.000000	0.000000	8
0.000006	0.000004	0.000002	0.000001	0.000001	0.000001	9
0.000029	0.000016	0.000010	0.000006	0.000004	0.000003	10
0.000098	0.000055	0.000033	0.000020	0.000013	0.000009	11
0.000255	0.000144	0.000086	0.000054	0.000035	0.000024	12
0.000555	0.000321	0.000194	0.000123	0.000080	0.000054	13
0.001057	0.000624	0.000384	0.000245	0.000162	0.000110	14
0.001817	0.001097	0.000687	0.000445	0.000296	0.000203	15
0.002881	0.001778	0.001134	0.000744	0.000502	0.000347	16
0.004286	0.002700	0.001752	0.001168	0.000797	0.000556	17
0.006056	0.003890	0.002567	0.001736	0.001200	0.000846	18
0.008203	0.005367	0.003600	0.002468	0.001727	0.001231	19
0.010733	0.007144	0.004864	0.003380	0.002393	0.001723	20
0.013641	0.009225	0.006371	0.004484	0.003210	0.002335	21
0.016917	0.011612	0.008127	0.005789	0.004189	0.003076	22
0.020544	0.014298	0.010133	0.007299	0.005336	0.003955	23
0.024505	0.017276	0.012388	0.009019	0.006658	0.004978	24
0.028778	0.020535	0.014887	0.010948	0.008156	0.006149	25
0.033340	0.024060	0.017624	0.013084	0.009833	0.007472	26
0.038170	0.027838	0.020590	0.015423	0.011686	0.008948	27
0.043242	0.031852	0.023775	0.017959	0.013714	0.010578	28
0.048535	0.036085	0.027167	0.020686	0.015914	0.012359	29
0.054026	0.040521	0.030756	0.023596	0.018280	0.014290	30
0.115665	0.092650	0.074781	0.060775	0.049703	0.040884	40
0.241719	0.207592	0.179013	0.154941	0.134560	0.117227	60
0.346803	0.308868	0.275817	0.246896	0.221493	0.199106	80
0.429764	0.391400	0.357150	0.326462	0.298881	0.274029	100
0.495425	0.458070	0.424165	0.393293	0.365111	0.339324	120
0.548177	0.512401	0.479541	0.449270	0.421319	0.395458	140
0.609966	0.576841	0.546023	0.517277	0.490407	0.465247	170
0.657190	0.626631	0.597943	0.570945	0.545489	0.521449	200
0.705043	0.677532	0.651483	0.626761	0.603257	0.580879	240
0.769627	0.746913	0.725172	0.704312	0.684271	0.664991	320
0.826734	0.808858	0.791594	0.774890	0.758703	0.742995	440
0.869838	0.855973	0.842497	0.829376	0.816581	0.804090	600
0.900736	0.889929	0.879380	0.869062	0.858958	0.849052	800
0.919787	0.910939	0.902280	0.893788	0.885451	0.877255	1000
1.000000	1.000000	1.000000	1.000000	1.000000	1.000000	INF

(*continued*)

Table 3 Lower Percentage Points of Wilks's Lambda (*continued*)
$[U_{m,a,b,\alpha}]$

$m = 8, \quad \alpha = 0.05$

b	a = 1	2	3	4	5	6
1	0.000000	0.000000	0.000000	0.000000	0.000000	0.000000
2	0.000000	0.000000	0.000000	0.000000	0.000000	0.000000
3	0.000000	0.000000	0.000000	0.000000	0.000000	0.000000
4	0.000000	0.000000	0.000000	0.000000	0.000000	0.000000
5	0.000000	0.000000	0.000000	0.000000	0.000000	0.000000
6	0.000000	0.000000	0.000000	0.000000	0.000000	0.000000
7	0.000138	0.000015	0.000004	0.000001	0.000001	0.000000
8	0.003295	0.000393	0.000090	0.000029	0.000012	0.000006
9	0.017079	0.002632	0.000659	0.000218	0.000087	0.000040
10	0.043574	0.008626	0.002458	0.000872	0.000361	0.000168
11	0.078039	0.019031	0.006148	0.002365	0.001032	0.000497
12	0.115676	0.033314	0.012011	0.004993	0.002304	0.001155
13	0.153630	0.050518	0.019990	0.008908	0.004335	0.002263
14	0.190453	0.069716	0.029839	0.014129	0.007216	0.003915
15	0.225477	0.090151	0.041241	0.020590	0.010980	0.006173
16	0.258443	0.111245	0.053875	0.028171	0.015610	0.009065
17	0.289300	0.132575	0.067447	0.036729	0.021061	0.012594
18	0.318105	0.153836	0.081699	0.046115	0.027265	0.016740
19	0.344966	0.174814	0.096415	0.056185	0.034144	0.021472
20	0.370015	0.195359	0.111416	0.066805	0.041616	0.026747
21	0.393387	0.215374	0.126559	0.077857	0.049601	0.032519
22	0.415217	0.234796	0.141726	0.089233	0.058021	0.038737
23	0.435632	0.253588	0.156826	0.100843	0.066804	0.045350
24	0.454749	0.271732	0.171785	0.112606	0.075884	0.052311
25	0.472677	0.289225	0.186549	0.124457	0.085199	0.059573
26	0.489514	0.306072	0.201075	C.136338	0.094698	0.067091
27	0.505352	0.322285	0.215331	0.148203	0.104332	0.074826
28	0.520271	0.337880	0.229293	0.160010	0.114060	0.082739
29	0.534345	0.352879	0.242945	0.171728	0.123844	0.090796
30	0.547639	0.367302	0.256277	0.183330	0.133653	0.098967
40	0.648630	0.484826	0.371902	0.289857	0.228618	0.182082
60	0.757690	0.627279	0.527185	0.447009	0.381482	0.327255
80	0.815243	0.708843	0.622840	0.550577	0.488795	0.435425
100	0.850742	0.761330	0.686819	0.622411	0.565838	0.515687
120	0.874811	0.797857	0.732425	0.674791	0.623251	0.576764
140	0.892201	0.824719	0.766516	0.714559	0.667497	0.624521
170	0.910793	0.853874	0.804039	0.758920	0.717493	0.679163
200	0.923918	0.874725	0.831204	0.791410	0.754525	0.720081
240	0.936396	0.894758	0.857556	0.823223	0.791114	0.760867
320	0.952108	0.920269	0.891472	0.864586	0.839159	0.814944
440	0.965057	0.941534	0.920045	0.899793	0.880463	0.861889
600	0.974316	0.956873	0.940825	0.925599	0.910972	0.896826
800	0.980707	0.967524	0.955337	0.943721	0.932512	0.921624
1000	0.984551	0.973956	0.964134	0.954746	0.945661	0.936815
INF	1.000000	1.000000	1.000000	1.000000	1.000000	1.000000

Table 3 (*continued*)

a						
7	8	9	10	11	12	*b*
0.000000	0.000000	0.000000	0.000000	0.000000	0.000000	1
0.000000	0.000000	0.000000	0.000000	0.000000	0.000000	2
0.000000	0.000000	0.000000	0.000000	0.000000	0.000000	3
0.000000	0.000000	0.000000	0.000000	0.000000	0.000000	4
0.000000	0.000000	0.000000	0.000000	0.000000	0.000000	5
0.000000	0.000000	0.000000	0.000000	0.000000	0.000000	6
0.000000	0.000000	0.000000	0.000000	0.000000	0.000000	7
0.000003	0.000002	0.000001	0.000001	0.000001	0.000000	8
0.000020	0.000011	0.000007	0.000004	0.000003	0.000002	9
0.000086	0.000047	0.000028	0.000017	0.000011	0.000008	10
0.000259	0.000144	0.000085	0.000052	0.000034	0.000023	11
0.000619	0.000351	0.000209	0.000130	0.000084	0.000056	12
0.001252	0.000727	0.000441	0.000278	0.000181	0.000122	13
0.002234	0.001331	0.000824	0.000527	0.000347	0.000235	14
0.003628	0.002215	0.001399	0.000910	0.000608	0.000416	15
0.005476	0.003422	0.002203	0.001457	0.000987	0.000683	16
0.007801	0.004982	0.003269	0.002197	0.001509	0.001057	17
0.010611	0.006915	0.004617	0.003151	0.002194	0.001555	18
0.013900	0.009228	0.006265	0.004339	0.003060	0.002194	19
0.017653	0.011923	0.008219	0.005771	0.004120	0.002987	20
0.021845	0.014991	0.010483	0.007456	0.005386	0.003946	21
0.026450	0.018419	0.013053	0.009397	0.006863	0.005078	22
0.031435	0.022192	0.015923	0.011593	0.008555	0.006390	23
0.036769	0.026287	0.019081	0.014041	0.010462	0.007885	24
0.042416	0.030685	0.022515	0.016733	0.012583	0.009565	25
0.048346	0.035361	0.026210	0.019663	0.014914	0.011428	26
0.054525	0.040293	0.030150	0.022818	0.017449	0.013472	27
0.060924	0.045457	0.034319	0.026189	0.020182	0.015694	28
0.067514	0.050831	0.038700	0.029764	0.023104	0.018089	29
0.074268	0.056394	0.043276	0.033529	0.026207	0.020651	30
0.146235	0.118316	0.096365	0.078964	0.065068	0.053897	40
0.281978	0.243910	0.211718	0.184362	0.161015	0.141011	60
0.388992	0.348380	0.312704	0.281253	0.253441	0.228779	80
0.470954	0.430871	0.394827	0.362322	0.332935	0.306310	100
0.534599	0.496197	0.461114	0.428982	0.399491	0.372376	120
0.585067	0.548712	0.515117	0.484002	0.455129	0.428296	140
0.643522	0.610267	0.579158	0.549999	0.522621	0.496881	170
0.687764	0.657345	0.628642	0.601508	0.575820	0.551470	200
0.732246	0.705079	0.679234	0.654605	0.631100	0.608645	240
0.791784	0.769570	0.748216	0.727659	0.707843	0.688722	320
0.843968	0.826629	0.809821	0.793502	0.777640	0.762209	440
0.883093	0.869724	0.856684	0.843948	0.831494	0.819306	600
0.911008	0.900629	0.890464	0.880494	0.870704	0.861084	800
0.928167	0.919691	0.911367	0.903183	0.895127	0.887192	1000
1.000000	1.000000	1.000000	1.000000	1.000000	1.000000	INF

Table 4 Upper Percentage Points of Roy's Greatest Basic Diagonal $[\theta_{p_1, p_2, p_3, \alpha}]$

$\alpha = 0.01 \quad p_1 = 2$

p_3	p_2 = 0	1	2	3	4	5	7	10	15
5	0.675	0.745	0.787	0.817	0.839	0.8568	0.8821	0.9066	0.9306
10	0.470	0.544	0.597	0.638	0.670	0.6970	0.7391	0.7834	0.8309
15	0.357	0.425	0.476	0.517	0.551	0.5803	0.6279	0.6810	0.7418
20	0.288	0.347	0.394	0.433	0.467	0.4951	0.5435	0.5998	0.6670
25	0.240	0.293	0.336	0.372	0.403	0.4309	0.4782	0.5347	0.6045
30	0.207	0.254	0.293	0.326	0.355	0.3812	0.4266	0.4819	0.5521
40	0.161	0.200	0.232	0.261	0.286	0.3094	0.3503	0.4017	0.4697
60	0.1114	0.140	0.165	0.186	0.206	0.2244	0.2576	0.3008	0.3608
80	0.0852	0.1080	0.1273	0.1448	0.1609	0.1759	0.2035	0.2402	0.2925
100	0.0692	0.0878	0.1038	0.1184	0.1319	0.1446	0.1682	0.1999	0.2458
130	0.0539	0.0685	0.0813	0.0930	0.1039	0.1142	0.1331	0.1595	0.1983
160	0.0441	0.0562	0.0668	0.0765	0.0857	0.09430	0.1105	0.1328	0.1662
200	0.0355	0.0453	0.0540	0.0619	0.0694	0.07653	0.08994	0.1085	0.1366
300	0.0239	0.0305	0.0365	0.0419	0.0471	0.05202	0.06137	0.07446	0.1076
500	0.0144	0.0185	0.0221	0.0255	0.0287	0.03192	0.03753	0.04574	0.07415
1000	0.00725	0.00930	0.01114	0.01285	0.01448	0.01605	0.01904	0.02328	0.03811

$\alpha = 0.05 \quad p_1 = 2$

p_3	p_2 = 0	1	2	3	4	5	7	10	15
5	0.565	0.651	0.706	0.746	0.776	0.7992	0.8337	0.8676	0.9011
10	0.374	0.455	0.514	0.561	0.598	0.6294	0.6787	0.7316	0.7889
15	0.278	0.348	0.402	0.445	0.483	0.5145	0.5671	0.6266	0.6954
20	0.222	0.281	0.329	0.369	0.403	0.4340	0.4855	0.5462	0.6197
25	0.183	0.236	0.278	0.314	0.346	0.3748	0.4239	0.4834	0.5580
30	0.157	0.203	0.241	0.274	0.303	0.3297	0.3760	0.4333	0.5070
40	0.121	0.158	0.190	0.218	0.243	0.2654	0.3064	0.3585	0.4282
60	0.0836	0.110	0.133	0.154	0.173	0.1909	0.2233	0.2661	0.3260
80	0.0638	0.0846	0.1027	0.1191	0.1345	0.1490	0.1756	0.2114	0.2630
100	0.0515	0.0686	0.0835	0.0972	0.1100	0.1221	0.1430	0.1753	0.2203
130	0.0400	0.0535	0.0652	0.0761	0.0864	0.09613	0.1141	0.1396	0.1771
160	0.0327	0.0473	0.0535	0.0626	0.0711	0.07925	0.09461	0.1159	0.1481
200	0.0263	0.0352	0.0432	0.0506	0.0576	0.06422	0.07687	0.09454	0.1215
300	0.0176	0.0237	0.0291	0.0342	0.0390	0.04356	0.05234	0.06471	0.09867
500	0.0106	0.0143	0.0176	0.0207	0.0237	0.02651	0.03194	0.03967	0.06110
1000	0.00535	0.00719	0.00888	0.01045	0.01195	0.01339	0.01618	0.02016	0.03547

From K. C. S. Pillai, *Statistical Tables for Tests of Multivariate Hypotheses*, The Statistical Center, University of the Philippines, 1960. Reproduced by permission of the author and publisher.

Table 4 (*continued*)

$\alpha = 0.01 \quad p_1 = 3$

p_3	p_2 = 0	1	2	3	4	5	7	10	15
5	0.75816	0.8040	0.8344	0.8564	0.8730	0.8894	0.0956	0.9247	0.9437
10	0.55857	0.6164	0.6590	0.6923	0.7192	0.7415	0.7767	0.8141	0.8544
15	0.43751	0.4936	0.5374	0.5730	0.6029	0.6285	0.6703	0.7172	0.7708
20	0.35856	0.4104	0.4519	0.4867	0.5166	0.5428	0.5866	0.6376	0.6985
25	0.30343	0.3506	0.3893	0.4223	0.4511	0.4707	0.5203	0.5726	0.6370
30	0.26286	0.3058	0.3416	0.3726	0.3999	0.4245	0.4670	0.5189	0.5847
40	0.20727	0.2434	0.2742	0.3012	0.3256	0.3477	0.3869	0.4350	0.5001
60	0.14555	0.17271	0.19631	0.21751	0.23692	0.2549	0.2874	0.3298	0.3883
80	0.11212	0.13378	0.15282	0.17011	0.18608	0.2010	0.2285	0.2662	0.3166
100	0.09118	0.10916	0.12508	0.13963	0.15317	0.1659	0.1895	0.2211	0.2670
130	0.07120	0.08552	0.09829	0.11004	0.12104	0.1314	0.1508	0.1772	0.2162
160	0.05842	0.07030	0.08096	0.09079	0.10003	0.1088	0.1253	0.1479	0.1816
200	0.04713	0.05681	0.06554	0.07362	0.08123	0.08849	0.1021	0.1222	0.1540
300	0.03178	0.03839	0.04441	0.04997	0.05526	0.06034	0.06991	0.08332	0.1038
500	0.01924	0.02329	0.02696	0.03043	0.03370	0.03686	0.04285	0.05130	0.06441
1000	0.00969	0.01174	0.01362	0.01538	0.01706	0.01868	0.02547	0.03011	0.03304

$\alpha = 0.05 \quad p_1 = 3$

p_3	p_2 = 0	1	2	3	4	5	7	10	15
5	0.66889	0.7292	0.7690	0.7994	0.8221	0.8456	0.8668	0.8934	0.9199
10	0.47178	0.5372	0.5862	0.6248	0.6564	0.6828	0.7246	0.7695	0.8185
15	0.36196	0.4218	0.4690	0.5078	0.5407	0.5690	0.6157	0.6687	0.7298
20	0.29310	0.3465	0.3898	0.4264	0.4582	0.4861	0.5333	0.5889	0.6559
25	0.24610	0.2937	0.3331	0.3671	0.3970	0.4237	0.4696	0.5251	0.5944
30	0.21201	0.2547	0.2907	0.3221	0.3500	0.3752	0.4192	0.4734	0.5429
40	0.16598	0.2013	0.2320	0.2584	0.2827	0.3050	0.3446	0.3950	0.4608
60	0.11566	0.14165	0.16440	0.18504	0.20400	0.2217	0.2538	0.2961	0.3550
80	0.08873	0.10925	0.12745	0.14404	0.15950	0.1740	0.2008	0.2382	0.2880
100	0.07199	0.08890	0.10403	0.11792	0.13091	0.1432	0.1660	0.1969	0.2421
130	0.05610	0.06951	0.08154	0.09268	0.10316	0.1131	0.1318	0.1574	0.1954
160	0.04594	0.05704	0.06705	0.07633	0.08510	0.09347	0.1092	0.1310	0.1637
200	0.03703	0.04604	0.05420	0.06182	0.06901	0.07591	0.08889	0.1079	0.1398
300	0.02491	0.03105	0.03665	0.04187	0.04684	0.05161	0.06072	0.07349	0.09332
500	0.01504	0.01882	0.02224	0.02546	0.02852	0.03149	0.03715	0.04517	0.05770
1000	0.00758	0.00947	0.01122	0.01286	0.01442	0.01594	0.01885	0.02741	0.02955

(*continued*)

Table 4 Upper Percentage Points of Roy's Greatest Basic Diagonal (*continued*)

$\alpha = 0.01 \quad p_1 = 4$

p_3	p_2 = 0	1	2	3	4
5	0.8110	0.8436	0.8662	0.8830	0.8959
10	0.6247	0.6708	0.7057	0.7334	0.7560
15	0.5016	0.5490	0.5867	0.6177	0.6439
20	0.4175	0.4627	0.4997	0.5309	0.5579
25	0.3570	0.3992	0.4343	0.4645	0.4910
30	0.3116	0.3507	0.3837	0.4125	0.4380
40	0.2483	0.2819	0.3108	0.3364	0.35960
60	0.1763	0.2021	0.22486	0.24541	0.26429
80	0.1367	0.1575	0.17603	0.19300	0.20872
100	0.1115	0.12900	0.14459	0.15899	0.17241
130	0.0874	0.10139	0.11402	0.12572	0.13670
160	0.0719	0.08353	0.09411	0.10396	0.11323
200	0.0581	0.06764	0.07634	0.08444	0.09213
300	0.03928	0.04583	0.05185	0.05749	0.06285
500	0.02383	0.02787	0.03159	0.03508	0.03842
1000	0.01202	0.01407	0.01547	0.01777	0.01948

$\alpha = 0.05 \quad p_1 = 4$

p_3	p_2 = 0	1	2	3	4
5	0.7387	0.7825	0.8131	0.8359	0.8537
10	0.5471	0.6004	0.6411	0.6736	0.7004
15	0.4307	0.4822	0.5235	0.5577	0.5869
20	0.3543	0.4017	0.4408	0.4741	0.5031
25	0.3006	0.3438	0.3802	0.4117	0.4395
30	0.2609	0.3004	0.3340	0.3635	0.3899
40	0.2063	0.2396	0.2685	0.2943	0.31767
60	0.1454	0.1704	0.19264	0.21284	0.23148
80	0.1122	0.1322	0.15014	0.16661	0.18196
100	0.0913	0.10795	0.12298	0.13686	0.14986
130	0.0714	0.08465	0.09672	0.10793	0.11849
160	0.0586	0.06963	0.07969	0.03909	0.09797
200	0.0473	0.05631	0.06454	0.07227	0.07959
300	0.03192	0.03808	0.04375	0.04909	0.05417
500	0.01934	0.02312	0.02661	0.02990	0.03306
1000	0.00974	0.01166	0.01344	0.01512	0.01674

Table 4 (*continued*)

$\alpha = 0.01$ $p_1 = 5$					
	p_2				
p_3	0	1	2	3	4
5	0.8478	0.8719	0.8892	0.9023	0.9126
10	0.6762	0.7136	0.7426	0.7659	0.7850
15	0.5544	0.5948	0.6274	0.6546	0.6777
20	0.4677	0.5074	0.5404	0.5685	0.5929
25	0.4038	0.4416	0.4735	0.5012	0.5255
30	0.3549	0.3904	0.4208	0.4475	0.4713
40	0.2854	0.3166	0.3438	0.3681	0.3963
60	0.2048	0.2293	0.2512	0.2710	0.2894
80	0.1597	0.1796	0.1977	0.2143	0.2296
100	0.1308	0.1476	0.1629	0.1771	0.1902
130	0.10284	0.11645	0.1289	0.1405	0.1514
160	0.08474	0.09615	0.10663	0.11641	0.12564
200	0.06862	0.07801	0.08665	0.09475	0.10244
300	0.04651	0.05300	0.05900	0.06466	0.07006
500	0.02828	0.03229	0.03602	0.03955	0.04292
1000	0.01429	0.01633	0.01825	0.02006	0.02180

$\alpha = 0.05$ $p_1 = 5$					
	p_2				
p_3	0	1	2	3	4
5	0.7882	0.8210	0.8447	0.8626	0.8768
10	0.6069	0.6507	0.6848	0.7125	0.7354
15	0.4882	0.5327	0.5689	0.5993	0.6252
20	0.4072	0.4494	0.4847	0.5150	0.5413
25	0.3488	0.3881	0.4215	0.4506	0.4764
30	0.3049	0.3412	0.3726	0.4002	0.4250
40	0.2433	0.2746	0.3021	0.3267	0.3490
60	0.1732	0.1973	0.2188	0.2385	0.2567
80	0.1344	0.1538	0.1714	0.1877	0.2028
100	0.1098	0.1261	0.1409	0.1547	0.1676
130	0.08612	0.09918	0.1112	0.1224	0.1329
160	0.07084	0.08175	0.09179	0.10120	0.11012
200	0.05729	0.06622	0.07448	0.08224	0.08962
300	0.03875	0.04490	0.05061	0.05600	0.06115
500	0.02353	0.02731	0.03084	0.03418	0.03739
10000	0.01187	0.01380	0.01560	0.01732	0.01897

(*continued*)

Table 4 Upper Percentage Points of Roy's Greatest Basic Diagonal (*continued*)

$\alpha = 0.01 \quad p_1 = 6$

	p_2				
p_3	0	1	2	3	4
5	0.8745	0.8929	0.9065	0.9169	0.9255
10	0.7173	0.7482	0.7724	0.7922	0.8086
15	0.5986	0.6334	0.6619	0.6858	0.7063
20	0.5111	0.5462	0.5757	0.6010	0.6231
25	0.4450	0.4790	0.5081	0.5335	0.5559
30	0.3936	0.4261	0.4542	0.4789	0.5011
40	0.3194	0.3484	0.3739	0.3969	0.4177
60	0.2315	0.2548	0.2757	0.2948	0.3215
80	0.1814	0.2006	0.2181	0.2342	0.2493
100	0.1491	0.1654	0.1803	0.1942	0.2072
130	0.11762	0.13091	0.14314	0.15457	0.16536
160	0.09713	0.10830	0.11901	0.12834	0.13754
200	0.07880	0.08803	0.09659	0.10466	0.11232
300	0.05355	0.05996	0.06594	0.07160	0.07701
500	0.03270	0.03661	0.04034	0.04388	0.04727
10000	0.01651	0.01855	0.02046	0.02229	0.02405

$\alpha = 0.05 \quad p_1 = 6$

	p_2				
p_3	0	1	2	3	4
5	0.8246	0.8499	0.8685	0.8830	0.8945
10	0.6552	0.6917	0.7206	0.7442	0.7639
15	0.5371	0.5758	0.6077	0.6346	0.6577
20	0.4535	0.4912	0.5231	0.5505	0.5746
25	0.3918	0.4276	0.4583	0.4852	0.5091
30	0.3447	0.3782	0.4074	0.4332	0.4564
40	0.2775	0.3069	0.3329	0.3563	0.3776
60	0.1995	0.2225	0.2433	0.2624	0.2801
80	0.1556	0.1745	0.1916	0.2075	0.2224
100	0.1275	0.1434	0.1580	0.1716	0.1843
130	0.10036	0.11319	0.12504	0.13615	0.14666
160	0.08272	0.09348	0.10388	0.11284	0.12175
200	0.06702	0.07586	0.08409	0.09186	0.09926
300	0.04545	0.05156	0.05728	0.06281	0.06790
500	0.02765	0.03143	0.03498	0.03835	0.04160
1000	0.01397	0.01590	0.01772	0.01946	0.02113

Index

O

P

Q

R

S